Roland Chin Ting-Chuen Pong (Eds.)

Computer Vision – ACCV'98

Third Asian Conference on Computer Vision
Hong Kong, China, January 8-10, 1998
Proceedings, Volume II

Springer

Series Editors

Gerhard Goos, Karlsruhe University, Germany

Juris Hartmanis, Cornell University, NY, USA

Jan van Leeuwen, Utrecht University, The Netherlands

Volume Editors

Roland Chin
Ting-Chuen Pong
Hong Kong University of Science and Technology
Computer Science Department
Clear Water Bay, Kowloon, Hong Kong, China
E-mail:(roland,tcpong)@cs.ust.hk

Cataloging-in-Publication data applied for

Die Deutsche Bibliothek - CIP-Einheitsaufnahme

Computer vision : proceedings / ACCV '98, Third Asian Conference on Computer Vision, Hong Kong, China, January 8 - 10, 1998. Roland Chin ; Ting-Chuen Pong (ed.). - Berlin ; Heidelberg ; New York ; Barcelona ; Budapest ; Hong Kong ; London ; Milan ; Paris ; Santa Clara ; Singapore ; Tokyo : Springer

Vol. 2 (1998)
(Lecture notes in computer science ; Vol. 1352)
ISBN 3-540-63931-4

CR Subject Classification (1991): I.3.5, I.5, I.2.9-10, I.4

ISSN 0302-9743
ISBN 3-540-63931-4 Springer-Verlag Berlin Heidelberg New York

This work is subject to copyright. All rights are reserved, whether the whole or part of the material is concerned, specifically the rights of translation, reprinting, re-use of illustrations, recitation, broadcasting, reproduction on microfilms or in any other way, and storage in data banks. Duplication of this publication or parts thereof is permitted only under the provisions of the German Copyright Law of September 9, 1965, in its current version, and permission for use must always be obtained from Springer-Verlag. Violations are liable for prosecution under the German Copyright Law.

© Springer-Verlag Berlin Heidelberg 1997
Printed in Germany

Typesetting: Camera-ready by author
SPIN 10661303   06/3142 – 5 4 3 2 1 0   Printed on acid-free paper

Preface

We are very pleased to have the opportunity to organize the 3rd Asian Conference on Computer Vision (ACCV'98). The conference is sponsored by the IEEE Hong Kong Section, Computer Chapter, the Sino Software Research Institute and the Department of Computer Science of the Hong Kong University of Science and Technology, and the Hong Kong Industry Department.

We received over 300 submissions of full papers (not including the invited papers for the special sessions) from 30 countries in April 1997. In order to provide a quality conference and quality proceedings, each paper was reviewed by at least three members of the program committee. The program committee selected and accepted 58 papers for oral presentation and 112 papers for poster presentation after the review process. Some of these papers were jointly submitted to ACCV'98 and ICCV'98 (to be held in Bombay in January 4–7, 1998) and they were reviewed in a coordinated effort. We must add that the program committee and the reviewers have done an excellent job within a tight schedule and we are very pleased with the quality of the papers.

Four eminent invited speakers, Professors Brian Funt of Simon Fraser University, Krishna Nathan of IBM, Eric Grimson of MIT, and Shoji Tominaga of Osaka Electro-Communication University, have contributed to the conference. We are grateful to them. In addition, we wish to thank Professors Jake Aggarwal, Shashi Buluswar, Yi-Ping Hung, Anil Jain, and Sharatchandra Pankanti for organizing the very high-quality special sessions. Last but not least, we would like to express our gratitude to all the contributors, reviewers, program committee and organizing committee members, and sponsors, without whom the conference would not have been possible.

Finally, we hope that you will benefit from these proceedings.

Roland T. Chin
Ting-Chuen Pong

January 1998

Conference Chair:
Helen Shen (Hong Kong U. of Science & Technology)
Song De Ma (Inst. of Automation, Beijing)

Program Co-Chairs:
Roland Chin (Hong Kong U. of Science & Technology)
T.C. Pong (Hong Kong U. of Science & Technology)
Saburo Tsuji (Wakayama U.)

Program Committee:
Jake Aggarwal (U. of Texas)
Narendra Ahuja (U. of Illinois)
Carlo Arcelli (Institute for Cybernetics, Italy)
Terry Caelli (Curtin U. of Technology)
Larry Davis (U. of Maryland)
Xiaoqing Ding (Tsinghua U.)
Charles Dyer (U. of Wisconsin)
Olivier Faugeras (INRIA)
Jun-ichi Hasegawa (Chyukyo U.)
Thomas Huang (U. of Illinois)
Katsushi Ikeuchi (U. of Tokyo)
Horace Ip (City U. of Hong Kong)
Anil Jain (Michigan State U.)
Ramesh Jain (U. of California, San Diego)
Ben Jang (IBM)
Ray Jarvis (Monash U.)
Rangachar Kasturi (Penn State U.)
Kwang Ik Kim (POSTECH)
Les Kitchen (U. of Melbourne)
Josef Kittler (U. of Surrey)
Kok Fung Lai (ITI, Singapore)
Louisa Lam (Hong Kong Inst. of Education)
Chung-Nim Lee (POSTECH)
Hsi-Jian Lee (Chiao Tung U.)
Seong-Whan Lee (Korea U.)
Takashi Matsuyama (Kyoto U.)
Dinesh Mital (Nanyang Technological U.)
Shree Nayar (Columbia U.)
Ram Nevatia (U. of Southern California)
Yuichi Ohta (Tsukuba U.)
Shmuel Peleg (Hebrew U.)
Brent Seales (U. of Kentucky)
Yoshiaki Shirai (Osaka U.)
Arnold Smeulders (U. of Amsterdam)
Ching Y. Suen (Concordia U.)
Michael Swain (U. of Chicago)
Eam KhwangTeoh (Nanyang Technological U.)
Baba Vemuri (U. of Florida)
Kazuhiko Yamamoto (Gifu U.)
Naokazu Yokoya (AIST-Nara)

Organising Committee:
Oscar Au (Hong Kong U. of Science & Technology)
Ronald Chung (Chinese U. of Hong Kong)
Horace Ip (City U. of Hong Kong)
Tong Lee (Chinese U. of Hong Kong)
Chiew Lan Tai (Hong Kong U. of Science & Technology)
H.T. Tsui (Chinese U. of Hong Kong)
Christopher Yang (U. of Hong Kong)

Sponsored by
IEEE Hong Kong Section, Computer Chapter
Sino Software Research Institute, Hong Kong U. of Science and Technology
Department of Computer Science, Hong Kong U. of Science and Technology
Hong Kong Industry Department

Contents of Volume II

Poster Session II

On Typical Implementations of Hough Transform for Improving Its Performances
Jun-ichiro Hayashi, Kunihito Kato, Toshio Endoh, Kazuhito Murakami, Takashi Toriu and Hiroyasu Koshimizu II-1

Hierarchical Segmentation and Representation with Dynamic Link Architecture Neural Network
Yunqiang Chen and SongDe Ma II-9

Perceptually Consistent Segmentation of Texture Using Multiple Channel Filter
Nan Zhang and Wee Kheng Leow II-17

Optimal Edge Detection under Difficult Imaging Conditions
Md. Shoaib Bhuiyan, Yuji Iwahori and Akira Iwata II-25

Restoring Image Quality Through Structure Preserving De-noising
Krishna Ratakonda and Narendra Ahuja II-33

Feature Saliency from Noise Variations in Invariants
Mark Jenkinson and Michael Brady II-41

Multiscale Image Representation and Edge Detection
Fang Chen and David Suter II-49

Rotation Invariant Texture Features from Gabor Filters
S.R. Fountain and T.N. Tan II-57

Euclidean Invariants of Linear Scale-Spaces
Alfons Salden II-65

Segmenting Objects at Multiple Scales : A Robust Approach
Farzin Mokhtarian II-73

Multi-grid Edge Models for Magnifying Digital Images
G. Qiu II-81

Scale and Rotation Invariant Recognition Method Using Higher-Order Local Autocorrelation Features of Log-Polar Image
Takio Kurita, Kazuhiro Hotta and Taketoshi Mishima II-89

Script and Language Identification from Document Images
G.S. Peake and T.N. Tan II-97

Document Categorization for Document Image Understanding
Hiroyuki Masai and Toyohide Watanabe II-105

Recognition of Various Bar-graph Structures Based on Layout Model
Naoko Yokokura and Toyohide Watanabe II-113

Word-Class Bigram Statistics Language Model for a Hand-Written Chinese Character Recognizer
Pak-Kwong Wong and Chorkin Chan II-121

Log Classification by Single X-ray Scans Using Texture Features from Growth Rings
Xinli Wang II-129

Precise and Fast Form Identification Method by Using Adaptive Base Lines for Matching
Hiroaki Takebe, Yutaka Katsuyama and Satoshi Naoi II-137

Combinatorial Coarse Classification Method for OLCCR
Jing Zheng, Xiaoqing Ding, Youshou Wu and Fanxia Guo II-145

Detecting Characters in Grey-Scale Scene Images
Yongmei Liu, Tsuyoshi Yamamura, Noboru Ohnishi and Noboru Sugie II-153

Conic Based Image Transfer for 2-D Objects: A Linear Algorithm
Akihiro Sugimoto II-161

Minimal Conditions on Intrinsic Paramenters for Euclidean Reconstruction
Anders Heyden and Kalle Åström II-169

Surface Based Hypothesis Verification in Intensity Images Using Geometric and Appearance Data
J.H.M. Byne and J.A.D.W. Anderson II-177

Next Best Viewpoint (NBV) Planning for Active Object Modeling Based on a Learning-by-Showing Approach
Hongbin Zha, Ken'ichi Morooka and Tsutomu Masegawa II-185

Object Recognition by Matching Symbolic Edge Graphs
Tino Lourens and Rolf P. Würtz II-193

Interpretation of Complex Scenes Using Bayesian Networks
Mark F. Westling and Larry S. Davis II-201

Recognition of Urban Scene Using Silhouette of Buildings and City Map Database
Peilin Liu, Wei Wu, Katsushi Ikeuchi and Masao Sakauchi II-209

A Cooperative Inference Mechanism for Extracting Road Information Automatically
Masakazu Nishijima and Toyohide Watanabe II-217

Model-Based Active Object Recognition Using MRF Matching and Sensor Planning
Tianrong Liu, Kap Luk Chan and Stan Ziqing Li II-225

Improved Image Classification Using Morphing
W. Brent Seales and Cheng Jiun Yuan II-233

Reconstruction of Non-manifold Objects from Two Orthographic Views
Chang-Hun Kim and Tae-Jung Suh II-241

3D Object Recognition Using Segment-Based Stereo Vision
Yasushi Sumi and Fumiaki Tomita II-249

Invited Talk

The State of Color Vision Research
Brian Funt II-257

Color Vision and Color Media Processing Research in Asia
Shoji Tominaga II-258

Session S1A: Recent Advances in Computer Vision

Recent Advances in Detection and Description of Buildings from Multiple Aerial Images
Sanjay Noronha and Ram Nevatia II-259

Visual Surveillance of Human Activity
Larry Davis, Sandor Fejes, David Harwood, Yaser Yacoob, Ismail Hariatoglu and Michael J. Black II-267

Bayesian Paradigm for Recognition of Objects - Innovative Applications
J. K. Aggarwal and Shishir Shah II-275

Toward Motion Picture Grammars
Ruud Bolle, Yiannis Aloimonos and Cornelia Fermüller II-283

Session S1B: Segmentation and Grouping

Hierarchical Texture Segmentation
P. Bajcsy and N. Ahuja II-291

Range Image Segmentation: Adaptive Grouping of Edges into Regions
Xiaoyi Jiang and Horst Bunke II-299

Optimising the Complete Image Feature Extraction Chain
M. Mirmehdi, P. L. Palmer and J. Kittler II-307

A Unified Framework for Salient Curves, Regions, and Junctions Inference
Mi-Suen Lee and Gérard Medioni II-315

Learning Multiscale Image Models of 2D Object Classes
Benoit Perrin, Narendra Ahuja and Narayan Srinivasa II-323

Session S2A: Computer Vision & Virtual Reality

3D Model Centered Framework for CV and VR
Michihiko Minoh II-332

Image-Based Geometrically-Correct Photorealistic Scene/Object Modeling(IBPhM): A Review
Zhengyou Zhang II-340

Measuring Object Surface Shape and Reflectance Properties
Yoichi Sato, Mark D. Wheeler, and Katsushi Ikeuchi II-350

Robust Image Composition Algorithms for Augmented Reality
Marie-Odile Berger and Gilles Simon II-360

Context-Based Recognition of Manipulative Hand Gestures for Human Computer Interaction
Kang-Hyun Jo, Yoshinori Kuno and Yoshiaki Shirai II-368

Session S2B: Motion Analysis

An Algorithm for Recursive Structure and Motion Recovery under Affine Projection
  Miroslav Trajković and Mark Hedley II-376

Relative Affine Depth: Structure from Motion by an Uncalibrated Camera
  Zhong-Ying Zhang and Hung-Tat Tsui II-384

The Eigenspace Method for Rigid Motion Recovery from less than Eight Point Correspondences
  Miroslav Trajković and Mark Hedley II-392

3D Shape and Motion Analysis from Image Blur and Smear: A Unified Approach
  Yuan-Fang Wang and Ping Liang II-400

3D Line's Extraction from 2D Spatio-temporal Image Created by Sine Slit
  Pingtao Wang, Katsushi Ikeuchi and Masao Sakauchi II-408

Toward Non-intrusive Motion Capture
  A. Bottino, A. Laurentini and P. Zuccone II-416

Session S3A: Object Recognition and Modeling

Appearance Based Visual Learning and Object Recognition with Illumination Invariance
  Kohtaro Ohba, Yoichi Sato and Katsushi Ikeuchi II-424

Evidence-Based Scene Interpretation Considering Subjective Certainty of Recognition
  Yasuhiro Taniguchi and Yoshiaki Shirai II-432

Robust Hypothesis Verification for Model Based Object Recognition Using Gaussian Error Model
  Frederic Jurie II-440

Shape Modeling from Multiple View Images Using GAs
  Satoshi Kirihara and Hideo Saito II-448

3-D Reconstruction of Multipart Self-Occluding Objects
  Nebojsa Jojic, Jin Gu, Helen C. Shen and Thomas S. Huang II-455

On Analysis of Cloth Drape Range Data
  Nebojsa Jojic and Thomas S. Huang II-463

Poster Session III

VR Models from Epipolar Images: An Approach to Minimize Errors in
Synthesized Images
  Mikio Shinya, Takafumi Saito, Takeaki Mori and Noriyoshi Osumi II-471

Shape and Pose Parameter Estimation of 3D Multi-Part Objects
  Satoshi Yonemoto, Naoyuki Tsuruta and Rin-ichiro Taniguchi II-479

Generating 3D Models of Objects Using Multiple Visual Cues in Image
Sequences
  Jiang Yu Zheng, Akio Murata and Norihiro Abe II-487

Strategical Tracking of Polyhedral Objects by Reactive Change of
Projection Pattern - Reactive Range Finder
  Takeshi Mita, Shinsaku Hiura, Hirokazu Kato and Seiji Inokuchi II-495

Autonomous Vision-Guided Robot Manipulation Control
  Wey-Shiuan Hwang and John (Juyang) Weng II-503

A New Adaptive Approach on Rapid Obstacle Detection in Range Image
  Qi Zhang, Weikang Gu and Xiuqing Ye II-511

Recognition of Shape Model for General Roads
  Keiichi Uchimura and Zhencheng Hu II-519

Visual Detection of Obstacles Assuming a Locally Planar Ground
  Manolis I.A. Lourakis and Stelios C. Orphanoudakis II-527

Potential-Based Modeling of 2D Regions Using Non-uniform Source
Distributions
  Jen-Hui Chuang, Chi-Hao Tsai, Wei-Hsin Tsai and Chuei-Yaw Yang II-535

A Linear Algorithm for Motion from Three Weak Perspective Images
Using Euler Angles
  Gang Xu and Noriko Sugimoto II-543

On Learning Spatio-Temporal Relational Structures in Two Different
Domains
  Adrian R. Pearce, Terry Caelli and Simon Goss..II-551

An Efficient Iterative Pose Estimation Algirithm
  S.H. Or, W.S. Luk, K.H. Wong and I. King II-559

A New Multistage Approach to Motion and Structure Estimation by
Gradually Enforcing Geometric Constraints
    Zhengyou Zhang II-567

Tracking a Person with Pre-recorded Image Database and a Pan, Tilt,
and Zoom Camera
    Yimimg Ye, John K. Tsotsos, Karen Bennet and Eric Harley II-575

Recovery of Motion and Structure from Optical Flow under Perspective
Projection by Solving Linear Simultaneous Equations
    Toshiharu Mukai and Noboru Ohnishi II-583

Vector Coherence Mapping: A Parallelizable Approach to Image Flow
Computation
    Francis K.H. Quek and Robert K. Bryll II-591

Robust Motion Segmentation Using Rank Ordering Estimators
    Alireza Bab-Hadiashar and David Suter II-599

Optical Flow in the Scale Space
    Qing Yang and SongDe Ma II-607

Motion Detection in Temporal Clutter
    Phillip M. Ngan II-615

A Novel Fast Three-Step Search Algorithm for Block-Matching Motion
Estimation
    William Booth, James M. Noras and Donglai Xu II-623

Moving Vehicle Detection and Tracking in Image Sequences
    Yi Lu, Jason Miller and Tie Qi Chen II-631

Gesture Recognition from Image Motion Based on Subspace Method and
HMM
    Yoshio Iwai, Tadashi Hata and Masahiko Yachida II-639

Identifying Faces under Varying Pose Using a Single Example View
    Dadet Pramadihanto, Yoshio Iwai, Masahiko Yachida and Haiyuan Wu II-647

Multiple Camera Based Human Motion Estimation
    Akira Utsumi, Hiroki Mori, Jun Ohya and Masahiko Yachida II-655

An Autonomous Facial Caricaturing Based on a Model of Visual Illusion-
Experimental Modeling of Visual Illusion
    Kazuhito Murakami, Mikiko Takai and Hiroyasu Koshimizu II-663

3D Estimation of Facial Muscle Parameter from the 2D Marker Movement
Using Neural Network
  *Takahiro Ishikawa, Hajime Sera, Shigeo Morishima and
  Demetri Terzopoulos* II-671

Appearance-Based Face Recognition under Large Head Rotations in Depth
  Shaogang Gong, Eng-Jon Ong and Peter J. Loft II-679

Skin-Color Modeling and Adaptation
  Jie Yang, Weier Lu and Alex Waibel..II-687

Human Information Retrieval by Face Extraction and Recognition on TV
News Images Using Subspace Method
  Yasuo Ariki, Noriyuki Ishikawa and Yoshiaki Sugiyama.............................II-695

Converting Facial Expressions Using Recognition-Based Analysis of
Image Sequences
  Takahiro Otsuka and Jun Ohya II-703

Muscle-Based Feature Models for Analyzing Facial Expressions
  Hiroshi Ohta, Hitoshi Saji and Hiromasa Nakatani....................................II-711

A Morphological Method for Moving Object Segmentation and Posture
Recognition
  Yi Li, Songde Ma and Hanqing Lu II-719

Detection of Glasses in Facial Images
  Xiaoyi Jiang, M. Binkert, B. Achermann and H. Bunke...............................II-726

Non-monotonic Continuous Dynamic Programming for Spotting Recognition
of Hesitated Gestures from Time-Varying Images
  T. Nishimura, T. Mukai and R. Oka..II-734

Face Recognition Using a Face-Only Database: A New Approach
  *Hong-Yuan Mark Liao, Chin-Chuan Han, Gwo-Jong Yu, Hsiao-Rong Tyan,
  Meng Chang Chen and Liang-Hua Chen*..II-742

Author Index..II-751

Contents of Volume I

Invited Talk

Pen Computing - An Overview
    Krishna Nathan I-1

Session T1A: Biometry I

Research Issues in Biometrics
    Ruud M. Bolle, Nalini K. Ratha, and S. Pankanti I-2

Automatic On-line Signature Verification
    Vishvjit S. Nalwa I-10

Integrating Faces and Fingerprints for Personal Identification
    Lin Hong and Anil Jain I-16

Automated Fingerprint Pattern Classification Error Analysis
    Weicheng Shen I-24

A High-Dimensional Indexing Scheme for Scalable Fingerprint-Based Identification
    Andrea Califano, Bob Germain, and Scott Colville I-32

Session T1B: Physics-Based Vision

Sign of Surface Curvature from Shading Images Using Neural Network
    Yuji Iwahori, Masamitsu Murakami, Robert J. Woodham and Naohiro Ishii I-40

On the Classification of Singular Points for the Global Shape from Shading Problem: A Study of the Constraints Imposed by Isophotes
    Takayuki Okatani and Koichiro Deguchi I-48

Determination of Sign of Gaussian Curvature of Surface Having General Reflectance Property
    Takayuki Okatani and Koichiro Deguchi I-56

Estimating Depth Through the Fusion of Photometric Stereo Images
    João L. Fernandes and José R. A. Torreão I-64

Out of the Dark: Using Shadows to Reconstruct 3D Surfaces
   M. Daum and G. Dudek..I-72

Session T2A: Color Vision I

Estimation of Reflection Parameters from a Color Image
   Shoji Tominaga..I-80

A Natural Norm for Color Processing
   Ron Kimmel..I-88

A Color Normalization Algorithm for Image Indexing
   In Kyu Park, Il Dong Yun and Sang Uk Lee..I-96

Adaptive Color-Image Embeddings for Database Navigation
   Yossi Rubner, Carlo Tomasi and Leonidas J. Guibas...................................I-104

A Large Capacity Steganography Using Color BMP Images
   Koichi Nozaki, Michiharu Niimi, Richard O. Eason and Eiji Kawaguchi......I-112

Session T2B: Robot Vision and Navigation

Dynamic Calibration of an Active Vision System to Compute the Ground Plane Transformation
   Fuxing Li and Michael Brady..I-120

Identification of 3D Reference Structures for Video-Based Localization
   Darius Burschka and Stefan A. Blum..I-128

Directing Robots with Visual Primitives for Navigation and Micro-manipulation
   W. B. Tong, S.K. Tso, S. Lang, G.Z. Lu and S.D. Ma...................................I-136

Combining Camera and Laser Radar for ALV Navigation
   Qi Zhang and Weikang Gu..I-144

Stereo Vision-Based Obstacle Detection for Partially Sighted People
   Stephen Se and Michael Brady..I-152

Session T3A: OCR and Applications

Evaluation and Application of Recognition Confidence in OCR
   Xiaofan Lin, Xiaoqing Ding, Youbin Chen, Jinhui Liu, and Youshou Wu........I-160

A New Nonlinear Shape Normalization Method for Off-line Handwritten
Chinese Character Recognition
    Youbin Chen, Xiaoqing Ding, Youshou Wu and Ming Chen I-168

A Novel Triangulation Procedure for Thinning Cursive Text
    Stanley S. Ipson, Muhammed Melhi and William Booth I-176

Digital Geometric Methods in Image Analysis and Compression
    Ari Gross and Longin Latecki I-184

Detection and Enhancement of Small Masses via Precision Multiscale
Analysis
    Dongwei Chen, Chun-Ming Chang and Andrew Laine I-192

A Method of Industrial Parts Surface Inspection Based on an Optics
Model
    Norifumi Katafuchi, Mutsuo Sano, Shuichi Ohara and Masashi Okudaira I-200

Poster Session I

Illumination Color from the Blurred Inter-reflection of a Reference Nose
    Mohamed Abdellatif, Yutaka Tanaka, Akio Gofuku and Isaku Nagai I-208

Shape Recovery from One Image under Multiple Light Sources
    Ying-li Tian, H.T. Tsui and S.Y. Yeung I-216

Spherical and Cylindrical Light Source Models for Shape Recovery
    Ying-li Tian, H.T. Tsui and S.Y. Yeung I-224

Polyhedral Shape Recovery Based on Interreflections
    Jun Yang, Dili Zhang, Noboru Ohnishi and Noboru Sugie I-232

Improved Supervised Color Constancy for Color Inspection
    Xuesheng Bai and Guangyou Xu I-240

Unsupervised Filtering of Munsell Spectra
    M. Hauta-Kasari, W. Wang, S. Toyooka, J. Parkkinen and R. Lenz I-248

Foveated Vision for Scene Exploration
    Naoki Oshiro, Atsushi Nishikawa, Noriaki Maru and Fumio Miyazaki I-256

Evolutionary Methods Applied to Binocular Disparity Estimation
    Carla L. Pagliari and Tim J. Dennis I-264

Robust Epipolar Geometry Estimation Using Genetic Algorithm
  Jinxiang Chai and SongDe Ma I-272

New Development of Stereo Vision: A Solution of Motion Stereo Correspondence
  M. Xie I-280

Acquisition of Three-Dimensional Information Using Omnidirectional Stereo Vision
  Atsushi Chaen, Kazumasa Yamazawa, Naokazu Yokoya and Haruo Takemura I-288

Error Analysis in Stereo Vision
  R.S. Ramakrishna and B. Vaidyanathan I-296

Detecting Targets in SAR Images: A Machine Learning Approach
  Qi Zhang, Zoran Duric and Ryszard S. Michalski I-305

Precise Matching by Robust Estimation of Deformation and Local Coherence
  Zhong-Dan Lan, Roger Mohr and Long Quan I-313

Active Viewpoint Control for Shape from Occluding Contours
  Takashi Akutsu, Kenichi Arakawa and Hiroshi Murase I-321

Point Selection: A New Comparison Scheme for Size Functions (With an Application to Monogram Recognition)
  Massimo Ferri, Patrizio Frosini, Alberto Lovato and Chiara Zambelli I-329

Sketch Up: Towards Qualitative Shape Data Management
  Costantino Collina, Massimo Ferri, Patrizio Frosini and Eleonora Porcellini I-338

Robust Matching and Hierarchical Recognition of 2-D Shapes Using "Chain of Circles"
  Jae-Moon Chung and Noboru Ohnishi I-346

Finding the Center of Rotational Symmetry from Noisy Forms
  Hyoung Seop Kim, Nachi Motomura and Seiji Ishikawa I-354

Recognition in Wavelet-Compressed Imagery
  Wei Hu and W. Brent Seales I-362

Fast Image Template and Dictionary Matching Algorithms
  Sung-Hyuk Cha I-370

Recognition of Planar Shapes Using Algebraic Invariants from Higher
Degree Implicit Polynomials
    Satish Kaveti, Eam Khwang Teoh, and Han Wang I-378

Object Recognition and Orientation via Zernike Moments
    Samer M. Abdallah, Eduardo M. Nebot and David C. Rye I-386

A Study of Zernike Moment Computing
    Simon X. Liao and Miroslaw Pawlak I-394

Query Expansion by Raw Image Features and Text Annotations in Image
Retrieval
    Kok F. Lai, Hong Zhou and Syin Chan I-402

Montage: An Image Database for the Fashion, Textile, and Clothing
Industry in Hong Kong
    Tak Kan Lau and Irwin King I-410

Auto Cameraman Via Collaborative Sensing Agents
    Qian Huang, Yuntao Cui, Supun Samarasekera and
    Michael Greiffenhagen I-418

Dynamic Adaptive Data Structures for Semantic Analysis and Synthesis
of Video Information
    V.V. Alexandrov, E.V. Laikov and B.E. Frenkel I-426

Recognition of Simple Curved Surfaces from 3D Surface Data
    Alan M. McIvor and Peter T. Waltenberg I-434

A Recursive Fitting-and-Splitting Algorithm for 3-D Object Modeling
Based on Superquadrics
    Hongbin Zha, Tsuyoshi Hoshide and Tsutomu Hasegawa I-442

Learning and Recognizing 3D Objects by Using Partial Planar Curve
Matching Method
    Jin Jia and Keiichi Abe I-450

Contour Matching Technique for 3D Object Recognition Using Kalman Filter
    M. Hanmandlu and V. Shantaram I-458

Kalman Filter Based Matching Technique for 3D Object Recognition
    M. Hanmandlu and V. Shantaram I-466

A Generating Method for 3-dimensional Knitting Cloth Shapes
    Tatsushi Funahashi, Tsuyoshi Miyazaki, Masashi Yamada,
    Hirohisa Seki and Hidenori Itoh I-474

A Fast Mesh Deformation Method to Build Spherical Representation Models
of 3D Objects
    Antonio Adán, Carlos Cerrada and Vicente Feliu I-482

Semi-automatic 3D Object Digitizing System Using Range Images
    C. Schütz, T. Jost and H. Hügli I-490

Invited Talk

Image Guided Surgical Systems
    Eric Grimson I-498

Session F1A: Biometry II

Technical Evaluation of Biometric Systems
    Brigitte Wirtz I-499

Face Recognition from Sequences Using Models of Identity
    Stephen J. McKenna and Shaogang Gong I-507

Enhancing Human Face Detection Using Motion and Active Contours
    Kin Choong Yow and Roberto Cipolla I-515

Learning Identity and Behaviour with Neural Networks
    A. Jonathan Howell and Hilary Buxton I-523

Open Sesame! Speech, Password or Key to Secure Your Door?
    Stéphane H. Maes and Homayoon S.M. Beigi I-531

Session F1B: Low-Level Processing

A Unified Framework for Image-Derived Invariants
    Yuan-Fang Wang and Ronald-Bryan O. Alferez I-542

Stereo Correspondences in Scale Space
    Christian Menard I-550

Fast Stereo Matching in Compressed Video
    Michael S. Brown and W. Brent Seales I-558

Robust Total least Squares Based Optic Flow Computation
    Alireza Bab-Hadiashar and David Suter I-566

Image Processing via the Beltrami Operator
  R. Kimmel, R. Malladi and N. Sochen I-574

Session F2A: Color Vision II

Efficient Contour Extraction in Color Images
  Aldo Cumani I-582

Color Edge Detection Using Orthogonal Polynomials
  R. Krishnamoorthi and P. Bhattacharyya I-590

Fast and Robust Segmentation of Natural Color Scenes
  Volker Rehrmann and Lutz Priese I-598

Segmentation and Tracking Using Color Mixture Models
  Yogesh Raja, Stephen J. McKenna and Shaogang Gong I-607

Object Tracking Using Adaptive Color Mixture Models
  Stephen J. McKenna, Yogesh Raja and Shaogang Gong I-615

Session F2B: Active Vision

A Learning Approach to Fixating on 3D Targets with Active Cameras
  Narayan Srinivasa and Narendra Ahuja I-623

Automatic Detection and Tracking of Human Heads Using an Active Stereo Vision System
  Cheng-Yuan Tang, Yi-Ping Hung, and Zen Chen I-632

Front Propagation and Level-Set Approach for Geodesic Active Stereovision
  Rachid Deriche, Christophe Bouvin and Olivier Faugeras I-640

A Bayes Nets-Based Prediction/Verification Scheme for Active Visual Reconstruction
  Éric Marchand and François Chaumette I-648

Actively Building Models with VIRTUE
  J. Lang and Michael R.M. Jenkin I-656

Session F3A: Face and Hand Posture Recognition

Using RBF Networks to Map GWT Ridge Images to Pose
  Alexandra Psarrou and Jonathan Tanner I-664

3-D Pose Estimation and Model Refinement of an Articulated Object from
a Monocular Image Sequence
    Nobutaka Shimada, Yoshiaki Shirai, Yoshinori Kuno and Jun Miura I-672

Face Synthesis with Arbitrary Pose and Expression from Several Images -
An Integration of Image-Based and Model-Based Approaches
    Yasuhiro Mukaigawa, Yuichi Nakamura and Yuichi Ohta I-680

Live Facial Expression Generation Based on Mixed Reality
    Hiromi T. Tanaka, Akira Ishizawa and Hiroaki Adachi I-688

Real-Time Tracking of Human Hands from a Sign-Language Image Sequence
    Kazuyuki Imagawa, Shan Lu and Seiji Igi I-698

The Model-Based Dynamic Hand Posture Identification Using Genetic
Algorithm
    Cheng-Chang Lien and Chung-Lin Huang I-706

Poster Session II

Parallel Implementation of Fractal Image Compression Using Multiple
Digital Signal Processors
    S.K. Chow, M. Gillies and S.L. Chan I-714

Comparison of Mean Field Annealing and Multiresolution Analysis in
Missing Data Estimation
    Hairong Qi, Wesley E. Snyder and Griff L. Bilbro I-722

Segmentation of MRF Based Image Using Hierarchical Genetic Algorithm
    Jin Wook Kim, Eun Yi Kim, Se Hyun Park and Hang Joon Kim I-730

Motion Compensated Color Video Classification Using Markov Random Fields
    Zoltan Kato, Ting-Chuen Pong and John Chung-Mong Lee I-738

Edge-Preserving Smoothing by Convex Minimization
    S.Z. Li, Y. H. Huang, J. S. Fu and K. L. Chan I-746

Author Index..I-755

On Typical Implementations of Hough Transform for Improving Its Performances

Jun-ichiro Hayashi[1], Kunihito Kato[2], Toshio Endoh[3], Kazuhito Murakami[1], Takashi Toriu[3] and Hiroyasu Koshimizu[1]

[1] School of Computer and Cognitive Sciences, Chukyo Univ., 101 Tokodate, Kaizu-cho, Toyota, 470-03 Japan
[2] Faculty of Engineering, Gifu Univ., 1-1 Yanagido, Gifu, 501-11 Japan
[3] Fujitsu Laboratories Ltd., 101 Tokodate, Kaizu-cho, Toyota, 470-03 Japan

Abstract

This paper proposes two typical implementations of Hough transform for improving Hough performances. RVHT(Randomized Voting Hough Transform) can provide higher detectability of shorter edge lines with lower computation cost than RHT(randomized HT) and PHT(probabilistic HT) algorithms.

Simultaneously, DTHT(Digital Template Hough Transform) can provide more higher detectability of shorter edges. Apart from the superiority of the direct line segment matching, since DTHT must prepare 4-dimensional parameter space for line segment detection, its computation must be reduced hereafter.

1 Introduction

Hough transform algorithm for detecting line segments from edge image has been and must be enhanced with respect to its performances such as the detectability and computation costs.
In order to reduce computation cost for voting many interesting algorithms had been proposed such as PHT(probabilistic Hough transform) and RHT(randomized Hough transform). First we introduced new randomized voting HT(RVHT) algorithm and clarify that RVHT is superior to these PHT and RVHT from each viewpoint of both the detectability of shorter edges and computation cost.

Basing on the principle that the edge line segment should be extracted by the direct matching between digital edge template and edge points in the image, we propose new algorithm DTHT(Digital Template Hough Transform) mainly to enforce the detectability for the shorter edge segments in the image.

In this paper we introduce these two typical implementations of Hough transform algorithms independently first, and then we discuss totally the performances of RVHT and DTHT algorithms. In chapter 2, we introduce RVHT algorithm and discuss on its property by comparing conventional RHT, and we introduce DTHT algorithm and investigate its performance in Chapter 3. It was clarified that RVHT can detect shorter edges with lower computation cost than RHT, and that DTHT can detect more shorter edges.

2 Implementation (1):RVHT algorithm

2.1 Introductions of RVHT and RHT

Fig.1 is a flow of RVHT and RHT.

First, let edge points be selected at random from the edge image. In RVHT algorithm, a locus of $\rho = x_i \cdot \cos\theta + y_i \cdot \sin\theta$ of the edge point (x_i, y_i) randomly selected is voted on the parameter space. In RHT algorithm, a vote is executed to the cell (ρ_k, θ_k) on the parameter space which is calculated from two edge points (x_i, y_i) and (x_j, y_j) randomly selected. Fig.2 shows the difference between the voting methods of these algorithms.

When one cell in the two dimensional array which represents the parameter space is greater than the threshold, the process of line detection is terminated and the cell is detected as a peak.

After this peak detection, all edge points on the line corresponding to this peak are deleted. Then, parameter space is cleared to zero, and the voting procedure is resumed. This procedure is repeated until an ending condition is satisfied.

The number of votes required to detect lines is reduced by the deleting procedure of edge points after

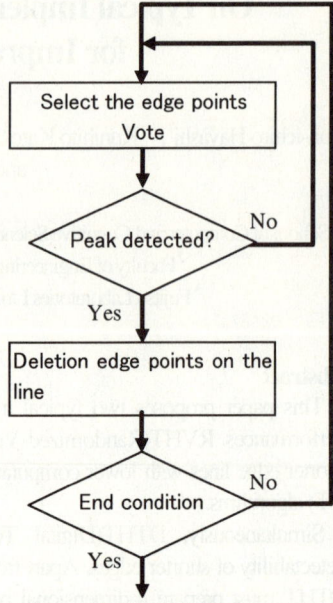

Fig.1. Flow of RVHT and RHT

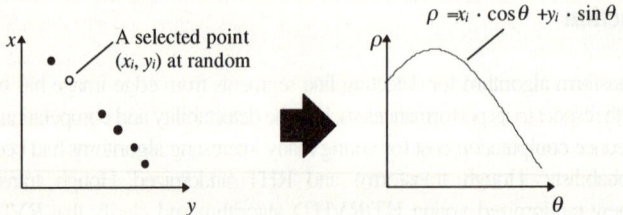

Voting method in RVHT algorithm

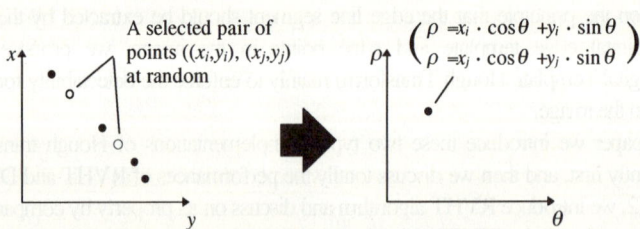

Voting method in RHT algorithm

Fig.2. Voting to the parameter space

the peak detection. And, the reduction of the number of votes improves the computation cost of the line detection. Moreover, the performance of the line detection is also improved by suppressing the suspicious peaks around the real peak, too.

2.2 Edge points deletion after line detection

After the peak detection process in the parameter space and line detection, the edge points neighboring the detected line in the input image are deleted as shown in Fig.3. If the parameter w in Fig.3 which determines the area for edge points deletion is defined larger in order to accelerate the computation cost, then the performance of line detection will be degraded because the edge points on the other lines are simultaneously deleted. On the other hand, if this parameter w is defined too small, we cannot improve computation cost because the suspicious peaks will be detected just in the same way of the normal Hough transform. Therefore, we have to set the value of this parameter w properly.

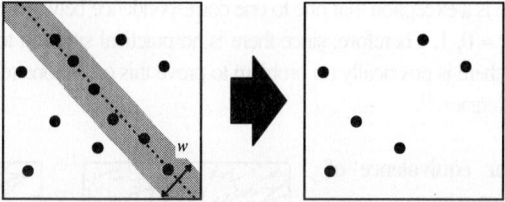

Fig.3. Edge deletion procedure after line detection

2.3 Postprocessing on parameter space

After the line detection, the renewal process of the parameter space is executed at the same time of edge points deletion. Original RHT algorithm clears all cell of the parameter space to 0. In this paper, we present three methods for parameter space renewal as follows. Fig.4 shows the difference among these methods.
1. Method 1 is to clear all cells in the parameter space to 0. This method is used in RHT as the renewal method.
2. Method 2 is to clear only the changes of the vote distribution generated by the edge points to 0 which were deleted in the edge points deletion process introduced in section 2.2.
3. Method 3 is to clear only the peak detected in parameter space to 0.

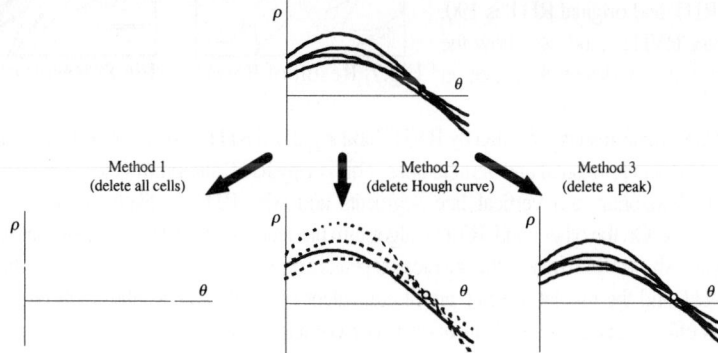

Fig.4. Renewal methods in parameter space

2.4 Relations between RVHT and RHT

Generally, the performance of RVHT and RHT are quite different. Nevertheless, the performances of RVHT and RHT become theoretically same only when the sequences of the edge points are strictly regulated as precisely given theorem as follows.

Theorem 1:

When the sequence of selection of the m edge points in RVHT is defined as $i_1, i_2, i_3, \cdots, i_m$ (provided that each edge point is not selected twice), and the sequence of selection of the pair of the edge points for RHT is defined as $(i_1, i_2) | (i_1, i_3), (i_2, i_3) | (i_1, i_4), (i_2, i_4), (i_3, i_4) | \cdots | (i_1, i_m), \cdots, (i_{m-1}, i_m)$, then line detection performance of these two algorithm becomes equivalent. We call this RHT algorithm with the specially regulated edge points as regulated RHT in this paper.

To prove this equivalence between these two algorithms, we have to prove the equivalence relation between two parameter spaces of each algorithms at the successive voting steps. Concretely, a cell in the parameter space whose value is t in RVHT has the voted value $s = {}_tC_2 = t(t-1)/2$ in the regulated RHT. There is a exception that one to one correspondence between two parameter spaces does not exist when $t = 0, 1$. Therefore, since there is no practical situation to detect a cell whose value $t = 1$ as a peak, there is practically no problem to prove this one to one relation. We explained this precisely in the reference [1].

2.5 Experiments for equivalence of detection performance

In this section, we present some experiments for comparing RVHT, the regulated RHT given in theorem 1 and RHT. We use parameter renewal methods 2 for RVHT and regulated RHT. Input image is shown in Fig. 5 which is an example of FRP microscopic image. Fig.6, 7, and 8 show the results of experiments, respectively. As for the experimental conditions, the resolution of the parameter space $\theta \times \rho$ is 180×284 for every algorithms, and, threshold for regulated RHT and original RHT is 190, and is 20 for RVHT. And, we show the calculation time to detect the lines in Table 1.

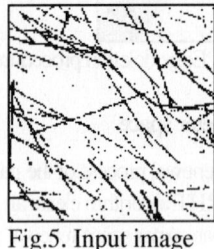
Fig.5. Input image

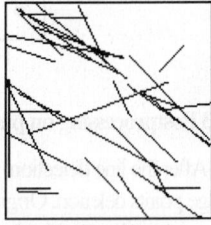
Fig.6. Result of regulated RHT

Fig.7. Result of RVHT

Fig.8. Result of RHT

We got the similar results provided by RVHT and regulated RHT. However, RHT algorithm has a tendency to detect horizontal and vertical lines. This is why RHT algorithm votes to the cells corresponding to horizontal and vertical line segments and why RHT is likely to be affected by digitization error. On the other hand, RVHT algorithm has tendency to detect some similar lines with RHT. This is why RVHT makes the suspicious peaks and why RVHT voting process draws sine curve. RVHT and the regulated RHT aren't equivalent completely under the digitized parameter space. This effect caused by the digitization is our coming subject to estimate by the quantitative analysis.

Table 1. Line detection time by each algorithms (*Sec.*)

	Regulated RHT	RVHT	Original RHT
Calculation time	10.9	1.2	13.1

3 Implementation (2):DTHT algorithm

In order to give a breakthrough to the problems of Hough transform[2],[3], a new Hough transform method called Digital Template Hough Transform(DTHT) is proposed[4]. DTHT can extract digital line segments directly from the digital image.

3.1 Basic idea

DTHT draws analog line segment that links a pair of the terminal point candidates in digital image, and the digital line segment should be extracted only when the pixel density of the digital line segment becomes greater than the threshold.
The basic procedure of DTHT can be summarized as follows:
(Step1): If the total number of edge points in the 8 neighbors is R, then basing on R, the attention pixels in the digital image are classified into (1) end point, (2) connecting point, (3) corner point, (4) branching point, (5) crossing point, and (6) inside point[4]. Every attention pixels will be classified into terminal and non-terminal pixels at the preprocessing of DTHT.
(Step1): After searching the attention pixels $b(x,y)$ from digital image, let the terminal point candidates $b_i(X_i,Y_i)$ and $b_j(X_j,Y_j)$ be selected, where $i \neq j$ and $i<j$.
(Step2): After calculating an analog line segment by eq.(1), let the digital line segment be generated as the template of the matching by analog line segment.

$$y = \frac{Y_j - Y_i}{X_j - X_i} x + Y_i - \frac{Y_j - Y_i}{X_j - X_i} X_i \qquad (1)$$

(Step3): Vote the pixel density of this digital line segment to the 4-dimensional parameter space, as shown in Fig.9.
(Step4): Detect the peaks which are greater than the threshold as the digital line segments, as shown in Fig.9.

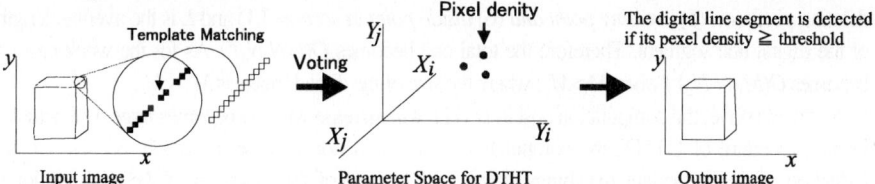

Fig.9. Template matching and voting to 4-dimensional parameter space and detecting digital line segment in DTHT(Step3~Step4)

3.2 The details of DTHT algorithm

Though we presented the basic procedure of DTHT at section 3.1, the concrete algorithms of the respective steps are given in this section.

3.2.1 Searching terminal point candidates

After being classified into terminal and non-terminal pixels, we extract the set $b(x,y)$ and, as shown in Fig.10(a), choose the terminal point candidates $b_i(X_i,Y_i)$ and $b_j(X_j,Y_j)$ ($i \neq j$, $i<j$). In order to give the efficiency, all attention pixels which appear just between $b_i(X_i,Y_i)$ and $b_j(X_j,Y_j)$ are omitted from the terminal point candidates[4].

3.2.2 Configuration of the digital template

As shown in Fig.10(b), the template of the digital line segment is introduced by eq.(1). Note that eq.(2) is used in place of eq.(1) when the slant of the analog line exceeds 45 degrees.

$$x = \frac{X_j - X_i}{Y_j - Y_i} y - \frac{X_j - X_i}{Y_j - Y_i} Y_i + X_i \qquad (2)$$

3.2.3 Digital template matching

Calculating the pixel density for the digital line segment, we can detect the digital line segment as the real segment when the pixel density exceeds the threshold (*for example*, 90%). For all digital line segments, this procedure is applied to all terminal point candidates in DTHT.

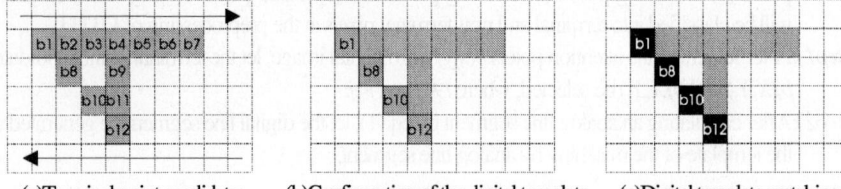

(a)Terminal point candidates   (b)Configuration of the digital template   (c)Digital template matching
Pixel density ≧ threshold

Fig.10. The details of DTHT

3.3 Order of Computation in DTHT

The order of computation for detecting all attention point candidates becomes $O((N-N_d)^2)$, and that for the template matching becomes $O(L)$, where N is the number of all pixels and N_d is the number of deleting pixels(*ex.* (5) *crossing point and* (6) *inside point in section* 3.1) and L is the average length of the digital line segment. Therefore the total one becomes $O(L(N-N_d)^2)$. As for the worst case, it becomes $O(M_x(N-N_d)^2)$ when $M_x>M_y$, where the size of the digital image is $M_x \times M_y$.

As stated above, the computation cost in DTHT will increase when N becomes large. But actually in the procedure of DTHT, the computation cost would not increase so rapidly because of the reduction of the redundant matchings stated in 3.2.1 and of the omission of the 4-dimensional parameter space.

3.4 Experimental Result

We show an example of experiments in Fig.'s 11-12. Fig.11(a) is an original image which is gray scale and of which size is 640 × 480. Fig.11(b) is an input image (*the number of edge point*=15981).

Fig.11(c) is an output image of Hough transform.Fig.11(d) is an output image of DTHT.

In the normal Hough transform, shorter lines are not detected. But in the output images of DTHT(*see* Fig.11(d)), shorter segments can be detected(*ex. Top lines of books*). And in the normal Hough transform, a curved line detection is very difficult. But in DTHT, it can be detected as a set of shorter line segments.

Table 2 shows the processing time and Table 3 the number of the detected lines in Fig.11(c)(d) and Fig.12(b)(c)(*by Sun SPARC Station*4). In Fig.11, processing time became longer than the Hough transform because the number of edge points were large. But in Fig12, the processing time of DTHT became almost the same as the Hough transform because the number of edge points were small. By the way, if edge points are not classified, the order of computation becomes $O(LN^2)$ and processing time become extremely longer than DTHT where the edge points were classified.

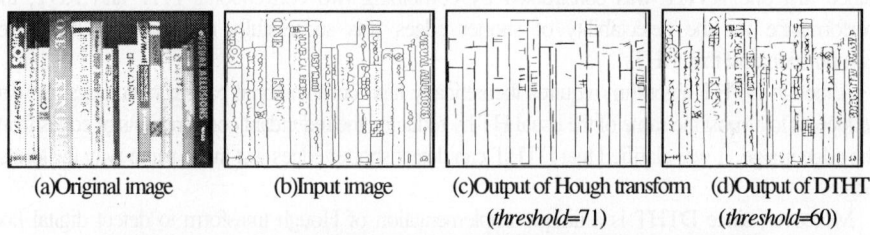

(a)Original image　　(b)Input image　　(c)Output of Hough transform　(d)Output of DTHT
　　　　　　　　　　　　　　　　　　　(*threshold=71*)　　　　(*threshold=60*)

Fig.11. Experimental results(*image size=640 × 480, attention pixel=15981*)

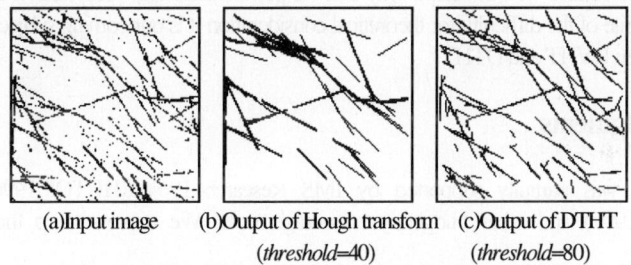

(a)Input image　　(b)Output of Hough transform　(c)Output of DTHT
　　　　　　　　　(*threshold=40*)　　　　(*threshold=80*)

Fig.12. Experimental results(*image size=128 × 128, attention pixel=2139*)

Table 2. Line detection time by each algorithms　(*Sec.*)

	Hough transform	DTHT	Not classified DTHT
Calculation time(Fig.11)	293	750	2697
Calculation time(Fig.12)	4	5	13

Table 3. Line detection segments by each algorithms (*Segments*)

	Hough transform	DTHT	Not classified DTHT
Detected segments(Fig.11)	550	1278	2024
Detected segments(Fig.12)	159	101	233

3.5 Conclusions of DTHT

We proposed a new Hough transform method, Digital Template Hough Transform DTHT, that extracts digital line segments directly from the edge image. As for the countermeasure to the threshold selection problem in Hough transform, we introduced a new pixel density measure between the end point and the another end point of the line segment. This pixel density measure enabled to evaluate fairly both shorter and longer line segments.

4 Discussion & Conclusions

In order to improve the performances of Hough transform, we introduced two typical algorithms. Since first one RVHT was constructed by combining two conventional PHT and RHT, the performance for the detectability of shorter edges was successfully realized with the lower computation cost than RHT.

Although DTHT algorithm requires theoretically high combinatorial pattern matchings, we can reduced it to almost the same of the usual HT by reducing the redundant combinations of edges from the edge image. It was clarified that DTHT can detect shorter edges just in the same way for longer edges.

Moreover, since DTHT is a unique implementation of Hough transform to detect digital line segments directly from edge image, it will be strongly expected to manage 4-dimensional DTHT Hough space to reduce well in such a way that the performance and computation cost should be balanced. In spite of the difficulty for theoretical consideration it is our coming subject to clarify the relation between RVHT and DTHT.

Acknowledgments

This paper was partially supported by IMS Research Project(HUTOP,1996-1997), IPA Project(1996-1997) and other budgetary encouragements. We express deep thanks to those supports.

References

[1] K. Kato, T. Endoh, K. Murakami, T. Toriu, H. Koshimizu. : "Theoretical & Experimental Considerations on the Detectability of Lines for Randomized Voting Hough Transform", Technical report of IEICE Vol. 96, No. 384, (PRMU96-87), pp.53-60(1996. 11)(in Japanese).
[2] T. Wada, T. Matsuyama. : "Frontiers of Object Detection by Hough Transform", J.IPS,36,3, pp.253-263(1995) (in Japanese).
[3] M. Morimoto, and H. Koshimizu. : "Hough Transform and New Pattern Sensing - Fundamentals -", J.SICE, vol.35, No.11, pp.869-877(1996.11) (in Japanese).
[4] J. Hayashi, K. Kato, K. Murakami, and H. Koshimizu. : "A Method for Line Segment Detection by Digital Template Hough Transform(DTHT)", Proc. of FCV'97, pp.39-46(Korea) (1997. 1).

Hierarchical Segmentation and Representation with Dynamic Link Architecture Neural Network

Yunqiang Chen , SongDe Ma

National Laboratory of Pattern Recognition, Chinese Academy of Science
Beijing P.O.Box 2728, 100080, P.R.China

Abstract. A segmentation scheme based on tracing objects through scale space is proposed. For analyzing the image structure in scale space, a hierarchical neural-oscillator network is proposed. Its intrinsic dynamics provides a natural and reliable representation of multiple segmented patterns. The pyramid-like structure and the local interactions between the neurons can easily integrate the local context information of pixels and global image structure. This model assumes that neural cells show oscillatory activities and the segmentation result is coded by synchronization of activities. This representation framework for neural network suggested by time correlation theory receives direct experimental support from neurophysiology experiment. Computer simulations demonstrate that the network can detect the topological character of images within several oscillatory cycles. Experiment on real image showed the model's ability of segmentation.

1 Introduction

Despite a wide variety of different segmentation techniques [1], no general theory of segmentation exists. Classical segmentation by marking boundaries is too limited, not being able to provide a natural basis for the utilization of segmentation. More potent segmentation systems can be based on the idea of tags: all parts of one segment are simply labeled with the same tag [2]. In this paper, segmentation with tags is explored in the biological context. We consider this task in the framework of the recently findings in cognitive science and the evidences from neurophysiology, then propose a new scheme based on Time Correlation Theory to segment and represent the segmentation result by synchronization of oscillators.

In cognitive science, many experiments on tachistoscopic perception of visual stimuli demonstrated that the visual system can extract the global topological properties although the detailed structure of the stimulus remains vague and amorphous [3]. These experiments suggest that the human vision system can achieve some kinds of topology detection even before edges are detected. These experimental facts are therefore relevant to Gibson's theory of invariance detection.

However, it has been demonstrated by Minsky from the perspective of the general theory of computation that the nature of topological invariant makes its computation difficult [4]. How can the visual system detects the topological structure in early vision even before the edges are extracted? Here we exploit a dynamic link architecture neural network to solve this problem. Computer simulation shows its ability to segment the scene into several connected regions in several oscillatory cycles.

Theoretical investigations of brain functions and perceptual organization suggest the mechanism of temporal correlation as a representational framework [5]. In particular, the correlation theory of von der Malsburg [5] asserts that an object is represented by the temporal correlation of the firing activities of the scattered cells coding different features of the object. A natural way of encoding temporal correlation is to use neural oscillations, whereby each oscillator encodes some features of an object. In this scheme, each segment is represented by a group of oscillators (points) that show synchrony (phase-locking with zero phase shift) of the oscillations, while different objects are represented by different groups whose oscillations are desynchronized from each other.

More recently, in neurophysiology evidences increasingly support this type of temporal coding mechanism for coherent object perception in early vision [6,7]. The discovery of synchronous oscillations in the visual cortex has triggered much interest in simulating the experimental results and in exploring oscillatory correlation to solve the problem of feature organization [8,9]. Most of these models rely on long-range (all-to-all) connections to achieve phase synchrony. While in our context, it is important to use local connection to achieve synchrony since long-range connections will inevitably discard the local connectivity information of the image [9]. Here we adopt the LEGION model presented by David Terman and DeLiang Wang [10] which uses locally excitatory connection to achieve synchrony in homogeneous region and global inhibitory connection to desynchronize different objects.

It has been demonstrated by David Terman that, for binary image, this kind of model can achieve global synchrony within connected regions and desynchrony between disconnected ones [10]. We extended this kind of model in three aspects to suit the need of hierarchical gray-level image segmentation. It can be seen later, within several cycles, dynamic segmentation result is attained.

2 Model Description

We now introduce the specific connectivity scheme needed for our network model. First we introduce the scale space theory. then we change the LEGION model to suit hierarchical gray-level image segmentation by modifying its connection schemes.

2.1 Scale Space Theory

In practice, the relevant details of images exist only over a restricted range of scale. A scene can be segmented into objects which are composed of several parts, which contain even small parts. Hence it is important to study the dependence of image structure on the level of resolution. It seems clear that visual perception treats images on several levels of resolution simultaneously. One nice way to interpolate between these different scales is to utilize the two-dimensional diffusion equation [11].

We can create a family of images $f(x,y,t)$, by setting $f(x,y;t=0)=f(x,y)$ and using the diffusion equation $\partial_t f(x,y;t) = [\partial_{xx}^2 + \partial_{yy}^2] f(x,y;t)$. The solution $f(x,y;t)$ of the diffusion equation at scale t can be obtained from $f(x,y;t=0)$ through convolution with the gaussian kernel

$$g(x, y) = 1/(4\pi t) \cdot \exp[-(x^2 + y^2)/(4t)]$$

It can be interpreted as sampling of input space by neurons with different sized receptive fields. During the process of resolution reduction by diffusion, local signal variations smooth out as "time" t proceeds. Extrema disappear one after the other, and finally only the strongest signal variations survive. So a hierarchical ordering of extrema is created.

Several schemes have been proposed to exploit the structure of scale space [12,13,14]. Most approaches deal with the behavior of edge in scale space or need a very dense sampling of scale space and are therefore computationally expensive. We propose here a simple neuronal mechanism for integrating discrete scale space data simultaneously: trace objects in several discrete scales by neuronal oscillators.

2.2 Neural Oscillator Modal

We develop our neural oscillator from the LEGION model [10] and extend it in three aspect. The basic neural oscillator i is defined in the simplest form as a feedback loop between an excitatory unit x_i and an inhibitory unit y_i:

$$x'_i = 3x_i - x_i^3 + 2 - y_i + \rho + P_i H(\sum_{k \in N(i)} \delta_{ik} - \theta_c) + S_i \tag{1a}$$

$$y'_i = \varepsilon \cdot (\gamma \cdot (1 + \tanh(\frac{x_i}{\beta})) - y_i) \tag{1b}$$

ρ denotes the amplitude of a Gaussian noise term. Si denotes coupling from other oscillators in the network. Without any coupling or noise, (1) corresponds to a standard relaxation oscillator. The x-nullcline of (1) is a cubic curve, while the y-nullcline is a sigmoid function, as shown in Fig. 1. If $P_i H(\sum \delta_{ik} - \theta_c) > 0$, (1) is oscillatory. The system displays limit-cycle behavior. The parameter γ control the rate of silent phase and active phase. If $P_i H(\sum \delta_{ik} - \theta_c) \leq 0$, (1) produces no oscillations (inactive), but it can be activated by excitatory connection from other oscillators. These two kinds of behavior are shown in Fig. 1. $P_i H(\bullet)$ is introduced to control the autonomous oscillation of neurons. P_i is the attention potential which control the extent (scale) to which the details are ignored. If we concentrate on the nth scale, $P_i=0.2>0$ for $i=n$, $P_i=0$ for $i \neq n$. H stands for heaviside step function, which is defined as $H(v)=1$ if $v \geq 0$, and $H(v)=0$ if $v<0$. $\delta_{ik}=1$ if the gray-level difference between i and k is less than the threshold θ_g, otherwise, $\delta_{ik}=0$. For the points around the edges, $H(\sum \delta_{ik} - \theta_c) = 0$. Then they are prohibited from oscillating autonomously. In this way, the noise and small regions are suppressed. Only for those points which are in the attention-directed level and at the same time are within homogeneous regions, $P_i H(\sum \delta_{ik} - \theta_c) > 0$. So they are oscillatory and can activate other points of the same gray-level by excitatory connection. The points around the edges and those in other scales can not be activated unless they receive excitatory connection from others. But no oscillator can retain active for a long time.

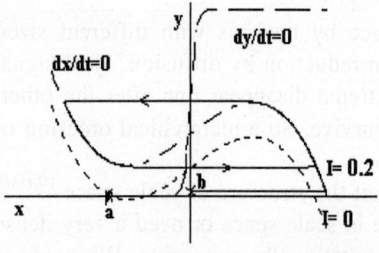

When $P_i*H>0$, the x-nullcline (dx/dt=0) intercept with y-nullcline (dy/dt=0) at one instable point, (3) is oscillatory. Starting at a random point such as b, it quickly converges to the stable trajectory of a limit cycle which alternates between the left branch and the right branch of the cubic.

When $P_i*H = 0$, the x-nullcline intercept with y-nullcline at one stable point "a" and one unstable point, so it produce no oscillation.
($I=P*H+S$)

Fig. 1 Nullclines and periodic orbit of a single oscillator

The oscillator model (1) may be interpreted as a model for the spiking behavior of a single neuron or the mean field approximation to a network of excitatory and inhibitory neurons.

2.3 Structure and Oscillatory Connections of Neural Network

We now present the structure of the neural network for the hierarchical segmentation. For the representation of scale space, we use a pyramidal data structure which looks like the quadtree graph structure. But the classic quadtree structure yields "blocky" segmentation [14]. In order to avoid these blocky artifacts, we introduce the intra-scale lateral connection. The structure is shown in Fig. 2. The connection scheme and parameters decide the segmentation effect.

As illustrated in Fig. 2, each oscillator is connected with its four nearest neighbors in the same scale by bi-directional excitatory connections. Neurons in coarser level are connected with four finer level neurons by unidirectional excitatory links. All the neurons give their output to the global inhibitor, and receive the inhibition from global inhibitor. The coupling term S_i in (1) is given by:

$$S_i = \sum_{k \in N(i)} W_{ik} S_\infty(x_k, \theta_x) - W_z S_\infty(z, \theta_{xz})$$

where $S_\infty(x, \theta) = \dfrac{1}{1 + \exp[-K(x - \theta)]}$

$W_{ik} = 6 \bullet (1 - |g_i - g_k| / \theta_g)$ is the excitatory connection (synaptic) weight from oscillator i to oscillator k, W_{ik} is much greater when the two points are of the same gray-level. θ is a threshold above which an oscillator can affect its neighbors. W_z is the weight of inhibition from the global inhibitor z, whose activity is defined as $dz/dt = \phi(\sigma_\infty - z)$. Here $\sigma_\infty=0$ if $x_i<\theta_{zz}$ for every oscillator, and $\sigma_\infty=1$ if $x_i \geq \theta_{zz}$ for at least one oscillator i. The activity of the global inhibitor can be described as follows. If the activity of every oscillator is below the threshold, then the global inhibitor will not receive any input. Then all the oscillators will not receive any inhibition from the inhibitor. Once an oscillator is in the active phase, it triggers the global inhibitor. At the same time, the active oscillator spreads its activation to its nearest neighbors, and from then to its further neighbors, which counter the inhibition from the global inhibitor. This will lead the whole connected region (and from coarser level down to finer levels) to be activated and inhibit other regions from oscillate autonomously, unless this region return to its inactive phase. Hence, we can describe this segmentation scheme as local cooperation and global competition.

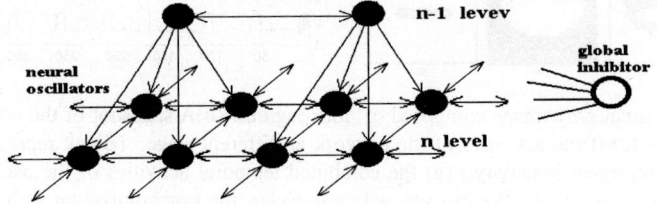

Fig. 2 Hierarchical Structure of the Neural Network

There are two kinds of basic oscillator behavior in the hierarchical network:
1. In the attention-directed scale, the neural oscillators in homogeneous region are oscillatory, when there are no activated oscillators, the first activated oscillator will lead all the oscillator in the region (till the finest level) to be activated.
2. Other oscillators will be excited if they are similar in gray-level with activated neighbors or its father. Because each oscillator can be activated only once in a cycle, it will be included in the most gray-similar region.

To ensure that each oscillator, whether it is in the interior or on the boundary of the network, has equal overall connection weights from its neighbors, Wang [9,15] proposed a mechanism called dynamic normalization. Simulations reveal that dynamic normalization will help the synchronization.

2.4 Computer Simulation

To illustrate how this model is used for segmentation, we simulate a one-level 32×32 network. We generated a three-objects image, shown in Fig. 3 (a): two blocks in a large ellipse and the background. The amplitude ρ of the Gaussian noise was set to 0.02. The parameter values in the differential equations were set as follows: $\varepsilon=0.02$, $\gamma=6.0$, $\beta=0.1$, $K=50$, $\theta_x=-0.5$, $\theta_{zf}=0.1$, $\theta_g=50$, $\phi=3.0$, $W_z=3.0$. These parameter decided the dynamics of the neurons. The total effective connections were normalized to 6.0. The initial status of all the oscillators were randomly initialized.

Fig. 3(b)-(f) show the instantaneous activities of the network at various stages of dynamic evolution. The activities were largely random at first (Fig. 3(b)), but after several cycles, the image was segmented into four desynchronized groups of oscillators. The patterns shown in Fig. 3(a) were completely segmented after just three oscillatory cycles. The four blocks of oscillators *popped out* one after another. As this simulation has shown, this activity distribution, in contrast to the conventional neural network paradigm, is not static, even for a constant input. The pattern of connectivity of the image was modulated to the timing of activities of the neurons.

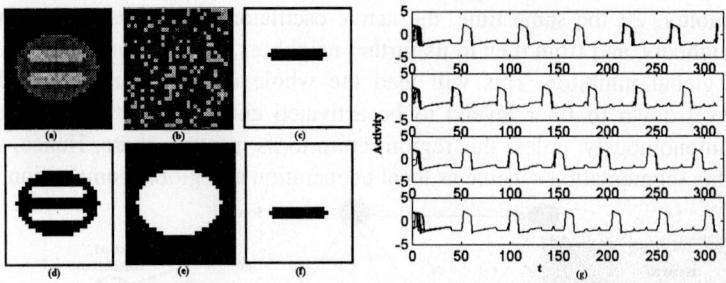

Fig. 3 (a) The input gray image composed of four regions. (b) A snapshot of the activities of the initial status (c)-(f) the activities of the network at different stage. (Black represents high activity, white represent inactivity.) (g) the combined temporal activities of the oscillators in the four regions, respectively. We can see, within 4 cycles, the synchronization within regions and desynchronization between different regions are achieved.

3 Experiment on Real Image

Computer simulation of the network on real image data is expensive. In order to test the segmentation properties of the whole neural network, we approximate the dynamics of the neural oscillators to their basic locking behavior. First, only the oscillator in homogeneous region can lead the whole region to be activated. Second, local coupling implies that two neighboring pixels are grouped into a region if their gray difference is below a threshold. Third, the oscillator which is connected with several desynchronized oscillators will synchronize with the most similar one.

Based on these simplifications, we segmented the gray-level image of a squirrel. The original image was smoothed by diffusion equation and reduced the resolution by half at each level. The five levels of the image were shown in Fig. 4.

Then we integrated the hierarchical data to segment this image. The attention potential decides to which extent we ignore the details. The parameters are set as following: $N(i)$ is set to be the nearest 8 points, θ_g at the attention-directed scale is 0.05, θ_g at the non-attention-directed scale is 0.08. We represented the segmentation result with different gray-levels. The pure black color means the oscillator corresponding to that point will never be excited at all. These never activated oscillators mean the abnormality. They could not be excited by their neighbors because the corresponding points' characters are not consistent with the region around them, and also they are so small to be considered as a region. Therefore the presence of black holes can be used to control further scrutiny of the scene. And the presence of the black holes, such as the eye in the squirrel body, may help the recognition progress in early level before we get the detail structure. Other gray values indicate different synchronized groups of oscillators. The finest level result was shown in Fig. 5. When attention was directed to the 1st level, the whole image was segmented into 3 groups of oscillators with black holes in them, the tail, the body and the background. When we noticed finer resolution level 2nd, the image was segmented into 5 parts. We got much more detail information of the image. We got the abdomen of the squirrel , and the tail was divided into two parts. When the attention was directed to finer scale, the image was segmented to many more parts. Hence we can get the hierarchical segmentation of the image.

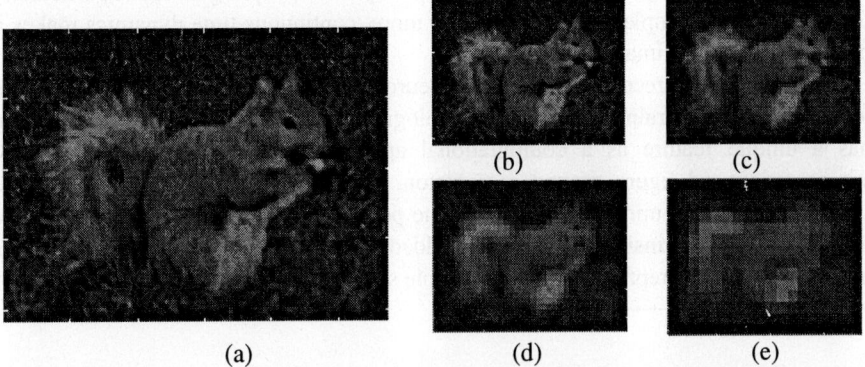

Fig. 4 The image is represented at several levels of resolution. The structure disappears gradually from finer to coarser level. Here, (a)-(e) represent level 5-1. Level 1 is the coarsest level in our experiment, level 5 is the original image (the finest level).

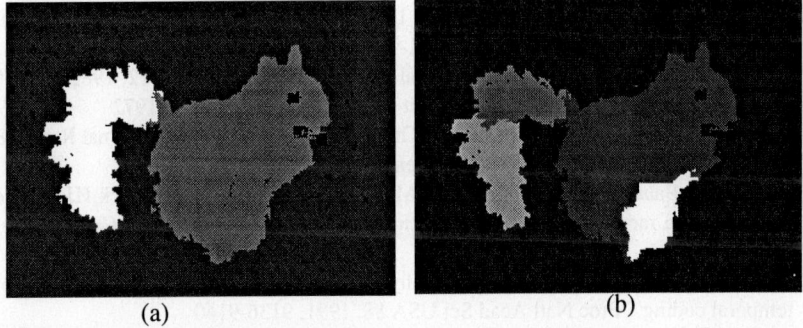

Fig. 5 (a) .The segmentation result when attention was directed to level 1. The image was segmented to 3 parts. The body of the squirrel, the tail, and the background. The black holes, such as the eye, denote that the oscillators there will not oscillate at all. (b). The result when attention was directed to level 2. It was segmented to 5 parts

From the experimental result, as we can see, we can get the precise boundary of large region and refrain the noise and small regions at the same time, because of the integration of multi-level data. And with attention potential, this framework can easily provide the hierarchical segmentation and representation of images.

4 Discussion

In this paper a new representation framework for neural network is introduced for segmentation of two-dimensional input data. It is based on the approximation of iso-intensity sheets in scale space by synchronizing neural oscillators. By integrating discrete scale space data, we refrain the noise and small detail without losing the precise boundary information.

Our method is performed by a population of active oscillators. This massively parallel network of dynamic systems with mainly local coupling makes it particularly feasible for VLSI implementation. Further more, continuous-time dynamics makes it possible to segment image in real time.

This model have received many direct neurophysiology support as a representation method in human brain. In addition to biological plausibility, oscillatory correlation has a unique feature as a computational approach to the engineering of scene segmentation and figure ground segregation. Due to the nature of oscillations, no single object can dominate and suppress the perception of the rest of the scene for a long time. The intrinsic dynamics embedded in oscillatory correlation provides a natural and reliable representation of multiple segmented pattern.

References

1. Haralick, R.M., Shapiro, L.G., "Survey: Image segmentation techniques." CVGIP 29, 1985, 100-132
2. Geman G, Geman S. Graffigne C, Dong P "Boundary detection by constraint optimization." IEEE PAMI 12, 1990, 721-741
3. Chen. L., "Topological Structure in Visual Perception" Science. 218. 12, 1982, 699-700
4. M. Minsky and S.Papert, Perceptrons, Mit Press, Cambridge, Mass., 1972
5. C. von der Malsburg. "The Correlation Theory of Brain Functions." Internal Report 81-2, Max-Planck-Institute for Biophysical Chemistry. Göttingen, FRG 1981
6. Eckhorn R, Bauer R, Jordan W, Brosch M, Kruse W, Munk M, Reitboeck HJ, "Coherent oscillations: a mechanism of feature linking in the visual cortex?" Biol Cybern 60, 1988 121-130
7. Engel AK, Konig P, Singer W "Direct physiological evidence for scene segmentation by temporal coding." Proc Natl Acad Sci USA 88, 1991, 9136-9140
8. P. König and T. B. Schillen, "Stimulus-dependent assembly formation of oscillatory responses: I. Synchronization." Neural Comput., 3, 1991, 155-166
9. D.L. Wang, "Emergent synchrony in locally coupled neural oscillators." IEEE Trans. Neural Networks, vol. 6, 1995, 941-949
10. David Terman, DeLiang Wang, "Global competition and local cooperation in a neetwork of neural oscillators", Physica 81, 1995, 148-176
11. Koenderinck, J.J "The Structure of Images." Biol. Cybern. 50, 1993, 363-370
12. Lifshitz, L.M., Pizer, S.M., "A Multiresolution Hierarchical Approach to Image Segmentation Based on Intensity Extrema." IEEE PAMI 12, 1990, 529-540
13. Lindeberg, T. "Detecting Salient Blob-Like Image Structures and Their Scales with a Scale-Space Primal Sketch: A Method for Focus-of-Attention." Int. J. Comp. Vis. 11:3, 1993, 283-318
14. C. Bouman, M. Sharpiro. "A multiscale random field model for bayesian image segmentation" IEEE Trans. On Image Processing. Vol. 3, No 2, 1994, 162-177
15. D.L. Wang, "Modeling global synchrony in the visual cortex by locally coupled neural oscillators." Proc. 15th Ann. Conf. Cognit. Sci. Soc., 1993, 1058-1063

Perceptually Consistent Segmentation of Texture Using Multiple Channel Filter*

Nan Zhang and Wee Kheng Leow

Department of Information Systems and Computer Science
National University of Singapore, Lower Kent Ridge Road, Singapore 119260
email: zhangnan, leowwk@iscs.nus.edu.sg

Abstract. Texture segmentation aims at dividing an image into perceptually uniform regions each containing a distinct texture. In images of natural scene, texture in a region can change gradually in scale and orientation due to perspective distortion. A naive segmentation method may erroneously group image patches with the same texture but slowly varying scales and orientations into distinct regions. This paper describes a novel segmentation method which takes into account the rate of change of texture scale and orientation. The method extracts scale and orientation information from the outputs of a set of Gabor filters, and use them to group image patches into perceptually uniform texture regions.

1 Introduction

Texture segmentation aims at dividing an image into perceptually uniform regions each containing a distinct texture. It is a very difficult task for images of natural scene. In these images, texture in a perceptually uniform region can change gradually in scale and orientation due to perspective distortion. Take Fig. 1 for example. This image contains several perceptually uniform regions each covered by one type of material: bricks or pebbles. Texture in the image decreases gradually in scale with increasing viewing distance from the bottom of the image to the top. Orientation of the brick texture also changes gradually from the left of the image to the right. Texture boundaries are perceived at locations where the type of texture differs (e.g., between bricks and pebbles) or when the texture scale or orientation changes abruptly (e.g., between the top and the bottom brick patterns).

A naive segmentation method may erroneously group image patches with the same texture but slowly varying scales and orientations into distinct regions. On the other hand, it may erroneously group neighboring patches with the same texture but sharp changes in scales or orientations into the same region. To correctly segment an image, a texture segmentation algorithm has to take into account the rate of change of texture scale and orientation. So far, existing works on image segmentation have not considered this issue. This paper presents a novel method that performs perceptually consistent segmentation of images containing natural texture.

---

* This research is supported by NUS Academic Research Grant RP950656 and NUS Research Scholarship HD950345.

Fig. 1. Image of floor containing several perceptually uniform regions each covered by a different type of texture: bricks or pebbles. Texture in a region can vary gradually in scale and orientation.

2 Related Works

Texture segmentation of images is performed based on texture features which can be divided into 4 categories: Markov random field, local statistics, Fourier spectrum, and Gabor filter magnitudes.

The Markov random field (MRF) model assumes that texture is formed by a stochastic process such that a pixel's intensity is given by a weighted sum of the intensities of neighboring pixels and a noise term [7, 10]. It is very difficult for MRF model to be scale- and orientation-invariant since it would require a different set of weights for every possible scales and orientations.

Local statistics such as the means, variances, and 3rd-order moments computed from the eigenvalues of covariance matrices have also been used as texture features [8, 9]. Although some of the features are orientation-invariant, they lack the information needed to determine whether the change of scale and orientation is gradual or abrupt.

Fourier spectrum can capture texture's frequency and orientation [4, 7]. Unfortunately, it is not localized in the spatial domain. It is impossible to extract a textured region's Fourier spectrum unless the region has already been segmented.

Gabor filters have the advantage of being optimally localized simultaneously in the spatial and the spatial frequency domains. Existing segmentation methods based on Gabor filters [1, 2, 3, 6] use only some of the filter channels, typically those with large output magnitudes. The invariant segmentation method described in this paper, however, uses all the filter channels so as to better estimate the rate of change of texture frequency and orientation. Existing methods handle only images containing *uniform texture*, i.e., texture that does not vary in scale and orientation within a perceptually uniform region. Typically, these images contain juxtaposed plane texture taken from the Brodatz album. Since these methods do not take into consideration gradual variation of texture scale and orientation, they cannot correctly segment images containing non-uniform texture. In contrast, the invariant method can segment images of natural scene that contain multiple non-uniform texture.

3 Multi-channel Texture Feature Extraction

The Gabor function $h(x,y)$ at image position (x,y) is a complex sinusoidal grating modulated by an oriented Gaussian function $g(x',y')$ [2]:

$$h(x,y) = g(x',y')\exp(2\pi j f x') \qquad (1)$$
$$g(x',y') = \frac{1}{2\pi\lambda\sigma^2}\exp\left[-\frac{(x'/\lambda)^2 + y'^2}{2\sigma^2}\right]$$

where $(x',y') = (x\cos\theta + y\sin\theta, -x\sin\theta + y\cos\theta)$ are rotated coordinates oriented at angle θ from the x-axis, λ is the aspect ratio, and σ is the scale parameter. The Gabor function has radial frequency f and orientation θ:

$$f = \sqrt{U^2 + V^2} \qquad \theta = \tan^{-1}\left(\frac{V}{U}\right) \qquad (2)$$

where U and V are spatial frequencies along the x- and the y-directions. Its frequency (octave) bandwidth B and orientation (radian) bandwidth Ω at half-peak are:

$$B = \log_2\left[\frac{\pi f \lambda \sigma + \alpha}{\pi f \lambda \sigma - \alpha}\right] \qquad \Omega = 2\tan^{-1}\left(\frac{\alpha}{\pi f \sigma}\right) \qquad (3)$$

where $\alpha = \sqrt{(\ln 2)/2}$. The range of spatial frequencies within the frequency and orientation bandwidths is called the *half-peak support*.

A multi-channel approach is adopted to represent texture. An input image $I(x,y)$ is filtered by a set of Gabor filters with different frequency f and orientation θ:

$$\begin{aligned} k_{c,f\theta}(x,y) &= h_{c,f\theta}(x,y) * I(x,y) \\ k_{s,f\theta}(x,y) &= h_{s,f\theta}(x,y) * I(x,y) \end{aligned} \qquad (4)$$

where $h_{c,f\theta}(x,y)$ and $h_{s,f\theta}(x,y)$ are the real and the imaginary components of Gabor function (Eq. 1). The magnitudes of the filters' outputs are given by:

$$k_{f\theta}(x,y) = \sqrt{k_{c,f\theta}^2(x,y) + k_{s,f\theta}^2(x,y)}. \qquad (5)$$

After Gabor filtering, the channels' outputs are smoothed by Gaussian filters to remove local variations introduced by the sinusoidal terms in the Gabor functions. The Gaussians have the same orientations and aspect ratios as the Gabors. Their scale parameters σ are set at four times those of the corresponding Gabors. After smoothing, the Gabor output magnitudes at each pixel location form the *texture feature vector* $\mathbf{k}(x,y)$ in the f-θ space.

Unlike existing methods, the Gabor filters' frequencies f_i and orientations θ_i are chosen such that there are some overlaps in the filters' half-peak supports:

$$f_i = \frac{f_m}{2^{i\beta B}} \qquad \theta_i = i\beta\Omega \qquad (6)$$

where f_m is the maximum spatial frequency and β determines the amount of overlap in the filters' supports. The amount of overlap is minimum when $\beta = 1$

and increases with decreasing β. In the current implementation, $f_m = 0.3$, $B = 0.75$ octave, $\Omega = 45°$, and $\beta = 0.5$. A total of 48 filters are used, with 6 spatial frequencies and 8 orientations.

In contrast to existing methods which use only some of the Gabor channels' outputs, the invariant segmentation method uses the *pattern* of outputs from all the channels as the texture feature. This method of representing texture feature is known as *distributed representation* in the Neural Networks literatures [5]. Distributed representation has the advantage of representing a large number of different texture types using a small number of channels' outputs. It also ensures that the output pattern will not be severely altered by slight changes in texture frequency and orientation, thereby facilitating a closer match between neighboring texture patches.

4 Invariant Texture Segmentation

After feature extraction, the texture feature vectors at each pixel location are grouped into regions in 2 stages: (1) region grouping, and (2) region merging.

4.1 Region Grouping

In the region grouping stage, similar texture feature vectors at neighboring locations are first grouped into *seed regions*. This operation reduces the number of feature vectors that will be involved in the next stage. Each seed region R_i is characterized by a *peak* vector $\mathbf{P}_i$ and a *mean* vector $\mathbf{M}_i$. Initially, one seed region R_1 is created which contains the feature vector $\mathbf{k}(0,0)$ at location $(0,0)$, and $\mathbf{P}_1$ and $\mathbf{M}_1$ are set to $\mathbf{k}(0,0)$. Subsequently, a feature vector $\mathbf{k}(x,y)$ is grouped into region R_i if

- it is near enough: there is a vector $\mathbf{k}'(x',y')$ in R_i that is a 4-neighbor of $\mathbf{k}(x,y)$ in the x-y space,
- it is similar enough: similarity $C(\mathbf{k},\mathbf{P}_i) > \Gamma_\mathrm{P}$, and $C(\mathbf{k},\mathbf{M}_i) > \Gamma_\mathrm{M}$,

where $\Gamma_\mathrm{P} = \cos 5°$ and $\Gamma_\mathrm{M} = \cos 3°$ are constant thresholds, and C is the cosine similarity:

$$C(\mathbf{k}_1, \mathbf{k}_2) = \frac{\mathbf{k}_1 \cdot \mathbf{k}_2}{\|\mathbf{k}_1\| \, \|\mathbf{k}_2\|} \, . \tag{7}$$

If a vector can be grouped into more than one region, then one of the regions is arbitrarily chosen to contain the vector. After grouping $\mathbf{k}$ into region R_i, the peak vector $\mathbf{P}_i$ and mean vector $\mathbf{M}_i$ are updated as follows:

$$\mathbf{P}_i = \max_{\mathbf{k}_j \in R_i} \mathbf{k}_j \qquad \mathbf{M}_i = \frac{1}{n} \sum_{\mathbf{k}_j \in R_i} \mathbf{k}_j \tag{8}$$

where n is the number of vectors in R_i.

4.2 Region Merging

After region grouping, seed regions are merged in the merging stage into larger regions that are more consistent with human's perception of the input image. Two regions that share a common boundary are merged if their feature vectors are similar enough. The similarity measure used in this stage is slightly more complex because it has to take into account gradual variations of texture scale and orientation. It can be shown that an isotropic scaling or a rotation of a texture image results in only a shift of the Gabor output pattern in the f-θ space. In an image of natural scene, however, changes in scale may not be isotropic due to perspective distortion. As a result, some parts of the Gabor output pattern may shift more than other parts. Therefore, to compute the similarity between two feature vectors, it is necessary to consider non-uniform shifting of Gabor output pattern.

The similarity measure between feature vectors are defined as follows. Suppose that feature vector $\mathbf{k}$ has a peak at (f, θ). Form a *local peak vector* by taking the feature vector components $k_{f'\theta'}$ in the 3×3 neighborhood of (f, θ). Now, consider two feature vectors $\mathbf{k}_1$ and $\mathbf{k}_2$ each having a set of local peak vectors $\{\mathbf{p}_i\}$ and $\{\mathbf{q}_j\}$. The *invariant similarity* $S(\mathbf{k}_1, \mathbf{k}_2)$ between $\mathbf{k}_1$ and $\mathbf{k}_2$ is defined as

$$S(\mathbf{k}_1, \mathbf{k}_2) = \tfrac{1}{2}[\zeta(\mathbf{k}_1, \mathbf{k}_2) + \zeta(\mathbf{k}_2, \mathbf{k}_1)] \tag{9}$$

where $\zeta(\mathbf{k}_1, \mathbf{k}_2)$ (and, similarly, $\zeta(\mathbf{k}_2, \mathbf{k}_1)$) is given as follows:

For each $\mathbf{p}_i$ of $\mathbf{k}_1$, find the matching $\hat{\mathbf{q}}_i$ such that

$$C_d(\mathbf{p}_i, \hat{\mathbf{q}}_i) = \max_l C_d(\mathbf{p}_i, \mathbf{q}_l) . \tag{10}$$

The similarity measure C_d is a distance-weighted vector dot-product:

$$C_d(\mathbf{p}, \mathbf{q}) = \frac{w(\mathbf{p}, \mathbf{q})\, \mathbf{p} \cdot \mathbf{q}}{\|\mathbf{p}\| \, \|\mathbf{q}\|} \tag{11}$$

where $w(\mathbf{p}, \mathbf{q})$ is a constant weighting factor that decreases with increasing Euclidean distance between $\mathbf{p}$ and $\mathbf{q}$ in the f-θ space. In other words, Eq. 10 finds the $\hat{\mathbf{q}}_i$ that is most similar and nearest (in f-θ space) to $\mathbf{p}_i$. Then, $\zeta(\mathbf{k}_1, \mathbf{k}_2)$ is computed as a normalized, weighted vector dot-product of all the local peak vectors $\mathbf{p}_i$ of $\mathbf{k}_1$ and their matching $\hat{\mathbf{q}}_i$ of $\mathbf{k}_2$:

$$\zeta(\mathbf{k}_1, \mathbf{k}_2) = \frac{\sum_i w(\mathbf{p}_i, \hat{\mathbf{q}}_i)\, \mathbf{p}_i \cdot \hat{\mathbf{q}}_i}{\sqrt{\sum_i \|\mathbf{p}_i\|^2 \sum_i \|\hat{\mathbf{q}}_i\|^2}} . \tag{12}$$

In the beginning of the merging process, each (seed) region R_i is characterized by its mean feature vector $\mathbf{M}_i$. As two regions are merged into one, both their mean vectors are collected in the merged region. It is necessary to collect the mean vectors instead of averaging them because, as a region gets larger in size, the mean vectors at its extreme ends may differ significantly even though they vary gradually over the entire region (e.g., Fig. 1). Two neighboring regions $\mathbf{R}_i$ and $\mathbf{R}_j$ can be merged if

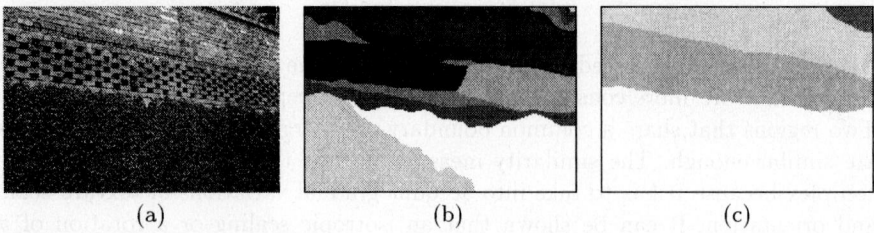

Fig. 2. (a) Image of a wall covered with leaves and two types of brick texture. (b) Non-invariant segmentation method splits a perceptually uniform region into several fragments (painted in different shades of gray). (c) Invariant segmentation method produces regions that are more consistent with human perception.

- they share common boundaries, and
- they are similar enough: $S(\mathbf{M}_i, \mathbf{M}_j) > \Gamma_S$

where $\mathbf{M}_i$ and $\mathbf{M}_j$ are the regions' mean vectors that are nearest to each other in the x-y space, and Γ_S is a constant threshold currently set at $\cos 23°$. At the end of the merging stage, the result is cleaned-up by merging very small regions into neighboring larger regions that are most similar in texture.

5 Test Results

This section illustrates some results of applying the texture segmentation method described in Section 4 on images of natural scenes. Figure 2 compares the performance of the invariant segment method with a typical non-invariant one. The non-invariant method used here is derived from the invariant method by replacing the invariant similarity measure S by a simpler cosine similarity C. The non-invariant method tends to split a perceptually uniform region into several fragments when there are significant differences between the scales or orientations of the texture patches (Fig. 2b). In contrast, the invariant method produces results that are more consistent with human perception (Fig. 2c).

Figure 3 illustrates another comparison between invariant and non-invariant methods. Due to perspective distortion, the texture on the metal cover in Fig. 3(a) changes gradually from very coarse scales to very fine scales. The non-invariant method segments the metal cover into several fragments each containing a texture of approximately the same scale (Fig. 3b). In contrast, the invariant method segments the cover into a single region (Fig. 3c). It also segments the other regions in the image well.

Figure 4 shows the result of segmenting Fig. 1. The two brick patterns at the top and the bottom are of the same type except for a rotation of 90°. The invariant method correctly identifies the texture boundary indicated by the abrupt change in orientation. It also groups image patches with the same texture but gradually varying scales and orientations into the same region. The final example illustrates the segmentation of boundaries formed by irregular texture, such as

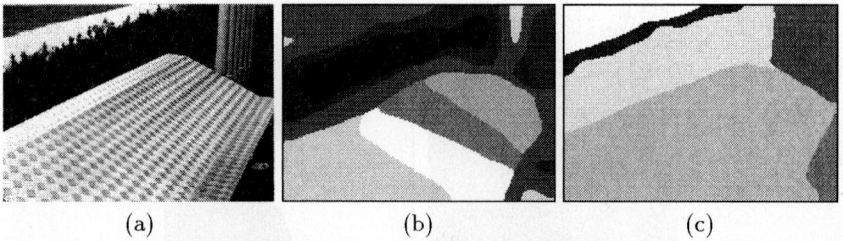

Fig. 3. (a) An image of a complex scene. (b) Non-invariant segmentation method splits the metal cover into several fragments according to the scale of the texture. (c) Invariant segmentation method identifies the patterns over the entire cover as the same texture despite gradual change in scale.

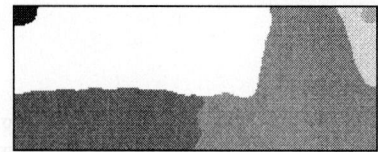

Fig. 4. Segmentation of Fig. 1. Invariant segmentation method correctly identifies the boundaries between the top and the bottom brick patterns that differ in orientations. It also segmented the pebble and the brick texture.

grass and rocks, that has rather ill-defined orientation. Figure 5 shows that the invariant segmentation method can identify the curvilinear boundary between the grass lawn and the rocks.

6 Conclusions

This paper has described a novel texture segmentation method that is invariant to gradual changes in texture scale and orientation. Two key features make the algorithm scale- and orientation-invariant. First, the method uses the pattern of outputs from all the Gabor channels as a distributed representation of texture feature. The channels have overlapping half-peak supports that reduce the variation of output pattern due to slight changes in texture scale and orientation. Second, the invariant similarity measure takes into account the non-uniform shifts of feature pattern due to perspective distortion, and performs individual matching of the local peaks in the feature pattern. The invariant segmentation method has been shown to perform well on images of natural scene containing multiple non-uniform texture patterns. In comparison with existing methods, it produces segmentation results that are more consistent with human perception.

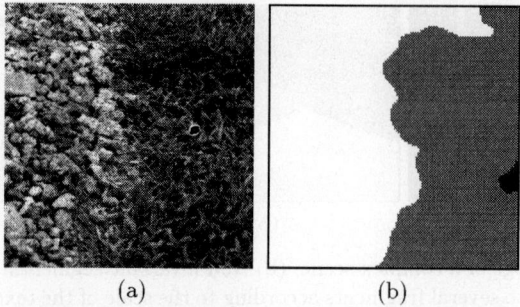

Fig. 5. (a) Image of ground covered with grass and rocks. (b) Invariant segmentation method can identify the curvilinear boundary between the grass lawn and the rocks.

References

1. J. Bigün and J. M. H. du Buf. N-folded symmetries by complex moments in Gabor space and their application to unspervised texture segmentation. *IEEE Transactions on Pattern Analysis and Machine Intelligence*, 16(1):80–87, 1994.
2. A. C. Bovik, M. Clark, and W. S. Geisler. Multichannel texture analysis using localized spatial filters. *IEEE Transactions on Pattern Analysis and Machine Intelligence*, 12(1):55–73, 1990.
3. D. F. Dunn, W. E. Higgins, and J. Wakeley. Texture segmentation using 2-D Gabor elementary functions. *IEEE Transactions on Pattern Analysis and Machine Intelligence*, 16(2):130–149, 1994.
4. J. M. Francos, A. Zvi Meiri, and B. Porat. A unified texture model based on a 2-D Wold like decomposition. *IEEE Transactions on Signal Processing*, pages 2665–2678, Aug. 1993.
5. G. E. Hinton, J. L. McClelland, and D. E. Rumelhart. Distributed representation. In David E. Rumelhart and James L. McClelland, editors, *Parallel Distributed Processing*. MIT Press, Cambridge, Massachusetts, 1986.
6. A. K. Jain and F. Farrokhnia. Unsupervised texture segmentation using Gabor filters. *Pattern Recognition*, 24(12):1167–1186, 1991.
7. F. Liu and R. W. Picard. Periodicity, directionality, and randomness: Wold features for image modeling and retrieval. *IEEE Transactions on Pattern Analysis and Machine Intelligence*, 18(7):722–733, 1996.
8. S. V. R. Madiraju, T. M. Caelli, and C.-C. Liu. On the covariance technique for robust and rotation invariant texture processing. In *ACCV'93 Asian Conference on Computer Vision*, pages 171–174, 1993.
9. S. V. R. Madiraju and C.-C. Liu. Rotation invariant texture classification using covariance. In *Proceedings of International Conference on Image Processing*, volume 2, pages 655–659, 1994.
10. D. K. Panjwani and G. Healey. Markov random field models for unsupervised segmentation of texture color images. *IEEE Transactions on Pattern Analysis and Machine Intelligence*, 17(10):939–954, 1995.

Optimal Edge Detection under Difficult Imaging Conditions

Md. Shoaib Bhuiyan[1], Yuji Iwahori[1] and Akira Iwata[2]

[1] Educational Center for Information Processing
[2] Dept. Electrical & Computer Engineering
Nagoya Institute of Technology, Showa, Nagoya, 466-8555, JAPAN
URL:- http://www.center.nitech.ac.jp/people/bhuiyan/pub.html

Abstract. This paper incrementally extends the energy minimization techniques for image analysis developed by [Koch et al. 1986]. Our application is edge extraction and we use the dual intensity and line processes introduced by [Geman and Geman, 1984]. The approach seeks to minimize a global energy functional that explicitly incorporates image properties to be minimized into weighted terms of the energy functional. Our specific contribution is modifying the weighting of terms in the energy functional that were previously independent of spatial gray level change to explicitly include spatial change in the weighting. We argue that the weighting used in previous implementations resulted in a reduced contribution from the edge components due to a dominance of the spatial intensity difference term as that spatial difference increases in size. Our specific modification compensates for this effect by scaling the edge process weighting factors by the spatial difference value (to the second order), thus, maintaining the same relative effect as the spatial difference increases. We found that the proposed algorithm works significantly better as compared to Koch et al. because of this modification.

1 Introduction

Edge detection is an indispensable procedure in both biological and machine vision. Consequently it has attracted a lot of research attention. Optimal edge detection using existing methods is not possible for images under difficult imaging conditions like non-uniform illumination, poor contrast, shadows and highlights. Here, optimal detection means marking the edge points while reducing the probability of falsely marking non-edge points.

[Geman and Geman, 1984] proposed a model which restores original image from degraded image (caused by mixture of noise) or from blurred image by employing a two-level Markov Random Field (MRF)— the high level MRF called *Line Process* represents discontinuities between the gray level of two neighboring pixels within an image (in other word, edge) while the low-level MRF called *Intensity Process* represents continuity of the image intensity. Edge detection is cast as a problem of finding an estimate of the line process that maximizes the *a posteriori* probability given an observed sample of the intensity process. However, the line process introduces local minima into the energy function, making the problem nonquadratic.

[Bhuiyan, 1996] reviews some work on edge and line detection schemes in which artificial neural networks (ANN) are either implicitly or explicitly involved. ANN's consist of neuron-like analog cells. Because the volume of data is large in vision, an analog ANN can play important role in preprocessing. This is because the lower accuracy of the computation supplied by the analog circuit is adequate for the preprocessing stages which often require more parallelism and more speed.

[Hopfield and Tank, 1985] showed that artificial neural networks (ANN) can be used in solving complex optimization problems and minimization problems. A Hopfield ANN is a single-layer, time iterative, feed-back network which has associative recall properties. This network has N processing elements (PE), each of which receives input from all the others. The input that a PE receives from itself is ignored. All the PE output signals are bipolar. The network has an energy function associated with it; whenever a PE changes state, this energy function always decreases. Starting at some initial position, the system's state vector simply moves downhill on the networks energy surface until it reaches a local minima of the energy function. However, Hopfield network deals only with quadratic problems.

[Koch et al., 1986] have shown how these networks can be generalized to solve the nonquadratic energy functionals of early vision by mapping the binary line processes into continuous variables bounded by 0 and 1. They also outlined one possibility for choosing an associated cost function. Earlier, we have shown that the coefficients in a cost function should not remain fixed for an image with inconsistent illumination and proposed a second-order variation for them [Bhuiyan et al., 1996]. We used a changing schedule of coefficients and compared it's performance for both synthetic and natural images with Sobel's operator, [Johnson, 1990] proposed "contrast based Sobel operator", [Marr-Hildreth's 1986] Laplacian-of-Gaussian (LoG) operator, and [Canny's 1986] operator in [Bhuiyan and Iwata, 1995]. We have also investigated it's noise immunity and compared with those of the above four methods. We have found the proposed technique to perform consistently better, especially for images where illumination varies widely. This work compares the performance of the proposed technique with that of Koch et al. original technique.

Experimentally, we found that the most difficult edges to extract from real world 8-bit digital images are those having spatial gray level difference of $1 \sim 20$ between adjacent pixels. This difficulty in feature extraction is especially compounded under difficult imaging conditions. We devised an empirical formula to assign higher weights for edges in $1 \sim 20$ region to be extracted. The formula is based upon our observation with natural and synthetic images, obtained from the image database (ftp://128.113.14.50/pub/image/still/usc/bgr/) of the Rensselaer Polytechnic Institute, USA and tested under various difficult imaging conditions. Mathematically written as follows:

$$C_{min} = \frac{l[1] \times 0.05 + l[2] \times 0.3 + l[3] \times 0.25 + l[4] \times 0.2}{M \times N \times 2} \qquad (1)$$

$l[1] = hlevel[1] + vlevel[1]$! 'no spatial difference in gray level
$l[2] = hlevel[2] + vlevel[2]$! 'spatial gray level difference of $1 \sim 20$
$l[3] = hlevel[3] + vlevel[3]$! 'gray level difference of $21 \sim 100$
$l[4] = hlevel[4] + vlevel[4]$! 'gray level difference of 101 or above

Where C_{min} is an image-dependent value to be used later in section 3.1. We are dealing with 8-bit digitized image (thereby having upto 256 gray levels) of M×N pixels. *hlevel* represents the horizontal adjacent gray level difference $|f_{i,j+1} - f_{i,j}|$ whereas *vlevel* represents the vertical adjacent gray level difference $|f_{i+1,j} - f_{i,j}|$. We observed that the best edge detection result is obtained when C_{min} value ranges from 0.05 for image region without any change in intensity values to 0.30 for image regions with rapid change in adjacent gray intensity value. By obtaining an working average for the particular image in hand, our algorithm can detect fine features otherwise undetected or would be difficult to detect.

2 Relaxation type Neural Network

In this neural network model, an image of M×N pixels has 3×M×N neurons. Figure 1(a) illustrates the two-dimensional lattice of the intensity and line processes. Three neurons are assigned to each pixel; the first shows the gray level intensity of the pixel (i,j) whose internal state variable is equal to it's output, represented by $f_{i,j}$. The second neuron decides whether there is any discontinuity or not between the two adjacent intensity levels $f_{i,j}$ and $f_{i,j+1}$. The output $h_{i,j}$ is given by a sigmoid function of the internal state variable $m_{i,j}$, i.e, $h_{i,j} = g_{i,j}(m_{i,j})$, where g(x) typically equals $(1+ e^{-2\lambda x})^{-1}$. Therefore, $h_{i,j}$ is expanded as follows. This neuron represents horizontal line process.

$$h_{i,j} = \frac{1}{1 + e^{-2\lambda m_{i,j}}} \qquad (2)$$

There is no discontinuity if $h_{i,j} \to 0$, but if it approaches 1 discontinuity does exist between the surrounding pixels. Similarly, the third neuron represents vertical line process. The output $v_{i,j}$ is given by a sigmoid function of the internal state variable $n_{i,j}$.

2.1 Energy Equations

[Koch et al., 1986] introduced energy function can be expanded as follows:

$$E = E_I + E_D + E_V + E_P + E_C + E_L + E_G \qquad (3)$$

This function has seven contributors. We have divided the line potential term E_L introduced by Koch et al. for simplicity into four sub-components E_V, E_P,

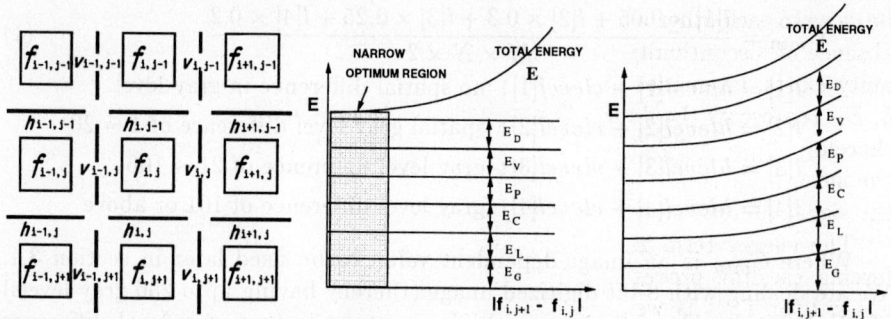

Fig. 1. (a) Intensity process $f_{i,j}$ and Line processes $h_{i,j}$, $v_{i,j}$. Each intensity value is enclosed on all four sides by a line process (b) Relationship of total energy E and it's components with Intensity Differential for fixed energy function coefficients (c) Proposed relationship of E and it's components with changing coefficients

E_C, and E_L, respectively. From now, E_L refers to the new, not the original E_L introduced by Koch et al. Each component of this function is described below and explanation follows:

$$E_I = \sum_{i,j}[(f_{i,j+1} - f_{i,j})^2(1 - h_{i,j}) + (f_{i+1,j} - f_{i,j})^2(1 - v_{i,j})] \quad (4)$$

$$E_D = C_D \sum_{i,j}(f_{i,j} - d_{i,j})^2 \quad (5)$$

$$E_V = C_V \sum_{i,j}[h_{i,j}(1 - h_{i,j}) + v_{i,j}(1 - v_{i,j})] \quad (6)$$

$$E_P = C_P \sum_{i,j}[h_{i,j}h_{i,j+1} + v_{i,j}v_{i+1,j}] \quad (7)$$

$$E_C = C_C \sum_{i,j}[h_{i,j} + v_{i,j}] \quad (8)$$

$$E_L = C_L \sum_{i,j}\{h_{i,j}[(1 - h_{i+1,j} - v_{i,j} - v_{i,j+1})^2 + (1 - h_{i-1,j} - v_{i-1,j} - v_{i-1,j+1})^2]$$
$$+ v_{i,j}[(1 - v_{i,j+1} - h_{i,j} - h_{i+1,j})^2 + (1 - v_{i,j-1} - h_{i,j-1} - h_{i+1,j-1})^2]\} \quad (9)$$

$$E_G = C_G \sum_{i,j}[\int_0^{h_{i,j}} g_{i,j}^{-1}(h_{i,j})dh_{i,j} + \int_0^{v_{i,j}} g_{i,j}^{-1}(v_{i,j})dv_{i,j}] \quad (10)$$

Where $f_{i,j}$, $v_{i,j}$, $h_{i,j}$, and $g_{i,j}^{-1}()$ correspond to the intensity process, the vertical line process, the horizontal line process, and the inverse sigmoid function, respectively. Note that the summation in E_D only includes nodes where measurements are available.

2.2 Energy Constraints

Each component of equation 3 makes up some constraint. The energy term E_I indicates that the intensities (gray levels) of the two neighboring pixels should be close to each other when line processes $h_{i,j}$ and $v_{i,j}$ equal to zero, i.e. meaning absence of discontinuity. But when $h_{i,j}$ or $v_{i,j}$ equals one, i.e. when a discontinuity exists, intensities of the neighboring pixels should be different. Equation 5, E_D represents the data reliability. $d_{i,j}$ is the intensity value of the pixel (i,j); therefore $(f_{i,j} - d_{i,j})^2$ describes the difference between the original data $d_{i,j}$ and the intensity process value $f_{i,j}$. This term is weighted by C_D, which depends on the signal-to-noise ratio. If $d_{i,j}$ is noise-free, then C_D might be $\gg 1$.

The energy term E_V of equation 6 makes up the boundary constraint by forcing the line processes to the corners of the hypercube—i.e., to approach either 0 or 1. It is weighted by C_V. E_P penalizes the formation of adjacent parallel lines and thus, makes up the single line constraint. It ensures that two adjacent line processes should not be activated simultaneously. E_C represents the *cost* that needs to be incurred for the introduction of each single line.

E_L ensures that the line processes should be active only along a line or a curve. Minimizing equation 9 requires $(h_{i,j}, v_{i,j}) \to 0$, and either one of $(h_{i+1,j}, v_{i,j}, v_{i,j+1})$ and $(v_{i,j-1}, h_{i,j-1}, h_{i+1,j-1}) \to 1$. Projecting this information to Figure 1(a) shows formation of a single continuous line. Thus E_L is an interaction term that favors continuous lines and penalizes both multiple-line interactions and discontinuous-line segments. The gain term E_G of equation 10 forces the line processes inside the hypercube $[0,1]^N$.

2.3 Network Dynamics

The following updated equations were introduced by Koch *et al.*:

$$\frac{df_{i,j}}{dt} = -\frac{\partial E}{\partial f_{i,j}}, \frac{dm_{i,j}}{dt} = -\frac{\partial E}{\partial h_{i,j}}, \text{ and } \frac{dn_{i,j}}{dt} = -\frac{\partial E}{\partial v_{i,j}} \qquad (11)$$

These three equations, when applied to determine the time-rate changes of the total energy term $E(f_{i,j}, h_{i,j}, v_{i,j})$, shows:

$$\begin{aligned}
\frac{dE}{dt} &= \sum_{i,j} \frac{\partial E}{\partial f_{i,j}} \frac{df_{i,j}}{dt} + \sum_{i,j} \frac{\partial E}{\partial m_{i,j}} \frac{dm_{i,j}}{dt} + \sum_{i,j} \frac{\partial E}{\partial n_{i,j}} \frac{dn_{i,j}}{dt} \\
&= \sum_{i,j} \frac{\partial E}{\partial f_{i,j}} \left(-\frac{\partial E}{\partial f_{i,j}}\right) + \sum_{i,j} \frac{\partial E}{\partial m_{i,j}} \left(-\frac{\partial E}{\partial h_{i,j}}\right) + \sum_{i,j} \frac{\partial E}{\partial n_{i,j}} \left(-\frac{\partial E}{\partial v_{i,j}}\right) \\
&= -\sum_{i,j} \left\{ \left(\frac{\partial E}{\partial f_{i,j}}\right)^2 + \left(\frac{\partial E}{\partial h_{i,j}}\right)^2 \frac{\partial h_{i,j}}{\partial m_{i,j}} + \left(\frac{\partial E}{\partial v_{i,j}}\right)^2 \frac{\partial v_{i,j}}{\partial n_{i,j}} \right\}
\end{aligned} \qquad (12)$$

From equation 2, $\frac{\partial h_{i,j}}{\partial m_{i,j}} = 2\lambda h_{i,j}(1 - h_{i,j})$, which is ≥ 0. Similarly it can be shown that $\frac{\partial v_{i,j}}{\partial n_{i,j}}$ is also ≥ 0. Therefore, none of the term in the above sum is negative, causing the rate of change of the total energy $\frac{dE}{dt}$ to be ≤ 0. So the total energy will always decrease. The system will evolve in such a manner as to find a minimum of E.

3 Why Energy Function Coefficients Should Change

Figure 1(b) illustrates the total energy E and it's component energies E_V, E_P, E_C, E_L and E_G— against intensity difference $|f_{i,j+1} - f_{i,j}|$. Note that the energy equations for these components containing line process terms i.e., equations (6)-(10) do not contain any $|f_{i,j+1} - f_{i,j}|$ terms. Therefore, as Figure 1(b) shows, the curve for total energy E tends to go upward as the intensity difference between pixels increases, though, the sum total of the above mentioned component energies remains unchanged if the corresponding energy coefficients remain unchanged. This causes the ratio of each of these energy components to change with the total energy E. Therefore, line processes become sensitive to the contrast if the energy function coefficients remain fixed in case of non-uniformly illuminated images.

We have had further evidence in [Bhuiyan et al., 1993], where we used a very simple model image having two different contrast regions— one being 10 times the other in difference of gray level between adjacent pixels. Then, false edges were obtained when the coefficients were set to detect the low contrast region and missed edges when they were set to detect the high contrast region, practically resulting only in a narrow optimum region in Figure 1(b) for edge detection. This clearly illustrates the inappropriateness of fixed value energy function coefficients, especially for images where contrast changes non-uniformly over the object. For such images, the energy-contrast relationship of figure 1(b) should be modified as proposed in Figure 1(c) to keep the ratio of E and it's component E_V, E_P, E_D, E_C, E_L, and E_G relatively close in order to accommodate the *whole region* for edge detection purpose.

3.1 Proposed Way of Changing the Coefficients

Since intensity changes over different range of scales in a natural scene, our sought after energy function coefficients C_* ($*$ represents either V, P, C, L, D or G) should ideally assume suitable values depending upon the particular range of scale of intensity within the image. As E_I varies as a second-order function of spatial intensity difference $|f_{i,j+1}-f_{i,j}|$, we propose that the component energies E_V, E_P, E_C, E_L, E_D, and E_G should also vary as a second-order function of $|f_{i,j+1} - f_{i,j}|$ as shown in Figure 1(c), in order to keep the rate of change of the component energies uniform against the total energy E. This can simply be done if our sought after coefficients C_* also vary in proportion to the square of $|f_{i,j+1} - f_{i,j}|$. Mathematically, we propose that

$$C_{(h_{i,j})} = C_{min}(f_{i,j+1}-f_{i,j})^2, \text{ and } \quad C_{(v_{i,j})} = C_{min}(f_{i+1,j}-f_{i,j})^2 \qquad (13)$$

$C_{(h_{i,j})}$ and $C_{(v_{i,j})}$ are directly proportional to $(f_{i,j+1}-f_{i,j})^2$; the image-dependent relating factor C_{min} is calculated from equation (1). It is the lowest limit below which the coefficients are not allowed to go, in order to avoid detecting spurious noisy edges when $|f_{i,j+1} - f_{i,j}|$ becomes very small.

Fig. 2. (a) Energy diagram of the proposed technique (b) Edge detection by the proposed technique, 70 iterations (c) Edge detection by Koch et al method, 50 iterations

4 Edge Detection

Initial choice of the parameters was made heuristically using cross-validation method against $C_I = 1.0$, $C_D = 2C_{min}$, $C_V = C_{min}$, $C_P = 10C_{min}$, $C_C = 2C_{min}$, $C_L = 8C_{min}$, $C_G = C_{min}$, $\lambda = 25$ and $C = 300C_{min}$. To have some control over the coefficients' updating during simulation runs, we maintained their relative weights. For any pixel (i,j), the proposed algorithm calculates $C_{(h_{i,j})}$ and $C_{(v_{i,j})}$, respectively for the horizontal and the vertical direction using equation (13). We also average these constants resulting from all seven adjacent pixels to take into account adjacent intensity differences as follows. These $C'_{(h_{i,j})}$ and $C'_{(v_{i,j})}$ values are updated every 10 iterations.

$$C'_{(h_{i,j})} = [C_{(h_{i,j})} + C_{(h_{i-1,j})} + C_{(h_{i+1,j})} + C_{(v_{i,j})} + C_{(v_{i,j+1})} + C_{(v_{i-1,j})} + C_{(v_{i-1,j+1})}] \div 7$$
$$C'_{(v_{i,j})} = [C_{(v_{i,j})} + C_{(v_{i,j+1})} + C_{(v_{i,j-1})} + C_{(h_{i,j})} + C_{(h_{i,j-1})} + C_{(h_{i+1,j})} + C_{(h_{i+1,j-1})}] \div 7$$
(14)

We follow a schedule where the original C value is lowered to half during the 5th, again during 20th, 40th, 70th and 100th iteration. The schedule is based upon our observation with images obtained from the image database of the Rensselaer Polytechnic Institute, USA and tested under various difficult imaging conditions. Edge detection is carried out as follows. For pixel (i,j), line processes $h_{i,j}$ and $v_{i,j}$ are initially set to 0.50, by setting internal state variables (m,n) to zero. All the energy function coefficients are set to their initial values. Intensity process and Line processes are activated. The intensity process $f_{i,j}$ approximates the original data $d_{i,j}$. Then the network computes the smoothest surface independent of all the line processes. This was done by subsequently updating the intensity process 100 times for every single update of the line process network, which makes the line process network appear virtually stationary, or substantially slower as compared to the intensity process. However, independent of the relative speed of the intensity and line processes, $E(f, h, v)$ is always a descending function. Figure 2(a) shows the energy diagram of the proposed algorithm for the Lena test image, Figures 2(b) and 2(c) show the corresponding edge maps in 70 and 50 iterations, respectively.

Consequently, $C(h_{i,j})$ and $C(v_{i,j})$ are determined for any pixel (i,j) using equation (13) and are averaged over seven adjacent pixels using equation (14). Care is taken so that $C'_{(h_{i,j})}$ and $C'_{(v_{i,j})}$ values do not go below the permitted minimum value, C_{min}. Initially C is set to $300 \times C_{min}$ for the first four iterations. This is meant to detect the most prominent edges i.e. where the differences of gray levels of adjacent pixels are quite high. C value is gradually lowered with increasing iteration. We imposed an upper limit of 100 iterations, though.

Once $C'_{(h_{i,j})}$ and $C'_{(v_{i,j})}$ values are determined from equation (14) and compared to the C value of the above schedule, the proposed algorithm computes the energy coefficient values C_V, C_P, C_C etc. by using the initial weighting pattern mentioned at the top of this section. Internal state variables (m,n) are calculated by applying equation (11) to equation (3). Their values, when applied to equation (4), provide $h_{i,j}$ and $v_{i,j}$. Energy component values are computed using equations (4)-(10), together with these values and the $f_{i,j}$ and $d_{i,j}$ values.

Edges are provided for sharp discontinuities in neighboring pixel intensity levels if any one of the line processes $h_{i,j}$ and/or $v_{i,j} \to 1$. Consequently, edges are detected gradually beginning from the most prominent toward less and less prominent ones. This results in a numerous number of short edges. The proposed neural net algorithm has one constraint (E_L) that works to connect these array of short edges; thus helping them grow to form long, continuous edges. Since real images contain multiple intensity gradients, this method helps to detect edges where contrast changes largely over the image by growing them from sharper regions towards dimmer regions.

References

[Geman and Geman, 1984] S. Geman and D. Geman, "Stochastic Relaxation, Gibb's Distributions and the Bayesian Restoration of Images," IEEE Trans., Pattern Anal. Machine Intell., Vol. PAMI-6, pp.721-741, 1984.
[Bhuiyan, 1996] Md. Shoaib Bhuiyan, "Contrast based Edge Detection using Artificial Neural Network," DEng Dissertation, Nagoya Institute of Technology, Nagoya, Japan, pp. 19-21, Jan 1996.
[Hopfield and Tank, 1985] J. J. Hopfield, and D. W. Tank, J. J. Hopfield, and D. W. Tank, "Neural" computation of decisions in optimization problems, *Biological Cybernetics*, vol. 52, pp. 141-152, 1985.
[Koch et al., 1986] C. Koch, J. Marroquin, and A. Yuille: Analog "neuronal" networks in early vision; in *Proc. Natl. Acad. Sci.*, USA, vol. 83. pp. 4263-4267, June 1986.
[Bhuiyan et al., 1993] M. S. Bhuiyan, M. Sato, H. Fujimoto, and A. Iwata "Edge detection by neural network with line process," in *Proc. Int'l. Joint Conf. on Neural Networks*, Nagoya, Japan, vol. 2, pp. 1223-1226, Oct. 25-29, 1993.
[Bhuiyan et al., 1996] M. S. Bhuiyan, H. Matsuo, A. Iwata, H. Fujimoto, and M. Sato, "Edge Detection using Neural Network for Non-uniformly illuminated Images," *IEICE Transactions on Information and Systems*, vol. E79D, no. 2, pp. 150-160, Feb 1996.
[Bhuiyan and Iwata, 1995] M. S. Bhuiyan and A. Iwata, "Performance Evaluation of a Neural Network based Edge Detector for high contrast images," in *World Congress on Neural Networks*, Washington, D.C., USA, Vol. 2, pp. 550-554, July 17-21, 1995.
[Marr and Hildreth, 1980] D. C. Marr and E. Hildreth, "Theory of Edge Detection," in Proc. Roy. Soc. London, vol. B207, pp. 187-217, 1980.
[Canny, 1986] J. Canny, "A Computational Approach to Edge Detection," *IEEE Trans., Pattern Anal. Machine Intell.*, vol. PAMI-8, no. 6. pp. 679-698, Nov. 1986.
[Johnson, 1990] R. P. Johnson, " Contrast based Edge detection," *Pattern Recognition*, pp. 311-318, 1990.

Restoring Image Quality Through Structure Preserving De-noising

Krishna Ratakonda and Narendra Ahuja
Department of Electrical and Computer Engineering
Beckman Institute, University of Illinois, Urbana, IL 61801.
krishna@stereo.ai.uiuc.edu

Abstract

In many image transmission and acquisition situations, the image may become corrupted by additive noise. De-noising refers to the process of removing the noise while maintaining good visual quality. This problem has assumed major significance with the increase in image related communication that has accompanied the exponential growth of the internet. Traditionally, image quality is measured in terms of PSNR (Peak Signal to Noise Ratio) which may have limited relation, at best, to the perceptual quality of the image. In this paper we present a novel de-noising scheme which results in significantly improved performance in terms of both perceptual quality and PSNR. Furthermore, we show that the de-noising framework that we propose encompasses the usual linear transform based de-noising schemes as special cases.

1 Introduction

Classical methods for de-noising images treat the problem no differently from its one-dimensional counterpart. There is no explicit way of preserving important structural information which is crucial for interpretation, recognition and other computer vision tasks. In fact most of the usual de-noising schemes perform poorly with respect to the preservation of structure. This paper proposes an approach to de-noising that improves PSNR while preserving structure. This is accomplished by employing structure sensitive constraints to guide the usual PSNR-enhancing de-noising schemes. Instead of making the usual assumption that the image is a smooth signal, we view the image as a multi-scale, piecewise-smooth, 2-D signal and use the PSNR based approach on each of these smooth parts defined by a multi-scale image segmentation. Clearly, the fidelity of image segmentation used would have a direct impact on the performance of the algorithm. To this end we use a multi-scale image segmentation algorithm that extracts image regions, which are in agreement with what is percieved, in the presence of significant geometric, topological and spectral complexity, and regardless of the number and parameters of the natural scales present in a given image. Since the image varies smoothly over each segmented region, the PSNR also improves significantly compared to that achieved by

---

[*] This research was supported by the Advanced Research Projects Agency under grant N00014-93-1-1167 administered by the Office of Naval Research and the NSF grant IRI 93-19038.

the same de-noising method applied globally over the entire image. Thus, the net effect is the integration of high-fidelity image segmentation and better than usual PSNR gains in each segment, leading to an improvement in perceptual quality in both optical and spectral senses.

The model for image estimation in the presence of additive noise may be written as $Y_{ij} = X_{ij} + \epsilon_{ij}$, where i, j are indices into the image represented as a matrix, Y is the noisy signal, X is the uncorrupted signal and ϵ represents the additive noise.

Many of the previously proposed de-noising methods with good performance use an orthonormal transform such as a Wavelet or a Fourier transform to map the noisy image Y into a domain where the energy of X is compactly represented. Note that we may consider the contributions due to X and ϵ indepedently due to the linearity of the transform. The signal compaction properties of the transforms ensure that most of the energy of X is concentrated into a few coefficients in the transform domain. Since the energy of X is not changed by an orthonormal linear transform (Parseval's relation) these few coefficients, which recieve significant contribution from X, have relatively large values.

If the noise ϵ is assumed to be identically and independently distributed, additive and white Gaussian (i.i.d. AWGN), the contribution to the transform coefficients due to noise is still distributed across the transform domain and their values retain the standard deviation characterizing the noise in the image domain. The contribution due to noise in each coefficient is small but effects all the coefficients. This distribution of the noise energy may also be explained by the fact that an orthonormal linear transform only results in a rotation of the coordinate system representing the image. Hence, the probability distributions of the noise coefficients, which are symmetric with respect to axis rotation due to i.i.d. assumption, are not affected by such a coordinate transformation.

Thus in the transform domain we get a representation of the noisy image Y in which the actual image X is typically concentrated in a few large values and the noise contributes small values to all transform coefficients. This discussion would suggest that applying a threshold on the transform coefficients and setting to zero all values below the threshold might be a good de-noising scheme. Donoho and Johnstone [1] analyzed the scenario under the i.i.d. AWGN assumption[1] and derived the optimal threshold to be applied when a linear, orthonormal transform is used. It is well known that linear transforms (including Wavelets) result in a distribution of the energy contributed by the edge areas in the image all over the transform domain in contrast to the energy contributed by the non-edge areas in the image which is localized to low frequencies. Since the transitions across edges contribute mainly to the small values, which are removed by the thresholding operation, we obtain distortion in edge areas.

Liu and Moulin [2] propose a more complicated approach which uses a lossy coder to code the image; the rate-distortion factor of the lossy coder being determined adaptively depending on the noise level corrupting the image. They use a Wavelet transform based lossy coder [3], which benifits mainly from the different representations of the noise and the signal in the Wavelet domain. The basic assumption is that the loss incurred in the coder is mostly due to the removal of noise which is valid since i.i.d. Gaussian noise is hard to compress. Their method works well for high SNRs and produces fewer artifacts than the simpler thresholding scheme of Donoho [1]. However edge distortion invariably occurs at low SNRs since the lossy coder does not perform as well for the i

---

[1] It may be noted that the AWGN assumption can be readily extended in a robust way to other probability distributions provided that the i.i.d. assumption is satisfied.

rate-distortion factors corresponding to low SNRs.

We introduce the new framework for structure mediated de-noising for i.i.d. AWGN in the next section and outline the key new ideas proposed. Our framework incorporates two major but independent components which are discussed in sections 3 and 4. Implementation details and a few illustrative results are presented in section 5.

2 Proposed Framework for De-noising

The key idea behind the proposed approach comes from the discussion in the previous section on the rate-distortion (Liu-Moulin) and hard thresholding (Donoho-Johnstone) schemes. Both these schemes inherently depend on the fact that the non-noisy signal is *compressible and hence compactly representable* in the Wavelet domain. The important point to note is that these schemes can be applied *independently on different portions of the image*, provided that the pixel population within each region is sufficiently large (we will define this more concretely later) to support the asymptotic assumptions involved. Thus we might think of the following processing: (a) split the image into disjoint sets of pixels and (b) apply the usual de-noising scheme over each set independently.

Provided that we split the image into disjoint collections of pixels, each collection characterized by grey value homogeniety, the Wavelet transform would be able to compress each of the regions individually to a high degree. This follows because the Wavelet transform (and in fact any other compacting linear transform) has difficulty in compacting signals with both smooth as well as high frequency components [4]. By segmenting the regions, say into regions characterized by gray level similarity, and choosing an individual region over which to apply the transform, we *effectively eliminate most of the high frequency coefficients* which cause loss in compaction, resulting in high compaction. This maximizes the performance of the schemes that we have discussed in Section 1.

A segmentation scheme which would robustly segment an image in the presence of Gaussian noise is needed. The presence of Gaussian noise is not likely to cause much degradation in the segmentation scheme relative to the case of the noise free image since: (a) segmentation involves pixel population analysis in order to find regions and (b) the noise is assumed to be i.i.d. and hence does not superimpose significant structure on the image. Recently, a new multi-scale segmentation scheme has been proposed that extracts image regions at all natural geometric and photometric scales present. At each scale, the regions extracted are consistent with perception for a wide range of the geometric, topological and spectral complexities [5]. We use this algorithm, in our present work, for segmentation.

The idea of exploiting the advantages of the Wavelet (and other linear) transforms over regions immediately leads to another problem: regions have arbitrary shapes and the Wavelets and other linear transforms cannot be applied over a region of arbitrary support. These transforms are defined only over square (or rectangular) supports. In fact, Wavelets may be applied, without computational tricks, only to supports which are powers of two. We propose a solution which not only extends the definition of the Wavelet transform (and any other discrete linear transform) to arbitrary supports but also leads to maximal compaction. Section 4 and a previous paper [6] provide the details.

To summarize, the framework that we propose involves three key steps:

1. Segment the image to obtain a multi-scale partition of the image into homogeneous regions and post process the results (section 3).

2. Convert each region into a form so that the linear transform can be applied (section 4).

3. Apply the transform based processing over each region individually, invert the linear transform and combine the results from different regions.

It is to be noted that the only restriction on the baseline de-noising scheme is that it is linear transform based. Most of the schemes which perform well in practice belong to this category. Furthermore, by considering the entire image as a single region we get back to the usual version of the baseline de-noising scheme. Thus, our framework includes as special cases all baseline de-noising schemes which are based on signal compaction in linear transform domains.

In the above discussion we have justified why we expect to obtain an advantage over the base-line processing scheme using our framework. Results show significant improvement in visual quality due to both PSNR increase and better structure preservation. The proposed framework can be viewed as a multi-representational paradigm which combines the advantages of linear schemes which generally work well with smooth signals, and the structural information provided by multi-scale segmentation which complements the strengths of the linear methods. Thus we benefit from the complementarity of representations.

3 Segmentation

In this paper, we will be using the segmentation information provided by the multi-scale segmentation transform described in [5]. As explained in the previous section, the performance of the segmentation algorithm is not expected to degrade very much due to i.i.d. AWGN. However, since we will be considering low SNRs, there might occur spurious structures in the image introduced by the noise which will be (correctly) detected by the low level segmentation method. Such noise artifacts will tend to be small and therefore could be removed by postprocessing the result of segmentation. Another reason for post processing is that the population of the pixels in each region should be above the minimum required to validate the asymptotic assumptions behind de-noising algorithms. Since the segmentation provided by the algorithm in [5] detects regions at all spatial scales, including those regions smaller than can be used, the small regions must be replaced by their sufficiently large parent regions at coarser scales, or if such a parent region cannot be found merged with a suitable neighbor, before applying the transform-based de-noising algorithms.

3.1 Post processing of segmentation data

We consider regions at the finest photometric scale for de-noising independently. For those regions that are too small, the path to the root of the segmentation tree is searched to find the nearest parent region having sufficiently large size. If no such region is found, that is if the finest scale region does not merge with any other region, then the region is deleted. This helps ensure that the asymptotic assumptions are followed and any spurious regions introduced by noise are disregarded.

To arrive at the size threshold, let X_i, $i = 1 \cdots n$ be a sequence of i.i.d. random variables. As n tends to ∞, the value of $\sum_{i=1}^{n}(X_i/n)$ approaches the expected value of the individual random variables. Under the assumption that the noise is zero mean, we formulate the following criterion to specify an acceptable level of the validity of the asymptotic results $E(|\sum_{i=1}^{n}(X_i/n))|^2) < 1$.

Since the expectation above decreases with increasing n, we obtain a threshold for the minimum size of the region. With Gaussian noise of standard deviation σ, we will thus need the number of pixels in a region is at least σ^2. This assumption will ensure that if we replace all the pixels within a region with the average value of the region (A), A will be a random variable with standard deviation smaller than 1 and mean equal to the the average of the same pixels in the uncorrupted image. Normally one would expect that artifacts would appear at region edges since we process each region independently of the others. The minimum size requirement on regions is intended to reduce this artifact by ensuring that the region mean in the noisy image is not very different from the region mean in the uncorrupted image.

Given the above threshold for the minimum size of a region, we obtain a simple algorithm. We apply the linear transform based de-noising to all the regions at finest scale that have a size larger than the threshold. For any region at the finest scale with size below the threshold, we perform the following steps:

1. If the region gets merged with another region at a coarser scale in the segmentation tree, since the regions at coarser scales are preserved at finer scale, find the smallest acceptable region along the path to the root and use it to replace all descendent leaves of the region.

2. If the region at a fine scale persists in all coarser scales of segmentation, merge it with the neighboring region which has minimum disparity in mean gray value from the current region.

Figure 1 shows the segmentation before and after post processing for the image Lenna. The output of the segmentation algorithm shows a number of small regions which are generated by isolated large values of noise. These regions are removed from the segmentation after post processing.

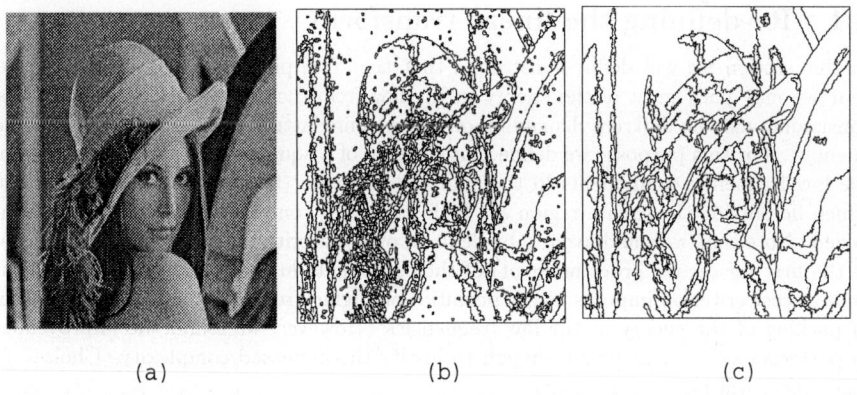

(a)   (b)   (c)

Figure 1: (a) Lenna corrupted by Gaussian noise of s.d. 15 (b) Segmentation at the finest scale and (c) Segmentation after post processing using multi-scale segmentation data.

4 Linear Transforms over Arbitrary Supports

Discrete linear transforms in two dimensions are in most cases defined over a rectangular support. The usual practice when we want to apply the transform over an arbitrarily shaped support is to fill out the rest of the support with zeros to make up the rectangle and then use the natural definition of the transform over a rectangle. This is an extension of the one dimensional case where we fill out an arbitrary length data set with zeros to form a data set of length 2^n either to increase the computational speed (through FFTs for Fourier Transforms) or to satisfy the definition of the transform (in the case of dyadic Wavelets). This, however, does not lead to a satisfactory definition of the linear transform in two or more dimensions for many applications.

The above discussion leads us to the following question: what should be the values attributed to the sample points which lie within the rectangle but not within the support of the function? With each possible choice of the values for the pixels which lie outside the support but within the rectangular region, we can associate a possible function-transform pair. We would like to define the free pixels (the pixels within the rectangular region but outside the support) so as to minimize the high frequency content in the transform domain. Thus we may view this as *minimizing a cost function which maximally shifts the energy in the low frequency values over all possible choices of the free pixels*. The choice of the free pixels should distribute the energy due to region boundary, as well as the energy within the rectangle but outside the region, to as low frequency values as possible.

In section 4.1, we provide a re-definition of the discrete linear transform that packs the energy into the low frequency areas. The solution in constructed within the projection on convex sets formalism. Since all the constraint sets are convex, this framework guarantees convergence to a solution. This kind of formulation has been treated by us in a more general context in [6]. If speed is an issue, it is to be noted that this algorithm may be reformulated to achieve quadratic convergence [7].

4.1 Re-defining the linear transform

In this section we will define constraints that the "free pixels" should satisfy, inorder that we may pack most of the energy in the low frequency areas of the image. These constraints may be different depending on the meaning attributed to the term "high frequency". For this purpose, we define ζ as the side of a square shaped region containing the low frequency components in the transform domain (figure 2(a)). All components which lie outside the square region are high frequency components and are to be minimized. Note that we could avoid ζ by defining our cost criterion to be the minimization of the first (or second order moment) of the coefficients in the transform domain. This kind of cost criterion imposes larger penalties on higher frequency terms, thus resulting in packing of the energy in the low frequencies. However, we found the improvement in performance not significant enough to justify the increased complexity. Choice of a value for ζ will be discussed later.

Now we have obtained two constraints that we would like the signal to satisfy:

1. In the transform domain the image should have minimum energy outside the square region of side ζ as defined above. That is all components outside the square should be as close to zero as possible in the L^2 norm sense.

2. In the spatial (or image) domain, the image should satisfy the known values of the function within its support.

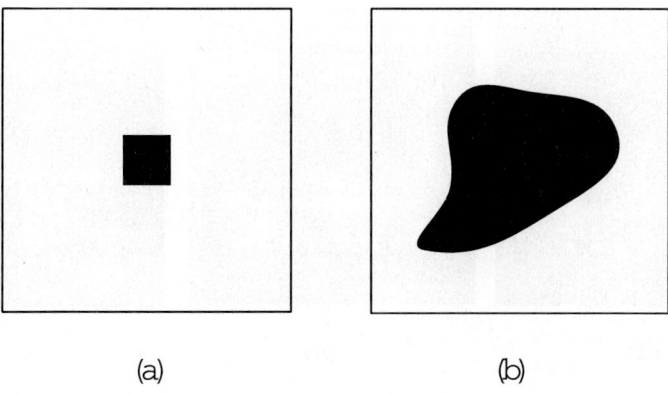

Figure 2: Constraints for the DFT. (a) DFT domain: Values outside the shaded square should be minimized (in L^2 sense). (b) Spatial (or image) domain: Known values of the function should not be disturbed.

These two constraints are illustrated in figure 2. That they may be put in terms of convex sets and a solution obtained using the Projection on Convex Sets algorithm is a straight forward exercise. The reader is refered to [8, 6] for more details.

What should be the value of ζ? We found that the results that we obtain (for this application in de-noising) do not vary much with ζ provided that we choose ζ not too conservatively. If we choose the value of ζ too conservatively, we might end up with too little variation in the free pixels and hence might not obtain optimal packing of the energy in the low frequencies. Since each region may be enclosed in squares of different sizes, we chose ζ to be proportional to the side length of the square on which the transform is being applied. As mentioned earlier, a less adhoc solution would be to avoid the choice of ζ altogether, and maximize the energy at low frequencies.

5 Implementation and Results

We use the Donoho-Johnstone hard thresholding criterion as our baseline scheme. As described in section 1, the optimal performance of this criterion requires maximal packing of the image energy in as few coefficients in the transform domain as possible. Since each of the individual regions processed with the baseline scheme already approximately satisfies this requirement for the optimal performance of the Donoho-Johnstone criterion, we expect only marginal changes in performance if we apply more complicated processing. From [1], we obtain the following threshold to be applied on the Wavelet coefficients for Gaussian noise of s.d. σ: $(Threshold)^2 = \sigma^2(2ln(N))$, where N is the number of pixels within a particular region. We use the Daubechies 4-tap Wavelet filters inorder to perform the simulations. The POCS algorithm for calculating the optimal transform over the arbitrary shaped region converges within 2-3 iterations to within 10% of its final value on the average. As was noted before, convergence can be speeded up even more by formulating the optimization problem in the formalism provided in [7].

Illustrative results are shown for the image Goldhill. Results for the Goldhill image for a Gaussian noise of standard deviation 15 are shown in figure 3. In figure 3(c), the edges are clearly better preserved than figure 3(b). This may be seen from the bars on

(a)   **(b)**   **(c)**

Figure 3: (a) Goldhill corrupted by i.i.d. AWGN of s.d. 15 (PSNR 24.6 dB) (b) Goldhill processed with Donoho-Johnstone criterion (PSNR 29.61 dB) (c) Goldhill processed with proposed approach (PSNR 31.70 dB)

the windows, the outline of the roofs and other portions of the image. Furthermore, the streaks that are found in figure 3(b) disappear in figure 3(c).

The results show that our algorithm preserves the structure as well as improves PSNR. Although results have been presented using the Donoho-Johnstone hard thresholding criterion as its baseline scheme, other transform based approaches could be easily used to de-noise the individual regions [2, 9].

References

[1] D. L. Donoho and I. M. Johnstone. Ideal spatial adaptation by Wavelet shrinkage. *Biometrika*, 81:425–455, 1994.

[2] Juan Liu and P. Moulin. Denoising of images using a complexity regularization criterion. *To be published in ICIP-97*.

[3] S. Servetto, K. Ramchandran, and M. T. Orchard. Morphological representation of wavelet data for image coding. *Proceedings ICASSP*, 1995.

[4] C. Herley, K. Ramchandran, et al. Time-varying orthonormal tilings of the time-frequency plane. *ICASSP*, 2:891–5, 1993.

[5] N. Ahuja. A transform for the detection of multi-scale structure. *IEEE Transactions PAMI*, December 1996.

[6] K. Ratakonda and N. Ahuja. Discrete multi-dimensional linear transforms over arbitrarily shaped supports. *To be published in ICASSP*, 1997.

[7] S. S. Dharanipragada and K. S. Arun. A quadratically convergent algorithm for convex set constrained signal recovery. *IEEE Transactions on Signal Processing*, 44(2):248–66, February 1995.

[8] H. H. Chen et al. A block transform coder for arbitrarily shaped image segments. *IEEE International Conference on Image Processing*, 1:85–89, 1994.

[9] J. Rissanen. Modeling by shortest data description. *Automatica*, 14:468–478, 1978.

Feature Saliency from Noise Variations in Invariants

Mark Jenkinson, Michael Brady

Department of Engineering Science, University of Oxford, Oxford OX1 3PJ

Abstract. Object localisation and recognition are fundamental problems in computer vision. The goal is to enable a mobile robot to locate general, non-polyhedral objects in complex settings. This requires considerable robustness and reliability, and so low level invariants are used as a robust starting point. In particular, a set of quantities is developed that are both *geometrically* and *photometrically* invariant. They are arranged as the components of a description vector, which are matched to locate model instances.

The paper analyses the variations of the invariant quantities. These arise in practice due to *image noise* and *spatial quantisation*: the case of image noise is treated here, quantisation being the subject of ongoing work. The noise models obtained show good agreement with experimental results.

A probability model for the variation of the description vectors is derived and used to define a saliency measure in the image. Combining this with a Non-Uniform selection strategy in a modified RANSAC (NU-RANSAC) scheme leads to a dramatic improvement in the probability of correctly matching points, which is the basis of localising the desired object.

1 Introduction

The motivation behind this work is to enable a mobile robot to localise, or recognise, general, non-polyhedral objects in a complex environment, such as a factory or stockyard in which neither clutter nor lighting can be controlled. More specifically, we seek to find deformable objects, such as sacks, that are lying on the ground and which the mobile robot should steer toward and pick up. Sacks are typically patterned with the name, or mark, of the manufacturer. The contrast with the background cannot, in general, be guaranteed.

Our initial approach was to apply active contour trackers [1], due to their ability to cope well with non-polyhedral objects. However, we found that there was an intrinsic lack of robustness, especially when the illumination changed, which prompted us to explore three fundamental issues: *invariance, selection* and *matching*.

Invariance. The idea is to compute a set of quantities that describe the desired object in a way that is independent of its position (geometric invariance) and the illumination conditions (photometric invariance). In this way the values themselves are already robust to geometric and photometric changes. Several different, independent invariants have been proposed, and can be treated as the components of a vector that describes a point and its local image patch. These description vectors are calculated at every point in an image.

The idea of using description vectors is developed in [8] for performing image database retrieval. Results reported there demonstrate the effectiveness of the description vector approach, though only geometric invariance is studied. In our case, photometric invariance is critical as non-photometrically invariant values are notoriously unreliable, especially when thresholded (see [6]).

Selection. It is also necessary to focus attention on the more interesting and informative patches by selection. Often this is done by thresholding the output of non-invariant edge or corner detectors. However, a measure of saliency is proposed here which defines how some points[1] are better suited for matching than others. The most salient points are then favoured when selecting candidate points.

Matching is the most critical aspect. The objective is to *match* an object in the scene with its model, which is assumed here to be an image of the object. This is done by matching small image patches, characterised by description vectors. Small patches have the advantage that only some will change when an object deforms, and that by using invariants, global geometric or photometric changes will have no effect at all.

A question that is fundamental to the approach is: when are two description vectors matched? Matching here means that the description vectors describe the same physical point on the object. In practice it is often assumed that if they are 'close', in terms of some constructed distance function, then they are matched. But how close is 'close'? Only by understanding how and why the vectors and their components vary can the results be made reliable and robust against changes in object pose and lighting. Often this question is ignored when constructing distance measures, and this considerably reduces their value.

2 Invariants

We assume in this paper that geometrical image transformations are limited to translations and rotations: that is, Euclidean transformations. This suffices for our application, because other transformations, such as scale changes and perspective distortions, can be removed by a pre-warping stage[5].

As non-trivial invariance is only a property that exists in continuous, noise-free images, we start by considering invariants in this domain. The quantities used in this paper that are (geometrically) invariant to Euclidean transformations, are:

$$L - \frac{\int_R L \, dS}{\int_R dS} \; ; \quad \sum_i L_{ii} \; ; \quad (\sum_i L_i L_i)^{1/2} \; ; \quad (\sum_{i,j} L_{ij} L_{ij})^{1/2} \; ; \quad (\sum_{i,j} L_i L_{ij} L_j)^{1/3} . \tag{1}$$

Here the notation denotes image intensity by L and employs index notation for the derivatives: e.g. $L_1 \equiv L_x \equiv \frac{\partial L}{\partial x}$; $\sum_i L_{ii} = L_{xx} + L_{yy} = \frac{\partial^2 L}{\partial x^2} + \frac{\partial^2 L}{\partial y^2}$

---
[1] equivalently image patches or description vectors in this context

etc. The first quantity also uses a surface integral over a circular region R, centred about the point of interest.

Note that all of the quantities are based entirely on local information at the point where they are calculated. Furthermore, they are closely related to familiar quantities: intensity deviation from the local average; the Laplacian; the magnitude of the gradient; the quadratic variation [3]; and the isophote image curvature.

Now consider photometric invariance. We assume that the photometric transformation is locally affine, that is: $L \rightarrow kL + c$ where k and c are locally constant. While this does not cover all possible photometric changes, most instances of changing scene illumination can be handled adequately within this framework.

Under such a transformation, all the quantities listed in (1) are unaffected by the offset, c, but scale linearly with k, and are therefore not invariant with respect to photometric scaling. However, by taking the quantities as components of a description vector, invariance is obtained simply by normalising the length. i.e. $\mathbf{V} = \mathbf{A}/\|\mathbf{A}\|$.

Euclidean invariants have been used in other work, especially that of Schmid and Mohr [8]. The notation used here is adapted from [8], however there are several differences in the definitions, notably the use of square roots, cube roots and demeaned local intensity. Also, no third derivatives are used as they have poor signal-to-noise ratios, which we have found to be unsuitable for the quality of images obtained from a mobile robot in uncontrolled lighting conditions. Furthermore we do not use the multi-scale approach taken in [8], but instead use a single, fixed size Gaussian convolution kernel. However, since there is no scale change in Euclidean transformations, a single scale approach is sufficient, preferably using a small Gaussian kernel to give good localisation.

3 Variations of the Invariants

There are two principle causes of variations for description vector components[2]: *pixelisation* and *image noise*. Pixelisation refers to the spatial discretisation of the image into pixels. Therefore discrete operators must be used, instead of continuous ones, and these are not invariant. As each cause is independent, they are treated separately herein.

3.1 Image Noise

Noise Model: In a CCD camera there is always some noise which corrupts the intensity readings. The causes of noise are investigated in Healey et. al. [4] who show that there are several different contributing processes. These have different distributions, including Gaussian, Poisson and Uniform. However, it is shown that the Gaussian component is usually dominant. The result is that the

---

[2] ignoring model variations

basic noise process, $\eta_0(x,y)$, in an isolated pixel can be modelled as a zero-mean Gaussian process with a variance of:

$$\sigma^2(x,y) = \sigma_1^2 L(x,y) + \sigma_2^2 \tag{2}$$

where $L(x,y)$ is the intensity at the pixel location (x,y), and σ_1^2 and σ_2^2 are constants.

This is the model of image noise that will be adopted here. However, there is also a leakage effect which correlates the noise on a local scale. This can be modelled as a set of leakage weights, ω_{ij}, such that the noise model becomes:

$$\eta(x,y) = \sum_{ij} \omega_{ij} \, \eta_0(x+i, y+j). \tag{3}$$

Note that the leakage is assumed to be position independent[3], and that the distribution remains Gaussian.

The covariance of the noise is then given by:

$$C_{ab} = E\{\eta(x_a, y_a)\eta(x_b, y_b)\} \tag{4}$$
$$= \sum_{ij} \omega_{ij} \, \omega_{i+\Delta x, j+\Delta y} (\sigma_1^2 L(x_a + i, y_a + j) + \sigma_2^2) \tag{5}$$

where $\Delta x = x_a - x_b$ and $\Delta y = y_a - y_b$. Note that care must also be taken with the limits of the summation, which are bounded locally.

Hence the noise model is *not* independent, identically distributed (i.i.d.) as is the common, but poor, assumption. Furthermore it has an intensity dependence.

The form of the noise can also be measured empirically (as was also done in [4]). Firstly, assume that the noise is not Markovian (i.e. its statistics do not change with time), which is reasonable for the steady-state case. Then the difference between two images of the same scene, taken in quick succession with a stationary camera, can be assumed to consist solely of noise. This allows empirical estimation of the noise-model parameters for the particular camera being used.

Noise Analysis: Discrete implementations are required in practice for all calculable quantities, such as those in (1). Typically this is achieved using convolutional masks. For example, a mask $m(u,v)$, generates an output:
$A(x,y) = \sum_{u,v} m(u,v) L(x+u, y+v)$.

Without loss of generality, take the output at $(x,y) = (0,0)$ and introduce a new index notation such that each location of the form (u,v) is denoted by a single index a - then the output is:
$A^{signal} = \sum_a m_a L_a$.

Furthermore, denote the noise similarly (as η_a) so that the observable output, in the presence of noise (i.e. $L \rightarrow L + \eta$) becomes:
$A = \sum_a m_a (L_a + \eta_a) = A^{signal} + A^{noise}$

---
[3] which is a good approximation except near the edges of the CCD array

Note that the noise processes, η_a, are different at each location and that the result of the convolution is to form a linear combination of the noise processes, which will be Gaussian.

In general, the aim of the noise analysis is to derive a model of the output noise, A^{noise}, in terms of a combination of the individual pixel-based noise processes, η_a. That is, to find the coefficients γ_a so that:
$$A^{noise} = \sum_a \gamma_a \eta_a.$$
The results of the analysis are summarised in the following table:

Case	A^{signal}	γ_a
Linear	$\sum_a p_a L_a$	p_a
Sum	$\sum_a p_a L_a + \sum_a q_a L_a$	$p_a + q_a$
Product*	$(\sum_a p_a L_a)(\sum_b q_b L_b)$	$(\sum_b q_b L_b) p_a + (\sum_b p_b L_b) q_a$ †
Square Root*	$(B^{signal})^{1/2}$	$\dfrac{\phi_a}{2(B^{signal})^{1/2}}$ ‡
Cube Root*	$(B^{signal})^{1/2}$	$\dfrac{\phi_a}{3(B^{signal})^{2/3}}$ ‡

Notes: † product terms are neglected as they are negligible for cases of interest
‡ uses the definition $B^{noise} = \sum_b \phi_b \eta_b$
⋆ an unbiased mean estimator exists for the non-linear cases (see [5])

Covariance: Applying these formulae allows the γ coefficients to be calculated for all the quantities listed in equation 1. In vector form, the noise component of the description vector is given by: $\mathbf{A}^{noise} = \Gamma \eta$ where Γ is a matrix formed from the γ_a terms, and η is a vector of pixel noise processes (η_a). Using this notation, the covariance is: $\mathcal{C}_A = \Gamma C \Gamma^T$ where C is the pixel-noise covariance (see equation 5). To a good approximation this gives the normalised description vector covariance as: $\mathcal{C}_V = \mathcal{C}_A / \|\mathbf{A}\|^2$.

Validation By using a non-linear optimisation method, the noise model parameters (i.e. ω_{ij}, σ_1^2 and σ_2^2) were estimated from a set of images of a test scene taken using our camera[4]. The resultant pixel-noise covariance matrix, C (equation 5), was then calculated and used to form an average description vector covariance matrix: $\mathcal{C}_A$.

This result, based upon the theoretical noise model, was then compared to the covariance measured statistically from the description vectors calculated using the same test images. The element values of the covariance matrices for the theoretical and experimental cases are plotted below for comparison. Good agreement was found, with the majority of the noise energy being well modelled.

3.2 Pixelisation

It was initially hoped that the variation due to pixelisation would be smaller than that due to noise, and only constitute a small correction. However, the

---

[4] note that similar results were obtained for other cameras

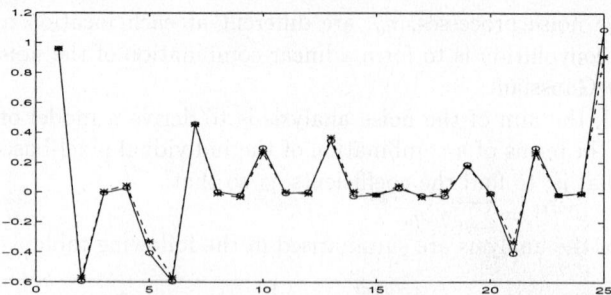

Fig. 1. A plot of the covariance matrix elements for the unnormalised description vector variations due to noise. The magnitude is plotted for both theoretical (stars on dashed line) and experimental (circles on solid line) estimations.

pixelisation variation is not negligible, even when minimised using methods such as those proposed in [7]. On the other hand, nor does it dominate.

Preliminary investigations into pixelisation variation show that it scales linearly with the standard deviation of the intensities in the local region, and that a Gaussian model is a reasonable first approximation. Therefore the total covariance due to both image noise and pixelisation is modelled here by: $C_{TOT} = C_{noise} + \sigma C_{exp}$ where σ is the standard deviation, C_{noise} the previously defined pixel-noise covariance (see (5)), and C_{exp} is the experimentally measured covariance (using a set of rotated images, and correcting for noise). Previous formulae remain valid if C_{TOT} is used instead of C: e.g. $C_A = \Gamma C_{TOT} \Gamma^\mathsf{T}$.

4 Saliency

Not all points carry equal information, and for matching purposes we want those that provide the strongest constraints. These are salient points. In terms of description vectors, the salient points are those that can be located easily and accurately. That is, they are points where the probability of mismatch is small. Therefore the measure of saliency is defined as the probability of mismatch, noting that this takes into account many possible mismatches. Hence it is a large scale view, as opposed to localisation which only looks at local mismatches.

Consider first a single pair of points in the model: the point of interest, $\mathbf{V_0}$, and another point, $\mathbf{V_i}$. In the scene image these description vectors will have different values, $\tilde{\mathbf{V}}_\mathbf{0}$ and $\tilde{\mathbf{V}}_\mathbf{i}$, due to the noise and pixelisation variations. Given the previously derived covariance describing this variation, the Mahalanobis distances (denoted by $\|\ldots\|_M$) can be calculated so that the probability of mismatch is:
$$p(\text{mismatch}) = p(\|\tilde{\mathbf{V}}_\mathbf{0} - \mathbf{V_0}\|_M > \|\tilde{\mathbf{V}}_\mathbf{i} - \mathbf{V_0}\|_M)$$
which can be found by analytical or numerical means. Therefore, for many pairs in an image section, the probability of mismatch, or saliency, is:
$$p(\text{mismatch}) = 1 - \prod_{i=1}^{n} p(\|\tilde{\mathbf{V}}_\mathbf{0} - \mathbf{V_0}\|_M < \|\tilde{\mathbf{V}}_\mathbf{i} - \mathbf{V_0}\|_M).$$

Results: Calculating the saliency measure described above gave the results shown graphically in figure 2. Note that the most salient points are typically those that are most 'interesting' to the eye.

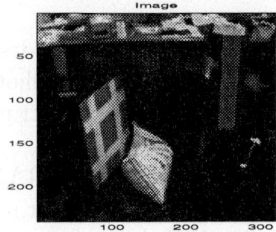
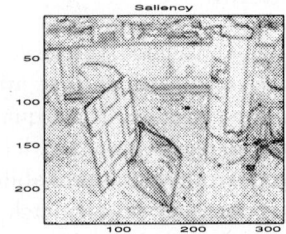

Fig. 2. Saliency values for an image. Left shows the original image. Right show the saliency measure at each point the with darkest points (lowest probability of mismatch) being most salient.

5 NU-RANSAC

Now consider the fundamental problem of finding, or localising, an object within the scene. One scheme that has been proposed for doing this is RANSAC [2], which works by randomly selecting points in one image and forming associations with points in the other. Each minimal set of associated pairs is then used to estimate the object's location and pose. Many such sets of estimates are made and they are evaluated by a verification stage to determine which is the best.

However, RANSAC treats all points equally; the saliency results show clearly that they are not. Therefore, a modification is proposed so that the selection is biased towards choosing highly salient points. As two pairs of matched points are required to estimate translation and rotation, four individual stages are proposed, each employing a biased, Non-Uniform (NU) selection probability.

In the first two stages only saliency and Mahalanobis distance are used to construct the biased probabilities. However in the latter two stages, once an initial associated pair has been selected, the image distance (in pixels) is also considered. This is done since small separations give bad transformation estimates, and the invariance of the Euclidean distance can be used to further reduce the search space.

These four stages select a single associated pair of points. In practice many such pairs need to be found and then tested via a verification stage. However, the biased selection probabilities substantially reduce the number of pairs that need to be looked at on average.

The preliminary results show that, by using saliency, the likelihood of an early mismatch occurring is at least an order of magnitude less than with normal RANSAC. As relatively few points ever need to be investigated (especially if the description vectors are arranged in hash tables), and each mismatch requires rejection from the verification stage, there is a significant benefit here. Furthermore, the increase in reliability of the initial matching can also be beneficial in other applications where matching needs to be done.

6 Conclusion

This paper has considered the issue of variations in invariant descriptors. In particular a set of quantities that are both geometrically *and* photometrically invariant has been studied. These quantities, arranged into a description vector, are used as the basis for an object localisation scheme.

Applications of the basic probability model are used in NU-RANSAC. This involves the probability of mismatch, defined as the measure for saliency. The saliency results show how promising areas are singled out for attention. This is exploited in NU-RANSAC resulting in a large increase in the probability of finding correct matches.

Future work will investigate the precise modelling of the variation due to pixelisation as it is essential that the total variation of the invariants is understood more thoroughly. This may lead to new quantities which are more stable with respect to pixelisation (see [7]).

Generally, this variation analysis provides an understanding of the way in which the measured values vary from the ideal values. The principle application of this is to facilitate *matching* by allowing the variation to be modelled. Therefore, any area where point matching is used can benefit from this analysis. Such areas not only include object recognition, but also finding stereo correspondences and tracking.

References

1. Andrew Blake, Michael Isard, and David Reynard. Learning to track the visual motion of contours. *Artificial Intelligence*, 8:101–133, 1995.
2. M.A. Fischler and R.C. Bolles. Random sample concensus: A paradigm for model fitting with applications to image analysis and automated cartography. *Comm. ACM*, 24(6):381–395, 1981.
3. W.E.L. Grimson. *From Images to Surfaces: A Computational Study of the Human Early Visual System*. MIT Press, Cambridge MA, 1981.
4. Glenn E. Healey and Raghava Kondepudy. Radiometric CCD camera calibration and noise estimation. *IEEE Transactions on Pattern Analysis and Machine Intelligence*, 16(3):267–276, 1994.
5. Mark Jenkinson. Object localisation using saliency. Technical Report OUEL 97, University of Oxford, 1997.
6. Peter Kovesi. *Invariant Measures of Image Features from Phase Information*. PhD thesis, University of Western Australia, 1996.
7. J.S.A. Merron and J.M. Brady. Isotropic gradient estimation. In *Conference on Computer Vision and Pattern Recognition*, pages 652–659, San Francisco, June 1996. IEEE Computer Society, IEEE Computer Society Press.
8. Cordelia Schmid and Roger Mohr. Object recognition using local characterization and semi-local constraints. *IEEE Transactions on Pattern Analysis and Machine Intelligence*, 19(5):530–534, 1997.

Multiscale Image Representation and Edge Detection

Fang Chen and David Suter

Department of Electrical and Computer Systems Engineering
Monash University, Victoria 3168, Australia
E-mail: [fchen,suter]@marvin.eng.monash.edu.au

Abstract. In this paper, we present an edge detection method based on the thin-plate spline with tension. Under regularization theory, the image is represented in a convolution form between the original image data and a two-dimensional kernel. This convolution kernel is derived from a PDE, which is related to a second order (thin-plate or bending) term and a first order (membrane or tension) term. This image representation involves two parameters: a smoothing parameter (or scale parameter) and a weighted smoothness parameter (which controls the degrees of the continuity of the reconstruction by placing a different weighting on the thin-plate and tension terms). By tuning these parameters, the image can be represented at different scales and with different smoothness requirements. Based on this convolution representation, the edges can be detected by differentiating the image to look for the zero-crossings of the Laplacian. By tuning the values of these two parameters, the edges at the different scales will be extracted.

1 Introduction

Detecting edges is a basic operation in image processing since the edges in a scene hold much of the important information of the image. The intuitive interpretation of the edge is that the edge occurs on the boundaries between two pixels when the intensity values of these two pixels are significantly different. This sharp intensity change will lead to peaks in the first derivative, or to zero crossings in the second derivative of the image intensity. Hence, a class of common methods for detecting edges uses a search for maxima in the slope of the surface or for zero crossings in the Laplacian of the surface. In addition, such methods also try to implement a low pass filtering (or smoothing) operation: to combat noise and to give a degree of scale selectivity. A well-known example of such an edge detector is the Marr-Hildreth's Laplacian of Gaussian filter [1], the bandwidth of the filter and the localization are determined by the variance of the Gaussian kernel.

We consider the problem of segmenting data (such as range depth data or other image data) into piece-wise smooth surfaces. Conceptually, the process involves fitting an interpolant (or approximation if the data is noisy) to the discrete data, and then segmenting the data by differentiating the interpolant to look for abrupt transitions. Variational techniques are used to reconstruct an image/surface under the thin-plate with tension smoothness constraint. In computer vision research, this type of approach has a long history: Grimson used a thin-plate type formulation [2], Terzopoulos [3] introduced a finite-element approximation of the spline and, a thin-plate spline with tension [4]. Various attempts have been made to improve the computational cost and the ability to preserve (and later detect) edges: including Graduated Non-convexity [5] and various Markov Random Field type characterizations [6]. More recently, Gokmen [7] has taken a thin-plate spline with tension approach and related it to a type of scale-space representation. He has also discussed approximations, including a one-dimensional filtering approach and shown a relationship to various edge detection filters. We extend this work to two-dimensional thin-plate spline with tension and related splines.

The main idea of the approach is to construct an image representation by minimizing an energy functional, which is associated with a thin-plate and a membrane. This new representation can then be expressed by convolving the original image data with a two-dimensional kernel, where the kernel depends on two parameters: one is a smoothing parameter λ, which controls the scale of the filter. Another is a weighted smoothness parameter ρ, which controls the trade-off in the second order continuity and the first order continuity. The edges can be extracted by differentiating the reconstruction to look for zero crossings in the Laplacian of the reconstructed image. By selecting the different λ and ρ, the edge will be detected at multiple scales and in response to different types of discontinuity. The main contribution of this paper is to establish a full 2-D approach (rather than the 1-D filter employed by Gokmen): we give explicit formulae for the filters and demonstrate their action on images.

The material is organized as follows: in section 2, we start with a thin-plate spline with tension minimization problem, then derive an image reconstruction operator, in convolution form, based on hybrid smoothness constraints. The explicit expression of the convolution kernel and its Laplacian in two dimensions, with different parameters, are derived in section 2.2. In section 3, an edge detector is developed based on the two-dimensional filter technique. The effect of the two parameters defined in this method is also discussed. In section 4, we demonstrate, with numerical examples, the edge detection at different scales and with different continuity requirements.

2 Thin-plate Spline with Tension

2.1 Image Reconstruction Based on Thin-plate with Tension

For modelling a smooth image/surface, we employ a smoothness functional constructed from the second order derivatives (thin-plate spline) and the first order derivative (membrane). In terms of physical explanation, a membrane defines a tension energy and the second order term is related to a bending energy. These two terms are combined together with a weight parameter so that the continuity of the interpolants derived from this model can be tuned by changing the value of the parameter. The minimization problem associated with the hybrid energy functional can be described as follows: for given original data $g(x,y)$ (may include noise), to find a smooth solution f to minimize the smoothness functional:

$$\int_{\Re^2} (f-g)^2 \, dx \, dy + \lambda \int_{\Re^2} \{\rho(f_x'^2 + f_y'^2) + (1-\rho)(f_{xx}''^2 + 2f_{xy}'^2 + f_{yy}''^2)\} dx \, dy. \quad (1)$$

The first term in (1) defines the residual error of the reconstruction from the original data. The second integral in (1) represents a measure of the roughness penalty of function f. The parameter $\lambda \geq 0$ is called a smoothing parameter or scale parameter, which controls the amount of the data smoothing with a large λ generating a coarse scale image representation, and vice verse. Another parameter $0 \leq \rho \leq 1$ is a weighted tension parameter to control the continuity of the solution.

The above minimization problem (1) is associated with an Euler-Lagrange equation:

$$f - \lambda \rho \Delta f + \lambda (1-\rho) \Delta^2 f = g \quad (2)$$

Let $\alpha = \lambda(1-\rho)$, $\beta = \lambda \rho$ and define a differential operator $P(D) = \alpha \Delta^2 - \beta \Delta + 1$, then f is a solution of the partial differential equation: $P(D)f = g$.

In the Fourier domain, f can be solved by a Fourier transform:

$$\mathcal{F}f = \frac{\mathcal{F}g}{P(\xi)} = (\mathcal{F}K) \cdot (\mathcal{F}g) = \mathcal{F}(K * g) \quad (3)$$

where $\xi \in \Re^2$, $K = \mathcal{F}^{-1}(\frac{1}{P(\xi)}) = \mathcal{F}^{-1}(\frac{1}{\alpha \xi^4 + \beta \xi^2 + 1})$.

The solution will be found in a convolution form by taking an inverse Fourier transform in equation (3):

$$f = K * g = \int K(x-y)g(y) \, dy. \quad (4)$$

Assume that g represents an original image signal, then the new image representation f is constructed by convolving the original data with a two-dimensional thin-plate with tension kernel K. The convolution kernel is influenced by a smoothing or scale parameter λ and a tension parameter ρ. Therefore, the new reconstruction (4) defines a multiscale image representation governing by λ and ρ. The major feature of this representation is that the scales and the degrees of the continuity of the reconstruction can be easily tuned by placing different values on λ and ρ.

The convolution operator in (4) works as a low-pass filter. As ρ decreases from 1 to 0, the smoothness constraints change from the membrane dominated to the thin-plate dominated, so that the degrees of the continuity of the reconstruction increase up to the second order smoothness. On the other hand, for a given ρ, it can be expected that the fine-scale details will disappear and images become more diffuse when the scale parameter λ increases, conversely reconstruction with small λ will focus on the fine features in the image.

It is noted that above thin-plate spline with tension model also contains two extreme cases: the membrane case with $\rho = 1$ and the thin-plate spline with $\rho = 0$.

2.2 Kernels and Laplacian of Kernels

In this section, we focus on deriving the explicit expressions of the kernel K for different selections of λ and ρ, then calculate the Laplacian of the kernel. In two dimensions, the kernel and its Laplacian operator are both expressed in the terms of a linear combination of the modified Bessel function K_0. Except for the true membrane case ($\rho = 1$), the kernel itself is a regular function, but the Laplacian of the kernel has a singularity at origin. In fact, this singularity is caused by the modified Bessel function. It is known [9] that the modified Bessel function is approximated by a $\log r$ term plus a constant when $r \to 0$. We use a local average operator [8] to remove this discontinuity.

Now we derive the explicit formula for kernel K and its Laplacian ΔK. From section 2.1, the kernel K can be calculated by an inverse Fourier transform:

$$K = \mathcal{F}^{-1}(\frac{1}{\alpha \xi^4 + \beta \xi^2 + 1}). \tag{5}$$

It is clear that the expression for K, as well as ΔK, depends on two coefficients α and β, or in turn, by λ and ρ. According to the different selection of λ and ρ, we identity the following four cases (we ignore the singular membrane case $\rho = 1$, which can be approximated by a limit setting $\rho \to 1$):

(i) **When $D = \beta^2 - 4\alpha > 0, \alpha \neq 0, \beta \neq 0$**

$$\begin{cases} K = \frac{1}{2\pi\alpha(b^2-a^2)}(K_0(ar) - K_0(br)) \\ \Delta K = \frac{1}{2\pi\alpha(b^2-a^2)}(a^2 K_0(ar) - b^2 K_0(br)) \end{cases}$$

where $a^2 = (\beta + \sqrt{D})/2\alpha$ and $b^2 = (\beta - \sqrt{D})/2\alpha$. K_0 is the zero order modified Bessel function of the second kind. $r = \sqrt{x^2 + y^2}$.

(ii) **When $D = \beta^2 - 4\alpha < 0, \alpha \neq 0, \beta \neq 0$**

$$\begin{cases} K = \frac{1}{2\pi\sqrt{-D}}\{2\theta + \sum_{n=1}^{\infty}[(\log \frac{\eta r}{2} + C_E - \sum_{k=1}^{n} \frac{1}{k})D_1 + \theta D_2]\frac{2r^{2n}}{4^n(n!)^2}\} \\ \Delta K = \frac{1}{2\pi\sqrt{-D}}\sum_{n=1}^{\infty}[(2 + 2n(\log \frac{\eta r}{2} + C_E) - 2n\sum_{k=1}^{n} \frac{1}{k})D_1 + 2n\theta D_2]\frac{4nr^{2n-2}}{4^n(n!)^2} \end{cases}$$

$$\theta = \frac{1}{2}\arctan\frac{\sqrt{-D}}{\beta}; \quad \eta = \alpha^{-1/4}; \quad u_1 = \eta\cos\theta; \quad u_2 = \eta\sin\theta; \quad C_E : \text{Euler constant};$$

$$D_1 = \sum_{k=1}^{n}(-1)^{k-1}\binom{2n}{2k-1}u_1^{2n-2k+1}u_2^{2k-1}; \quad D_2 = \sum_{k=0}^{n}(-1)^k\binom{2n}{2k}u_1^{2n-2k}u_2^{2k}.$$

(iii) **When $D = \beta^2 - 4\alpha = 0, \alpha \neq 0, \beta \neq 0$**

$$\begin{cases} K = \frac{r}{4\pi\alpha c}K_1(cr), \quad (K_1 \text{ is the first order modified Bessel function}) \\ \Delta K = \frac{1}{4\pi\alpha}\{1 + 2E + \sum_{n=1}^{\infty}[1 + 2(n+1)E - 2(n+1)\sum_{k=1}^{n}\frac{1}{k}]\frac{(\frac{1}{4}(cr)^2)^n}{(n!)^2}\} \end{cases}$$

where $E = \log\frac{cr}{2} + C_E$ and $c = \sqrt{\frac{\beta}{2\alpha}}$.

(iv) Thin-plate case: $\rho = 0$, that is, $\alpha > 0$, and $\beta = 0$

$$\begin{cases} K = \frac{1}{4\pi\sqrt{\alpha}}\{2\theta + \sum_{n=1}^{\infty}[(\log\frac{\eta r}{2} + C_E - \sum_{k=1}^{n}\frac{1}{k})D_1 + \theta D_2]\frac{2r^{2n}}{4^n(n!)^2}\} \\ \Delta K = \frac{1}{4\pi\sqrt{\alpha}}\sum_{n=1}^{\infty}[(2 + 2n(\log\frac{\eta r}{2} + C_E) - 2n\sum_{k=1}^{n}\frac{1}{k})D_1 + 2n\theta D_2]\frac{4nr^{2n-2}}{4^n(n!)^2} \end{cases}$$

where η, D_1, D_2 and u_1, u_2 are defined same as in case 2 except $\theta = \pi/4$.

3 Edge Detection with Laplacian Filter

In order to detect discontinuities or edges, a natural approach is to look for maxima (peaks) in the slope of the surface, or for zero crossings in the Laplacian of the surface. As discussed in section 2.1, the new image representation is obtained in a convolution form: $f = K * g$, where g is an original image, K is a two-dimensional thin-plate spline with tension kernel. Thus an explicit method of edge detection can be produced by directly calculating the Laplacian of the reconstructed image, then looking for the edge points in zero-crossings. Using the derivative rule for the convolution operator, we have: $\Delta f = \Delta(K * g) = \Delta K * g$. This edge detector has bandpass filter characteristics, where the bandwidth is determined by parameters λ and ρ.

The performance of edge detector is greatly effected by the two parameters λ and ρ. When the tension parameter ρ is relatively large, the kernel is less smooth, and the shape of the filter is sharp and narrow. Otherwise, the kernel is more smooth and the filter becomes wide. The ability of the filter to *localize edges* of different types is controlled by tension parameter ρ, with a small ρ implying a good localization of edge detector - as demonstrated in the next section.

On the other hand, the scale parameter λ, with fixed ρ, is capable of controlling the edge detector so as to extract edges at *different scales*. The filter with a large scale (large λ) will detect the coarse edges, while the filter with a small scale will extract much finer features in the image. In particular, for noisy images, one can enforce a larger λ in the filter to suppress the noise.

In summary, the edge detector we have developed provides an efficient tool for general purpose edge detection. It has the capability of controlling the scales of the filter (λ) as well as degrees of discontinuity (ρ).

4 Experimental Results

In this section, we demonstrate the capabilities of the edge detector described above with examples obtained on synthetic images and real image. We also demonstrate how the parameters can be optimized to achieve the trade-off between localization of edges and noise suppression in noisy image edge detection. For these purpose, three examples are included to illustrate the localization with selected types of discontinuity, noise suppression, and multiple scale edge detection, respectively.

Figure 1 shows two different synthetic black and white bar images and the edge detection associated with different tension parameter settings (ρ) for a given

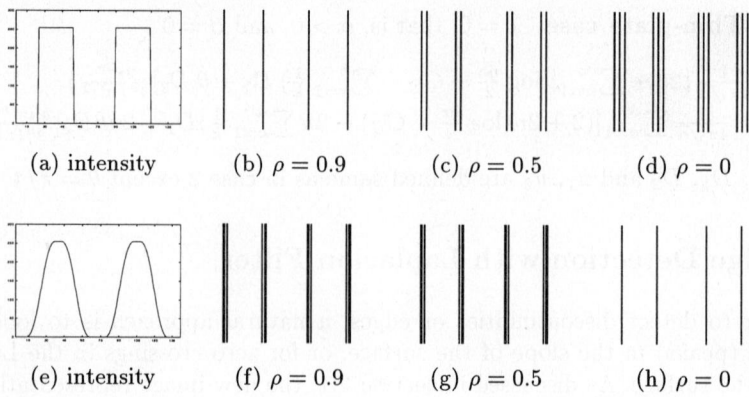

Fig. 1. Synthetic bar images edge detection with different degrees of continuity: Top one shows that the kernel enforcing low order of continuity has better localization of this type of discontinuity, while bottom one shows that the filter derived using $\rho = 0$ performs better at locating edges with higher continuity (for both cases, $\lambda = 1$).

scale parameter $\lambda = 1$. Results in the top row (case 1) are obtained from a sharp two-value bar image, while results in the bottom row (case 2) are derived from a blurred bar image. In this figure, the first column shows the 1-D profiles of the intensity for two images, a strict step function for the case 1 and a smoother function for the case 2. Edges shown in the second, third and last columns are detected based on the membrane dominated model ($\rho = 0.9$), intermediate model ($\rho = 0.5$) and the true thin-plate spline model ($\rho = 0$), respectively. As expected, case 1 shows that the filter with a low order of continuity has superior localization performance for this type of step discontinuity. But in case 2, the higher degrees of smoothness filter shows a good localization for such types of continuity since the change in intensity near the boundaries has a smoother transition.

In Figure 2, a noisy synthetic checkboard image is used to demonstrate the noise suppression of the edge detector. As discussed above, a smoother edge detector is expected to be of use for combating the noise and producing the correct edges. In this figure, (a) shows the original synthetic noisy image. (b), (c), (d) show the edges detected with different values of λ (setting $\rho = 0$). As we can seen, the edge detector with a relatively small scale ($\lambda = 0.1$) preserves lots of finer features, such as the noise, while a large scale ($\lambda = 10$) edge detector captures only the coarse scale changes and provides a much clear picture of the edges.

Figure 3 gives an example of multiple scale real image edge detection: (a) shows the original real image, (b), (c) and (d) show the edges detected at different scales based on the thin-plate spline model ($\rho = 0$). It can be seen that the finer features are preserved at the small scale, such as in (b), while for a large λ in (d), a more coarser scale edges are obtained.

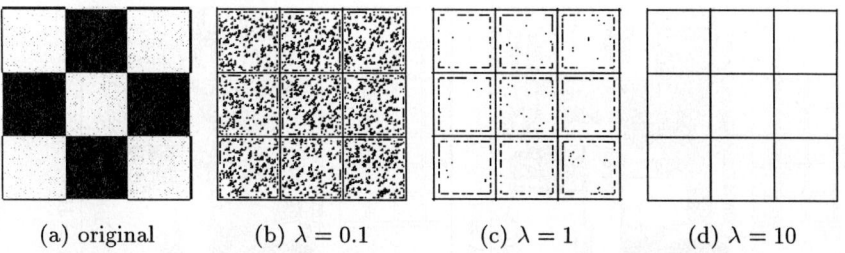

(a) original   (b) $\lambda = 0.1$ (c) $\lambda = 1$ (d) $\lambda = 10$

Fig. 2. Synthetic checkboard image (with random noise) edge detection associated with the different scale parameter λ (for given $\rho = 0$) shows the larger scale filter has a good noise suppression performance since the finer scale features, such as noise, have been discarded during the smooth procedure.

5 Conclusion

We have extended the work of Gokmen to give an explicit method of detecting edges using a true 2-D convolution filter based on the thin-plate spline with a membrane term. We have also illustrated how the derived method is capable of extracting edges at multiple scales. Moreover, unlike the related Marr-Hildreth type operator, our operator can be tuned to select edges with various degrees of continuity.

References

1. David Marr and E. Hildreth. Theory of edge detection. *Proc. R. Soc. Lond. B*, 207:187–217, 1980.
2. W. E. L. Grimson. Implementation of a computational theory of visual surface interpolation. *Computer Vision, Graphics, and Image Processing*, 22:39–69, 1983.
3. D. Terzopoulos. Multilevel computational processes for visual reconstruction. *Computer Graphics, Vision, and Image Processing*, 24:52–96, 1983.
4. D. Terzopoulos. Regularization of inverse visual problems involving discontinuities. *IEEE Trans. Pattern Analysis and Machine Intelligence*, 8(4):413–424, July 1986.
5. A. Blake and A Zisserman. *Visual Reconstruction.* MIT Press, Cambridge MA, 1987.
6. S. Geman and D. Geman. Stochastic relaxations, Gibbs distributions, and the Bayesian restoration of images. *IEEE Trans. on Pattern Analysis and Machine Intelligence*, 6:731–741, 1984.
7. M. Gokmen and A. K. Jain. $\lambda\tau$-space representation of images and generalized edge detector. *Proceedings CVPR'96*, pages 764–769, New York, June 1996.
8. M. Benbourhim . Splines elliptiques et interpolation par moyennes locales. *Congres National d'Analyse Numerique*, 1985.
9. Milton Abramowitz and Irene A. Stegun. Handbook of mathematical functions. *Dover publications, INC. New York*, 1964.

Fig. 3. Real image edge detection at different scales based on the thin-plate spline model ($\rho = 0$).

Rotation Invariant Texture Features from Gabor Filters

S. R. Fountain and T. N. Tan

Dept. of Computer Science, The University of Reading, Reading RG6 6AY, UK
Email: S.R.Fountain@reading.ac.uk, T.Tan@reading.ac.uk

Abstract: This paper presents a thorough investigation into the use of Gabor filters for the extraction of rotation invariant texture features. Numerous experiments have been conducted to discover the effect of different parameter settings on classification results. The optimum parameter settings are established and tested by classification and content based image retrieval experiments on a large database of randomly rotated Brodatz texture images. Resistance of the method to Gaussian noise is also examined. The issues studied in this paper are of great importance for practical applications but have not been adequately addressed by existing work on texture analysis.

1 Introduction

Texture analysis is important for many applications such as content based image retrieval and scene analysis. Research on the subject has been active for decades. The vast majority of techniques developed to date (see [10] for a recent review) assume that textures are acquired from the same viewpoint. This is an unrealistic assumption for most practical applications. For example, if images are obtained from scanning photographs they are often subject to random skew angles. Texture analysis methods should ideally be invariant to viewpoints in such applications. Obtaining viewpoint invariant texture features is a very difficult task [1]. Rotation (e.g. due to skew angles) and scale (e.g. change in focal length) invariance are important aspects of the general viewpoint invariance problem.

This paper focuses on rotation invariance (however the method used is readily extendible to scale invariance). Following on from our previous work [2] the paper presents a comprehensive study on the use of Gabor filters for rotation invariant texture analysis. Previous studies into rotation invariant texture analysis have proven computationally expensive to execute (e.g. [6]), or train (e.g. [5]), others require a restricted data set (e.g. [17]). In contrast Gabor filters are comparatively computationally inexpensive (highly suited to parallel implementation) and require no data set restrictions rendering them ideal for most applications involving texture analysis (e.g. content based image retrieval).

Gabor filters are highly desirable for texture analysis as they give maximum resolution in both the frequency and space domains. There is also evidence that Gabor filters provide good models for the response profiles of many cortical cells in the visual cortex [14] thus further increasing their validity. Pairs of symmetric and antisymmetric filters are often used to model visual cortical cells [13]. Tan [9] gives a detailed description of this application.

A detailed investigation in to the optimum parameter settings for the Gabor method and its resistance to noise is presented. Such studies are of great practical importance but have not been addressed in previous work (e.g. [2-4, 11]). Many papers on the subject fail to include extensive testing on large databases (e.g. [7]), or do not use random rotation angles (e.g. [6]). These two issues are also major concerns of this paper. Classification and content based image retrieval results on a database of 1320 images from 44 different Brodatz texture classes are presented.

The paper is organised as follows: a general introduction to Gabor filters is followed by the description of filter design for the work in this paper. Experiments concentrating on the input parameters of the system are presented along with the corresponding results. Classification and image retrieval on a large database then follows. Finally the method's resistance to noise is addressed.

2 Rotation Invariant Feature Extraction Using Gabor Filters

From a signal analysis point of view signals can either be described as functions of time or frequency content. The uncertainty relation $\Delta t \Delta f \geq k$ (Δt is the uncertainty of location in time domain, Δf is the uncertainty of signal in frequency domain and k is a constant) states that it is not possible to achieve simultaneous optimal localisation in both domains [12]. If a signal is perfectly defined in time (i.e. $\Delta t=0$) it is undefined in the frequency domain (i.e. $\Delta f=\infty$). The converse is also true. A family of functions which achieves the minimum joint uncertainty (i.e. $\Delta t \Delta f = k$) was found by Gabor [12]. Such optimal functions are now commonly called the Gabor functions and are the product of a Gaussian function and a complex sinusoid. A 2-dimensional isotropic Gabor function h(x,y) centred at frequency (u_o, v_o) is given as follows:

$$h(x,y) = g(x,y) e^{-2\pi j(u_o x + v_o y)} \tag{1}$$

where g(x,y) is an isotropic Gaussian of the form:

$$g(x,y) = e^{-\frac{1}{2}\left[\frac{x^2+y^2}{\sigma^2}\right]} \tag{2}$$

This function can be decomposed in to an even ($h_e(x,y)$) and odd ($h_o(x,y)$) part, also known as the symmetric and antisymmetric filter respectively:

$$h_e(x,y) = g(x,y) \cos(2\pi f(x \cos\theta + y \sin\theta))$$
$$h_o(x,y) = g(x,y) \sin(2\pi f(x \cos\theta + y \sin\theta)) \tag{3}$$

where $f = \sqrt{u_o^2 + v_o^2}$ and $\theta = \arctan(v_o/u_o)$.

2.1 Filter design for rotation invariant texture analysis

Gabor filtering is performed in the frequency domain for increased efficiency. The Fourier transforms of the equations in (3) are given as follows:

$$H_e(u,v) = \left(e^{s[(u-u_o)^2+(v-v_o)^2]} + e^{s[(u+u_o)^2+(v+v_o)^2]}\right)/2$$
$$H_o(u,v) = \left(e^{s[(u-u_o)^2+(v-v_o)^2]} - e^{s[(u+u_o)^2+(v+v_o)^2]}\right)/2j \tag{4}$$

where $s = -2\pi^2\sigma^2$ and $j = \sqrt{-1}$. The output images of the two filters $q_e(x,y)$ and $q_o(x,y)$ are obtained in the frequency domain via FFT, for example, q_e is obtained by:

$$q_e(x,y) = FFT^{-1}[P(u,v) \times H_e(u,v)] \tag{5}$$

where P(u,v) is the Fourier transform of the input image p(x,y). The outputs of the two filters are combined (see [9,13] for combination justification) at each pixel by:

$$q(x,y) = \sqrt{q_e^2(x,y) + q_o^2(x,y)} \tag{6}$$

The radial frequency (f) and orientation (θ) with respect to the horizontal axis are required to position a Gabor filter. For each radial frequency f, multiple filters are obtained by sampling around the circle of radius f with sampling interval Δθ. A total of 180/Δθ filters are required as conjugate symmetry is exploited.

The resulting sequence of 180/Δθ filtered images at each frequency are used to obtain rotation invariant texture features. The energies of the filtered images form a periodic function of θ with period π. The magnitudes of the periodic function's Fourier coefficients are invariant to image rotations (since a rotation of the image corresponds to a translation of the periodic function and Fourier magnitudes are invariant to translations). The first n magnitudes are used as n rotation invariant texture features. Thus for a set of M radial frequencies, we obtain a total of $n*M$ rotation invariant texture features.

3 Experimentation

A database of ten Brodatz texture classes [8], shown with bold italic labels in Figure 1, was used as a test bed for initial experimentation. Each 512*512 image was rotated by an arbitrary angle and randomly cropped to 128*128; 45 images were obtained from each texture class in this way. The resulting 450 images were subjected to histogram equalisation to prevent bias towards images possessing similar grey levels.

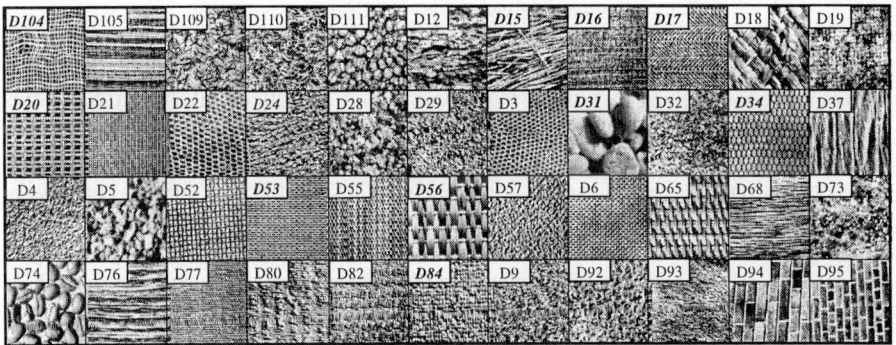

Figure 1 The texture classes used for initial testing (italic bold labels) and final testing

Numerous experiments were performed in order to discover the optimum parameter settings in terms of accuracy and efficiency. The minimum number of features, optimum sampling interval (Δθ) and frequency combinations were investigated. The first six powers of two (i.e. 2,4,8,16,32,64) were selected as radial frequencies for the frequency analysis in which all possible combinations were examined.

The robustness of the method to noise was tested by adding Gaussian noise (with standard deviation ranging from 0 to 90) to each image in the database. Finally the method was tested on a large database containing textures of all 44 classes shown in Figure 1. Classification and content based image retrieval (CBR) experiments were conducted on both clean and noisy images. In all classification experiments the

Euclidean classifier was used. During classification half of the database is used as training data and the other half as test data.

3.1 Frequency Combinations

Each combination of the six radial frequencies was studied via classification experiments. The best frequency combinations and relative effectiveness of features derived from each frequency were identified. In these experiments a sampling angle of 10° was used and three features (n=3) per radial frequency were retained. Thirteen of the frequency combinations (specified in Table 2) gave a 100% correct classification rate. The accuracy of the remaining combinations is shown in Figure 2.

52.3%	89.1%	96.3%	97.8%
64	4,32	2,8,64	4,16
4	2,4,64 / 2,32	2,16	2,4,16 / 2,4,32
2	8,16,64 / 8,64	2,8,32	4,8,16
4,64	32,64	2,4,8,64	8,16 / 2,16,64 / 4,16,64
2,64	4,8,64 / 2,32,64	4,8,32,64 / 2,8,32,64	2,8,16 / 2,4,8,16 / 2,4,16,64
16	2,4,32,64	2,8 / 2,4,8	16,32
32	4,32,64	2,4,8,32	16,32,64
8	2,4	8,32,64 / 2,4,8,32,64	2,8,16,64 / 4,8,16,64 / 2,4,8,16,64
4,8	8,32 / 4,8,32	16,64	4,16,32
			↓
			99.8%

Figure 2 The accuracy of different frequency combinations excluding those which give a 100% classification rate

It can be seen that a frequency of 64 alone gives the worst recognition rate of 52.3%. This is closely followed by single frequencies 4 and 2. Notice that these frequencies lie towards the edges of the frequency space. In comparison the most successful single frequency result was 8 which gave an 87% classification accuracy. The other two central frequencies 16 and 32 followed with an accuracy of 86.2% and 86.5% respectively. Notice that these single frequencies give better results than pairs of very high and low frequencies i.e. (2,64) and (4,64). The most successful frequency pair was (16, 32) which obtained an outstanding 99.4% accuracy and is present in each of the thirteen top combinations (see Table 2).

The gain in recognition rate of adding an extra frequency to each existing combination was measured. For example, the success of frequency combination (2,16,64) was compared to (2,4,16,64), (2,8,16,64) and (2,16,32,64) and the classification success noted. Table 1 shows the mean gains of this operation over all frequency combinations. It can be seen that increasing from one frequency to two is the most advantageous with a maximum gain of 45.3%. The mean gain decreases as more frequencies are added; in some cases frequency addition has an adverse effect.

3.2 Number of Features

For a given set of frequencies, the other factor which affects classification accuracy is the number of features used. Similar experiments were conducted to study the effects of using less features on the performance of the 13 frequency combinations shown in Table 2 (i.e. those frequency combinations which give perfect classification results when 3 features per frequency were used). It is found that reducing the number of features per frequency from 3 to 2 has no effect on performance. However, when only one feature per frequency was used, the recognition rates for frequency combinations (8,16,32), (2,16,32) and (4,16,32,64) dropped to 99.8, 99.6 and 99.8 respectively (all other combinations still achieved perfect results).

The above results show that the minimum number of features required to achieve perfect classification is 4 (e.g. one feature per frequency from combination (2,16,32,64) etc.). The use of 6 features (i.e. two features per frequency from combinations (2,16,32) and (8,16,32)) also gives a perfect recognition rate and is computationally more efficient (in series) since one less frequency iteration is required during the programs execution.

3.3 Sampling Interval

The sampling interval $\Delta\theta$ also has an impact on classification accuracy. The optimal values for $\Delta\theta$ were found as follows. Classification results were collected at sampling angles between 5° and 45° in 5° steps; an angle of 2° was also tested. Frequency combinations (2,16,32) and (8,16,32) were used with 3 features per radial frequency. Figure 3 shows the results of the study. It can be seen that angles between 5° and 15° give the optimum classification results. If speed is a particular concern an angle of 40° seems to be acceptable. It is unclear at present why there is a notch at $\Delta\theta=25°$.

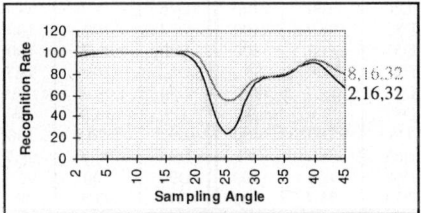

Figure 3 The effects of the sampling interval on the recognition rate

Freq Change	Ave.	Max.	Min.
1 → 2	18.1	45.3	2.6
2 → 3	4.1	19.2	-6.7
3 → 4	1.6	7.8	-5.2

Table 1 The effects on the recognition rate of adding an extra frequency (%)

3.4 Classification on a Large Database

The results discussed so far were obtained from a database of 450 images generated from 10 Brodatz texture classes. In this section, the effectiveness of the rotation invariant features is further examined by using a much larger database of 44 Brodatz texture classes (as shown in Figure 1).

Each texture was randomly rotated, cropped and histogram equalised to prevent bias towards images with similar grey levels. This resulted in a database of 1320 images. The optimum frequency combinations discovered in Section 3.1 were used during classification experiments. A sampling angle of 10° was applied and 3 features per radial frequency were employed. Table 2 shows the classification results. All frequency combinations used in the experiments gave perfect classification rates on the small database. For the large database, the best result (94.4%) is obtained via the use of all 6 radial frequencies.

Number of Frequencies	Frequency Combination	1st Place Recog. Rate	Number of Frequencies	Frequency Combination	1st Place Recog. Rate
3	2,16,32	84.4	5	2,4,16,32,64	91.1
	8,16,32	86.9		2,8,16,32	92.5
4	4,16,32,64	88.1		2,4,8,16,32	93.2
	2,16,32,64	88.3		4,8,16,32,64	93.4
	2,4,16,32	88.6		2,8,16,32,64	93.6
	8,16,32,64	89.8	6	2,4,8,16,32,64	94.4
	4,8,16,32	92.0			

Table 2 Classification rates on the large database

The addition of extra frequencies gives better results, but the relative strength of frequency combinations with the same number of elements remains similar to the small test database results. For example, (8,16,32) was a stronger combination than (2,16,32) on the small database (see Figure 3). This is also reflected in the recognition rates using the large database. Table 3 shows the average classification accuracy for all texture classes using 3 features for each of the 6 radial frequencies. A total of 94.4% of all textures are correctly classified.

Text.	Class.	CBR	Text.	Class.	CBR	Text.	Class.	CBR	Text.	Class.	CBR
D104	100	94.8	D20	93.3	95.3	D4	90	96	D74	96.7	98
D105	73.3	91	D21	100	100	D5	100	100	D76	76.7	94.7
D109	90	96	D22	100	100	D52	96.7	100	D77	100	100
D110	90	96	D24	80	94	D53	100	100	D80	96.7	100
D111	100	100	D28	100	98	D55	100	99.3	D82	100	98.7
D12	73.3	96.7	D29	86.7	96	D56	96.7	96.7	D84	93.3	92
D15	96.7	99.3	D3	100	100	D57	90	94	D9	83.3	96
D16	100	97.3	D31	100	100	D6	100	100	D92	90	100
D17	100	100	D32	100	96.7	D65	96.7	98	D93	96.7	98.7
D18	96.7	99.3	D34	100	100	D68	100	100	D94	100	99.3
D19	100	100	D37	73.3	97.3	D73	96.7	100	D95	93.3	96

Table 3 Classification and content based retrieval rates (%) for the large database

3.5 Content Based Image Retrieval

Content based retrieval was used in order to apply the method to a practical application. A query image is presented to a retrieval system [15] which searches the database and returns the n closest images according to a similarity criterion. In this experiment each of the 1320 images was presented to the retrieval system in turn. The similarity measure used is the Euclidean distance between the Gabor feature vectors of the query and database images. The parameter settings are identical to those used in the classification experiments on the large database (cf. Section 3.4). The five closest images are returned per search.

Table 3 shows the probability (%) that each image returned from a search is of the same texture class as the query image. For example if a sample of texture D104 is presented to the system an average of 94.8% of the images returned from the search belongs to class D104. It is interesting to note that in some cases classes with perfect classification rates give imperfect retrieval rates and vice versa. 14 classes give the same (not necessarily perfect) results in both situations, 19 classes are more successful during retrieval and 11 classes are more successful during classification.

An average of 98% of all the textures returned by the retrieval system were of the same texture class as the query image. This compares favourably to a 94.4% average overall classification rate.

3.6 Resistance to Gaussian Noise

The images used in the previous sections are all clean noise free images. In practice images are often contaminated with noise and it is of great importance that the performance of the rotation invariant features under noise conditions is properly characterised [16]. For this purpose various levels of Gaussian noise (σ=10,30,50,70,90) were added to each image in both databases. The large database now contains 6600 (1320*5) images and the small database 2250 (450*5). Figure 4 shows texture D104 with various levels of noise added to it.

Classification experiments were performed on both databases. The noisy images were classified to the original noise free exemplar class vectors. Content based retrieval was also performed using each of the 6600 noisy images to query the original noise free database. The parameters are the same as in previous experiments. The success rate of each experiment is shown in Figure 5.

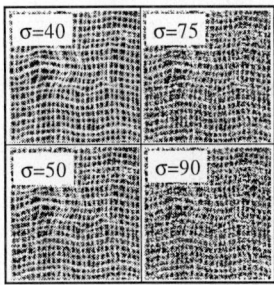

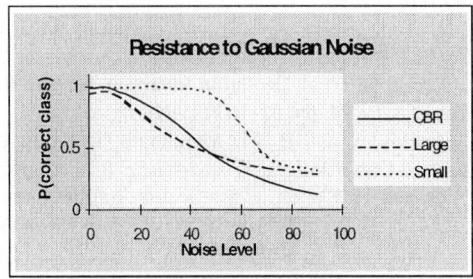

Figure 4 The addition of Gaussian noise to texture D104

Figure 5 Results for CBR on the large database and classification on both databases

On the small database the classification rate remains at 100% until $\sigma\approx40$ at which point it rapidly deteriorates. In contrast classification results on the large database show an almost immediate but less vigorous deterioration. Increasing the noise level has less effect at each step; the curve gradually shallows. At $\sigma\approx90$ the classification rates of the large and small databases almost converge.

It can be seen from Figure 5 that image retrieval is more resistant to low levels of noise than classification, at $\sigma\approx50$ the two curves cross and the converse is true. The CBR rate steadily falls whereas the classification rate levels out. It is interesting to note that for an average retrieval rate of one correct texture class per search the method is successful until $\sigma\approx75$. Example images containing the aforementioned levels of noise are presented in Figure 4 (Figure 1 shows the original).

4 Conclusions

The Gabor method for rotation invariant texture classification and retrieval has been thoroughly investigated. Extensive experimental evidence has been presented. The effects on classification accuracy of the sampling angle, the number of features, and the number and combinations of radial frequencies have been examined. The noise robustness of the method has also been studied.

It was found that combinations of central radial frequencies are the most successful. Adverse effects due to the curse of dimensionality become apparent as the number of frequencies increase. A sampling angle between 5° and 15° was found to be optimum. Classification experiments on a database of 44 Brodatz texture classes containing 1320 images at random orientations were performed. An average classification rate of **94.4%** was obtained.

The method has also been applied to content based image retrieval where a query image is presented to a retrieval system and the 5 closest images are returned from the database. An average of **98%** of all images returned by the retrieval engine were of the same texture class as the query image. A noise level of $\sigma \approx 75$ is reached before the average return is one correct image per search.

5 References

[1] T. N. Tan, Geometric Transform Invariant Texture Analysis, *Proc. of SPIE*, Vol. 2488, pp475-485 (1995).
[2] T. N. Tan, Noise Robust and Rotation Invariant Texture Classification, *Proc. of EUSIPCO-94*, pp1377-1380 (1994).
[3] H. Greenspan et. al., Rotation Invariant Texture Recognition using a Steerable Pyramid, *Proc. of ICPR94*, pp162-167 (1994).
[4] G. M. Hayley and B. M. Manjunath, Rotation Invariant Texture Classification using Modified Gabor Filters, *Proc. of IEEE ICIP95*, pp262-265 (1994).
[5] J. You and H. Cohen, Classification and Segmentation of Rotated and Scaled Texture Images using Tuned Masks, *Pattern Recognition*, Vol.26, No.2, pp245-258, (1993).
[6] R. Kashyap and A. Khotanzad, A Model Based Method For Rotation Invariant Texture Classification, *IEEE Transactions on Pattern Analysis and Machine Intelligence*, PAMI-8(4), pp. 786-804 (1986).
[7] S. Madiraju et al, On The Covariance Technique for Robust and Rotation Invariant Texture Processing, *Proc. of ACCV '93*, (1993).
[8] P. Brodatz, Textures: A Photographic Album for Artists and Designer, NY (1966).
[9] T. Tan, Texture Feature Extraction via Cortical Channel Modelling, *Proc. 11th IAPR Inter. Conf. Pattern Recognition*, IEEE Computer Society Press, C607-C610, (1992).
[10] T. Reed and J. du Buf, A Recent Review of Texture Segmentation and Feature Extraction Techniques, *CVGIP: Image Understanding*, Vol. 57, pp359-372, (1993).
[11] M. Leung and A. M. Peterson, Multiple Channel Neural Network Model for Texture Classification and Segmentation, *Proc. of IEEE Inter. Conf. on Acoustics, Speech and Signal Processing*, pp2677-2680 Toronto, Ontario, Canada, (1991).
[12] D Gabor, Theory of Communications, *J. Inst. Elec. Engng*, Vol. 93, pp429-459, (1946).
[13] D. Pollen and S. Ronner, Visual Cortical Neurons as Localised Spatial Frequency Filters, *IEEE Trans. SMC*, Vol. 13, pp907-916, (1983).
[14] S. Marcelja, Mathematical Description of The Responses of the Simple Cortical Cells, *J. Opt. Soc. Am.*, Vol. 70, pp1297-1300, (1980).
[15] S. Fountain and T. Tan, Rotation Invariant Retrieval and Annotation of Image Databases, *BMVC*, Vol. 2, pp390-399,(1997).
[16] R. Haralick, Performance Characterisation in Computer Vision, *BMVC*, pp1-8, (1992).
[17] G. Eichmann and T. Kasparis, Topologically Invariant Texture Descriptors, *CVGIP*, Vol. 41, pp267-281, (1988).

Euclidean Invariants of Linear Scale-Spaces

Alfons Salden*

INRIA, 2004 route des Lucioles, BP 93,
F-06902 Sophia-Antipolis Cedex, France

Abstract. The similarity jet of a linear scale-space is described in its most concise set of local and multi-local Euclidean invariants. The stability and (partial) equivalence of topologies on these invariants regardless additive uniform Gaussian noise is demonstrated.

1 Introduction

In this paper Hilbert's method [2] is applied to find a solution to the equivalence problem for the similarity jet of a linear scale-space of a two-dimensional grey-valued image [3]. Furthermore, a partially well-ordered topology induced on the found set of local and multi-local invariants is shown to be stable and (partially) equivalent above an effective scale related to that of additive uniform Gaussian noise [3]. The found set of invariants becomes important in describing magnetic resonance images or setting up of affine invariant frames in case of weak perspective imaging [3]. Its solution is also relevant as input to modern geometry and topology to quantify image formation and to construct dynamic scale-space theories [3].

The paper is organised as follows. In section 2 the equivalence problem is defined, in section 3 Hilbert's method is briefly treated and in section 4 a partially well-ordered topology of the invariants is shown to be stable and (partial) equivalent for two input images that are slightly perturbed versions of each other.

2 The Equivalence Problem

On the basis of a conservation law for the exchange of energy between a region and its surrounding on a two-dimensional Euclidean space E^2 and application of the divergence theorem a linear scale-space of a two-dimensional grey-valued input image satisfies an isotropic linear diffusion equation with suitable initial-boundary value conditions [3]. The solution space to this initial-boundary value problem is a linear scale-space which has a particular jet-structure:

---

* This work is supported by the Netherlands Organisation of Scientific Research, grant nr. 910-408-09-1, and by the European Communities, H.C.M. grant nr. ERBCHBGCT940511

Definition 1. The jet of a linear scale-space is defined by: $j^\infty(L_0) = \{x, s, L_\mathbf{n} | \mathbf{n} \in \mathbb{Z}_0^+ \times \mathbb{Z}_0^+\}$, with $L_\mathbf{n} = L_0 * G_\mathbf{n}$; $G_\mathbf{n} = \frac{\partial^n G}{\partial x^{i_1}...\partial x^{i_n}}$, $G_0 = G$, where x are the spatial coordinates, s is the scale parameter, L_0 the input image and G the Green's function corresponding to the initial-boundary value problem.

Requiring invariance under the similarity group [3] yields a similarity jet:

Definition 2. The similarity jet of (Definition 1) is defined by: $j^\infty(\Lambda_0) = \{\xi, \Lambda_\mathbf{n} | \xi = \frac{x}{\sqrt{s}}, \Lambda_\mathbf{n} = s^{\frac{n+2}{2}} L_0 * G_\mathbf{n}, \mathbf{n} \in \mathbb{Z}_0^+ \times \mathbb{Z}_0^+\}$, where ξ labels Euclidean distances and $\Lambda_\mathbf{n}$ have the same physical dimension and are differential energies.

As our image domain coincides with Euclidean space E^2 the fundamental properties of (Definition 2) are invariants under the group of spatially homogeneously Euclidean movements $E(2, \mathbb{R})$.

Definition 3. A function I on (2) is called a Euclidean invariant or simply an invariant, if and only if, $I(g(j^\infty)\Lambda_0) = det^w(J(g))I(j^\infty \Lambda_0)$, $\forall g \in E(2, \mathbb{R})$, where J is the Jacobian of the Euclidean group action and $det^w(J(g)) = 1$, $\forall g \in E(2, \mathbb{R})$ is the weight of the invariant I.

Now (Definition 2) and (Definition 3) provide a natural basis for stating the equivalence problem.

Definition 4. The equivalence problem is the problem of finding a complete and irreducible set of invariants necessary and sufficient to describe any property I of (Definition 2) irrespective any action of $E(2, \mathbb{R})$.

3 Hilbert's Solution Method

The lenght measures ξ^i and the differential energies $\Lambda_\mathbf{r}$ can locally coordinatise so-called binary forms.

Definition 5. A binary form Q of n-th order is a homogeneous polynomial in ξ with coefficients $\Lambda_{i_1...i_r}$: $Q(\xi) = \frac{1}{n!} \sum_{i_k=1}^{2} \Lambda_{i_1...i_n} \xi^{i_1} ... \xi^{i_n}$.

In the context of (Definition 4) energy can be approximated by a truncated Taylor series of binary forms up to r-th order.

The above mentioned complete and irreducible set of invariants will primarily be generated by considering the resultant of two binary forms of equal order (see section (3.1) and (3.2)). In order to define such a resultant first of all the transvection of two binary forms P and Q of arbitrary orders, the polar of a binary form and the Cayley-form of two binary forms of equal order have to be introduced [1]:

Definition 6. The k-th order transvection $[.,.]^k$ of two binary forms P and Q of arbitrary orders is defined by: $[P, Q]^k(\xi) = \lim_{\xi \to \eta} \prod_{l=1}^{k} \epsilon_{i_l j_l} \frac{\partial}{\partial \xi^{i_l}} \frac{\partial}{\partial \eta^{j_l}} P(\xi)Q(\eta)$, in which ϵ_{ij} is the parity of the ordered pair (ij).

Definition 7. The κ-th order polar Q_{η^κ} of a binary form Q of order n is defined by: $Q_{\eta^\kappa} = \binom{n}{\kappa}^{-1} \prod_{p=1}^{\kappa} \eta^{i_p} \frac{\partial Q(\xi)}{\partial \xi^{i_p}}$.

Definition 8. The Cayley-from F of two binary forms Q^1 and Q^2 both of order n is defined by: $F = \frac{Q^1 Q_{\eta^n}^2 - Q_{\eta^n}^1 Q^2}{(\xi \eta)}, (\xi \eta) = \xi^1 \eta^2 - \xi^2 \eta^1$.

Using $Q^i Q_{\eta^n}^j = \sum_k \frac{\binom{n}{k}\binom{n}{k}}{\binom{2n-k+1}{k}} [Q^i, Q^j]_{\eta^{n-k}}^k (\xi \eta)^k$ it appears that the Cayley-form F can be written as: $F = \sum_{i=0}^{n-1} \sum_{j=0}^{n-1} c_{ij} (\xi^1)^i (\xi^2)^{n-1-i} (\eta^1)^j (\eta^2)^{n-1-j}$ with c_{ij} a symmetric matrix function c on the coefficients $Q_{i_1...i_n}^1$ and $Q_{j_1...j_n}^2$ of the binary forms Q^1 and Q^2, respectively. On the basis of these definitions the resultant can be defined as the determinant of the matrix c [1]:

Definition 9. The resultant R of two binary forms Q^1 and Q^2, respectively, both of order n is defined by: $R(Q^1, Q^2) = det(c)$.

Because our solution method hinges on that of Hilbert it is necessary to know which systems of binary forms to consider. Hilbert searches for the invariants of forms under actions of the general linear group $Gl(n, \mathbb{R})$, whereas we are interested in the invariants under $E(2, \mathbb{R})$. Felix Klein, however, pointed out that equivalence problems for forms under special transformation groups can be conceived as particular projective ones [5]:

Theorem 10. *A complete set of integral and rational invariants of the system of (binary) forms $Q^1, \ldots, Q^r$ is equivalent to a complete set of projective invariants of the system of equations: $q^i = Q^i(\tilde{\xi}, 1) = 0$, $\tilde{\xi} = \frac{\xi^1}{\xi^2} \in \mathbb{RP}^1$ adjoined to equation: $\sigma(\tilde{\xi}) = \frac{1}{2}(\tilde{\xi}^2 + 1) = 0$ under the fractional action $\tilde{\xi} \to \frac{p\tilde{\xi}+q}{r\tilde{\xi}+s}$ on the projective line $\mathbb{RP}^1$ induced by the linear group $GL(2, \mathbb{R})$ on $\mathbb{R}^2$.*

Therefore, we first solve the affine equivalence problem by means of Hilbert's method before tackling the Euclidean one.

3.1 The Solution of the Affine Equivalence Problem

A local binary form On the basis of (Theorem 10) the affine equivalence problem is a special case of a projective problem for binary forms. For this projective problem there always exists a complete system of a finite number of invariants [5]. If such a system exists, then the next problem is finding such a system. In this context Hilbert addressed the problem of finding a complete set of integrally and algebraically independent invariants for a system of (binary) forms that is invariant under the general linear group $GL(n, \mathbb{R})$. His construction method for such a set is based on the concept of null forms [2].

Definition 11. A null form is a (binary) form all of which invariants vanish.

In the case of a single binary form the question arises how such a null form looks like and which invariants have to vanish in order it to be a null form. To this end Hilbert proves the following [2]:

Theorem 12. *If all the invariants of a binary form of order $n = 2h + 1$, respectively $n = 2h$ are zero, then the binary form possesses an $(h + 1)$-fold linear factor-and conversely, if it possesses an $(h + 1)$-fold linear factor, then all invariants are equal to zero.*

Thus a binary form of order $n = 2h + 1$ or $n = 2h$ is a (canonical) null form if it has a $(h+1)$-fold linear factor and therefore is parametrised by $n - h$ coefficients.

Definition 13. A canonical null form N_n of order $n = 2h + 1$ or $n = 2h$ is defined as a product of a $(h + 1)$-fold linear factor ξ_1^{h+1} and a binary form $q_{n-(h+1)}$ of order $(n - (h + 1))$: $N_n(\xi) = \xi_1^{h+1} q_{n-(h+1)}(\xi)$.

Now the proof of (Theorem 12) in [2] supplies us with a method for constructing a set of invariants of a single binary form, through which all others can be expressed as an integral algebraic function of that set. Let's follow closely and elucidate the first part of the proof presented by Hilbert [2].

In order to explicitly state which invariants of a single binary form Q_n of order $n = 2h + 1$, respectively $n = 2h$, have to vanish such that it is a null form Hilbert starts off with the construction of the set Z_n of transvectants $[Q_n, Q_n]^k$:
$$Z_n \equiv \{Q_n, [Q_n, Q_n]^2, \ldots, [Q_n, Q_n]^{2g}\} \text{ where } g = \begin{cases} h & \text{if } n = 2h+1 \\ h-1 & \text{if } n = 2h \end{cases}. \text{ Next}$$
he uses the following auxiliary theorem to retrieve his invariants [2,1,4]:

Theorem 14. *The necessary and sufficient condition for a binary form Q_n of order $n = 2h + 1$, respectively $n = 2h$, to have a root of multiplicity $h + 1$ is equivalent to requiring the set of transvectants, Z_n, to have one linear factor in common.*

The latter condition on the set of transvectants, Z_n, is now formulated in terms of the vanishing of the resultant of two linear independent combinations U and V of powers of the transvectants that have the same order, namely M. Here M is defined to be the least common multiple of the numbers $n, \ldots, 2(n - 2g)$, such that the powers $m_0, m_1, \ldots, m_g$ of the transvectants are related to this least common multiple M and the order n of the binary form Q_n as follows: $M = m_0 n = 2m_1(n - 2) = \ldots = 2m_g(n - 2g)$, where g is h and $h - 1$, respectively, depending on whether $n = 2h + 1$ and $n = 2h$, respectively. On the basis of these numbers the two forms U and V with indeterminate parameters $u = u_0, u_1, \ldots, u_g$ and $v = v_0, v_1, \ldots, v_g$, are given by: $U = \sum_{k=0}^{g} u_k([Q_n, Q_n]^k)^{m_k}, V = \sum_{k=0}^{g} v_k([Q_n, Q_n]^k)^{m_k}$. Now (Theorem 14) holds, if and only if, the resultant R of the forms U and V vanishes: $R(U, V) = \sum_\mu J_\mu P_\mu = 0$, where each P_ν is a product of powers of the parameters u and v above and J_ν are the sought invariants for odd order n and J_μ and $[Q_n, Q_n]^n$ those for even order n.

A local system of binary forms Now the search for invariants boils down to deriving the conditions for a system of binary forms $(Q_1, \ldots, Q_r)$ to be a system of canonical null forms $(N_1, \ldots, N_r)$. The set of simultaneous invariants come subsequently about upon application of the following theorem [3]:

Theorem 15. *A complete and irreducible set of simultaneous integrally and algebraically independent invariants of a system of two binary forms (Q_p, Q_q) of orders say p and q, respectively, is generated firstly by determining the least common multiple $M_{p,q}$ of the orders p and q of the binary forms and secondly by computing the simultaneous invariants of the new system of two binary forms $\left(Q_p^{\frac{M_{p,q}}{p}}, Q_q^{\frac{M_{p,q}}{q}}\right)$ through calculation of the integrally and algebraically independent invariants of a linear combination of this new pair of binary forms of order $M_{p,q}$.*

Calculating the invariant of each binary form Q_n and the simultaneous invariants of pairs of binary forms (Q_p, Q_q) of system $(Q_1, \ldots Q_r)$ is necessary and sufficient to solve the affine equivalence problem for (Definition 2).

Multi-local systems of binary forms Let's conclude by solving the affine equivalence problem for multi-local systems of binary forms $\{(Q_1, \ldots, Q_r) \| Q\}$. The solution is an extension of the solution for the local problem for a system of binary forms $(Q_1, \ldots, Q_r)$ to a bi-local problem concerning two systems of binary forms $\{(Q_1, \ldots, Q_r), (P_1, \ldots, P_r)\}$, to a tri-local, etc. The solution is obtained by consecutively adjoining forms P to the system $(Q_1, \ldots, Q_r)$ and solving the problem for the extended problem analogous that for one local system of binary forms treated above.

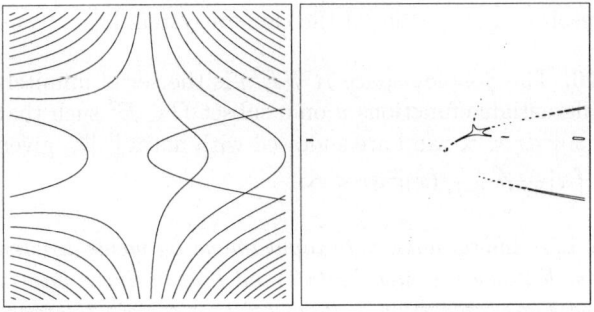

Fig. 1. To the left level contours of $z(x,y) = x^2 + xy^2$ of a surface with a cusp at the origin together with its parabolic curve projected onto the (x,y)-plane. To the right level contours on which invariant $R_{1,1}$ is approximately zero. The disconnected curve segments are obviously close to the projected parabolic curve.

Example The systems of linear forms $\{(Q_1),(P_1)\}$ can be put in the system of linear forms $\{N_1(Q), N_1(P)\}$, if and only if, the following invariant vanishes: $R_{1,1} = [Q_1, P_1]$. This image property is just the resultant of both linear forms of the systems of the corresponding linear forms. Such an invariant can be useful in establishing an affine invariant frame. Here, however, this invariant is used to find the parabolic curves on a surface S. Let us consider the invariant $R_{1,1}$ on the basis of the image gradient fields ∇I and $\nabla I'$ of two orthographic images I and I', respectively, of a Lambertian surface illuminated from two different source directions $\hat{s}$ and $\hat{s}'$ at the same point in both orthographic images. It's shown [3] that the above image gradient fields have to be parallel along the projected parabolic curves. Consequently, the invariant $R_{1,1}$ has to vanish. Thus one finds as photometric invariant objects of the pair of images the projected parabolic curves of the surface (see figure 1).

3.2 The Solution of the Equivalence Problem

The solution of (Definition 4) using the projective formulation in (Theorem 10) comes about by adjoining the quadratic binary $\Sigma = \frac{1}{2}\xi^i\xi^i$ to the systems of the previous section (3.1) and extend Hilbert's method to the extended systems.

4 Stability and Partial Equivalence

As the energies $\Lambda_\mathbf{n}$ satisfy a maximum principle [3] numerical stability under the linear filtering is definitively guaranteed. The stability of (Definition 2) or its invariants I under addition of uniform Gaussian noise ν with zero mean and fixed variance comes about by conceiving $j^\infty(\bar{A})$ of $\bar{L}_0 = L_0 + \nu$ as a by scale s parametrised subset of the Sobolev-space $\mathcal{H}^\infty(E^2)$ and fixing tolerances on the basis of the resolution properties of the camera system [3].

Definition 16. The Sobolev-space $\mathcal{H}^\infty(E^2)$ is the set of infinitely many times continuous differentiable functions u on a subset $\Omega \subset E^2$ such that $u \in L^2(E^2)$: $\|u\|_0^2 = \int_\Omega |u(x)|^2 dx < \infty$, and are endowed with norm $\| \; \|_\infty$ given by: $\|u\|_\infty^2 = \sum_{n=0}^\infty \sum_{i_n=1}^2 \int_\Omega |\frac{\partial^n u}{\partial x^{i_1}\ldots\partial x^{i_n}}(x)|^2 dx < \infty$.

Proposition 17. *Adding noise ν to input image L_0 yields differences in square integrable sense between the invariants I and $\bar{I}$ upon linear filtering that can be neglected above some scale s_n, i.e. there exists a certain tolerance value $\delta > 0$ and scale s_n such that:* $\|\bar{I} - I\|_{\infty^2(s)} < \delta, \; \forall \; s \geq s_n$.

Besides stability in the above sense one also expects some (partial) equivalence of the topology induced by the invariants I of (Definition 2). Partial in the sense that certain subsets have to be omitted or added to realise equivalence. In order to demonstrate stability and (partial) equivalence we first create the full topology $\mathcal{O}$ of the field of invariants I derived in the previous section.

Definition 18. The topology $\mathcal{O}$ on a field of invariants I is a set of open subsets S of this field satisfying $\emptyset \in \mathcal{O} \wedge S \in \mathcal{O}$, $(A \in \mathcal{O} \wedge B \in \mathcal{O}) \Rightarrow A \cap B \in \mathcal{O}$ and $A_i \in \mathcal{O} \Rightarrow \cup_i A_i \in \mathcal{O}$.

This topology $\mathcal{O}$ can be endowed by means of s and the attained levels $\epsilon(I)$ of the invariants I with an ordering relation $\prec$ yielding an ordered topology $(\mathcal{O}, \prec)$.

Definition 19. An ordering relation $\prec$ on the topology (18) is a relation that satisfies transitivity $v \prec w \wedge w \prec x$ if $v \prec x$; $v, w, x \in S$, and trichotomy, $\forall v, w \in S || v = w \vee v \prec w \vee w \prec v$, respectively.

Definition 20. An ordered topology $(\mathcal{O}, \prec)$ is the topology (18) endowed with and ordering relation (19).

On the basis of s and $\epsilon(I)$ one can subsequently endow (20) with a partial well-ordering yielding a partial well-ordered topology.

Definition 21. An ordered topology (20) is partially well-ordered if and only if $\forall T \subset S | \exists w \in T | w \preceq w'$, $\forall w' \in S$.

Definition 22. A partial well-ordered topology $(\mathcal{O}, \prec_w)$ is the ordered topology (20) satisfying the well-ordering relation (21).

Let's make explicit the partially well-ordered topology (22) induced by s and $\epsilon(I)$ attained on the set of critical points and the set of critical zero-crossings.

Definition 23. The set $\partial K_x(I)$ of critical points of I is defined by: $\partial K_x(I) = \{(x, s) | \nabla I(x, s) = 0\}$.

Definition 24. The set $\partial K_s(I)$ of zero-crossings of I is defined by: $\partial K_s(I) = \{(x, s) | \partial_s I(x, s) = 0\}$.

Definition 25. The set $\partial K_{x,s}(I)$ of critical zero-crossings of I is defined by: $\partial K_{x,s}(I) = \partial K_x(I) \cap \partial K_s(I)$.

The partially well-ordering by s comes about by extending the scale ranges such that they include the previous ones. The additional partial well-ordering of the field of invariants on each subset of the critical sets (23) and (25) at one scale s comes about by extending analogously the ranges of attained levels $\epsilon(I)$. These considerations lead us to inflict the following topology on the field of invariants I in order to assess stability and (partial) equivalence of $\bar{L}_0$ and L_0.

Proposition 26. *The partially well-ordered topology (22) of the field of invariants I on the critical sets (23) and (25) is proposed to be generated by well-ordering all possible subsets of those critical sets firstly on the basis of the scale parameter s and secondly on the basis of the levels $\epsilon(I)$ at one scale s.*

Note that metrical relations on the subsets of the critical sets can be added to refine the well-ordering [3]. In figure 2 the stability and (partial) equivalence are visually demonstrated. Above a certain scale-width the d_3-levels at the critical sets and these sets themselves of the two input images become partially indistinguishable. Realise that the set-theoretic difference operator has to be used in order to eventually establish (partial) equivalence.

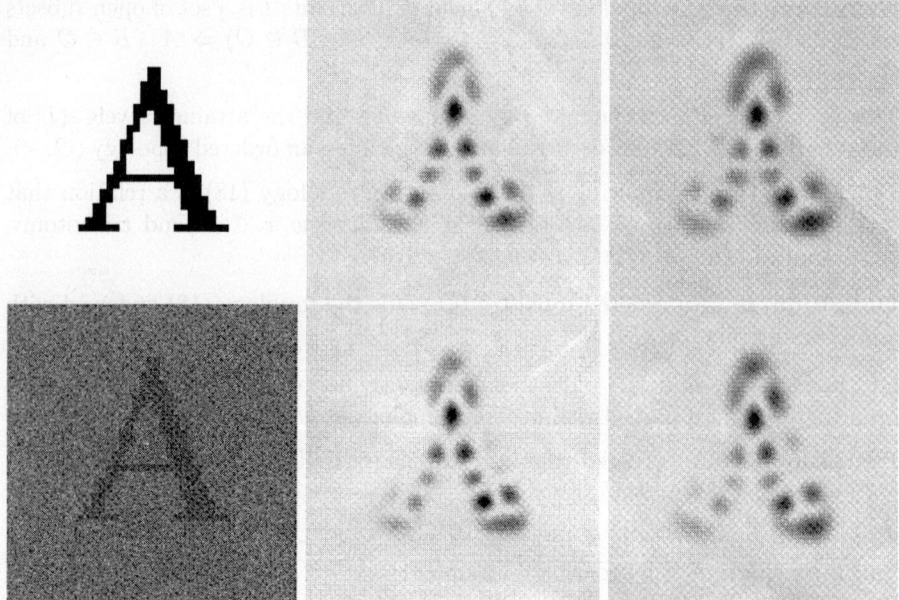

Fig. 2. First row from left to right: the 256×256 input image of a capital letter "A" of dynamic range of 200 for the background and 0 for the letter itself and its third order discriminant d_3 at scale-width of $\sqrt{2s} = 15$ and $\sqrt{2s} = 20$ pixel-distances, respectively. Second row from left to right: the 256×256 input image of the capital letter "A" from the first row with additive uncorrelated Gaussian noise with mean zero and variance of $5 * 10^5$ and its third order discriminant d_3 at the same scale-widths as the figures in the two other figures on the first row.

References

1. Gordan, P.: Ueber die Bildung der Resultante zweier Gleichungen. Math. Annalen, **50** (1871) 355–414
2. Hilbert, D.: Ueber die vollen Invariantensystemen. Math. Annalen, **42** (1893) 313–373
3. Salden, A.H.: *Dynamic Scale-Space Paradigms*. PhD thesis, Utrecht University, The Netherlands, 1996
4. van der Waerden, B.L.: *Moderne Algebra*. Springer-Verlag, Berlin, 1940.
5. Weitzenböck, R.: *Invariantentheorie*. P. Noordhoff, Groningen, 1923.

Segmenting Objects at Multiple Scales: A Robust Approach

Farzin Mokhtarian

Centre for Vision, Speech, and Signal Processing
Department of Electronic and Electrical Engineering
University of Surrey, Guildford, England GU2 5XH, UK
E-mail: *F.Mokhtarian@ee.surrey.ac.uk*

Abstract. Effective local segmentation of contours is an important problem which arises in occluded object recognition as well as other areas. For any recognition system to perform successfully, the segmentation procedure used must be robust in presence of noise and local distortions of shape. Furthermore, it should be based on geometric invariants so that the segmentation will not be affected by arbitrary choices. This paper proposes a new multi-scale segmentation routine for planar contours which is based on the curvature scale space representation. Curvature zero-crossing segments extracted from a continuum of scales are utilized for robust segmentation of planar curves. Two algorithms for efficient termination of the multi-scale segmentation process have also been developed. This approach is more robust than techniques which try to recover features/segments from a *stable* scale and, as a result, risk over- or under-segmentation of the input contour.

1 Introduction

Robust segmentation of a planar contour is an important problem which occurs in occluded object recognition among other areas. For such an object recognition system to perform reliably, the segmentation algorithm must be robust with respect to noise and local variations in object shape. A large amount of work has been carried out on contour segmentation. Much of the earlier work focused on polygon approximation of contour data [1]. Various criteria have been utilized to determine the number as well as the location of polygon vertices. A sequence of polygons with improving fit to the contour data can be obtained by gradually reducing the acceptable error of fit. This polygon sequence can be considered a multi-scale representation of the contour data. However, the locations of the chosen vertices tend to be arbitrary and as a result, this technique can not be considered robust with respect to noise and local shape distortions.

A newer class of algorithms segment an input contour by first smoothing it to remove some noise and then extracting the points where the absolute value of the curvature function has a local maximum. These local maxima are also referred to as *corners*. An attempt is then made to identify a *stable* scale at which the corresponding corners can be extracted [10, 11]. This group of algorithms is generally more robust than the polygon fitting algorithms described earlier.

However an important problem arises since feature points are always extracted from a single scale whereas features can in general occur at multiple scales on free-form contours. As a result, the segmentation can still suffer from noise or from loss of structure due to over-smoothing.

This paper proposes a new multi-scale segmentation procedure for 2-D contours which is based on the curvature scale space (CSS) image. The CSS image is an organization of curvature zero-crossing points extracted from an input contour at a continuum of scales. Section 2 presents a brief overview of the CSS technique. Section 3 describes the proposed multi-scale segmentation method. That segmentation method was used in conjunction with an occluded object recognition system presented in [5]. Two criteria for efficient termination of the multi-scale segmentation process are proposed in section 4. The role of these criteria is to remove very long segments which do not contribute to the recognition system. Section 5 presents the results of the object recognition system based on the multi-scale segmentation method. Section 6 contains the concluding remarks.

2 The curvature scale space representation

A curvature scale space representation is a multi-scale organization of the invariant geometric features (curvature zero-crossing points and/or extrema) of a planar curve. The CSS representation of a planar curve represents that curve uniquely modulo scaling and a rigid motion [3]. To compute it, the curve Γ is first parametrized by the arc length parameter u:

$$\Gamma(u) = (x(u), y(u)).$$

An *evolved version* Γ_σ of Γ can then be computed. Γ_σ is defined by:

$$\Gamma_\sigma = (X(u,\sigma), Y(u,\sigma))$$

where

$$\mathcal{X}(u,\sigma) = x(u) \otimes g(u,\sigma) \quad \mathcal{Y}(u,\sigma) = y(u) \otimes g(u,\sigma)$$

where $\otimes$ is the convolution operator and $g(u,\sigma)$ denotes a Gaussian of width σ. Note that σ is also referred to as the *scale* parameter. The process of generating evolved versions of Γ as σ increases from 0 to ∞ is referred to as the *evolution* of Γ. This technique is suitable for removing noise from a planar curve. Evolving contours can be considered an early form of active contours (snakes) [2] since they are similar in behaviour to snakes without any external constraints. The CSS representation contains curvature zero-crossings or extrema extracted from evolved versions of the input curve. In order to find such points, one needs to compute curvature accurately and directly on an evolved version Γ_σ of a planar curve. Curvature κ on Γ_σ is given by [8]:

$$\kappa(u,\sigma) = \frac{\mathcal{X}_u(u,\sigma)\mathcal{Y}_{uu}(u,\sigma) - \mathcal{X}_{uu}(u,\sigma)\mathcal{Y}_u(u,\sigma)}{(\mathcal{X}_u(u,\sigma)^2 + \mathcal{Y}_u(u,\sigma)^2)^{1.5}}$$

where
$$\mathcal{X}_u(u,\sigma) = \frac{\partial}{\partial u}(x(u) \otimes g(u,\sigma)) = x(u) \otimes g_u(u,\sigma)$$
$$\mathcal{X}_{uu}(u,\sigma) = \frac{\partial^2}{\partial u^2}(x(u) \otimes g(u,\sigma)) = x(u) \otimes g_{uu}(u,\sigma)$$
and
$$\mathcal{Y}_u(u,\sigma) = y(u) \otimes g_u(u,\sigma) \qquad \mathcal{Y}_{uu}(u,\sigma) = y(u) \otimes g_{uu}(u,\sigma).$$

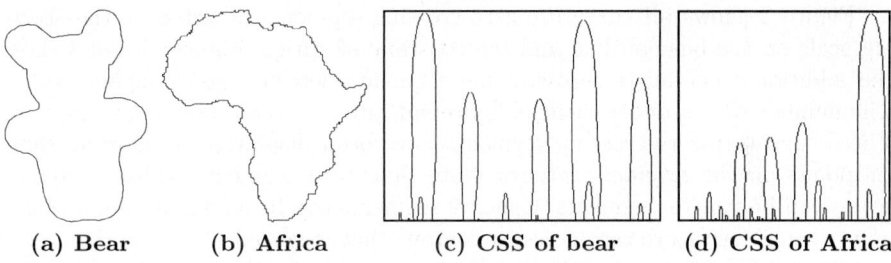

(a) Bear (b) Africa (c) CSS of bear (d) CSS of Africa

Fig. 1. Two input contours and their CSS images

The function defined implicitly by $\kappa(u,\sigma) = 0$ is the CSS image of Γ. For all values of σ larger than a σ_c, evolved curves Γ_σ will be simple and convex. This suggests that the computation can stop as soon as σ_c is reached [9, 4]. Figure 1 shows two input contours and their CSS images. For further examples of CSS images, see [8].

3 Multi-Scale Segmentation of Planar Curves

The purpose of the multi-scale segmentation procedure described here is to divide a 2-D contour into basic segments to be used by a local matching algorithm for occluded object recognition. Curvature zero-crossing points are used as inherent feature points to segment the contour since their locations are invariant to rotation, uniform scaling, and translation of the contour. (Note, however, that corners (or maxima of the absolute value of curvature) are also inherent feature points which can be combined with curvature zero-crossing points, if necessary, to obtain a richer segmentation.) Since curvature zero-crossing points can be extracted from a contour at different scales, the choice of an appropriate scale to be used for feature detection is a crucial one. On free-form contours, features exist at different scales and therefore attempts to discover a *natural* scale to be used for feature extraction will be misguided. Some heuristics may be used to select a specific scale but if the scale chosen is too small, the segmentation will be affected by noise and local distortions of shape, and if it is too large, important structure on the contour may be lost.

The solution used here was motivated by the main underlying concept of the curvature scale space representation: utilize information from multiple scales rather than prefer a single scale. The adaptation of that concept to the problem considered here necessitates the extraction of curvature zero-crossing segments from different scales. This process would ensure robustness to noise and local shape distortions as well as small features that may be part of the model object but missing from a similar (but physically distinct) input object. The multi-scale segmentation procedure essentially constructs a union of curvature zero-crossing segments found at multiple scales by detecting and eliminating repeated segments. For a detailed description, see [6].

Figure 2 shows all curvature zero-crossing segments detected at the starting scale on the bear outline and the coastline of Africa. Figures 3 and 4 show the additional curvature zero-crossing segments detected at all higher scales. The number shown under each subfigure indicates the corresponding value of σ. These curvature zero-crossing segments have been displayed by marking their endpoints on the *original* contours. Note that most segments added at higher scales will be convex segments. Concave segments can be added at higher scales only when a curvature zero-crossing contour that is nested inside another zero-crossing contour disappears. This example demonstrates that the initial segmentation of the input contours is not satisfactory since parts of those contours are oversegmented but that useful segments are added at higher scales. Starting the segmentation at a higher scale may have removed some useful segments as well as noise.

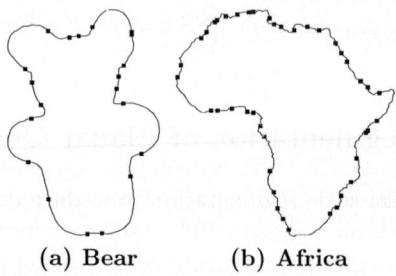

(a) Bear (b) Africa

Fig. 2. Curvature zero-crossing segments at the initial scale

4 Efficient termination of multi-scale segmentation

The algorithm described in the previous section recovers curvature zero-crossing segments from an input contour at multiple scales until the final scale of the corresponding CSS image is reached at which the number of curvature zero-crossings found drops to zero. So the termination of the multi-scale segmentation

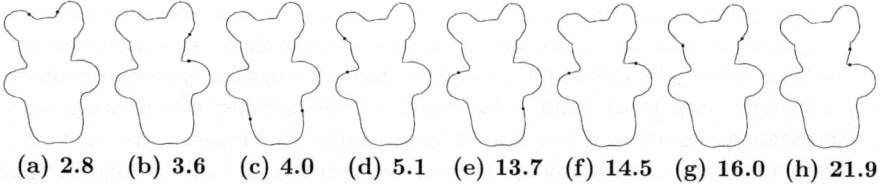

(a) 2.8   (b) 3.6   (c) 4.0   (d) 5.1   (e) 13.7   (f) 14.5   (g) 16.0   (h) 21.9

Fig. 3. Curvature zero-crossing segments of Bear at higher scales

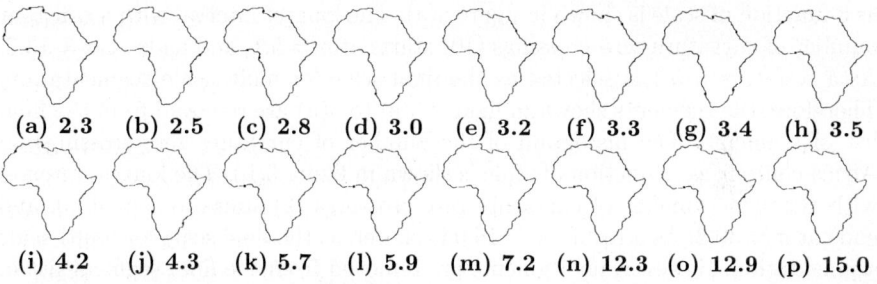

(a) 2.3   (b) 2.5   (c) 2.8   (d) 3.0   (e) 3.2   (f) 3.3   (g) 3.4   (h) 3.5

(i) 4.2   (j) 4.3   (k) 5.7   (l) 5.9   (m) 7.2   (n) 12.3   (o) 12.9   (p) 15.0

Fig. 4. Curvature zero-crossing segments of Africa at higher scales

process is well-defined. Note however that some of the final segments recovered can be very long and can in fact cover most of the input contour. Examples are segments shown in figures 3h and 4p. Very long segments do not contribute to object recognition systems since the presence of occluding objects makes them redundant. Moreover, they reduce the efficiency of the recognition system since it will have to process more segments. This section proposes two algorithms for efficient termination of the multi-scale segmentation process so that only the small to medium scales are covered. The termination criteria are determined automatically, eliminating the need to use arbitrary thresholds. The following is a description of each algorithm:

- The underlying idea of the first algorithm is to determine the longest scale interval in which the number of curvature zero-crossing points remains constant. The starting scale of that interval is then defined as the final scale used for the multi-scale segmentation process. An initial pass through the CSS image is required in order to build a histogram of the number of curvature zero-crossing points found at each scale. The longest interval with a constant number of curvature zero-crossing points is then recovered from that histogram.

- The second algorithm incorporates the idea that only curvature zero-crossing segments whose lengths are below a reasonable ratio (when divided by the length of the input contour) should be accepted. The multi-scale segmentation process described in the previous section is first applied to recover segments at all scales of the CSS image. For each of those segments, a length-ratio is defined by dividing its length by the full length of the input contour. The next step is to build a histogram of the number of curvature zero-crossing segments as a function of length-ratio. The longest interval with a constant number of segments is then recovered from that histogram. The starting value of length-ratio corresponding to that interval is then utilized as the cutoff ratio.

The histogram of the number of curvature zero-crossings on the bear contour as a function of scale is shown in figure 5(a). The longest interval with a constant number of curvature zero-crossings (10) starts at $\sigma = 5.1$, and ends at $\sigma = 13.7$. As a result, $\sigma = 5.1$ is selected as the final scale for multi-scale segmentation. Therefore, the segments shown in figures 3(e) to 3(h) are removed from the final list of segments. The histogram of the number of curvature zero-crossings on Africa contour as a function of scale is shown in figure 5(b). The longest interval with the same number of curvature zero-crossings (2) starts at $\sigma = 15.0$, and ends at $\sigma = 30.2$. As a result, $\sigma = 15.0$ is chosen as the final scale for multi-scale segmentation. However, no segments are removed from the final segment list in this case.

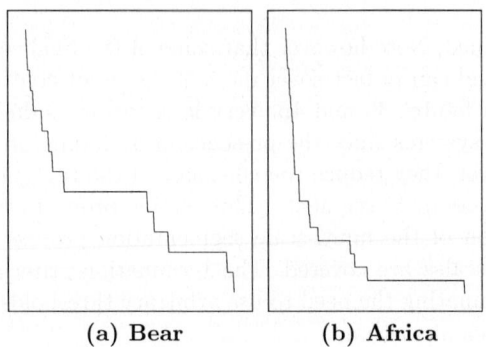

(a) Bear   (b) Africa

Fig. 5. Number of curvature zero-crossings vs. scale

The histogram of the number of curvature zero-crossing segments on the bear contour as a function of the ratio r of their lengths to the length of the whole contour is shown in figure 6(a). The longest interval with the same number of curvature zero-crossing segments (25) starts at $r = 0.55$ and ends at $r = 0.95$. As a result, $r = 0.55$ is selected as the cutoff length ratio. It can be said that, in this case, this procedure has eliminated curvature zero-crossing segments that are longer than about half the length of the input contour. This results in the removal of the segment shown in figure 3(h) from the final list of segments. The histogram of the number of curvature zero-crossing segments on Africa contour as a function of the ratio r of their lengths to the length of the whole contour is shown in figure 6(b). The longest interval with a constant number of curvature zero-crossing segments (58) starts at $r = 0.6$ and ends at $r = 0.85$. Hence, $r = 0.6$ is chosen as the cutoff length ratio. The result is the removal of the segment shown in figure 4(p) from the final segment list.

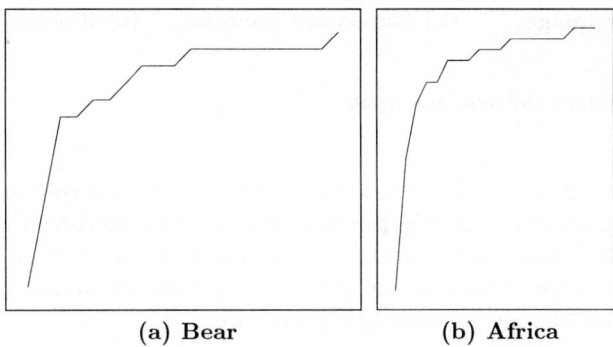

(a) Bear   (b) Africa

Fig. 6. Number of curvature zero-crossing segments vs. length ratio

Based on these two examples, it appears that the second algorithm performs better in terms of removing very long segments but preserving the ones detected at small to medium scales.

5 Results and discussion

The multi-scale contour segmentation algorithm demonstrated here was used to segment a number of model and image contours used as input to an object recognition with occlusion system [7]. Figure 7(a) shows one of the images on which the system was tested. Figure 7(b) shows the outermost contour recovered from 7(a) and figure 7(c) shows the recognition result. Model objects have been registered with the image data. The challenging nature of the recognition problem is highlighted by the fact that actual 3-D objects were used for the experiments which gave rise to some perspective distortions.

6 Conclusions

This paper proposed a new multi-scale segmentation procedure for 2-D curves based on the curvature scale space representation. It also proposed automatic procedures for the termination of that multi-scale segmentation process in order to enhance efficiency. Curvature zero-crossing segments were extracted from

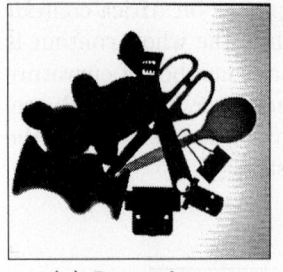 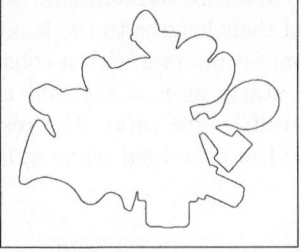 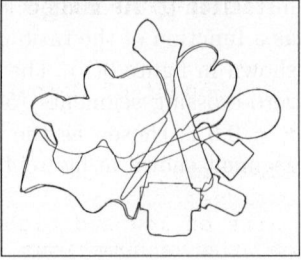

(a) Input image   (b) Outermost contour   (c) Recognition result

Fig. 7. Input image and system output

multiple scales and used for robust segmentation of free-form contours. This segmentation procedure is robust in presence of noise and local distortions of shape, and because it is based on geometric invariants, it is not affected by arbitrary choices. It was shown that reliable occluded object recognition can be achieved through this multi-scale segmentation procedure.

References

1. N. Ayache and O. D. Faugeras. Hyper: A new approach for the recognition and positioning of 2-d objects. *IEEE Trans Pattern Analysis and Machine Intelligence*, 8(1):44–54, 1986.
2. M. Kass, A. Witkin, and D. Terzopoulos. Snakes: active contour models. In *Proc International Conference on Computer Vision*, pages 259–268, 1987.
3. F. Mokhtarian. Fingerprint theorems for curvature and torsion zero-crossings. In *Proc IEEE Conference on Computer Vision and Pattern Recognition*, pages 269–275, San Diego, CA, 1989.
4. F. Mokhtarian. Zero-crossings of curvature and torsion in the limit. In *Proc Asian Conference on Computer Vision*, pages III:457–461, Singapore, 1995.
5. F. Mokhtarian. Silhouette-based object recognition with occlusion through curvature scale space. In *Proc European Conference on Computer Vision*, pages I:566–578, Cambridge, England, 1996.
6. F. Mokhtarian. Multi-scale contour segmentation. In *Proc International Conference on Scale-Space Theory in Computer Vision*, Utrecht, The Netherlands, 1997.
7. F. Mokhtarian. Silhouette-based occluded object recognition through curvature scale space. *Machine Vision and Applications*, 10:87–97, 1997.
8. F. Mokhtarian and A. K. Mackworth. A theory of multi-scale, curvature-based shape representation for planar curves. *IEEE Trans Pattern Analysis and Machine Intelligence*, 14(8):789–805, 1992.
9. F. Mokhtarian and S. Naito. Scale properties of curvature and torsion zero-crossings. In *Proc Asian Conference on Computer Vision*, pages 303–308, Osaka, Japan, 1993.
10. A. Rattarangsi and R. T. Chin. Scale-based detection of corners of planar curves. *IEEE Trans Pattern Analysis and Machine Intelligence*, 14(4):430–449, 1992.
11. P. L. Rosin. Representing curves at their natural scales. *Pattern Recognition*, 25:1315–1325, 1992.

Multi-grid Edge Models for Magnifying Digital Images

G Qiu

School of Computing & Mathematics, University of Derby
Derby DE22 1GB, United Kingdom

Abstract: A set of discrete ideal step edge models have been designed for small windows of pixels. These models consist of pairs of low and high resolution ideal step edge patterns. Based on the assumptions that within a small image region, the underlying edge structure and the average pixel intensity of the low and high resolution image samples should be identical, a small window of low resolution pixels are mapped to a high resolution lattice. The method involves mainly table look-up operation and is therefore computationally efficient. The models have been applied to the reconstruction of high resolution images. Simulation results show that the images expanded by the new method have sharper edges and lower reconstruction errors than those produced by traditional nearest neighbor, bilinear and bicubic interpolation methods.

1. Introduction

Image resolution expansion is used in many aspects of image processing and computer vision. Standard methods use interpolation which fits the original data with a continuous model and then re-sample the interpolated two dimensional signal on a new sample grid [1-3]. The simplest method for image resolution expansion is the nearest neighbor interpolation [1, 2]. It is very fast and easy to implement, but the resultant images usually appear blocky. Smoother images can be obtained by using bilinear interpolation [1, 2], or bicubic spline interpolation [1, 3]. However, these methods smooth the image data in flat as well as detail areas, as a result, the expanded images often appear rather blurry.

Recently, a nonlinear interpolation method based on a continuous spatial edge fitting model has been developed for image resolution expansion [4]. Compared with traditional methods, the edge fitting method can generate high resolution images with sharper edges and reduced reconstruction errors.

In this paper, we develop a discrete edge model matching (EMM) method for image resolution enhancement. The method is based on the assumptions that within a sufficiently small image region, the edge structure can be modeled as ideal step edge and that the underlying structure of the edge remains the same in high and low resolution grids. Three criteria which ensure the visual integrity of the images are developed and used for the mapping process. Within a small region, it is assumed that, 1) the orientation of the step edge will remain unchanged in both high and low resolution lattice; 2) the amplitude of the gradient calculated in low and high resolution lattice are the same; and, 3) the average intensities calculated for the low and high resolution samples should be identical.

The new method involves mainly table look-up operations and is computationally efficient. Its complexity is lower than that of the continuous edge model method of [4]. Extensive computer simulations show the new method outperforms the traditional methods in the sense that the images expanded by the new method have sharper edges and lower reconstruction errors than those produced by traditional methods.

2. Multi-grid Discrete Edge Models

The method of [4] maps a small neighborhood about each low-resolution pixel to a best fitting continuous space step edge. Edge fitting provides, for each low-resolution neighborhood, the location of the best fitting edge in the continuous space, and two associated intensity values lying on the two sides of the boundary. A higher resolution sampling grid is then super-imposed over the resulting region of the edge fit. The values of the higher resolution pixels are approximated as being one of the two intensity values [4]. According to the authors, such a nonlinear interpolation method was able to produce high resolution images with sharper edges as well as lower reconstruction mean square errors than those produced by linear techniques.

One of the major tasks in edge fitting is to find the edge boundary, i.e. the line that divides the small testing region into two almost homogeneous regions. The method of [4] is based on a continuous spatial edge fitting model; even though edge fitting may be accurate, the implementation is quite complicated. Furthermore, when working with digital images, and expanding the image by a small factor[1], the method maybe unnecessarily too complicated.

In discrete domain, the line parameters can be quantized without affecting the accuracy of edge fitting. This

---

[1] Although in theory, the method can be used to increase the resolution of an image by a factor of any number, large size expansion violates the pre-assumption of the model and is unlikely to give good performance.

observation has led to the development of a discrete multiresolution edge matching model (EMM). Because the step edge model assumption will only be valid within a small region, the edge patterns are created for a 2 x 2 pixels region in the low resolution lattice and 3 x 3 pixels in the high resolution lattice, as shown in Fig. 1

The approach adopted in this work is to first quantize the orientation parameter, then for each quantized orientation, we quantized the line[2] such that each will specify a different edge pattern within the 3 x 3 window. With reference to Fig. 1, let's first fix $\rho = 0$, for $\theta = 0$ the line specifies one edge pattern. To find the next orientation which will specify another different pattern, the line is rotated. As can be seen from Fig. 1 so long as the rotated angles fall within the range, $0° < \theta < 45°$, the new lines specify the same new pattern. Based on this observation, a comprise may be found by setting the new orientation half-way through the range, i.e., θ is quantized to 22.5 degree interval. Based on this principle, sixteen quantized orientations are used.

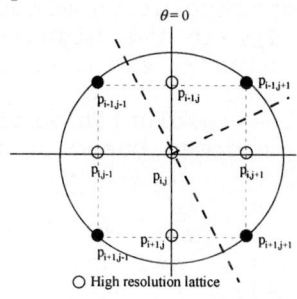

○ High resolution lattice
● Low resolution lattice

Fig. 1 Edge fitting window

For each quantized orientation, we fix the orientation value and try to quantize the line such that each new line will specify a distinctive low and high resolution edge patterns. If two lines give exactly the same high or low resolution patterns, they are merged[3] to create only one pair of patterns. In this way, for a given orientation, an edge pattern (2 x 2) in the low resolution grid will have exactly one corresponding edge pattern (3 x 3) in the high resolution grid. Following this principle, the corresponding edge patterns in the low and high resolutions can be determined for each quantized orientation.

---

[2] The line is described as $x\cos\theta + y\sin\theta = \rho$
[3] The reason is that in our current application, the high and low resolution patterns must have a one to one mapping, otherwise there will be an ill-posed one to many mapping situation.

A set of pairs of 2 x 2 and 3 x 3 step edge patterns generated are listed in Appendix A.

3. High Resolution Image Reconstruction

Using the set of pairs of step edge patterns, we have developed a method for increasing the spatial resolution of digital images. Within a small region, it is assumed that, 1) the orientation of the step edge will remain unchanged in both high and low resolution lattice; 2) the amplitude of the gradient calculated in low and high resolution lattice are the same; and, 3) the average intensities calculated for the low and high resolution samples should be identical.

Based on these assumptions and the edge models described in the last section, a procedure for expanding image spatial resolution is described as follows: Each 2 x 2 pixel window, $w[m, n]$, in the low resolution image is mapped to a 3 x 3 pixel window, $W[m, n]$, at the high resolution grid. The low resolution windows overlap each other, so do the high resolution windows. Using a method similar to that of [4], in the high resolution grid, the border pixels of the windows are averaged.

Mapping of a 2 x 2 low resolution window to a 3 x 3 high resolution grid is performed based on the following three steps.

Step 1, calculating the mean, M of the 2 x 2 window of low resolution samples, and then the mean value is removed from this pixels,

$$M = \frac{1}{4}\sum_{i=0}^{1}\sum_{j=0}^{1} w[i,j] \qquad w'[i,j] = w[i,j] - M \qquad \forall i,j \qquad (1)$$

Step 2, the orientation of the 2 x 2 window at the low resolution grid is calculated. The Prewitt like operators [1] defined as follow are used.

$$h_x = \frac{1}{2}\begin{bmatrix} -1 & 1 \\ -1 & 1 \end{bmatrix} \qquad h_y = \frac{1}{2}\begin{bmatrix} 1 & 1 \\ -1 & -1 \end{bmatrix}$$

The gradients, the orientation and amplitude are calculated as

$$g_x = \langle w', h_x \rangle \quad g_y = \langle w', h_y \rangle \quad \theta = \tan^{-1}\left(\frac{g_y}{g_x}\right), \quad g = \sqrt{g_x^2 + g_y^2} \qquad (2)$$

θ is then quantized to the nearest value of the set of 16 angles. If the orientation of the quantized value has only one pattern, then this window is mapped to the 3 x 3 pattern in the list. However, if multiple mapping patterns exist for the orientation, then 2 x 2 masks are used to decided which pattern should be used. The masks are generated by replacing a with 1 and b with -1 in the list. The inner product between w'[m, n] and these masks are calculated and the window is mapped to the 3 x 3 pattern which corresponds to the mask producing the largest inner product value.

Step 3, the final step is to decide the values of A and B in the 3 x 3 pattern. This is done by setting the mean and the amplitude of the gradient of the 3 x 3 pattern equal to that of the 2 x 2 window of pixels.

4. Experimental Results

To evaluate the performance of the proposed edge matching method for enhancement of spatial resolution of images, we have performed many computer simulations. We use a set of six images randomly chosen to represent different type of nature images. They are Mandrill, Goldhill, Lena, MIT, Mobile, and Textures. Some such as Mandrill, Goldhill and Lena are well known, MIT is a sub-image taken from the HDTV video sequence of the same name, Mobile is a sub-image taken from the Mobile Calendar video sequence. Textures consists of a composition of four different Brodatz textures.

All the original images are 512 x 512, 8 bits per pixel gray scale image. The image is first sub-sampled by a factor q as follows

$$F_q(i,j) = \frac{1}{q \times q} \sum_{m=0}^{q-1} \sum_{n=0}^{q-1} F_o(qi+m, qj+n) \quad i,j = 0,1,\cdots,\frac{N}{q}-1 \quad (3)$$

where F_o is the original image, N is the dimension of the original image (N = 512). F_q is then expanded back to the original resolution. In this way, mean square errors computing the difference between the original image and the expanded image can be used to evaluate the performance of different image expansion techniques. Apart from the new method, three well known image interpolation method, nearest neighbor, bilinear and bicubic are also implemented. We adopt the SNR measure used in [5] for quantitative comparison of different image expansion methods. The improved SNR is defined as

$$\Delta_{SNR} = 10\log_{10}\left(\frac{\|F_0 - F_{nb}\|^2}{\|F_0 - F_e\|^2}\right) \text{dB} \quad (4)$$

where F_{nb} is the nearest neighbor expanded image, F_e represents the expanded image by one of the following methods, bilinear, bicubic or the new edge model matching

Table 1, Comparison of image expansion methods, $q = 2$

Images	Δ_{SNR} (dB)		
	EMM	Bicubic	Bilinear
Mobile	2.42	0.07	0.79
MIT	2.72	0.02	0.83
Lena	2.16	-0.80	-0.05
Baboon	0.45	-0.44	-0.05
Texture	3.89	0.44	1.46
Goldhill	2.61	-0.36	0.42

Table 2, Comparison of image expansion methods, $q = 4$

Images	Δ_{SNR} (dB)		
	EMM	Bicubic	Bilinear
Mobile	2.28	0.39	0.63
MIT	2.22	0.38	0.58
Lena	2.39	-0.36	0.05
Baboon	0.45	-0.12	0.07
Texture	3.69	0.66	1.07
Goldhill	2.82	0.10	0.41

(EMM) method. Table 1 shows the improved SNR when the images are reduced by a factor of $q = 2$ in both dimensions and expanded back to the original resolution. Table 2 shows the improved SNR when the images are reduced by a factor of $q = 4$ in both dimensions and expanded back to the original resolution. It is seen the new method significantly outperforms the traditional interpolation methods. The images expanded by EMM also have sharper edges than those produced by the interpolation methods. In [4], better SNR performance of the edge fitting method was also reported, but no numerical result was available. Informal comparison of the images show similar visual quality. Informal comparison with that of multi-frame technique [5] also show the new method is competitive. The new method is computationally simpler than that of [4] and much easier to implement.

5. Concluding Remarks

A new method for image resolution enhancement based on discrete edge pattern modeling has been presented. Compared with other schemes, the new method has the advantages that it is computationally relatively simple and its performance is superior in terms of producing sharper images and lower reconstruction error. In the current model, only step edge models have been developed, no formal model is applied to textures. Our future work will try to develop texture models in combination with the models described in this paper. Other edge models such as line pattern are also under investigation.

Appendix A, Step Edge Model Mapping List

In these pattern, it is assumed that a > b and A > B. Another half of the set of pairs of patterns can be found easily by swapping the position of a and b, and A and B of their supplementary orientations patterns. Notice that for convenience, ramp edges are assumed for the horizontal and vertical orientations.

$\theta = 0°$

Low Resolution Patterns	b a
	b a
High Resolution Patterns	B (A+B)/2 A B (A+B)/2 A B (A+B)/2 A

$\theta = 22.5°$

Low Resolution Patterns	b a	b a	a a
	b b	b a	b a
High Resolution Patterns	B B A B B A B B B	B A A B A A B B A	A A A B A A B A A

$\theta = 45°$

Low Resolution Patterns	b a	a a
	b b	b a
High Resolution Patterns	B A A B B A B B B	A A A B A A B B A

$\theta = 67.5°$

Low Resolution Patterns	b a	a a	a a
	b b	b b	b a
High Resolution Patterns	B A A B B B B B B	A A A B A A B B B	A A A A A A B B A

$\theta = 90°$

Low Resolution Patterns	a a
	b b
High Resolution Patterns	A A A (A+B)/2 (A+B)/2 (A+B)/2 B B B

$\theta = 112.5°$

Low Resolution Patterns	a b	a a	a a
	b b	b b	a b
High Resolution Patterns	A A B B B B B B B	A A A A A B B B B	A A A A A A A B B

$\theta = 135°$

Low Resolution Patterns	a a	a b
	a b	b b
High Resolution Patterns	B A A B B A B B B	A A A B A A B B A

$\theta = 157.5°$

Low Resolution Patterns	a b	a b	a a
	b b	a b	a b
High Resolution Patterns	A B B A B B B B B	A A B A A B A B B	A A A A A B A A B

References
1. W. P. Pratt, **Digital Image Processing**, New York: Wiley, 1978
2. A. K. Jain, **Fundamentals of Digital Image Processing**, Prentice Hall, 1989
3. H. S. Hou and H. C. Andrews, "Cubic spline for image interpolation and digital filtering", IEEE Trans. ASSP, Vol. 26, pp. 508 - 517, 1978
4. K. Jensen and D. Anastassiou, "Subpixel edge localization and the interpolation of still images", IEEE Trans. on Image Processing, vol. 4, pp. 285 - 295, 1995
5. R. R. Shultz and R. L. Stevenson, "Extraction of high-resolution frames from video sequence", IEEE Trans. Image Processing, vol. 5, pp. 996 - 1011, 1996

Scale and Rotation Invariant Recognition Method Using Higher-Order Local Autocorrelation Features of Log-Polar Image

Takio KURITA[1], Kazuhiro HOTTA[2], and Taketoshi MISHIMA[2]

[1] Electrotechnical Laboratory, 1-1-4, Umezono, Tsukuba City, 305, JAPAN
[2] Saitama University, 255, Shimo-Okubo, Urawa City, 338, JAPAN

Abstract. This paper proposes a scale and rotation invariant recognition method which uses higher-order local autocorrelation (HLAC) features of log-polar image. Linear scalings and rotations are represented as shifts in the log-polar image which is obtained by re-sampling of the input image. HLAC features of log-polar image become robust to the linear scalings and rotations of a target because HLAC features are shift-invariant. By combining these features with a simple classifier which uses linear discriminant analysis, we can design a scale and rotation invariant recognition system. Robustness to the scalings and rotations are confirmed by experiments on 2D shapes and face recognition. Robustness to the changes of backgrounds is also confirmed by experiments on face recognition.

1 Introduction

When people watch a target, they capture the target with the center of the visual field by eye movement. The density of the structure of human's vision system increases toward the center of the visual field and decreases from the center toward the periphery. The center of retina is called "fovea". Fovea has rich eyesight and its visual angle is only five degrees.

In the field of computer vision, the interest in active vision which acquires necessary information by moving the cameras is increasing[1, 2]. The one of the reasons of the interest is due to the developments of small, light, and fast camera which can be controlled by computer. It is known that space variant sensor like biological vision is effective in active vision[3, 4] . The mechanism which always keeps the target at the center of the visual field is called "gaze control". That is one of the important elements of active vision system.

Shift invariance is important for recognition system which uses usual static camera because the target changes the position within the image frame. On the other hand, scale and rotation invariance becomes more important in recognition system which uses active vision head with gaze control mechanism, because active vision system can maintain the target with the center of camera. In this paper we consider the second case, where we can use active vision head.

One of the simplest models of space variant sensor is log-polar transformation. In the log-polar image, higher weights are given for the central region of the original image than peripheral region. This property of the log-polar images is good for target recognition because the peripheral region often includes unnecessary information such as background. There is another important property. In the log-polar coordinate, linear scaling and rotation are represented as shifts if the origin of the polar plane does not move. Therefore, shift-invariant features extracted from log-polar images become invariant to scalings and rotations of the target.

In this paper, we use higher-order local autocorrelation (HLAC) features [5] which are inherently shift-invariant, additive, and computationally inexpensive. Kurita et al.[6, 7] developed a real-time face recognition system based on computation of HLAC features. Goudail et al.[8] obtained a peak recognition rate of 99.9% by using a large database of 11,600 images of 116 different faces. These attempts assume the images taken by the usual static camera. Here we propose the recognition method for images which are taken by active vision system. Since HLAC features of log-polar images become invariant to scalings and rotations, we can design a scale and rotation invariant recognition system by combining these features with a simple classifier which uses linear discriminant analysis.

In section 2 we describe the property of log-polar transformation which is one of the simplest models of space variant sensor. Then HLAC features of log-polar image are introduced in section 3. In section 4 experimental results of 2D shapes and face images recognition.

2 Log-polar image

Input image is generally represented as a collection of pixel points on the Cartesian coordinate. Log-polar image can be constructed by the following transformations of the coordinates. At first, the point (x, y) on the Cartesian coordinate is transformed to the point $(\rho = \sqrt{(x^2 + y^2)}, \theta = \arctan(\frac{y}{x}))$ on the polar coordinate. Then the point on the polar coordinate is transformed to the point $(z = \log(\rho), \theta)$ on the log-polar coordinate by taking the logarithm of the scale ρ. Figure 1 (a) and (b) show Cartesian coordinate and log-polar coordinate.

In this paper, we use the re-sampling method by the reverse transformation to obtain a log-polar image from a input image. To obtain the pixel value at the point (z_i, θ_j) on the log-polar image, the point is reversely transformed to the point $(\exp(z_i)\cos(\theta_j), \exp(z_i)\sin(\theta_j))$ on the Cartesian plane. Then the value of the point (z_i, θ_j) is estimated as the average intensity value of the neighboring points of the back-projected point $(\exp(z_i)\cos(\theta_j), \exp(z_i)\sin(\theta_j))$ on the input image. We can obtain a log-polar image by performing this estimation for all points on the log-polar image. Figure 1 (c) and (d) show sampling point of input image. It is noticed that sampling density increases toward the center of image and decreases from the center toward the periphery. The number of sampling points depends on the resolution of re-sampling to obtain the log-polar image.

Scalings of a target are represented as shifts along z $(= \log(\rho))$ axis on the log-polar image (coordinate). Rotations of a target are also represented as shifts along θ axis but the upper end and the lower end are connected because 0 and 2π are equivalent.

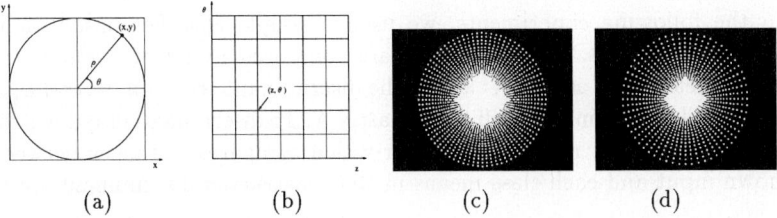

Fig. 1. Sampling points of input image to construct the log-polar image. (a) Cartesian coordinate. (b) Log-Polar coordinate. (c) The resolution of input image is 320×240. The resolution of log-polar image is 80×80. (d) The resolution of log-polar image is 60×60.

3 Higher-order local autocorrelation features of log-polar image

To obtain scale and rotation invariant features, we have to extract shift invariant features from log-polar image because the scalings and rotations are represented as shifts in log-polar image.

It is well known that autocorrelation function is shift-invariant. Its extension to higher orders is higher-order autocorrelation function. The Nth-order autocorrelation functions with N displacements $(\boldsymbol{a}_1, \boldsymbol{a}_2, \cdots, \boldsymbol{a}_N)$ from the reference point $\boldsymbol{r}$ are defined by

$$x^N(\boldsymbol{a}_1, \boldsymbol{a}_2, \cdots, \boldsymbol{a}_N) = \int I(\boldsymbol{r})I(\boldsymbol{r}+\boldsymbol{a}_1)\cdots I(\boldsymbol{r}+\boldsymbol{a}_N)d\boldsymbol{r} \quad (1)$$

where function $I(\boldsymbol{r})$ denotes a log-polar image and $\boldsymbol{r} = (z = log(\rho), \theta)$. The number of these autocorrelation functions obtained by the combination of the displacements over the image are enormous. Here we reduce them for practical application. At first, we restrict the order N up to the second ($N = 0,1,2$). Then, we also restrict the range of displacements to within a local 3×3 window, because the correlation within local region is much higher than the correlation between far points. By eliminating the displacements which are equivalent by the shift, the number of the patterns of the displacements are reduced to 35. Since these features are obviously invariant to shift, HLAC features extracted from log-polar images become robust to linear scalings and rotations of a target in the original image. By combining these features with a simple classifier which uses linear discriminant analysis, we can design a scale and rotation invariant recognition system.

4 Experiments

To show effectiveness of the proposed method, we have performed experiments using images with different scales and rotations of targets. Recognition targets are 2D shapes and human faces.

In the following experiments, we used a simple classifier based on linear discriminant analysis. New features y are computed by linear combinations of the HLAC features x as $y = A^T x + b$. The discriminant criterion $J = \text{tr}(\hat{\Sigma}_W^{-1} \hat{\Sigma}_B)$ is maximized to obtain the coefficient matrix A. To discriminate unknown image, we can use a nearest neighbor classifier which compares distances between the unknown input and each class means in the constructed discriminant space.

4.1 2D shapes recognition

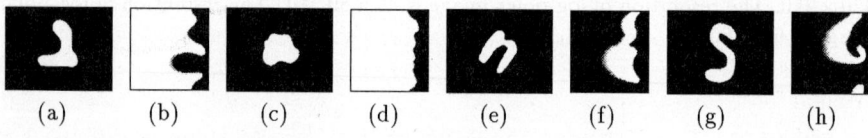

Fig. 2. Examples of 2D shape images and their log-polar images. (a) Shape 1. (b) Log-Polar image of (a). (c) Shape 2. (d) Log-Polar image of (c). (e) Shape 3. (f) Log-Polar image of (e). (g) Shape 4. (h) Log-Polar image of (g).

Figure 2 shows 2D shapes [9, 10] which are used in the experiment and their log-polar image. In this experiment, we collected images with different sizes (6 sizes) and rotations (12 angles) for each shape in Figure 2. The different sizes are obtained by changing the distance between the camera and the object. The rotation were changed about $\pi/6$ rad at a time by hand over the range from 0 to 2π rad. The number of images used in this experiment is 288 (4 classes × 6 sizes × 12 rotations). The resolution of these images is 320 × 240 pixels and each of images is binalized by Otsu's thresholding method [11, 12]. Since the resolution of input images and the resolution of log-polar images effect to the recognition rate, we measured them with different resolution of input images (320 × 240, 160 × 120, 80 × 60) and log-polar images (120 × 120, 80 × 80, 60 × 60, 40 × 40, 30 × 30, 20 × 20).

In this experiment, all of the images were used to construct discriminant space and the recognition rate were estimated by leave-one-out method. The recognition rates were 100% for all combinations of resolutions of input images and log-polar images. From these results, we can say that the proposed method has ability to recognize 2D shapes with different sizes and rotations.

Next, to show the robustness to scalings, we have performed the experiment in which images with only one size (4 classes × 1 size × 12 rotations) are used in learning and then the rest (4 classes × 5 sizes × 12 rotations) are used for evaluation of the recognition rate. For the comparison, we also calculated the recognition rates by using HLAC features extracted from original images directly. In this experiment, the resolution of input images was set to 160 × 120 pixels. The results are shown in Table 1 (a) and (b). The recognition rates obtained by HLAC features extracted from log-polar images are about 99%, while the best recognition rates obtained by HLAC features extracted from input images directly are at most 50 %.

Figure 3 (a) and (b) shows discriminant spaces constructed in this experiment. In Figure 3 (a) and (b), each of four classes is plotted with different symbols (Each symbol corresponds to each class of Figure 2). Figure 3 (a) and (b) shows discriminant space constructed from HLAC features extracted from log-polar image and HLAC features extracted from input images, respectively. Since there is the case which include noises when all eigenvalues are used, we restricted the dimension by a constraint in which cumulative proportion becomes more than 99%. Consequently, the dimension of discriminant space constructed from HLAC features extracted from log-polar image became three. However, the dimension of discriminant space of HLAC features of input image became two. It is noticed that Figure 3 (a) consists four group which corresponds to the original four classes, while each class is separated into exactly five groups which correspond to five scales of a target in Figure 3 (b). From these results, the proposed method becomes more robust to scalings of targets by using log-polar transformation.

Similarly, to show the robustness to rotations, we have performed the experiment in which only one rotation (4 classes × 6 sizes × 1 rotation) is used in learning and the rest (4 classes × 6 sizes × 11 rotations) are used for evaluation of the recognition rate. Again, for the comparison, we also calculated the recognition rates by using HLAC features extracted from original images directly. In this experiment, the size of input images was also set to 160 × 120 pixels. The results are shown in Table 1 (c) and (d). The recognition rate for HLAC features extracted from log-polar images is about 95 % except the case in which the resolution of the log-polar image is 40 × 40, but the best recognition rate obtained by HLAC features extracted from input images is only 55.68 %. From these experiments, the proposed method is very robust to the changes of scalings and rotations of 2D shapes.

HLAC features extracted from log-polar image are scale and rotation invariant but then do not shift invariant. To evaluate the sensitivity of the recognition rate to shift of a target, we have measured the recognition rates by changing distance between the center of the image and the center of gravity of the target. The resolution of input images is set to 160 × 120 pixels and the resolution of log-polar images is 20 × 20 pixels. The images without shifts were used in learning and recognition rates were estimated by right shifted images, respectively. The result is shown in Figure 3 (c). The horizontal axis shows distance between the center of image and the center of gravity of the target. The vertical axis shows the recognition rate. During the distance between center of image and the center of gravity of a target is less than 7 pixels, the recognition rate achieves more than

Table 1. Generalization to scalings and rotations (The resolution of input image is 160 × 120)

(a) HLAC features extracted from log-polar image

The resolution of log-polar image	Rate (%)
40 × 40	98.75
30 × 30	99.58
20 × 20	100.00

(b) HLAC features extracted from original image

The resolution of input image	Rate (%)
80 × 60	47.92
40 × 30	50.00
20 × 15	37.08

(c) HLAC features extracted from log-polar image

The resolution of log-polar image	Rate (%)
40 × 40	76.14
30 × 30	95.83
20 × 20	96.59

(d) HLAC features extracted from input image

The resolution of input image	Rate (%)
80 × 60	37.12
40 × 30	55.68
20 × 15	26.89

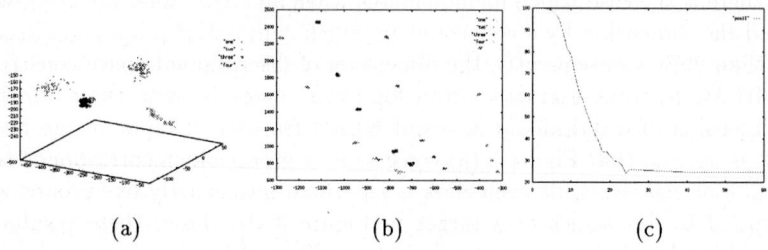

(a)　　　　　　　　(b)　　　　　　　　(c)

Fig. 3. Generalization to scalings:　　(a) HLAC features extracted from log-polar image.(The resolution of log-polar image is 20 × 20. The dimension of the discriminant space is three.)　(b) HLAC features extracted from input image.(The resolution of input image is 40 × 30. The dimension of the discriminant space is two.)　(c) Effect by changing distance from center.

about 95%. But, as the distance becomes more far, the recognition rate becomes suddenly very low. This means the recognition rate of the proposed method is very sensitive to the shift of the target. However, this sensitivity of shift of the target is not harmful for the recognition system combined with active vision. It is useful to identify the position of the target because we can do matching more accurately by checking the distances in discriminant space constructed by linear discriminant analysis. Currently, we are developing a face detection method using this property.

4.2 Face recognition

In the previous experiments, binary images are used but the proposed method can be applied to gray scale images (face images) without any changes. Examples of face images with different sizes and their log-polar images are shown in Figure 4 from (a) to (f). The face images were taken under fluorescent lights, namely no special lighting is used. We collected the face images whose scalings were

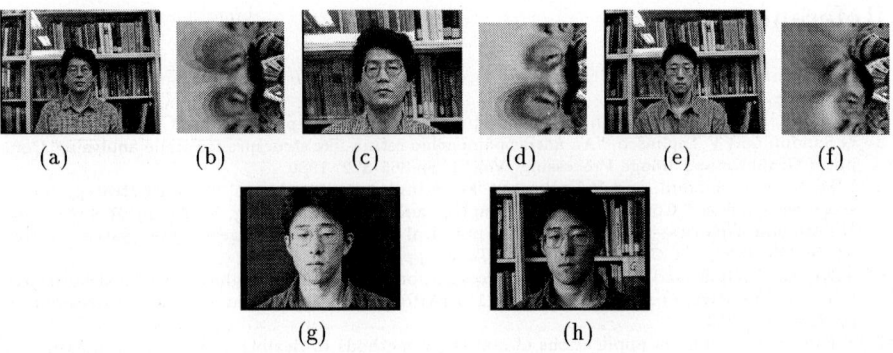

Fig. 4. Examples of face images, their log-polar images, and face images with different background. (a) Face image 1. (b) Log-polar image of (a). (c) Face image 2. (d) Log-polar image of (c). (e) Face image 3. (f) Log-polar image of (e). (g) Face image 4. (h) Face image 5.

changed 7 steps by changing zoom parameter of camera, which can be controled by computer. The number of images are 1000 (5 persons × 200 images) from 5 persons. The resolution of original images is 320 × 240 pixels. Since both the resolution of input images and the resolution of log-polar images influence the recognition rates, we have performed experiment with changing the both resolutions. At first, all the images were used in learning and the recognition rates were estimated by leave-one-out method. We obtained good results. For some combinations of the resolution of input images and the resolution of log-polar images, the best recognition rates achieved over about 99.70%. Even the worst recognition rate achieved 97.90%. This means that the proposed method has ability to recognize face images with different sizes. In this experiment, the recognition rate was highest when the resolution of log-polar images was 60 × 60 pixels. The resolution of input images also affect to the recognition rate. In this experiment, highest recognition rate 99.70% was obtained when the resolution of input images was 160 × 120 pixels.

Since log-polar image gives higher weights for center region, namely face region, than peripheral region, namely background or clothes, it is expected that the recognition rate is also less influenced by the changes of background. To investigate such effect of different background, we have performed face image recognition with different background. Examples of face images with different background are shown in Figure 4 (g) and (h). The 1400 images (20 images × 7 scales × 2 background × 5 persons) are gathered. All the images were used in learning and the recognition rate was estimated by leave-one-out method. Almost all recognition rates were 100%. As expected, this shows that the proposed method is robust to the changes of backgrounds.

References

1. J.Y.Aloimonos and I.Weiss, "Active vision," International Journal of Computer Vision, pp.333-356, 1988.
2. D.H.Ballard, "Animate vision," Artificial Intelligence, Vol.48, pp.57-86, 1991.
3. G.Sandini and V.Tagliasco, "An anthropomorphic retina-like structure for scene analysis," Computer Graphics and Image Processing, Vol.14, pp.365-372, 1980.
4. L.Massone, G.Sandini, and V.Tagliasco, "Form-invariant: Topological mapping strategy for 2D shape recognition," Computer Vision, Graphics and Image Processing, Vol.30, pp.169-188, 1985.
5. N.Otsu and T.Kurita, "A new scheme for practical flexible and intelligent vision systems," *Proc. IAPR Workshop on Computer Vision, Tokyo*, pp. 431-435 , 1988.
6. T.Kurita, N.Otsu, and T.Sato, "A face recognition method using higher order local autocorrelation and multivariate analysis," Proc. 11th IAPR International Conf. on Pattern Recognition, pp.213-216, 1992.
7. T.Kurita ,"A study on applications of statistical methods to flexible information processing (in Japanese) ,"Researches of the Electrotechnical Laboratory, No.957, November,1993.
8. F.Goudail, E.Lange, T.Iwamoto, K.Kyuma, and N.Otsu, "Face recognition system using local autocorrelations and multiscale integration," IEEE Trans. on Pattern Analysis and Machine Intelligence, Vol.18, No.10, 1996.
9. I.Sekita, T.Kurita, and N.Otsu, "Complex autoregressive model for shape recognition," IEEE Trans. Pattern Analysis and Machine Intelligence, Vol.PAMI-14, No.4, pp.489-496, 1992.
10. T.Kurita, I.Sekita, and N.Otsu, "Invariant distance measures for planar shapes based on complex autoregressive model," Pattern Recognition, Vol.27, No.7, pp.903-911, 1994.
11. N.Otsu, "A threshold selection method from gray-level histograms," IEEE Trans. Systems, Man, and Cybernetics, Vol.SMC-9, No.1, pp.62-66, 1979.
12. N.Otsu ,"Mathematical studies on feature extraction in pattern recognition (in Japanese)," Researches of the Electrotechnical Laboratory,No.818, July, 1981.
13. T.Kurita , K.Hotta , T.Mishima, "Scale and rotation invariant recognition of 2D shapes and face images using higher-order local autocorrelation features of log-polar image (in Japanese)," Techinical report of IEICE,PRMU96-212,pp.151-158,1997.
14. T.Kurita , K.Hotta , T.Mishima, "Scale and rotation invariant recognition of Face images Using Higher Order Local Autocorrelation Features of Log-Polar Image (in Japanese)," Trans. of IEICE, Informatoin and Communication Engineers D-II, Vol.J80-D-II No.8, pp.2209-2217,1997.
15. K.Hotta, T.Kurita, T.Mishima, "Scale and rotation invariant recognition of 2-D shape using higher-order local autocorrelation features of log-polar image (in Japanese),"Proceedings of the 1997 IEICE general conference,D-12-229,pp.436,1997.
16. O.Hasegawa, K.Itou, T.Kurita, S.Hayamizu, K.Tanaka, K.Yamamoto, N.Otsu, "Active agent oriented multimodal interface system," Proc. of IJCAI'95, pp.82-87, 1995.
17. M.Kirby , L.Sirovich, "Application of the Karhunen-Loeve procedure for the characterization of human faces," IEEE Trans. Patt. Anal. and Mach. Intell., Vol.12, pp.103-108,1990
18. N.Kita ,"Active vision systems using human vision as inspiration ", Trans. of Information Processing Society of Japan , Vol.36, No.3, pp.264-270,1995
19. Y.Kuniyoshi, N.Kita, S.Rougeaux, and T.Suehiro, "Active stereo vision system with foveated wide angle lenses," Proc. of 2nd Asian Conf. on Computer Vision, Vol.I, pp.359-363, 1995.
20. L.Berthouze, S.Rougeaux, F.Chavand, and Y.Kuniyoshi, "Calibration of a foveated wide-angle lens on an active vision head," Proc. of CVPR'96, pp.183-188, 1993.
21. H.Yamamoto, Y.Yeshurun, Martin D.Levine, "Foveated vision system with attentional mechanisms (in Japanese) ,"Trans.IEICE D-II, Vol.J77-D-II,, No.1, pp.119-129, 1994.
22. J.Van der Spiegel, G.Kreider, C.Claeys, I.Debusschere, G.Sandini, P.Dario, F.Fantini, P.Bellutti, G. Soncini, "A foveated retina-like sensor using ccd technology,"Analog VLSI and Neural Network Implementations, Boston,1989
23. M.Tistarelli and G.Sandini, "On the advantages of polar and log-polar mapping for direct estimation of time-to-impact from optical flow," IEEE Trans. on Pattern Analysis and Machine Intelligence, Vol.15, NO.4, pp.401-410, 1993.
24. M.Kreutz, B.Völpel, and H.Janβen, "Scale-invariant image recognition based on higher-order autocorrelation features," Pattern Recognition, Vol.29, No.1, 1996.

Script and Language Identification from Document Images

G.S.Peake and T.N.Tan
Department of Computer Science, University of Reading
England, RG6 6AY
E-mail: G.S.Peake@reading.ac.uk

Abstract

In this paper we present a review of current script and language identification techniques. The main criticism of the existing techniques is that most of them rely on either connected component analysis or character segmentation. We go on to present a new method based on texture analysis for script identification which does not require character segmentation. A uniform text block on which texture analysis can be performed is produced from a document image via simple processing. Multiple channel (Gabor) filters and grey level co-occurrence matrices are used in independent experiments in order to extract texture features. Classification of test documents is made based on the features of training documents using the K-NN classifier. Initial results of over 95% accuracy on the classification of 105 test documents from 7 scripts are very promising. The method shows robustness with respect to noise, the presence of foreign characters or numerals, and can be applied to very small amounts of text.

1 Introduction

The world we live in is becoming increasingly multilingual and, at the same time, increasingly automated. In the optical character recognition (OCR) and document image processing (DIP) communities this is beginning to present a problem. Almost all existing work on OCR makes an important implicit assumption that the language of the document to be processed is known beforehand. Individual OCR tools have been developed to deal best with only one specific language. In an automated environment such document processing systems relying on OCR would clearly need human intervention to select the appropriate OCR package, which is obviously not desirable. A pre-OCR language identification system would enable the correct OCR system to be selected in order to achieve the best character interpretation of the document. This area has not been very widely researched to date, despite its growing importance to the document image processing community and the progression towards the "paperless office".

In the first part of the paper, we present a review of the existing work. We discuss the principles, merits and weaknesses of each approach. Our review of existing work has motivated us to develop a novel approach to script identification. The new approach is described in the second part of the paper. Experimental results are included to illustrate the performance of the new method.

2 Previous Work

In this section, we present a review of existing work on script and language identification from document images. Each of the research groups currently involved in this area has taken a different approach. These approaches are discussed in the following.

2.1 Text Symbol Templates

Hochberg et al. present a method for automatically identifying script from a binary document image using cluster-based text symbol templates [1]. Text symbols are rescaled to 30×30 pixels. Symbols are grouped into clusters based on Hamming distances. The centroid (template) of each cluster is calculated. Classification requires N textual symbols from the test image which are compared (using the Hamming distance) with all template symbols with a reliability of at least $R\%$. The mean matching score for each script is calculated and the script with the best score chosen as the script of the document. A very high level of accuracy (up to 98%) can be obtained for both test images and "challenge" images (containing graphics and foreign characters).

Misclassifications generally arise due to test images containing markedly different fonts to the training images. This problem may be solved by including training images with a wider variety of fonts. It is claimed that the method is insensitive to page skew angles of up to 10°. In the dataset used, all the images were scanned at a resolution of 200 dpi or higher. In images scanned at lower resolutions, the text symbols will be much coarser which may present problems.

2.2 Projection Profiles

Wood et al. present some observations about the characteristics of various scripts, with particular reference to the effect such characteristics have on the horizontal and vertical projections of the document image [2]. They comment that other parts of the text (e.g. foreign characters), which have different characteristics, interfere with the definitive profiles. In order to remove these interfering effects and enhance the desired characteristics (and increase robustness with respect to skew), they apply a number of filtering methods to the original image including medial axis transforms, erosion, dilation and run length filtering. However, it is not clear how, (or if) the projection profiles can be analyzed automatically to determine the script, and no general system is suggested. The method would require no character segmentation, may be insensitive to point size, and would cope with a small amount of page skew. It is possible that particularly flamboyant fonts and italicized text may present some problems.

2.3 Upward Concavities

Spitz proposes a method for distinguishing between Asian and European languages by examining the upward concavities of connected components [3]. Upward concavities are described as [3]:

"Where two runs of black pixels appear on a single scan line of the raster image, if there is a run on the line below which spans the distance between these two runs, an upward concavity is formed on the line."

The vertical distribution of the upward concavities is shown to be markedly different between Asian and European languages. Gross script identification is performed by analyzing the variance of the distribution. In [5] this analysis is described as a "simple heuristically determined value of variance".

The work is then continued to discriminate between Chinese, Japanese and Korean if the script is classified as Asian. An optical density function is computed whereby the total number of black pixels in each character (its density) is tabulated in reading order across the text. Study of the distributions of these optical density functions shows distinct differences between Korean, Chinese and Japanese, although there are some problems discussed regarding actually performing classification in some cases. Sibun and Spitz go on to perform the classification of European languages on the basis of character shape codes in [4]. Character shape codes relate to the dimensions of characters rather than the actual characters themselves. Language identification of several Roman alphabet languages is performed by statistical analysis of frequent combinations of character shape codes in the languages investigated. The work in [3] and [4] is combined in [5].

2.4 Neural Networks

Lee and Kim use a self-organizing neural network in order to determine not the script of the entire document, but the script of individual characters within the document [6]. The only scripts they actually discuss are Chinese, Korean and English.

Initially, a non-linear normalization of character shapes (based on character density and character dimension information) is performed. There is no discussion of how the characters themselves are obtained. Zero, first, second and third order features are calculated using a Mesh feature system, overlapping contour direction codes, or Kirsch masks (which are explained in depth in [6]). There are then two classification stages, a coarse classifier (using a self organizing feature map which clusters characters into groups of all English, all Chinese, all Korean or mixed characters), followed by a fine classifier which classifies the characters in the mixed groups and presumably performs actual character identification.

Different combinations of these classification methods are combined with the different methods of obtaining features. The results they present show above 95% accuracy for all experiments, with as high as 98.27% being achieved for classification based on mesh features.

These results are obviously limited to English, Korean, Chinese, numerals and a handful of special characters. Also, the method does seem to be perhaps overly complicated and not ideally suited to the problem of script classification of a whole document image. It would, however, be very useful in recognizing small areas of foreign/special characters in a larger document.

2.5 Remarks

Some of the above approaches rely on accurate character segmentation or connected component analysis. The problem of character segmentation presents a paradox similar to that presented by OCR, namely that character segmentation can best be performed when the script of the document is known. Some scripts, such as Chinese, have the characters laid out in a regular array, making character segmentation a relatively simple matter. Korean and Japanese tend not to have overlapping characters, but, at the other extreme, contain horizontally disjoint characters. Spitz has developed a method for dealing with these scripts in [7]. Mono-spaced Roman fonts present little problem, but the use of proportional fonts, italics and kerning produces characters which are conjoined (such as the f joining to the i in "fi"), and which persistently overlap in terms of character bounding box space. The Arabic scripts are even more difficult to segment due to the deliberate overlapping and conjoining of cursive characters during the typesetting process. A method has been proposed, however, by Hashemi et al. for the segmentation and recognition of Persian and Arabic characters [8]. The process of scanning can also contribute to these artifacts, and furthermore can cause characters to become unintentionally split.

As can be seen from the above discussion, there is a need to apply different processing methods depending on the script of the document so as to achieve the best results possible from character segmentation. Performing character segmentation before the script of the document is known may prove to be inefficient. What we propose here is a method for identifying the script or language family of a document, without requiring character segmentation (or even connected component analysis), or placing any emphasis on the information provided by individual characters themselves.

Fig. 1: Examples of typical documents used: a) Chinese, b) English, c) Greek, d) Korean, e) Malayalam, f) Persian, g) Russian.

3. The New Algorithm

The new algorithm is inspired by the simple observation that a uniform block of text (where the line and word spacing is normalized), written in any language, can clearly be seen to have a texture [12]. Different scripts produce different textures. The texture differences are due to the variations in character density and stroke orientation. The previous researchers have commented on the spatial characteristics of different scripts, but do not really exploit them: Hochberg's templates rely on the general differences in character shapes, Spitz's upward concavities take advantage of different character densities and stroke positions, Lee and Kim use the orientations of contours, but Wood is the only one so far to regard the text as a whole rather than examining the individual characters.

Texture analysis is quite difficult to apply directly to documents - it is quite rare to find uniform blocks of text occurring in documents - there are usually interfering aspects such as variable word spacing in fully justified documents, and variable line spacing between paragraphs. We have found, however, that with some simple processing, a uniform block of text can be extracted from document images, to which established texture analysis methods can be applied. This presents a major extension of our previous work [12], where Gabor multichannel filtering is used to produce rotation invariant features to which texture classification is applied. The text blocks used there, however, are unrealistic as they require uniform spacing. Here we have extended the language set from 6 to 7 languages (now including Korean). We have used multiple frequencies instead of 1 and removed the rotation invariance requirement as this is irrelevant if the text has been skew-compensated. We have also made a comparison between the Gabor filter method and the use of grey level co-occurrence matrices (GLCMs).

3.1 Creating a Uniform Block of Text

The input to this stage is a binary image of a section of a document which has been skew-compensated [9] and from which graphics and pictures have been removed (at present the removal of non-textual information is performed manually, though page segmentation algorithms (e.g. [13]) could have readily been employed to perform this automatically). The text may contain lines with different point sizes and variable spaces between lines, words and characters. Punctuation symbols and foreign characters (such as Arabic numerals in Chinese text) may appear.

There are four main steps to obtain a uniform block of text from an arbitrarily formatted document:

Text Line Location

The horizontal projection profile (HPP) of the document is computed and smoothed. The peaks correspond to the centre of the text lines, and the valleys correspond to the blank areas between lines. Only a limited amount of smoothing over a small window can be applied here, as too much smoothing will cause the required peaks and valleys to be lost due to the generally very close spacing of peaks.

The height of each text line is then taken as being the width of the corresponding peak, up to the point on either side of the peak maximum where the curvature of the HPP either smooths out (in the case where there is a large white space above or below a line), or the lowest point in a sharp valley between two peaks. Text lines are then checked to ensure they do not overlap.

Outsize Text Line Removal

At present there is no method implemented for standardizing the height of the text lines, so all lines with a height much greater or much smaller than the mean line height are removed. The necessary thresholds are computed statistically in the following way: the mean and the standard deviation of the text line heights are calculated. All text lines which fall outside the range defined by *mean ± std. dev.* are removed. This process is repeated (including calculating the new mean and standard deviation), providing that the repetition will not cause all remaining text lines to be removed.

Spacing Normalization

This process comprises 3 steps:
1. Eliminate white space between lines: the vertical bounds of each line are known, so it is a simple matter to adjust the lines so that all line spacings are set to be the same predefined value.
2. Left justification: this step is important with respect to the *padding* stage. The x (horizontal) position of each text line is updated so that the leftmost point of the first character on the line is at x=0.
3. Normalize inter-word spacing to a maximum of 5 pixels: the pixels of each text line are projected vertically and examined from left to right. All runs of white pixels greater than 5 pixels wide are reduced to 5. This step forces the maximum gap between two characters to be 5 pixels. Gaps smaller than this are allowed to remain, because they may well, depending on the script, be gaps between characters rather than words.

Padding

The text is then padded to a block of a predefined size (here 128×128 pixels) in 2 stages. First the space between the last filled pixel on each text line and the right hand side of the image is calculated. An inter-word space of 5 pixels is subtracted from the gap size. This number of pixels is then copied from the start of each scan line in the text line to the space. The gap between the end of the existing text and the beginning of the copied text is always known to be 5 pixels because of the left-justification stage. Padding may be performed in this way because the texture is content-independent. Secondly the block is padded to the required height (if necessary) in the same manner by copying the appropriate number of scan lines from the top of the image. The block is then extracted from the top left hand corner of the image and texture features may then be extracted from the block.

3.2 Feature Extraction

In principle any texture analysis technique (see [14] for a recent survey) can be applied to the uniform text blocks created in the way discussed above. Here, two established methods are implemented to obtain texture features: Gabor filtering, and GLCMs. The former is becoming very popular and the latter widely recognized as the benchmark technique in texture analysis.

Gabor Filtering

Gabor filters have been shown to be a good model of the processing that takes place in the human visual cortex, and have been used successfully in both texture segmentation [10] and texture classification [11]. The mathematical details of Gabor filters can be found in [15] and elsewhere (e.g. [16]).

A Gabor filter requires as input: an $N \times N$ pixel image (where N is a power of 2), an angle (θ) and a central frequency (f). Parameters θ and f specify the location of the Gabor filter on the frequency plane. The filtering is performed in the frequency domain using FFT. Commonly used frequencies are powers of 2. In [11] it has been shown that, for an image of size $N \times N$, the important frequency components are likely to be found within $N/4$ cycles/degree, so here we are using frequencies at 4, 8, 16, and 32 cycles/degree. The width of the filter is determined by sigma (σ) of the modulating Gaussian function (assumed to be isotropic), which varies inversely proportionally to f.

For each central frequency, filtering is performed at 0°, 45°, 90° and 135°. This results in 16 output images (4 from each frequency) from which the texture features are extracted as follows: the mean and the standard deviation of each output image are calculated, giving 32 features per input image. Testing was conducted using all 32 features and various subsets of the features (e.g. the 16 mean features, the 8 features from a single frequency etc.).

Grey Level Co-occurrence Matrices

GLCMs are in general very expensive to compute due to the requirement that the size of each matrix is $N \times N$, where N is the number of grey levels in the image. In our case, however, because there are only 2 grey levels, it is reasonable to use GLCMs.

GLCMs were constructed for five distances ($d=1..5$) and four directions 0°, 45°, 90° and 135°. This gives, for each input image, 20 matrices of dimension 2×2. In other applications, where the size of the GLCM is prohibitively large to allow the use of the matrix element values directly, measures such as energy, entropy, correlation and so on are computed from the matrix and used as features [17]. Here, however, there are only 4 elements in each matrix, and, due to diagonal symmetry, 2 of those values are identical, giving 3 independent values from each matrix. Now we have 60 features per input image. Again, testing was conducted using various subsets of these features.

4 Experimental Results

The images used were scanned from newspapers, magazines, journals and books at 150 dpi greyscale. 25 examples were chosen for each of: Chinese, English, Greek, Korean, Malayalam, Persian and Russian. These languages represent most of the major scripts of the world (other and more languages could have been included were they available). Each image was selected from an entire page scan so that it would contain no graphics, and would resemble reasonably closely the output from a document segmentation system (although sections with multiple paragraphs were included because these are easily dealt with by the uniform block system, and images with multiple columns were used). Foreign characters, numerals and italicized text were present in many of the images. A small amount of page skew was inevitably introduced during the scanning process. This was compensated for using the method outlined in [9]. Images were scaled to have approximately the same average text height so that, in the absence of point size normalization, vastly different point sizes between documents would not affect the texture. Fig. 1 shows some examples of typical document images used in the experiments.

The images were divided first into 15 training and 10 test images per script (Set A), followed by 10 training and 15 test images (Set B). Images in the training sets did not appear in the test sets. Testing was conducted using different combinations of the features. All classification was performed using the K-NN classifier, with K=5. Table 1 shows the results from the Gabor filter method.

These results show that certain combinations of features produce a very high level of accuracy in the classification process. The single most useful frequency is 32, but this combined with a frequency of 16 produces the best results. For the shaded box at "*f=16&32 (means only)*" in Set A, the total number of features used was 8: there were four output images at *f*=16 and another four at *f*=32 and only the mean from each output image was used in this test. The shaded cells highlight the highest score for each set of tests. For Set A (15 training, 10 test images per language, 70 test images in total) 95.71% (=67/70) of the images were classified correctly. For Set B (10 training, 15 test images per language, 105 test images in total) 95.23% (=100/105) images were classified correctly. Table 1 also shows that in general the use of more training samples leads to higher classification accuracy.

The best result obtained for the GLCMs was only 77.14% accuracy (=54/70 images classified correctly from Set A), using all 60 texture features (5 values of d)×(4 directions)×(3 matrix values). Tests were not performed on Set B using the GLCM features because the Set A results (which should be more accurate than the Set B results) were so poor.

features	all	means only	std. devs. only	all at f=4	all at f=8	all at f=16	all at f=32	all at f=16 &32	f=16 & 32 (means only)
Set A	94.29	94.29	84.29	50.00	77.14	78.57	94.29	94.29	95.71
Set B	90.48	90.48	80.95	48.57	63.81	72.38	92.24	95.23	92.24

Table 1: Results of Gabor Filter Method - % of documents classified correctly.

5 Conclusions

We have reviewed the current work in the area of document script and language identification, and concluded that, although there is progress being made in the field, much of it relies on character segmentation, which we would like to avoid at the pre-language-determination stage. We have presented a novel method for script identification based on texture analysis which does not require character segmentation. Two texture analysis methods have been implemented: Gabor filters and grey level co-occurrence matrices. In tests conducted on exactly the same sets of data (documents taken from 7 languages), the Gabor filters proved to be far more accurate than the GLCMs, producing results which are over 95% accurate (comparable to results obtained in existing work). The key points of this method are:

- No character segmentation or connected component analysis is required and the presence of small areas of foreign characters, numerals and italicized text in the document does not affect the overall texture of the block extracted (hence no effect on the method).
- It is robust with respect to noise (use of an unsophisticated thresholding method caused some of the resulting binary images to contain noise).
- The point size of each character does not have to be normalized. (Although the documents were scaled so that they all had approximately the same average point size, there was no exact point size measurement). In future, the individual text lines, not characters, will be scaled (without character segmentation). This will eliminate the need for text line removal.

- Due to the padding process only a small amount of text is required. It may be possible to apply this method to single words if enough of the script characteristics are present.
- The method is not complicated - simple, established texture classification techniques have been employed.

Future work will involve the extensions outlined above, plus research into the area of discriminating between languages which are written in the same script, such as the family of Roman alphabet languages (e.g. English, French, Spanish and so on).

References

[1] J. Hochberg, L. Kerns, P. Kelly and T. Thomas, Automatic Script Identification from Images Using Cluster-based Templates, IEEE PAMI, Vol. 19, No. 2, February 1997, pp. 176-181.

[2] S. L. Wood, X. Yao, K, Krishnamurthi, L. Dang, Language Identification For Printed Text Independent of Segmentation, Proc. of IEEE ICIP 95, pp. 428-431.

[3] A. L. Spitz, Script and Language Determination from Document Images, Proceedings of the Third Annual Symposium on Document Analysis and Information Retrieval, 11-13 April 1994, pp. 229-235.

[4] P. Sibun and A. L. Spitz, Language Determination: Natural Language Processing from Scanned Document Images, Proc. of ANLP '94, pp. 15-21.

[5] A. L. Spitz, Determination of the Script and Language Content of Document Images, IEEE PAMI, Vol. 19, No. 3, March 1997, pp 235-245.

[6] S-W. Lee and J.-S. Kim, Multi-lingual, Multi-font, Multi-size Large-set Character Recognition using Self-Organizing Neural Network, Proc. of IDCAR '95, pp. 23-33.

[7] A. L. Spitz, Text Characterization by Connected Component Transformations, SPIE Proceedings, Vol. 2181, 1994, pp. 97-105.

[8] M. R. Hashemi, O. Fatemi and R. Safavi, Persian Cursive Script Recognition, Proc. of IDCAR '95, pp. 869-873.

[9] G. S. Peake and T. N. Tan, A General Algorithm For Document Skew Angle Estimation, Proc. of IEEE ICIP '97 (in press).

[10] T. N. Tan, Texture Edge Detection by Modelling Visual Cortical Channels, Pattern Recognition, Vol. 28, No. 9, 1995, pp. 1283-1298.

[11] T. N. Tan, Texture Feature Extraction via Visual Cortical Channel Modelling, Proc. 11th IAPR Inter. Conf. Pattern Recognition, Vol. III, 1992, pp. 607-610.

[12] T. N. Tan, Written Language Recognition Based on Texture Analysis, Proc. of ICIP '96, Lausanne, Switz., September 1996, Vol. 2, pp. 185-188.

[13] A. K. Jain and Y. Zhong, Page Segmentation using Texture Analysis, Pattern Recognition, Vol. 29, 1996, pp. 743-770.

[14] T. Reed and J. M. Hans Du Buf, A review of recent texture segmentation and feature extraction techniques, CVGIP: Image Understanding, Vol.57, 1993, pp. 358-372.

[15] D. Gabor, Theory of Communication, J. Inst. Elec. Engng. 93, 1946, pp. 429-459.

[16] J. G. Daugman, Uncertainty Relation for Resolution in Space, Spatial Frequency, and Orientation Optimized by Two-Dimensional Visual Cortical Filters, J. Opt. Soc. Am. A, Vol. 2, 1985, pp. 1160-1169.

[17] R. M. Haralick, Statistical and Structural Approaches to Texture, Proc. of IEEE, Vol. 67, 1979, pp.786-804.

Document Categorization for Document Image Understanding

Hiroyuki Masai[1,2] and Toyohide Watanabe[1]

[1] Department of Information Engineering,
Graduate School of Engineering, Nagoya University
Furo-cho, Chikusa-ku, Nagoya 464-01, Japan
[2] R & D Group, Media Lab., OKI Electronic Industry Co.,Ltd.
550-5 Higashiasakawa-cho, Hachioji, Tokyo 193, Japan(current address)

Abstract. In the knowledge-based document image understanding, it is important to distinguish the layout structures of individual documents exactly with a view to making use of adaptable document model. At least, the document models which are characterized heuristically by the application-specific layout structures are not always applicable to every document. In this paper, we propose a categorization method of various kinds of documents. Our categorization method on the basis of the classification and verification paradigm divides various kinds of documents into appropriate document types stepwisely. First, the classification procedure divides the given documents using rough features about documents, and then the verification procedure is applied to the globally categorized document sets, using the detail features.

1 Introduction

Today, though many methods/approaches about document image understanding have been proposed/developed, these are not always applicable to the analysis/recognition of other documents besides some documents of targets. In particular, the knowledge-based document image understanding system which makes use of document models makes the recognition ability clear from a logical point of view, but is dependent on the representation/specification manners of document models[1], which are derived heuristically from application-specific features, because different kinds of documents are associated with different logical/layout features such as structures, composition rules, data formats, relationships among data items, etc. As the first step, it is necessary to distinguish the types of individual documents effectually[4].

In this paper, we address a categorization method to divide various kinds of documents into the document types with respect to the domain-dependent features. We define the document type as a document set which can be distinguished in layout features. Our categorization method is constructed stepwisely on the basis of classification and verification paradigm[6]: the first is the division of document groups, and the second is the separation of document types. Here, the document group is a set of document types with the same or similar layout features, and is introduced with respect to the categorization efficiency. In the division phase of document groups the classification procedure first analyzes various kinds of documents with the features for pixel distribution, and then the verification procedure is applied to this analyzed/divided document. Next, in the separation phase of document types the classification procedure first analyzes the divided document group with the approximate features of layout structures, dependent on individual document types, and then the verification procedure distinguishes each document type with the detail features of layout structures. Then, in our classification and verification paradigm the classification makes use of the global features of individual document sets and also the verification works with the local features of individual document sets. Also, the verification procedure takes a role of re-classification.

2 Document Structure

Generally, the document structure consists of logical structure and layout structure. The logical structure is a tree structure which represents semantically inclusive relationships among items. Items are semantical components of documents. While, the layout structure is a tree structure which represents spatially inclusive/connective relationships among item areas. Item areas are partial areas which spatially include item data in document images.

Furthermore, we divide the layout structure into physical and logical layout structures concerning the definition of document sets[4]. These structures are decided according to the specification of item areas. When we pay our attentions to the allocation of item data in a 2-dimensional sheet in case of table-form documents in Fig.1(a), item data are allocated to areas which are surrounded by line segments as shown in Fig.1(b). For all table-form documents, item data are allocated to areas which are previously decided as Fig.1(c). In this case, we call the layout structure the physical layout structure. On the other hand, in case of name cards in Fig.1(a), item data are allocated according to geometrical relationships among them as shown in Figs.1(b) and (c). Geometrical relationships are relationships such as indentation, centering, variation of sizes of white spaces and so on among item data. In this case, item areas are specified on the basis of geometrical relationships among item data, and we call this layout structure the logical layout structure.

These layout structures are not mutually independent. Namely, some documents may have both layout structures. For example, in newspapers the composition of columns to be separated by line segments is dependent on the physical layout structure, and the composition of item areas based on geometrical features between sub-headlines and news items in columns is derived from the logical layout structure.

3 Outline of Categorization

Generally, many of researches about document image understanding have been performed in case of the same logical structure and different layout structure. If we can define document sets for the difference of logical structures and categorize document images into them, the document image understanding can be performed individually by appropriate methods. When we categorize document images, because it is impossible to categorize into an appropriate document type at one try. In this way, it is reasonable to define document sets on the basis of document structures.

Our categorization method consists of classification/verification process, and categorizes document images into document groups and document types in the order, shown in Fig.2. Our classification/verification process accomplishes the efficient categorization, because this applies the verification procedure to document sets distinguished appropriately in the classification process, but does not perform excessive verification process. Also, this accomplishes the effective categorization which recovers erroneous classification.

4 Document Set

For the purpose of understanding some kinds of documents, it is necessary first to determine target documents previously, and then apply the procedure of document image understanding to those documents. We consider six kinds of documents such as report-form document, newspaper, name card, book directory, table-form document and check as target documents, because their approaches/methods have been already developed[4]. Fig.3 shows examples of these documents.

4.1 Document group

Document groups are defined according to the difference among specifications of item areas, with respect to physical and/or logical layout structures as shown in Table 1. The document group 1 has both physical and logical layout structures, and report-form documents, newspapers and so on belong to this group. For report-form documents in Fig.3(a), item areas are specified by both columns and geometrical features such as indentations, variations of sizes of white spaces and so on[3]. Columns are previously specified areas, and item data such as titles, contents of chapters/sections and so on are allocated according to geometrical relationships in those columns. While, the document group 2 has only logical layout structure, and name cards, library cataloging cards, book directories and so on belong to this group. For name cards in Fig.3(b), geometrical features are found among item data such as company's name, person's name and so on, and it is possible to specify item areas according to them[1]. On the other hand, the document group 3 has only physical layout structure, and table-form documents, checks and so on belong to this group. For table-form documents in Fig.3(d), their item areas can be specified by line segments[2].

4.2 Document type

Document types are defined according to the difference among logical structures, and we show them in Table 2. In name card of Fig.3(b), there are items such as company's name, person's name, address and so on. While, in book directory of Fig.3(c), items such as title, author, publisher and so on exist. Since logical structures are different in individual applications, document types are defined in every application. Also, document types belong to document groups and inherit the layout features of document groups. For example, name card type and book directory type belong to the document group 2, because their item areas are specified by geometrical features though their logical structures are different.

5 Categorization of Document Group

We categorize document images into document groups as the first step. The classification process is performed on the basis of pixel features of documents, and the verification process is done by extracting layout features of individual document groups.

Classification Pixel features relate to the distribution of black pixels in document images. The extraction of them is to get the ratio of black/white pixels. We introduce two indexes: the ratios of line elements and black pixels, and we analyze document images according to those indexes as shown in Table 3.

The ratio of black pixels in the document group 1 is high because this group generally consists of sentence-form text data. Also, the ratio of line elements is low even if there exist line segments which organize columns. For the document group 2, since item data are word-form text data, the length of item data is short. Also, item areas are specified on the basis of geometrical features. Therefore, the ratio of black pixels is low. The ratio of line elements in the document group 3 is high because item areas are enclosed by line segments. Also, black pixels which construct line segments make the ratio of black pixels high.

This classification is first to analyze document images by the ratio of line elements, and then to analyze them by the ratio of black pixels as shown in Fig.4. The first analysis classifies document images into the document group 3 and others, and the second analysis classifies the others into the document groups 1 and 2. Thresholds in Fig.4 are determined through pre-experiments. Furthermore, in the second analysis we perform swelling procedure before getting the ratio of black pixels certainly. Document images in Fig.4 are extraction images of line elements and swelling images.

Verification We verify the classified document groups according to the layout features in Table 4. Layout features are related to physical and logical separators so that item areas are specified independently.

In the document group 1, physical separators such as line segments and white spaces organize columns. Also, geometrical features such as indentation, large space, centering and so on are found in columns, and can be regarded as logical separators because their existences represent geometrical relationships. Namely, item areas are specified by physical and logical separators. Therefore, the verification of document group 1 is performed by checking whether item areas are specified by physical separators and geometrical features. In the document group 2, item areas are specified by geometrical features. Therefore, the verification of document group 2 is performed by checking the existences of geometrical features. In the document group 3, vertical/horizontal line segments, which are physical separators, specify item areas. Therefore, the verification of document group 3 is performed by checking whether item areas are enclosed by physical separators such as line segments.

6 Categorization of Document Type

We categorize document groups into document types as the second step. We use layout features at the classification process and logical features at the verification process.

Classification Layout features in this process relate to the number of item areas, and they are shown in Table 5. For report-form document and newspaper types which belong to the document group 1, columns/rows are specified by physical separators, and the item data are done by logical separators. Generally, in this document type the number of item areas is undeterministic. However, the number of columns or rows is about 2 in report-form documents, and about 10 in newspapers. While, for name card and book directory types, the logical structure is fixed, and items shown in Table 2 exist. The number of them is about 6 in name cards, and about 12 in book directories. Since these document types belong to the document group 2, the number of items is the same as the number of item areas which are specified by logical separators. For table-form document and check types, line segments are physical separators, and item areas are specified by enclosing them. The number of such item areas is 2 or more in table-form documents, and 1 in checks.

In this way, the number of item areas shown in Table 5 is sufficient for classification of document types. It is necessary to extract item areas in a rough-and-ready manner. We accomplish this with a method, proposed by Y.Ishitani et al.[5].

Verification In the verification of document types, since it is impossible and unnecessary to extract all items, we extract a few items which are sufficient for identification of some document types which belong to the same document group. Here, we empirically judge which items are appropriate, and verify individual document types by extracting items shown in Table 6.

For example, in the name card type the name group which includes person's name has a special geometrical feature. If there is the feature in document images, those images have an item of name group. Therefore, the verification of name card type is to check whether such an item exists or not.

7 Experiments

We show experimental results about our categorization method. Sample document images are 50 sheets of report-form documents, name cards, book directories and table-form documents, individually. These images are digitalized by the image scanner with 200 dpi and 256 gray levels.

Table 7 shows results about categorization of document groups and document types on the basis of each feature. The numbers show correct results for classification/verification process. For example, in case of book directory, 44 sheets of document images are correctly categorized into the document group 2, and 48 sheets are correctly done by the verification process. Furthermore, 42 sheets of document images are correctly categorized into book directory type, and 45 sheets of them are correctly done by the verification process. When we pay our attentions to the ratios of correct results of each categorization, we find that the verification process grows up the ratio of correct categorization. Namely, the verification process recovers erroneous classification.

8 Conclusion

In this paper, we defined two document sets: document group and document type, and proposed the categorization method based on the classification/verification process. Furthermore, we showed which document sets target documents belong to, and reported the categorization method of documents based on some features such as pixel distribution, layout and logical structures. Also, we showed that our categorization method was useful through our experiments.

Our future work is to develop an integrated system by connecting the existing document image understanding system for individual documents with our categorization mechanism.

Acknowledgments We are very grateful to Prof. T. Fukumura of Chukyo University, and Prof. Y. Inagaki and Prof. J. Toriwaki of Nagoya University for their perspective remarks, and also wish to thank Dr. Y. Sagawa, Mr. K. Asakura , Mr. T. Sobue and our research members for their many discussions.

References

1. T. Watanabe, Q. Luo, Y. Yoshida, and Y. Inagaki: "A Stepwise Recognition Method of Library Cataloging Cards on the Basis of Various Kinds of Knowledge", *Proc. of 10th IPCCC*, pp.821-827(1991).
2. T. Watanabe, H. Naruse, Q. Luo and N. Sugie: "Structure Analysis of Table-form Documents on the Basis of the Recognition of Vertical and Horizontal Line Segments", *Proc. of 1st ICDAR*, pp.638-646(1991).
3. Q. Luo, T. Watanabe and N. Sugie: "A Structure Recognition Method for Japanese Newspapers", *Proc. of 1st SDAIR*, pp.217-234(1992).
4. T. Watanabe, Q. Luo and N. Sugie: "Structure Recognition Methods for Various Types of Documents", *Int'l Journal of MVA*, Vol.6,pp.163-176(1993).
5. Y. Ishitani: "Document Layout Analysis Based on Emergent Computation", *MIRU'96*, Vol.I,pp.343-348(1996)(in Japanese).
6. H. Masai and T. Watanabe: "Identification of Document Types from Various Kinds of Document Images Based on Physical and Layout Features", *Proc. of MVA'96*, pp.369-372.

Table 1. Document group

document group	examples of documents	logical layout structure	physical layout structure
1	report-form document newspaper	○	○
2	name card book directory	○	
3	table-form document check		○

Table 2. Document type

document type	items
report-form document	chapter, section (title, content), figure, table, etc.
newspaper	headline, sub-headline, news item, figure, photograph, etc.
name card	logo, company's name, title, person's name, address, etc.
book directory	title, subtitle, author, publisher, etc.
table-form document	many kinds of items, etc.
check	sum of money, signature, recipient, etc.

Table 3. Pixel features of document groups

document group	ratio of line elements	ratio of black pixels
1	low	high
2	low	low
3	high	middle

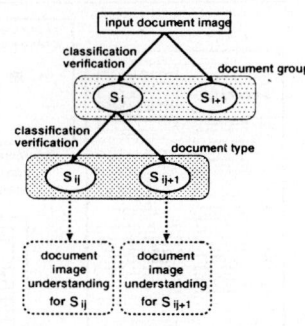

Fig. 2. Categorization method

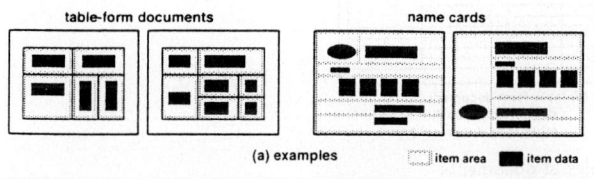

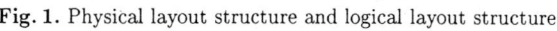

Fig. 1. Physical layout structure and logical layout structure

Table 4. Layout features of document groups

document group	examples of documents	layout features
1	report-form document, newspaper	·item areas specified by physical separators ·geometrical features in such item areas
2	name card, book directory	·geometrical features
3	table-form document, check	·item areas specified by physical separators

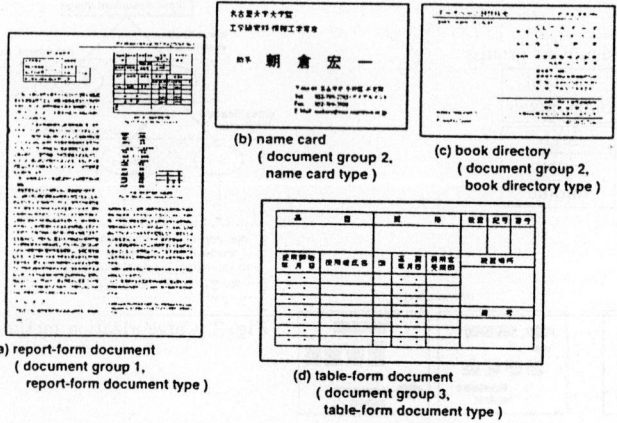

(a) report-form document (document group 1, report-form document type)
(b) name card (document group 2, name card type)
(c) book directory (document group 2, book directory type)
(d) table-form document (document group 3, table-form document type)

Fig. 3. Examples of documents

Table 5. Layout features of document types

document types	layout features
report-form document	about 2 columns
newspaper	about 10 rows
name card	about 6 item areas
book directory	about 12 item areas
table-form document	more than equal 2 item areas
check	1 item area

Table 6. Logical features of document types

document type	items
report-form document	chapter/section title
newspaper	sub-headline
name card	name group
book directory	title
table-form document	many item areas
check	sum of money

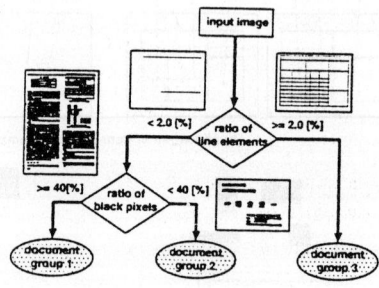

Fig. 4. Classification of document groups

Table 7. Experimental results

document group	document type	input	categorization of document group		categorization of document type	
			classification	verification	classification	verification
1	report-form document	50	47	46	46	42
2	name card	50	50	48	41	44
	book directory	50	44	48	42	45
3	table-form document	50	41	47	46	46
	total	200	182	189	175	177
	ratio [%]	--	91.0	94.5	87.5	88.5

Recognition of Various Bar-graph Structures Based on Layout Model

Naoko YOKOKURA and Toyohide WATANABE

Department of Information Engineering,
Graduate School of Engineering, Nagoya University
Furo-cho, Chikusa-ku, Nagoya 464-01, JAPAN
E-mail: {yokokura,watanabe}@watanabe.nuie.nagoya-u.ac.jp

Abstract. The structure recognition of bar-graphs is one of interesting subjects in the research fields of document understanding and drawing interpretation. In this paper, we address the recognition method of bar-graphs under the assumption that bar-graph has its own layout structure. The bar-graphs are very simple in comparison with other graphs, but the structures are too complicated: we observe various structures or many different representations in our environment. Our method is constructed flexibly so as to analyze such complicated structures successfully: our layout network, which represents the knowledge of layout structure of bar-graphs, is composed of three subnets with a view to establishing the concise specification of variant and invariant structures.

1 Introduction

Business-graphs as a kind of drawings are one of presentation means which can represent the corresponding information contents visually/intentionally. Bar-graphs are more basic type than others, and have been used in many cases. In the bar-graph, the graph primitives may be arranged generally under some composition rules. However, it is not always easy to distinguish these primitives individually because they are intersected/connected complicatedly or the geometrical/spatial configurations are varied according to applications in addition to the representation ways. Traditionally, the bottom-up-oriented image processing approaches have been applied to the analysis of such diagrams[1, 2, 3]. Especially, M.H.Lee et al. reported the recognition method on the basis of the knowledge about hierarchical structure of logical objects in the line-graph[4]. Also, we attacked at the analysis issue of bar-graph in the bottom-up and top-down manners under the assumption that bar-graphs have their own layout structures[5]. These approaches were successful to analyze simple structures, but the abilities are limited because the structures are variously complicated.

In this paper, we address an advanced recognition method of bar-graphs under the assumption that bar-graphs are associated inherently with their own layout structures. Our recognition method is an enhanced version of the previously proposed method[5]. Of course, we make use of layout-based approach as well as the previous method. In this version, our idea is to separate the layout network, which is specified as the representation knowledge of layout structure, into several parts so as to be adaptable to the variations of positional relationships among composite elements and different representation means.

2 Outline of Bar-graph Structure Recognition

2.1 Problems in our proposed method

We have already proposed a method for recognizing the structures of bar-graphs by using the layout knowledge[5]. In this method, the layout knowledge was expressed as the network-form structure, which is called as a layout network. Our proposed method has been composed of two processing phases: the first is an extraction of composite elements of bar-graph, and the second is an identification of graph primitives by interpreting the layout network. The layout network represents all possible bar-graph structures and the topological structure becomes complicated. Therefore, rules were used to make a suitable interpretation of bar-graphs possible, attended with the layout network. However, there are various kinds of bar-graphs and this method was not always successful.

Figure 1 shows some typical examples of bar-graphs. In our previously proposed method, indexes on horizontal axis are defined in the layout network as "indexes depend on bars or divisions", and in the corresponding rule "if indexes depend on all bars, then the indexes depend on bars one by one". Therefore, bar-graphs as shown in Figures 1(c) and (d) were recognized erroneously. Another error was derived from the extraction of composite elements. Figure 2 shows an example of erroneous extraction. This figure denotes that some composite elements were erroneously extracted as one element. In this case, this erroneously extracted element will be also erroneously identified. These errors can be avoided by adding the rule like "indexes may depend on one or more bars", but this addition makes the network and rules complex.

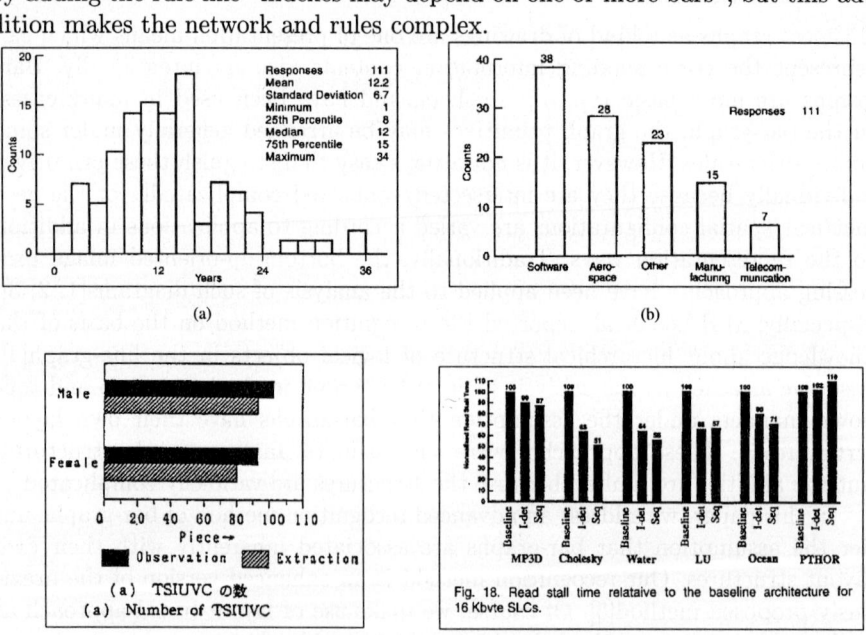

Fig.1. Example of various bar-graphs

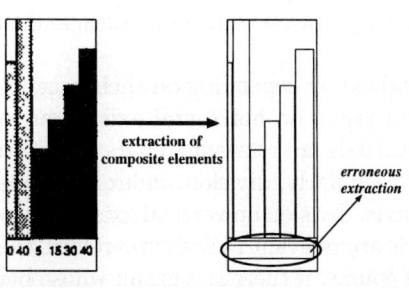

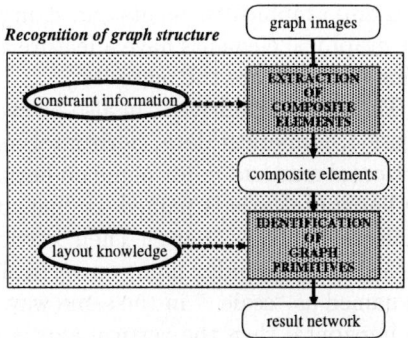

Fig.2. Example of erroneous extraction

Fig.3. Processing flow

2.2 Improvement of our method

To treat various kinds of bar-graphs, we improve each phase in the above method: an extraction phase and an identification phase. Our improved processing flow is shown in Figure 3.

Extraction of composite elements The layout structures of bar-graphs are constructed by relationships among graph primitives. First, it is necessary to distinguish composite elements in bar-graphs explicitly. In the previous method, graphical elements depend mainly on histograms of bar-graph images, and string elements are constructed by merging neighboring characters. However, it is not effective to use only the distance as criterion of merging characters. Thus, in our improvement method characters are not merged directly.

After the distinguishment of composite elements, graphical elements are checked. Since graphical elements are connected to the axis, they are labeled at the distinguishment phase, and examined by using the constraint information about graphical elements: divisions are allocated so that they are arranged at regular intervals. Then, composite elements with named bars are divided into some groups if some of them are connecting with each other or nearer than others. This is because an index of bar is assigned to one or more bars.

Identification of graph primitives In our previous method, the layout network represented all possible graph structures at once(e.g. both vertical bars and horizontal bars): the network was complex because all possible relationships among graph primitives were specified together. However, bars are identified in the vertical or horizontal direction, and all possible relationships must not be represented in the layout network: if there are vertical bars, it was not checked well that the relationships among bars and indexes are on vertical axis. Thus, we divide the layout network into four parts; each part specifies the partial layout structure in which the graph primitives have strong relationships with each other.

3 Layout Structure of Bar-graph

3.1 Graph primitives

The bar graph is composed from various kinds of graph primitives. Figure 4 shows a typical bar-graph and its primitives. The graph primitives can be categorized into graphical elements and string elements, depending on their attributes. The graphical elements have a feature that they always connect to axes, and the string elements are the opposite.

The graph primitives can also be categorized by depending on their usage. In Figure 4, numerical data, horizontal axis, division on horizontal axis, index on horizontal axis and index name on horizontal axis are relevant to bars: therefore, the horizontal axis is named as "bar axis"; similarly, division, index and index name are named as "bar..". Then, vertical axis, division on vertical axis, index on vertical axis and index name on vertical axis are relevant to scale: therefore, they are named as "scale.." in the same way. Of course, if there is a graph whose bars are horizontal then the vertical axis is named as "bar axis", and the horizontal axis is named as "scale axis".

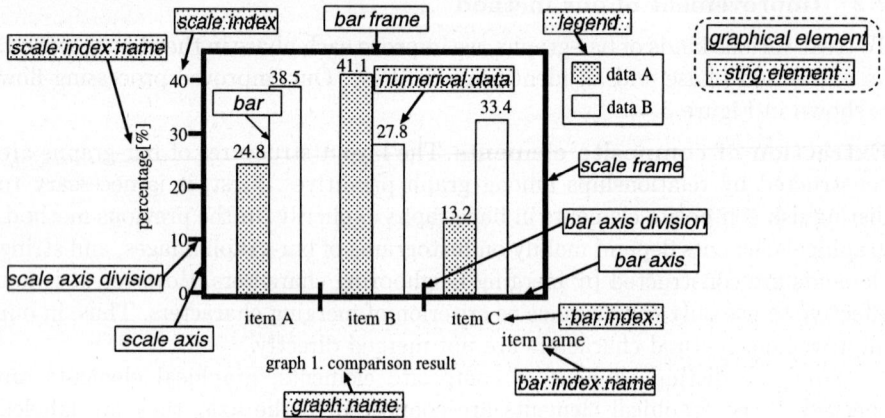

Fig. 4. Examples of graph primitives

3.2 Layout network for bar-graph structure

The graph primitives are arranged under the mutual relationships. For example, the bar axis is arranged under vertical bars or on the left side of horizontal bars, and each scale index is arranged under or on the left side of scale divisions.

We regard the layout structure of bar-graphs as a set of these relationships, and express them by using the network-form structure shown in Figure 5. We call this the layout network. This network is divided into three subnets, and each subnet is composed of primitives which have strong relationships with each other: the data subnet, the scale subnet, and the addition subnets. The data subnet is composed of elements relevant to bars, the scale subnet is composed of elements relevant to scale. The addition subnet is composed of other elements. In this network, a node indicates graphical primitive and an edge indicates a locational relationship among two connected nodes. Strings attached to "connecting/neighboring" edges represent a direction of connection from parent node to child node(here, "above" means a direction from "*bar axis*" to *bar*"). Strings expressed with italic characters represent a restrictive extent for the existence of child node. For example, Figure 6 shows how to interpret nodes with the edge of the restrictive extent in the graph image.

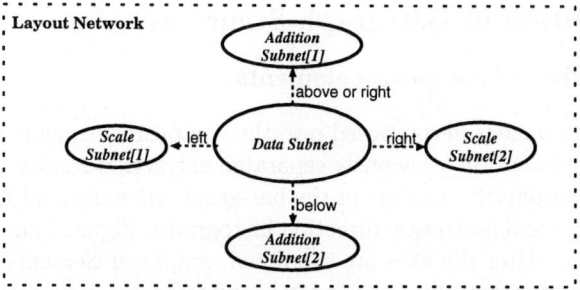

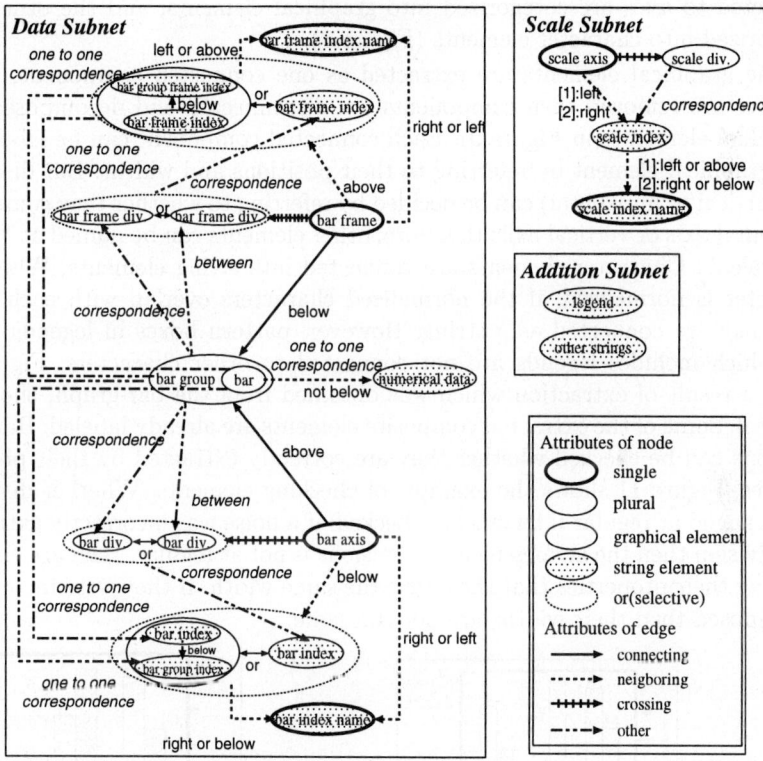

Fig. 5. Layout network

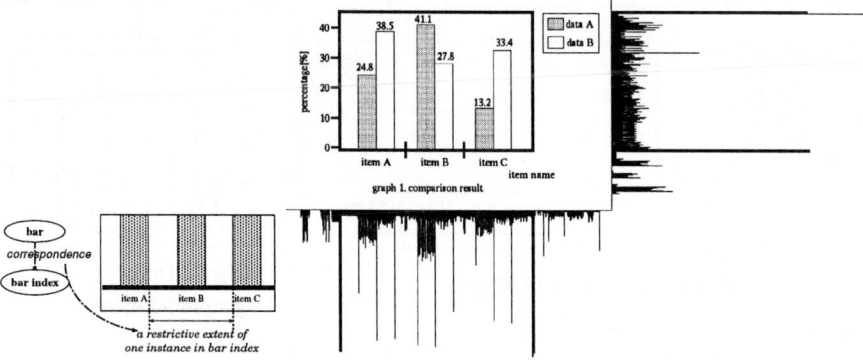

Fig.6. Restrictive extent   **Fig.7.** Histogram in graph image

4 Recognition of Bar-graph Structure

4.1 Extraction of composite elements

The bar-graph image is decomposed into the composite elements stepwisely one by one on the basis of the previously separated extraction results. First, the axes which are the landmark elements in the bar-graph are extracted from the image by using vertical and horizontal direction histograms. Figure 7 shows histograms for graph image. After the axes are extracted, graphical elements and character elements are categorized by referring to connected components: the elements connected to axes are categorized into graphical elements, and the others are categorized into character elements (Figure 8).

The graphical elements are extracted as one connected component, but if the axes are removed then components are disconnected and decomposed into individual elements (in Figure 9). Each connected component can be labeled as each graphical element by referring to their positions and widths. Bar direction (i.e. vertical or horizontal) can be decided by referring to whether they connect to horizontal axis or vertical axis: therefore, other elements can be named as "bar.." or "scale..". Character elements are connected into string elements. When one character is normalized, if the normalized characters overlap with each other then they are connected as a string. However, pattern boxes in legends and a box which includes legends are not connected to other characters. Figure 10 shows a result of extraction which was obtained from the bar-graph, shown in Figure 4. Some of the extracted composite elements are already labeled. Bars and divisions can be checked whether they are correctly extracted by their physical features. Figure 11 shows the example of checking elements. Whether divisions are arranged at regular intervals are checked: if a noise was incorrectly extracted as a division then the arrangement of divisions is not at regular. Bars are checked by using their properties that they have the same width. If they were incorrectly decomposed then their widths are not the same.

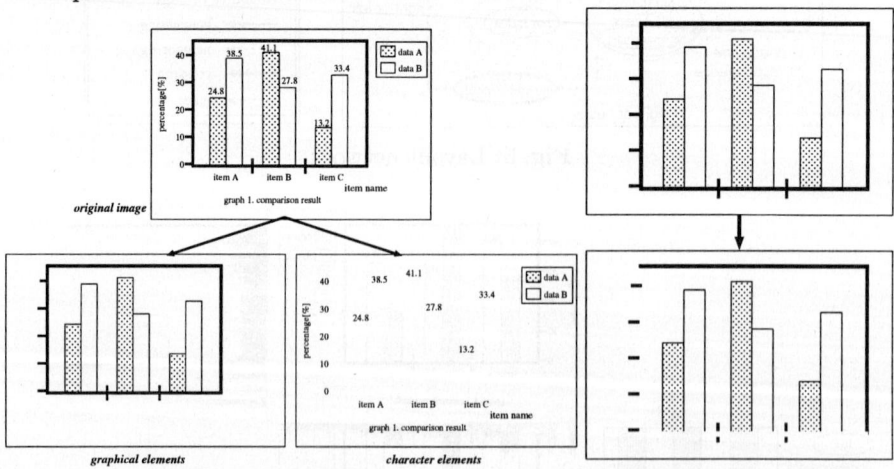

Fig. 8. Categorized elements

Fig. 9. Decomposed graphical elements

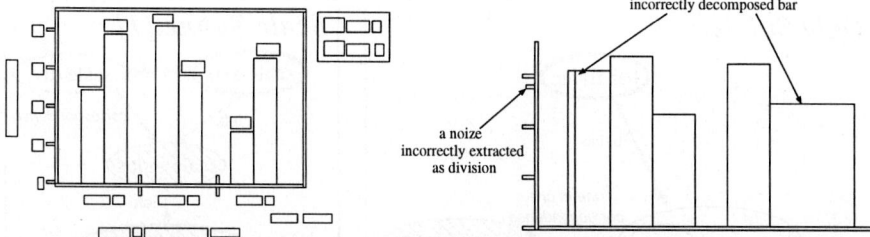

Fig.10. Extracted composite elements   **Fig.11.** Checking incorrectly extracted elements

4.2 Identification of graphical primitives

In this phase, the previously extracted composite elements are identified as graph primitives by using the layout knowledge of bar-graph and a result network is generated as the recognized result. This result network is regarded as an instance of the layout network. The processing steps are as follows:

Step1 Search each node in the layout network where parent nodes of searching node were already searched.

Step2 Extract graph primitives which correspond to the parents of searching node from the result network. If the primitive does not exist in searching composite elements, then extract graph primitives which correspond to grand-parent nodes of searching node.

Step3 Select only composite elements which satisfy constraint conditions, associated to graph primitives extracted in Step2: the searching node and edges connected to searching node(in Figure 12).

Step4 Label composite elements selected in Step3 with properties of searching node as graph primitives, and add the primitives to result network.

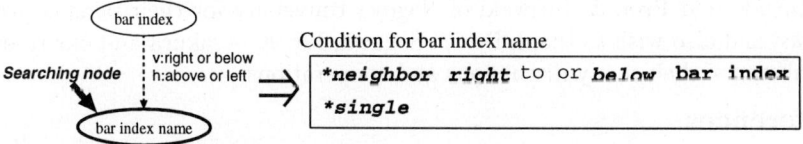

Fig. 12. Condition in searching node

Figure 13 shows a result network which was obtained from the bar-graph shown in Figure 4 (see and compare with the layout network shown in Figure 5).

5 Conclusion

In this paper, we proposed a recognition method of various bar-graph structures using the layout structure knowledge of bar-graphs. We consider the layout structure of bar-graphs as relationships among composite elements, and represented these relationships by using a network structure, called the layout network. This layout network is constructed from three subnets. Each subnet has primitives which have strong relationships with each other. Our proposed method is composed from two processings: the first is the extraction of composite elements,

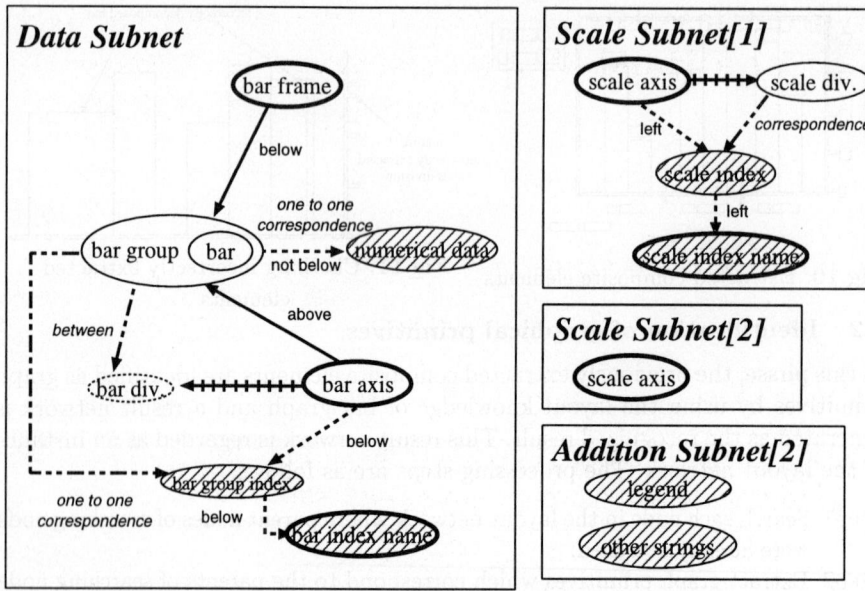

Fig. 13. Result network

and the second is the identification of graph primitives. In the extraction step, bar-graph composite elements are distinguished from a bar-graph image. In the identification phase, extracted composite elements are identified as graph primitives by using the layout network.

Acknowledgments

We are very grateful to Prof. T. Fukumura of Chukyo University, and Prof. Y. Inagaki and Prof. J. Toriwaki of Nagoya University for their perspective remarks, and also wish to thank Dr. Y. Sagawa, Mr. K. Asakura and our research members for their many discussions and cooperations.

References

1. R. Kasturi and et al.: "A System for Recognition and Description of Graphics", *Proc. of ICPR'88*, Vol. 1, pp. 255–259 (1988).
2. R. P. Futrelle and et al.: "Understanding Diagrams in Technical Documents", *IEEE Computer*, Vol. 25, No. 7, pp. 75–78 (1992).
3. Y. Nakamura and R. Furukawa: "Integration of Pattern Information and Natural Language Information toward Diagram Understanding", *Trans. on IPSJ*, Vol. 36, No. 1, pp. 196–205 (1995) [in Japanese].
4. M. H. Lee, N. Babaguchi and T. Kitahashi: "Symbolization and Presentation of Graph Images for Intelligent Communication of Document Images", *Proc. of ACCV'95*, Vol. 3, pp. 680–684 (1995).
5. N. Yokokura and T. Watanabe: "Recognition of Composite Elements in Bar Graphs", *Proc. of MVA'96*, pp. 348–351 (1996).

Word-Class Bigram Statistics Language Model for a Hand-Written Chinese Character Recognizer

Pak-Kwong Wong and Chorkin Chan

Department of Computer Science, The University of Hong Kong,
Pokfulam Road, Hong Kong.
{pkwong, cchan}@cs.hku.hk

Abstract. A word-class bigram statistics language model supported by a huge Chinese lexicon of 87,326 word entries has been investigated on its effectiveness in upgrading the accuracy of a hand-written Chinese character recognizer. The concept of the homogeneity of a word-class is introduced in classifying words into word-classes. On the average, the bigram statistics language model upgrades the recognition rate by 12.4% so that the overall system performance reaches 89.6%.

1 Introduction

An image recognizer of a line of unknown characters can be asked to propose an n-degree lattice of character candidates that are most likely to reveal the true content of the line. The correspondence between a sequence of character candidates and a sequence of words is usually ambiguous because of ambiguous segmentation of the characters into words. A language model as a post-processor, can help selecting among the candidates by evaluating their respective soundness in forming a natural line-of-text of the language because the linguistic information of the characters can provide a useful basis for improving the recognition rate [1]. In this study, a character recognizer [2] is employed to test a bigram statistics language model as a post-processor of the recognizer. The character recognizer supports a vocabulary of 4,616 characters and accepts writer independent off-line hand-written character images (Chinese characters, alphanumeric and punctuation symbols) from a scanner. It outputs a user-specified number n of candidates for each character image forming a lattice. Because Chinese, unlike western languages in which words are separated by blanks, has no word markers except the punctuation symbols, if there are m character images lying between a pair of punctuation symbols, the number of possible candidate sequences is m^n which can be extremely large for large m and n. Inevitably, many words can be formed in the lattice just coincidentally. This paper investigates how to select the "best" candidate sequence out of this large number of possible choices efficiently and accurately by means of a post-processing statistical language model.

2 The Lexicon

A lexicon named WORDDATA [3] is acquired from the Institute of Information Science, Academia Sinica, Taiwan. There are 78,410 word entries in WORDDATA covering most, if not all, of the Chinese words actively used in modern texts such as journals, newspapers, and literature in Taiwan. Each word entry is associated with a usage frequency and the membership of at least one syntactic/semantic word-class. A text corpus named "The Selection of Hundred Kinds of the Press in 1994", of over 63 million characters of news lines is acquired from Beicheng Co. Ltd., Beijing, China, as the testbed of the language model. Due to cultural differences between Taiwan and China, many words encountered in the corpus cannot be found in the lexicon. The lexicon must therefore be enriched before it can be applied. The first step towards this end is to merge a lexicon published in China into WORDDATA, increasing the number of word entries to 85,855. This updated lexicon is then applied to segment, by means of maximum matching with word binding forces [4], the text corpus into words for two purposes: 1. to discover words not in the lexicon so that the latter can be further enriched; 2. to collect the relevant statistics.

3 Word Segmentation by Maximum Matching

Segmentation of a line of text into words by maximum matching [5] favours long words and is a greedy algorithm in nature, hence, sub-optimal. Segmentation may start from either end of the line without any difference in the segmentation results. In this study, the forward direction is adopted. Words in the lexicon are divided into 5 groups according to their word lengths. They correspond to words of 1, 2, 3, 4, and more than 4 characters with group sizes equal to 7025, 53532, 12939, 11269, and 1090 respectively.

A data structure of bins and tries exploiting the unique characteristics of Chinese words has been adopted to organize the lexicon allowing an extremely effective search for words. With such a lexicon, an algorithm known as 'Maximum Matching Based on Word Binding Forces' is then applied to segment the huge text corpus.

Being a greedy algorithm, maximum matching word segmentation often suffers from selecting a long word at the expense of the word that follows. Errors are usually associated with single-character words after such segmentation. If the first character of a line is identified as a single-character word, what it means is that there is no multi-character word entry in the lexicon that starts with such a character. In that case, there is not much one can do except accepting it as a valid word. On the other hand, when a character is identified as a single-character word γ preceded by another word β in the line, one cannot help wondering whether the sole character composing γ should not be combined with the character(s) at the end of β to form a multi-character word instead, even that means changing β into another word of a shorter length. In that case, every possible character sequence alternative starting with a sub-sequence from the end of β, involving γ and possibly other characters beyond γ down the line should be evaluated according to the product of its constituent word binding forces. The binding force of a word is a measure of how strongly the characters composing the word are bound together as a single unit. This force is often equated to the usage frequency of the word but it is sometimes adjusted to another value to improve the word segmentation accuracy. In this respect, the proposed algorithm is a structural as well as statistical approach. It is as efficient as the maximum matching segmentation method because word binding forces are utilized only in exceptional cases. However, much of the word ambiguities are eliminated, leading to a very high word identification accuracy. Segmentation errors associated with multi-character words can be reduced by adding words to or deleting words from the lexicon as well as adjusting the word binding forces.

The CPU time used for segmenting a text of 1.2 million characters randomly selected over the text corpus is only 5.7 seconds on an IBM RISC System/6000 3BT computer. The hand-verified average segmentation accuracy over randomly sampled texts of 200,000 characters in total is 99.7%. Most of these errors are caused by proper nouns not included in the lexicon. They are hard to avoid unless these proper nouns become popular enough to be included into the lexicon. Consequently, the lexicon is expanded to contain 87,326 word entries. Each added entry is assigned a word binding force and the membership of a syntactic/semantic word-class. The text corpus of 63 million characters is then word segmented with this enhanced lexicon by maximum matching with word binding force.

4 A Language Model of Word Class Bigram Statistics

The limitation of word segmentation by maximum matching as a language model is its failure to capture the inter-dependence of words in a line of text. The use of bigram statistics in a language model [6] is a step towards overcoming this shortcoming. Since there are almost 90,000 words in the lexicon, the number of parameters in such a language model will be astronomical. A common practice is to employ the bigram statistics between word-classes [7] instead. If a sequence of character images $o_1, ..., o_T$ is segmented into $o_1^{w_1}, ..., o_{k_1}^{w_1}, o_1^{w_2}, ..., o_{k_2}^{w_2}, ..., o_1^{w_h}, ..., o_{k_h}^{w_h}$ correpsonding to a word sequence of $w_1, ..., w_h$ which in turn, belonging to word-classes $s_1, ..., s_h$ respectively, the soundness of the segmentation can be measured in terms of:

$$L = p(s_0 \mid s_h) \prod_{i=1}^{h} p(s_i \mid s_{i-1}) p(o_1^{w_i}, ..., o_{k_i}^{w_i} \mid s_i) \qquad (1)$$

where s_0 is a word-class of punctuation symbols appearing before and after the sequence of character images. $p(s_i \mid s_{i-1})$ and $p(s_0 \mid s_h)$ can be collected from the segmented text corpus while the suitability of $o_1^{w_i}, ..., o_{k_i}^{w_i}$ forming a word w_i in s_i is defined as:

$$p(\mathbf{o}_1^{w_i}, ..., \mathbf{o}_{k_i}^{w_i} \mid s_i) = p(w_i \mid s_i) \prod_{j=1}^{k_i} p(\mathbf{o}_j^{w_i} \mid c_j^{w_i}) \qquad (2)$$

Here, word w_i is a character sequence $c_1^{w_i}, ..., c_{k_i}^{w_i}$. $p(w_i \mid s_i)$ is computed from the segmented text corpus. $p(\mathbf{o}_j^{w_i} \mid c_j^{w_i})$ is a measure of similarity between the observed image $\mathbf{o}_j$ and the character $c_j^{w_i}$ of word w_i supplied by the image recognizer. The principle of dynamic programming is employed to determine the optimal segmentation of the character images into words. 500 pages of 400 characters each are hand-written by 300 writers. Out of these samples, 100 pages are used to determine the number of candidates, n, generated by the recognizer for each image while the remaining 400 pages are used to test the performance of the language model. Figure 1 illustrates the effect of a bigram statistics language model of 192 word-classes adopted by WORDDATA in upgrading the recognition rates of the 100 language model tuning pages as a function of the number of candidates n for each image. $n = 6$ is chosen for the bigram statistics language model. One can observe that the recognition rate is actually degraded by the application of bigram statistics of word-classes. This suggests the possibility of improper word clustering in WORDDATA which will be investigated in a later section.

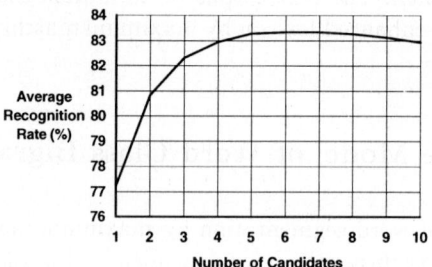

Fig. 1. Average recognition rates with a bigram statistics language model of 192 word-classes

5 Homogeneity of Syntactic/Semantic Classes in Lexicon

Words in WORDDATA are grouped into 192 word-classes with each word belonging to mostly one but up to four word-classes. In this investigation, each word is assigned only the membership of the most important class indicated in WORDDATA. A natural and objective criterion in measuring the soundness of any clustering is that all members within a cluster should have a similar pattern of associations with all clusters. From the text corpus, the probability of observing word w_j of s_i placed before a word of class s_q, for all q, can be computed. Associated with word w_j of s_i, there is therefore a probability vector $\mathbf{p}_j^{s_i}$ of 192 components, viz., $p_{j_k}^{s_i}$ for $k = 1, 192$. $p_{j_k}^{s_i}$ is the probability of seeing

w_j of class s_i before any word of class s_k in a line of the corpus. Since each word belongs to one class only in this investigation, there is no ambiguity if we drop the superscript s_i in $\mathbf{p}_j^{s_i}$ and its components. These vectors are normalized so that they lie on the surface of a unit hyper-sphere.

The centroid $\mathbf{C_i}$ of class s_i is defined as a unit vector along the direction of the average probability vector of all the words (weighted by the prior probability of the word) of the class. With this concept in mind, the homogeneity of class s_i, a word-class of M_i words, can be defined as:

$$H_i = \sum_{j=1}^{M_i} P(w_j) \mathbf{C}_i \cdot \mathbf{p}_j^{s_i} \quad (3)$$

Figure 2 gives the homogeneity histograms when the lexicon is divided into 192 and 470 word-classes respectively. Obviously, according to the histogram corresponding to 192 word-classes, the clustering of the words has a lot to be desired as far as homogeneity is concerned because of the existence of classes of very low homogeneity.

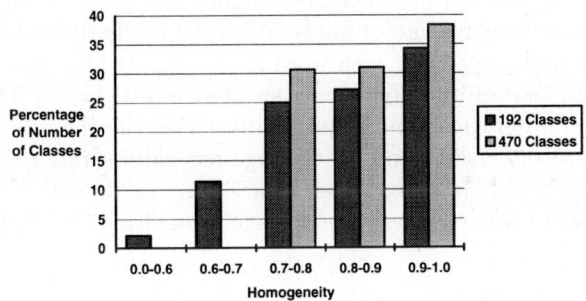

Fig. 2. Homogeneity histogram of 192 and 470 word-classes

Various thresholds are chosen over a number of iterations so that any word-class with a homogeneity below it will be split into two as in ISODATA, except that the feature space is confined to a unit hyper-sphere surface. A newly formed word-class with a homogeneity still below the threshold will be further split repeatedly. Finally, 470 word-classes are formed and bigram statistics between them are collected from the text corpus. At the end of an iteration corresponding to a particular homogeneity threshold, the effect of the word-class bigram statistics language model on the recognition of the 100 earmarked pages is measured and the results are illustrated in Figure 3. One can also notice from Figure 2 that there is no more word-class of very low homogeneity in the histogram corresponding to 470 classes.

The 400 pages of hand-written character samples reserved for final testing are then used to test the language model. The average recognition rate upgrades of the language model is 9.7% from 77.2% when no language model is applied.

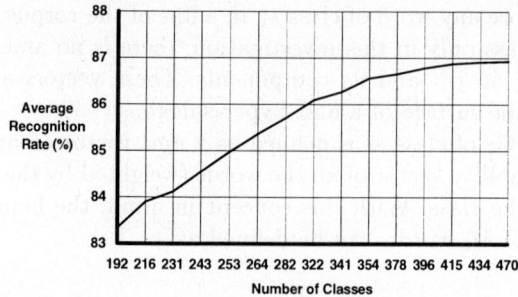

Fig. 3. Average recognition rates with a bigram statistics language model of various number of word-classes

The word-class splitting process discussed above is hierarchical. To mitigate the ill effect caused by any mis-classification of words in WORDDATA, after the number of classes has stabilized at 470, each word is re-assigned to a word-class whose centroid has the minimum inner product with the probability vector of the word. As soon as a word has been re-assigned, the centroids of the two word-classes affected are updated accordingly. With the newly defined word memberships, the probability vector of each word is re-computed by going over the text corpus again and so are the homogeneities of all word-classes. This process repeats over several iterations and Figure 4 illustrates the changes in the average homogeneity over all classes and the average recognition rate of the image recognizer with the word-class bigram statistics language model as the process of word-class reassignment iterates. The average recognition rate upgrade is 10.2%.

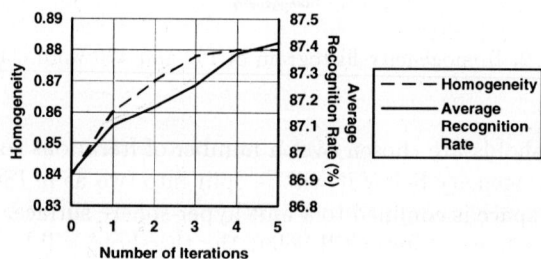

Fig. 4. Average homogeneity of 470 word-classes and average recognition rate as a function of word-class reassignment iterations

Figure 5 illustrates the improvement effected by the language model as a function of the recognition rate of the image recognizer when no language model is applied.

The recognition rate corresponding to "top 6" in the legend gives the probability of a character image with the true identity included amongst the top 6

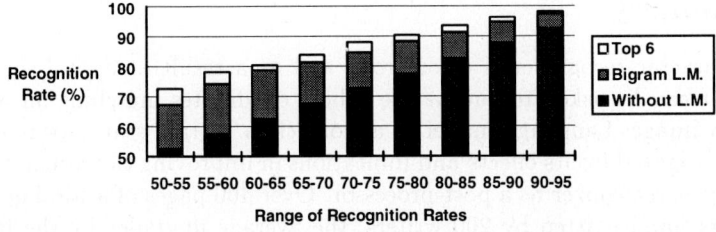

Fig. 5. Recognition rate improvement by the language model

candidates. It is the maximum recognition rate any language model can do with the image recognizer if the candidates of any image are limited to those in the top 6 choices proposed by the image recognizer. One can see that there is not much room for further improvement by any language model unless one can find means of increasing the number of candidates without the associated high risk of coincidental word formations. In the next section, a way of expanding the list of candidates with this objective will be discussed.

6 Confusion Set of a Character

A character is often mis-identified by the image recognizer for images of other characters that bear some similarity to it. The collection of such "similar" characters together with the character itself can therefore be used to define its "confusion set". When a_k^j, the kth image of character c_j is matched against a character template/model c_i of the vocabulary, a similarity measure $L(a_k^j|c_i)$ is computed by the image recognizer. As a consequence of matching every training image sample to all characters of the vocabulary, a "similarity" measure of c_j to c_i can be defined as:

$$S_{i,j} = \sum_{k=1}^{M_j} L(a_k^j|c_i)/M_j \tag{4}$$

where M_j is number of training samples of character c_j. The confusion set of c_i can be defined as those c_j's with $S_{i,j}$ ranked within the top N on the descending list of $S_{i,j}$. In the following experiment, N is chosen to be 20.

After an unknown line of images having been segmented into words, any image o failing to be a component of a multi-character word has n candidates for consideration whose confusion sets may overlap. An expanded set of candidates for image o can be constructed by adding to the original set of candidates those characters which appear in more than m of these n confusion sets. In this experiment, m is chosen to be 3, i.e., half of n. Then, the same bigram statistics language model above is applied again and the recognition rate is upgraded by 12.4% so that the overall system performance reaches 89.6%.

7 Summary

A character recognizer is error-prone and as a result of digital signal processing, it can be asked to output the n-best candidates matching an observed unknown image. Language model of a word-class bigram statistics nature has been investigated for its effects and limitations in improving the recognition rate of the image recognizer as a post-processor. Over 400 pages of a total of 160,000 characters hand-written by 200 writers, the average upgrades by the language model is 12.4%.

As by-products of this study of statistical language model for Chinese, a large lexicon of 87,326 words each assigned a word binding force and a wordclass membership has been constructed. In addition, bigram statistics between word-classes (470 strong) have been computed. These by-products are extremely useful for other linguistics applications.

ACKNOWLEDGEMENTS

This investigation has been supported by the Hong Kong Research Grant Council as well as the Industrial Support Funds of the Hong Kong Industry Department.

References

1. K.T. Lua, "From Character to Word – An Application of Information Theory", *Computer Processing of Chinese and Oriental Languages*, Vol. 4, No. 4, pp. 304-313, March 1990.
2. S.L. Leung, P.C. Chee, Q. Huo and C. Chan; "Contextual Vector Quantization Modeling of Hand-printed Chinese Character Recognition"; *Procs. of IEEE International Conference on Image Processing*, pp. 432-435, Washington, D.C., October 1995.
3. "WORDDATA", Chinese Knowledge Information Processing Group, Technical Report No. 93-05, Institute of Information Science, Academic Sinica, Taiwan.
4. P.K. Wong and C. Chan, "Chinese Word Segmentation based on Maximum Matching and Word Binding Force", *Procs. of COLING'96*, Vol. 1, pp. 200-203, Copenhagen, August 1996.
5. Y. Liu, Q. Tan and K.X. Shen, "The Word Segmentation Rules and Automatic Word Segmentation Methods for Chinese Information Processing (in Chinese)", *Tsinghua University Press and Guangxi Science and Technology Press*, page 36, 1994.
6. W. Eckert, F. Gallwitz and H. Niemann; "Combining Stochastic and Linguistic Language Models for Recognition of Spontaneous Speech"; *Procs. of IEEE International Conference on Acoustics, Speech and Signal Processing*, pp. 423-426, Atlanta, May 1996.
7. L.S. Lee et al; "Golden Mandarin (II) - An Intelligent Mandarin Dictation Machine for Chinese Character Input with Adaptation/Learning Functions", *Procs. of IEEE Int. Sym. on Speech, Image Processing and Neural Networks*, pp.155-159, Hong Kong, 1994.

Log Classification by Single X-ray Scans Using Texture Features from Growth Rings

Xinli Wang

The Norwegian Pulp and Paper Research Institute
P.O. Box 24, Blindern, N-0313 Oslo, Norway
Xinli.Wang@pfi.no

Abstract. An automatic log classification system is described. The classification was based on texture features obtained from measurements of growth rings from a single X-ray scan along each log. The speed of the log feeding restricted the possibility of full-scale scanning and image reconstruction. A preliminary analysis to study the projection of growth rings was performed by developing a simulated X-ray system. Then the projection images of 347 logs obtained in a medical scanner were used in the classification. It was not possible to measure the rings exactly due to the low resolution of the scanner and the high moisture content in sapwood. Instead, texture features in the central part of the logs were used in the classification.

Image processing methods were used to locate the knot-free areas and to segment the growth rings automatically. These methods were both simple and fast enough that they could be performed in parallel with the log scanning. A grey-level-gap-length method was used for texture feature extraction and a back propagation neural network was utilised for log classification.

The system was able to sort the logs into two different ring width classes with an 89 % correct classification. The results show that automatic log sorting is feasible by X-ray scanning and image processing.

1 Introduction

Sorting logs into homogeneous quality classes is a good way to improve the stability and efficiency of wood processing, such that the values of the raw material can be better utilised. Wood quality can be characterized by several properties depending on the end products, but the mean width of growth rings seems to be a useful parameter for both saw-logs and pulp-logs [5] [2] [11].

Up to now, no automatic sorting system exists which is able to classify the logs according to their ring width simply because the structures are too difficult to detect. Growth rings can vary from less than 0.1 mm to more than 10 mm in width. The diameter of the logs is normally in a range from 100 mm to 500 mm. In a Norwegian sawmill, logs pass through the grading unit one after another at a speed of 1 - 2 m / sec. The dimension, the speed and the resolution together make a high demand on the scanning and processing capability of the system.

Optical scanning is sensitive to the surface unevenness, decays and cracks on the log end. The maximum contrast between earlywood and latewood is about 18% in absolute light reflectance [11]. This contrast becomes much less when the rings are very tight. Spatial resolution and processing speed are also problems for optical scanning system [4].

Since earlywood fibres have lower density than latewood fibres, non-destructive detection using X-ray scanning is an alternative. X-ray imaging is a density-based measurement that can reveal the inner structure of the logs. But traditional computer tomography (CT) methods need a full range of scanning around the object to reconstruct the images. It normally takes more than one hour in a medical scanner to scan a 5-meter long log with one image per 10 mm of log length. Another problem is that the growth rings in sapwood are almost undetectable because of the high moisture content. Besides, the spatial resolution of the X-ray scanners is about 0.5 mm / pixel, which is not sufficient to detect all of the rings.

However, growth rings are typical texture features and the width of the rings is normally larger in heartwood than in sapwood. Measuring the ring textures only in heartwood may reduce the problems caused by high moisture content and low spatial resolution [10]. Since growth rings are more or less symmetric, extracting features directly from one projection instead of reconstructed images will largely reduce the data dimension and the processing time. The remaining questions are if the features from one projection are representative for the growth rings, and how wide the rings must be for the result to be reliable.

The objective of this study was to examine the feasibility of log sorting using one-pass X-ray scanning. In a pre-study, a computer simulated fan-beam system was developed and tested on an image of artificial rings to study the problems caused by the limited spatial resolution and the superposition of objects. Then texture features from projection images of 347 logs were extracted after image processing. The features were further fed to a back-propagation neural network that classified the logs into two growth ring classes.

2 Preliminary Analysis

Substances of different densities will attenuate X-rays differently. For an object with a given density and chemical composition, the attenuation coefficient μ is linearly related to the object density ρ by

$$\mu = k \cdot \rho \quad (1)$$

where k is decided by the chemical composition of the object as well as the scanning energy [3] [7]. By measuring the intensity of the transmitted photons, the attenuation coefficient and further the density of the object can be estimated.

In order to study the projection profile of growth rings under different scanning resolutions, a computer simulation of a fan-beam system was developed. The system consists of an X-ray source and a line of detector elements, as shown in Fig. 1. The spatial resolution of the scanner depends on the number of detector elements, the dimension of the detector, as well as the distance between the X-ray source and the image centre.

In principle, the projection of a ray-beam detected on a detector element can be obtained by performing integration along the ray over the object. A profile of the

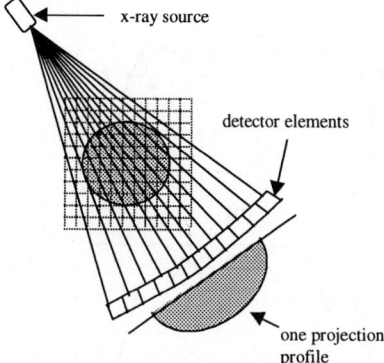

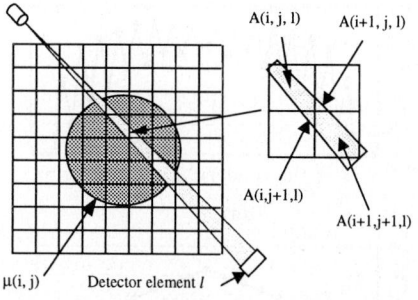

Fig. 1. A simulated fan beam scanning system.

Fig. 2. An illustration of a beam of rays passing through an object µ(i, j) and detected on a detector element l. Here A(i, j, l) is the fractional area of the pixel (i, j) covered by the beam.

projection is a set of such integrations for all of the detector elements. Assuming an object µ(i, j) is a two-dimensional distribution of linear attenuation coefficient, and the fractional area of pixel (i, j) covered by the beam to detector l (1<l<J) is A(i, j, l), as shown in Fig. 2, then the projection profile P(l) is

$$P(l) = \sum_{(i,j) \in b(i,j,l)} A(i,j,l) \cdot \mu(i,j) \tag{2}$$

where b(i, j, l) is the set of pixels traversed by the beam to detector l, and J is the total number of detector elements.

Given an 8-year old log having a cylindrical shape and evenly distributed rings, a cross-section image of the log will look like the one in Fig. 3a. Here, each "growth ring" consists of one black ring of "earlywood" (4 pixels) and one white ring of

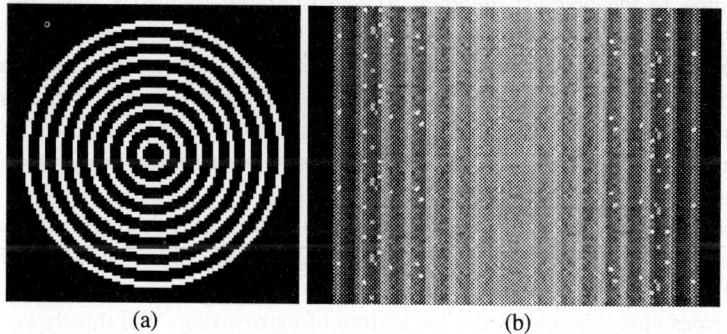

(a)   (b)

Fig. 3. (a) An image of 8 artificial growth rings. Black rings are earlywood and white rings are latewood.
(b) The image of projection profiles detected on 116 detector elements when scanning around the rings.

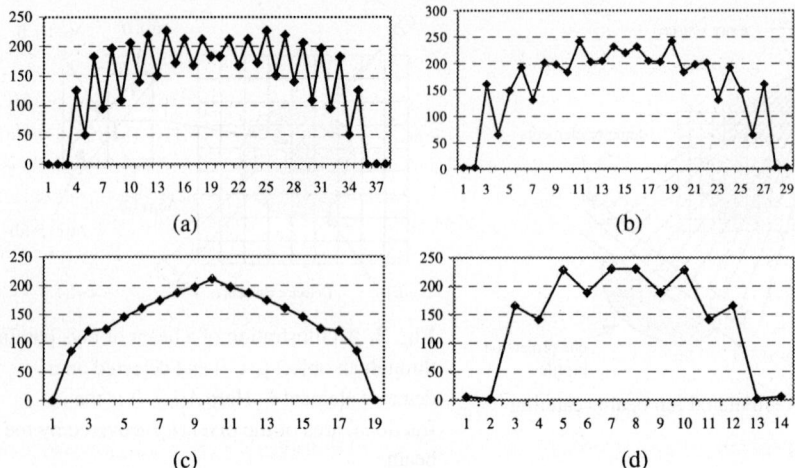

Fig. 4. The figures in (a), (b), (c) and (d) show the projection profiles of Fig. 3a scanned at 2, 1.5, 1, and 0.7 pixels per ring, respectively.

"latewood" (2 pixels). Rotating the scanning system around the rings produces projections as shown in Fig. 3b. Each line in the image represents a projection in a specific direction. The white spots in Fig. 3b are noise due to the discrete representation of the rings. Since the projections are the same for all directions, a scan along the log will produce the same profiles according to the cylindrical assumption.

Fig. 3b shows that the superposition of the rings in the projection causes a gradual rise of the grey levels and a reduction of the contrast towards the centre of the rings. Even so, 8 x 2 accumulated stripes from the "latewood" are still visible. The edges between earlywood and latewood are smoothed and the width of the projected rings depends mainly on the number of detector elements. In Fig. 3b the resolution is about 6 pixels per growth ring.

According to the Nyquist theorem [8], the sampling frequency must be at least twice the highest frequency of the signal in the direction perpendicular to the projection. So the scanner must have a spatial resolution of at least two pixels per growth ring. When the resolution is insufficient, aliasing occurs. Examples of the profiles scanned at 2, 1.5, 1, and 0.7 pixels per ring can be found in Fig. 4. The figures show that when the resolution is less than 2 pixels per ring the profile turns into a signal with lower frequencies and a reduced contrast.

The pre-analysis shows that it is possible to detect growth rings from a single projection when the ring width is sufficiently large. If the ring width is less than two pixels, most of the rings will be filtered out. The texture of wide rings will appear to be thin stripes along the log, while the texture of narrow rings will mostly be blurred. This property may be used to differentiate wide-ring logs from narrow-ring logs.

3 Experiments

In a large experiment, scoutview images (edge-enhanced projection images) taken from the butt part of 347 logs were analysed and classified. These images were obtained using a single scan along each log in a medical X-ray scanner. The maximum diameter of the logs is 540 mm and the minimum is 160 mm. The average ring width is about 1.4 mm. The spatial resolution of the images is 1 mm per pixel. Therefore, only the growth rings wider than 2 mm may be detectable. Fig. 5a and Fig. 5c show the images of two logs with wide and narrow rings, respectively.

In order to analyse the scoutview images, a set of image processing methods were developed, and texture features were extracted. The classification was performed using a neural network classifier.

3.1 Image Processing

A global thresholding method [6] was used to segment the logs from the background. Then a binary morphologic closing operation with a 5 x 5 element was carried out to smooth the outline of the logs. The average diameter of the logs was measured.

Since knots normally contain high-density wood fibre and more water than knot-free areas, knot areas will have high grey levels. The mean grey level for each

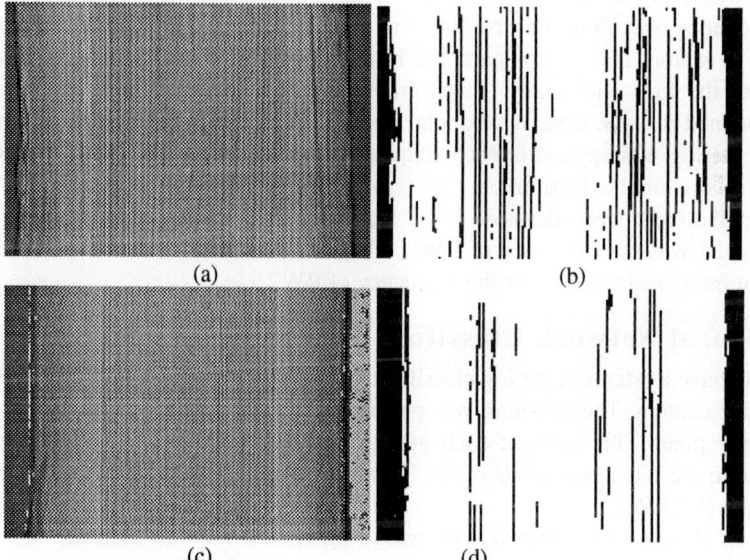

Fig. 5 (a) The scoutview of a log with wide growth rings. The mean growth ring width is 2.2mm.
(b) The resulting image of (a) after image processing.
(c) The scoutview of a log with narrow growth rings. The mean growth ring width is 1mm.
(d) The resulting image of (c) after image processing.

projection line was calculated in order to locate knot-free areas.

At each knot-free area, a local thresholding method [12] was used to segment the earlywood from the latewood. The threshold t at pixel (i, j) was determined by

$$t = \mu(i,j) + k \cdot \sigma(i,j) \qquad (3)$$

where μ and σ were the local mean and standard deviation respectively, and k was a constant. A window of 5 x 1 pixels was used to capture the vertical stripes of growth rings. After segmentation, a decision-based method was applied to remove black spots less than 17 pixels in height and 5 pixels in width.

Fig. 5b and Fig. 5d are the results of Fig. 5a and Fig. 5c after image processing. The mean growth ring width (MGRW) of the log is 2.2 mm in Fig. 5a and 1.0 mm in Fig. 5c. It can be seen that the log with narrow rings will have a few stripes left in the resulting image.

3.2 Feature Extraction

Texture features within the inner two thirds of the log radius were measured based on a grey-level-gap-length method [9] from the processed images. A gap is a peak or a valley between two pixels having the same grey level. A concatenated white gap and black gap in a projection was counted as a 'ring' width, although it might not represent a real growth ring.

The counting was done line by line and the frequency of each 'ring' width was stored in a vector. This vector gave the histogram of the 'ring' widths. In order to distinguish the logs with many 'rings' from those with a few 'rings', a weighting factor obtained through dividing the total number of counting by the total number of pixels in the log was applied to each frequency. Finally, a mean filtering of width 3 was used to smooth the distribution.

Four parameters were defined: RW1 was the 'ring' width with the maximum frequency; F1 was the frequency of RW1; RW2 was the 'ring' width with the second maximum frequency and F2 was the frequency of RW2.

3.3 Neural Network Classifier

In order to have a reference for log classification, all logs were scanned again optically using a line camera. The scanning was performed along a radius with a resolution of 0.1 mm per pixel. The width of each growth ring was registered. The average ring width within the inner two thirds of the log radius was defined as the Centre Growth Ring Width (CGRW).

Two log classes were defined. The wide ring class (WR) included the logs with CGRW larger than 1.5 mm, and the narrow ring class (NR) contained the logs with CGRW less than 1.5 mm. Among the 347 logs, 206 logs belonged to the wide ring class and 141 logs to the narrow ring class. The average CGRW was 2.2 mm for wide ring class and 1.2 mm for the narrow ring class.

A back-propagation neural network [1] was used for log classification. This neural network had five input neurodes, three hidden neurodes and two output neurodes. In addition to the texture parameters RW1, RW2, F1 and F2, the log diameter (D) was used as input. The wide-ring class (WR) and the narrow-ring class (NR) were the

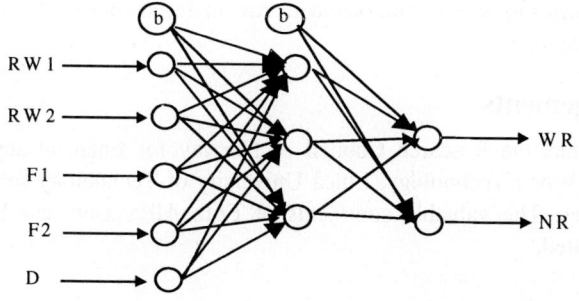

Fig. 6 A back-propagation neural network with the texture parameters RW1, RW2, F1, F2, and the log diameter (D) as inputs, and the wide-ring class (WR) and the narrow-ring class (NR) as outputs.

output neurodes. An illustration of the neural network is shown in Fig. 6. The bias neurodes, which were labelled *b*, served as threshold units for the layers to which they were connected.

The weights of the network were randomly initiated between −0.3 and 0.3. After each iteration, all weights were adjusted according to the error values back propagated from the output neurodes. The neural network would stop training when the average mean-squared error was less than 0.02 or when the number of iterations exceeded 2000. The weights obtained from the training process were used in final testing.

80 logs from the wide-ring class and 80 logs from the narrow-ring class were used as a training set for the neural network. Then the remaining 187 logs were used as a test set, which included 126 logs from the wide-ring class and 61 logs from the narrow-ring class. An output value larger or less than 0.5 was used to determine whether a log belonged to a given class or not.

4 Results and Discussion

During the training process, an 8 % error in the classification was obtained after 2000 iteration. The testing on the extra 187 logs gave an 89 % correct classification. Because the classes were defined according to the average ring width, a log in the narrow-ring class that had both very wide rings and very narrow rings might be classified as a wide-ring log. The result will be better if the spatial resolution increases.

Since most of the measurements were done line by line, the image processing and feature extraction could be performed almost in parallel with the scanning. An X-ray scanner can produce more than 360 projections per second. If the projection is performed at every 4 mm in the log length direction, the logs can be measured at a speed of 1.5 m / s, which is sufficient for the requirements of log grading today.

The results show that log classification according to their growth rings is feasible by X-ray scanning and image processing. Such a sorting system will enable the wood and paper industries to receive homogeneous raw materials in terms of construction or papermaking quality.

Acknowledgements

The author thanks the Research Council of Norway for financial support, and the Department of Wood Technology, Luleå University of Technology for the supply of data and images. The valuable comments of Fritz Albregtsen and Bent Foyn are greatly appreciated.

References

1. R. C. Eberhart and R. W. Dobbins (Ed.), *Neural Network PC Tools*, Acad. Press, 1990.
2. A. Grönlund and S. Grundberg, *Relations between Annual Ring Width and Different Properties for Individual Logs*, 2nd IUFLO Workshop, Berg-en-Dal, South Africa, 1996.
3. G. T. Herman, *Image Reconstuction from Projections*, Academic Press, New York, 1980.
4. L. Jonsson, *Recording Annual Ring Development Using Image Analysis*, Workshop on Scanning Tech. and Image Proc. on Wood, Skellefteå Campus, Sweden, 1992.
5. L. A. Jozsa and G. R. Middleton, *A Discussion of Wood Quality Attributes and Their Practical Implications*, Foringtek Canada Corp. No. SP-34, 1994.
6. J. Kittler and J. Illingworth, *Minimum Error Thresholding*, Pattern Recognition, 19, 41-47, 1986.
7. L. O. Lindgren, Medical CAT-scanning, *X-ray absorption coefficients, CT-numbers and their relation to wood density*, Wood Sci. Technol. 25, 341-349, 1991.
8. H. Nyquist, *Certain Topics in Telegraph Transmission Theory*, AIEE Trans., 617-644, 1928.
9. X. Wang, F. Albregtsen and B. Foyn, *Texture Analysis Using Grey Level Gap Length Matrix*, in G. Borgefors (Ed.): "Theory and applications of image analysis II, Selected Papers from 9th SCIA", World Scientific Publishing, Singapore, 65-78, 1995.
10. X. Wang, O. Hagman and S. Grundberg, *Sorting pulpwood by X-ray Scanning*, Proc. International Mechanical Pulping Conference, Sweden, 395-399, 1997.
11. X. Wang and K. R. Braaten, *Growth rings and spruce pulpwood sorting*, Nordic Pulp & Paper Research J., Vol. 3, 1997.
12. J. White and G. Rohrer, *Image Thresholding for Optical Character Recognition and Other Applications Requiring Character Image Extraction*, IBM J. of Research and Development, 27, 400-411, 1983.

Precise and Fast Form Identification Method by Using Adaptive Base Lines for Matching

Hiroaki Takebe Yutaka Katsuyama and Satoshi Naoi

Fujitsu Laboratories Ltd.
4-1-1 Kamikodanaka, Nakahara-Ku, Kawasaki 211, Japan

Abstract. Conventional form identification methods have been based on the normalization of an input image. So, if the base for normalization is different from that of the true model, it is difficult to identify its form. In this paper, we propose a form identification method, which prevents the difference from spreading throughout the process. In the method, the local ruled line structures are analyzed exhaustively by varying a pair of base lines of an input image and a model.The process is realized efficiently by generating the correspondence possibilities between ruled lines, and grouping these possibilities. We registered 100 models with a dictionary, and experimented on form identification under the various conditions. The result shows that the method has high accuracy and practical processing speed.

1 Introduction

Many form documents, such as slips and fixed business forms, are in circulation now. In order to file these documents electronically, it is necessary to identify their forms automatically. Some identification methods had been reported [1][2][3]. However, the problems with these method are that they depend too much upon the result of feature extraction. Afterwards, two methods which are robust against image fluctuations have been proposed. In Asano's method [4], a form document is represented as a set of center points of cells. Identification is considered to be a point-correspondence problem, and they use the geometric-hashing method in matching. In Ishitani's method [5][6], a form document is represented as a set of ruled lines. The correspondence between ruled line structures is acquired by extracting the largest maximal clique from an association graph. Here, a node of the association graph represents correspondence of ruled lines, and compatiblle nodes are connected with each other by an arc.

Both methods are based on the normalization of an input image so as to fit a model in a dictionary. The base for normalization, such as the origin of a table in an image, is variable to be extracted according to its fluctuation caused by noise and faintness, as shown by Fig.1. But they don't deal well with the situation when failures occur in fitting an input image to the true model because there is a difference between the base features of an input image and those of a model. Asano deals with the situation by entering multiple patterns for a model whose bases for normalization are varied. However, the problem with this method is its disability to cover all combinations of bases and reducing the identification ability on account of the increase in similar models. In Ishitani's method, it seems that the situation can be dealt with by easing the generation condition of nodes. But the method has two problems. The first, the time to generate an association graph and extract the largest maximal clique from it increases greatly, because the scale of an association graph increases. The second problem is that the generation of many nodes reduce the ability of identify, because the condition to connect nodes by an arc depends only on the qualitative relationship between nodes. In this paper, we propose a form identification method where the local ruled line structures are exhaustively and efficiently analyzed by varying a pair of base lines of an input image and a model. This results in a form identification method, which is robust in the sense that it is not seriously affected by a difference between the base for normalization of an input image and that of a model, and which is highly accurate and fast. In section 2 and section 3, we give the details of the algorithm used. Following that, we present experimental results that demonstrate the effectiveness of the method in section 4.

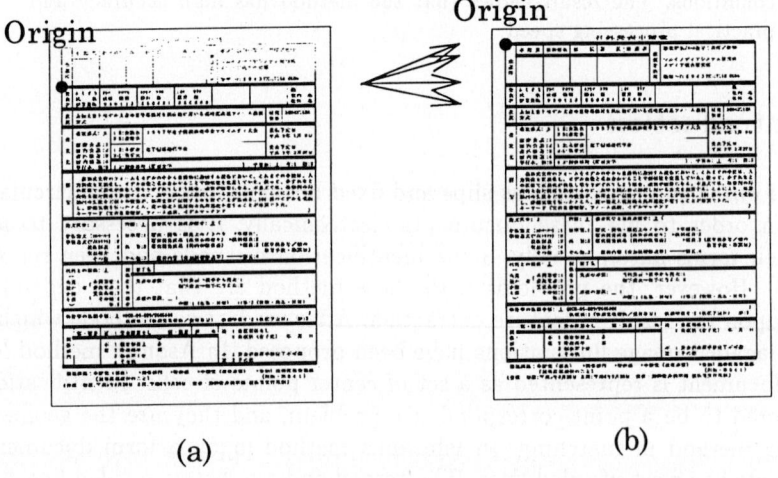

Fig. 1. Example results from extraction of ruled lines, when (a) the top part is faint, and (b) the extraction was succesful, and variety of pairs of base lines.

2 Calculation of Feature Quantities

First, we extract ruled lines from an input image. Then we calculate feature quantities about the ruled line structure.

2.1 Global Features

We calculate the global feature of a ruled line, which is represented by a set of three integers. A global feature shows where the ruled line takes its position in the table and how long its length is for the size of the table. The base width, W, base height, H, and origin (x_0, y_0) of a table in an image are calculated in advance. Let the length of a ruled line be expressed as $r1$, and its center coordinates be as (x_1, y_1). The global feature (DL, DX, DY) of a ruled line is calculated as follows. $DL = \frac{r_1 \times 100}{W}$, $DX = \frac{(x_1 - x_0) \times 100}{W}$, $DY = \frac{(y_1 - y_0) \times 100}{H}$.

2.2 Local Features

We calculate local feature of a pair of ruled lines, which is represented by a set of real numbers. A local feature shows their relative relationship about their lengths and positions. Let the length of one ruled line be expressed as r_1 and its center coordinates as (x_1, y_1), and let the length of the other's be r_2 and its center coordinates be (x_2, y_2). We define the local feature (dl, dx, dy) of a pair of ruled lines as follows. $dl = \frac{r_2}{r_1}$, $dx = \frac{x_2 - x_1}{r_1}$, $dy = \frac{x_2 - x_1}{r_1}$.

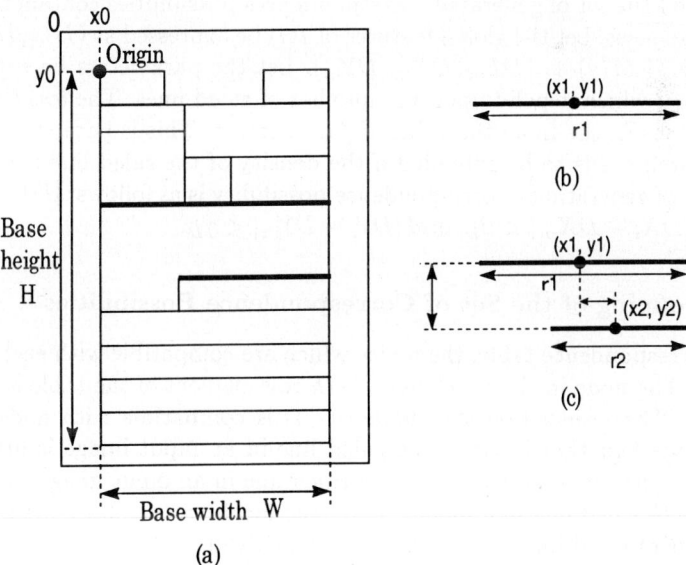

Fig. 2. (a) Base features (b) global features of ruled line and (c) local features of pair of ruled lines.

3 Model Matching

The ruled line structure of an input image is matched to that of a model in a dictionary by using their global features and local features twice in the horizontal case and in the vertical case. As a result, the similarity score of a form of the input image for a model is calculated.

3.1 Generation of Correspondence Possibilities

We express the number of ruled lines in an input image as p, and the number of the ruled lines in a model as m. First, the table of size $p \times m$ is made. The $i-th$ line $j-th$ row element of the table represents the data linking the $i-th$ ruled line in an input image and the $j-th$ ruled line in a model. We call this the correspondence table. Next, the possibility for correspondence between the $i-th$ ruled line $I(i)$ in an input image and the $j-th$ ruled line $M(j)$ in a model is judged by their global features. If it is judged that it is possible for $I(i)$ to correspond to $M(j)$, a node is placed in the $i-th$ line $j-th$ row element. Thus, the set of correspondence possibilities is represented by the correspondence table as shown by Fig.3.

The condition of generating a correspondence possibility is easy enough to permit multiple ruled lines to correspond to a ruled line. So, even if the base features of an input image are different from those of the true model, it may well be that the set of generated correspondences possibilities contain the correct correspondences. Let the global features of $I(i)$ be expressed as (DL_i, DX_i, DY_i), and those of $M(j)$ as (DL_m, DX_m, DY_m). Let the parameters be expressed as $\alpha_D, \beta_D, \gamma_D$, which depend upon the number of ruled lines. The less the number of ruled lines is, the more the value of parameter is. This is because the domain of the search needs to be extended if the density of the ruled line is rough. The condition of generating a correspondence possibility is as follows. $|DL_i - DL_m| < \alpha_D$, and $|DX_i - DX_m| < \beta_D$, and $|DY_i - DY_m| < \gamma_D$.

3.2 Grouping of the Set of Correspondence Possibilities

In the correspondence table, the nodes which are compatible with each other are grouped. The node in the $i-th$ line $j-th$ row element of the table is expressed as $n(i,j)$. The proposition that node $n(i,j)$ is compatible with node $n(k,l)$ is equal to the fact that if the $i-th$ ruled line in an input image is fitted to the $j-th$ ruled line in a model, the $k-th$ ruled line in an input image can be fitted to the $l-th$ ruled line in a model. Here, the local feature of a node means that of the pair of ruled lines which the node is made of.

First, a node is regarded as the base node, and the pair of ruled lines which the base node is made of are regarded as the base lines. Nodes compatible with the base node are searched for. Here, the domain of the search is represented as a part of the correspondence table. In the concrete, using a constant θ, the domain of the search for the base node $n(i,j)$ is as follows. $\{(x,y)|\ i < x < i + \theta,\ and\ j < y\}$. The target nodes of the search are the nodes which are not yet grouped in the domain. They don't contain the nodes which are theoretically incompatible with the base node. This restriction of the target node for the search makes the process efficient. We express the local feature of the base node as (dl_0, dx_0, dy_0), and the local feature of the target node as (dl_1, dx_1, dy_1). We express the fixed thresholds as $\alpha_d, \beta_d, \gamma_d$. The conditions of grouping are as follows. $|dl_1 - dl_0| < \alpha_d$, and $|dx_1 - dx_0| < \beta_d$, and $|dy_1 - dy_0| < \gamma_d$.

Next, if the node which is compatible with the base node is found, the node is made to join the group which the base node belongs to, and is regarded as the new base node instead. Then a new search starts. If there is no such node in the domain, the search ends. A node which is not grouped yet is regarded as the new base node instead, and a new search starts. In Fig.3, the base node, the domain of the search for it, and the target nodes of the search for it are shown. The nodes connected with ea ch other by lines belong to the same group.

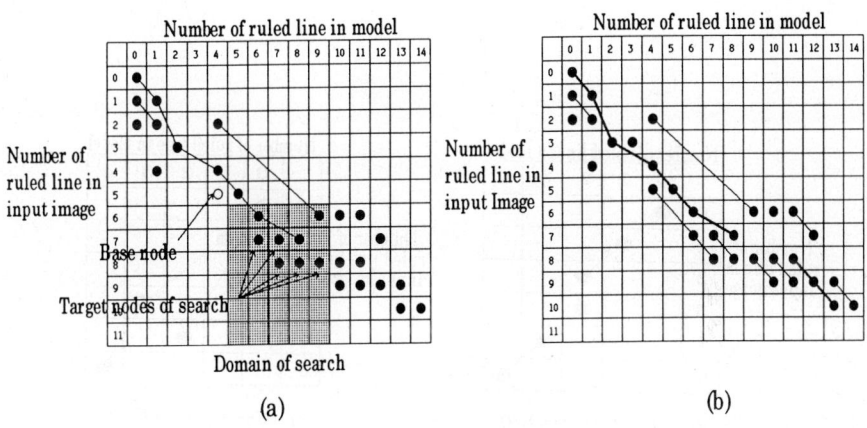

Fig. 3. (a) Grouping of nodes (b) Result of the grouping, and the optimum combination of groups

3.3 Search of Optimum Combination of Groups

Moreover we investigate combinations of groups which are compatible with each other. The proposition that two groups are compatible with each other is equal to the fact that the sets of ruled lines, which the nodes in the groups consist of, do not overlap with each other. There are two kinds of cases that the sets do overlap with each other. The same ruled line overlaps or the order of the ruled lines is inverted, as shown by Fig.4. In the set of the combinations, the combination which has the most number of nodes is regarded as the optimum combination of groups. In Fig.3, the optimum combination of groups is two groups where the nodes are connected with the heavy line.

3.4 Calculation of Detail Similarity Score

We express the number of the horizontal ruled lines in an input image be as ih, that in a model as mh, and the number of nodes of the optimum combination of groups as $maxh$. The detail similarity score of the structure of the horizontal ruled lines of an input image for that of a model, which is expressed as SH, is defined as follows. $SH = \frac{maxh}{ih} + \frac{maxh}{mh}$.

In the same way as the horizontal case, the similarity of the structure of the vertical ruled lines for that of a model, expressed as SV, is defined by using iv, mv, and $maxv$ as follows. $SV = \frac{maxv}{iv} + \frac{maxv}{mv}$.

Finally, the detail similarity score of the structure of ruled lines of an input image for that of a model, which is expressed as S, is defined as follows. $S = SH + SV$.

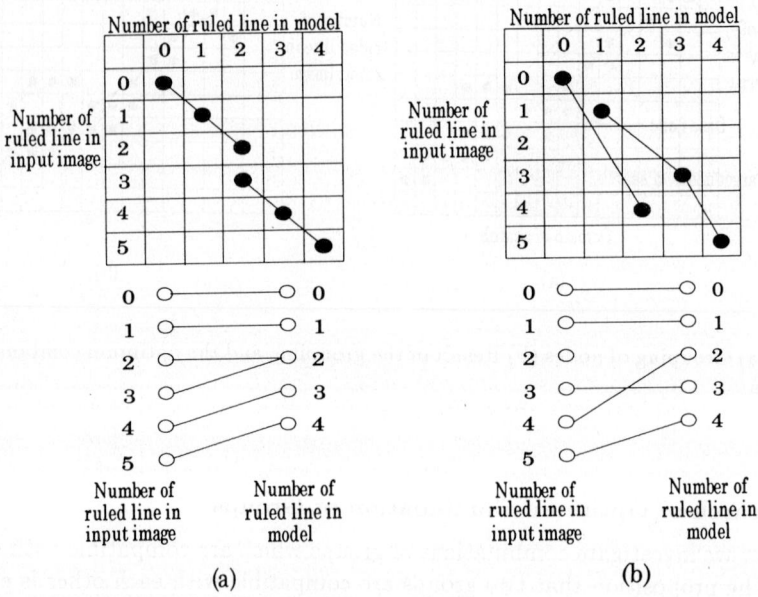

Fig. 4. Examples of groups which are incompatible with eath other. (a) the same reuled line overlaps, and (b) the order of the ruled lines is inverted.

4 Experimental Result

4.1 Experimntal Condition

The subjects of the experiment are the 100 kinds of A4-size business documents.They are scanned by a 400 dpi flatbed image scanner, and converted into image data. They are binarized by the Ohtsu's method [7]. The ruled lines are extracted, and the features are entered into a dictionary with the identification number of the model automatically. The fluctuated images are prepared as follows.

1. The threshold of binarization is changed. We tried the case in which 50 was added to the threshold and that in which 20 was subtracted from the threshold.The values of 50 and 20 are the limits for a document before it becomes extremely faint and deep, according to the result of our experiment.

2. The documents are scanned under three resolutions. The binalization threshold is fixed. The image scanner used is different from that used above, and scanning method is ADF.

3. The documents are inputted from a fax with a resolution of 200 × 100dpi. They are printed out, and scanned by the same image scanner as (b). The resolution is 400dpi, and binalization threshold is fixed. Again, scanning is by ADF.

We have tested 100 sheets of eath type. We use a 166 MHz Pentium processor, and Windows NT 3.51.

4.2 Results

The results of the correction rate are shown in Table 1, and the mean processing time is about 64 ms per sheet. The identification accuracy is quite high, and the processing speed is practical. The errors when inputting from a fax are caused by the fluctuation in the ratio between the horizontal and vertical sizes of an image.The correct correspondences are generated in the correspondence table, but are not grouped by the fluctuation.

Experimental Documents	1st Correction Rate
Alteration of Threshold	100%
400dpi	100%
300dpi	100%
200dpi	100%
FAX(200 × 100 dpi)	98%

Table 1. Result of 1st correction rates

5 Conclusion

In this paper, we have described the form identification method using adaptive base lines for matching. In the method, the local relationship between the ruled line structures are exhaustively and efficiently analyzed by varying base lines for matching. This results in a form identification method which is robust, and which has high accuracy and high speed. In fact, the form identification experiment using a dictionary with 100 models shows that the method has high accuracy, even for images whose scanning conditions are varied, and that the processing speed is 64 ms per sheet. In the near future, we extend targets of the method to a document which is characterized by underline such as a questionnaire and a document which is characterized by arrangement of character strings.

References

1. Q. Luo, T. Watanabe, and N. Sugie: Automatic Acquistion of Layout Knowledge for the Structure Recognition of Table-Form Documents. *Trans. IEICE Japan*, vol. J76-D-2, No. 3, pp. 534–546, 1983
2. Q. Luo, T. Watanabe, and N. Sugie: Structure Recognition of Various Kinds of Table-Form Documents. *Trans. IEICE Japan*, vol. J76-D-2, No. 10, pp. 2165–2176, 1993
3. T. Watanabe, Q. Luo, and N. Sugie: Structure Recognition Methods for Various Types of Documents. *Machine Vision and Applications*, vol. 6, No. 3, pp. 163–176, 1993
4. M. Asano, S. Shimotsuji: Form Document Identification Using Cell Structures. *Trans. IEICE Japan*, vol. J80-D-2, No. 1, pp.131–138, 1997
5. Y. Ishitani: A Model Matching Method for Form Image Understanding. *Technical Report of IEICE*, PRU94-34, pp.57–64, 1994
6. Y. Ishitani: Model Matching Based on Association Graph for Form Image Understanding. *Proc. of the 2nd ICDAR*, pp.287–292, 1995
7. N. Ohtsu: An Automatic Threshold Selection Method Based on Discriminant and Least Squares Criteria. *Trans. IEICE Japan*, vol. J63-D, No. 4, pp. 349–356, 1980

Combinatorial Coarse Classification Method for OLCCR

Jing Zheng, Xiaoqing Ding, Youshou Wu and Fanxia Guo
Image Processing Group, Department of Electronic Engineering, Tsinghua University
Beijing, 100084, P.R.China
Email: zj@ocrserv.ee.tsinghua.edu.cn

Abstract

This paper presents a new coarse classification method based on combination of two independent approaches for on-line Chinese character recognition (OLCCR). Through introducing a feedback control provided by the fine classifier, we are able to reach a good trade-off between precision and speed. In the paper, we make some theoretical analysis on the form of combination, and compare it with competing methods. Both theoretical analysis and experimental results show that our method has the characteristics of high correct rate, good recognition speed, being easy to control and adaptable to various needs for practical application through parameter adjustment.

1 Introduction

In the technique of on-line Chinese character recognition (OLCCR), in order to obtain good trade-off between time consumption and recognition rate, two-stage classification methods [1][2][3] are often used. That is, a fast but less accurate algorithm is employed first, to select a subset that is assumed to contain the characters being similar with the input pattern. The subset being much smaller than original alphabet is called candidate set. After that, a precise but slow algorithm is utilized to select the final result from the candidate set. By this way, the whole recognition time can be greatly reduced only with tiny recognition rate loss compared with the case that only the precise algorithm is used. The former step, which produces the candidate set, is called coarse classification.

There are basically two categories of methods for coarse classification: one utilizes global features, such as stroke number, segment number and some statistical features of character[4]. Another takes advantage of local features, such as shape and sequence of several first and last written strokes of a character[5].

If only regularly script needs handling, each of the aforementioned methods can be effective enough. However, now a good recognition algorithm must be able to handle relatively free script where variations such as stroke connection, stroke order variation, stroke shape distortion occur frequently. These variations have great influence on the efficiency of coarse classification. For example, stroke connection changes the number of strokes and segments; stroke order variation makes local feature hard to get. Practical experience tells us simple methods based on single aspect of features is not competent to perform good recognition on low limited script.

According to this, we present a combinatorial coarse classification method that utilizes both global and local features and makes use of the feedback of fine classification. In the following sections we will see that the discussed method is indeed better than common ones.

2 Framework of combinatorial coarse classification method

In our recognition system, the recognition alphabet is 6763 characters in large. And according to the computation complexity of our fine classifier, we hope the candidate set of coarse classification to be smaller than 50 characters, and the algorithm of coarse classification must be able to handle both stroke connection and stroke order variation. This is a very difficult requirement for any simple method being commonly utilized. So we present a scheme that combines two algorithms in a serial form where the second takes the output of the first as its input, and we hope it can give much smaller candidate set than any single algorithms. As we know, the second algorithm must be independent of the first one, otherwise, the combination will be meaningless. Obviously, the method based on global statistical features is totally independent of that based on local structural features, and we select these two kinds of algorithms to construct our combinatorial coarse classification.

We select the first radical and the last radical of a character as the local feature to perform coarse classification. A radical refers to a group of strokes that construct a sub-structure shared by many Chinese characters. The first and last radical refer to the first and last written radical of a character respectively. The advantage of this feature is that they are easy to extract and less influenced by stroke order comparing with other local features.

The global feature we select is so called directional segment feature that will be discussed later. Because this kind of feature has good describing ability, and is able to utilize the characteristics of on-line recognition's specialty, it has very good discriminating power when recognizing regularly script.

Two classifiers have been designed based on aforementioned two kinds of features receptively. According to the characteristic of each classifier, we select the former one (radical based classifier) to be the first step classifier.

Because the aforementioned two classifiers are connected in a series-wound form, the problem of error accumulating exists. So long as one classifier fails to select the correct character in its output, the whole method fails. According to this case, we present a new method that utilizes a supplementary classifier controlled by a feedback that fine classifier provides to improve correct rate. The principle is as follows:

The fine classifier generates a matching score for each recognition result, and the value of the score ranges from 0 to 100, reflect the degree of matching: the higher the value, the better the matching. If the score is low, it means that a recognition error probably occurs. The recognition errors are often caused by two reasons: one lies in fine classification, the other lies in coarse classification. We have found out that when the matching score is low, the second category of errors occur very frequently. Hence, we set a threshold that when the matching score is lower than it, a supplementary coarse classifier is called.

We observe that the errors of coarse classification are mainly caused by the first-step classifier, because the second-step classifier is very accurate. So the supplementary classifier only needs to be independent of the first-step classifier. We directly use the algorithm of the second-step classifier to perform supplementary coarse classification. However, this time, the input of supplementary classifier is the whole recognition alphabet set that is 6763 Chinese characters in large, and the output is enlarged to 200 characters. These 200 characters are sent to the fine classifier to perform a second recognition process which determines the final result.

Experiments show that, the supplementary coarse classification has conspicuous effect in improving correct rate, and is not very time-consuming. We will show this point in the following sections.

The framework of the whole coarse classification method is illustrated as fig. 1.

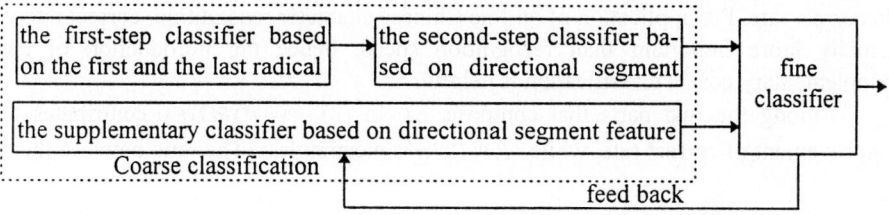

Fig.1. combinatorial coarse classification framework

3 Influence of the supplementary classification

What role does the supplementary classification play in the whole method? In this section, we will make a theoretic analysis on above question.

Suppose that correct rate of the first-step classifier is P_1, and the threshold of matching score is T. When the first-step classification is wrong, the probability of matching score being less than T denotes $P(T|E)$; when the first-step classification is correct, the corresponding probability denotes $P(T|C)$. Suppose that the correct rate of the supplementary classifier is P_2'. It is easy to know that the contribution of the supplementary classifier to correct rate of coarse classification is:

$$\Delta P = (1-P_1)P(T|E)P_2' \quad (1)$$

Suppose that the time consumption of supplementary classification and corespondent fine classification is τ. Then the average processing time caused by supplementary classification is:

$$\Delta T = \left[(1-P_1)P(T|E) + P_1 P(T|C)\right] \cdot \tau \quad (2)$$

In our coarse classification method, P_1 is about 0.97 and P_2' is about 0.90 for average qualified writing. When T evaluates 50, $P(T|E)$ equals 0.4 while $P(T|C)$ equals 0.03 approximately. Thus, according to equation 1, the accretion of correct rate is about 1.1 percent. Considering that P_1 is already as high as 97 percent, the improvement is rather appreciable.

If a user's writing habit is not very normal, correct rate of the first-step classifier based on local structural feature is probably affected to some extent, for example, say it drops to 0.90. Because the supplementary classifier is stroke order free, its performance will not change. In this case, ΔP is about 3.6 percent, and that is very conspicuous.

As for extra procession time brought about by the supplementary classification, when P_1 equals 0.97, ΔT is about 0.04τ; When P_1 equals 0.90, ΔT is about 0.07τ. In our method, τ is about 3 times as much as the average time of recognizing a character. That is to say, 12%~21% extra processing time brings about 1.1~3.6 point accretion in correct rate. For application of on-line Chinese character recognition, correct rate is usually more important than recognition speed, hence the introduction of the supplementary coarse classification is helpful.

Among the two parts that compose ΔT in (2), $(1-P_1)P(T|E)\tau$ contributes to improvement of correct rate, while $P_1 P(T|C)\tau$ is meaningless. The ratio between them is:

$$\frac{(1-P_1)}{P_1} \bigg/ \frac{P(T|E)}{P(T|C)} \tag{3}$$

We can see that, from the aspect of time consuming, when P_1 is comparatively low while $P(T|E)/P(T|C)$ is relatively high, the supplementary classification is of good efficiency. From the aspect of correct rate, the higher the $P(T|E)$, the larger the ΔP. Because P_1 and P_2' is determined by the quality of input writing, the selection of T is crucial for efficiency of coarse classification. According to definition, $P(T|E)$ and $P(T|C)$ changes monotonously with T, and satisfy:

$$\begin{cases} P(T|E) = P(T|C) = 0 & T = 0 \\ P(T|E) = P(T|C) = 1 & T = 100 \end{cases} \tag{4}$$

We have observed that the curve of $P(T|E)$ and $P(T|C)$ according to some practical samples, its shape is illustrated by fig 2 :

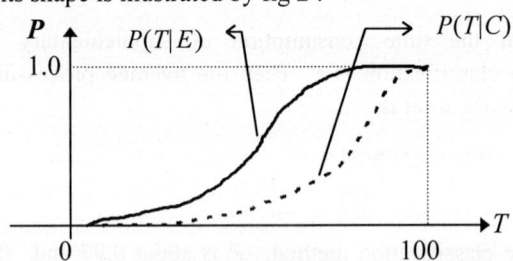

Fig.2. curve of $P(T|E)$ and $P(T|C)$

As we can see, in a large scope of T, $P(T|E)$ is far larger than $P(T|C)$, this is mainly because our fine classifier is able to give reasonable matching score. Researcher can select an optimal T that makes both the correct rate and recognition speed reaches a good trade-off.

4 Introduction of methods of the two of classifiers

4.1 Classifier based on the first and the last radical

To extract head and tail radicals from handwriting, we employ dynamic programming (DP) technique in two directions: from head to tail and from tail to head[6]. Although several kinds of writing styles have been summerized, handing different stroke order is still the main weak point of this method

Altogether we define 75 types of first radicals and 88 types of last radicals, and build the recognition dictionary for them. All 6763 Chinese characters are organized as a classification tree, where those characters having the same first and last radicals share the same parent node.

For each input character, 16 possible first and last radicals are extracted respectively. Then characters with these first and last radicals are sorted according to matching distance, and the first $N_1 = 256$ characters are taken as the candidate set of first-step coarse classification.

4.2 Classifier based on directional segment feature

In on-line character recognition technique, trajectory of pen stylus is easy to be described as series of line-segments. It has been reported that the direction of these line segments has powerful discrimination ability[2]. We can use statistical method describing a character with a feature vector based on directions and its locations of these small segments. We call this kind of feature the directional segment feature.

In our method, taking speed and stability of diverse direction in to consideration, we only deal with two kinds of segment directions: horizontal and vertical. By adopting projection method, a 16-dimensional feature fector is obtained to describe an individual character..

We use 75 sets of samples to train our classifier, each set has 6763 characters. An improved competitive learning strategy is employed to generate a recognition dictionary with 28245 vectors that is 479KB in large.

The purpose of the second-step classifier is select $N_2 = 50$ characters as the candidate set from the output of the first-step classifier and an algorithm based on NN (Nearest Neighbor) method is adopted. Experiment shows that when the first-first step classifier succeeds, the second-step classifier is able to reach a correct rate of 99.8% for average qualified writing.

The supplementary classifier is designed to select 200 characters from the whole recognition alphabet set that is 6763 characters in large. To accelerate its speed, we employ SOM (Self Organized Mapping) to cluster the 28245 feature vectors into 256 categories, thus a two-layer classification tree is built. For odinary quality script, it is able to reach a recognition rate of 92%.

5 Compare with other forms of combination

Because coarse classification is much less time-wasting than final classification, for the coarse classification method, the main factor that influences performance of the whole system lies in the correct rate and the size of candidate set.. As shown in fig.

3, we list three conceivable schemes: scheme 1 is our method, while scheme 2 and scheme 3 take the intersection and union of the two classifiers' output as their candidate set.

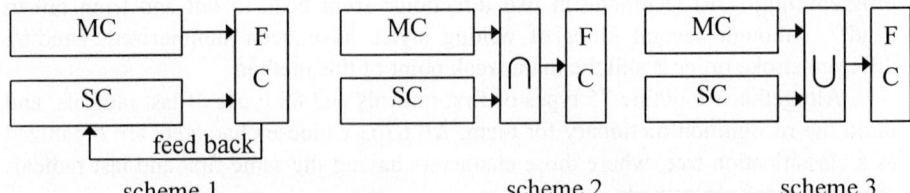

scheme 1　　　　　　　　　scheme 2　　　　　　　　scheme 3
Fig.3.　different forms of combination

In section 3, we have discussed scheme 1 in detail, according eq. 1 and 2, the correct rate of it is:

$$P^{(1)}(C) = P_1 P_2 + (1 - P_1) P(T|E) P_2' \qquad (5)$$

Where P_2 is the correct rate of the second-step classifier, and $P_1 P$ is correct rate of the main classifier.

The average size of candidate set for scheme 1 is:

$$N^{(1)} = 50 + \left[(1 - P_1) P(T|E) + P_1 P(T|C)\right] \times 200 \qquad (6)$$

As for scheme 2, obviously it has the smallest candidate set, however its correct rate is

$$P^{(2)}(C) = P_1 P_2 P_2' \qquad (7)$$

As mentioned before, P_2' is only 92% approximately, so $P^{(2)}(C)$ is by no means larger than 92% which is insufficient for coarse classification

As for scheme 3, it has the highest correct rate evidently. Because the two algorithms are independent of each other, and $P_2 \approx 1$, the gross correct rate for scheme 3 is:

$$P^{(3)}(C) \approx P_1 P_2 + (1 - P_1 P_2) P_2' \qquad (8)$$

And the size of its candidate set satisfy:

$$200 \leq N^{(3)} \leq 250 \qquad (9)$$

When T is set an appropriate value, that is, $P(T|C)$ being small while $P(T|E)$ being relatively large, because P_1 is high, $\left[(1 - P_1) P(T|E) + P_1 P(T|C)\right]$ can be of very little value, say, less than 0.08. At this time, $N^{(1)} < 66$, is less than one third of $N^{(3)}$. For $P_2 \approx 1$, we can see that:

$$P^{(3)}(C) - P^{(1)}(C) \approx (1 - P_1)(1 - P(T|E)) P_2' \qquad (10)$$

As shown in fig.2, in a large scope of T, $P(T|E)$ is far larger than $P(T|C)$. Accordingly, we can select a T, which makes a small $P(T|C)$ and a large $P(T|E)$, then scheme 1 has a good correct rate that approximates that of scheme 3, while time cost can be far less than that of scheme 3. In fact, according to eq.4, when $T=100$, $P(T|E) = P(T|C) = 1.0$. At this time, $P^{(3)}(C) - P^{(1)}(C) = 0$. It can be said that, scheme 3 is an extreme case of scheme 1 when $T=100$.

From above analysis we can see that, because of introducing feedback from the fine classifier, scheme 1 is pretty flexible that it can reach good correct rate as well as fine recognition speed. Through careful selecting T, designer can obtain a good trade-off between precision and velocity. In sum, scheme 1 has indisputable advantage over the other two.

6 Experiments and results

To verify the effectiveness of our method, we select 3 sets of samples that each is 6763 characters in size to test our method. All these samples have not been trained. According to writing quality, these samples are divided into 3 categories: excellent, good and middle.. We test the correct rate of coarse classification, total recognition rate of OLCCR system and the recognition speed for several kinds of combination form (referred in previous section) and different value of T. The testing machine is pentinum-120 running Win95 operation system. The results are listed in talbe1~3.\

Table 1 testing index (the sample's quality : excellent , number : 6763)

method	correct rate of coarse classification (%)	recognition rate (%)	recognition speed (characters / s)
scheme 1 , T=0	98.31	97.84	4.2
scheme 1 , T=50	99.22	98.61	3.7
scheme 1 , T=75	99.51	98.85	2.8
scheme 2	92.15	91.90	6.7
scheme 3	99..68	98.91	1.1

Table 2 testing index (the sample's quality : good , number : 6763)

method	correct rate of coarse classification (%)	recognition rate (%)	recognition speed (characters / s)
scheme 1 , T=0	95.38	94.01	4.0
scheme 1 , T=50	97.40	95.85	3.3
scheme 1 , T=75	98.29	96.66	2.4
scheme 2	84.22	83.32	6.4
scheme 3	98.40	96.72	1.0

Table 3 testing index (the sample's quality : middle , number : 6763)

method	correct rate of coarse classification (%)	recognition rate (%)	recognition speed (characters / s)
scheme 1 , T=0	90.01	87.75	3.9
scheme 1 , T=50	92.67	90.30	2.8
scheme 1 , T=75	94.11	91.51	1.5
scheme 2	75.44	75.41	5.9
scheme 3	94.18	91.53	0.9

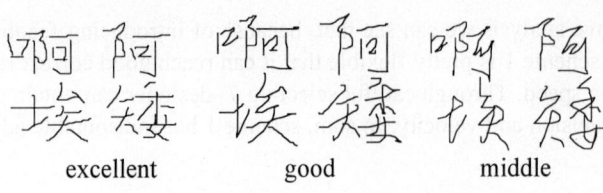

      excellent   good   middle

Fig.4   examples of different quality

Fig.4 shows some examples of excellent, good and middle quality samples. From the testing result, we can see that, scheme 2 is the fastest method among the all, while it has the worst recognition rate. Especially when the quality drops, the recognition rate is influenced badly, because P_2' drops down rapidly. In comparison, scheme 1 is far more quick than scheme 3, and far more accurate than scheme 2. When $T=75$, its recognition rate approximates that of scheme 3, while its speed is twice as fast. Perhaps people notice that while sample's quality drops, the recognition speed for all methods of scheme 1 degrades. This is mainly because the matching score given by fine classifier decrease, thus cause the accretion of $P(T|E)$ and $P(T|C)$, and more chances for the supplementary classification to be executed.

7 Conclusion

According to experimental results, we can see that, our method that combines two totally independent algorithms is very effective that it is able to find a 50-characters candidate set from 6763 categories of Chinese character with very high correct rate. More important, through introducing a feedback from the fine classifier, our method has conspicuous advantage in performance and flexibility over those simple combination forms who just calculate the intersection or union of the results of different classifiers.

Reference

[1] Y.S Wu, X.Q.Ding, Chinese character recognition---principle, method and implement, Advanced education publishing house, 1992 (In Chinese)
[2] C.C.Tappert, "The state of the art in on-line handwriting recognition", IEEE Trans. Patt. Anal. Machine. Intell., Vol.12, No.8, pp.787-808, 1990
[3] F.Nouboud and R.Plamondon, "On-line recognition of handprinted characters:suvery and beta tests", Pattern Recognition, Vol.23, No.9, pp.1031-1044,1990
[4] Y.Xia,S.P.Ma,Z.H.Yang,J.P.Gong and Y.Yang, "Method and system for on-line recognizing constraint free Chinese character", The fifth Conference on Oriental Language, pp.85-89 (in Chinese)
[5] T-Z.Lin and K-C.Fan, "Coarse classification of on-line chinese characeers via structure feature-based method", Pattern Recognition, Vol.27, No.10, pp.1365-1377, 1994
[6] Y-T.Tsay and W-H.Tsai, "Attribute string matching by split-and-merge for on-line chinese character recongnition", IEEE Trans. Patt. Anal. Mach. Intell., Vol. 15, No.2, pp.180-185, 1993

Detecting Characters in Grey-Scale Scene Images

Yongmei Liu,[1] Tsuyoshi Yamamura,[1] Noboru Ohnishi[2] and Noboru Sugie[3]
{liu,yamamura,ohnishi}@ohnishi.nuie.nagoya-u.ac.jp

[1] Department of Information Engineering, Nagoya University,
Furo-cho, Chikusa-ku, Nagoya 464-01, Japan
[2] Bio-Mimetic Control Research Center, RIKEN
[3] Faculty of Science and Technology, Meijo University

Abstract. This paper proposes a method for detecting characters in grey-scale scene images for navigating vision-based mobile robot by character information. First, we extract subregions with high spatial frequency and great variance in grey-level from an input scene image as candidates of character components. Then, we select characters by using several heuristics such as constraints of size and shape, bimodality of an intensity histogram, alignment and proximity of characters. We conducted an experiment using 20 indoor and 40 outdoor images. As a result, character lines are detected with a high rate of 80%.

1 Introduction

Character information plays an important role in our everyday life. There is much character information such as signboards and traffic (road) signs around us. We humans can separate these characters from other objects, get necessary information from them immediately, and use the information to reach our destination. Lately, with the development of optical character recognition (OCR) technology, characters ranging from printed alphabetical letters to handwritten Chinese characters can be recognized at a fairly high rate. Currently, however, most of these OCR systems are focused on extracting and recognizing characters in document images only[1]. If characters in scene images could also be recognized, it will be very valuable for mobile robot or autonomous vehicle navigation. A robot that can recognize a traffic (road) sign as a landmark in its map of the environment can then use this information to locatize itself in its environment.

Research on extracting characters in scene images includes: minimization of a cost function, which reflects some features specific to Japanese printed characters[2]; use of color segmentation and spatial variance to locate text in CD and book cover images[3]; and extraction of character pattern candidates using position and grey-level information[4].

Here, we propose a method for detecting character lines in scenes from their grey-scale images. The proposed method is based on the general features of both characters and character lines. First, we extract subregions with high spatial frequency and great variance in grey level as candidates of character components.

Next, from the candidates, characters are selected according to the size and shape of the rectangle surrounding a candidate region and the bimodality of an intensity histogram in the region. Finally, character lines are determined using arrangement features such as proximity of characters and alignment of characters.

We explain the detail of detecting characters and extracting character lines in Section 2. We show experimental results along with discussions in Section 3 and conclude in Section 4.

2 Detection of Characters in Scene Images

2.1 Characters in Scene Images

The method presented here has been primarily designed to detect Japanese characters, written on signboards and traffic (road) signs, in grey-scale scene images. Compared with other objects in scene images, Japanese characters (Japanese Kana letters and Chinese characters) written on signboards and traffic (road) signs have the following features.

F1. The spatial frequency of character components is high.
F2. The grey-level contrast between character components and their backgrounds is high.
F3. The aspect ratio of Japanese characters is almost constant.
F4. Characters in the same character line are adjacent to each other.
F5. Characters in the same character line are of similar size.
F6. Characters are horizontally or vertically aligned.

2.2 Outline of Algorithm

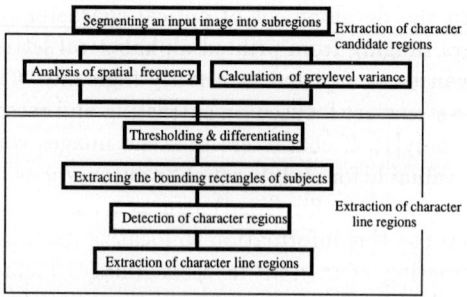

Fig. 1. Processing flow

Here, we use the features of characters stated in Section 2.1 to extract character components. Figure 1 shows the algorithm for detecting characters in grey-scale scene images.

Input images acquired by a CCD camera are digitized into 640 (width)×480 (height) pixels of 256 grey levels. To extract regions with locally high spatial frequency, we divide an input image into subregions of P×Q pixels during pre-processing. According to the result of preliminary experiments using different subregion sizes such as P=Q=4, P=Q=8, P=Q=16, P=Q=32 and P=Q=64, we adopted P=Q=16 as the optimum size of a subregion. The extraction of character candidates and threshold processing are done in each subregion.

2.3 Extraction of Character Candidate Regions

Generally, compared with noncharacter components in scene images, character components have high spatial frequency and high grey-level contrast. We compute the two-dimensional Fourier transform and the mean and standard deviation of grey level for each subregion.

Analysis of Spatial Frequency Most textures can be characterized based on the local spatial frequency. Here, we use the two-dimensional Fourier transform to evaluate localization properties, and then calculate the distribution of spatial frequency in subregions and extract subregions of high spatial frequency[5].

A 2-D Fourier transform is represented as:

$$\mathbf{F}(k,l) = \sum_{i=0}^{P-1}\sum_{j=0}^{Q-1} f(i,j) \cdot \exp\{-J2\pi(\frac{k(i-1)}{P} + \frac{l(j-1)}{Q})\} \quad (1)$$

$$k = 0,1,...,P-1; l = 0,1,...,Q-1; J = \sqrt{-1}$$

where (i,j) is the local coordinate in a subregion, and f(i,j) is the grey-level value of point(i,j).

Using $\mathbf{F}(k,l)$, we can compute the average value of frequency weighted by its amplitude. This can be represented as:

$$\omega_{sub} = 2\pi \sum_{k=0}^{P/2}\sum_{l=0}^{Q/2} \| \mathbf{F}(k,l) \| \sqrt{k^2 + l^2} \quad (2)$$

where $\| \mathbf{F}(k,l) \|$ is the Fourier amplitude spectrum:

$$\| \mathbf{F}(k,l) \| = \sqrt{\{ReF(k,l)\}^2 + \{ImF(k,l)\}^2} \quad (3)$$

As shown in Eq. (2), the value of ω_{sub} reflects the proportion of high frequency components in each subregion. We calculate the average value $\bar{\omega}$ and standard deviation σ of ω_{sub} over the whole image, and extract subregions with relatively high frequency components by the following equation:

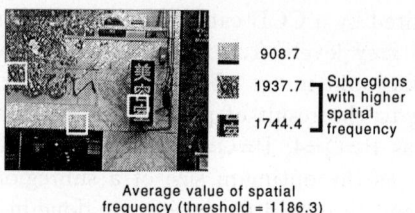

Fig. 2. ω_{sub} of subregions

$$\omega_{sub} \begin{cases} \geq \bar{\omega} - \sigma/8, & \Longrightarrow \text{Character candidates} \\ < \bar{\omega} - \sigma/8, & \Longrightarrow \text{Noncharacter} \end{cases} \tag{4}$$

Here the threshold value $\sigma/8$ is chosen by examining the results of a preliminary experiment using 20 scene images.

Figure 2 shows the ω_{sub} values of three subregions with different texture. As we can see from the figure, subregions with more line components, such as characters and tree leaves, have more high frequency components.

Evaluation of Grey-Level Contrast In order to discriminate characters from other objects, we evaluate grey-level contrast in subregions using the standard deviation of the grey-level value. Subregions with large standard deviations are chosen as character candidates by thresholding.

2.4 Detection of Characters

Some character candidates selected by analysis of spatial frequency and evaluation of grey-level contrast are noncharacter regions such as leaves of trees. The next crucial step, therefore, is to select the correct components that correspond to characters only.

Finding the Bounding Rectangle of Each Object In order to detect specific objects from a scene image, sometimes it is necessary to find each object and then select the target object from all these objects. Here, in order to detect characters from character candidates, first, we convert the input grey images into binary images using the method of local thresholding[6], which is a kind of adaptive thresholding method utilized for image segmentation. The input image is divided into the same subregions as in Section 2.2. We compute the local threshold of each subregion by the method of Otsu[7]. To cope with the background noise, the threshold of a target subregion is determined by a linearly weighted sum of thresholds of the target subregion itself and its 8-neighbor subregions. By using the interpolated thresholds, we can get smooth binary images. We then detect edge components using a 3×3 Sobel filter and obtain the boundary of each object. By calculating the minimum and maximum values of x and y coordinates of the boundary, we can find the bounding rectangle of each object.

Many Chinese characters and Japanese Kana characters are composed of more than one segment. In order to detect these multisegment characters, we combine rectangles considering their positional relation (including relation, crossing relation and neighboring relation), and features of shape and size. Examples of the relations are shown in Figure 3.

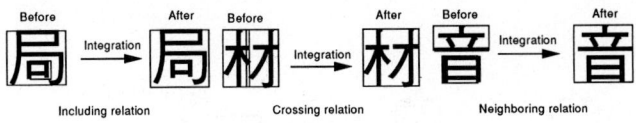

Fig. 3. Merging bounding rectangles

Selecting Character Components We use the following three conditions to remove noncharacter components.

- Aspect ratio of the bounding rectangle (width /height)
  Generally, for Japanese Kana characters and Chinese characters, the ratios of width and height of their bounding rectangles are approximately the same. We choose rectangles with aspect ratios ranging from $\frac{2}{3}$ to $\frac{3}{2}$ as characters.
- Bimodality of intensity histogram
  In the bounding rectangles, the grey-level contrast between characters and their backgrounds is relatively high, and the grey-level histograms of character regions have two remarkable peaks (bimodality). We compute the grey-level histogram of each bounding rectangle and examine the bimodality of the histogram. Rectangles of which the grey-level histograms have two remarkable peaks are selected as characters.
- The size of bounding rectangle
  Bounding rectangles of characters must be larger than 25×25 pixels and smaller than 120×120 pixels.

Detecting Character Lines In scene images, characters are usually in the form of character lines. Therefore the characteristics of character arrangement are also important in detecting character components. We use the following conditions:

- Similarity of size
  We restrict the area ratio of two adjacent rectangles in a same character line to be between 0.5 to 2.0.

- Proximity of characters
  We limit the space between two adjacent characters in the same character line by the following equation:

$$(a+b)/2 < c < 3(a+b)/2 \tag{5}$$

,where a, b, and c are shown in Figure 4(a).
- Alignment of arrangement
  As shown in Figure 4(b), if the horizontal (vertical) center line of rectangle 1 passes through rectangle 2, and the horizontal (vertical) center line of rectangle 2 passes through rectangle 1, we say that they are in alignment.

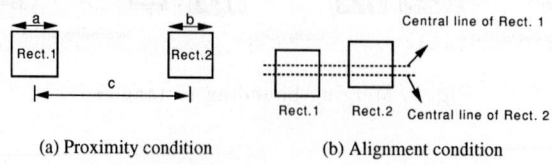

(a) Proximity condition   (b) Alignment condition

Fig. 4. Character arrangement

If more than three bounding rectangles satisfy all the three conditions, we select them as a character line.

3 Experimental Results

As mentioned in Section 1, the algorithm was designed primarily to detect character components in scene images. We used images of both indoor and outdoor real scenes that contain character components such as signboards and traffic signs. Input images were digitized into 256 grey levels and 640×480 pixels. We conducted experiments on 40 outdoor and 20 indoor scene images. The results are shown in Tables 1. Figure 5 shows two examples of experimental results.

Table 1. Experimental Result

	Number	Number of extraction	Rate of extraction (%)	Error
Char.	397	351	88.4	23
Char. line	77	63 (71)	81.8 (92.2)	12

In the table, "Error" indicates the number of noncharacters (lines) misextracted by the method, and values in parentheses represent the number (rate) of character lines, the characters of which are fully or partially extracted. About 90% of characters and 80% of character lines in these 60 scene images were detected correctly. The detection rate of character lines will increase to 90% if we include character lines whose characters are partially detected. The processing time differs from image to image. Generally it takes 15 to 20 seconds to process an image on an ONYX workstation.

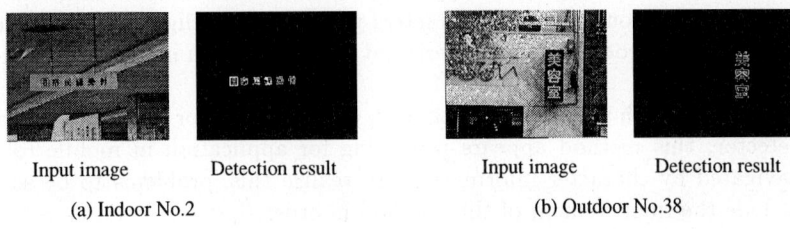

Input image   Detection result   Input image   Detection result

(a) Indoor No.2   (b) Outdoor No.38

Fig. 5. Experimental Results

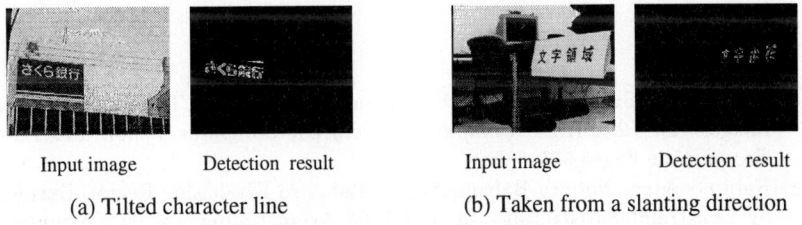

Input image   Detection result   Input image   Detection result

(a) Tilted character line   (b) Taken from a slanting direction

Fig. 6. Experimental Results

Our algorithm can also extract character lines that is slightly tilted and character lines that taken from a slightly slanting direction. Figure 6(a) shows an example of an outdoor scene with a slightly tilted character line and its detection result, and Figure 6(b) shows the input and detection result of an indoor scene taken by the Sony CCD camera on the top of our mobile robot at a slanting direction.

The proposed algorithm could detect most of the characters in these 60 scene images well. However, we failed to find characters that have a relatively low grey-level contrast and special simple characters such as "1", "一", "三" and "川".

4 Conclusion

A method for extracting characters from grey-scale scene images was described in this paper. This method was developed based on the general features of characters and can be used to detect characters with different font, size, and grey level. By analyzing the spatial frequency and evaluating grey-level contrast to extract character candidates, the calculation load can be lightened.

The main shortcoming of this approach is that it cannot detect characters of low grey-level constrast and characters which are not aligned. Nevertheless, the approach produces good experiment results for both indoor and outdoor scene images.

If combined with a character recognizer, distance detector and robot obstacle detector, this method appears promising for application in mobile robots navigated by character information. To realize this, problems to be solved include the improvement of this method in order to detect characters taken from a slanting direction and to decide the direction of the camera.

Acknowledgments

We thank Dr. T.Tanaka, Assistant Professor of Computation Center, Nagoya University, for his constructive and helpful comments to this work.

References

1. Anil K. Jain, Sushil K. Bhattacharjee, and Yao Chen. On Texture In Document Images. In *1992 IEEE Computer Society On Computer Vision and Pattern Recognition*, Pages 677–680.
2. Keiji Gyohten, Noboru Babaguchi,and Tadahiro Kitahashi. Region Extraction by Constraint Satisfaction. In *ACCV'93 Asian Conference on Computer Vision*, pages 846–849, Osaka, Japan, November 1993.
3. Yu Zhong, Kalle Karu, and Anil K.Jain. Locating Text in Complex Color Images. *Pattern Recognition*,28(10):1523–1535, October 1995.
4. Jun Ohya, Akio Shio, and Shigeru Akamatsu. Recognizing Characters in Scene Images. *IEEE Transactions on pattern analysis and machine intelligence*, 16(2):214–220, February 1994.
5. Rafael C. Gonzalez, and Richard E. Woods. *Digital Image Processing*. Addison-Wesley Publishing Company, 1992.
6. Akio Shio. A Dynamic Thresholding Method for Character Detection in Complex Scene Images. In *Trans. IEICE Japan*, J71-D,no.6 pages 863–873. (In Japanese)
7. Nobuyuki Otsu. An Automatic Threshold Selection Method Based on Discriminant and Least Squares Criteria In *Trans. IEICE Japan*, J63-D, no.3, pages 349–356. (In Japanese)

Conic Based Image Transfer for 2-D Objects: A Linear Algorithm

Akihiro Sugimoto

Advanced Research Laboratory, Hitachi, Ltd.
Hatoyama, Saitama 350-03, Japan
e-mail: sugimoto@harℓ.hitachi.co.jp

Abstract. This paper presents a study, based on conic correspondences, on the relationship between two uncalibrated images. We show that, for a pair of corresponding conics, the parameters representing the conics satisfy a linear constraint. We also present a linear algorithm for uniquely determining from the coefficients of this constraint the transformation between corresponding points or lines in the two images up to a scale factor. Accordingly, conic correspondences enable us to easily handle both points and lines in uncalibrated images of a planar object.

1 Introduction

One of the fundamental difficulties in recognizing objects from images is how to deal with a number of images obtained from the same object. Clarifying the relationship between images of the same object is thus of fundamental importance. Knowledge of the relationship between images provides us with many advantages in handling images of the same object. For example, it allows us to know the number of stored images required to establish a geometric model of an object. Moreover, we have only to transmit the stored images of an object to a recipient, who can then synthesize an image of the object from any viewpoint. Thus, the relationship between images of the same object plays a key role in important problems in computer vision and multimedia.

A conic is one of the most important image features. This is because many man-made objects have circular parts, and circles are perspectively projected onto conics. Furthermore, the conic is a more compact primitive than points or lines and can be more robustly and more exactly extracted from images. Hence, developing algorithms dealing with conics is significant. Nevertheless, there are fewer articles (for instance, [7, 8, 15]) dealing with conics to clarify the relationship between images while the multilinear constraints on multiple images of points or lines were derived [2, 3, 5, 6, 9, 10, 11, 13]. Quan [7, 8] showed that two polynomial constraints are satisfied by corresponding conics in two uncalibrated images and addressed a geometric invariant and conic reconstruction from two images. However, it has not yet established the uniqueness of the solution due to the nonlinearity of the derived polynomial constraints. Methods dealing with conics are developed based on nonlinear constraints on conics and, therefore, the procedures require complex nonlinear computation and do not ensure the uniqueness of the solution.

This paper presents a study, based on conic correspondences, on the relationship between two perspective images acquired by an uncalibrated camera. We

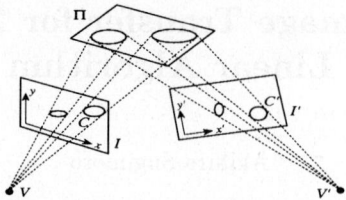

Fig. 1. Two different images of conics in plane Π.

show that for a pair of corresponding conics, the parameters representing a conic in one image are transformed by a projective transformation of degree 5 to the parameters representing the corresponding conic in the other image. We also present a linear algorithm for uniquely determining from this transformation, the transformation between corresponding points or lines in the two images up to a scale factor.

2 Plane Transformation

Let $\mathcal{P}^n$ be the n-dimensional projective space over the real number field **R**. When we observe coplanar points in 3-D, introducing the homogeneous coordinates allows us to express the transformation between their two uncalibrated images as a plane projective transformation. That is, letting the homogeneous coordinates of the corresponding points in the images be $\boldsymbol{x}$ and $\boldsymbol{x}'$, we have $\boldsymbol{x}' \simeq M^{-1}\boldsymbol{x}$, where M^{-1} is a 3×3 nonsingular matrix and where $\simeq$ implies the equality sign up to a scale factor. We call M a *pointwise transformation*.

The relationship between two images with respect to points or lines is aggregated into a pointwise transformation. For example, when a pointwise transformation between two images is given, we can predict the coordinates of points or lines in the second image from those in the first image.

3 Conic-based Relationship between Two Images

3.1 Conicwise Transformation

A conic in $\mathcal{P}^2$ is expressed in the form of a quadratic homogeneous equation:
$$\boldsymbol{x}^{\mathrm{T}} C \boldsymbol{x} = 0, \tag{3.1}$$
where C is the 3×3 symmetric matrix given by $C \simeq \begin{pmatrix} a & f & e \\ f & b & d \\ e & d & c \end{pmatrix}$. We see that a conic bijectively corresponds to a point in $\mathcal{P}^5$. Hereafter, for conic (3.1), we refer to $\boldsymbol{q} = (a, b, c, d, e, f)^{\mathrm{T}}$ ($\in \mathcal{P}^5$) as a *conic vector*.

Conics are transformed into conics under projective transformations. Denote by $\boldsymbol{q}$ and $\boldsymbol{q}'$ the conic vectors for the corresponding conics, both of which are obtained by observing a conic in plane Π (see Fig. 1). Then, $\boldsymbol{q}$ is transformed to $\boldsymbol{q}'$ by a projective transformation of degree 5. That is, when we let S be a 6×6 matrix, we have
$$\boldsymbol{q}' \simeq S\boldsymbol{q}. \tag{3.2}$$

(3.2) gives the linear constraint on corresponding conics in two images when we observe a conic in space. We call S a *conicwise transformation*. When we observe coplanar conics from two different viewpoints, a conicwise transformation relates their two images. Every time we observe a conic in Π and are given the correspondence between its two images, we have five independent linear constraints. We can thereby uniquely determine the conicwise transformation up to a scale factor if we observe seven conics in a *general position* in Π. Here seven conics are said to be in a *general position* when the seven points in $\mathcal{P}^5$ that correspond to the seven conics can be a projective coordinate frame for $\mathcal{P}^5$.

Proposition 1. *Suppose that we observe a conic in plane Π and that we are given the correspondence of its images in planes I and I'. Then, we can uniquely determine the conicwise transformation between I and I' up to a scale factor if we observe at least seven conics in a general position in Π.*

3.2 Conicwise Transformation and Pointwise Transformation

When points are subject to a pointwise transformation M, C is transformed to

$$C' \simeq M^\mathrm{T} C M. \tag{3.3}$$

It is easy to see that each entry of S is quadratic homogeneous in the entries of M. In fact, if we define $Q_{ijk\ell} := M_{ki} M_{\ell j}$ $(i,j,k,\ell \in \{1,2,3\})$ and then rewrite (3.3) in terms of the entries of the conic vectors, it follows that

$$C'_{(ij)} \simeq \sum_{(k\,\ell)} \sigma(k,\ell)\, Q_{(ij)(k\,\ell)}\, C_{(k\,\ell)}, \tag{3.4}$$

where $C_{(k\,\ell)}$ denotes the entry of the conic vector that is deduced from $C_{k\ell}$ ($C'_{(ij)}$ is defined in the same manner). The notation $(\cdots)$ implies the symmetrization of the indices inside parentheses; we define that the symmetrized indices are aligned in such a way as $(1\,1),(2\,2),(3\,3),(2\,3),(3\,1),(1\,2)$. Moreover, function σ is defined by $\sigma(k,\ell) := 1$ (if $k = \ell$), 2 (otherwise).

From (3.2) and (3.4), we can establish the link between a conicwise transformation S and its corresponding pointwise transformation M:

$$S_{(ij)(k\,\ell)} \simeq \sigma(k,\ell)\, M_{(k \atop (i} M_{\ell) \atop j)}. \tag{3.5}$$

Therefore, a conicwise transformation is determined from its corresponding pointwise transformation. Then, is the converse true? Namely, can we determine the corresponding pointwise transformation from a conicwise transformation? In the next section, we answer this question affirmatively.

4 Decomposing a Conicwise Transformation to a Pointwise Transformation

Our aim is, for a given $\widehat{S} \simeq S D^{-1}$ where $D = \mathrm{diag}(1,1,1,2,2,2)$, to uniquely determine M up to a scale factor. The problem is reduced to solving an overdetermined system of nonlinear equations. Here, we develop a method that makes full use of the redundancy involved in $\widehat{S}$ and which only requires linear computation. We remark that for an unknown real number ρ,

$$\widehat{S}_{(ij)(k\,\ell)} = \rho\, M_{(k \atop (i} M_{\ell) \atop j)}. \tag{4.1}$$

4.1 Column/Row-based Decomposition

Denote by $u_{(ij)}$ the (ij)th row vector of $\widehat{S}$ $((ij) = (11),(22),\ldots,(12))$ and also denote by c_i the ith column vector of M ($i = 1,2,3$). Defining a mapping from a vector in $\mathbf{R}^6$ to a symmetric 3×3 matrix,

$$\text{Mtx} : (x_1, x_2, x_3, x_4, x_5, x_6)^{\text{T}} \mapsto \begin{pmatrix} x_1 & x_6 & x_5 \\ x_6 & x_2 & x_4 \\ x_5 & x_4 & x_3 \end{pmatrix},$$

we have $\text{Mtx}(u_{(ij)}) = \frac{\rho}{2}(c_i c_j^{\text{T}} + c_j c_i^{\text{T}})$, which yields

$$\text{Mtx}(u_{(ii)} + u_{(jj)} - 2u_{(ij)}) = \rho(c_i - c_j)(c_i - c_j)^{\text{T}}.$$

Proposition 2. *Let $i \neq j$. $\text{Mtx}(u_{(ii)} + u_{(jj)} - 2u_{(ij)})$ is a matrix of rank 1, and its eigenvalue λ_{ij}^{c} that has the maximum absolute value and eigenvector c_{ij} (its Euclidean norm is 1, i.e., $\|c_{ij}\| = 1$) associated with λ_{ij}^{c} are given by*

$$\lambda_{ij}^{\text{c}} = \text{sgn}(\rho)\left[\sqrt{|\rho|}\,\|c_i - c_j\|\right]^2, \quad c_{ij} \simeq c_i - c_j, \tag{4.2}$$

where $\text{sgn}(\rho) = 1$ *(if $\rho > 0$)*, -1 *(if $\rho < 0$)*.

Remark. When noise exists, the rank of $\text{Mtx}(u_{(ii)} + u_{(jj)} - 2u_{(ij)})$ is greater than 2. In this case, two eigenvalues other than λ_{ij}^{c} are near each other but are far from λ_{ij}^{c}. That is, from the viewpoint of magnitude, the two eigenvalues are grouped together, whereas λ_{ij}^{c} belongs to another group. λ_{ij}^{c} and c_{ij} are therefore computed robustly [1, 4]. □

Expressing c_i in terms of λ_{ij}^{c} and c_{ij}, we obtain the following column-based decomposition of M (c_{ij} and $c_i - c_j$ do not necessarily have the same sign):

$$c_i - c_j = \mu_{ij}^{\text{c}} \gamma \sqrt{|\lambda_{ij}^{\text{c}}|}\, c_{ij}, \tag{4.3}$$

where $\mu_{ij}^{\text{c}} \in \{1,-1\}$ and $\gamma = 1/\sqrt{|\rho|}$. (4.3) gives a system of linear homogeneous equations in the entries of c_i and γ. Only six of the nine equations are independent. We can easily and robustly determine μ_{ij}^{c} by only judging which is the greater of given two values. See Sugimoto [12] for details.

In the same way, we obtain the row-based decomposition of M:

$$r_k - r_\ell = \mu_{k\ell}^{\text{r}} \gamma \sqrt{|\lambda_{k\ell}^{\text{r}}|}\, r_{k\ell}, \tag{4.4}$$

where $\mu_{k\ell}^{\text{r}} \in \{1,-1\}$, and $v_{(k\ell)}$ and r_k are, respectively, the $(k\ell)$th row vector of $\widehat{S}$ and the kth row vector of M. (4.4) therefore gives a system of linear homogeneous equations in the entries of r_k and γ, only six of which are independent.

4.2 Parametrizing the Pointwise Transformation

The column-based decomposition of M and the row-based decomposition of M are obtained independently. In other words, μ_{ij}^{c} and $\mu_{k\ell}^{\text{r}}$ together do not necessarily ensure the consistency of signs when we regard (4.3) and (4.4) as one system. Originally, μ_{ij}^{c} and $\mu_{k\ell}^{\text{r}}$ are closely related to each other since c_i and r_k express the same M in different ways. The consistency of signs between (4.3) and (4.4) is ensured by a pair of $\{\mu_{ij}^{\text{c}}\}$ and $\{\mu_{k\ell}^{\text{r}}\}$ or by a pair of $\{\mu_{ij}^{\text{c}}\}$ and $\{-\mu_{k\ell}^{\text{r}}\}$.

Proposition 3 allows us to easily determine which pair ensures this consistency. That is, we put $\kappa \in \{1, -1\}$ and choose a κ that nullifies

$$\sum_{(ij),\, i \neq j} \left(\text{the } \pi(i,j)\text{th component of } [\mu_{ij}^c \boldsymbol{c}_{ij} + \kappa \mu_{ij}^r \boldsymbol{r}_{ij}] \right).$$

Here, for $i \neq j$ $(i, j \in \{1, 2, 3\})$, we define $\pi(i, j) := \{1, 2, 3\} - \{i, j\}$. When noise exists, we choose a κ that minimizes the absolute value of the above equation.

Proposition 3. *The column vectors $\boldsymbol{c}_i$ and the row vectors $\boldsymbol{r}_i$ of M satisfy*

$$\sum_{(ij),\, i \neq j} \left(\text{the } \pi(i,j)\text{th component of } [(\boldsymbol{c}_i - \boldsymbol{c}_j) \pm (\boldsymbol{r}_i - \boldsymbol{r}_j)] \right) \overline{\not=} 0.$$

Now we can modify (4.4) so that its signs are consistent with those of (4.3). That is, we have only to replace $\mu_{k\ell}^r$ with $\kappa \mu_{k\ell}^r$ on the right-hand side of (4.4). Consequently, we can regard all the equations together as one system of linear homogeneous equations in M and γ. We have 18 constraints, only eight of which are independent.

Proposition 4. *Let* $U := \begin{pmatrix} 1 & 1 & 1 \\ 1 & 1 & 1 \\ 1 & 1 & 1 \end{pmatrix}$, $V := \begin{pmatrix} m_{11} & m_{12} & m_{13} \\ m_{21} & m_{22} & m_{23} \\ m_{31} & m_{32} & 0 \end{pmatrix}$ *(where $m_{ij} \in \mathbf{R}$), and put $M = sU + V$ ($s \in \mathbf{R}$). The system of linear homogeneous equations that is derived from (4.3) and the modified (4.4) (i.e., with $\mu_{k\ell}^r$ replaced by $\kappa \mu_{k\ell}^r$) is expressed in the form of*

$$A \boldsymbol{f} = 0 \qquad (\operatorname{rank} A = 8;\ \boldsymbol{f} \neq 0), \tag{4.5}$$

where A is an 18×9 coefficient matrix and $\boldsymbol{f} = (m_{11}, m_{12}, \ldots, m_{32}, \gamma)^{\mathrm{T}} \in \mathbf{R}^9$.

Since we know the rank of A, linear computation enables us to uniquely determine the least-squares solution up to a scale factor. Denote by $\widehat{\boldsymbol{f}} = (\widehat{m}_{11}, \widehat{m}_{12}, \ldots, \widehat{m}_{32}, \widehat{\gamma})^{\mathrm{T}}$ ($\widehat{\gamma} = 1$) the solution obtained by setting γ to be 1. Putting $\widehat{V}$ to be V determined by $\widehat{m}_{ij}$, we can express M with two parameters s and t as follows:

$$M = sU + t\widehat{V} \qquad (s, t \in \mathbf{R};\ t \neq 0). \tag{4.6}$$

4.3 Determining the Ratio s/t

From (4.6), we express (4.1) using s and t, from which we obtain

$$s^2 + \left(\frac{\widehat{m}_{ki}\widehat{m}_{\ell j} + \widehat{m}_{kj}\widehat{m}_{\ell i}}{2} - \operatorname{sgn}(\lambda_{ij}^c)\, \widehat{S}_{(ij)(k\ell)} \right) t^2$$
$$+ \frac{\widehat{m}_{ki} + \widehat{m}_{\ell j} + \widehat{m}_{kj} + \widehat{m}_{\ell i}}{2} st = 0. \tag{4.7}$$

This is a system of quadratic homogeneous equations in s and t.

Solving (4.7) with respect to s and t is equivalent to solving (4.7) with respect to s^2, t^2 and st (these are regarded as independent parameters) under the constraint $\det \begin{pmatrix} s^2 & st \\ st & t^2 \end{pmatrix} = 0$. We have two steps in obtaining a solution: we first solve a given system ignoring the constraint and then modify the solution so that it satisfies the constraint. This modification is performed so that the difference between the original and modified solutions is minimized under a certain norm. The reason we have these two steps is that they simplify the procedure for obtaining the solution.

Putting $g = (s^2, t^2, st)^T$, we can then express (4.7) in the form of

$$Bg = 0 \quad (\text{rank} B = 2; \; g \neq 0). \tag{4.8}$$

Here B is a 36×3 coefficient matrix. If we regard s^2, t^2 and st as independent parameters, (4.8) gives a system of linear homogeneous equations. Linear computation thereby enables us to uniquely determine the least-squares solution of (4.8) up to a scale factor. Denote by $\widehat{g} = (\widehat{s^2}, \widehat{t^2}, \widehat{st})^T$ ($\widehat{st} = 1$) the solution obtained by setting st to be 1 and define $W := \begin{pmatrix} \widehat{s^2} & \widehat{st} \\ \widehat{st} & \widehat{t^2} \end{pmatrix}$. Letting $W = \psi_1 w_1 w_1^T + \psi_2 w_2 w_2^T$ ($\psi_1 \geq \psi_2 \geq 0$) be the spectrum resolution of W and putting $W' := \psi_1 w_1 w_1^T$, we see that W' has the following property. The proof is similar to that of Tsai–Huang [14].

Proposition 5. *Of the 2×2 symmetric matrices that satisfy* $\det \begin{pmatrix} s^2 & st \\ st & t^2 \end{pmatrix} = 0$, *$W'$ is nearest from W under the Frobenius norm.*

Hence, putting $w_1 = (w_1, w_2)^T$, we see that w_1/w_2 gives s/t. By returning to (4.6), we can uniquely determine M up to a scale factor. (The scale factor is determined by setting the criteria that the determinant of M is positive and that M is normalized under the Frobenius norm.)

Remark. Whereas we have $\psi_2 = 0$ in the case where no noise exists, we have $\psi_2 > 0$ in the presence of noise. Even in the presence of noise, however, we have $\psi_1 \gg \psi_2$. This indicates that the computation of w_1 is robust. □

5 Experimental Results

On the basis of the decomposition procedure above, we present our experimental results and show that the outputs of the algorithm are stable and robust with respect to noise.

We created a 3×3 matrix M, each entry of which was randomly and independently generated within the interval of $[-50.00, 50.00]$. We used this M to compute $\sigma(k, \ell) M_{(k} M_{\ell)}$. We then computed S' by multiplying an unknown scale to it, where the scale factor was randomly generated within the interval of $[-10.00, 10.00]$. Next, we added Gaussian noise to each entry s'_{ij} of S' ($i, j \in \{1, 2, \ldots, 6\}$), independently. The mean of Gaussian distribution was set to be 0.0 and its standard deviation was set to be $r \times |s'_{ij}|$, where r represents a noise level and r was changed by 0.01 from 0.0 (noiseless) to 0.1 (10% noise). Here we evaluate the accuracy of each entry of a computed conicwise transformation in terms of noise levels. By applying our algorithm to S, we estimated M. For each noise level, we iterated the procedure "compute S by adding noise to S' and apply our algorithm to S to estimate M" 50 times. The mean values of estimated M over the 50 times are shown in Fig. 2. To evaluate the deviation

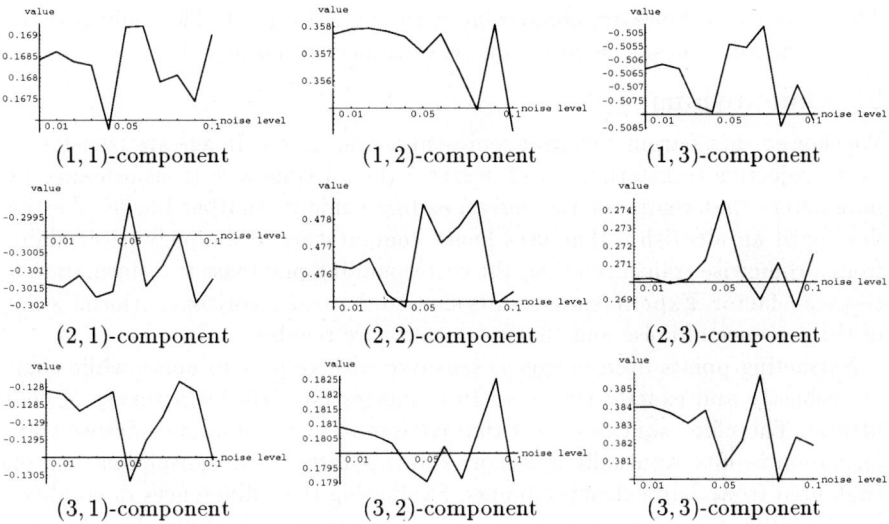

Fig. 2. Means under several noise levels.

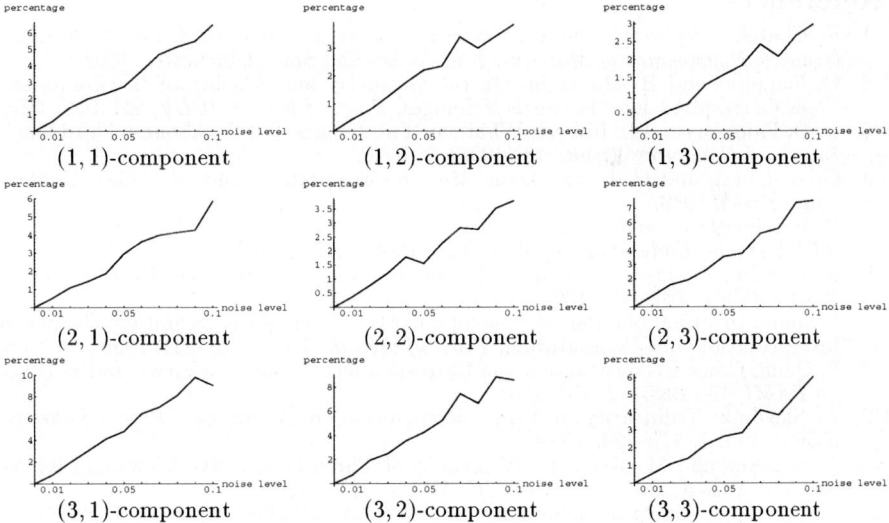

Fig. 3. Coefficients of variations under several noise levels.

of the estimated M over the 50 times, we calculated coefficients of variations, i.e., the ratio of the standard deviation to the mean, which are shown in Fig. 3.

Figure 2 shows that each entry of M is accurately estimated up to the second digit of precision under all noise levels. Depending on the noise level, the third digit of precision changes, but the change is limited to at most 2% of each value. We can thus conclude that M is accurately estimated even in the presence of noise and our algorithm is robust. On the other hand, Fig. 3 shows that the coefficients of variations (almost) monotonically increase as the noise level increases.

This increase is, however, compatible with the noise level. This indicates that the estimated values of M are stable with respect to noise.

6 Conclusions

We showed that parameters that represent a conic in one image are transformed by a projective transformation of degree 5 (i.e., a conicwise transformation) to parameters that represent the corresponding conic in another image. We then developed an algorithm that uses linear computation in uniquely determining, from a conicwise transformation, the corresponding pointwise transformation up to a scale factor. Experimental results showed that the algorithm performs stably in the presence of noise and that its outputs are reliable.

Extracting points from images is sensitive with respect to noise, while conics are robustly and exactly extracted from images (see [16] for a survey of conic fitting). Therefore, we can expect that computation based on a conicwise transformation results eventually in a more stable pointwise transformation than one computed from only extracted points. Evaluating their differences in stablity is left for future research. Also left for future investigation is the evaluation of the practical efficiency of the algorithm in real situations.

References

1. F. Chatelin: *Valeurs propres de matrices*, Masson, Paris, 1988. (W. Ledermann (trans.): *Eigenvalues of Matrices*, John-Wiley and Sons, Chichester, 1993.)
2. O. Faugeras and B. Mourrain: On the Geometry and Algebra of the Points and Lines Correspondences Between N Images, *Proc. of the 5th ICCV*, 951–956, 1995.
3. O. D. Faugeras and L. Robert: What can Two Images Tell Us About a Third One?, *Int. J. of Computer Vision*, 18 (1996), 1, 5–19.
4. G. H. Golub and C. F. van Loan: *Matrix Computation*, 2nd ed., Johns Hopkins Univ. Press, 1989.
5. R. Hartley: Lines and Points in Three Views — An Integrated Approach, *Proc. of ARPA Image Understanding Workshop*, 1009–1016, 1994.
6. R. Hartley: A Linear Method for Reconstruction from Lines and Points, *Proc. of the 5th ICCV*, 882–887, 1995.
7. L. Quan: Invariant of a Pair of Non-coplanar Conics in Space: Definition, Geometric Interpretation and Computation, *Proc. of the 5th ICCV*, 926–931, 1995.
8. L. Quan: Conic Reconstruction and Correspondence from Two Views, *IEEE Trans. on PAMI*, 18 (1996), 2, 151–160.
9. A. Shashua: Trilinearity in Visual Recognition by Alignment, *Proc. of the 3rd ECCV*, Vol. 1, 479–484, 1994.
10. A. Shashua and M. Werman: Trilinearity of Three Perspective Views and its Associated Tensor, *Proc. of the 5th ICCV*, 920–925, 1995.
11. A. Shashua: Algebraic Functions for Recognition, *IEEE Trans. on PAMI*, 17 (1995), 8, 779–789.
12. A. Sugimoto: *Image Transfer of 2-D Objects Based on Conic Correspondences—A Linear Algorithm Using Redundancy—*, ARL Research Report, 96-003, 1996.
13. B. Triggs: Matching Constraints and the Joint Image, *Proc. of the 5th ICCV*, 338–343, 1995.
14. R. Y. Tsai and T. S. Huang: Uniqueness and Estimation of Three Dimensional Motions Parameters of Rigid Objects with Curved Surfaces, *IEEE Trans. on PAMI*, 6 (1984), 1, 13–27.
15. I. Weiss: 3-D Curve Reconstruction from Uncalibrated Cameras, *Proc. of ICPR*, Vol. 1, 323–327, 1996.
16. Z. Zhang: Parameter Estimation Techniques: A Tutorial with Application to Conic Fitting, *Image and Vision Computing*, 15 (1997), 1, 59–76.

Minimal Conditions on Intrinsic Parameters for Euclidean Reconstruction *

Anders Heyden, Kalle Åström

Dept of Mathematics, Lund University
Box 118, S-221 00 Lund, Sweden
email: {heyden,kalle}@maths.lth.se

Abstract In this paper will be investigated what constraints on the intrinsic parameters that are needed in order to reconstruct an unknown scene from a number of its projective images. Two such minimal cases are studied in detail. Firstly, it is shown that it is sufficient to know the skew parameter, even if all other parameters are unknown and varying, to obtain a Euclidean reconstruction. Secondly, the same thing can be done for known aspect ratio, again when all other intrinsic parameters are unknown and varying. In fact, we show that it is sufficient to know any of the 5 intrinsic parameters to make Euclidean reconstruction.

An algorithm, based upon bundle adjustment techniques, to obtain Euclidean reconstruction in the above mentioned cases are presented. Experiments are shown on the slightly simpler case of both known aspect ratio and skew.

1 Introduction

During the last years there has been an intensive research on the possibility to obtain reconstructions up to an unknown similarity transformation (often called *Euclidean reconstruction*), without using fully calibrated cameras. It is a well-known fact that it is only possible to make reconstruction up to an unknown projective transformation (often called *projective reconstruction*) when nothing about the intrinsic parameters, extrinsic parameters or the object is known. Thus it is necessary to have some additional information about either the intrinsic parameters, the extrinsic parameters or the object in order to obtain the desired Euclidean reconstruction.

A priori information about the object can be used in a fairly straight-forward manner, see [3]. Several researchers have dealt with the problem of using additional information about the extrinsic parameters, i.e., the camera orbit, see for example [12].

One common situation is when the intrinsic parameters are constant during the whole (or a part) of the image sequence. This approach leads to the well-known Kruppa equations. These equations are highly nonlinear and difficult to solve numerically. Several attempts to solve this problem have been made, see [9], [2]. In [5] the same problem is solved by a global optimisation technique, where a lot of smaller optimisation problems have to be solved in order to get a starting point for the last optimisation.

---

* This work has been done within then ESPRIT Reactive LTR project 21914, CUMULI and the Swedish Research Council for Engineering Sciences (TFR), project 95-64-222

Another constraint, called the *modulus constraint* have been used in [11], to obtain Euclidean reconstruction from constant intrinsic parameters. This formalism has been extended to the case when the focal length is varying between the different imaging instants, see [10]. The practical implications of this result is questionable since when the focal length varies, by zooming, the principal point varies also.

The results presented in this paper is motivated by the fact that when a CCD-camera is used in order to capture an image sequence and the zoom is used, as in active vision, both the focal length and the principal point varies. However, it is often the case that the aspect ratio and/or the skew are constant during the whole sequence (at least if the same camera is used), since they are determined by the geometry of the light sensitive array. It is furthermore possible to measure, once and for all, the aspect ratio and/or skew for a camera and then use these values for all image sequences captured by that camera. When these quantities are known, they can be compensated for and we may assume that the aspect ratio is equal to 1 and/or the skew is equal to 0.

It has been shown, see [7] that it is possible to make Euclidean reconstruction when both the skew and aspect ratio are known. In this paper we will extend that result and show theoretically that Euclidean reconstruction is possible in two minimal cases. Firstly, when the aspect ratio is known and secondly, when the skew is known. In both cases are all other intrinsic parameters unknown and allowed to vary between the different imaging instants. Although of less practical importance, it is also shown that it is possible to make Euclidean reconstruction if only one coordinate of the principal point is known or if the focal length is known. In all these cases, not only a Euclidean reconstruction of the object are obtained, but also the intrinsic camera parameters for each camera, i.e. auto-calibration is performed. The proof is based on the assumption of generic camera motion. The theoretical result is verified by experiments on simulated data.

2 The Camera Model

The image formation system (the camera) is modeled by the equation

$$\lambda \begin{bmatrix} x \\ y \\ 1 \end{bmatrix} = \begin{bmatrix} \gamma f & sf & x_0 \\ 0 & f & y_0 \\ 0 & 0 & 1 \end{bmatrix} [R \mid -Rt] \begin{bmatrix} X \\ Y \\ Z \\ 1 \end{bmatrix} \Leftrightarrow \lambda \mathbf{x} = K[R \mid -Rt]\mathbf{X} = P\mathbf{X} \ . \quad (1)$$

Here $\mathbf{X} = [X\, Y\, Z\, 1]^T$ denotes object coordinates in extended form and $\mathbf{x} = [x\, y\, 1]^T$ denotes extended image coordinates. The scale factor λ, called the **depth**, accounts for perspective effects and (R, t) represent a rigid transformation of the object, i.e. R denotes a 3×3 rotation matrix and t a 3×1 translation vector. Finally, the parameters in K represent intrinsic properties of the image formation system: f represents focal length, γ represents the **aspect ratio**, s represents the skew and (x_0, y_0) is called the principal point. The parameters in R and t are called **extrinsic parameters** and the parameters in K are called the **intrinsic parameters**. In this paper we will mostly deal with cameras where $s = 0$ and/or $\gamma = 1$.

Definition 1. A camera that can be modeled as in (1), with $s = 0$ is called a **non-skew camera**. When $\gamma = 1$ it is called an **aspect-free camera** and when both $s = 0$ and $\gamma = 1$ it is called a **camera with Euclidean image plane**. Internal calibration matrices K of type

$$K_{\text{ns}} = \begin{bmatrix} \gamma f & 0 & x_0 \\ 0 & f & y_0 \\ 0 & 0 & 1 \end{bmatrix}, \quad K_{\text{af}} = \begin{bmatrix} f & s & x_0 \\ 0 & f & y_0 \\ 0 & 0 & 1 \end{bmatrix} \quad \text{and} \quad K_{\text{Eip}} = \begin{bmatrix} f & 0 & x_0 \\ 0 & f & y_0 \\ 0 & 0 & 1 \end{bmatrix} \quad (2)$$

are called **non-skew calibration matrices**, **aspect-free calibration matrices** and **Euclidean calibration matrices** respectively.

Observe that it is not necessary that $\gamma = 1$ and/or $s = 0$ in order to use the subsequent results. It is sufficient that they are known, since then the image coordinates can be transformed to new ones obeying the constraints on γ and/or s, for details see [8].

The following results, one of them shown in [6] and one shown in [4], will be needed later. For a proof, see [8].

Lemma 2. *A camera matrix* $P = \begin{bmatrix} u^T & \\ v^T & t \\ w^T & \end{bmatrix}$, *normalised such that* $w.w = 1$, *represents a non-skew camera if and only if*

$$(u \times w).(v \times w) = 0 \quad (3)$$

an aspect-free camera if and only if

$$|(u \times w) \times (u \times w)| = (v \times w).(v \times w) \quad (4)$$

and a camera with Euclidean image plane if and only if

$$(u \times w).(v \times w) = 0 \quad \text{and} \quad (u \times w).(u \times w) = (v \times w).(v \times w) \ . \quad (5)$$

Here $a.b$ *denotes the scalar product of* a *and* b.

Observe that the condition $w.w = 1$ can easily be fulfilled by multiplying the camera matrix by a suitable constant, since a camera matrix is only defined up to scale.

It is possible to state similar conditions for the cases of known f, x_0 and y_0. By making coordinate changes in the images similar to the previous cases it is no restriction to assume that $f = 1$, $x_0 = 0$ and $y_0 = 0$ respectively. The conditions on the camera matrices are in turn $|v \times w| = w.w$, $u.w = 0$ and $v.w = 0$ respectively.

Now we have the necessary tools to prove that it is possible to obtain a Euclidean reconstruction, when sufficiently many point correspondences are given in a sufficient number of images.

3 Euclidean Reconstruction is Possible

For a moment, we do not take into account the special form of the camera matrices, (2), for cameras with known skew and/or aspect ration, and instead work with totally uncalibrated cameras, as in (1). Then it is possible to make reconstruction up to an unknown projective transformation. This means that it is possible to calculate camera matrices $P_i, i = 1, \ldots, m$ that fulfills

$$\lambda_i \mathbf{x}_i = P_i \mathbf{X}, \quad i = 1, \ldots, m , \qquad (6)$$

where $\mathbf{x}_i$ denotes extended image coordinates in image i and λ_i denotes the corresponding depth in image i. It can easily be seen from (6) that given one such sequence of camera matrices, $P_i, i = 1, \ldots, m$, and a reconstruction, $\mathbf{X}$, also $P_i H, i = 1, \ldots, m$ and $H^{-1}\mathbf{X}$ is a possible choice of camera matrices and reconstruction, where H denotes a nonsingular 4×4 matrix. Multiplication of $\mathbf{X}$ by such a matrix corresponds to projective transformations of the object. In our case H can not be chosen arbitrarily since every camera matrix has to obey the conditions in Lemma 2.

The next step is to show that given a sequence of camera matrices that solves the projective reconstruction problem and represents cameras with known skew and/or aspect ratio, i.e. fulfills one of the conditions in Lemma 2, then the only possible transformations H that preserve these conditions are the ones representing similarity transformations. In order to show this some notations will be introduced.

Denote by $\mathcal{M}_P$ the manifold of all 3×4 projection matrices, i.e., the set of all 3×4 matrices defined up to scale. Denote by $\mathcal{M}_{ns}$, $\mathcal{M}_{af}$ and $\mathcal{M}_{Eip}$ the manifold of all camera matrices that represents non-skew cameras, aspect-free cameras and cameras with Euclidean image planes, respectively, i.e., all 3×4 matrices that can be written as in (2), and thus obeying one of the conditions in Lemma 2. Denote the group of all projective transformations, represented by 4×4 matrices, by $\mathcal{G}_P$. The subclass of transformations that preserves the properties in Lemma 2 is denoted by $\mathcal{G}_{ns}$, $\mathcal{G}_{af}$ and $\mathcal{G}_{Eip}$ respectively, e.g.

$$\mathcal{G}_{ns} = \{ H \in \mathcal{G}_P \,|\, (P \in \mathcal{M}_{ns}) \Rightarrow PH \in \mathcal{M}_{ns} \} .$$

Finally, the group of all similarity transformations will be denoted by $\mathcal{G}_S$ and will be represented by

$$\mathcal{G}_S = \{ H = \begin{bmatrix} \lambda R & t \\ 0 & 1 \end{bmatrix} \,|\, RR^T = I, 0 \neq \lambda \in \mathbb{R} \} . \qquad (7)$$

It can easily be seen that the group of similarity transformations is contained in $\mathcal{G}_{ns}$ as well as in $\mathcal{G}_{af}$ and $\mathcal{G}_{Eip}$. Thus $\mathcal{G}_S \subseteq \mathcal{G}_{ns} \subseteq \mathcal{G}_P$, $\mathcal{G}_S \subseteq \mathcal{G}_{af} \subseteq \mathcal{G}_P$ and $\mathcal{G}_S \subseteq \mathcal{G}_{Eip} \subseteq \mathcal{G}_P$. Note that $\mathcal{G}_{ns}$, $\mathcal{G}_{af}$ and $\mathcal{G}_{Eip}$, are precisely the transformation groups that groups of interest. If, for example, $\mathcal{G}_{ns} = \mathcal{G}_P$ then it is only possible to make reconstruction up to a projective transformation in the non-skew camera case. If $\mathcal{G}_{ns} = \mathcal{G}_S$ then it is possible to make reconstruction up to a similarity transformation.

Theorem 3. *Let $\mathcal{G}_{ns}$, $\mathcal{G}_{af}$ and $\mathcal{G}_{Eip}$ respectively denote the class of transformations in 3D-space that preserves the conditions in Lemma 2 and $\mathcal{G}_S$ the group of similarity transformations in 3D-space. Then*

$$\mathcal{G}_{ns} = \mathcal{G}_{af} = \mathcal{G}_{Eip} = \mathcal{G}_S .$$

Proof. Consider first the case of a aspect-free camera. From the discussion above we have $\mathcal{G}_S \subseteq \mathcal{G}_{\text{Eip}}$.

Observe that the constraints on the camera matrices in Lemma 2 only involve the first 3×3 submatrix. Use the notation $H = \begin{bmatrix} A & b \\ c & d \end{bmatrix}$, where A is a 3×3 matrix. Assume that P represents an aspect-free camera, H a projective transformation and

$$PH = K[R|t]\begin{bmatrix} A & b \\ c & d \end{bmatrix} = [K(RA+tc)\,|\,K(Rb+td)] \in \mathcal{M}_{\text{af}} \ . \tag{8}$$

Then $K(RA+tc)$ can be factorised $K(RA+tc) = K'R'$ where K is an aspect-free calibration matrix and R' denotes an orthogonal matrix. Since (8) is valid for any P that represents a camera matrix, i.e., for any K, R and t, we first choose $t = 0$.

Assume that A has the property that for every aspect-free calibration matrix K and orthogonal R, it is possible to factorise KRA according to $KRA = K'R'$, for some aspect-free calibration matrix K' and orthogonal R'. Then also UAV has this property for every pair of orthogonal matrices U and V, since

$$KRUAV = KR''AV = K'R'''V = K'R' \ ,$$

where R'' and R''' denote orthogonal matrices. Now, using the singular value decomposition of A we can write

$$D_1 = U_1 A V_1 = \begin{bmatrix} a & 0 & 0 \\ 0 & b & 0 \\ 0 & 0 & c \end{bmatrix} \quad \text{and} \quad D_2 = U_2 A V_2 = \begin{bmatrix} b & 0 & 0 \\ 0 & c & 0 \\ 0 & 0 & a \end{bmatrix} \ .$$

where D_2 is obtained by a simple permutation of the rows and columns in U_1 and V_1 respectively. Replacing A by D_1 and choosing $R = I$ in (8), Lemma 2 gives $a = b$ and replacing A by D_2 gives $b = c$. Thus all singular values of A are equal, which means that A is a multiple of an orthogonal matrix.

Consider now the case, where $t \neq 0$, and the condition that for every aspect-free calibration matrix K, every orthogonal R and every t, $K(RA+tc)$ can be factorised as $K(RA+tc) = K'R'$ for some aspect-free calibration matrix K' and orthogonal R'. If $RA+tc$ can be factorised in this way then so can $(RA+tc)V$ for every orthogonal matrix V. Choose V such that $cV = [s\,0\,0]$, then choose $R = (AV)^{-1}$ and $t = [1\,0\,0]^T$. These choices give

$$(RA+tc)V = RAV + tcV = \begin{bmatrix} 1 & 0 & 0 \\ 0 & 1 & 0 \\ 0 & 0 & 1 \end{bmatrix} + \begin{bmatrix} 1 \\ 0 \\ 0 \end{bmatrix}[s\,0\,0] = \begin{bmatrix} 1+s & 0 & 0 \\ 0 & 1 & 0 \\ 0 & 0 & 1 \end{bmatrix} \ ,$$

and according to Lemma 2, $s = 0$, which in turn implies $c = [0\,0\,0]$.

Summing up, H is of the form $H = \begin{bmatrix} \lambda R & b \\ 0 & d \end{bmatrix}$, where λ is a scalar and R an orthogonal matrix. Dividing by d gives $\frac{1}{d}H \in \mathcal{G}_S$. Thus $\mathcal{G}_{\text{af}} \subseteq \mathcal{G}_S$ from which the first part of the theorem follows.

Consider now the case of a non-skew camera. Again from the discussion above we have $\mathcal{G}_S \subseteq \mathcal{G}_{\text{Eip}}$. The proof is analogous to the above until $D_1 = U_1AV_1 =$

diag(a, b, c) is obtained. Then according to (8)

$$R = \frac{1}{\sqrt{2}} \begin{bmatrix} 1 & 1 & 0 \\ -1 & 1 & 0 \\ 0 & 0 & \sqrt{2} \end{bmatrix} \implies RD_2 = \frac{1}{\sqrt{2}} \begin{bmatrix} a & b & 0 \\ -a & b & 0 \\ 0 & 0 & \sqrt{2}c \end{bmatrix} .$$

Now, according to Lemma 2, RD_2 is a skew-free calibration matrix if and only if

$$((a, b, 0) \times (0, 0, c)) \cdot ((-a, b, 0) \times (0, 0, c)) = (bc, -ac, 0) \cdot (bc, ac, 0) = c^2(b^2 - a^2) = 0 ,$$

i.e. if and only if $a = b$. Using a permutation of the singular values as before gives $a = b = c$ and then we can proceed as in the aspect-free case again until the choice of cV. This time we chose $cV = [0\, s\, 0]$, which gives

$$(RA + tc)V = \begin{bmatrix} 1 & s & 0 \\ 0 & 1 & 0 \\ 0 & 0 & 1 \end{bmatrix} ,$$

which according to Lemma 2 gives $s = 0$ and so on.

Finally, the case of a camera with Euclidean image plane follows from either of the two cases above and this completes the proof.

A glimpse at the proof gives that the three remaining minimal cases can be proven similarly:

- If only the focal length, f is known (assumed to be equal to 1), it is possible to make Euclidean reconstruction. (A similar proof as in the aspect-free case.)
- If only the x-coordinate or the y-coordinate of the principal point is known, it is possible to make Euclidean reconstruction. (Similar to the proof in the skew-free case, but use

$$R = \frac{1}{\sqrt{2}} \begin{bmatrix} 1 & 0 & 1 \\ 0 & \sqrt{2} & 0 \\ -1 & 0 & 1 \end{bmatrix} \quad \text{and} \quad cV = [0\, 0\, s]$$

in the case of $x_0 = 0$ and similarly in the case of $y_0 = 0$.

Thus reconstruction up to a similarity transformation is possible if any of the internal calibration parameters f, s, γ, x_0 or y_0 are known.

Observe that this theorem is valid only under the assumption that the camera motion is sufficiently general. This fact is used implicit in the formulation of the theorem and in the proof, by requiring that $P = K[\,R\,|\,-Rt\,]$ can be chosen arbitrarily.

Finally, it can be shown that these reconstruction problems can be formulated as polynomial equations in the intrinsic parameters in the first image and the parameters describing the plane at infinity, see [6]. The polynomial equations are exactly the ones in Lemma 2, i.e. two equations per image for the case of Euclidean image planes and one equation per image in the other cases. Since the equations arises from different images they are in general independent. This means that in the case of Euclidean image planes at least 4 images are needed since we have 6 parameters (3 for the plane at infinity and 3 for the unknown intrinsic parameters) and 8 equations (6 equations will not give a unique solution). In the other cases we have 7 parameters and one equation per images, which means that at least 8 images are needed in general.

4 Experiments

The method was tested on simulated data in the case of a camera with Euclidean image plane. The numerical computations has been made using a so called *bundle adjustment technique*. Briefly parameters are introduced for all object coordinates and camera matrices. Then the squared difference between the image coordinates obtained from these parameters and the true image coordinates are minimised with respect to the parameters. Apart from this least squares solution also an estimate of the accuracy of the parameters can be obtained. For further details see [8]. Note that the calculated parameters not only gives a Euclidean reconstruction, but also the intrinsic parameters for all cameras.

First Simulation. First an experiment was performed with 10 points in 15 images. The points were taken as random points with coordinates between -300 and $+300$ units. The camera positions were chosen at random approximately 1000 units away. The standard deviation σ together with the focal length f and the position (x_0, y_0) of the principal point of the first camera and the RMS of reconstructed object positions in percent of overall scale are presented in Table 1 for different levels of noise, $\sigma[e]$.

Second Simulation. Second an experiment was performed with 50 points in 20 images. The points were taken as random points with coordinates between -500 and $+500$ units. The camera positions were chosen at random approximately 1000 units away. The standard deviation σ together with the focal length f and the position (x_0, y_0) of the principal point of the first camera and the RMS of reconstructed object positions in percent of overall scale are presented in Table 1 for different levels of noise, $\sigma[e]$.

σ	f	x_0	y_0	Δ
0	2112.191	25.433	8.250	0.000
0.1	2096.894	33.395	6.853	0.371
0.2	2107.966	43.571	5.061	2.193
0.5	2143.423	56.123	31.375	1.727
1	1982.302	9.773	-16.357	3.611
2	2057.016	352.815	-22.979	11.247
5	1974.814	314.814	32.671	18.755

σ	f_1	x_0	y_0	Δ
0.0	1010.752	4.435	1.355	0.000
0.1	1010.787	4.460	1.385	0.017
0.2	1012.072	4.723	1.271	0.135
0.5	1008.164	4.416	1.959	0.225
1	1010.795	5.023	2.970	0.251
2	1014.648	7.878	2.285	0.357
5	1007.924	12.647	-1.364	0.669
10	1020.446	-6.559	-7.934	2.033

Table 1: Some estimated parameters and the reconstruction error in the first and second simulation respectively.

It is important to note that many points are needed in many images since there are so many unknown parameters. The first simulation with 10 points in 15 images with 300 equations and 158 unknown degrees of freedom is much less stable than the second simulation with 50 points in 20 images, (2000 equations and 323 unknown degrees of freedom).

5 Conclusions

In this paper we have shown that it is possible to reconstruct an unknown object from a number of its projective images up to similarity transformations, i.e. angles and ratios of lengths can be calculated. This is possible even when the focal distance and the principal point change between the different imaging instants. The only thing we need to know about the cameras is the aspect ratio and/or the skew. These parameters are defined by the geometry of the light sensitive area and need only be measured once for each camera. In many cases it is reasonable to assume that the skew is 0 and the aspect ratio is 1. This is called a camera with Euclidean image plane. The other two minimal cases, aspect ratio equal to 1, called an aspect-free camera, and skew equal to 0, called a non-skew camera, are also treated and shown to give reconstructions up to an unknown similarity transformation. Although of less practical importance, it is also shown that it is possible to make Euclidean reconstruction if only one coordinate of the principal point is known or if the focal length is known.

The paper contains a theoretical proof of these facts as well as an experimental validation using simulated data in the case of a camera with Euclidean image planes. In these experiments a bundle adjustment technique has been used to estimate all undetermined parameters, i.e. the reconstructed object, the relative position of the cameras and the intrinsic parameters at the different imaging instants.

References

1. American Society for Photogrammetry, *Manual of Photogrammetry*, Ed. C.C. Slama, 4th edition, 1984.
2. Faugeras, O. D., Luong, Q.-T., Maybank, S. J., Camera Self-Calibration: Theory and Experiments, *ECCV'92, Lecture notes in Computer Science, Vol 588. Ed. G. Sandini, Springer-Verlag*, 1992, pp. 321-334.
3. Faugeras, O., D., Stratification of three-dimensional projective, affine and metric representations, *Journal of the Optical Society of America A.*, vol. 12, No. 3/March 1995, pp. 465-484.
4. Faugeras, O., D., *Three-Dimensional Computer Vision*, MIT Press, Cambridge, Mass., 1993.
5. Hartley, R., I., Euclidean Reconstruction from Uncalibrated Views, *Applications of Invariance in Computer Vision, Lecture notes in Computer Science, Vol 825. Ed. Joseph L. Mundy, Andrew Zisserman and David Forsyth, Springer-Verlag*, 1994, pp. 237-256.
6. Heyden, A., Geometry and Algebra of Multiple Projective Transformations, *Doctoral Thesis, CODEN:LUFTD2/TFMA--95/5002--SE, ISBN 91-628-1784-1, Lund, Sweden*, 1995.
7. Heyden, A., Åström, K., Euclidean Reconstruction from Image Sequences with Varying and Unknown Focal Length and Principal Point, *Proc. CVPR'97, IEEE Computer Society Press*, 1997, pp. 438–443.
8. Heyden, A., Åström, K., Minimal Conditions on Intrinsic Parameters for Autocalibration and Euclidean Reconstruction, *Internal Report, Dept. of Mathematics, Lund University, CODEN:LUFTD2/TFMA--97/????--SE, Lund, Sweden*, 1997.
9. Luong, Q.-T., Matrice Fondamentale et Calibration Visuelle sur l'Environnement-Vers une plus grande autonomie des systèmes robotiques, *PhD-thesis, Université de Paris-Sud, Centre d'Orsay*, 1992.
10. Pollefeys, M., Van Gool, L., Oosterlinck, A., Euclidean Reconstruction from Image Sequences with Variable Focal Length, *ECCV'96, Lecture notes in Computer Science, Vol 1064. Ed. B. Buxton, R. Cipolla, Springer-Verlag*, 1996, pp. 31-44.
11. Pollefeys, M., Van Gool, L., Oosterlinck, A., The modulus constraint: a new constraint for self-calibration, *Proc. ICPR'96, IEEE Computer Society Press*, vol. 1, 1996, pp. 349-353.
12. Zisserman, A., Beardsley, P., Reid, I., Metric Calibration of a Stereo Rig, *Proc. IEEE Workshop on Representation of Visual Scenes*, 1995.

Surface Based Hypothesis Verification in Intensity Images Using Geometric and Appearance Data

J.H.M. Byne and J.A.D.W. Anderson

Dept. of Computer Science, University of Reading, Berks RG6 6AY, UK.
email: J.H.M.Byne@reading.ac.uk

Abstract. In this paper we discuss current work concerning Appearance-based and CAD-based vision; two opposing vision strategies. CAD-based vision is geometry based, reliant on having complete object centred models. Appearance-based vision builds view dependent models from training images. Existing CAD-based vision systems that work with intensity images have all used one and zero dimensional features, for example lines, arcs, points and corners. We describe a system we have developed for combining these two strategies. Geometric models are extracted from a commercial CAD library of industry standard parts. Surface appearance characteristics are then learnt automatically by observing actual object instances. This information is combined with geometric information and is used in hypothesis evaluation. This augmented description improves the systems robustness to texture, specularities and other artifacts which are hard to model with geometry alone, whilst maintaining the advantages of a geometric description.

1 Introduction

In this paper we describe our current work on combining two important, but opposing vision strategies: CAD-based vision [3] and Appearance-based vision [14]. CAD-based vision relies on having available a complete 3D object centred representation of the objects for recognition from which search strategies are derived. Appearance based matching does not assume geometric models are available, but builds view dependent models for matching from example images. Appearance based matching seems to work well with objects that are difficult to handle with pure geometric methods, for example objects containing many specularities and textures. However, appearance-based matching has some problems coping with partial object occlusion, and objects at distances or orientations not present in the training images, which are better handled by CAD-based vision systems. Mundy, *et al*, [13] have reported an experimental comparison between the two approaches and give results showing that both methods have their own strengths and weaknesses. Thus combining these two approaches would appear to offer a more general and robust solution.

In this paper we present a technique for extracting models from a commercial CAD system. These models contain all the geometric information stored in the CAD package, and symbolic labels for the types of materials and surface finishes represented. The system then builds statistical models of these surface types by observing instances of the model from a number of different views. Using the geometric and appearance models, vision programs are compiled that perform hypothesis verification and pose refinement.

In section 2 we describe how models are extracted from CAD. Building statistical descriptions of surface appearance is discussed in section 3. We describe hypothesis verification in sections 4 and 5. In the remaining sections we present results and conclusions.

2 Extracting Models from CAD

In the past, the geometrical models used in recognition programs have often been constructed in a very *ad-hoc* fashion. In this section we discuss a method we have developed for extracting models from a commercial CAD system for use with both our existing edge based vision programs, reported elsewhere [4, 20, 21, 12], and for the new techniques discussed in this paper that use both geometric and appearance based information.

Using CAD modelling software has significant advantages over existing *ad-hoc* construction methods. The ease and speed with which CAD models can be constructed facilitates testing of our existing model based vision software with a large range of objects, and makes feasible the testing of those that contain many complex curved surfaces, which were previously very tedious to model. Also there is a wide range of pre-existing CAD models that have been constructed for other purposes, for example industrial design, that may be used.

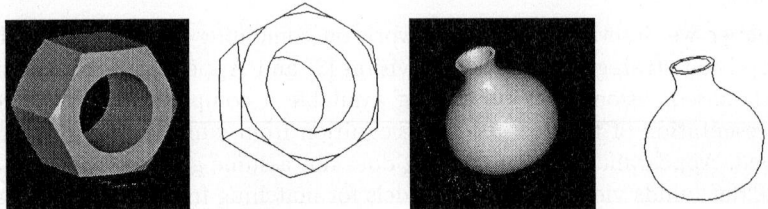

Fig. 1. Examples of CAD Models, a JIS Hex Nut and a bottle with free form surfaces, and their polyhedral approximations showing predicted image discontinuities.

The system we have implemented converts arbitrary CAD models from a commercial CAD system (SDRC Master Series [19]), which may contain complex free form surfaces, to models which can be used with existing model-based vision programs at Reading. The CAD models are approximated using planar patches. By investigating the underlying CAD geometry and surface information for each planar patch, the edges in the approximate model are recorded as Fold, Extremal or Invisible. Figure 1 shows examples of the CAD rendered models, a JIS (Japanese Industrial Standard) Hex Nut, taken from the SDRC

[19] library of standard fasteners, and a bottle containing free form surfaces, and their polyhedral approximations with expected image discontinuities marked.

The CAD systems is used to fit a planar mesh to the model using its FEA procedures. This mesh is then extracted and stored as a winged edge polyhedral (WEP) model [2]. The analytical CAD geometry is then extracted and associated with each face and edge in the WEP model. The FEA meshing produces multiple elements on planar surfaces which is unnecessary for our purposes, so the WEP model is processed to combine these elements. Each edge in the WEP model is examined in turn, if the two faces associated with it belong to the same CAD surface and are coplanar, the edge is removed, merging the faces. Then each vertex in the WEP model is examined, if it contains only two edges and they are colinear the vertex is deleted, merging the two edges. When the WEP model is written out in our primitive file format [11] for use by the existing model-based vision software, concave polygons are tessellated into convex polygons. The extracted models contain only convex planar faces allowing very fast projection to facilitate the matching processes discussed below.

In order to facilitate the surface based matching, additional information is also extracted from the CAD database. In the polyhedral approximation, each polygon records the surface type and the analytical geometry of the CAD face it represents. The geometry includes a numerical description and a category for the CAD surface – for example sets of polygons may be labelled as representing planar, spherical, cylindrical or conic surfaces. The surface type is a symbolic label which denotes the face material – for example yellow plastic, wood, coarse steel, or similar. When many polygons have been used to represent a single CAD face, they are labelled to make this explicit, so that they may be treated as a homogeneous region for matching.

3 Learning Surface Appearance

Although the geometric models extracted from CAD contain symbolic information describing the types of materials and surfaces they contain, there is no explicit information about their appearance. In this section we discuss how this gap is bridged automatically by observing instances of modelled objects and building statistical models of their appearance.

A calibrated robotic system [6], that provides control over camera position and orientation, is used to record a set of training images for each object. The geometric model is fitted by hand to one of the images from the training set. The model is fitted to other images in the training set by applying the known translations that were applied to move the camera. To overcome kinematic errors in the robot positioning, the pose of the model is refined by aligning edges extracted from the geometric model with image discontinuities using existing techniques [5].

When the models have been fitted to training images, a correspondence between the various CAD surfaces and the image regions representing them is obtained. This allows statistical information about each of the various surface types to be recovered. A number of local operators are applied to the image region and feature vectors for each surface type are recorded. The operators include normalised colour features as in [8], and texture features [18, 16, 1].

The recovered feature vectors are used to construct a probability density function using a mixture of Gaussians model, P, giving $P(\mathbf{f} \mid s_i)$, the probability of $\mathbf{f}$ occurring given surface s_i.

As new feature extractors are developed and implemented, they can be incorporated into the system to enhance its discriminatory power. However, if the system has a large feature library, the cost of computing all of them will become high. While this is acceptable for off-line training, it is not acceptable for on-line recognition. In practice a given vision task may require only a small subset of the available features for adequate discrimination. When a vision program is required to recognise a particular set of surface types, we use a feature selection algorithm to chose the most appropriate feature subset. We use the branch and bound algorithm [15] as it provides optimal subset selection despite the higher computational overhead compared to other methods, such as the floating search technique [17]. The training and feature selection phases of the system are viewed as off-line compilation, and computational time spent here that results in better on-line performance is well spent.

4 Labelling Images

When an image containing unknown objects is presented to the system the first step is to apply a low level probabilistic labelling to each pixel. For a given scene the system must a have a set of models, M, which contains all the models that might be presented to the system. Each model, $m \in M$, will contain a number of different surface types, we denote this set surf(m). Thus for any given scene the set of surface types, S, that may occur is

$$S = s_0 \cup \bigcup_{m \in M} \text{surf}(m),$$

where s_0 represents something other than the model surfaces. It is convenient to regard s_0 as the background. For each pixel i, in the input image I, a feature vector $\mathbf{f}_i$ is computed. Using the probabilities as computed above, the probability of i belong to surface $s \in S$ may by calculated, as shown below, using Bayes rule assuming equal prior probabilities.

$$P(s_j \mid \mathbf{f}_i) = \frac{P(\mathbf{f} \mid s_j)}{\sum_{s \in S} P(\mathbf{f} \mid s)}$$

As objects are recognised and located in the image, labels are assigned to each pixel. We use l_i to denote the label given to pixel i. Initially when no objects have been hypothesised, $l_i = s_0 \, \forall i \in I$.

5 Evaluating Hypotheses

When an object is hypothesised to exist at a specific location, a surface template, T, is constructed, labelling which pixels are expected to belong to which surface type. This surface template can be constructed using well documented graphics algorithms, see for example [7]. Using a depth buffering algorithm, T may be constructed in time $O(s)$, where s is the number of surfaces contained within the object. Also many desktop machines have hardware support for such operations, permitting extremely efficient construction of T. In contrast, edge based hypothesis verification which requires contour predicting algorithms are typically $O(s^2)$ and are not as well supported in hardware.

Given the surface template T, and the probabilistic image labelling as computed above, a goodness of fit metric can readily be computed as follows. For each pixel label in the template T we have available the probability of it being correct. To evaluate the hypothesis we need to combine the information from each label in T. An appropriate means of combining the probabilities is to use the chi-squared distribution to compute a combination of tests ([10] pp. 35). Thus with k tests, that is labels in T, the statistic a is distributed as χ^2 with $v = 2k$ degrees of freedom:

$$a = \sum_{i=1}^{k} \log p_i$$

Unfortunately the chi-squared distribution involves an incomplete Gamma function and is therefore expensive to compute. However, a suitable approximation exists ([9] pp. 166-167). Thus for sufficiently large number of degrees of freedom v, say $v = 30$, the variable z is an approximately normal variable with zero mean and unit standard deviation:

$$z = \sqrt{2\chi^2} - \sqrt{2v - 1}$$

Substituting a in z gives:

$$z = \sqrt{-4 \sum_{i=1}^{k} \log p_i} - \sqrt{4k - 1}$$

Setting $g = \frac{z}{2}$ and ignoring the small -1 term gives the goodness of fit metric as:

$$g = \sqrt{-\sum_{i=1}^{k} \log p_i} - \sqrt{k}$$

Notice that the smaller the value of g the better the fit.

While this goodness of fit score may be used to evaluate how well a single object fits the image data, it may also be used to measure how well the whole image interpretation explains the image data. When a set of models have been hypothesised, a surface template t may be computed for the whole image and g may be computed as above. If the set of hypothesised models contain partial or total occlusions this will be accommodated in the construction of T, using the depth buffering algorithm. The goodness of fit of small image regions may also be computed allowing the system to decide where the interpretation is poor and further processing is required.

6 Experiments and Results

In this section we discuss the performance of surface based evaluation, and compare the results with our existing edge based schemes. In our experiments we have used CAD models of JIS nuts and bolts from the SDRC fasteners library and CAD models of toy blocks, a toy lorry, a door knob and a model alligator.

One of the main advantages of the surface based approached discussed here is that when models are misaligned there is still some overlap between the surface patches. This is not true for the edge based methods, where when models are significantly misaligned there is no correspondence between projected edges and image discontinuities. The graphs in figure 6 show the much smoother evaluation space for the surface based technique.

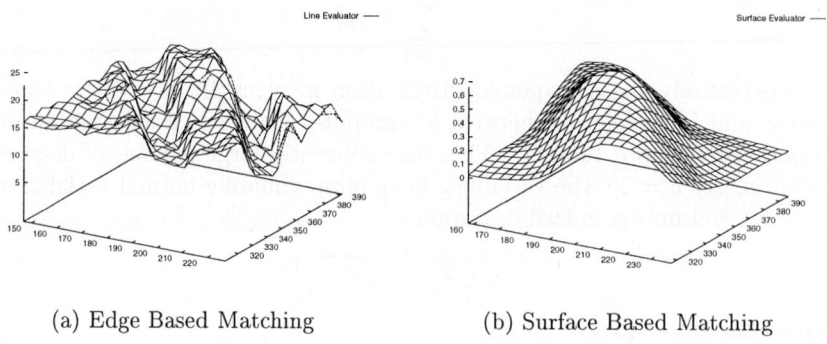

(a) Edge Based Matching (b) Surface Based Matching

Fig. 2. Graphs showing Edge and Surface based evaluation function as the model position is varied in space.

To demonstrate some of the advantages of the smoother evaluation space we performed the following experiment. Objects were assume to lie on a work surface giving them three degrees of freedom: two translations on the surface and one rotation around an axis orthogonal to it. An algorithm for pose refinement was then implemented that performed a simplex search over these parameters to optimise the match score. Examples of perturbed positions from which the technique converged to the correct position using the reported surface based evaluation are shown in figure 3.

To compare the performance of edge and surface based evaluation we perturbed the models randomly from their true position and recorded the number of search iterations required for the models to converge to their true positions. The graph in figure 2 shows the percentage of the randomly perturbed starting positions that converged to the true pose against the number of evaluations performed during the simplex search, for surface, edge and a combined evaluation function.

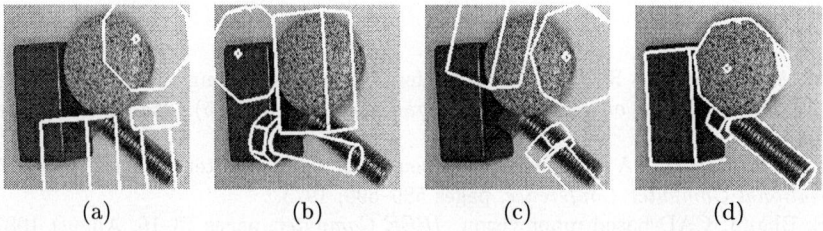

Fig. 3. Examples of perturbed positions (a,b,c) and the position to which they converge (d) using the surface based evaluation technique with simplex search.

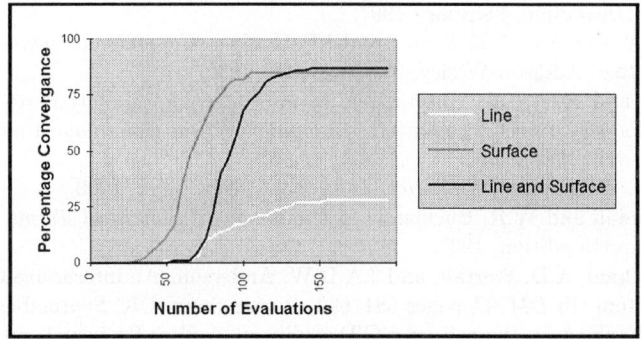

Fig. 4. Convergence rate using simplex search.

The graph clearly shows that the convergence with the surface based technique is faster and more reliable than with the edge based method. The convergence of the combined technique is slightly more reliable than the surface based method but convergence is slower and evaluation more expensive. In some situations, where adjacent surface types are confusable the combined method is superior.

7 Conclusions

Effective surface based techniques for matching CAD models to intensity images have been presented. The techniques have two significant advantages over existing edge based methods. The matching score is more tolerant to misalignment between model and object, providing a much smoother evaluation surface. We have shown that this results in a three fold improvement in pose refinement. Also the method can be used to measure the quality of the entire scene description, or parts of it, to decide where further recognition efforts should be directed. We have also noted that both techniques are complimentary and can be used together to provide greater robustness.

References

1. M. Amadasun and R. King. Textural features corresponding to textural properties. *IEEE Trans. on Systems, Man, and Cybernetics*, 19(5):1264–1274, September/October 1989.
2. B.G. Baumgart. A polyhedral representation for computer vision. In *AFIPS National Computer Conference*, pages 589–596, 1975.
3. B. Bhanu. CAD-based robot vision. *IEEE Computer*, pages 13–16, August 1987.
4. K. Brisdon. *Hypothesis Verification using Iconic Matching*. PhD thesis, University of Reading, UK, November 1990.
5. K. Brisdon, G.D. Sullivan, and K.D. Baker. Feature aggregation in iconic model evaluation. In *Alvey Vision Conf.*, pages 19–24, Manchester, August 1989.
6. J.H.M. Byne and J.A.D.W. Anderson. CAD-based vision. In *IEE Colloquium on Industrial Inspection*, February 1997.
7. J. Foley, A. van Dam, S.K. Feiner, and J.F. Hughes. *Computer Graphics-Principles And Practice*. Addison-Wesley, Reading, MA, 1990.
8. L. Grewe and A.C. Kak. Interactive learning of a multiple-attribute hash table classifer for fast object recognition. *Computer Vision and Image Understanding*, 61(3):387–416, May 1995.
9. P.G. Hoel. *ELementary Statistics*. John Wiley, New York, 1960.
10. M.G. Kendall and W.R. Buckland. *A Dictionary of Statistical Terms*. Longman, London, fourth edition, 1982.
11. B. M-Rouhani, A.D. Worrall, and J.A.D.W. Anderson. An interactive CAD-based vision system. In *BMVC*, pages 681–690, Birmingham, UK, September 1995.
12. S.J. Maybank, A.D. Worrall, and G.D. Sullivan. A filter for visual tracking based on a stochastic model of driver behaviour. In *ECCV*, Cambridge, UK, April 1996.
13. J. Mundy, A. Liu, N. Pillow, A. Zisserman, S. Abdallah, S. Utcke, S. Nayar, and C. Rothwell. An experimental comparision of appearance and geometric model based recognition. Number 1144 in Lecture Notes in Computer Science, pages 247–269, 1996.
14. H. Murase and S.K. Nayar. Visual learning and recognition of 3-D objects from appearance. *International Journal of Computer Vision*, 14(1):5–24, 1995.
15. P. Narendra and K. Fukunaga. A branch and bound algorithm for feature subset selection. *IEEE Trans. on Computers*, 26(9):917 – 922, September 1977.
16. A.P. Pentland. Fractal-based description of natural scenes. *IEEE Trans. of Pattern Analysis and Machine Intelligence*, 6(6):661–673, November 1984.
17. P. Pudil, J. Novovičová, and J. Kittler. Floating search methods in feature selection. *Pattern Recognition Letters*, pages 1119–1125, November 1994.
18. K.V. Ramana and B. Ramamoorthy. Statistical methods to compare the texture features of machined surfaces. *Pattern Recognition*, 29(9):1447–1459, 1996.
19. SDRC I-DEAS master series. 2000 Eastman Drive, Milford, Ohio 45150, USA.
20. A.D. Worrall, J.M. Ferryman, G.D. Sullivan, and K.D. Baker. Pose and struture recovery using active models. In *BMVC*, pages 137–146, Birmingham, UK, September 1995.
21. A.D. Worrall, F.R. Marslin, G.D. Sullivan, and K.D. Baker. Model based tracking. In *BMVC*, pages 310–318, Glasgow, UK, September 1991.

Next Best Viewpoint (NBV) Planning for Active Object Modeling Based on a Learning-by-Showing Approach

Hongbin Zha, Ken'ichi Morooka and Tsutomu Hasegawa

Graduate School of Inf. Sci. & Elec. Eng., Kyushu Univ., Fukuoka, Japan

Abstract. The paper presents a new method of creating a complete model of a curved object from a sequence of range images acquired by a fixed range finder. A robot arm is used in order to change poses of the object. To accomplish the modeling fast and accurately in an optimal manner, we propose a new on-line viewpoint planning algorithm to choose the Next Best Viewpoint (NBV) based on already obtained the partial model.

1 Introduction

Automatic generation of a complete boundary description for any 3-D object has become a popular topic in vision, robotics and CAGD researches. There are several approaches to the 3-D modeling problem. For a simple object, we can decompose it manually into some primitives. For a complex one, however, we have to utilize some active methods that take a lot of images of it by 3-D sensors and integrate the images automatically into a unified description ([6], [9]). The latter belongs to the so-called *learning-by-showing* paradigm and has demonstrated a great adaptability to shape variations.

In general, the following three steps are necessary in an active modeling approach ([5], [3], [10]): 1)data acquisition; 2)registration; 3)integration of views. While many useful techniques have been proposed for the last two steps, relatively little attention was paid to the first.

To acquire sufficient image data for the whole modeling, we have to move the sensor or the object to accomplish a complete sensor coverage. Two issues should be taken into consideration here. First, image acquisition and registration are quite time-consuming, and thus the number of images should be as small as possible. Second, one main source of errors in range finder is the sharp shape changes on the object surface. That is, the accuracy of acquired images is variant with the geometry of the objects to be modeled, and thus we have to arrange the image acquisition order carefully according to a certain prediction on the shape changes.

One of the straightforward solutions to the problems is to introduce a planning process on the *Next Best Viewpoint (NBV)* prior to taking ([2]). In the paper, we propose a new method for the NBV planning purpose.

2 Active object modeling

In the section, we give an overview of our active object modeling method. Fig.1(a) shows the setup of the system we utilize to carry out the whole modeling procedure. It comprises the following two major implements in addition

(a) system setup (b) range finder

Fig. 1. Setup of the modeling system.

to a workstation performing image processing and planning computation: a 6-DOF PUMA robot arm, and a stereo range finder with a controllable laser-slit scanner, which is illustrated in Fig.1(b).

As th first step, a range image of a given object to be modeled is taken to generate an initial (partial) object model. The current NBV is computed, and then by using the robot arm to move the object to the corresponding pose, the expected image is taken and registered with the partial model. This procedure is repeated and terminates when the whole object surface is scanned. This whole procedure can be partitioned into the following four computational processes: *triangulation of image, image registration, integration,* and *NBV determination.* Detailed explanations of triangulation and integration process can be found in [7] and [11], and here we give a brief explanation on the registration process and the partial model.

Image registration :   To create a complete 3-D object model, in the integration phase, it is necessary to know the precise motion between a new range image and the already generated partial model. We use a modified version of the *ICP algorithm* ([1], [3], [12]) here to estimate the motion. The registration algorithm is an iterative process minimizing errors resulting from pseudo data correspondences.

In the registration process, we define the *control points* as smooth surface points locating at the visible regions from the new viewpoint. However, they should be further selected randomly to produce a control point set whose size is below a certain number.

Partial model :   As new images are integrated, the partial model grows in its surface coverage by incrementally gathering new triangular patches. Here we call all patches in the partial model as *scanned patches*, which are further classified into *smooth patches, occluding patches, unreliable patches,* and *regular patches.* A smooth patch is one containing at least one control point as its vertices, while an occluding patch one containing at least one boundary edge of the partial model. A unreliable patch is determined by the angles between its normal and all of the previously selected NBVs. If the angles are all larger than a specified threshold, the large obliqueness of the patches makes the reliability of data uncertain. Patches not belonging to the above three types are regular patches.

Moreover, as in [8], we also define rectangular *void patches* which are attached to the edges of the occluding patches as virtual surfaces of the occluded parts of the object. The normal n_{pv} of a void patch is determined by

$$\cos^{-1}(n_{pv} \cdot (-r_{pv})) = \Theta_v, \qquad (1)$$

where r_{pv} is the direction of the occluding ray associated with the neighboring occluding patch. Θ_v is the breakdown angle of the used range finder, and in our case $\Theta_v = 60°$.

3 Viewpoint planning algorithm

In the section, we present an new algorithm for solving the NBV problem by means of analytic or heuristic evaluations on suitability of viewpoints as the NBV. The evaluations are based on computation of factors such as extending, resampling, and smoothness constraints on the data registration and integration.

Our algorithm is most similar in some aspects with the methods proposed in [4] and [8]. In [4], however, the rating function for the viewpoint selection only considered constraints on configurations of a robot for modeling an indoor space. In the method in [8], although the possibility of data merging and overlapping is both taken into accounts, the problem is largely simplified there because the author used a range scanner movable only in a 2-D circle trajectory.

At first we define several fundamental visibility functions in the viewpoint space before detailed algorithms are explained.

3.1 Viewpoint space

Given an object to be modeled, we define the viewpoint space as a set of 3-D locations where the used sensor can be set to take range images. In general, the viewpoint space is formed by regular discrete points on the surface of a sphere with the object at its center and a radius large enough relative to the object. In our approach, owing to the use of the robot arm, such a 3-D placement can be realized except for a small number of singular configurations. The sphere surface is discreted here by a widely used *triangulated geodesic tessellation* ([7]),

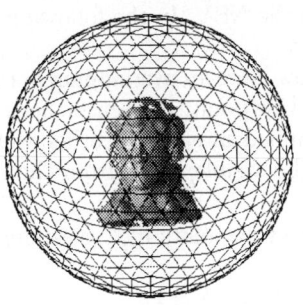
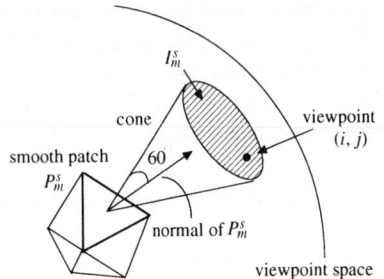

(a) viewpoint space (b) a smooth patch and its visible region in viewpoint space

Fig. 2. Viewpoint space.

as shown in Fig.2(a). Since the sensor is always posed facing straight at the object at the center, the viewpoint space can be represented as a 2-D space (i,j) where i and j correspond to semiregular latitudinal and longitudinal dimensions.

To evaluate the suitability of each viewpoint as the NBV, we have to define some measures for it with the viewpoint space as the domain. The factors to be evaluated include the visibility, voidness, reliability, smoothness of patches in the partial model. In the following, as an example, we give the definition of the visibility function of smooth patches for a viewpoint.

Given a smooth patch P_m^s, we say that it is visible from a viewpoint (i,j) if the angle of its normal and the viewpoint direction of (i,j) is smaller than Θ_v given in eq.(1). The viewpoints from which P_m^s is visible form a circular region $I_m^s(i,j)$ as illustrated in Fig.2(b). Then, we define a binary function

$$A_m^s(i,j) = \begin{cases} 1, & (i,j) \in I_m^s \\ 0, & otherwise \end{cases}. \quad (2)$$

For all smooth patches in the current partial model, the visibility of them as a whole can be computed by

$$N_s(i,j) = \sum_m A_m^s(i,j). \quad (3)$$

As is clear from the definition, $N_s(i,j)$ represents a "voting" process of all smooth patches on the superiority of viewpoints, and $N_s(i,j)$ is simply the votes the viewpoint (i,j) collects. By the same way, we can define the visibility functions $N_o(i,j)$, $N_u(i,j)$ and $N_v(i,j)$ for the occluding patches, unreliable patches, and void patches, respectively.

It is noticed that although all of the visibility functions have to be re-evaluated once a new image is merged into the partial model, the complexity of the computation can be kept small by an iterative algorithm ([7]). At each iteration, only the votes from the patches in the newly merged image need to be modified.

3.2 Potential candidates of NBV

Our NBV planning algorithm is composed of the following major steps: *(i)*select potential candidates of the NBV; *(ii)*evaluate the suitability of them using a rating function and find the most suitable one as the NBV. In the following, we explain briefly how to select viewpoint candidates.

The accuracy of image registration relies on how large an overlapping portion exists between the two data sets ([1]). At a new viewpoint, the sensor needs to resample a part of the object that has been scanned already. In other words, some triangulated patches of the object should be visible again from the new viewpoint.

In our method, we select as the candidates the viewpoints $v_{ij}^{(k)}$ satisfying

$$0° \leq \cos^{-1}(v_{ij}^{(k)} \cdot v_{NBV}^{(l)}) \leq \Theta_c (l = 0, \cdots, k-1) \quad (4)$$

where k denotes the number of iterations of pose planning, $v_{NBV}^{(l)}$ the l-th previous NBVs, and $\Theta_c = 120°$. The decision is based on a rough assumption that an overlapping portion exists between the partial model and the image taken at a viewpoint not far away from the previously selected NBVs by an angel 120°. This is enough since the candidate election process is just a rough guess on the NBV for computational saving.

To reduce the candidates furthermore, we also exclude those viewpoints in a neighborhood of previous NBVs $v_{NBV}^{(l)}$.

3.3 Rating function for NBV evaluation

We evaluate the suitability of the potential NBV candidates by using a rating function

$$f^{(k)}(i,j) = w_e^{(k)} f_e^{(k)}(i,j) + w_s^{(k)} f_s^{(k)}(i,j) \tag{5}$$

where $f_e^{(k)}, f_s^{(k)}$ are factor functions at the kth iteration of planning, and $w_e^{(k)}, w_s^{(k)}$ weighting coefficients. The factor functions rate the candidates in terms of some physical or heuristic constraints. The viewpoint of the largest value of $f^{(k)}(i,j)$ will be chosen as the NBV. The definitions of $f_e^{(k)}, f_s^{(k)}$ and $w_e^{(k)}, w_s^{(k)}$ are given in the following.

$f_e(i,j)$: **Extending and resampling constraints** The first and primary requirement for NBVs is that the sensor set at the NBV should sample into the partial model a large and new portion of the object surface in order that the partial model extends as fast as possible. The new portions includes two kinds of patches when a new image is taken and integrated. The first is the void portions representing surfaces unscanned. This is approximated in our method by the void patches in the partial model. The second is the portions that need to be re-sampled to refine the acquired data. This kind of regions are mainly composed of occluding patches and unreliable patches. Then, we define $f_e(i,j)$ as

$$f_e(i,j) = c_v N_v(i,j) + c_{ou}(N_o(i,j) + N_u(i,j)) \tag{6}$$

where c_u, c_{ou} are coefficients accounting for the confidence in estimating $N_u(i,j)$, $N_o(i,j), N_u(i,j)$. The coefficients are determined empirically in our experiments. It is noted that, to simplify expressions, the superscripts (k) are omitted here and in the following equations.

$f_s(i,j)$: **Smoothness constraints** As stated in Section 2, at the level of point matching, the registration process is based on correspondence between control points. The control points are selected from smooth regions in the overlapping area since they are more stable than those at sharply changing regions. To make sure the constraint is satisfied, a certain number of control points should appear in the new image, and thus, an image having a larger smooth area in the overlapping portions is desirable. Consequently, $f_s(i,j)$ is computed directly by the visibility function for the smooth patches as

$$f_s(i,j) = N_s(i,j). \tag{7}$$

w_e, w_s : **Weighting coefficients** The principal strategy for controlling the weight coefficients is as follows. If the new image is registered and integrated in a good accuracy, we select the current NBV to sample a large new portions into the partial model. Otherwise, we select a near viewpoint to increase the opportunity of scanning the partial model once again. That is, we should choose a large w_e in the former case while a large w_s in the latter.

In our implementation, the cases are detected by the error between the partial model and the image integrated to the model just now. Then, we define

$$w_e = \frac{\alpha}{E_r}; \quad w_s = 1 - \frac{\alpha}{E_r} \qquad (8)$$

where E_r is the final error of the registration algorithm used in the preceding image registration, and α a constant to be empirically determined.

4 Results of experiments

We have made a experiment to demonstrate the applicability of our method for real images. The experiments are performed in the environment shown in Fig.1, The range errors of the range finder are about $1.5[mm]$ in the depth of $2[m]$.

The experiment reported here is carried out to show the performances of the NBV planning algorithm. We use a mini-statue shown in Fig.3(a) as the object to be modeled. At first we acquired the initial range image of the statue at the initial viewpoint $v^{(0)}$ as indicated in Fig.3(b). Fig.4(a) is the picture illustrating the pose of the object in the arm's end effector. Fig.4(b) and (c) show, respectively, the range image and the triangulated mesh representation for the image. As shown here, there are several holes on the surface due to occlusions and errors in data acquisition.

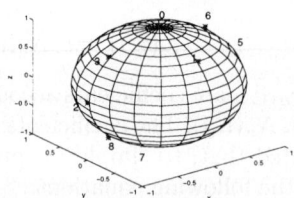

(a) the object : a mini-statue (b) NBVs at the viewpoint sphere

Fig. 3. The object to be modeled and selected NBVs

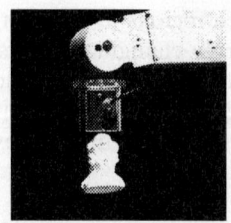

(a) initial (0th) pose   (b) image for the pose in (a)   (c) triangulated meshes for (a)

Fig. 4. Range images acquired from the initial viewpoint.

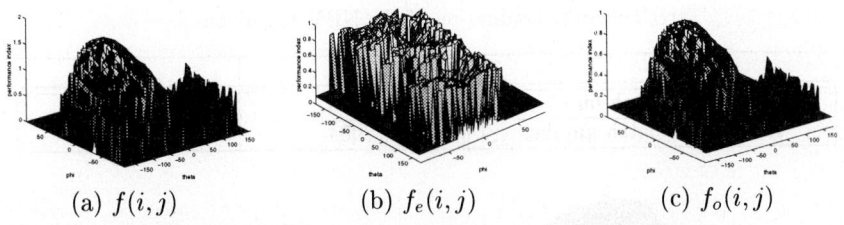

Fig. 5. Rating function computed for the 1st NBV.

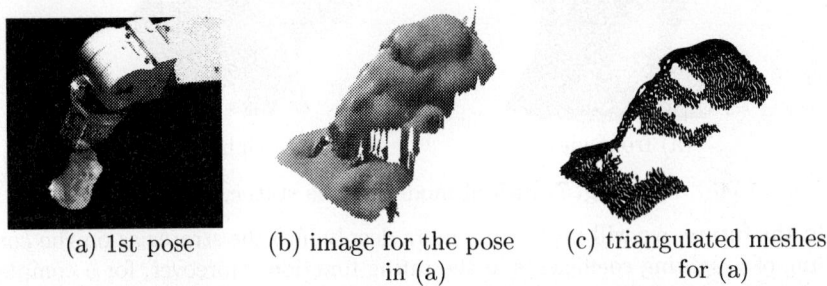

Fig. 6. Range images acquired from the first NBV.

Using the image, we start our viewpoint planning algorithm in order to decide the 1st NBV. The graphs of the rating function and each factor in it at $v^{(0)}$ are shown in Fig.5. They are represented in the world coordinates used in creating the viewpoint space. Moreover, the weighting coefficients in eq.(5) and (8) in the rating function are chosen as $c_v = 10.0$, $c_{ou} = 1.0$, $\alpha = 0.5$.

In the graph of $f(i,j)$, the maximum is located at the viewpoint whose direction vector is $v_{MAX} = (-0.0616, 0.7071, 0.7044)^T$ in the world coordinate system, which is selected as the $v_{NBV}^{(1)}$ as shown in Fig.3(b). The computational time for determining it on a microSPARCII (70 [MHz]) was 475 CPU seconds. In practical implementation, the viewpoint is transformed back into object pose then achieved by the arm manipulation (Fig.6(a)). The 1st image and the triangulated meshes are shown in Fig.6(b) and (c). From the two images, we create a partial model of the object, which is illustrated in Fig.7.

Here we give a little more detailed investigation on the rationality of our NBV algorithm. Table.1 lists, respectively, the numbers of void patches N_v and smooth patches N_s which are both computed as patch number estimations in the planning algorithm or counted in the real image. While N_s in the both cases are nearly the same, N_v in the real image is about double that from the computation. This may result from the fact that a too small number of void patches are used here in the void volume approximation.

5 Conclusions

In this paper, we presented a new approach of generating 3-D models with emphasis on planning on the NBVs. The proposed algorithm is based on analytic or heuristic evaluations of the suitability of candidate viewpoints as the NBV.

Results of our experiment show it is applicable and practical in real situations.

Table 1. Evaluation of the NBV Algorithm

	N_v	N_s
Estimated patch number	3360	1733
Real image patch number	7084	1844

(a) front view (b) right view

Fig. 7. Partial model for the statue.

In the future, we will investigate more deeply into the strategies on the controlling of weighting coefficients in the rating function. Moreover, for a complex object as those used in our experiments, we have also to solve problems concerned with planning on arm configurations as well as changes of grasping locations.

References

1. P. Besl, N. Mckay, "A Method for Registration of 3-D Shapes", *IEEE Trans. PAMI*, vol.14, no.2, pp.239–256, 1992.
2. C.I. Connolly, "The Determination of Next Best Views", *Proc. ICRA*, pp.432–435, 1985.
3. Y. Chen, G. Medioni, "Object Modeling by Registration of Multiple Range Images", *Image and Vision Computing*, vol.10, no.3, pp.145–155, 1992.
4. E. Kruse, R. Gutsche, F.M. Wahl, "Efficient, Iterative Sensor Based 3-D Map Building Using Rating Functions in Configuration Space", *Proc. ICRA*, pp.1067–1072, 1996.
5. A. Hilton, *On Reliable Surface Reconstruction from Multiple Range Images*, Technical Report VSSP-TR-5/95, Dept. EEE, Univ. of Surrey, 1995.
6. K. Higuchi, M. Hebert, K. Ikeuchi, "Building 3-D Models from Unregistered Range Images", *Proc. ICRA*, pp.2248–2253, 1994.
7. K. Morooka, *Automatic Modeling of 3-D Curved Objects on the Basis of Viewpoint Planning*, M.S. Thesis, Dept. CSCE, Kyushu University (in Japanese), 1997
8. R. Pito, "A Sensor-Based Solution to the 'Next Best View' Problem", *Proc. ICPR*, pp.941–945, 1996.
9. J.R.Stenstrom, C.I.Connolly, "Constructing Object Models from Multiple Images ", *Int. Journal Computer Vision*, vol.9, no.3, pp.185–212, 1992.
10. M. Soucy, D. Laurendeau, "A General Surface Approach to the Integration of a Set of Range Views", *IEEE Trans. PAMI*, vol.17, no.4, pp.344–359, 1995.
11. H. Zha, K. Morooka, T. Hasegawa, T. Nagata, "Active Modeling of 3-D Objects: Planning on the Next Best Pose(NBP) for Acquiring Range Images", *Proc. Int. Conf. Recent Advances in 3-D Digital Imaging and Modeling*, pp.68–75, 1997.
12. Z. Zhang, "Iterative Point Matching for Registration of Free-Form Curves and Surfaces", *Int. Journal Computer Vision*, vol.13, no.2, pp.119–152, 1994.

Object Recognition by Matching Symbolic Edge Graphs

Tino Lourens[1] and Rolf P. Würtz[2]

[1] Computing Science, University of Groningen, The Netherlands,
http://www.cs.rug.nl/~tino/tino.html
[2] Institute for Neurocomputing, Ruhr-University Bochum, Germany,
http://www.neuroinformatik.ruhr-uni-bochum.de/ini/PEOPLE/rolf/

Abstract. We present an object recognition system based on symbolic graphs with object corners as vertices and outlines as edges. Corners are determined in a robust way by a multiscale combination of an operator modeling cortical end-stopped cells. Graphs are constructed by line-following between corners. Model matching is then done by finding subgraph isomorphisms in the image graph. The complexity is reduced by adding labels to corners and edges. The choice of labels makes the recognition system invariant under translation, rotation, and scaling.

1 Introduction

Labeled graph matching is a method used successfully for, e.g., invariant face recognition [1]. In that system, vertices are assigned local texture elements and edges carry the geometrical information about the relative locations. Applicability of that method is limited to richly structured or textured objects like faces. Objects with homogeneous surfaces do not provide the sort of vertex labels required there and can only be matched using their outlines. In this paper we present a somehow complementary matching scheme that matches corners and connecting edges with the idea that a combination of both approaches will yield a fairly general recognition method.

Images as well as stored models are represented as graphs whose vertices correspond to object corners and whose edges code for edges connecting corners in the image. The method is invariant under translation, rotation, and scaling and robust under changes in background, limited changes in perspective, and small distortions, but can currently not handle significant occlusion.

For model matching it is assumed that corners can only be matched onto corners and connecting edges must match edges in the image. This makes the process equivalent to finding subgraph isomorphisms, which is NP-complete in the number of vertices. In our approach, the complexity is drastically reduced by demanding that the labels of matched vertices and edges, respectively, must be roughly equal. We will show experimentally that the problem is so reduced to a tractable size.

The corner extraction method, which is described here only briefly, is based on a model for end-stopped cells in the visual cortex and thus models some aspect of human vision. Recently, models start to evolve that present neuronal algorithms for the graph extraction method [2]. A neuronal model for graph matching was proposed in [1]. Although the evidence is still sketchy, it can be expected that the whole algorithm allows a neural implementation and thus potentially models aspects of human object recognition, for which corners and edges play an important role. In [3], e.g., it is shown that partial contour deletion only impedes object recognition if it is accompanied by altering corner attributes.

2 Symbolic edge graphs

We will describe objects and scenes as graphs with corners as vertices and outlines as edges. The most important prerequisite for such a method is a robust corner detector.

2.1 Robust corner detection

Our method for detecting corners yields position, sharpness, size and color and grey-scale distribution (contrast). The subtended angle can be determined a posteriori by following the line segments that constitute the corner (see section 2.2). It is based on a model of cortical end-stopped cells [4]. These model cells are a nonlinear combination of the amplitudes of well-known Gabor functions [1] applied to an image. We denote those Gabor responses by $\mathcal{C}_{\sigma\theta}$ at scale σ and orientation θ.

Sharp corners are characterized by strong responses over a wide frequency range. If only high frequency cells respond, the feature is likely to be noise or texture rather than a corner. We found that averaging the responses over a range of frequencies yields a much more robust corner detection. The whole algorithm is described in [5]. Here, it suffices to mention that that algorithm yields the corner positions with sufficient accuracy and reliability. With a slight and biologically justified extension of the concept of complex cells, line and corner detection can be extended to color channels, which is also described in detail in [5].

2.2 Line following

A simple method to get the outlines of objects are binarized responses of appropriate edge detectors. Those, however, have turned out to be very sensitive to changes in local contrast and size of thresholds. We are using a method that starts from corners and collects evidence for a line to connect this corner with another one. Thus, the resulting edge graphs can be called "symbolic". The edge detector consists of the Gabor moduli $\mathcal{C}_{\sigma\theta}$. The complete graph extraction algorithm is shown in figure 1.

Orientation selection A corner (or line-end) contains implicit information that one or more lines start from this position. Therefore we start searching for

```
1    Procedure ExtractGraph (C, I, G)
2        C' := Corner cluster elimination C
3        forall c ∈ C'
4            O := all orientations at distance d from c where a segment is found
5            forall o ∈ O
6                repeat
7                    follow segment with intial orientation o
8                until stop criterion fulfilled
9                if stop criterion is another corner
10                   store detected segment
11       optimize detected segments
12       G := represent detected segments as a 2-D planar graph
```

Fig. 1. Algorithm for graph extraction.

possible lines using a circle with the corner at the center and a certain radius d. On this circle a number of equidistant samples are taken and the response of the $\mathcal{C}_\sigma$-operator, which is the average of $\mathcal{C}_{\sigma\theta}$ for all θ, is determined. In order to assure that the algorithm indeed follows a line, the operator is also averaged over all scales, yielding the operator $\mathcal{A}$. A line segment in orientation ϕ is selected if the following conditions are satisfied:

1. The response of the $\mathcal{A}$-operator is above a threshold T.
2. The response is 20% higher than the weakest response from the samples of the $\mathcal{A}$-operator on the circle.
3. It is a local maximum among the samples taken.
4. The angle θ_0 where the response of $\mathcal{C}_{\sigma\theta}$ is maximal lies closer to ϕ than the sampling stepsize of the orientations. This angle is called the *principal orientation* $\mathcal{O}_\sigma$ at the current image location.

Following a line The subsequent steps of line following are simpler than the starting one, because the line must roughly continue in the same direction. If a line diverts too much from that heuristic the corner detector must find a corner there. Thus, the circles around the corner are replaced by arcs of twice the sampling stepsize around the current line orientation.

Stop criteria At every step the following five stop criteria are checked, and following the current line terminates if one of them is satisfied.

1. Another corner is found.
2. One of the samples falls outside the image.
3. There is no sample where the output of the $\mathcal{A}$-operator is above the threshold T and the principal image orientation falls inside the possible range.
4. The orientation at coordinate (x_i, y_i) in the line does not correspond with the preferred orientation $\mathcal{O}_\sigma$.

5. The response of the $\mathcal{A}$-operator differs too much from the response at the previous step.
6. The length of the line exceeds twice the image size.

Line optimization The line following has a certain step size which may not be chosen too large. Consequently, straight lines or lines of small curvature get represented by too many intermediate points.

During the process of line following a point $(x_i, y_i),) \leq i \leq n$, is calculated at every step. (x_0, y_0) is the position of the corner where the line starts from and (x_n, y_n) is the end-point of the line, which is usually another corner. If a straight line can be drawn between the two corners, all coordinates $(x_1, y_1), \ldots, (x_{n-1}, y_{n-1})$ can be dropped. In general, if we have three coordinates $(x_{i-1}, y_{i-1}), (x_i, y_i)$, and (x_{i+1}, y_{i+1}) then (x_i, y_i) can be eliminated if the angle at (x_i, y_i) is smaller than δ'. In the simulations we used $\delta' = 5°$. When the point with index i is eliminated the same procedure is repeated for points with indices $i-1, i+1$, and $i+2$. If coordinate (x_i, y_i) is not removed, the procedure is repeated for coordinates with indices $i, i+1$, and $i+2$.

We apply the line following algorithm separately at every scale. This minimizes the chance for missing a line, but now, a line may be detected more than once. When this is the case, or more precisely, if several lines connect the same pair of corners and the average distance between them becomes smaller than a threshold, all but the shortest one are eliminated.

2.3 Graph labels

After all these steps we end up with a graph that has corners as vertices and curve segments described by a series of points as its edges. Once stored models and the image to be analyzed are represented in this way, model matching can be done by finding a copy of the model graph in the image graph.

Each corner is labeled with the angles between all pairs of adjacent line segments starting from it. The edges are labeled with the relative length of the line segments (i.e. the ratio of the length to the length of the longest line segment in the whole graph). This choice of labels automatically yields invariance under translation, rotation and changes in size.

3 Graph matching

A brute force approach to test if two graphs are isomorphic is to exhaustively test every one-to-one mapping between the vertices of the graphs, which implies that every possible permutation is tested. We are using a modified version of the algorithm for subgraph isomorphism is from Ullman [6] based on tree search with backtracking. To cut down evaluation expenses the above mentioned labels are assigned to vertices and edges. To cut down evaluation expenses in graph matching often only the best matching copy is searched. This is not acceptable

1	**Procedure** MatchGraph (G, MG)
2	stack $:= \emptyset$
3	**forall** $v \in V(G)$
4	$L_0 := v$ /* list of parsed vertices of image graph */
5	$ML_0 :=$ first model vertex /* list of parsed vertices of model graph */
6	$mv := 1$ /*number of matched vertices*/
7	$cv :=$ first model vertex /* vertex being evaluated */
8	$cva :=$ first angle of cv /*angle of cv to be evaluated */
9	$ED := 0.0$ /* accumulated edge difference */
10	$AD := 0.0$ /* accumulated angle difference */
11	Push $(mv, cv, cva, L, ML, ED, AD)$
12	**while** stack $\neq \emptyset$
13	Pop $(mv, cv, cva, L, ML, ED, AD)$
14	**if** $mv = cv = \#V(MG)$ /* Match found */
15	Evaluate maximum and average relative length differences
16	**else**
17	**if** Angle cva of vertex cv can be evaluated
18	**if** Evaluation accepts angle and ratio
19	**if** $cva =$ last angle of cv
20	$cv :=$ next (cv)
21	$cva :=$ first angle of cv
22	**else**
23	$cva :=$ next (cva)
24	Push $(mv, cv, cva, L, ML, ED, AD)$
25	**else** /* Add a missing vertex */
26	$ML_{mv} :=$ missing-vertex
27	**forall** $v \in V(G) - L$ /* Unused vertices only */
28	$L_{mv} := v$
29	**if** ProperVertex $(EC, \max ED - ED)$
30	Push $(mv + 1, cv, cva, L, ML, ED + EC, AD)$

Fig. 2. Algorithm for graph matching.

here, because the same model may appear several times in the image graph. Consequently, we are interested in all "copies" of a model graph G_m in the given image graph G.

We assume that the model graphs are either constructed by hand or extracted from "clean" images. Thus, they contain *all* edges, wheras in the image graph some may be missing. Our matching process allows for this by allowing every non existing edge between two different vertices to be added at a cost of 1, the total cost being limited by a parameter, which has been set to 2 in the experiments. Thus, our graph matching model finds both exact and inexact copies of the models in the image graph.

The algorithm for graph matching is illustrated in figure 2. It can find all copies of a model graph G_m in an image graph G. Lines 2–11 are the initial stage of the algorithm, we start with an empty stack and push all vertices of the image graph on the stack one after another, since each of them can, in principle, be matched with the first vertex of the model graph. Lines 12–30 constitute the matching proper. In line 13 we take a possible partial solution from the top of the stack and check if we have a complete match (line 14). If the match is complete, the maximum and average relative length differences to the model are calculated, if both are below a certain threshold the match is accepted and displayed. If we do not have a complete match we go to line 17. Here we check if the current angle and ratio can already be evaluated. If we can not evaluate because one or both vertices to form the angle are still missing, then we parse the missing model vertex by adding it to the list (line 26) and find all possible vertices in the image graph (lines 27–30). A vertex v is added if it is not matched yet and if the cost EC of adding edge (L_{cv}, v) is smaller than the allowed cost.

The speed of the algorithm depends mainly on the condition of line 18. If angle and ratio differences are chosen properly most of the partial matches will be rejected here, and the path is rejected by not pushing it back on top of the stack. During matching, the difference in ratio δr between two pairs of edges is obtained by scaling the edge pairs in such a way that the first edge of both pairs is one, then the ratio is the size of the rescaled second edge of the first pair: the size of the rescaled second edge of the second pair. The average length difference is evaluated by using the relative lengths of both model and found match in the image graph, as described earlier in this chapter. The absolute difference of the model edge with its corresponding image edge is taken, when there was no edge between v_b and v_b in the image graph we used the relative length of $dist(v_a, v_b)$. The average of all edges is taken to represent the average relative length difference.

4 Results

Figure 3 shows (what used to be) a color image and the extracted image graph. We have used the model graph illustrated in figure 3c) to find all the markers in the image.

The result is illustrated in figure 3d). We allowed at most two out of the seven edges in the model graph to be added and a maximal δr of 5. We tolerated an angle difference of at most 36° and also an average angle difference of at most 36°, since there are 10 angles in the model graph this means that the summed angular difference should not exceed 360 degrees. When a match is found we tolerate a maximum relative length difference of 50% and an average relative length difference of 10%. The matching time for the image graph is are approximately 1 second.

5 Discussion

We have presented a graph matching scheme for object recognition based on corners and outlines of objects. We have used a robust and biologically motivated

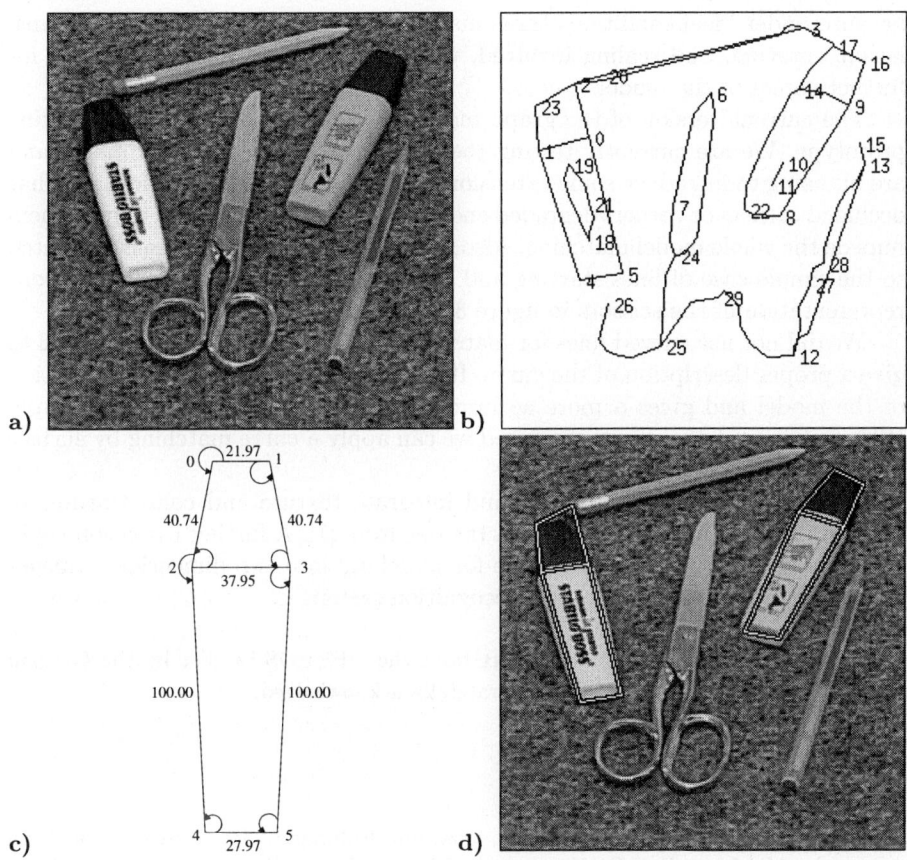

Fig. 3. a) Input image. b) Image graph extracted from a), with numbered vertices. c) A model graph with implicit attributes, such a the (relative) length of an edge and the angle between two edges. d) Found matches of the markers. The used parameters are: maximum five edges to be added, ratio between two edge pairs $\delta r = 5$, average angle tolerance of 10%, maximum angle tolerance of 10%, maximum length tolerance of 50%, and an average length tolerance of 10%.

operator to detect the corners. A relatively sophisticated algorithm has been used to follow lines between corners. This makes these graphs true symbolic information.

The NP-complete problem of subgraph matching has been greatly simplified by assigning angles between edges and relative sizes of edges as labels to the corners. Although we currently can not make formal statements about the resulting complexity we have shown that the time requirements can be cut down to reasonable amounts for realistic problem sizes. The choice of labels makes the matching invariant under translations, rotations and scaling. When two isomorphic but different models are found, the similarity of the labels is used as as selection criterion. We conjecture that the complexity of the algorithm is $O(N^4)$, because under ideal conditions three matching point pairs determine the translation, rotation, and scaling involved, while the fourth is needed to check for further copies of the model.

The current version of the graph matching system is only the first working prototype. We are currently testing the robustness on many more images and are planning to introduce some extensions. The most serious limitations are that occluded corners or corners degraded enough for the corner detector to miss them impede the whole model matching. Also, the line following algorithm is restricted to the simple case of lines starting and ending at corners. This causes the poor representation of the scissors in figure 3b).

We did not use curved lines for matching but help vertices which are used to give a proper description of the curve. It can be used to get a better description of the model and gives a more accurate cost of a match in the image graph. When additional coordinates are used we can apply a curve matching by surface difference algorithm.

Further developments will try and integrate texture and color features in order to combine this approach with the one from [1]. A further extension could use three-dimensional model graphs for matching into two-dimensional images and could yield a truly 3D object recognition system.

Acknowledgments. Financial support from the NEUROS-Project by the German Research Minister (01 IN 504 E 9) is gratefully acknowledged.

References

1. Martin Lades, Jan C. Vorbrüggen, Joachim Buhmann, Jörg Lange, Christoph von der Malsburg, Rolf P. Würtz, and Wolfgang Konen. Distortion invariant object recognition in the dynamic link architecture. *IEEE Transactions on Computers*, 42(3):300–311, 1993.
2. Lothar Weitzel, Klaus Kopecz, Claus Spengler, Reinhard Eckhorn, and H.J. Reitboeck. Contour segmentation with recurrent neural networks of pulse-coding neurons. In G.Sommer and J.J. Koenderink, editors, *Proceedings of the 7'th International Conference on Computer Analysis of Images and Patterns,Kiel, Germany, September 10-12, 1997*, 1997.
3. Irving Biedermann. Recognition-by-components: A theory of human image understanding. *Psychological Review*, 94(2):115–147, 1987.
4. Friedrich Heitger, Lukas Rosenthaler, Rüdiger von der Heydt, Esther Peterhans, and Olaf Kübler. Simulation of neural contour mechanisms: from simple to end-stopped cells. *Vision Research*, 32(5):963–981, 1992.
5. Rolf P. Würtz and Tino Lourens. Corner detection in color images by multiscale combination of end-stopped cortical cells. In *Proceedings of the International Conference on Artificial Neural Networks, Lausanne, Switzerland, October 1997*, Lecture Notes in Computer Science, pages 901–906, Berlin Heidelberg New York, 1997. Springer Verlag.
6. J. R. Ullman. An algorithm for subgraph isomorphism. *Journal of the Association for Computing Machinery*, 23(1):31–42, 1976.

Interpretation of Complex Scenes Using Bayesian Networks*

Mark F. Westling[1] and Larry S. Davis[2]

[1] Perceptus Technologies, 7500 Woodmont Ave. Suite 1007,
Bethesda MD 20814, USA
[2] Computer Vision Laboratory, Center for Automation Research,
University of Maryland, College Park MD 20742, USA

Abstract. In most object recognition systems, interactions between objects in a scene are ignored and the best interpretation is considered to be the set of hypothesized objects that matches the greatest number of image features. Visual and physical interactions, however, provide a rich source of information: occlusion explains why features might be undetected, and physical constraints ensure a realisable interpretation. We show how these interations can be easily modeled using a Bayesian network, and how the problem of interpretation can be cast as finding the most likely explanation for such a network.

1 Introduction

The recognition of multiple objects in complex scenes is usually performed without consideration of object interactions. When an image contains multiple objects, each object is hypothesized independently and the best interpretation of the scene is considered to be the set of hypothesized objects that matches the largest number of image features. Interactions, however provide a rich source of information in understanding a scene. The recognition of visual interactions, i.e., occlusion, can account for missing features and add evidence to hypotheses. The verification of physical interactions, such as spatial interference between objects (i.e., overlap), can prevent physically unrealisable interpretations.

These kinds of interactions between objects and features can be represented and reasoned about using a Bayesian network. Bayesian networks provide a natural method for modeling the relationships between hypothesized objects when the goal is to determine the best set of hypotheses for interpreting an image. A further advantage is that background information, such as the prior probabilities of the presence of certain types of objects, can be easily incorporated.

We present a method for automatically generating Bayesian networks for modeling interactions between 3-D objects in complex scenes, and show how computing the most likely composite hypothesis is equivalent to finding the best

---

* The partial support of the Defense Research Projects Agency (ARPA Order No. C635) and the Office of Naval Research (grant N000149510521) is gratefully acknowledged.

interpretation. To address the problem of how to determine the actual probabilities required by a network, we show that to compute the best interpretation of a scene, only relative probabilities are important. We show that with certain restrictions, computing the joint probability of a set of hypotheses is equivalent to a feature counting strategy. An investigation into these restrictions reveals weaknesses in counting strategies and shows how a Bayesian network can be effective with only a few assumptions.

2 Related Work

Bayesian networks, in combination with geometric reasoning systems, have been used in 3-D object understanding to relate model components to predicted appearances and to control the process ([9], [3], [8], [2], [13], [12], [14]).

Verification of hypotheses by counting features can be viewed as collecting positive evidence. This notion is extended by [6] and [7] to include negative evidence, e.g., for unlikely conditions such as image edges that pass through model edges at large angles. Similarly, [16] derives a statistical match function that penalizes unmatched image features while [1] includes a penalty for unmatched model features. Occlusions are specifically considered by [15] as an explanation for unmatched model features.

3 Problem Definition

We assume grayscale or color images containing multiple rigid, opaque objects. These can either be orthographic projections of 2-D objects (e.g., laminar parts) or perspective projections of 3-D objects restricted to three (or fewer) degrees of freedom. (This includes, for example, objects translated across a ground plane and rotated about a vertical axis.) Our goal is to find the best interpretation of the scene, where we define an interpretation as a collection of hypothesized objects (hereafter referred to simply as hypotheses), each consisting of a model and a pose that locates and orients the model in the scene. We assume that we have some method of generating a set of hypotheses from an image and a collection of models; in the implementation, we use a memory-based technique described in [17].

4 Network Design

A Bayesian network is a directed acyclic graph (DAG) in which nodes represent random variables and arcs represent influence, quantified by conditional probabilities. Nodes can be organized into two general groups: evidence nodes, which can be "fixed" to an observed values, and hidden nodes, whose values are inferred from evidence. We represent hypotheses as true/false hidden nodes, where the probability of true represents confidence or belief in the hypothesis. Associated with each hypothesis node is a prior probability, which for convenience may be set to 0.5 to represent no prior knowledge. Image features are represented as evidence nodes. Each hypothesis predicts which features, if any, should be detected at various parts of the image. In the case of occluding or overlapping hypotheses, there may be several different predictions at a single location. We assign two possible values to a feature node: one representing the feature that was actually detected, and the other representing the absence of that feature (or, the presence of a different feature at the same location). Each feature node is then observed to be the value representing what was detected.

In the general case, a feature may be influenced by more than one parent hypothesis. Figure 1 illustrates an example where two different hypotheses influence the same feature. There are two cases here: either the hypotheses predict the same feature, or they predict different features at this location. In the first case, the hypotheses are competing, and if we attribute feature f_2 to, say, hypothesis h_1, then f_2 should give no evidential credit to hypothesis h_2. The second case is more complicated. Suppose that h_1 partially occludes h_2 at the region that contains f_2. If h_1 is present, then f_2 should appear as it is predicted by h_1, not as it is predicted by h_2. Both hypotheses are mutually compatible, however, and in this situation the presence of h_1 actually supports the presence of h_2, by explaining why f_2 was not matched by h_2. The assumption here is that is a predicted feature is more likely to be missing if it is occluded than if there is no explanation for why it was not detected.

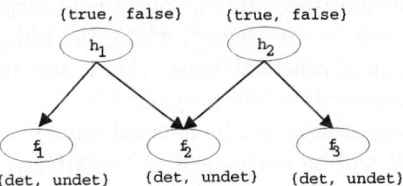

Fig. 1. Bayesian network representing visual interactions between two hypotheses.

The links between hypotheses and features are quantified using conditional probabilities. For each feature, we formulate a conditional probability that relates the presence (or absence) of the feature to the presence (or absence) of its parents. For example, in Figure 1, we provide $P(f_1|h_1)$, $P(f_2|h_1, h_2)$, and $P(f_2|h_2)$, along with the priors $P(h_1)$ and $P(h_2)$.

4.1 Relationship to Feature Counting.

This formulation implies independence between the features predicted by a hypothesis. This is also implied by feature counting, as noted by Breuel ([4]). Breuel formulates the interpretation of a single-object scene as a statistical decision problem: the best object is the one that maximizes the probability of all features, conditioned on that object.

This formulation can be extended easily to multiple objects by conditioning the set of features on a set of hypotheses. We are interested in in finding an interpretation $I = \{h_i\}$ that maximizes:

$$P(I) = \prod_i P(h_i|f_j^i) \qquad (1)$$

where f_j^i is the set of features influenced by h_i. Using Bayes' theorem and converting the notation to odds and likelihood, we can rewrite these terms as:

$$O(h_i|f_j) = O(h_i) \prod_j L(f_j|h_i), \qquad (2)$$

and by taking the logarithms of both sides, obtain:

$$\log O(I) = \sum_i \log O(h_i) + \sum_i \sum_j \log L(f_j|h_i). \qquad (3)$$

The log-odds of an interpretation is the sum of the logs of the prior odds of each hypothesis plus the sum of the logs of the likelihood ratios of the features given their parent hypotheses. Thus, if the probability of each hypothesis is 0.5, i.e., neutral, then the sum of the log-odds of the priors disappears, and if all features are considered equally likely, the joint probability of an interpretation is equivalent to counting features.

A feature counting system, like our formulation, assumes conditional independence of features. Since features are obviously not independent, our formulation is what is often termed "naive Bayes", where the independence assumption is knowingly violated. In a practical sense, this is not too bad: although errors introduced by this assumption will propagate to the posterior probabilities for the hypotheses, all hypotheses will be affected more or less equally, and the calculation of the MPE will be unchanged. Alternatively, one could augment the formulation with higher-order terms that capture interactions ([4]).

This representation suggests two ideas not usually considered in feature counting systems. First, most systems typically assume equiprobable hypotheses; casting the problem as a Bayesian network, however, forces one to consider whether priors add more information to an interpretation. Second, the weight of each feature in a counting system clearly depends on the log-likelihood of the feature given its parent. To be meaningful, two conditions must hold:

$$\log L(f_i^j|h_i) > 0 \equiv P(f_i^j|h_i) > P(f_i^j|\overline{h}_j), \qquad (4)$$

which states that a detected feature is more likely to arise from a true hypothesis than from a false hypothesis, and

$$\log L(\overline{f_i^j}|h_i) \leq 0 \equiv P(\overline{f_i^j}|h_i) \leq P(\overline{f_i^j}|\overline{h}_i), \qquad (5)$$

which states that an undetected feature is more likely to arise from a false hypothesis than from a true hypothesis. Thus, detected features add evidence to hypotheses, and undetected features either add nothing or subtract evidence from hypotheses.

4.2 2-D Example

As an example, consider the situation in Figure 2(a). The object h_1 partially occludes the second object h_2, creating three groups of features, f_1, f_2, and f_3. The Bayesian network in Figure 2(b) can be used to model this image. As stated previously, this requires that we supply the conditional probabilities $P(f_1|h_1)$, $P(f_2|h_1, h_2)$, and $P(f_2|h_2)$, along with the priors $P(h_1)$ and $P(h_2)$. Specification of $P(f_1|h_1)$ (and $P(f_2|h_2)$) requires a 2 × 2 table containing probabilities for all combinations of f_1 and h_1 (and f_2 and h_2). Full specification of $P(f_2|h_1, h_2)$ would require eight values, but observe that due to occlusion, knowing that h_1 is true makes the value of h_2 irrelevant. Thus, we need only six values, the two from $P(f_2|h_1) = T$ and the four from $P(f_2|h_1 = F, h_2)$.

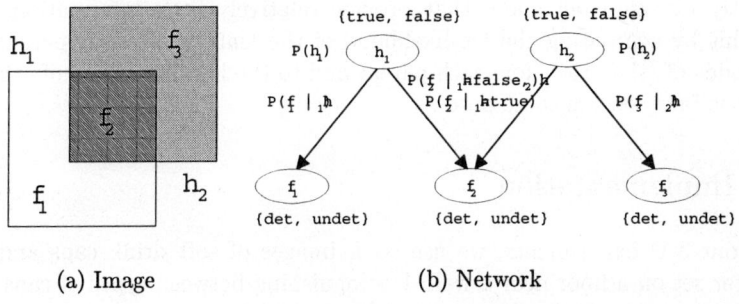

Fig. 2. Example image and network for simple occlusion.

4.3 Constraints Between Objects

The most common constraint between objects is interference, i.e., spatial overlap. To be physically realisable, no objects in an interpretation can overlap. This is complicated by uncertainty in object pose determination: true interference can be asserted only when poses are known exactly.

The simplest way of representing a set of mutually exclusive hypotheses is to use a single, multivalued node, with a distinct value representing each hypothesis and an extra value representing the absence of all of the given hypotheses. This representation works when interference can be determined precisely, but in some cases there may be uncertainty. Moderate interference might be explained by error in pose estimate, especially for abutting objects. In such situations, one may wish to represent mutual exclusion probabilistically: the presence of both objects may be unlikely but not impossible. This can be represented by a true-false dummy node whose value is false with strong confidence only when both objects are present. We then observe the node with value set to true. Then, any evidence supporting one of the hypotheses will drive down the belief in the other hypothesis.

4.4 Simplifying the Network

A network created as described in the previous section can be greatly simplified through a few steps that take advantage of the fact that all evidence is known when the network is created.

Step 1: Cluster features. Feature nodes that have the same detection characteristics (either all detected or all undetected) and that share the same parental relationships can be merged into single nodes.

Step 2: Move evidence into priors. Evidence from single-parent feature nodes can be moved into the parent hypothesis node. A feature node with a single parent does not capture any interaction and therefore can be "factored out" by combining it with the prior distribution of the parent, using a simple application of Bayes' Theorem.

Step 3: Discard minor interactions. The network can often be simplified greatly by removing nodes that provide relatively little information. We can do this by comparing the log-likelihood of the feature given its parents to the log-odds of (the priors for) each parent and to the log-likelihoods of other interactions between the same parents.

5 Implementation

For our 3-D experiments, we use color images of soft drink cans and various clutter set on a floor near a wall. Distinguishing between types of cans requires an appearance-based approach, since the only differences between can types are the complex graphics that appear on labels. Images features are small (10×10 pixel) rectangular regions containing one of sixteen different colors, extracted by reducing the resolution of the image and then mapping the color to a palette. This gives a rough approximation to the dominant color at each local neighborhood across the entire image. These features are mapped to lists of matching objects and poses using a look-up table that is computed off-line, and these lists are accumulated using a generalized Hough transform. Raw hypotheses are extracted from peaks in the accumulator array. The entire process is very fast due to the trade-off of space for time ([17]).

There are many methods and tools for representing and computing on Bayesian networks. We chose the symbolic probabilistic inference (SPI) architecture ([11]). SPI allows Bayyesian networks to be expressed using a local expression language ([5]) that captures only the relevant components of individual probability distributions. This ability to use partial distributions is ideal for expressing occlusion relationships, as described in Section 5.2. An efficient algorithm for computing the MPE of an SPE network is given in [10].

6 Example

To demonstrate how our method works on 3-D scenes, we use color blocks as image features. Figure 3 shows an example image with extracted colors. In this case, the 320 × 240 pixel image was reduced in resolution such that 10 × 10 pixel regions were reduced to a single color that was then mapped into a 16-color palette using the built-in image manipulation functions of the Apple Macintosh. These colors provide indexes to feature-pose maps for the three different types of cans and the baseboard of the wall.

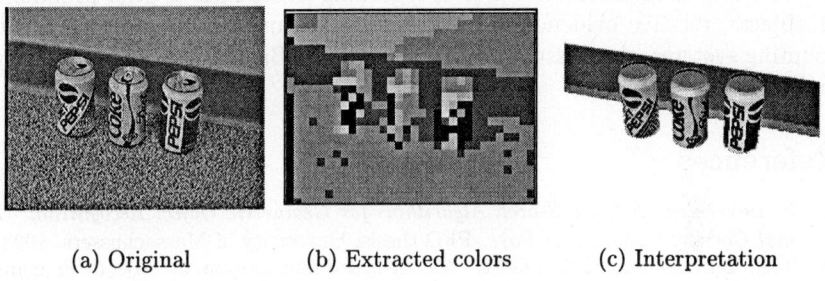

(a) Original (b) Extracted colors (c) Interpretation

Fig. 3. Example image with extracted colors and synthesized interpretation.

The ten best hypotheses were collected from each of these four object types. These 40 hypotheses were then reduced to a set of 18 by finding clusters of neighboring poses for a single model and interpolating the surrounding neighborhoods.

These interpolated hypotheses and their associated features were used to automatically create the Bayesian network shown in Figure 4. This network contains four nodes representing clusters of mutually exclusive hypotheses and ten nodes representing clusters of features.

A synthetic image created from the best interpretation is shown in Figure 3(c). This interpretation directly matches most predicted individual image features and also accounts for the majority of missing features from the wall that are occluded by the cans. Although it differs slightly from the actual image for various reasons (mainly errors in camera calibration and pose interpolation), it is quite close to the original image.

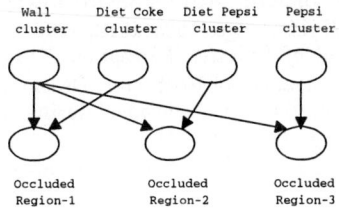

Fig. 4. Bayesian network created from the hypotheses and features.

7 Conclusions

We have shown how physical and visual interactions between objects can provide a information useful for interpreting complex scenes. These interactions can be represented using a Bayesian network that relates hypotheses to image features. Finding the best interpretation of the modeled scene is equivalent to finding the MPE of a Bayesian network. The Bayesian network formalism forces one to consider and to make explicit choices regarding issues such as prior probabilities of objects, relative evidence of features, and feature independence. In feature counting systems, these issues are hidden, but the Bayesian network formulation makes them explicit.

References

1. R. Beveridge. *A Local Search Algorithms for Geometric Object Recognition: Optimal Correspondence and Pose.* PhD thesis, University of Massachussets, 1993.
2. T.O. Binford and T.S. Levitt. Model-based recognition of objects in complex scenes. In *1994 ARPA Image Understanding Workshop*, November 1994.
3. T.O. Binford, T.S. Levitt, and W.B. Mann. Bayesian inference in model-based machine vision. *Uncertainty in AI*, 3:73–94, 1989.
4. T.M. Breuel. Higher-order statistics in object recognition. In *CVPR93*, pages 707–708. IEEE, 1993.
5. B. D'Ambrosio. SPI in large BN2O networks. In *Proceedings of the Tenth Annual Conference on Uncertainty in Artificial Intelligence (UAI-94)*, pages 128–135, Seattle, WA, 1994.
6. C. Hansen and T. Henderson. CAGD-based computer vision. *PAMI*, 11(11):1181–1193, 1989.
7. D.P. Huttenlocher and S. Ullman. Recognizing solid objects by alignment with an image. *IJCV*, 5(2):195–212, November 1990.
8. T.S. Levitt, J.M. Agosta, and T.O. Binford. Model-based influence diagrams for machine vision. *Uncertainty in AI*, 5:233–244, 1990.
9. T.S. Levitt, T.O. Binford, G.J. Ettinger, and P. Gelband. Probability-based control for computer vision. In *DARPA89*, pages 355–369, 1989.
10. B. Li, Z.and D'Ambrosio. A framework for ordering composite beliefs. *IEEE Transactions on Systems, Man, and Cybernetics*, 25(2):243–255, 1995.
11. Z. Li and B. D'Ambrosio. Efficient inference in Bayes nets as a combinatorial optimization problem. *Intl Journal of Approximate Reasoning*, 11(1):55–81, 1994.
12. J. Liang, F. Jensen, and H. Christensen. A framework for generic object recognition with Bayesian networks. In *Proceedings of the First International Symposium on Soft Computing for Pattern Recognition*, Reading, U.K., March 1996.
13. W.B. Mann and T.O. Binford. SUCCESSOR: Interpretation overview and constraint system. In *ARPA96*, pages 1505–1518, 1996.
14. R. D. Rimey. *Control of Selective Perception Using Bayes Nets and Decision Theory.* PhD thesis, University of Rochester, December 1993.
15. C. Rothwell. Reasoning about occlusions during hypothesis verification. In *ECCV96*, volume 1, pages 599–609, 1996.
16. W. Wells. *Statistical Pattern Recognition.* PhD thesis, Massachussetts Institute of Technology, 1993.
17. M. Westling and L. Davis. Object recognition by fast hypothesis generation and reasoning about object interactions. In *ICPR96*, pages 148–153, Vienna, 1996.

Recognition of Urban Scene Using Silhouette of Buildings and City Map Database

Peilin Liu, Wei Wu, Katsushi Ikeuchi, Masao Sakauchi

University of Tokyo, 7-22-1 Roppongi Minato-ku Tokyo 106,JPN

Abstract. This paper proposes an approach for understanding urban scenes by recognizing key buildings in silhouette using model-based object recognition scheme. A city map-database is used to build a world model, based on the information of location and orientation of the video camera measured by GPS. The general correspondence hypotheses between the line vectors extracted from the silhouette of ground objects and the world model are established by using dynamic programming technique. The general correspondence hypotheses are verified and modified according to the local position relations. The approach was successfully tested in recognition of the buildings in real scenes.

1 INTRODUCTION

Model-based object recognition scheme is widely applied to the varied applications of computer vision such as autonomous navigation, pose estimation, surveillance. The robot's position and orientation are estimated by establishing correspondence between the line features extracted from the images acquired by the robot and the model features[1][2]. Most model-based vision systems employ a CAD model of objects in terms of geometric components, and use multisensory information to search the correspondences between the CAD-model and images in order to identify specified objects for industrial processes, outdoor scene interpretation. However these model-based object recognition systems usually have difficulties in recognition of objects in real urban scene due to the complicated modeling process . Geographic Information Systems(GIS) is the one of the most effective ways to integrate and manipulate the spatial data in the real world[3]. In a Geographic Information System, map database is used to visualize spatial data and to reveal the relations, therefore can also help to build a world model. In our research, a city map-database with 3-D descriptions of the rooftops of the buildings is used to build a world model. Considering of the urban scenes, the buildings which are very high or in special shapes are very prominent. We select features of the prominent buildings in the view for corresponding between the model and images. The parameters of view can be obtained with the sensors for location and orientation. The feature correspondences between images and the model are established using the dynamic programming technique. Incorrect feature correspondences caused by sensor uncertainty and image clutters are modified by computing cost with a similarity evaluation method. With the correspondence of building features between images and the model, other objects in the viewing field of the world model can be projected to the image, so that the contents of the scene can be understood extensively.

In the following sections, we discuss the details of the approach of understanding the urban scenes by recognizing the key buildings in the silhouette. Section 2 describes features used in searching the image-model correspondences, and presents how to build the model for correspondence searching with the features, and how to extract features in the input image for matching with the model. Section 3 discusses the matching algorithm using dynamic programming technique to get the correspondence hypotheses between the model and the input image, and introduces a method of verification and modification of the correspondence hypotheses. Section 4 presents the experimental results of the proposed approach using the real world images of urban scene. Section 5 discusses the conclusions and future research directions.

2 FEATURES FOR MATCHING

Model-based object recognition consists of matching features between an image and a pre-stored object model. The types of features required and number of features used depends on the objects for recognizing and the model description. The types selected should make the correspondence searching between the input image and the model easy. Considering this, we use the vectors composing the key buildings' rooftops as correspondence features.

2.1 Extracting features from the input image

We notice that high buildings are very prominent in the urban scenes. The silhouette of ground objects almost consists of the contours of the buildings in the urban environment. The high line segments in the silhouette are made by high buildings. If the line segments of the silhouette can be extracted from the image, these line segments might be considered to be the contours of the high buildings. Extracting the silhouette from image is very easy. The canny edge detector detects the edges from the input image. The pixels which are reached first when scanning from the top to the bottom of the image construct the silhouette. The pixels of silhouette are chained and segmented into line vectors. We extract horizontal vectors which is higher than the threshold defined as the candidates for corresponding with model. The features extracted from the input image are represented as:

$$I = \{I_i | i = 1, 2, ..., M\} \qquad (1)$$

here I_i is one of the vectors in the silhouette of the input image.

2.2 Features Using in the Model Description

We use a city map-database with the 3-D descriptions of the rooftops of the buildings to construct the model for matching. With the parameters of the video camera, the 3-D descriptions of the rooftops of the buildings can be projected to 2-D descriptions in the term of the line vectors easily. The parameters of the camera can be obtained by the GPS and gyros. With the information of the location and orientation of the camera, the viewing field can be defined. Considering the uncertainty of the sensors, we define the viewing field wider than the one defined by data of sensors. In order to decrease the number of features, we further select the buildings which are higher than the threshold defined to construct the model for matching. We call these buildings as key buildings which will be searched from the input image. So the model can be represented as:

$$M = \{M_j | j = 1, 2, ..., N\} \quad (2)$$

Here, the M_j is one of the vectors of the key buildings' rooftops.

3 MATCHING ALGORITHM

Dynamic programming solves an N-stage decision process as N single-stage processes. This reduces the computational complexity to the logarithm of the original combinatorial one[4]. The DP technique requires that the two sequences used for matching should be monotonous and continuous and satisfy the initial conditions[5]. So far several methods of using Dynamic Programming technique to solve the stereo matching problem have been proposed [6]. They showed satisfactory results of appling DP technique to stereo matching problem. However in the case of model-image correspondence, the requirements of DP can not be satisfied in some degree. As the position order of buildings in the urban environment and the map-database are fixed, the vector sequences of the input image and the model can be assumed to be monotonous and continuous. The remainder of this section will present our matching algorithm for solving the model-image correspondence problem using the DP method.

3.1 Searching Correspondences Using DP

Finding the Optimal Path. We get two vector sequences from the input image and the model as shown in equation (1),(2). The problem of obtaining correspondences between the vectors of the input and the model can be solved as a path finding problem on a 2-D plane shown as figure 1 . A path has a vertex at the intersection (i,j) when the vector I_i and vector M_j are matched. The image vector and the model vector matched represented as I_{o_m}, and M_{o_m}, respectively. This correspondence is represented as (I_{o_m}, M_{o_m}). Here, a primitive path corresponds to matching of the relative distance and the relative height of the vectors. Thus the cost of a path is defined as:

$$D(I_i, M_j) = |(h_{I_i} - h_{M_j}) + (d_{I_i} - d_{M_j})| \quad (3)$$

where $0 \leq i \leq M, 0 \leq j \leq N$

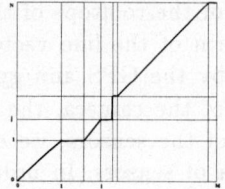

Fig. 1. 2-D search plane

h is the relative height of the vectors' midpoint, d is the relative distance of the vector's midpoint. I_i and M_j are the indexes of them. The DP process actually is an iterative process. The iteration starts at the (0,0) and computes $D(I_i, M_j)$ for each intersection. At each intersection the primitive path that gives the minimal cost is recorded. The sequence of primitive paths which gives $D(I_M, M_N)$ at the intersection (M,N) is the optimal path. By finding the optimal path from this 2-D search plane, the correspondence hypotheses between sequences of the input image and the model can be established. The correspondence hypotheses are represented as

$$Corr. = \{(I_{o_i}, M_{o_i}) | i = 1, 2, ..., m\} \quad (4)$$

where m is the number of the correspondence.

The problems of searching correspondences using DP. The dynamic programming searches the correspondences between the input image and the model by finding an optimal path locally. Therefore it can not deal with the situation when there are errors caused by the sensor uncertainty and the image clutters very well. Being able to get reliable and accurate correspondences, we propose an estimation method described below to verify the correspondence hypotheses and modify the incorrect correspondences.

3.2 Verifying the correspondence hypotheses

We verify the correspondence hypotheses by computing the local similarity degree for each correspondence. We first give the definitions of the ratio of the image vector $r_{I_{o_i}}$, the ratio of the model vector $r_{M_{o_i}}$, and the similarity degree of a correspondence between the vectors of the image and the model s_i.

The ratio of the image vector:

$$r_{I_{o_i}} = d_{I_{o_i}} / L_I \quad (5)$$

The ratio of the model vector:

$$r_{M_{o_i}} = d_{M_{o_i}} / L_M \quad (6)$$

The similarity degree of a correspondence

$$s_i = |r_{I_{o_i}} - r_{M_{o_i}}| \quad (7)$$

Here, the L_I and the L_M are the width from the beginning vector's midpoint to the ending vector's midpoint in the image vector sequence and the model vector sequence respectively. $d_{I_{o_i}}$ and $d_{M_{o_i}}$ represent the relative distance of the vectors' midpoint of the matched image vectors and the matched model vectors respectively.

Since the buildings are rigid objects, when the value of the similarity between the image vector and the model vector s_i is below a threshold defined, the correspondence is assume to be correct. Otherwise the correspondence of I_{o_i} and M_{o_i} is assumed to be incorrect.

3.3 Modifying the incorrect corresponding

Several kinds of modification operation can be used to modify the incorrect correspondences: connection, deletion, and separation. The modification costs of these operations are first computed by the equation (8).

$$cost_i = \begin{cases} -1 & if\ s(r_{I_{o_i}}, r_{M_{o_i}}) < \theta \\ (s_i - s_{old_i})/(s_i + s_{old_i}) & if\ s(r_{I_{o_i}}, r_{M_{o_i}}) \geq \theta \end{cases} \quad (8)$$

Here, s_i is the similarity after modification and the s_{old_i} is the one before modification. The cost value is getting close to -1 when the similarity degree is getting less. Therefore, according to the cost value, a suitable modification can be selected to obtain the most reliable and accurate correspondences.

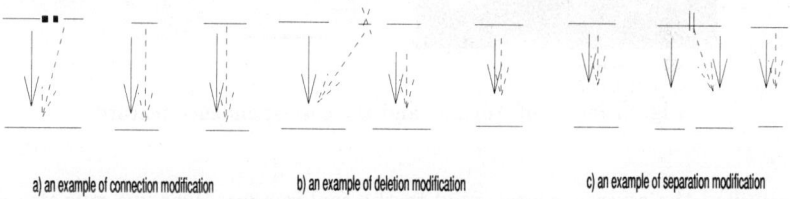

a) an example of connection modification   b) an example of deletion modification   c) an example of separation modification

Fig. 2. The modification operation

Figure 2 illustrates how the modification operations work well in obtaining the more reliable and accurate correspondences. In the Figure 2, the dotline brushes show the correspondence obtained by the dynamic programming, and the solid ones shows the correspondences after modification. The upper vectors are vectors of the input image, and the lower are the vectors of the model. The connection operation is used to modify the fragmentation of the vectors caused by noise in the input image. As shown in figure 2.a, a dotline connects the two separated vectors. The vector is only connected with the neighboring ones. The DP process needs not be carried out during the connection operation. The deletion operation can be used to deal with the additional vectors caused by noise in the input image as shown in figure 2.b. The vector marked × is deleted, the and DP process is executed again with the new vector sequences in which the mismatched vectors are not included. The separation operation can used to cope with the object occlusions in the input image shown in the figure 2.c. The vector marked with || are separated into the two vectors and the DP process is carried out again with the new vector sequences.

The connection operation is given highest priority, since the DP process needs not be carried out during the connection operation. The connection cost of all vectors which needs to be modified is calculated. The one with the most minimal modification cost will be carried out. If the connection cost is too big, the deletion or the separation operation are considered. After the modification, the results are verified. If the similarities of the vector are in the extent allowed, the correspondences are assumed to be established. Otherwise the correspondences are considered to be incorrect with the endpoint predicated. In this case, the endpoint for corresponding is changed to the other vector if possible, and then the searching process with the new endpoint for corresponding is carried out again. If there is no suitable vector for the endpoint, we select the correspondences with the most minimal cost and end the matching process.

4 Experimental Result

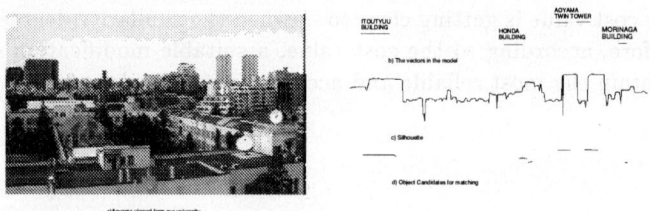

Fig. 3. Image of Aoyama and the correspondence features

We applied the approach presented to the real images. Here, we give two examples to illustrate how our approach for understanding urban scene works. One is Aoyama viewed from the rooftop of one building in our university which is about 21 meters high. The viewing direction is about 45 °from north. The image is shown in figure 3.a. The other one is the Shinjuku viewed from Tower about 333 meters high. The viewing direction is about 60 °from north, shown in figure 6.a.

For recognizing the buildings in the image of Aoyama, we got five objects which might consist of the silhouette in the image of Aoyama from map database. They are Aoyama twin tower, Honda building, ect. The five objects are laid in the order from left to right according to the geographical position to form the matching model, as shown in the figure 3.b. We extracted silhouette from the input image and segmented them into vectors shown in figure 3.c. Eight vectors used to corresponding with the model were extracted shown in figure 3.d.

The 2-D search plane of Aoyama is shown in the figure 4.

Five vectors are searched from eight vectors as the correspondences between the model. The correspondences are then verified. The correspondences which

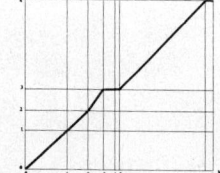 

Fig. 4. 2-D search plane of the image of Aoyama and the correspondence result

similarity degree is larger than the threshold are modified. Table 5 shows the similarity degree of the vectors and the modification cost. From the table 5, we can see that the last two vectors of the Aoyama image is connected. After connection, the similarity degree and cost are getting small. Since the similarity degree after connection is smaller than the threshold, the result resumes to be right. The image-model corresponding result is shown in figure 4.

Coor No	Similarity degree Before modification		Similarity degree after modification		Modification Operation		Modification Cost Before		Modification Cost After	
	Image1	Image 2	Image1	Image 2	Image1	Image 2	Image1	Image 2	Image1	Image 2
0	0.0008	0.0954	0.0008	0.0954			-1	-1	-1	-1
1	0.0006	0.1054	0.0006	0.1054			-1	-1	-1	-1
2	0.0278	0.0868	0.0278	0.0868			-1	-1	-1	-1
3	0.0378	0.0723	0.0235	0.0723	Connection		0	-1	-0.23	-1
4		0.2369		0.0206		Deletion		0		-0.84
Average	0.0186	0.1194								
Variance	0.0164	0.0598								
Threshold	0.0332	0.1791								

Fig. 5. The result of verification and modification

For recognizing the key buildings in the image of shijuku viewed from Tokyo Tower, we got 6 key objects which might consist of the silhouette in the image of Aoyama based the information of the map. The six objects are laid according to the geographical position form the matching model, as shown in the figure 6.b. We extracted silhouette from the input image and segmented them into vectors shown in figure 6.c. Seventeen vectors used to corresponding with the model were extracted shown in figure 6.d.

 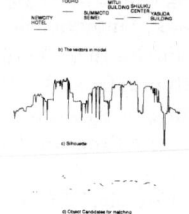

Fig. 6. Image of Shijuku and the correspondence features

In the image of Shinjuku, the third vector are deleted and the DP process is carried out again. The 2-D search plan before deletion and the one after deletion are shown in 7. The corresponding result is also shown in figure 7.

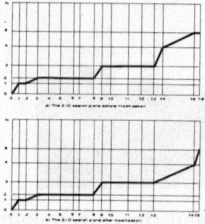 

Fig. 7. 2-D search plane for the image of Shinjuku and correspondence result

5 CONCLUSION

This paper describes an approach to understanding of urban scenes by recognizing key buildings in silhouette using model-based object recognition scheme. The silhouette of ground objects is extracted from the input image and segmented into line vectors. The correspondence hypotheses between the line vectors extracted from the silhouette of ground objects and the world model are obtained by using the dynamic programming technique. The correspondence hypotheses are then verified to ensure the correspondences to be reliable and accurate. After the key buildings are recognized through the correspondence process, the other objects in the viewing field of the world model can be projected to image, so the contents of the scene can be understood extensively. This approach proposed is suitable for the case of taking video from the high places. Future work includes understanding of urban scene using model-based object recognition scheme for the case of taking video while walking in the street.

References

1. Rai Talluri,J.K.Aggarwal,"Mobile Robot Self-location Using Model-Image Feature Correspondence" IEEE Robotics And Automation, **12**(1996) No.1,Feb.1996.
2. T.Tsubouchi and S.Yuta,"Map assisted vision system of mobile robots for reckoning in a building environment, "in Proc. IEEE Int. Conf. Robotics Automat.(Raleigh),Mar,1987,pp1978-1984.
3. Menno-Jan Kraak, "Integrating Multimedia in Geographical Information Systems", IEEE Multimedia Journal, Autumn,1996, 59-65
4. A.V.Aho, J.E.Hopcroft, and J.D.Ulman, The Design an Analysis of Computer Algorithms. Reading, MA: Addison Wesley,1974
5. R.Bellman:Dynamic Programming, Princeton University Press, Princeton, New Jersey,1957
6. Y.Ohta and T Kanade," Stereo by intra- and interscanline search using dynamic programming", IEEE Trans. on Pattern Analysis and Machine Intelligence, **7**(1985) No.2,pp139-154.

A Cooperative Inference Mechanism for Extracting Road Information Automatically

Masakazu Nishijima and Toyohide Watanabe

Department of Information Engineering,
Graduate School of Engineering, Nagoya University
Furo-cho, Chikusa-ku, Nagoya 464-01, Japan
Phone: +81-52-789-4409, Fax: +81-52-789-3808
E-mail: watanabe@nuie.nagoya-u.ac.jp

Abstract. The subject for extracting road information automatically from map images has been regarded as effectual means to construct GIS. Until today, many different methods/approaches on this subject have been reported, but it is not easy to extract road information successfully because various kinds of map elements are overlayed and intersected. This paper addresses an advanced approach to identify road information on the basis of cooperative inference mechanism. This mechanism generates hypotheses for un-extracted roads simultaneously and interprets them. Namely, roads, disjointed by the existences of other map elements or not extracted as road fragments, are connected as a result of checking up the conformity, consistency and adequacy among hypotheses or the adjustment between hypotheses and other map elements.

1 Introduction

The issue about the automatic extraction of road information from map images is one of important subjects to compose the basic resource data in constructing GIS(Geographical Information System). Until today, many methods/approaches have been reported, concerning this issue[1, 2]. For example, the method for tracking parallel and continuous pixels[3], the skip-scan method[4] and so on were typically developed. However, since these methods were based on only the bottom-up processing, the processing capabilities are limited.

While, the methods based on the top-down processing also were proposed: the integration method of bottom-up and top-down processings[5], and the cooperation method among bottom-up and top-down processings[6]. In the integration method, the bottom-up processing is first applied to urban map images in order to extract road information locally and then the top-down processing is used to refine the extracted road information globally, using the road network as a road model. Here, the road network is a topological graph for expressing road information[7]. While, in the cooperation method the bottom-up processing and top-down processing are controlled repeatedly to refine road information.

However, it is not easy to distinguish roads from urban map images completely even if these methods were applied. In the refinement process, hypotheses are generated by checking the adjustment between the heuristic knowledge about road configurations and then the locally extracted road information one by one. Also, the verification process of hypotheses is sequentially performed. Namely, the hypothesis is generated from only one disjointed point in the road network and only one check-point is corresponded to constraint conditions. The interrelations between individually disjointed points are ignored.

This paper addresses an advanced approach which refines road information on the basis of cooperative inference mechanism. In our mechanism, hypotheses for un-extracted roads are generated simultaneously, and interpreted cooperatively.

2 Approach

The cooperative inference mechanism is composed of two processes: cooperative interpretation of hypotheses[8] and complementary adjustment between hypotheses and other map elements[9]. Here, hypothesis means that road fragments should be inferable from disjointed points. Of course, the inferred facts that existing road fragments may be inconsistent to the heuristic knowledge are also hypotheses.

In the cooperative interpretation process, the hypotheses are first generated from several disjointed points simultaneously; second, the conformity, consistency and adequacy among these hypotheses are checked; and finally the reasonable hypotheses are selected. Of course, it takes much time to perform this process among all disjointed points, and it is not effective to apply this process to hypotheses which are not interrelated at all. The concept of cooperative region is introduced with a view to defining effective hypotheses. Fig.1 shows the cooperation among related disjointed points. In Fig.1, hypotheses are illustrated as arrows extended from disjointed points.

On the other hand, in the complementary adjustment process other map elements are first found out; second, the extracted map elements are searched from the disjointed points; third, links are generated to indicate the correspondences between the disjointed points and the extracted map elements; and finally disjointed points linked to the same map elements are connected through common edges. Fig.2 represents the cooperation between a character region and disjointed points. In Fig.2, links are illustrated as arrows set between disjointed points and a character region.

3 Outline of System

Our system is composed of three phases: identification of map elements, construction of road network and refinement of road network, as shown in Fig.3.

Identification of map elements We make use of color and shape information in order to distinguish individual map elements successfully. Our current extraction system identifies the following map elements.
- characters: rectangular blocks(almost square shapes) for parts of character strings such as city names, building names, street names and so on
- bus-routes: single sequential lines
- roads: pairs of parallel line segments

Construction of road network The road network is constructed by searching out pairs of successive parallel line segments[5]. In this case, we regard intersections as the beginning points of searching roads because the intersection is the most confident object. Of course, intersections can be identified in advance[5]. This construction procedure works from the intersection toward the direction of connective roads until all of connective roads are checked.

Refinement of road network All of road fragments are not yet extracted completely because the construction procedure is performed based on the bottom-up processing and depends on input images too strongly. In this phase, the road network is refined under the cooperative inference mechanism, in addition to the generation/verification mechanism of single hypotheses[6]. The cooperative inference mechanism is composed of two processes: the cooperation among disjointed points and the cooperation between disjointed points and characters.

4 Cooperation among Disjointed Points

We call incomplete road network, which was composed in the construction phase, an initial road network. Generally, many disjointed points exist in the initial road network. The hypotheses for un-extracted roads are generated from these disjointed points, checked up with respect to the cooperative regions and verified whether inferable roads exist among disjointed points. Then, the corresponding disjointed points are connected in the modification procedure.

Extraction of cooperative regions It is not easy to infer geographical shapes of roads between disjointed points. If an edge connecting two different disjointed points is assumed and also this assumed edge does not cross any other edges, two disjointed points are connective. The cooperative region is defined as a rectangular region, in which disjointed points are enclosed, and is extracted in the following steps:

1. Extract pairs of connective disjointed points.
2. Collect a set of connective disjointed points. For example, as for three disjointed points i, j and k, three pairs (k, i), (i, j) and (j, k) are extracted.
3. Identify a minimum rectangular region, which encloses all connective disjointed points of this collected set inclusively.

Fig.4 shows the generation process of a cooperative region.

Generation of hypotheses The generation of hypothesis is to presume the road which leads to the tip of disjointed point. The edge, which we call VE(Virtual Edge), is generated toward an extensible direction of road from each disjointed point. The direction of VE is calculated with respect to that of the existing edge which is already connected to the disjointed point. Also, the length of VE is set up according to that of cooperative region. We treat the following characteristic points as disjointed points:

TypeA: terminal points in road network

TypeB: connective points in road network in case that distance between two road fragments is large

TypeC: side points in road fragment which are not connected to other

Interpretation of hypotheses VE is a hypothesis for representing the presumed road. First, a pair of VE's, regarded as the same continuous road, is selected from a set of them. Two different VE_i and VE_j indicate the same road when they are satisfied with either of the following relations, and also construct a pair of nodes which connect to each disjointed point, based on VE_i or VE_j.

- Relation1 "opposition": When two VE's are extensible to the opponent's disjointed point mutually, the relation between them is "opposition".
  [Condition for opposition]
  The angle between two VE's: $\theta < \theta_0$. Here, θ_0 is a threshold value.
- Relation2 "intersection": When two VE's are caught as a line segment and intersected mutually, the relation between them is "intersection".

However, each disjointed point is not corresponded by 1-to-1 relation in the area where two or more roads interrelate. In such an area, it is appropriate to treat pairs of related disjointed points together and grasp the geographical shape. We regard this area as the cooperative region defined in Section 4.1. In order to interpret mutual relationships among several hypotheses in the cooperative region, a relation table is constructed, as shown in Fig.5(b).

The road network is refined, based on this relation table. First, pairs of disjointed points related to "opposition" are connected. And then, pairs related to "intersection" are connected according to the result of "opposition" pairs. Fig.6 represents the refinement process of road network with respect to the interpretation of hypotheses.

Verification of hypotheses Hypotheses generated in the interpretation procedure are verified with empty regions. The empty region is a closed region, in which any other map elements do not exist. Before the verification process, empty regions are extracted, using color information. In the verification process, if hypothesis for roads goes through the empty region, this hypothesis is judged as a mistaken one and then deleted.

5 Cooperation between Disjointed Points and Characters

The road network is refined complementarily using other map elements such as characters and bus-routes. This refinement process is performed, based on the positional and semantical relationships among map elements.

Positional and semantical relationships The positional relationship is defined between two different map elements, which are overlayed mutually on one map sheet. While, the semantical relationship is established between two related map elements under the condition that the existence of a map element depends on the existence of another map element. For example, our map elements under these relationships are paired: (1) road and character for positional relationship, and (2) road and bus-route for semantical relationship.

On one map sheet, characters are always printed over roads. It is possible to infer a road among disjointed points if characters overlayed on roads could be desirably distinguished. On the other hand, whenever bus-routes are printed, the roads are printed along them. It is possible to complement roads if the bus-routes, parallel to roads, could be recognized.

Refinement process using other map elements This refinement process works as (1) construction of bus network, using bus-routes, (2) recognition of intersections or T-junctions and (3) searching of character regions from disjointed points, and connection of them with links.

We explain the cooperation between disjointed points and characters in the step (3) precisely. First, the searching edge of l_0(a threshold value) in length is generated toward an extensible direction of road from each disjointed point. If the searching edge is kept within a character region, the link which connects between the disjointed point and a character region is generated. In case that the searching edge is within two or more character regions, the link is generated between the disjointed point and the nearest character region. Fig.7 shows a generation result of link. Next, these links are interpreted. If only one disjointed point is connected to a character with a single link, it is sure that this point was disjointed by a character or became a normal terminal point after the recognition of T-junctions(in the step (2)). If several disjointed points are connected to a character, the roads should be hidden under the character. Finally, the road network is refined based on interpretation of links. Fig.8 shows the refinement process of road network, based on character regions. In case that the number of links is single, the edge neighboring to the disjointed point is extended until contacting to the border line of a character region, and then the recognition of T-junction is performed for the extended edge. Of course, if T-junction cannot be recognized, the disjointed points should be a normal terminal point and the extended edge is deleted. While, in case that the number of links is two or more, a new node is generated at the center of gravity among disjointed points. And, each disjointed point is connected to it.

6 Experiment

The original urban map of scale 1:10000 is digitalized by the image scanner with 72 dpi and 256 RGB levels. Fig.9 is a binarized urban map image(600x600 pixels). Fig.10 shows the initial road network in the bottom-up processing and Fig.11 shows the road network in generation/verification mechanism of single hypotheses. Fig.12 is the finally refined result in our approach. When we compare Fig.11 and Fig.12, we can find out 12 newly detected road fragments. The mark "A" represents the road fragments distinguished between disjointed points and characters, B does ones detected among disjointed points, and C does ones identified between roads and bus-routes.

Table 1 shows the recognition result for other map images. The numeral values in each field indicate the numbers of road fragments(or edges). And, the numeral values in each parenthesis indicate the recognition ratios based individually on the bottom-up processing, generation/verification mechanism of single hypotheses and our proposed approach. By applying our approach, the recognition ratio raised up from 80.0% to 93.7%, in average.

7 Conclusion

In this paper, we proposed the cooperative inference mechanism to extract road information from urban map images automatically. This mechanism makes it possible that inferable/interpretative disjointed points, which were not connected in the traditional methods/approaches, are connected by the adjustments among disjointed points and/or between map elements. Also, our experiment made it clear that our mechanism is applicable and the extraction ratio is high.

Acknowledgements We are very grateful to Prof.T.Fukumura of Chukyo University, and Prof.Y.Inagaki and Prof.J.Toriwaki of Nagoya University for their perspective remarks, and also wish to thank Dr.Y.Sagawa, Mr.K.Asakura and our research members for their many discussions and cooperations.

References

1. G.Maderlechner and H. Mayer:"Conversion of High Level Information from Scanned Maps into Geographic Information Systems", *Proc.of ICDAR'95 Vol.1*, pp.253-256.
2. C.Nakajima and T.Yazawa:"Automatic Recognition of Facility Drawings and Street Maps Utilizing the Facility Management Database", *Proc.of ICDAR'95 Vol.1*, pp.516-519.
3. T.Miyataki, H.Matsushima and M.Ejiri:"Extraction of Roads from Topographical Maps Using a Parallel Line Extraction Algorithm", *Trans.on IEICE, Vol.J68-D, No.2*, pp.153-160(1985) [in Japanese].
4. T.Nagao, T.Agui and M.Nakajima:"Automatic Extraction of Roads Denoted by Parallel Lines from 1/25,000 Scaled Maps Utilizing Skip-scan Method", *Trans.on IEICE, Vol.J72-D-II, No.10*, pp.1627-1634(1989) [in Japanese].
5. T.Hayakawa, T.Watanabe, Y.Yoshida and K.Kawaguchi:"Recognition of Roads in an Urban Map by Using the Topological Road-network", *Proc.of MVA'90*, pp.215-218.
6. T.Watanabe, T.Hayakawa and N.Sugie:"A Cooperative Integration Approach of Bottom-up and Top-down Methods for Road Extraction of Urban Maps", *Proc.of ICARCV'92*, pp.61-65.
7. T.Watanabe and T.Fukumura:"Towards an Architectural Framework of Map Recognition", *Proc.of ACCV'95 Vol.3*, pp.617-622.
8. M.Nishijima and T.Watanabe:"An Automatic Extraction Approach of Road Information on the Basis of Recognition of Character Regions", *Proc.of ICSC'95*, pp.173-180.
9. M.Nishijima and T.Watanabe:"An Automatic Extraction of Road Information on the Basis of Cooperative Hypotheses Interpretation Mechanism", *Proc.of MVA'96*, pp.147-150.

Table 1. Experimental result

	original	bottom-up	single	our approach
map1	132	75(56.8)	91(68.9)	122(92.4)
map2	157	102(65.0)	125(80.0)	152(96.8)
map3	155	139(89.7)	139(89.7)	149(96.1)
map4	164	103(62.8)	113(68.9)	155(94.5)
map5	124	101(81.4)	103(83.1)	112(90.3)
map6	193	156(80.8)	164(85.0)	177(91.7)
sum	925	676(73.1)	735(80.0)	867(93.7)

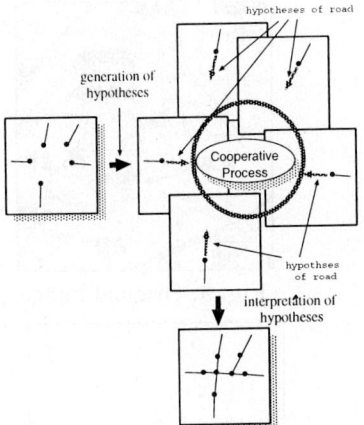

Fig. 1. Cooperation among disjointed points

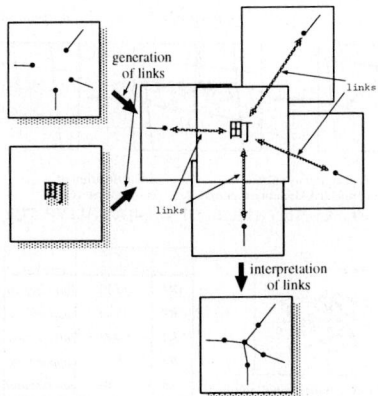

Fig. 2. Cooperation among disjointed points and character region

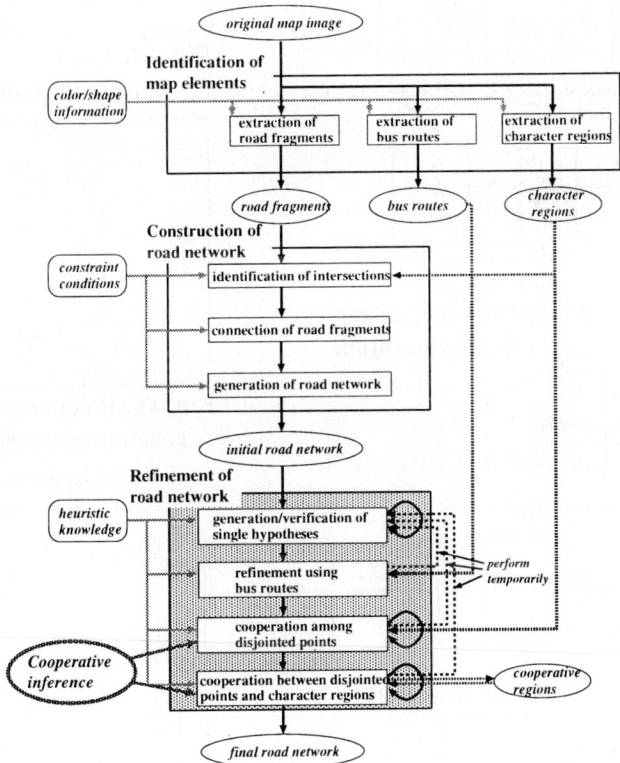

Fig. 3. Processing flow

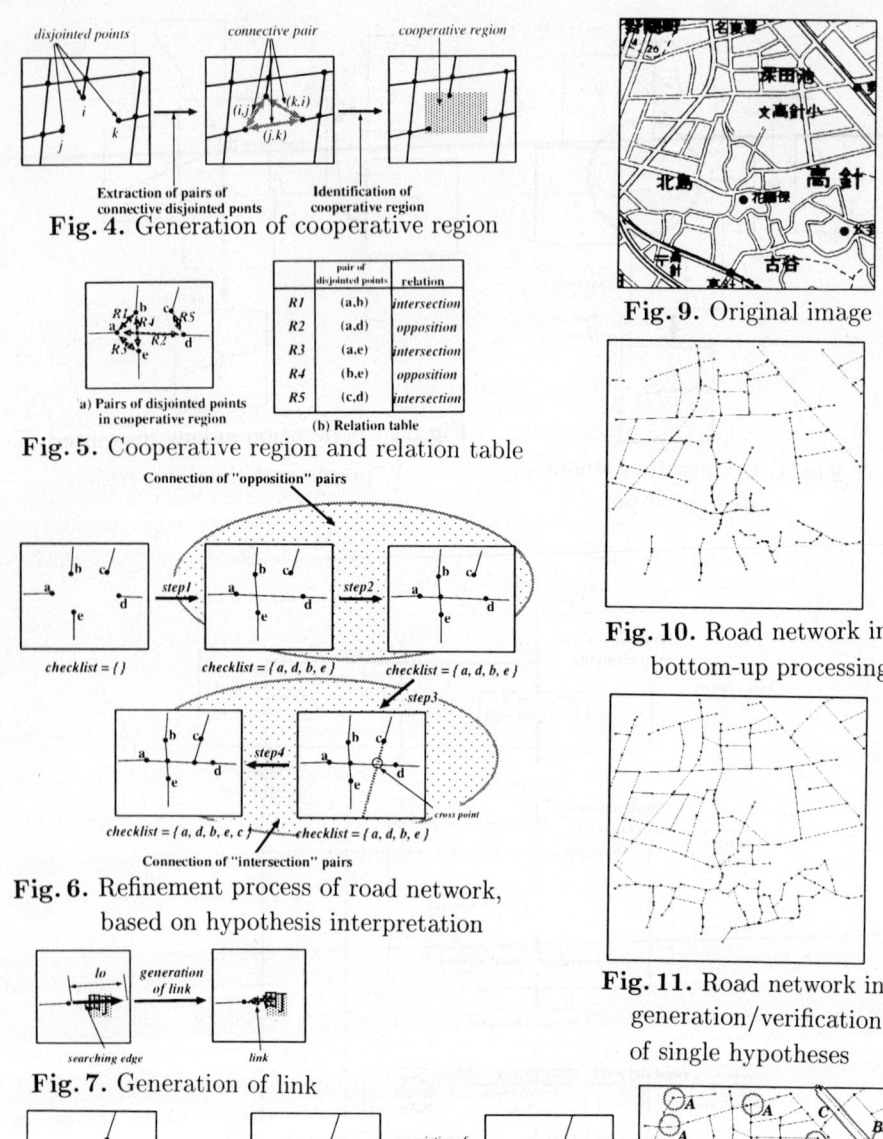

Fig. 4. Generation of cooperative region

Fig. 5. Cooperative region and relation table

Fig. 6. Refinement process of road network, based on hypothesis interpretation

Fig. 7. Generation of link

Fig. 8. Refinement process of road network, based on character regions

Fig. 9. Original image

Fig. 10. Road network in bottom-up processing

Fig. 11. Road network in generation/verification of single hypotheses

Fig. 12. Road network in our approach

Model-Based Active Object Recognition Using MRF Matching and Sensor Planning

Tianrong Liu, Kap Luk Chan[†], Stan Ziqing Li

School of Electrical and Electronic Engineering
Nanyang Technological University
Nanyang Avenue, Singapore 639798
[†]E-mail: eklchan@ntu.edu.sg

Abstract. This paper presents an active object recognition algorithm using MAP-MRF matching and sensor planning strategy. The matching between the sensed and model object is based on surface properties. A new measure of surface distinguishability is defined for sensor planning. MAP-MRF framework is used for generating matching label set. A measure of confidence of correct match is determined based on the posterior energy. An active object recognition algorithm is used for determining the next viewpoint of the camera if ambiguity exists in the matching result. The next viewpoint is chosen based the surface with highest distinguishability. Experimental results on images under perfect and imperfect segmentation are presented.

1. Introduction

Most computer vision algorithms perform object recognition based on the information contained in a single image, typically a set of features is extracted and matched against model features, with the best match determining the result. In recognizing a 3D object from a 2D intensity image from a single viewpoint, ambiguity may occur because the features extracted are not always sufficient for the recognition task [1]. One approach to overcoming this problem is to utilize multiple sensor observations obtained from different viewpoints, to acquire more information in order to perform unambiguous object recognition [1-6].

An active object recognition algorithm for recognizing 3D objects from multiple 2D intensity images is presented in this paper. Three major factors affecting the capability and performance of such an active recognition system are considered: object representation, matching methodology and sensor planning strategy. Matching between the sensed and model objects is based on the surface properties. A new surface measure called the *surface distinguishability* is defined using some of the surface properties to represent the distinctiveness of a model surface. The Markov Random Field (MRF) matching framework[7] is used for establishing correspondences between the visible surfaces in the scene object and those in the models. A posterior energy function is used to measure the similarity between sensed and model surfaces. Given an image, a matching label set is created through matching. A measure of the confidence of correct match is calculated based on minimized energy functions. When recognition ambiguity exists in the label set according to our defined termination criterion, the next viewpoint of the camera is determined using the normalized surface distinguishability among unmatched surfaces of the hypothesized matching object in order to capture more helpful information for reliable recognition.

2. Object Representation

An object-centered representation scheme based on surface properties for object representation is adopted in our work. Since the features extracted from an image are represented by the viewer-centered coordinate system, a transformation must be performed to realign the coordinate system before the observed features can be matched to the model. Note that an invariant representation is undesirable in here because the pose of the object is to be recovered for sensor planning.

2.1 Surface properties for object matching

The surface properties used for matching are estimated 3D surface area, 3D surface compactness and surface orientation. These surface properties are derived from 2D intensity images. Our empirical models developed in [8] are used for estimating these object-centered features to remove the foreshortening effect.

The 3D Surface Area for model surface is just the surface area in 3D. For sensed surface in a 2D intensity image, the actual 3D surface area is estimated using empirical model developed in [8]. The 3D Surface Compactness is used to represent the shape of the surface. The actual 3D surface compactness is again estimated from the surface compactness in 2D image by our empirical model. For planar surfaces, the surface orientation for a model surface is the unit surface normal for a certain pose of the model object. The surface normal can be uniquely determined if we know the 3D position of four points belonging to a planar surface. The 3D coordinates of some corner points of a planar surface are determined by the triangulation method from [9] using a pair of images in the image sequence captured during sensor movement.

2.2 Defining the Surface Distinguishability for Sensor Planning

A new measure for the model surfaces, called the *surface distinguishability*, is defined to express how distinctive a model surface is when compared to other model surfaces in the model database and it is defined for a model surface S_i as:

$$D(S_i) = \left[\sum_{\substack{j=1 \\ j \neq i}}^{M} \exp\left(-\sum_{k=1}^{F} \left| \frac{f_{ik} - f_{jk}}{f_{ik} + f_{jk}} \right| \right) \right]^{-1} \quad (1)$$

where F is the number of features used,
 f_{ik} and f_{jk} represent the kth feature value for surface i and surface j respectively, and
 M is the total number of model surfaces in the database.

In calculating $D(S_i)$, only the surface area and the compactness features are used and hence $F=2$. The surface orientation is not adopted as a feature in defining the surface distinguishability because such a feature cannot make a surface more distinctive than others.

For some model objects with high symmetry, all the identical surfaces are therefore grouped together and the normalized surface distinguishability for a group is defined as:

$$D_n(S_i) = \frac{D(S_i)}{\sum_{i=1}^{N} D(S_i)} \qquad (2)$$

where N is the number of surface groups in a model object.

The normalized surface distinguishability, $D_n(S_i)$, expresses the confidence of recognizing an object through this surface and this property is used in our sensor planning strategy for choosing the next viewpoint of the camera.

3. Object Matching based on Markov Random Field Framework

Under the MRF theory, object matching is posed as labeling problems in which the solution to a problem is a set of labels assigned to image pixels or features. In our work, the m sensed surfaces are represented in a set $S = \{1, ..., m\}$ associated with a vector $d_1(i)$ composed of two *unary properties,* the estimated 3D surface area and compactness. Two different sensed surfaces are related to each other by a vector $d_2(i,i')$ composed of one *binary (bilateral) relations,* the angle between the surface normals of two sensed surfaces. The labeling of a scene in terms of a model object is denoted by $f = \{f_i \in \Lambda^+ | i \in S\}$ where Λ^+ denotes the actual label set in matching the sensed surface to the model surface plus a NULL label [7], i.e. $\Lambda^+ = \Lambda \cup \{0\}$ where $\Lambda = \{1,..., M\}$. The object matching problem according to the MRF matching scheme developed in [7] is thus considered as finding an optimal label f^* for each model object by minimizing the following posterior energy:

$$U(f|d) = U(f) + U(d|f) \qquad (3)$$

Detailed description for $U(f)$ and $U(d|f)$ in equation (6) is explained in [7].

In our work, since there are L model objects, then L MAP solutions, $f^{(1)}, ..., f^{(L)}$, are obtained after matching the scene to each of the models in turn. Each configuration $f^{(l)}$, where $l=1,..., L$, corresponds to the label with the highest confidence for model object l. The optimal label f^* for each model object is determined by searching for the global minimum solution of the posterior energy function. Since the number of model and sensed surfaces are not too big in our experiments, an exhaustive search is performed. For a large number of model objects and sensed surfaces, more efficient optimization methods can be employed [7].

The *confidence of correct match* for each model object is then defined as:

$$\Psi(f^{(l)}) = \frac{1/U(f^{(l)}|d)}{\sum_{k=1}^{L} 1/U(f^{(k)}|d)} \quad (4)$$

where $U(f^{(k)}|d)$ is the minimum *posterior energy* for model object K.

The model object with the highest confidence of correct match is hypothesized as the recognized object.

4. Active Recognition using a Sensor Planning Strategy

Reliable recognition is achieved by evaluating the ambiguities existing in the current matching results and in this case is an evaluation of the termination condition. The termination condition is determined based on the comparison among the confidence of correct match for each model object. Assume the nth model object has the maximum confidence of correct matching, that is

$$\Psi(f^{(n)}) = Max(\Psi(f^{(l)})), \quad l = 1,...,L \text{ and } 1 \leq n \leq L \quad (5)$$

thus the termination condition is expressed as:

$$\Psi(f^{(n)}) \geq K \times \Psi(f^{(l)}), \quad l = 1,...,L \text{ and } l \neq n \quad (6)$$

This means that the maximum confidence of correct match for one model should be K times higher than that of all the other model objects where K is a constant. The bigger K is, the lower possible ambiguity may exist in the recognition result but more views may be needed to achieve this. In our experiments K is selected to be 2.

If no new evidence is available, i.e. no more new surfaces can be seen from a new view, the active recognition algorithm is also stopped and the current label that has the highest confidence of correct match in the current label set is determined to be the recognition result.

If the label set is generated and the termination condition is not met, the next viewpoint of the camera needs to be determined in order to capture more information for reliable recognition. The next viewpoint is chosen based on the normalized surface distinguishability defined in section 2. The next viewpoint of the camera is determined to be the position from which the surface with the highest surface distinguishability among the unmatched surfaces of the hypothesized correct matching model should be visible in the new image. The advantages of determining the next viewpoint of the camera in this way is that: (i) the surface with high distinguishability is more helpful in recognizing the model object from other model objects; (ii) the credibility of matching a sensed surface to a model surface is largely affected by the visible area of the sensed surface, the more surface area is visible, the more accurate the 3D features could be estimated [8]. A new matching label set for all the sensed surfaces is produced if there are newly sensed surfaces detected in the new image. The NULL label in the MRF matching framework permits partial object. If there is no new information found in the image taken from the desired position, i.e. no new surface appeared in the new image, and there is a NULL label in this current label set for the hypothesized matching object, the next viewpoint is still determined based on this object. In this case, the surface with the next highest distinguishability in the current hypothesized matching object is used for planning the next view. If there is no NULL label included in the currently hypothesized matching object, and no new surface found in the new image, the label for the current model object is considered incorrect. Thus this label is removed from the label set. The label for another model object that has the next lowest posterior energy is used for choosing the next viewpoint. The active recognition algorithm described above is illustrated in figure 1.

Begin {*active recognition algorithm* }
    Capture an image from initial viewpoint and perform MRF matching.
    Choose the label for the object with highest confidence of correct match from current label
        set.
    Repeat
        Select an unmatched surface with highest distinguishability on the object.
        Repeat
            Determine the camera position, **C**, to see this surface.
            Move sensor to position **C**.
            If *a new surface seen from* **C**, **then**
                generate new match label set.
            Else If *the object with highest confidence of correct match contains NULL*
                matches, **then**
                Select the unmatched surface with next highest distinguishability from
                the object.
            End {if}
        End (if-else}
        Until *a new surface seen* or *all unmatched surface from the object have been explored*
            or *the object does not contain NULL match*.
        If no new match label set generated, **then**
            Choose the label for the object with next highest confidence of correct match.
        End {if}
    Until *all objects have been explored* or *a new match label set is generated*.
End.

Fig.1 Active recognition algorithm based on MRF framework.

5. Experimental Results

In this section, the experimental results on real images are presented. Figure 2 shows six model objects in our experiment with real images. Their surfaces are assigned labels from *1* to *37* (see Appendix). The surface groups for each model object are listed in Table 1.

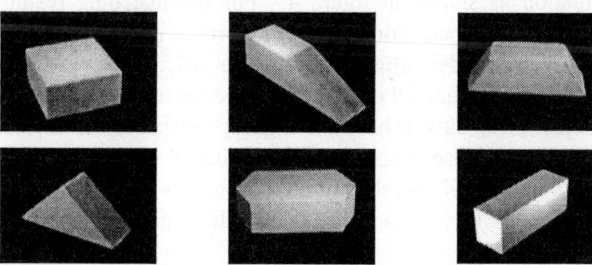

Fig.2. Model object 1 to 6 for experiment with real images (from top to bottom, left to right.)

Table 1 The surface groups for model object in real image

Model object	1	2	3	4	5	6
Surface group	{5,6} {1,2,3,4}	{10},{12} {7,9}, {11}, {8}	{14,16} {13,15} {17},{18}	{19,21} {20,22} {23}	{30,31} {25,26,28,29} {24,27}	{33,35}, {32,34,36, 37}

The matching result of recognizing model object 6 based on MRF framework is shown in table 2. From the matching of the first view, model object 3 has the highest confidence of correct match. Since the termination condition is not met, view 2 is determined in order to capture model surface 14, which has the highest surface distinguishability of model object 3, into the new image. After combining the information from these two views, model object 2 has the highest confidence of correct match. View 3 is then taken in order to take model surface 9 into the image. After combining the information from these three images, model object 6 has the highest confidence of correct match and the matching algorithm is also stopped as the termination condition is now met.

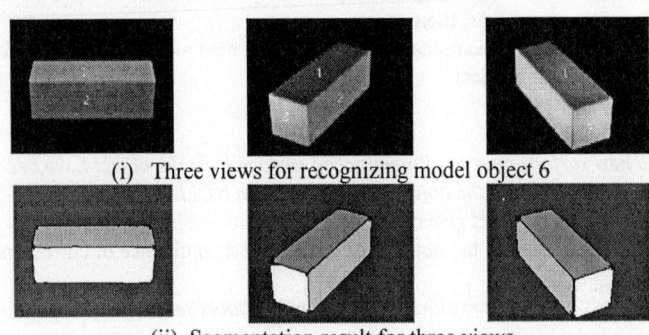

(i) Three views for recognizing model object 6

(ii) Segmentation result for three views

Fig.3. Three views for recognizing model object 6

Table 2. Matching result for fig. 3 (confidence of correct match)

Object	1	2	3	4	5	6
View 1	0.02975	0.249	0.3526	0.087	0.0358	0.2457
View 2	0.033	0.3986	0.094	0.038	0.043	0.1155
View 3	0.0793	0.1355	0.1066	0.077	0.3929	0.486

Table 3 lists the first 3 matching result for recognizing model object 6 under imperfect segmentation as shown in figure 4. For the matching result of view 1, model object 2 has the highest confidence of correct match. According to the unmatched surfaces of model object 2, surface *10*, has the highest surface distinguishability. Model surface *10* is therefore to be captured in the next image. The label for model object 2 remains to have the highest confidence of correct after view 2. The third view is taken also according to the label of model object 2. This time the label for model object 2 still has the highest confidence of correct match. Thus the next image is the same as view 1 in order to take the model surface *8* into the new image. No new surface is visible in this view and there is no NULL assigned to the sensed surface in the label, therefore the label for model object 2 is discarded. According to our matching algorithm, the label that has the second highest confidence of correct match is used to determine the next view. This time the label for model object 6 has the highest confidence of correct matching and the next viewpoint is determined to be the position of view 3 in figure 4, again no new surface is obtained and the label of model object 6 is also discarded. This procedure is continued in the remaining label set and all the labels are eventually discarded. Thus there is no result obtained for this experiment.

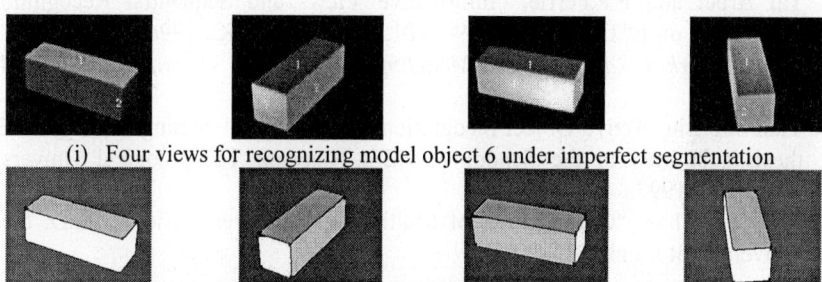

(i) Four views for recognizing model object 6 under imperfect segmentation

(ii) Segmentation result for 4 views

Fig. 4 Four views for recognizing model object 6 under imperfect segmentation

Table 3. Matching result for figure 4 (confidence of correct match)

Object	1	2	3	4	5	6
View 1	0.1326	0.2	0.189	0.156	0.156	0.1655
View 2	0.088	0.269	0.184	0.1317	0.113	0.214
View 3	0.089	0.251	0.208	0.114	0.099	0.2398

6. Conclusion

An active object recognition algorithm using MRF matching and sensor planning strategy is presented. The matching between the sensed and model object is based on two unary surface properties and one binary surface property. Experiments have been conducted on real images for cases with good image segmentation and poor image segmentation. For perfectly segmented images, the algorithm performs well. For imperfectly segmentation, the algorithm will terminate without recognizing any object, but this is still better than recognizing the wrong object. Further work needs to be conducted to improve this matching algorithm making it more robust to the imperfect segmentation. Current work considers only planar surfaces. Future work can extend into matching non-planar surfaces and multiple objects scene with occlusion.

Reference
1. Keith D. Greban et al, "Planning Multiple Observations for Object Recognition", International Journal of Computer Vision, 12:2/3, pp.137-172, 1994

2. Seth A.Hutchinson and Avinash C.Kak, "Planning Sensing Strategies in a Robot Work Cell with Multi-Sensor Capabilities", IEEE Transactions on Robotics and Automation, Vol.5, No.6, pp.765-783, 1989
3. Jasna Maver and Ruzena Bajcsy, "Occlusions as a Guide for Planning the Next View", IEEE Transactions on PAMI, Vol. 15, No. 5, 1993
4. HuiQun Liu et al, "Model-based Next View Planning by Using Rules—Automatic Feature Prediction and Detection", Proceeding of ICCVPR, pp.773-776, 1994
5. Robert B. Fisher, *From Surfaces to Objects—Computer Vision and Three Dimensional Scene Analysis*, John Wiley & Sons, 1989
6. Tal Arbel and F.P.Ferrie, "Informative Views and Sequential Recognition", Proceeding of ECCV, Vol A, 469—481, Cambridge, UK., 1996
7. S.Z.Li, *Markov Random Field Modeling in Computer Vision*, Springer-Verlag, 1995
8. Tianrong Liu, Active Object recognition using Sensor Planning Strategy, MEng thesis submitted to the School of EEE, Nanyang Technological University, Singapore, 1997.
9. Kap Luk Chan, "Optimization of Multiple View Stereo Vision", Ph.D. thesis, University of London, UK, 1991.

Appendix : Label assignment of model objects

Figure A1 to Figure A7 show the label assignment for model object 1 to 6 in experiment:

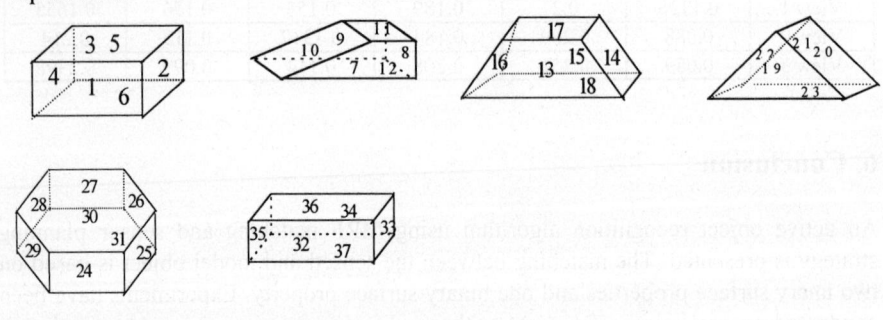

Improved Image Classification Using Morphing

W. Brent Seales and Cheng Jiun Yuan

Computer Science Department
University of Kentucky
Lexington, Kentucky 40506, USA

Abstract. Principal component methods for classifying images have received broad attention and application. For objects with varying appearance, such as three-dimensional objects, increasing the number of object poses represented in the training set is the primary method for improving classification rate. In this paper we show how to improve the performance of this kind of an appearance-based image recognition system. The improvement is obtained by adding new views to the training set which have been generated from existing training data via a morphing algorithm. We show that adding morphed views to the training set increases recognition rate over the same data without morphed views.

Keywords: object recognition, morphing, principal components, appearance models

1 Introduction

The problem of recognizing objects is made more difficult by the change in appearance of 3-D objects as a function of pose. Appearance-based methods for object recognition, such as principal component methods, rely on many different views of an object in order to classify previously unseen images [11].

In this paper we show that recognition rates can be improved by adding images generated from a morphing algorithm to the database of views of an object. The goal is to improve recognition performance using a model that is not based on the tedious acquisition of a large number of views [4]. The algorithm to generate the new, synthetic views is critical, however, and must create views according to realistic constraints over how objects appear. One approach is to generate new views from a complete 3-D model, although this requires 3-D model acquisition, which is a difficult problem. Another approach is simply to obtain many views of an object over large numbers of lighting and pose conditions [6]. It is not always practical, however, to obtain many images of objects in a controlled setting. Our methods provide a way to improve classification performance without using a 3-D model and without requiring the acquisition of a large number of views of an object.

The next section presents a short summary of the principal component method, which we use for image classification. Section 3 reviews the key aspects of the morphing algorithm which make it suitable for generating data for this approach. Experimental results in Section 4 show our performance gains from a prototype system, and the paper concludes with a summary of our findings.

2 Multi-View Classification

The recognition approach at the heart of our algorithms is the method of principal components. Principal component analysis (PCA) allows a large amount of data to be represented with a smaller, lower-dimensional subspace. This is a type of compression based on the Hotelling transform, which is the method of principal components for discrete variables and is known in the continuous case as the Karhunen-Loeve (KL) transform [2]. We base our recognition strategy on PCA, but improve several shortcomings by incorporating image view-morphing. The Hotelling and KL transforms have been applied to object recognition [11, 7, 6, 8].

2.1 Principal Components

We develop the Hotelling transform [3] by defining the input as a set of k images $\{\mathbf{f}_1, \mathbf{f}_2, \ldots, \mathbf{f}_k\}$, represented as vectors, where each image of dimension $n \times m$ has $N = n \times m$ pixels. We treat an image as a vector by placing the pixel values in scanline order. First we subtract from each input vector the average value

$$\mathbf{m} = \frac{1}{k} \sum_{i=1}^{k} \mathbf{f}_i \tag{1}$$

of the input vector set.

Now we compute the covariance matrix $\mathbf{C}$ of the mean-adjusted input vectors $\mathbf{f}$:

$$\mathbf{C}_{N \times N} = \mathbf{U}\mathbf{U}^T \tag{2}$$

where the elements of $\mathbf{U}$ are written as

$$[\mathbf{f}_1 - \mathbf{m}, \ \mathbf{f}_2 - \mathbf{m}, \ \ldots, \ \mathbf{f}_k - \mathbf{m}]_{N \times k} \tag{3}$$

and the elements of $\mathbf{U}^T$ are written as

$$\begin{bmatrix} (\mathbf{f}_1 - \mathbf{m})^T \\ (\mathbf{f}_2 - \mathbf{m})^T \\ \vdots \\ (\mathbf{f}_k - \mathbf{m})^T \end{bmatrix}_{k \times N} \tag{4}$$

The Hotelling transform is obtained from the eigenvectors and eigenvalues of $\mathbf{C}$:

$$\lambda_i \Lambda_i = \mathbf{C}\, \Lambda_i \tag{5}$$

The matrix of ordered eigenvectors Λ is the Hotelling transform, and can be viewed as a matrix that projects input vectors, adjusted by the mean vector $\mathbf{m}$, into eigenspace:

$$\mathbf{p} = \Lambda(\mathbf{f} - \mathbf{m}) \tag{6}$$

Because Λ is orthogonal, an original input vector $\mathbf{f}$ can be recovered exactly from its projection $\mathbf{p}$, the matrix Λ and the mean vector $\mathbf{m}$ by the relation

$$\mathbf{f} = \Lambda^T \mathbf{p} + \mathbf{m} \tag{7}$$

All k eigenvectors are needed in order for the equality to hold. But one advantageous property of principal component methods is that the eigenvectors are ordered. This ordering allows the original input vectors to be approximated by the first w eigenvectors $\{\Lambda_1, \Lambda_2, \ldots, \Lambda_w\}$ with $w \ll k$:

$$\mathbf{f} \approx \Lambda^T_{w \times N} \mathbf{p} + \mathbf{m} \tag{8}$$

Given two vectors that are projected into eigenspace, the closer the projections in eigenspace, the more highly correlated the original vectors [6]. Furthermore, the Hotelling transformation is distance-preserving, which means that

$$\|\mathbf{f} - \mathbf{g}\|^2 = \|\mathbf{p} - \mathbf{q}\|^2 \tag{9}$$

where $\mathbf{p}$ and $\mathbf{q}$ are the projections of $\mathbf{f}$ and $\mathbf{g}$ in eigenspace. When fewer than k eigenvectors are used, the equality becomes an approximation:

$$\|\mathbf{f} - \mathbf{g}\|^2 \approx \|\mathbf{p} - \mathbf{q}\|^2 \tag{10}$$

Thus Eq. (9) shows that distance between input vectors is preserved under the Hotelling transform.

2.2 The Classification Algorithm

Classification of an unknown image is performed using the matrix of ordered eigenvectors Λ. The unknown image is projected into eigenspace and the distance of its projection to the other projected points in the set is the basis for the classification. Note that the entire space is constructed upon the assumption that the new view will be somehow registered with the views which have been used to form the eigenspace. A new image of an object which has been shifted significantly will not align with other images in which the object is centered, for example. Fig. 1 illustrates significant feature shifts induced by a normal change in pose. This registration problem is typically solved by a normalization process, which attempts to center the object of interest and correct for scale and intensity changes. When registration is not done, the classification performance degrades substantially.

3 Improved Recognition Via Morphing

There are (at least) two major problems in using the PCA approach for 3-D object recognition. First, PCA is appearance-based, using 2-D views. Objects which are 3-D change in appearance as a function of pose. Many views are required to represent their appearance. Second, objects may not be registered because of translations and changes in scale. Objects which are not registered cause a degradation in the performance of recognition based on PCA. We use view-morphing to solve the problem of needing many different views in the training set, and the problem of mis-registration.

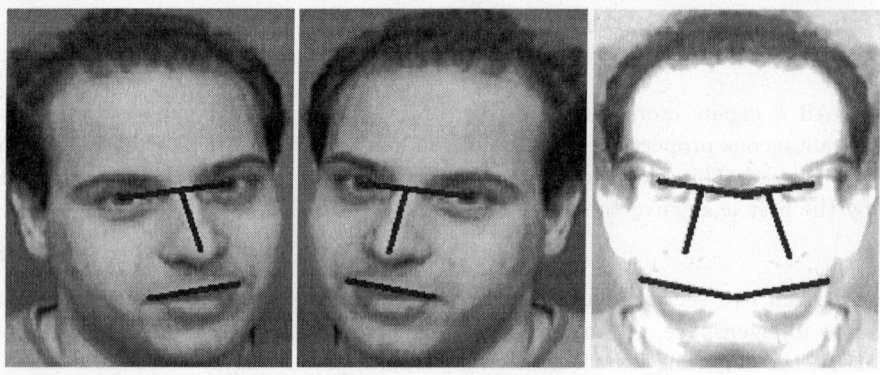

Fig. 1. These two poses of one subject illustrates that large feature shifts are caused from normal pose variation.

3.1 The Morphing Algorithm

The view-morphing algorithm generates accurate views of image data from basis images and calibration information. Calibration can come from truly calibrated cameras, or from point sets which form a relative basis for reconstruction. In general, morph algorithms are appearance-based, which makes them good candidates for improving recognition in the context of the principal component method. For 3-D objects, however, the synthetically-created view must closely approximate object appearance for new poses. The algorithm for morphing that we use specifically addresses how image appearance will change as a function of a changing virtual viewpoint [10].

The view morphing algorithm uses a virtual camera and affine relationships between image planes. These affine relationships, derived from the epipolar geometry of the image set, allows points which are in correspondence to be used to construct intermediate views of a scene. The morphing algorithm is strongly geometric and avoids any 3-D computation. Rather, from point correspondences alone, the necessary affine transformations from one image plane to another generate intermediate image points, which are the image points which would have been generated by a virtual camera viewing the scene. Clearly several constraints must be satisfied in order for a morphed, synthetic view to accurately represent the projection of the 3-D object. The algorithm is completely described in [10].

3.2 Adding Morphed Views to Training Data

The key aspect of improving image classification via morphing is the ability to generate realistic 2-D views from basis views. These new 2-D views are then added to the training set, improving the ability to classify previously unseen

Fig. 2. Two poses of each of 40 distinct subjects were used in this classification experiment. The two subjects above were selected to be morphed.

poses of objects to be recognized. In terms of the appearance model we are using for the principal component classification, the morphing algorithm gives an ideal approach to generating similar vectors for the database.

The goal is to find a small basis set of images for a particular 3-D object from which all other views can be accurately morphed. With such a basis set, new views can be generated via morphing and added to the training set automatically. Currently, we select basis images by hand and morph via the Seitz-Dyer algorithm [10]. We are studying automatic algorithms for selecting basis poses which minimize the number of basis poses while maximizing the number of views which can be realistically generated via view-morphing from the basis set.

View-morphed images lessen the impact of mis-registration by interpolating between basis images which exhibit translations and scale changes. While this does not completely eliminate the need for registration of images, the morphed views more closely approximate the image content over a larger range of scale and translation changes. It is also possible to use the morphing algorithm to generate large numbers of images at varying scales and translations, in anticipation of new views which will vary across multiple scales. The feature shifts shown in Fig. 1 are interpolated by the morph so that features track continuously in the "tween" views from one position to another. These new morphed views in the database will more closely register with a pose which does not align directly with either basis image.

4 Experimental Results

Our prototype system is implemented in C++ on a Unix platform. The eigenspace decomposition is performed using a shared-memory multiprocessor and the "lapack" numerical algorithms package [1]. All our experiments are actually performed on JPEG-compressed data [9] where we see a large speedup in computation time without a degradation in classification rate.

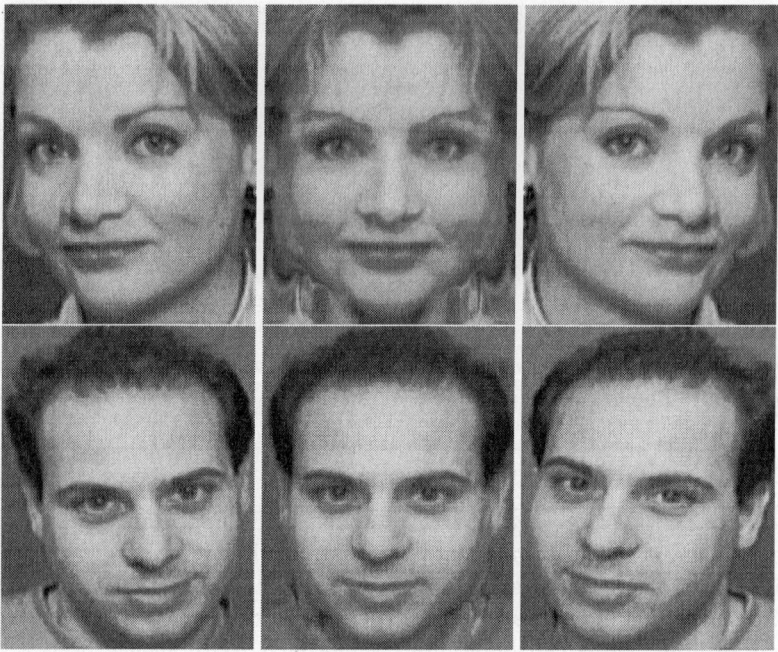

Fig. 3. The middle view for each subject above is a "tween" view from the morphing algorithm [10]. The "tween" views for these two subjects are added into the training set, and this improves the recognition rate for those subjects.

Fig. 2 shows two subjects from the face database we used for experimentation. The database was obtained from The Olivetti Research Laboratory [5]. Two poses from each of 40 subjects were used to create a training set. The remaining 8 real poses of each subject were classified using the eigenspace method and the two-pose, 40-subject training set. The classification results were not good (249/320 correct, which is 78%). The reason is that two poses are typically not enough views to cover the change in appearance over the many poses of each subject. In order to improve this rate, the two subjects shown in Fig. 2 were morphed to create synthetic views, such as a frontal view. These morphed views were added to the original two-pose, 40-subject training set and the classification was again performed. Fig. 3 shows one "tween" frame for each of the two subjects. Arbitrarily many tween frames can be generated, to create a very smooth transition between the two basis images. We selected 3 tweens for each subject and added them to the training set. In the case of both morphed subjects, the classification rate improved substantially.

In particular, with morphed views added, all poses of one subject are recog-

Fig. 4. This shows the similarity between one of the morphed views and a different pose of each subject. The left image for each subject is a real pose; the right image is a morph. By selecting appropriate basis images for morphing, real poses can be closely approximated.

nized correctly (Fig. 2, left). This represents a change in classification rate for this subject from 62% without morphed views to 100% with morphed views included. The second morphed subject ((Fig. 2, right) improves from 25% to 50%. Another significant result was that adding the morphed views did not change any of the other subjects' classification rates. There was improvement for the two morphed subjects and no change in the classification rate for all other subjects.

Fig. 4 shows the similarity between the synthetic morphed views and other real poses in the data set. This similarity accounts for the improvement in the classification rate. The synthetic views more closely match the new poses than do any original poses in the database from any subject. Notice also that the view-morphing causes mis-registered features to more closely align, improving classification rates.

5 Conclusion

In this paper we have shown that view-morphing is a powerful tool which can generate training set data to improve the classification of objects in an object recognition system based on principal component analysis. Using the morph algorithm to generate new views substantially decreases mis-classification rates in our experiments. In particular, two subjects from a large database were morphed, and the morphed views were added to the training data. With this new training set, image classification improved for the two subjects by 38% and 25% respectively. The classification rate of other images was not affected by adding morphed views. This technique has application in many problems where it is not always practical, possible or even necessary to obtain a large number of views of an object.

Acknowledgments

We gratefully acknowledge the support of the National Science Foundation for this research under grants IRI-9308415, CDA-9320179 and CDA-9502645.

References

1. E. Anderson, Z. Bai, and C. Bischof. *LAPACK Users' Guide, Second Edition*. Society for Industrial and Applied Mathematics (SIAM), Philadelphia, PA, 1995.
2. K. Fukunaga. *Introduction to Statistical Recognition*. Academic Press, 1990.
3. R.C. Gonzales and R.E. Woods. *Digital Image Processing*. Addison-Wesley, 1993.
4. J. Krumm. Object detection with vector quantizated binary features. In *Proceedings CVPR '97, San Juan, Puerto Rico*, pages 179–185. IEEE, June 1997.
5. Olivetti Research Laboratory. The ORL database of faces, April 1992. Images available at URL: *http://www.cam-orl.co.uk/facedatabase.html*.
6. H. Murase and S. Nayar. Visual learning and recognition of 3-D objects from appearance. *International Journal of Computer Vision*, 14:5–24, 1995.
7. A. Pentland, R.W. Picard, and S. Sclaroff. Photobook: Content-based manipulation of image databases. *Int. Journal of Computer Vision*, to appear, 1996.
8. W.B. Seales, M.D. Cutts, C.J. Yuan, and W. Hu. Content analysis of compressed video. Technical Report 265-96, Computer Science Dept., University of Kentucky, Lexington, Kentucky, 1996.
9. W.B. Seales, M.D. Cutts, C.J. Yuan, and W. Hu. Object recognition in compressed imagery. *Image and Vision Computing*, 1998.
10. S. Seitz and C. Dyer. View morphing. In *Proc. SIGGRAPH*, 1996.
11. M. Turk and A. Pentland. Eigenfaces for recognition. *Journal of Cognitive Neuroscience*, 3(1), 1991.

Reconstruction of Non-manifold Objects from Two Orthographic Views

Chang-Hun Kim[1] and Tae-Jung Suh[2]

[1] Korea University, Seoul 136-701, Korea
[2] Joong-Kyoung Junior Technical College, DaeJeon 300-100, Korea

Abstract. Since the two-view orthographic representation of a 3D object is ambiguous, it requires a numerous amount of combinatorial searches in the process of reconstruction. This paper presents an efficient algorithm for reconstructing polyhedral 3D objects from two-view drawings. The main feature of the algorithm is to improve the reconstruction process speed. First, the partially constructed objects are reconstructed from the restricted candidate faces corresponding to each area on the two-view drawings. Then the complete objects are obtained from the partially constructed objects by adding perpendicular faces with geometrical validity. By limiting the number of candidate faces corresponding to areas only, the combinatorial search space can be considerably reduced. In addition, the reconstruction finds the most plausible 3D object that human observers are most likely to select first among the given multiple solutions. Several examples from a working implementation are given to demonstrate the completeness of the algorithm.

1 Introduction

Engineering drawings have been used to describe the design of products. A lot of works have been done on reconstruction of 3D objects from orthogonal projections[3, 4, 8, 9, 11]. The early works proposed before 1986 are summarized in Nagendra and Gujar[10] Characteristics of relevant papers after 1986 are given by You and Yang[13]. Also, Lequette[6], Dutta and Srinivas[2] introduced different types of algorithms which are capable of reconstructing cylindrical and conical faces as well, and Kim etc.[5] introduced heuristics to speed up combinatorial search process. Their applications, however, are limited to three orthographic views.

The reconstruction of a 3D model from two orthographic views has also been tried [2, 6, 12, 13]. However, it requires more combinatorial searches in the reconstruction process compared with three orthogonal views. Also, multiple solutions that are not what the designers intended to achieve would be another problem. Dutta and Srinivas[2] made an assumption that every solid has a planar rectangular base. However, they did not generate all possible solutions. You and Yang[13] generated only manifold objects and did not consider the most plausible object.

In this paper, a two phase algorithm will be described in order to generate all possible non-manifold polyhedral 3D objects from two-view drawing. The algorithm reduces the search space and improves the search speed by classifying the candidate faces into ones corresponding to each area on the two-view drawings and others. Then, the algorithm finds the most plausible 3D object that human observers are most likely to select first among the given multiple solutions[7]. To find more plausible solutions prior to unusual ones, it evaluates the face states according to the heuristic rules. Heuristic rules are dynamically applied in the combinatorial search process so that the way of interpretation follows the human understanding. Thus, the algorithm generates all valid 3D objects and finds the most plausible solution first.

2 Problem Statement

2.1 Two-view drawings and 3D object

An object projected in the two orthographic views can be defined by a set of points P and a set of lines L. The set of lines L represents the connectivity between the points in the set P. An object is bounded by planar faces. A face is a directed finite area in a plane and bounded by a closed sequences of edges lying on the same plane.

For the simplicity, let the set of the area, the object, the face, the edge, and the vertex be A, O, F, E, and V, respectively[9]. The symbols of operators used in this study are defined by :

$[A \wedge B]$: *A subset of set A which consists of set B for the given sets A and B.*
$[A \vee B]$: *A subset of set A which provides element B for the given set A and element B.*
$[A \sim B]$: *A subset of set A obtained from the projection of set B on each view for the given 2D components set A and 3D components set B. The set B is reconstructed from the set A.*

2.2 Partially constructed 3D object and complete 3D object

If two orthographic views are given, some faces have no area in the views, they correspond to segment in each view. We denote these faces as *perpendicular face* and edges of *perpendicular face* as *perpendicular edge*. The algorithm consists of the first phase partially constructed objects generation and the second phase complete objects reconstruction. In order to describe the partially constructed object, we use the following definitions.

[Definition 1]
$F_p = \{f | f \in F, n(A \sim f) = 0\}$: F_p *is a set of perpendicular faces.*
$E_p = E \wedge F_p$: E_p *is a set of perpendicular edges.*
$O_p = \{f | f \in ((F \wedge O_c) - (F_p \wedge O_c))\}$: O_p *is a partial object without $F_{(p)}$.*
$O_c = \{f | f \in ((F \wedge O_p) \cup (F_p \wedge O_c))\}$: O_c *is a complete object with F_p.*

2.3 Most plausible solution

In the case of 3D object reconstruction from drawings, if the data input is small, it is assumed that the small amount of input data would reduce the effort and the time for completing the drawing and managing the data. However, the input data used to complete the drawing might bring a difficulty in the 3D object reconstruction process. Since two-view drawing representation may have ambiguity, multiple number of solutions that is not what the designer intended to achieve would be a problem. To solve this problem, we device heuristic rules to generate the sequence number for the solutions from two-view drawing. Note that the three-view drawing has less ambiguity than the case of two-view drawing.

3 Extraction of 3D Components

The algorithms to generate 3D components from the geometrical information given in a set of drawings were proposed by Markowsky and Wesley[8, 9], Haralick and Queeney[3]. However, the papers considered only complete sets of three orthographic views for the application.

3.1 Wireframe generation

A feature represented on the top and the front views has the same X-coordinate on the third-angle projection drawing. Therefore, a candidate vertex can be registered by searching a pair of matching points which lie on the same line between the top and the front views. In addition, a 3D edge can be represented by a line or a point on the drawing. Hence, a 3D edge can be extracted by connecting the two registered candidate vertices. In the case of using only two-view drawings, there are possibilities of generating redundant or pathological edges due to the lack of the side view which can tell the validity of the generated 3D vertices. One of the effective methods for eliminating the pathological edges was given by You etc.[13]. Fig. 1 shows an example of the reconstructed wireframe.

3.2 Generation of candidate faces

All possible surfaces can be created from information of two successive 3D edges. Some pruning operations are performed to remove pathological cases[13]. Pruning operations are consecutively performed until no edges need to be removed. In this study, a special projection technique is employed to simplify the process of searching the candidate faces. At first, the 3D edges on the plane are projected into one of orthographic views. It can be noted that the edges on the original plane and the projection plane has one-to-one relationship. Thus, we use a simple process for finding a closed area from the set of edges on the projection plane. The candidate faces are then obtained by extracting the array of edges. For example, Fig. 1 shows the reconstructed candidate faces.

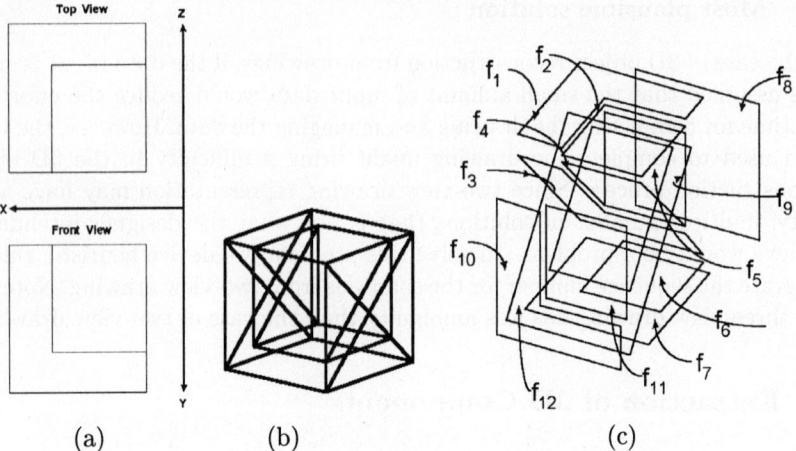

Fig. 1. 3D wireframe and faces reconstructed from two orthographic views: (a) two orthographic views. (b) wireframe. (c) candidate faces.

4 Reconstruction Algorithm

The two phase algorithm generates the partially constructed object first, then later, a complete object is accomplished by generating the matching perpendicular faces. A set of decision rules is developed to enhance the speed of eliminating the pathological elements. In addition, a heuristic approach is also employed to generate most plausible object first.

4.1 Two phase reconstruction algorithm

The reconstruction of 3D objects can be regarded as a combinatorial search problem of the candidate faces which form the objects. Therefore, the conditions for two phase reconstruction algorithm are developed and used to reduce the search space by searching limited candidate faces satisfying the given conditions. The conditions are developed based on the geometrical and topological characteristics of the object. There are conditions for each stage.

[Conditions for partial object(O_p)]

1. $\forall e \in (E \wedge O_p), n((F \wedge O_p) \vee e) \geq 1$
2. $\forall f \in (F \wedge O_p), n(Cross((F \wedge Op), f)) = 0$; *(Cross(F, f) is a subset of F intersecting wity f)*
3. $\forall l \in L, n((E \wedge O_p) \vee l) \geq 1$
4. $L = L \sim (E \wedge O_p)$
5. *There is no face which has an edge inside of the object when the face is projected as a solid line in a view.*
6. *There is at least one faces which has an edge inside of the object when the face is projected as a dashed line in a view.*

[Conditions for complete object(O_c)]
1. The same conditions as the given conditions for the partial object (O_p) are applied again except the first condition.
2. $\forall e \in (E \wedge O_c), n((F \wedge O_c) \vee e) \geq 2m$ (m is a positive number)

Fig. 2, shows the reconstruction process, in the figure, the partially constructed object(Fig. 2(a)) has been generated from the first phase and it does not contain perpendicular faces. The completed object(Fig. 2(c)) has been obtained by generating the perpendicular faces(Fig. 2(b)) in the second phase.

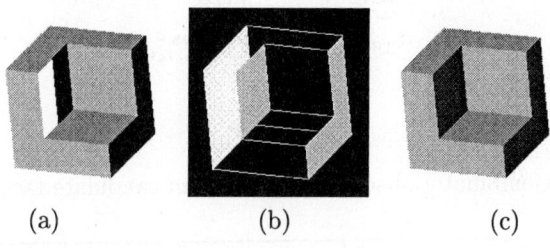

(a) (b) (c)

Fig. 2. Reconstruction process: (a) partially constructed 3D object. (b) patched perpendicular faces. (c) complete 3D object

4.2 Combinatorial search process

In fact, the reconstruction of an object can be simplified as a process of filtering the pathological elements of the candidate faces. The rules determine the states of candidate faces in the combinatorial search process and detect the pathological elements[5]. Fig. 3 shows the combinatorial search process of the candidate faces shown in Fig. 1 and Fig. 4 shows the objects reconstructed from Fig. 3.

4.3 Heuristic rules

In this study, a heuristic approach is employed to increase the search speed. The approach utilizes the human experience in drawing 3D objects such that the selected candidate face would be a true face with a high probability. The approach consists of the two steps. The first stage is to select a series of faces from the candidate faces so that it allows the decision rules to determine the face states effectively. The second stage is to generate the most geometrically plausible objects first[5]. Applying the heuristic rules, Fig. 5 shows an example of the sequence for solutions.

5 Experiments

In this study, C++ language and a Silicon Graphics Indigo 2 workstation were used for the experiment. Furthermore, Motif and OpenInventor were also used for

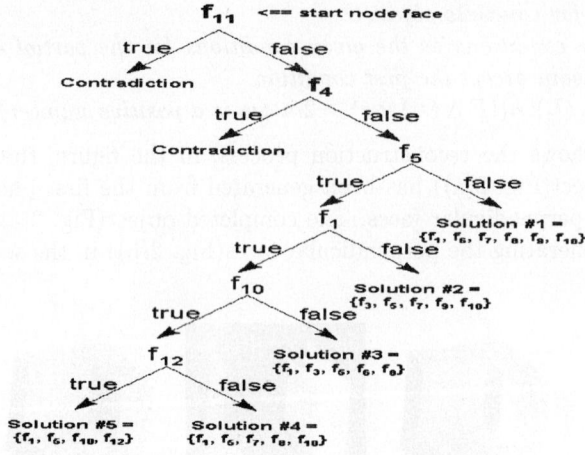

Fig. 3. Combinatorial search process from candidate faces of Fig. 1

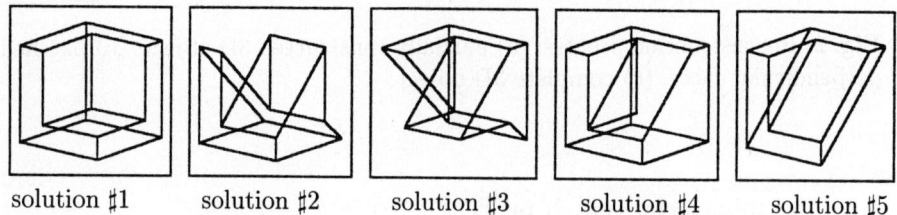

Fig. 4. Examples of multiple manifold solutions

a 3D user interface. Fig. 6(a) - (c) show the various examples of the most plausible solutions of all possible 3D objects generated from two-view drawings and the corresponding side views. If the two-view drawings have much multiplicity[2], the number of perpendicular faces is rapidly increased. But, our algorithm only generates minimal faces that have consistency of the given drawings. For Fig. 6(a), (b), and (c), our algorithm generates 24, 23, and 18 candidate faces, but You and Yang[13] generates 38, 131, and 72 faces respectively.

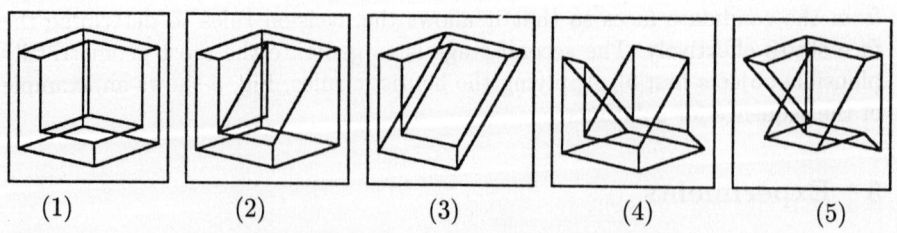

Fig. 5. Sequence of the solutions in Fig. 4

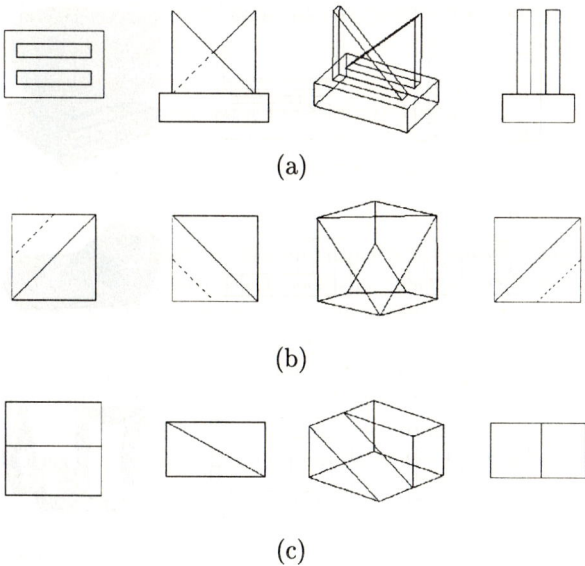

Fig. 6. Examples of the most plausible solution of all possible 3D objects generated from two-view drawings and the corresponding side views

Fig. 7 shows two-view drawings of the practical engineering drawings and the most plausible solution obtained applying for the algorithm.

6 Conclusion

In this study, an algorithm for analyzing and reconstructing 3D objects from two-view drawings has been presented. The two phase algorithm has been used to limit the search space and improve the searching process by classifying the candidate faces into ones corresponding to each area on the two-view drawings and others. Furthermore, this study has used the decision and heuristic rules to generate the most plausible object first among the multiple solutions.

The future work to be applied to the practical objects with curved surfaces. Further research also is required to attempt to find the optimal object configuration using techniques other than heuristic search. We believe that the algorithm presented here will be able to cope with these problems.

Acknowledgments

This paper was supported by NON DIRECTED RESEARCH FUND, Korea Research Foundation, 1996.

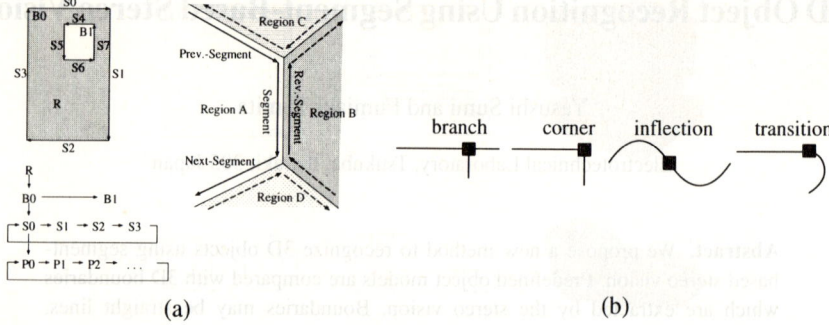

Fig. 1. Boundary representation of an image. (a) Data structure; (b) Segmentation of the boundary.

the VVV project is to develop a versatile vision system that can be used for various applications such as hand-eye robots and autonomous vehicles. The rest of this paper describes the reconstruction of 3D information in the VVV system, the modeling of objects, and the object recognition algorithms. Experimental results are finally shown to demonstrate the effectiveness of this method.

2 Reconstruction of 3D data

2.1 Boundary representation of an image

In the VVV system, we adopt *B-rep*, boundary representation of an image, as an intermediate description which describes both the 2D structure of an image and 3D structure of a scene. Each of stereo images is converted into the B-rep data structure by the image segmentation and the boundary segmentation procedures [6, 7].

Fig. 1 (a) illustrates the data structure of the B-rep which consists of the four layers: *region* (R), *boundary* (B), *segment* (S), and *point* (P). The boundaries are segmented into straight, convex, or concave segments at the points shown in Fig. 1 (b). The inflection points and the transition points are particularly useful for free-form boundaries. In this paper, we call a pair of B-rep forms converted from stereo images *a stereo B-rep*.

2.2 Segment-based stereo vision

3D data of a scene is reconstructed by the segment-based stereo vision system [8]. We call the B-rep point with its 3D position *data point* and the B-rep containing the data points *3D B-rep*. Fig. 2 shows examples of stereo images (a), stereo B-rep (b) and reconstructed 3D B-rep (c).

2.3 Geometrical features

Geometrical features for recognition are generated from the 3D B-rep. We define two types of geometrical features, *data vertexes* and *data arcs*. They are generated by fitting

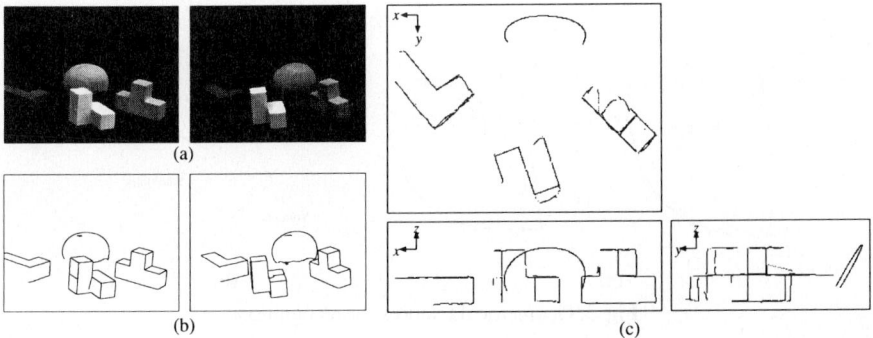

Fig. 2. (a) Stereo images (640×480 pixels, 256 gray levels); (b) Stereo B-rep; (c) 3D B-rep in three orthographic views.

lines and/or circles to the B-rep segments. The data vertex consists of its 3D position and two tangent vectors and the data arc consists of the position of the end point of the segment, and two vectors given by the circle. The geometrical features are generated by the following procedures (see Fig. 3 (a)):

1. A line or circle is fitted in 3D space for data points on each of the B-rep segments.
2. If the fitting error is large, the data points are bisected and the fitting is done for each part of the data point recursively. As a result, the B-rep segment is approximately expressed by a line, a circular arc, or the combination of multiple lines and arcs.
3. When the segment is a single arc, the arc can be used as a data arc.
4. When the segment is expressed by a combination of lines and arcs and an arc is fitted at the end part of the segment, the arc is used as a data arc.
5. Two tangential lines are determined by lines or arcs at the ends of connected two B-rep segments. A data vertex is defined by the two tangential vectors at the midpoint of a line segment which is the shortest distance between the two tangential lines.

Fig. 3 (b) shows geometrical features (vertexes) generated from the 3D B-rep in Fig. 2.

3 Object model

Object models consist of *model vertexes*, *model arcs* and *model points*, as shown in Fig. 4 (a). The model vertexes and arcs are geometrical features and the data structures are the same as the data vertexes and arcs. The model points reflect the whole shape of the objects and sample points on the B-rep segments at equal intervals. Each of the model points has a 3D position and a normal vector.

We have developed both sensor-based and CAD-based modeling systems. The sensor-based system uses the stereo vision system as a sensor. Fig. 4 (b) is an example of CAD-based object models which is shown by connecting adjacent model points.

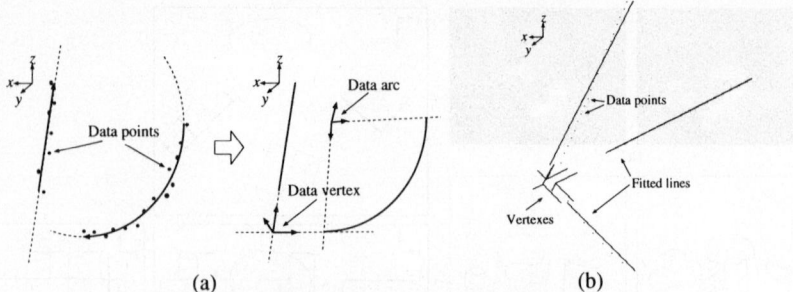

Fig. 3. Geometrical features for recognition.

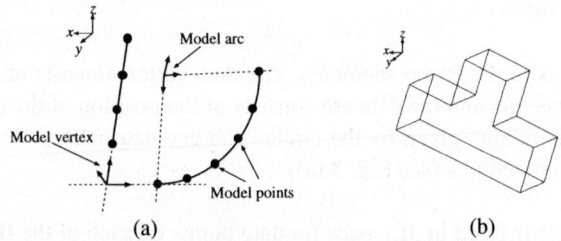

Fig. 4. Object model.

4 Recognition algorithm

The position and orientation of an object are expressed as a 4×4 transformation matrix $T = \begin{bmatrix} R & t \\ 0\,0\,0 & 1 \end{bmatrix}$, where R is a 3×3 rotation matrix and t is a 3D translation vector which move an object model. In other words, the recognition algorithm is a procedure for calculating T by comparing an object model and scene data which is reconstructed by the stereo vision. An object is recognized in the following two phases: the *initial matching* and the *fine adjustment*.

4.1 Initial matching

In the initial matching phase, the candidates of T are roughly calculated by comparing model features (vertexes and arcs) with data features. However, correct correspondence between a model feature and a data feature is not known in advance, so every data feature which is similar to the model feature is a candidate for the correspondence.

When a model vertex V_M moves to the position and orientation of a data vertex V_D as shown in Fig. 5, t is calculated by the 3D coordinates of V_M and V_D, and R is also calculated by two vectors of V_M and V_D.

If the angle θ_M between two vectors of the model vertex is largely different from θ_D of the data vertex, we can assume that the model vertex does not correspond to the data vertex. That is, if $V_M(i)$ $(i=1,\ldots,m)$ and $V_D(j)$ $(j=1,\ldots,n)$ satisfy

$$|\theta_M(i) - \theta_D(j)| < \Theta, \tag{1}$$

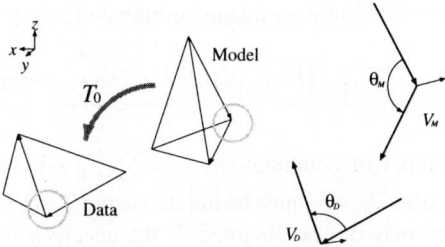

Fig. 5. Initial matching with vertexes.

a candidate of the transformation matrix, $T_{ij}(0)$, is calculated, where m and n are the numbers of vertexes in the model and the scene. Θ is a threshold value. In the case of arcs, the radii are used for θ.

4.2 Fine adjustment

The fine adjustment process tests if the candidates selected in the initial matching phase are correct and makes the transformation matrix $T_{ij}(0)$ more accurate. The procedures are as follows:

1. Model point $\bar{P}(\bar{k})$ ($\bar{k} = 1, \ldots, \bar{p}$) is moved to $\bar{P}'(\bar{k})$ by $T_{ij}(0)$, where $\bar{p}$ is the number of model points. 3D positions $\bar{P}(\bar{k})$ and unit normal vectors $\bar{N}(\bar{k})$ of $\bar{P}(\bar{k})$ are transferred to

$$\bar{P}'(\bar{k}) = R_{ij}(0)\bar{P}(\bar{k}) + t_{ij}(0), \\ \bar{N}'(\bar{k}) = R_{ij}(0)\bar{N}(\bar{k}). \quad (2)$$

2. If

$$\cos^{-1}\left(\frac{\bar{S}'^{t}(\bar{k}) \cdot \bar{N}'^{t}(\bar{k})}{|\bar{S}'(\bar{k})|}\right) > \frac{\pi}{2}, \quad (3)$$

$\bar{P}'(\bar{k})$ is observable from a camera position C, where

$$\bar{S}'(\bar{k}) = \bar{P}'(\bar{k}) + t_{ij}(0) - C \quad (4)$$

is a vector that expresses the direction of observing $\bar{P}'(\bar{k})$. Let $P(k)$ ($k = 1, \ldots, p; p \le \bar{p}$) and $P(k)$ denote the observable model point and its 3D coordinate.
3. If a data point $D(l)$ ($l = 1, \ldots, q$) exists in the vicinity of $P(k)$, let (k, l) be the combination of $P(k)$ and $D(l)$, where q is the number of data points.
4. The transformation matrix T'_{ij} which moves $P(k)$ to $D(l)$ are given by the least squares method which minimizes

$$J = \sum_{(k,l)} |R'_{ij}P(k) + t'_{ij} - D(l)|^2. \quad (5)$$

5. $P(k)$ can be transformed into an image coordinate $[col_k \ row_k]$. If a mean square error

$$\epsilon^2 = \frac{\sum_{(k,l)}\{(col_k - col_l)^2 + (row_k - row_l)^2\}}{r} \qquad (6)$$

is large or the number of combinations $r = \sum_{(k,l)} 1$ is few compared to p, the above processing from 2. to 5. must be iterated using $T(u) = T'T(u-1)$, because sufficient accuracy may not be obtained. If the accuracy does not converge after many iterations, the candidate should be considered incorrect, so we exclude the candidate.

6. After all the candidates have been processed, the transformation matrix $T_{ij}(u)$ which has the largest r is selected as the final recognition result.

A candidate $T_{ij}(0)$ may not be accurate enough, even if the candidate is correct, because local features are used in the initial matching phase. Therefore, the fine adjustment procedures are applied by the following two steps.

Initial adjustment: The position and orientation of the object is roughly adjusted by using only the model points near the geometrical feature.

Main adjustment: The position and orientation is further refined using all the model points.

Fig. 6 shows an example of both the initial adjustment and the main adjustment. This shows that both the initial adjustment and the main adjustment were iterated three times. Although the corresponding data points were found only in the vicinity of the geometrical feature at first, they were finally found over the entire model points.

5 Experiments

Fig. 7 shows example experimental results. The model was moved according to the recognition result and straight lines connecting two adjoining model points were projected and displayed on a left stereo image respectively.

In Fig. 7(a), an object corresponding to the model in Fig. 4 was correctly found from the recognition data shown in Fig. 3. (b) shows another result for the same object. The object was recognized from a cluttered scene in which there were many objects of partly similar shape. In (c), a result for a partially occluded object is shown. (d) shows a result for an object consisting of only a free-form boundary.

The computational time depends primarily on the complexity of the model and the scene. In the case of Fig. 7(a), the recognition took about 6 seconds*. For the scene of (b) in which there were roughly twice the number of data vertexes, it took about 10 seconds. The computational times for (c) and (d) were approximately one second because there were far fewer vertexes or arcs corresponding between the data and model. The recognition error in 3D space usually does not exceed 2 mm** and it is no problem to pick up the recognized object by a robot manipulator.

---

* SuperSPARC, 40MHz
** Errors which arise in the stereo vision measurement are excluded.

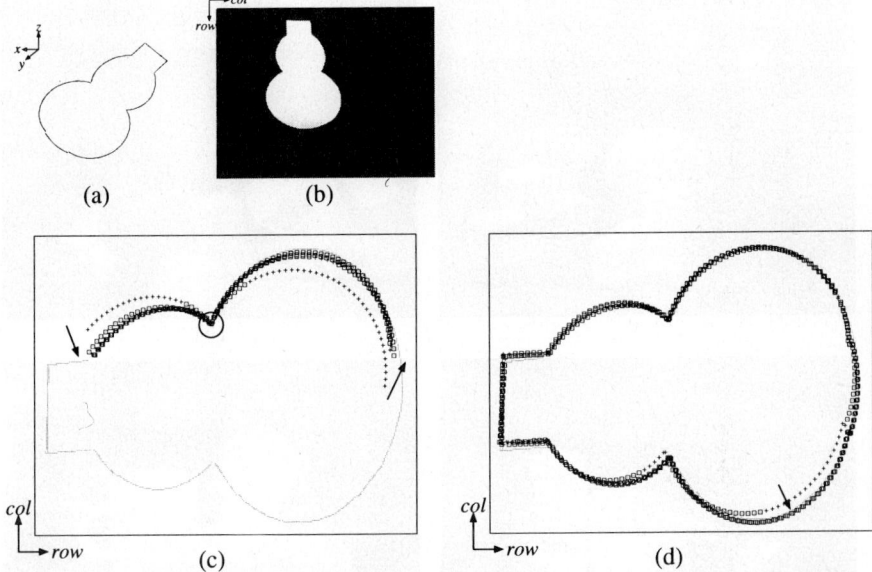

Fig. 6. (a) Object model built by the sensor-based modeler (a piece of cardboard); (b) Stereo image (left); (c) Initial adjustment; (d) Main adjustment; · (dot): data point, ▫: model point that has a data point in the vicinity, +: model point that has no data point, ○: the position of the geometrical feature (a vertex, in this case) used in the initial matching phase.

6 Conclusion

In this paper, we have presented a method to recognize the 3D position and orientation of an object. We consider that the method is useful for the vision of intelligent robot systems because of the following advantages:

- Based on stereo vision. No special sensing devices, such as a laser range finder, are required.
- The targets are not only polyhedra but also objects with free-form boundaries.
- Robust for occlusion as shown by the result in Figure 7(c), because the local features of an object are used for the initial estimation of the object position.

This method is, however, only effective for objects including some fixed edges and does not cope with apparent boundaries of curved objects. We have already developed a recognition method for the apparent contours that is an extension of the method proposed in this paper [9]. We can now combine these two methods to recognize any rigid object.

Acknowledgments

We would like to express our thanks to the members of the Computer Vision Section, ETL and the VVV working group for helpful discussions.

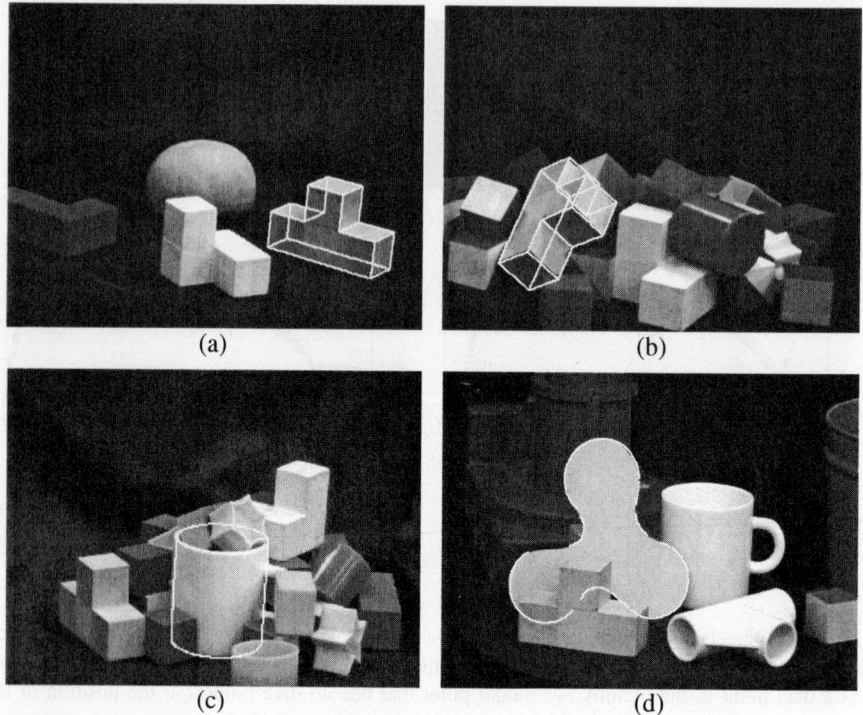

Fig. 7. Experimental results. (a) a toy block; (b) the toy block in a cluttered scene; (c) a partially occluded mug; (d) a piece of cardboard which consists of a free-form boundary.

References

1. Kriegman, D. J., Ponce, J.: On recognizing and positioning curved 3-D objects from image contours. *IEEE Trans. on PAMI*, **12**, 12, 1127–1137, 1990
2. Wong, K. C., Kittler, J.: Recognizing polyhedral objects from a single perspective view. *Image and Vision Computing*, **11**, 4, 211–220, 1993
3. Arman, F., Aggarwal, J. K.: Model-based object recognition in dense-range images — A review. *ACM Computing Surveys*, **25**, 1, 5–43, 1993
4. Porrill, J., et al.: TINA: A 3D vision system for pick and place. *Image and Vision Computing*, **6**, 2, 91–99, 1988
5. Tomita, F.: Toward practical 3D vision. *J. Robotics Society of Japan*, **12**, 8, 1124–1127, 1994 (in Japanese)
6. Tomita, F., Tsuji, S.: *Computer Analysis of Visual Textures*, chapter 3. Kluwer Academic Publishers, 1990
7. Sugimoto, K., Tomita, F.: Boundary segmentation by detection of corner, inflection and transition points. *Proc. IEEE Workshop on Visualization and Machine Vision*, 13–17, 1994
8. Kawai, Y., et al.: Search for stereo correspondence based on connectivity of segments. Technical Report PRMU96–135, IEICE, January 1997 (in Japanese)
9. Sumi, Y., Kawai, Y., Yoshimi, T., Tomita, F.: Recognition of 3D free-form objects using segment-based stereo vision. *Proc. ICCV98*, January 1998 (to appear)

The State of Color Vision Research

Brian Funt

School of Computing Science
Simon Fraser University
Vancouver, British Columbia
Canada V5A 1S6

ABSTRACT

The members of the Computational Vision Laboratory at Simon Fraser University have been studying color for over a decade. I will discuss some of the main color issues, the progress we have made in understanding them and the application of our methods to color-based object recognition and digital photography.

To explain color perception, we must explain how it is that we see colors as relatively stable despite changes in the incident illumination. I make the assumption that color, like the rest of visual perception, is there to give us information about the world, the surface properties of objects in particular, and so the stability and reliability of the information is important. The problem of color stability arises because the light reaching our eyes from an object is the product of the object's surface reflectance and the spectrum of the light illuminating the object. We do not have direct access to the properties of the incident light, so somehow we must estimate them from the light we receive from the object. To make matters worse, our eyes only measure the spectrum at extremely low resolution.

Clearly a stable representation of object color would be useful in computer vision. It is equally important for digital photography since we do not want images to produce images whose color balance depends on the qualities of the ambient scene illumination. Until recently the growth of digital photography was limited less by camera technology than by the lack of economical digital printers with photographic quality, a situation that has changed dramatically in the last six months with the introduction of some new ink jet printer technology (e.g. the Hewlett-Packard Photo Smart printer). Digital photography is providing new impetus to the search for better models of color perception.

Comparative tests of many different color constancy methods show that two work most reliably: one based on a neural network to estimate the illumination properties from a color histogram, and a second based on the constraints provided by the image gamut.

Color Vision and Color Media Processing Research in Asia

Shoji Tominaga

Osaka Electro-Communication University
Osaka, Japan

ABSTRACT

I will survey the state of color vision and color media processing research in Asia from a computation viewpoint. First, I will discuss briefly the principles of color vision and color measurement. Then I describe why color constancy is important and difficult to be realized in machine vision. The goal of computational color constancy is to recover the physical properties of illuminant and surface from photosensor responses. I introduce a vision system for color constancy and our current results.

Next, I will discuss algorithms for understanding complex color images. Real images exhibit rich and complex structure, whose nature is determined by the physical and geometric properties of illumination, reflection, and imaging. I suggest some reflection models adequate for describing light reflection of a variety of materials. Then approaches are introduced for estimating scene parameters based on the reflection models.

Finally, I will discuss the issue of color management. The concept of device-independent color reproduction of image has received wide spread attention since the advent of desk-top publishing systems. Color reproduction requires color conversion between the color signals, depending on a device, and the standard color coordinates, representing color appearance. Mapping methods are introduced for solving the color conversion problem for color printers.

Recent Advances in Detection and Description of Buildings from Multiple Aerial Images

Sanjay Noronha and Ram Nevatia*
Institute for Robotics and Intelligent Systems
University of Southern California
Los Angeles, California 90089-0273
Keywords: Stereo, Building Detection, Multiple Images

Abstract

A brief description of a method for detection and description of rectangular buildings from two or more registered aerial intensity images is provided. A new interactive editing module that can use partial results of the automated process to efficiently correct errors and derive complete descriptions is also described.The automated system operates by grouping features hierarchically to form roof hypotheses which are then verified by using wall and shadow evidence. Grouping and matching steps are interleaved and multiple descriptions are preserved when clear choices are not available. Some recent results are given.

1 Introduction

3-D models of buildings in urban environments are important for a variety of applications and several systems to extract them have been developed (e.g. [1]). This is a highly challenging task. The desired object boundaries are typically highly fragmented due to low contrast, occlusion caused by nearby vegetation and by smaller structures on the roofs, and need to be grouped to yield the desired objects. In our work, we limit the buildings shapes to be rectilinear (i.e. rectangular or compositions of rectangular shapes) to aid the task of organization. However, many other structures such as roads, sidewalks and parking lots can also give rise to rectilinear organizations and need to be distinguished from the building structures. The second major problem is to infer 3-D shapes of the objects, while this is possible with single images (e.g. [4]), the task is made easier by using multiple images. We now need to make correspondences between the features in different images. This task too is difficult in the aerial image domain. Area correlation methods are likely to have difficulty as the viewpoints can be widely separated, the images may be taken at different times and the building roofs have limited texture.

Our approach is to first form hypotheses for building roofs and to verify the hypotheses by using evidence from cast shadows and visible walls (if any). We use a hierarchy

---

* This research was supported, in part, by the Defense Advanced Research Projects Agency of the United States Department of Defense under grant No. DACA76-97K-0001, monitored by the Topographic Engineering Center and in part by a subgrant from Purdue University under Army Research Office grant No. DAAH04-96-10444.

of feature levels (consisting of lines, parallels, "U" shapes and parallelograms). Matching at one level is used to form group hypotheses at the next level. We maintain multiple matches at each level and resolve them only when sufficient information becomes available at the higher levels. We briefly describe our system and show some recent results; more details may be found in [5]. We also describe an interactive editing module of our system that uses partial results from automatic analysis to help correct the errors in a highly efficient way.

2 Description of the Automatic System

Our system can be considered to have three major stages: first, features are extracted and grouped to form roof *hypotheses*, stronger candidates are then *selected* from this set, and finally further reduction is made by *verifying* hypotheses by examining other evidence such as presence of walls and shadow. These three steps are described briefly below.

2.1 Grouping Features to form Roof Hypotheses

Initially, lines are detected using the Canny edge-detector and grouped on the basis of colinearity and proximity. These lines are matched across all views; multiple matches are retained. The constraint used in matching is that the matching lines fall in a quadrilateral defined by the epipolar lines from ends of one line and the allowed disparity range. Figure 1 shows an example with images and matched lines.

Figure 1 Images with matched lines overlaid

Next, binary junctions (formed by the intersection of exactly two lines) are formed and matched. The matching junctions should be on a corresponding epipolar line, the lines forming the junctions should match across the views and be orthogonal in 3-D and the computed height should be within the allowed range. In addition, if a third view is available, the usual trinocular constraint is applied.

Next, we search for parallels and their matches. Parallels are formed between pairs of lines in the same view that are separated by the less than the maximum width of a building. For two parallels to match, each line forming a parallel in one view should match with a line forming the matching parallel in the other view. While the task domain causes a large number of parallels in each image because of the alignment of buildings, roads, parking lots and shadows, the number of parallel matches is typically lower than the number of lines in any image.

Then, U's are formed by an alignment of two junctions, or by a parallel that has closure evidence near one of its ends. In the case of U match being formed from a parallel match coupled with evidence of closure, no other constraint are applied. In the case that a U match is formed from two aligned junctions, these junction matches should be approximately coplanar.

Finally, parallelograms are formed as a basis for roof hypotheses. To hypothesize a 3D rectangle the minimal requirement is a U match (a match may be in 2 or more images), with additional evidence arising from an "opposing" U match, an "opposing" U in a single image or a parallel match. The components of a parallelogram match (line matches and junction matches) must be coplanar in 3D.

2.2 Selection of Building Hypotheses

The parallelogram matches serve as roof hypotheses, and are equivalent to having a 3D model of the buildings. They satisfy the constraints of being rectangular in 3D and nearly coplanar. In addition, the height and orientation with respect to the ground is known. However, additional processing is still necessary to distinguish building parts from rectangular areas on the ground. This selection is done by using the actual line and junction evidence that is present in forming the roof hypotheses. Some kinds of lines and junctions can actually reduce the confidence in the hypotheses. The selected hypotheses for an example are shown in Figure 2.

2.3 Verification and Combination of Building Hypotheses

It may be noted that so far the evidence that was used was concerning the roof only. The presence of lighting causes shadows to be cast. When the view is oblique, some vertical sides of the walls of the building may be visible. These cues are used to verify the selected hypothesis.,The combined score from the wall evidence and shadow evidence is thresholded to obtain rectangular building (or building component, in the case of non-rectangular rectilinear buildings) hypotheses.

Rectilinear buildings can be decomposed into rectangular components. Verified rectangular hypotheses are examined for combination according to two criteria of proximity, and overlap. Results after this step of processing are shown in Figure 3.

3 Some Results of Automated Processing

We show results on two windows from images taken over Ft. Hood, Texas, and Fort Benning, Georgia. These sites have building shapes that are largely restricted to being rectilinear but images are otherwise very complex. These datasets are commonly used for evaluation of building detection systems.

Figure 4 shows the results obtained on a complex area of Fort Hood. 17 buildings

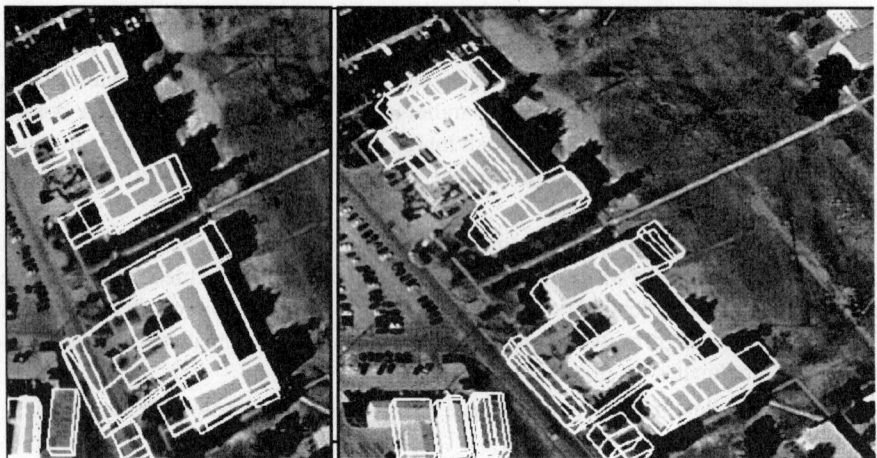

Figure 2 Selected hypotheses from Figure 1

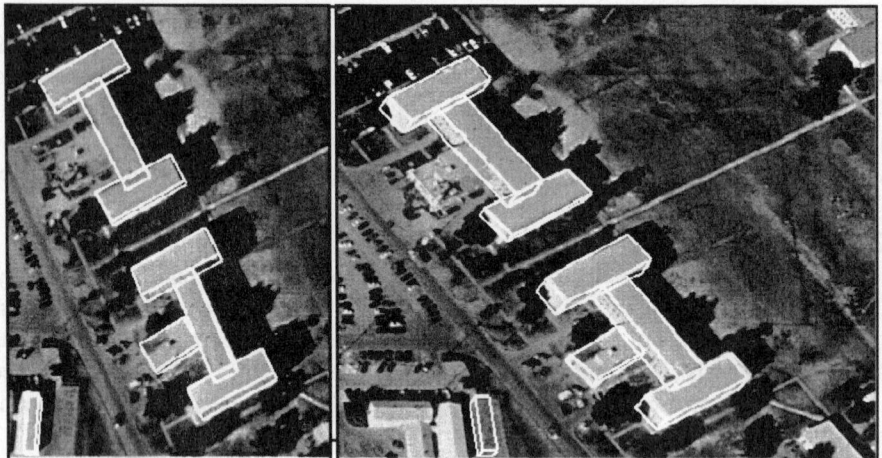

Figure 3 Results are verification and combination

were correctly detected, 3 buildings were not detected, and 4 rectangular building fragments were missed. The buildings in the image are not very high (typically less than 10m tall), and there are numerous instances buildings occluding themselves. This phenomenon caused the 2 rectilinear fragments labeled A and B, in Figure 4 to be missed completely. The building labeled C in Figure 4 was missed because it was too low to verify based on the height evidence alone, and is partially occluded by surrounding foliage. The building labeled D in Figure 4 was missed because of its complex shape that cannot be conveniently decomposed into rectangular fragments.

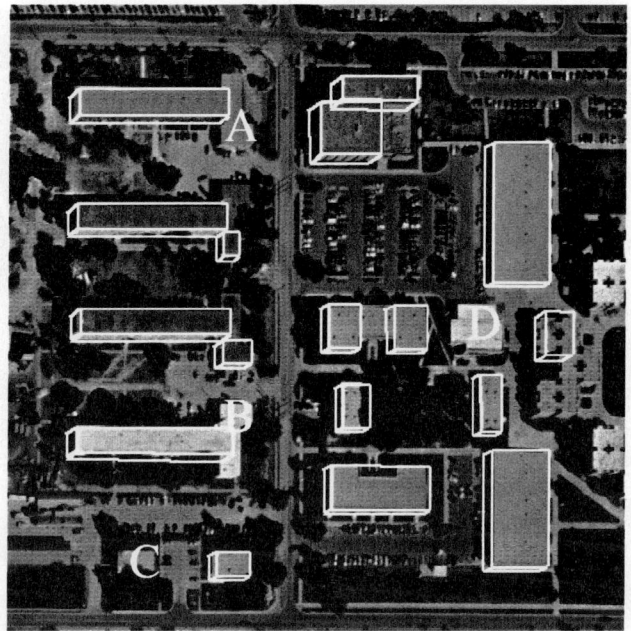

Figure 4 Results on a complex area of Fort Hood, Texas

Figure 5 and Figure 6 show results obtained on the Fort Benning site. In total, 20 buildings (or partial buildings) were correctly detected, with no buildings missed, and 1 false alarm. Buildings labeled A, B, and C in Figure 5 are examples of partial hypotheses. The reason these hypotheses were not completed is that the bounding roof walls are raised above the rest of the roof, causing shadows to be cast on the roof itself. This is a special case of self-occlusion that the system has no knowledge of, currently. A similar problem is encountered in the multi-level complex building in Figure 6 in the area labeled D. However, the system correctly detects and describes other fragments of the same complex building. This building is particularly hard because it has many levels, coupled with gabled roofs. The false alarm is in Figure 6 and is labeled E. This false alarm is very deceptive, and looks like a sloping part of a building even to a human observer. The only cue that it is at ground level is the vegetation that occurs in a part of it.

Fort Benning is a fairly difficult site, with a variety of buildings. The gabled roofs have sub-structures on them. The pathways between the buildings have the same texture as the many of the rooftops. In addition there are confusing details like vegetation and vehicles. Considering the difficulties, the automatic system does a very good job.

4 Interactive Corrections

While the performance of our automatic system, we believe, is an advance over previously available techniques, the building models need to be refined further to meet the needs of most application tasks. For this purpose, we have developed an interactive

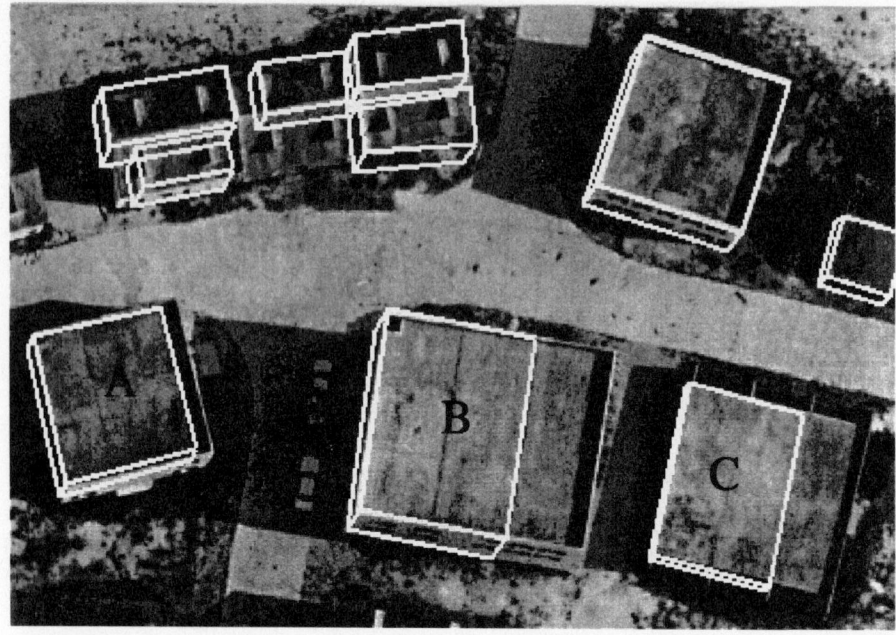

Figure 5 Results on an area of Fort Benning

methodology that can use partial results of the automated analysis. Our method follows the one described in [2], but is made easier due to availability of multiple images.

There are two distinct tools available to the user to interactively construct or refine a model. The first tool requires only an approximate location for a building; it proceeds as follows:

- the user clicks once within the area contained by each building the user would like to model.
- the system searches for and returns the best hypothesis that contains this point. This hypothesis is taken from the set of selected hypotheses from the automated analysis.

The second tool requires a user to specify approximate junctions. It is divided into two sub-methods: one that deals specifically with roofs that are parallel to the ground plane; the other deals with roofs that slope with respect to the ground. The method that deals with roofs parallel to the ground plane operates as follows:

- The user specifies a junction by clicking near one corner of a building. This causes the system to look for junctions in the neighborhood. If one or more junctions is found, the system attempts to construct hypotheses from all the junctions found, by looking for U-completions and lines parallel to the lines forming the junctions. The system chooses the best hypothesis, and then determines the 3D height of the hypothesis by matching the 2D hypothesis with the line evidence in all the other available images. In many cases, this is sufficient to construct a complete 3D hypothesis. If a hypothesis is formed it is displayed that the user may accept or choose to provide further hints.

Figure 6 Results on an area of Fort Benning

- If no hypothesis is formed or the user is not satisfied with the generated hypothesis from the first interaction described above, the user needs to specify another junction by clicking near another corner of the desired building. The process of constructing a 2D hypothesis and extending to 3D, is repeated, with the additional constraint of using the line defined by the two junctions near the user specified locations. This interaction usually suffices to complete most buildings that are not modeled after the first interaction.
- If no hypothesis has been generated at this point, or if the generated hypothesis is not satisfactory, the user can choose to click again on a third corner of the desired building. Three points are sufficient to form a roof hypothesis (a parallelogram in the image); the plane of the roof can be determined from just a single view by employing the constraint that the roof hypothesis must form a rectangle in 3-D. Two choices are possible, however. The system chooses the one which forms the smaller angle with the ground plane.
- The system determines the height of the building by searching for a value that maximizes the score for the building hypothesis. Should the height of the hypothesis be unacceptable to the user, it can be corrected by a single click on any "ground" point. The system figures out which "ground" point is being pointed to and reconstructs the 3D hypothesis with this new input. In our experience, this correction has rarely been necessary.

The method that deals with rooftops that are inclined with respect to the ground plane in 3D (but are still rectangular) requires exactly 3 mouse clicks. The user clicks near three of the corners of the building being modeled. The method then proceeds in a manner similar to the three junction case described above. Again, two choices are possible for the roof orientation and we select the one that is the least sloping.

In the example shown in Figure 7 below the model was constructed using this tool. While the main purpose of the tool is to augment and correct the model of the site that is generated by the automatic system, we demonstrate its effectiveness in the following example by constructing the model from scratch. To construct the model, the user needed an average of 1.5 clicks near a building junction. The computed heights were very accurate, and required no correction.

Figure 7 Results of the smart interactive system on Fort Benning

Bibliography

[1] A. Gruen and M. Baltsavias, Editors, "Automatic Extraction of Man-Made Objects from Aerial and Space Images", Birkhauser Verlag, 1997.

[2] S. Heuel, R. Nevatia, "Including Interaction in an Automated Modeling System", Symposium on Computer Vision, Miami Beach, FL, November 1995(383-388)

[3] M. Ito and A. Ishii, "Three-view stereo analysis", IEEE Transactions on Pattern Analysis and Machine Intelligence, 8:524-532, 1986.

[4] C. Lin, and R. Nevatia, "3D Descriptions of Buildings from an Oblique View Aerial Image", IEEE International Symposium of Computer Vision, 377-382, 1995.

[5] S. Noronha and R. Nevatia, "Detection and Description of Buildings from Multiple Aerial Images", IEEE Computer Vision and Pattern Recognition, pp. 588-594, San Juan, PR, June, 1997.

Visual Surveillance of Human Activity

Larry Davis[1] Sandor Fejes[1] David Harwood[1]
Yaser Yacoob[1] Ismail Hariatoglu[1] Michael J. Black[2]

[1] Computer Vision Laboratory, University of Maryland, College Park, MD 20742
[2] Xerox Palo Alto Research Center, 3333 Coyote Hill Road, Palo Alto, CA 94304
lsd, fejes, harwood, yaser, hismail@umiacs.umd.edu,
black@parc.xerox.com

1 Introduction

In this paper we provide an overview of recent research conducted at the University of Maryland's Computer Vision Laboratory on problems related to surveillance of human activities. Our research is motivated by considerations of a ground-based mobile surveillance system that monitors an extended area for human activity. During motion, the surveillance system must detect other moving objects and identify them as humans, animals, vehicles. When one or more persons are detected, their movements need to be analyzed to recognize the activities that they are involved in. Ideally, the surveillance system would be able to accomplish this even while continuing to move; alternatively, the system could stop and stare at that part of the scene containing people.

In Section 1 we describe a novel approach to the problem of detecting independently moving objects from a moving ground camera, and illustrate the approach on sequences taken in very cluttered environments. Current research focuses on the problem of classifying those independently moving objects as people based on a combination of their appearance and movement. In Section 2 we describe a system that can track multiple moving people using sequences taken from a stationary camera. This system of algorithms, which has been implemented on a PC and can process 10-30 frames per second (depending on the number of people within the field of view and the resolution of the imagery) uses a hierarchy of tracking modules to identify and follow people's heads, torsos, feet, ... Finally, in Section 4 we explain how the recovered motion of these people can be classified into various activity classes using a principal component model of the time variation of motion of the body parts.

2 Detecting independent motion using directional motion estimation

This section briefly describes an application of the theory developed in [2] to the problem of detecting independent motion in long image sequences. The ap-

---

*** The support of the Defense Advanced Research Projects Agency (ARPA Order No. C635), the Office of Naval Research (grant N000149510521) is gratefully acknowledged.

proach, which is based on two simple geometric observations about directional components of flow fields, allows general camera motion, a large camera Field Of View (FOV), and scenes with large depth variation; no point correspondences are required. Due to the projection method the original problem of detecting independent motion is reduced to a combination of robust line fitting and one-dimensional search. More details about the method can be found in [1, 3].

2.1 Properties of directional components of visual displacement fields

Given an optical flow field, we construct a scalar field by projecting the optical flow vectors onto a given projection direction. Our approach to motion estimation is then based on analyzing cross sections of this projected flow field; in particular, cross-sections both parallel and orthogonal to the chosen projection direction. This analysis leads to recovering the projections of the camera motion, which we call the directional motion parameters. In the simple case of a narrow-FOV camera the rotational projected flow is constant along the parallel cross-sections, and varies linearly along the orthogonal cross-sections; we call this the *linearity property*. Since the projected translational flow is zero at the projection of the FOE on any parallel cross-section, the second observation leads to what we call the *divergence property*: points to the "left" of the projected FOE in a parallel cross-section have projected flow less than the flow at the projected FOE, and points to the "right" have greater projected flow. Since we do not know, a priori, what the projected rotational flow is, we estimate at *each point* along every parallel cross-section that flow value which best satisfies the divergence property. This is, essentially, equivalent to estimating a flow value that minimizes negative depth [6] along that parallel cross-section. Orthogonal cross-sections of these new projected flow values are then constructed. Finally, using the linearity property, the projected rotation parameters are estimated by finding that orthogonal cross-section on which the projected (new) flow values are best fit by a linear model. Extensions of the algorithm to large FOVs (accomplished by embedding it into a recursive derotation framework) or very small FOVs (in which the divergence property of the projected flow cannot be used) can be easily achieved. In any event, once the projected motion parameters are estimated we know by the epipolar constraint that along any parallel cross-section, all points to the left of the projected FOE should have projected motion less than the final estimate of rotation at the projected FOE obtained from the linear model, and all points to the right should have greater projected flow.

2.2 The detection algorithm

The ability to verify the epipolar constraint for arbitrary flow fields using only low-dimensional projections of the original flow field provides a simple basis for detecting independently moving objects. For this purpose we need to incorporate only one or a small collection of directional components of the flow field. The image locations where the linearity and the divergence constraints of projections are violated are considered as regions with independent motion.

Fig. 1. Detection of moving people (top and bottom) and a moving vehicle (middle) from a hand-carried camera. Each row illustrates two (non-consecutive) frames of long image sequences taken from several seconds apart.

In practice, one must take into account the fact that the linear fitting process used to estimate the projected rotation parameters must be robust to both measurement error in flow estimation and errors introduced by the presence of independently moving points; and, in the detection of independently moving points, one must take into account measurement error in flow estimation. The first problem is addressed using robust line fitting, in which the parallel cross-sections corrupted by independent motion are eliminated from the fit by a repeated-median-based robust line-estimator [7]. The second problem is addressed by first assuming that the parallel cross sections that are included in the robust line fit ("inliers") in fact do not include any independently moving points. This allows us to identify detection thresholds that adapt to changes in imaging conditions. Intuitively, for each parallel inlier cross-section we find the projected flow vector which most violates the assumption that no pixel along an inlier cross-section is moving independently. We then consider the worst violator amongst all the inlier cross-sections, and use the magnitude of the difference in flow between that pixel and the flow at the projected FOE on that cross-section as our adaptive threshold. This threshold is then applied to the remaining "outlier" cross sections. This simple automatic adaptive thresholding procedure provides a good trade-off between sensitive detection and low false alarm rate, and is a significant improvement of the detection algorithm over those which apply fixed thresholds.

In order to further improve the reliability and robustness of the algorithm, frame-by-frame-based instantaneous detections need to be integrated over both space and time. We employ temporal integration over motion trajectories using tracking to verify detections and eliminate short-term drop-outs. Finally, a spatial integration provides grouping of independently moving pixels that pass the temporal analysis based on coherence in location and velocity.

Figure 1 shows three examples of detecting independent motion from a hand-carried camera. The camera FOV is relative large (55°) while the scenes contain different degree of depth variation. In all examples the primary input for our algorithm was the simple normal flow as a particular directional component of the flow field. In each frame dark pixels indicate local detections verified by the temporal filter. The high-lighted bounding boxes represent groupings of these detections using spatial and velocity coherence. Current research focuses on characterizing the appearance and motion of independently moving objects to classify them as people, vehicles, etc.

3 The W^4 System

W^4 is a real time system for tracking people and their body parts in monochromatic imagery. It constructs dynamic models of people's movements to answer questions about what they are doing, and where and when they act. It constructs appearance models of the people it tracks so that it can track people (who?) through occlusion events in the imagery. In this section we describe the computational models employed by W^4 to detect and track people and their parts. These models are designed to overcome the inevitable errors and ambiguities that arise in dynamic image analysis. These problems include instability in segmentation processes over time, splitting of objects due to coincidental alignment of objects parts with similarly colored background regions, etc.

W^4 has been designed to work with only monochromatic video sources, either visible or infrared. While most previous work on detection and tracking of people has relied heavily on color cues, W^4 is designed for outdoor surveillance tasks, and particularly for night-time or other low light level situations. In such cases, color will not be available, and people need to be detected and tracked based on weaker appearance and motion cues. W^4 is a real time system. It currently is implemented on a dual processor Pentium PC and can process between 10-30 frames per second depending on the image resolution (typically lower for IR sensors than video sensors) and the number of people in its field of view. In the long run, W^4 will be extended with models to recognize the actions of the people it tracks. Specifically, we are interested in interactions between people and objects - e.g., people exchanging objects, leaving objects in the scene, taking objects from the scene. The descriptions of people - their global motions and the motions of their "parts" - developed by W^4 are designed to support such activity recognition.

W^4 currently operates on video taken from a stationary camera, and many of its image analysis algorithms would not generalize easily to images taken from a moving camera. Other ongoing research in our laboratory attempts to develop both appearance and motion cues from a moving sensor that might alert a system to the presence of people in its field of regard [4].

W^4 consists of five computational components: background modeling, foreground object detection, motion estimation of foreground objects, object tracking and labeling, and locating and tracking human body parts. The background scene is statically modeled by the minimum and maximum intensity values and temporal derivative for each pixel recorded over some period, and is updated periodically. For each frame in the video sequence, foreground objects are detected by frame difference thresholding, connected component analysis, and morphological analysis. These foreground objects are tracked and labeled by a forward matching process from previously detected objects to currently detected objects. Motion models, which are based on matching silhouette edges of foreground objects in two successive frames and a recursive least square method, are used during object tracking to estimate the expected location of objects in future frames. A cardboard human model of a person in a standard upright pose is used to model the human body and to locate human body parts (head, torso, hands, legs and feet). Those parts are tracked using dynamic template matching methods. Figure 2 illustrates some results of the W^4 system.

Fig. 2. Examples of using the cardboard model to locate body parts in different situations: four people meet and talk (first line), a person sits on a bench (second line), two people meet (third line).

4 Activity Modeling and Recognition

Activity representation and recognition are central to the interpretation of human movement. There are several issues that affect the development of models of activities and matching of observations to these models,

- Repeated performances of the same activity by the same human vary even when all other factors are kept unchanged.

- Similar activities are performed by different individuals in slightly different ways.
- Delineation of onset and ending of an activity can sometimes be challenging.
- Similar activities can be of different temporal durations.
- Different activities may have significantly different temporal durations.

There are also imaging issues that affect the modeling and recognition of activities

- Occlusions and self occlusions of body parts during activity performance.
- The projection of movement trajectories of body parts depend on the observation viewpoint.
- The distance between the camera and the human affect image-based measurements due to the projection of the activity.

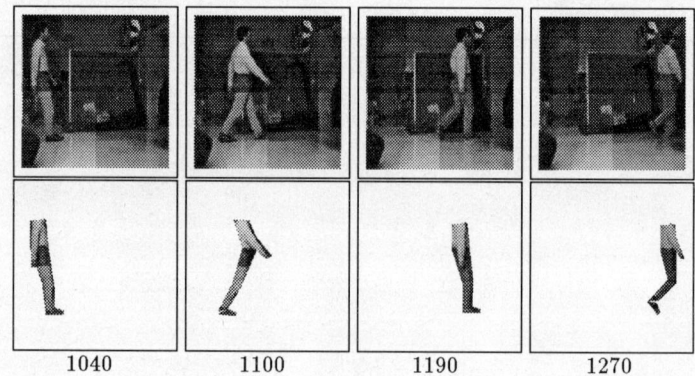

Fig. 3. Image sequence of "walking" and the five-part tracking

An observed activity can be viewed as a vector of measurements over the temporal axis. Consider as an example Figure 3, which shows both selected frames from an image sequence of a person walking in front of a camera and the model-based tracking of five body parts (i.e., arm, torso, thigh, calf and foot). We developed (see [8] for details) a method for modeling and recognition of these temporal measurements while accounting for some of the above variations in activity execution. This method is based on the hypothesis that a reduced dimensionality model of activities such as "walking" can be constructed using principal component analysis (PCA, or an eigenspace representation) of example signals ("exemplars"). Recognition of such activities is then posed as matching between principal component representation of the observed activity ("observation") to these learned models that are subjected to "activity-preserving" transformations (e.g., change of execution duration, small change in viewpoint, change of performer, etc.).

4.1 Experimental Results

We employ a recently proposed approach for tracking human motion using parameterized optical flow [5]. This approach assumes that an initial segmentation of the body into parts is given and tracks the motion of each part using a chain-like model that exploits the attachments between parts to achieve tracking of body parts in the presence of non-rigid deformations of clothing that cover the parts.

A set of 44 sequences of people walking in different directions were used for testing. The model of multi-view walking was constructed from the walking pattern of one individual while the testing involved eight subjects. The first six activity bases were used. The confusion matrix for the recognition of 44 instances of walking-directions are shown in Table 1. Each column shows the best matches for each sequence. The walkers had different paces and stylistic variations, some of which where recovered well by the affine transformation. Also, time shifts were common since only coarse temporal registration was employed prior to recognition.

Walking Direction	Parallel	Diag.	Away	Forward
Parallel	11	2		
Diagonal	3	14		1
Perp. Away			6	
Perp. Forw.	1	1	1	4
Total	15	17	7	5

Table 1. Confusion matrix for recognition of walking direction

Next, we illustrate the modeling and recognition of a set of activities that we consider challenging for recognition. We chose four activities that are overall quite close in performance: *walking, marching, line-walking*[2], and *kicking while walking*. Each cycle of these four activities lasts approximately 1.5 seconds.

We acquired tens of sequences of subjects performing these four activities as observed from a single view-point. Temporal and stylistic variabilities in the performance of these activities are common. Clothing and lighting variations also affected the accuracy of the recovery of motion measurements from these image sequences.

Table 2 shows the confusion matrix for recognition of a set of 66 test activities. These activities were performed by some of the same people who were used for model construction as well as new performers. Variations in performance were accounted for by the affine transformation. Up to 30% speed-up or slow-down as well as up to 15 frames of temporal shift were accounted for by the affine transformation used in the matching.

---

[2] A form of walking in which the two feet step on a straight line and spatially touch when both are on the ground.

Activity	Walk	Line-Walk	Walk. to Kick	March
Walk	11	3		3
Line-Walk	3	24		1
Walk to Kick			12	
March	1	1		7
Total	15	28	12	11

Table 2. Confusion matrix for recognition results

References

1. S. Fejes and L.S. Davis. Detection of independent motion using directional motion estimation. Technical Report CS-TR-3815, CAR-TR-866, University of Maryland, 1997.
2. S. Fejes and L.S. Davis. Direction-selective filters for egomotion estimation. Technical Report CS-TR-3814,CAR-TR-865, University of Maryland at College Park, 1997.
3. S. Fejes and L.S. Davis. What can projections of flow fields tell us about the visual motion. To appear in *Proceedings of the International Conference on Computer Vision*, Bombay, India, January 1998.
4. S. Fejes, L.S. Davis "Exploring Visual Motion Using Projections of Flow Fields" *Proc. of the DARPA Image Understanding Workshop*, New Orleans, LA,1997
5. S. X. Ju, M. Black, and Y. Yacoob. Cardboard people: A parameterized model of articulated image motion. *Proc. Int. Conference on Face and Gesture*, Vermont, 1996, 561-567.
6. S. Negahdaripour and B.K.P. Horn. Using depth-is-positive constraint to recover translational motion. *Proceedings of the Workshop on Computer Vision*, pages 138–144, 1987.
7. A.F. Siegel. Robust regression using repeated medians. *Biometrika*, 69:242–244, 1982.
8. Y. Yacoob and M. Black Parameterized Modeling and Recognition of Activities. *To appear in ICCV-98*.

Bayesian Paradigm for Recognition of Objects — Innovative Applications

J. K. Aggarwal and Shishir Shah

Computer and Vision Research Center
Department of Electrical and Computer Engineering
The University of Texas at Austin
Austin, Texas 78712-1084, U.S.A.

Abstract. This paper describes three innovative uses of the Bayesian paradigm for recognition of objects. A brief overview of the recognition problem and the use of the statistical approach are provided, along with the various stages for solving a problem. In addition, the paper presents formulations and results obtained by using Bayesian approaches in recent applications: human motion tracking, texture segmentation, and target recognition.

1 Introduction

The recognition of an object may be thought of as classifying the object into one of M classes based upon the observed features $X = \{x_1, x_2, \ldots, x_d\}$. This problem of classification may be approached in a number of distinct ways. For example, one may use a probabilistic framework, a neural network based classification strategy, or a rule-based paradigm driven by compiled knowledge. The choice of technique depends upon a number of issues including the availability of previous knowledge. If one is given N samples with corresponding features and the observed objects, then the problem may be solved in any of the three domains. For the above problem, each technique has its advantages and disadvantages; and each technique has its own rules of the game - the recognition process.

In the case of statistical analysis, one estimates the *a priori* probabilities and conditional probability density functions from the given data. Using Bayes Rule one estimates the *a posteriori* probability functions and accomplishes the task of recognition for a given feature vector and the corresponding object class. There are many issues and refinements depending upon whether or not one assumes the probability density functions are Gaussian. Life is easier with a Gaussian distribution, but in real life a Gaussian distribution rarely occurs.

The neural network framework poses the problem as the iterative computation of weights of a decision function using either supervised or unsupervised learning techniques. The simplest case consists of a single perceptron where the weights are computed for a two-class discrimination problem. The classification

---

* This research was supported by ARO DAAH-94-G-0417 and DAAH 049510494.

problem is formulated as $W(n)^T X(n) > \theta$, where the weights W are determined iteratively at each step n using the given data set $\mathbf{X} = \{X(1), \ldots, X(N)\}$. Again the above process may be generalized in a variety of ways depending upon the complexity of the network. Another important observation is that if the underlying distributions are Gaussian, the result of the Bayesian paradigm and the result of the single layer perceptron-based decision surfaces, given sufficient training data, will ultimately yield the same result [9, 1].

The rule-based strategy consists of estimating the rules in an abstract form. The knowledge-base in the rule-based system is generally represented either as a set of procedures or rules in a declarative fashion. The most expressive hypotheses employed involve the representation of the output function as a set of IF-THEN rules that jointly define the function. One way to learn a set of rules from given data is to first form a decision tree, then translate the tree into an equivalent set of rules, with one rule for each leaf node in the tree [9]. Rule-based approaches are used for object recognition due to their emergence from inductive learning and explanation-based reasoning. Rule-based paradigms provide a logical and understandable manner for using symbolic knowledge or domain knowledge in performing complex and heuristic tasks.

These three paradigms differ in their philosophies and their underlying mathematics. We have used all three in our research efforts to address problems of recognition. However, in this paper, we present recent results on three applications based upon the use of Bayesian strategy. The rest of the paper is organized as follows: section 2 presents an overview of the Bayesian paradigm for object classification. Section 3 presents our work in tracking of humans in an indoor environment; section 4 presents another application of Bayesian techniques derived from our work on the segmentation of textured scenes to identify region of interest to detect targets in video imagery. The final application, derived from our work on thermal imagery to recognize targets, is presented in section 5. These three applications are diverse and the use of the Bayesian paradigm yields interesting and meaningful results. Finally, section 6 presents some concluding remarks.

2 Bayesian Object Classification

Bayesian methods provide a formal means to reason about partial beliefs under conditions of uncertainty [8, 5]. Bayesian statistics have been used at various stages of the object recognition process to provide a firm theoretical footing as well as to improve performance. Bayesian approaches also provide error estimates with their decisions, which give another perspective for analyzing systems. Bayesian statistics have been used in the object recognition paradigm for indexing, model matching and incorporating neighborhood relations under different contexts with significant degree of success. Bayes' theorem indicates that the overall strength of belief in a hypothesis should be based on our previous knowledge and the observed evidence, and is based on these two factors. Other attractive features of the Bayes theorem are: (1) its ability to pool evidence from different sources while making a hypothesis, and (2) its amenability to recursive and incremental computation schemes, especially when evidence is accumulated in a sequential manner.

Since the beginning of image analysis, object models have been used for recognition purposes. Statistical techniques allow for automatic computation of object models using implicit features at certain locations or explicit structural models. Object models are represented by parameterized or non-parameterized density functions of their features. The calculation of the involved parameters is often solved by direct maximum likelihood estimations or the Expectation Maximization (EM) algorithm [4]. In general, M different objects are represented by density functions, the types and parameters of which may vary between objects and applications. Any Bayesian object classification system provides three main stages; (a) *localization stage,* where the class conditional data $\{X_\eta | 1 \leq \eta \leq N\}$ is localized for the M object classes, (b) *training stage,* where the parameters $\alpha_m, 1 \leq m \leq M$, of the model density functions have to be estimated from a sample set $\{X_\eta | 1 \leq \eta \leq N\}$ of data, and (c) *classification stage,* where the class number M of the observed object is determined.

Considering the case of recognition between two objects, M_1 and M_2, let the prior information of the objects be given by $P(M_1)$ and $P(M_2)$. Now consider that given the object data, we are able to extract a relevant feature for recognition, X. Thus the recognition problem can be posed as the identification of object M_i, given only the feature X. From a set of training samples, we can estimate the density function that represents each of the objects given by $p(X|M_1)$ and $p(X|M_2)$. The posterior probability of recognizing the objects is given by the inversion formula,

$$P(M_i|X) = \frac{p(X|M_i)P(M_i)}{p(X)}, \quad i = 1, 2 \quad (1)$$

The denominator $p(X)$, given by $p(X|M_1)P(M_1)+p(X|M_2)P(M_2))$, is a normalizing constant. Thus the recognition is based on deciding object M_1 if $P(M_1|X) > P(M_2|X)$ and vice versa.

3 Human Motion Tracking

A Bayesian framework for tracking moving humans in an indoor environment from a sequence of synchronized monocular grayscale images captured from multiple fixed cameras is developed in [3]. The tracking system consists of three main modules: Single View Tracking, Multiple View Transition Tracking, and Automatic Camera Switching. Bayesian classification schemes based on motion analysis of human features are used to track (spatially and temporally) a subject image of interest between consecutive frames.

Tracking from a single view includes segmentation of moving objects from the still background, distinguishing human subjects from the other segmented objects, extracting features from the segmented human subjects, and matching the subject image in successive frames by applying a Bayesian classifier. Segmentation is performed through dynamic recovery of the background image [2] and human classification is performed using moment invariant Zernike features to reduce the ambiguity of matching during tracking. Since the velocity of hands and legs can be considered to converge to that of the head and trunk over several frames, only the upper human body is considered for the purpose of classification and tracking.

2D points belonging to the medial axis of the segmented human upper body image are considered as features for matching. Three types of features are considered for tracking: location, intensity, and geometric features. The location of N feature points on the medial axis of a upper body at time t forms a feature vector $\mathbf{X}_t = \{x_{1t}, x_{2t}, \ldots, x_{Nt}\}^T$ where $x = \{u, v\}$, the image location. The intensity feature is an N dimensional feature vector $\mathbf{Y}_t = \{y_{1t}, y_{2t}, \ldots, y_{Nt}\}^T$, in which y_{mt} is the average intensity of the neighborhood of the mth feature points with $1 \leq m \leq N$. In addition, the subject height ratio between consecutive frames is used as the geometric feature g_t. To locate the most likely match of the subject image in a consecutive frame, a prior probability function, $P(\Theta)$, is postulated, where Θ is the feature parameters of the subject to track. For simplicity of computation without loss of generality, $P(\Theta)$ is assumed to be uniformly distributed. The class conditional density under independent feature assumption is postulated as:

$$p(\mathbf{Z}_t|\Theta) = p_x^{w_x}(\mathbf{X}_t|\Theta_x) p_y^{w_y}(\mathbf{Y}_t|\Theta_y) p_g^{w_g}(g_t|\theta_g), \qquad (2)$$

where w_x, w_y, and w_g are the the weights associated with $p_x(.), p_y(.)$, and $p_g(.)$, indicating the relative contribution of the corresponding feature in matching. Finally, a discriminant function $P(\Theta|\mathbf{Z}_t)$ is formulated as maximizing the likelihood and thus minimizing $D_t = -\log p(\mathbf{Z}_t|\Theta)$. The densities $p_x(.), p_y(.)$, and $p_g(.)$ are assumed to be normally distributed and the N points along the medial axis are sparsely sampled to address the non-rigidity of the human form. In formulating the likelihood function for location features, the velocity direction of the 3D points that are projected as the feature points is assumed to be constant for three consecutive frames (about 1 second under our sampling rate), a reasonable assumption given the speed of normal human walking.

To track the subject of interest across the views of multiple pre-calibrated fixed cameras, location and intensity features are used. Features are projected to a common camera coordinate system that is determined at the calibration stage. The correspondence model is developed by assuming the class-conditional distributions to be multivariate Gaussians. The likelihood function for $p_y(.)$ is similar to the earlier formulation, except that the intensity value of each pixel is normalized by the ratio of the average intensities of various cameras. Feature correspondence between location parameters from different camera views is obtained through spatial matching and spatial-temporal matching. Feature correspondence between frames captured from different cameras at the same time instant is based on the correspondence between a 2D point and its corresponding epipolar line. Spatial-temporal matching of an image point is achieved by estimating the projection of a 3D point in the view of camera i at time t given the projections in the previous frame. Correspondence between the projections is obtained through stereo estimation of the 3D point at time $t-1$, and the established matching of the subject image at time $t-1$ and t captured by C_j provides the information on depth change of the same point [2].

Automatic camera switching consists of prediction and optimal camera selection. Prediction is based on location, subject height, and tracking confidence. Tracking confidence is a measure of robust matching between consecutive frames. Since the weighted sum of *Mahalanobis* distances is used as the key to finding the most likely match between two consecutive frames, tracking confidence and optimal camera selection is also derived using the same. The system is successful in tracking people in indoor scenes and can tolerate occlusions up to three frames.

4 Texture Segmentation

The texture segmentation method developed in [10] consists of a Bayesian formulation for labeling similar regions. Gabor Wavelets are used to extract relevant features from the image. The features obtained are coarse clustered to obtain the approximate region labelings. Each cluster is considered to be suboptimal, with missing data, and thus the parameters are estimated using the EM algorithm. Final segmentation is then performed in a recursive manner while maximizing the *a posteriori* class probability for each region label.

A multi-scale filtering approach is employed, using Gabor Wavelets to extract texture information from images. Sixteen texture features for each pixel are used and a measure of similarity in their distribution characterizes the similarity of pixels in the image. The grouping of such similar pixels would result in a uniform region within the image. The goal is to partition the image space into clusters that have similar measures in the feature space. We have to discriminate among K possible texture categories to form similar clusters, and we do this by using the Nearest Neighbor(NN) algorithm.

Given the coarse clusters or the initial labeling of data, the segmentation problem is reduced to estimating the distribution parameters and the feature labelings to represent homogeneous regions. The image is defined as Y, where each pixel y is sampled from the space of features S. The distribution of the image pixels is given as the class conditional density function $p(y|x)$, where x comes from the true segmented regions X, with *a priori* distribution of $p(x)$. A Bayesian formulation in conjunction with the Markov Random Field model is considered to introduce dependencies between adjacent scales. In such a framework, the decisions are made by evaluating the *a posteriori* probability for each class and choosing the one with the highest probability as the true class. The feature space is a set of textures, each coming from a different scale of the image, thus, the segmentation/labeling of the pixels at a scale n can be given by X^n. Further, using properties of the Markov chain, true labeling is obtained when the pixels in X^n are conditionally dependent on the pixels at the next coarser level X^{n+1}. Thus, to find the true segmentation X, we start from the highest level. The labeling of each pixel in $X^n_{i,j}$ at location (i,j) depends only on the fixed neighborhood, $(\delta i, \delta j)$, of the next coarser level $X^{n+1}_{(\delta i, \delta j)}$. The neighborhood is defined by the spread of the Gabor filter. Thus the conditional density function for class labels at each scale is given as:

$$P(X^n = x^n | X^l = x^l, l > n) = p(x^n | x^{n+1}) \tag{3}$$

True segmentation is obtained by minimizing the Bayes Risk which is equivalent to maximizing the posterior probabilities over the images, and the problem can be considered as that of Maximum A Posteriori(MAP) estimation given by:

$$x^* = \arg \max_{x \in X} p(y|x) P(x) \tag{4}$$

The prior probability of region labeling, $P(x)$, is estimated from the distribution derived from coarse clustering as a product of Gaussian kernels using the Parzen window approach.

In trying to get the MAP estimate, MRF parameters, (θ_x, θ_y) are introduced for the posterior probabilities over the image, where θ_y is the mean(μ) and covariance(Σ) of the assumed multivariate Gaussian distribution for each texture type. Generalizing for each level of calculation, the MAP estimate is given by:

$$\theta_x^{t+1} = \arg \max_{\theta_x} \ln P(x^t | \theta_x) \tag{5}$$

$$x^{t+1} = \arg \max_x \ln P(x|\theta_x^{t+1}) + \ln p(y|x, \theta_y)$$

One of the main tasks in any MAP formulation is the estimation of parameters for the various probability distribution functions considered. This can be solved by maximizing the likelihood through the use of EM algorithm. Results of the developed methodology are presented for texture images and, as an application of object segmentation, on images with tactical vehicles in [10].

5 Target Recognition

A system for detection and recognition of targets in second generation Forward Looking InfraRed (FLIR) images is presented in [6, 7]. Recognition is based on a methodology for target recognition by parts. No information about the 3D structure of the object is available; the only *a priori* information is a database containing images of the object viewed at different angles (aspects) and distances. A hierarchical, modular architecture based on Bayesian statistics for the recognition and pose estimation of the targets is developed.

Recognition involves deciding the type of the target in the image from a set of known targets (T_i, $i = 1, \ldots, K$) and its pose (or aspect) β_l, $l = 1, \ldots L$ using information from the image ($\mathbf{X}$). In Bayesian theory, this information can be determined by finding the T_i and β_l that maximize the *a posteriori* probability $P(\beta_l, T_i | \mathbf{X})$. The posterior probability can be derived from Bayes theorem. More precisely (dropping the subscripts for clarity),

$$P(\beta, T | \mathbf{X}) = P(\beta | T, \mathbf{X}) P(T | \mathbf{X}). \tag{6}$$

The maximum *a posteriori* probability (MAP) estimates of the target and pose $(\widehat{T,\beta})$ is obtained as:

$$\widehat{T,\beta} = \arg\max_{T,\beta} P(\beta,T|\mathbf{X})$$

$$\equiv \arg\max_T[(\arg\max_\beta P(\beta|T,\mathbf{X})) + P(T|\mathbf{X})]$$

$$\equiv (P(\beta_l|T_i,\mathbf{X}) + P(T_i|\mathbf{X})) > (P(\beta_k|T_j,\mathbf{X})] + P(T_j|\mathbf{X})), \quad (7)$$

$\forall j \neq i$, where, $\beta_l = \arg\max_\beta P(\beta|T_i,\mathbf{X})$.

The developed system uses a hierarchical modular structure (HMS) for parts-based target recognition. In the most general form, the lowest level in the hierarchy identifies the class of the vehicle (e.g., tanks vs. trucks), while the higher levels use the information from the lower levels, as well as features extracted from the original target parts, for classification within each class (e.g., a M-60 tank vs. a M-1 tank). At each level, targets are recognized using their parts and thus each target classifier is made up of modules, each of which is trained on a specific part of the target. Each modular classifier is trained to recognize the part under different viewing angles and transformations (translation, scaling and rotation). A single layer of the hierarchy computes $P(T_i|\mathbf{X})$ and $P(\beta_l|T_i,\mathbf{X})$ for all known target to use equation (7). Each target T_i is considered to be made up of p_i parts. The presence of a specific target in the image is decided by accumulating evidence for that target. The inputs to the HMS are features extracted from the different parts of a target that have to be recognized. These inputs are presented sequentially to the system. X_t represents the input feature vector of the tth part of the object under consideration and $\mathbf{X} = \{X_1, X_2, \ldots, X_t, \ldots, X_n\}$ represents the entire feature vector set, where n represents the number of parts in the object to be recognized. Since the input parts are seen sequentially, this posterior probability can be incrementally updated using the recursive Bayesian updating rule [8]:

$$P(T_i|\mathbf{X_{t-1}}, X_t) = P(T_i|\mathbf{X_{t-1}}) \frac{P(X_t|T_i)}{P(X_t)}. \quad (8)$$

Each new part feature X_t is assumed to be independent of the previous data $\mathbf{X_{t-1}}$. The classifier modules can be considered to represent the conditional probability density functions for the input feature X_t given a specific part R_{ij} of a specific object T_i. $P(\beta|T,\mathbf{X})$ can also be determined within each expert module using similar reasoning.

The conditional density functions associated with each classifier module for a specific part of the object is computed as a mixture of Gaussian densities, the parameters of which are obtained by using the EM algorithm. The prior probabilities for each part classifier are based on *frequency* and *saliency*. Frequency specifies how many times the part is expected to be seen during the recognition experiments, while *saliency* specifies the relative importance of one part over another in recognizing a target or determining its pose.

Once the parts of the target are identified, each part is normalized for translation and rotation and then represented using Zernike moments. For the experimental results, two distinct sets of data were used, one for training the system and the other for testing the system. The **training set** consisted of six targets from three classes (tanks, trucks and armored personnel carriers). The **testing set** consisted of images of the same vehicles used for training but obtained under different viewing conditions and varying segmentation outputs. The overall recognition rate was 90.05%.

6 Remarks

We have presented a summary on the use of Bayesian methodology for solving problems in recognition of objects. Bayesian formulation is presented for three applications, whose results show that such a mathematical framework is suitable for 2D and 3D vision problems. Further research should concentrate on suitable models of encoding prior information and more efficient parameter estimation techniques or model learning procedures. The consideration of statistical dependencies and the explicit modeling of sensor and object properties will also increase the robustness of object recognition systems.

References

1. C. M. Bishop. *Neural Networks for Pattern Recognition*. Oxford University Press, New York, 1995.
2. Q. Cai and J. K. Aggarwal. Tracking human motion using multiple cameras. In *Proc. of Intl. Conf. on Pattern Recognition*, pages 68–72, Vienna, Austria, August 1996.
3. Q. Cai and J. K. Aggarwal. Automatic tracking of human motion in indoor scenes across multiple synchronized video streams. to appear, *Proceedings of the International Conference on Computer Vision*, Bombay, India, January 1998.
4. A. P. Dempster, N. M. Laird, and D. B. Rubin. Maximum likelihood from incomplete data via the EM algorithm. *Journal of the Royal Statistical Society*, 39-B:1–38, 1977.
5. R. O. Duda and P. E. Hart. *Pattern Classification and Scene Analysis*. A Wiley-Interscience Publication, 1973.
6. D. Nair and J. K. Aggarwal. Hierarchical, modular architectures for object recognition by parts. In *13th International Conference on Pattern Recognition*, volume 1, pages 601–606, Vienna, Austria, August 1996.
7. D. Nair and J. K. Aggarwal. Robust automatic target recognition in 2nd generation flir images. In *3rd IEEE Workshop on Applications of Computer Vision*, pages 194–201, Sarasota, Florida, December 1996.
8. J. Pearl. *Probabilistic Reasoning in Intelligent Systems: Networks of Plausible Inference*. Morgan Kaufmann Publishers, Inc. San Mateo, California, 1988.
9. R. Schalkoff. *Pattern Recognition: Statistical, structural and neural approaches*. John Wiley, 1992.
10. Shishir Shah and J. K. Aggarwal. A Bayesian segmentation framework for textured visual images. In *Proceedings of Computer Vision and Pattern Recognition*, pages 1014–1020, Puerto Rico, 1997.

Toward Motion Picture Grammars

Ruud Bolle[1], Yiannis Aloimonos[2], and Cornelia Fermüller[2]

[1] Exploratory Computer Vision Group, IBM T.J. Watson Research Center,
Yorktown Heights, NY 10598, USA
[2] Computer Vision Laboratory, Center for Automation Research,
Institute for Advanced Computer Studies, Computer Science Department,
University of Maryland, College Park, MD 20742-3275, USA

Abstract. We are interested in processing video data for the purpose of solving a variety of problems in video search, analysis, indexing, browsing and compression. Instead of concentrating on a particular problem, in this paper we present a framework for developing video applications. Our basic thesis is that video data can be represented at a higher level of abstraction as a string generated by a grammar, termed motion picture grammar. The rules of that grammar relate different spatiotemporal representations of the video content and, in particular, representations of action.

1 Introduction

The goal of this paper is to provide the steps toward a formal framework for dealing with video search, analysis, indexing and compression. The underlying idea is to develop a number of representations of the video content that refer to the shape and motion of objects as well as to the actions that can be found in a video clip. Most previous work on video search has concentrated on simple segmentation of video-clips into shots, finding critical frames to represent the shots and grouping the shots based on low-level information. Thereby the problem of video search is reduced to that of still-image search, but even here the features used to classify the still-images are often too low-level and do not match the way a typical user would think of a typical image.

The basic, elementary part of a video clip is the shot. Closely related shots combine to tell a story or display a scene. Each shot is composed of frames or images. Segmentation into shots is usually achieved via simple low-level feature descriptions such as color histograms or texture patterns. One can also segment using a simple distance metric that compares raw light intensity levels of corresponding pixels for successive images after a simple algorithm has been applied to compute correspondence [1–8]. These simple minded, low-level, information-based segmentations are usually quite adequate. Each shot then can be represented using one or more key or mosaicked frames [1, 9–15]. A key frame is just a single frame selected out of the sequence of frames forming a shot. If there is motion, many key frames may be needed to represent a shot or many frames may be pasted together into a mosaic.

In any case, the key or mosaicked frame representation is essentially static. We can now apply static video query technology (e.g., [16]), but even though we necessarily must have computed some motion information in order to register the different frames of a mosaic, we are using an all too static representation. It will be difficult to make and answer queries about certain motion features or motion between shots. If we want to search for scenes involving running or jumping (to take a simple example), we will have difficulties. We should use a more dynamic representation.

Whatever we do at the shot level, there are too many shots in a typical two hour film. We need to group the shots into meaningful units and perhaps recognize scenes and scene discontinuities. This is not easy to do without domain-specific knowledge, but we can group shots using the same kind of similarity metrics used to segment a film into shots. This clustering must also take into account that two shots far apart in time probably should not be in the same equivalence class [17]. We might have not one but several different groupings using different similarity metrics. Then, based on the pattern of sequences of equivalence classes of shots, we may be able to identify and classify scenes and, for example, recognize a possible conversation by the occurrence of an A-B-A-B-A-B pattern [14].

This clustering is dependent on low-level static shot attributes such as global color histograms or statistical texture features. We suggest here that 2D and 3D shape and motion information is also interesting.

Ideally, we would also like to use very high-level or semantic information to index video. For a domain-specific application such as television news, it may be possible to construct an a priori model that will help identify some simple semantic information such as the occurrence of a news story as opposed to a weather report or commercial break [18,19].

Of course, doing real semantics is too difficult, but we imagine the user asking queries like, "Find scenes where Bill Clinton is jogging." Much could be accomplished if we could hire humans to annotate every shot of every video clip in the database. Unfortunately this manual annotation is too tedious and time-consuming. Automatic annotation is desirable. Perfect automatic annotation with semantic descriptors is too much to hope for, but we may be able to learn (from examples) lower-level structural descriptions that correlate with semantic descriptions. The correlation will not be perfect, but might work much of the time. The lower-level representations should not be too low-level: Computer vision researchers know something about detecting information that is more high-level than texture statistics or color histograms of images compressed using the discrete cosine transform, and this information is useful.

2 Representations

There are several kinds of representations of interest to us: 3D object, 2D object, 3D motion, 2D motion, 4D motion-object, image feature. There are many ways each of these can be used at many scales of resolution with many different similarity metrics to facilitate video indexing, compression and search. We would prefer to use many different representations. Different representations are better for different purposes on different kinds of video databases.

Objects A two (or three) dimensional object representation is a map $f : F \to 2^{(O \times R^n)}$. Here F is the set of frames, O is the set of possible object shapes, $n = 2$ or $n = 3$ and thus R^n is the set of locations. $f(\mathcal{F})$ for $\mathcal{F} \in F$ is a map from $O \times R^n$ to 2. An element (o, l) of $O \times R^n$ is an ordered pair consisting of an object o and location l. As usual 2 is the set with two members 0 and 1 and if $f(\mathcal{F})(o, l) = 1$, then the object o is found at location l in frame $\mathcal{F}$. Since for most possible pairs (o, l) and most frames $\mathcal{F}$, we have $f(\mathcal{F})(o, l) = 0$, it might be simplest to implement object representations by listing for each frame the set of its objects and their locations. There are many ways of representing shapes $o \in O$. In general, O might even be partial. If we never see the back side of the moon, it suffices to describe the part of the shape we do see. A description might be obtained by providing a set of points P with integer coordinates such that $P \in O$, or we might use a simple spline approximation to the boundary of the object [20, 21]. In the case of 3D objects, our representation might be either object-centered, and thus invariant, or it might be viewpoint dependent. An example of the latter is a 2 1/2 D description that associates a depth to each pair (i, j) such that (i, j) is the 2D frame (= image) location to which real-world point $P \in O$ projects and i and j are integers. For more 3D representations see [22].

Motions A 2D or 3D motion representation is a sequence of maps $m : M \to R^n$ for each $M \in F$ with M a frame, $n = 2$ or 3 or 6. If $n = 2$ or 3, we are associating with each point in each frame an optical flow or real-world motion vector. If $n = 6$, we are assuming approximate local rigidity and associating with each point three parameters of translation and three of rotation. For non-rigid motion, more parameters might be needed [23]. Then for each frame M, $m(P)$ is only defined at a few points $p \in M$. It is difficult to provide a dense, accurate, optical flow map estimate. The motion representation might be sparse.

Actions We can combine object and motion representation and obtain a 3D or 4D action representation. A 4D action representation is intended to describe the motions of objects through sequences of frames. A video then may be considered to be a set of actions (including actions where there is no motion, because remaining still is just motion in which all motion parameters are zero). An action (in 4D) is a pair consisting of an action shape and a spatiotemporal location (i.e., a point in R^4, a place and a time). An action shape describes an action using a coordinate system in which the action begins at time $t = 0$ and at $t = 0$, the center of mass of the region where the action is taking place is the origin. One way to describe an action shape is to list for each time $0 \leq x \leq t_{\max}$, for each point P that at t is part of the action, the motion at P at time t. Here $t_{\max}$

is the duration of the action. An important action shape is global rigid motion. (Every point is moving with the same rigid motion.) This usually represents camera motion. Action representations are the foundation of a non-frame, non-shot based decomposition of a video sequence. With actions as with motion and object representations, we may need to approximate and compute information at multiple scales.

Image features The image feature representations are what video researchers have hitherto most often chosen to investigate. They associate to each pixel or to each window of an image either a simple feature or list of simple features such as light intensity, edge density, predominant texture orientation, the leading coefficient of a wavelet transform, or a histogram of features such as a color histogram. If windows are used, it must be very simple to find the sizes and locations of the windows. The windows might be overlapping. Again image feature representations might be approximate or sparse. It is important to note that among the features used can be two or three-dimensional motion features.

If we remind ourselves that all the representations we have mentioned can and should be given at multiple scales of resolution, we can find another way of thinking of feature representations: They are really object representations and summaries of object representations. We also need to remember that there are other ways in which descriptions can be given at varying levels of detail. (It might suffice to say that a certain object is a triangle without giving the lengths of its sides and its location need not be very precisely known. It might suffice to just say that motion is nonrigid in a certain region without giving the parameters of translation and rotation.)

3 Grammars

Using the multiple representations available to us we can encode much of the accessible information. We can encode that a conversation is not just an A-B-A-B pattern of shots but also that the same object recurs in all the A (and maybe in the B). This object is large and well-lighted and has a face or person shape.

In fact, once we humans know something about a certain category of video clip (e.g., movie drama, television documentary) or video scene (e.g., a play in a sports game, a conversation, a surprise meeting on the street between two people) we can construct a simple grammar to represent our knowledge. The grammar should be context-free, if possible, to simplify parsing.

It is too difficult to take a raw video and parse it using some grammar to obtain the high-level meaning of video scenes and segments. It is easier to start with one of our representations (action, object, motion, etc.) and parse. Even this is not quite right. Not all information is best handled using a grammar formalism. Algebraic constraints relating parameters may not be sufficiently structural information to best be treated via grammars.

What we need are attribute grammars [24–26]. Instead of just having a rule $X \to YZ$, we might also have associated with Z an attribute or parameter and the set of possible values it might take. Furthermore, there might be universal constraints on these attributes, for example, attribute A and B must always have the same sign. Thus for a certain kind of film, we might have a rule Conversation $\rightarrow$ Conversation Type 3 and a rule Conversation Type 3 $\rightarrow$ Conversation of Type 3 AB where A represents the shot of one actor and B the shot of another actor. Associated with A and B are attributes defining how well-lit and how central the actors are in the shot. There can be algebraic relations between the lighting attributes of A and B and lighting attributes of Conversation Type 3. Also, rather than having a separate rule for each actor, we might have a general rule Conversation Type 3 $\rightarrow$ Conversation Type 3 AB that works for all pairs of actors and we may force the same pair of actors to appear in successive shots by putting an algebraic constraint on attributes associated with A, B, and Conversation Type 3. The attributes discriminate between different actors (e.g., discriminate by roundness of face) and then we have associated with Conversation Type 3 a pair of roundness values and with each of A, B a pair of values and these must match. Much more complex algebraic constraints, relating attribute values, are possible.

Using attributed grammars, we might be able to parse one of the representations we suggest and produce a very high-level description, but these attributed grammars need to be stochastic to allow for the fact that all rules of video have exceptions, and so we can represent the fact that even if there are many ways of shooting a certain kind of scene, certain ways are much more frequently employed than others. Stochastic, context-free grammars [27] and hidden Markov models [28] are closely related. Non-terminals in the grammar formalism act like hidden variables (not directly observable variables) in the hidden Markov formalism.

We have suggested stochastic, attributed, context-free grammars as a useful formalism for representing the high-level structure of videos. To make sense of this suggestion it is necessary to be more concrete.

The first question that arises is what should be our alphabet of terminal vocabulary. We have already declared that we do not wish to use our grammar to generate data at the pixel level (at any scale of resolution). Also, if we were really taking seriously the multidimensional geometric character of images, we would have to deal with the additional complexities present in array grammars or the pattern grammars of Grenander [29]. We do not wish the sentences our grammar generates to live in multidimensional space; instead they should be strings. The strings are strings of actions (with their attributes). The question is, exactly which strings of actions.

The set of well-formed strings of actions might be something that is application specific and defined by the system designer in consultation with representative domain experts. Or, more generally, we might specify a universal language of action strings. The action strings might represent the significant actions that occur in a segment of a video, or only those actions that start during the segment in question, or the string might represent action that starts during the segment in question, and also other actions that make significant changes in their attributes during the segment in question. Order of actions in the string should essentially represent temporal order: The question is, which action began first or had significant changes in its parameters first?

Actions can be described at many levels of resolution and detail, so we have to make many choices in order to pick an adequate terminal vocabulary. It might be necessary to leave these choices to the domain experts.

In any case, we can analyze a video and obtain the set of most likely parsings or the most likely values of hidden Markov nodes. This provides higher-level descriptors that can be used to annotate and index video. Now we can search for higher-order information in the video. The grammars we need should best be supplied by hand with the advice of domain experts.

Because relatively little work has been done on learning stochastic grammars or attribute grammars [30, 31] and what little work has been done has focused on relatively simple grammars with few rules, we might have great difficulty actually learning a good grammar from examples [32, 33], and we should start under the assumption that the grammar is going to be very simple with few rules and the action string descriptions we will be working with are short and the actions are not described in very much detail.

4 Conclusions

Video has both syntax and semantics, just as language does. Video, however, has an additional structure encapsulated in spatiotemporal and action representations obeying certain photometric and geometric constraints. This paper has suggested how these representations can be related through a grammar to semantics.

References

1. Arman, F., Hsu, A., Chiu, M.: Feature management for large video databases. SPIE 1908, Storage and Retrieval for Image and Video Databases (1993) 2–12
2. Liu, H. C., Zick, G. L.: Scene decomposition of mpeg compressed video. SPIE 2419, Digital Video Compression: Algorithms and Technologies (1995) 26–37
3. Otsuji, K., Tonomura, Y., Ohba, Y.: Video browsing using brightness data. SPIE 1606, Visual Communications and Image Processing (1991) 980–989
4. Sethi, I. K., Patel, N.: A statistical approach to scene change detection. SPIE 2420, Storage and Retrieval for Image and Video Databases III (1995) 329–338
5. Shahraray, B.: Scene change detection and content-based sampling of video sequences. SPIE 2419, Digital Video Compression: Algorithms and Technologies (1995) 2–13
6. Swain, M. J., Ballard, D. H.: Color indexing. International Journal of Computer Vision **7** (1991) 11–32
7. Zhang, H., Kankanhalli, A., Smoliar, S. W.: Automatic partitioning of full motion video. Multimedia Systems **1** (1993) 10–28
8. Zhang, H. J., Low, C. Y., Smoliar, S. W.: Video parsing and browsing using compressed data. Multimedia Tools and Applications **1** (1995) 89–111

9. Mann, S., Picard, R. W.: Virtual bellows: Constructing high quality still from video. International Conference on Image Processing, volume 1 (1994) 363–367
10. Sawhney, H. S., Ayer, S., Gorkani, M.: Model based 2D & 3D dominant motion estimation for mosaicking and video representation. Technical report, IBM Almaden Research Laboratory (December 1994)
11. Szeliski, R.: Image mosaicking for telereality applications. Technical Report CRL 94/2, DEC Cambridge Research Laboratory (1994)
12. Teodosio, L., Bender, W.: Salient video stills: Content and context preserved. Proceedings, Multimedia '93, ACM (1993) 39–46
13. Tonomura, Y., Akutsu, A., Otsuji, K., Sadakata, T.: Video map and video space icon: Tools for anatomizing video content. INTERCHI '93 Conference on Human Factors in Computing Systems, ACM (1993) 131–136
14. Yeung, M. M., Yeo, B. L.: Data modelling of videos with temporal events and its applications. Technical Report TR-EE-ISS-YM9603, Princeton University (April 1996)
15. Yow, K. D., Yeo, B. L., Yeung, M. M., Liu, B.: Analysis and presentation of soccer highlights from digital video. Second Asian Conference on Computer Vision, volume 2 (1995) 499–503
16. Hibino, S., Steiner, E. A. R.: A visual query language for identifying temporal trends in video data. International Workshop on Multi-media Database Management Systems (1995) 74–81
17. Yeung, M., Yeo, B.: Time-constrained clustering for segmentation of video into story units. ICPR '96, volume 6 (August 1996) 375–380
18. Swanberg, D., Shu, C. F., Jain, R.: Knowledge-guided parsing in video databases. SPIE 1908, Storage and Retrieval for Image and Video Databases (1993) 13–25
19. Zhang, H. J., Gong, Y. H., Smoliar, S. W., Yan, S. Y.: Automatic parsing of news video. International Conference on Multimedia Computing and Systems (1994) 45–54
20. Dierckx, P.: Curve and Surface Fitting with Splines. Clarendon: Oxford (1993)
21. Tiller, W.: Rational b-splines for curve and surface representation. IEEE CGA **3** (1983) 61–69
22. Faugeras, O. D.: Three-Dimensional Computer Vision. Cambridge, MA: MIT Press (1992)
23. Shulman, D., Aloimonos, J. Y.: (non-)rigid motion interpretation: a regularized approach. Proc. Royal Society, London B **233** (1988) 217–234
24. Fu, K. S.: Syntactic Pattern Recognition and Applications. Englewood Cliffs, NJ: Prentice Hall (1982)
25. Lee, K. H., Eom, K. B., Kashyap, R. L.: Character recognition based on attribute-dependent programmed grammar. IEEE Transactions on Pattern Analysis and Machine Intelligence **14** (1992)
26. Zhao, M.: Two-dimensional extended attribute grammar method for the recognition of hand-printed chinese characters. Pattern Recognition **23** 1990
27. Charniak, E.: Statistical Language Learning. MIT Press (1993)
28. Huang, X. D., Ariki, Y., Jack, M. A.: Hidden Markov models for Speech Recognition. Edinburgh University Press (1990)
29. Grenander, U.: Elements of pattern theory. Johns Hopkins, Baltimore (1996)
30. Abney, S.: Stochastic attribute valued grammars. Currently working at AT & T Labs Research in Florsham Park, NJ
31. Keller, B., Lutz, R.: Learning stochastic context-free grammars from corpora using a genetic algorithm. ICANNGA (1997)

32. Johnson, M.: Attribute valued logic and the theory of grammar. CSLI Lecture Notes, volume 16. CSLI (1988)
33. Torenvliet, L., Trautwein, M.: A note on the complexity of restricted attribute valued grammars. Computational Linguistics in the Netherlands, Meeting at Twente (1995)

Hierarchical Texture Segmentation

P. Bajcsy and N. Ahuja

Beckman Institute, 405 N. Mathews Ave., Urbana, IL 61801
Tel: (503)620-6601, (217)333-1837
E-mail: peter@stereo.ai.uiuc.edu, ahuja@vision.ai.uiuc.edu

Abstract: *We present a new hierarchical texture segmentation method that partitions an image into textured regions. A textured region is viewed as a set of uniformly distributed primitives. A primitive is a region with constant gray values. Gray values within a primitive can be corrupted by noise. Any noisy primitive contains gray values from a δ-wide interval (δ-homogeneous primitive). The noisy primitive is described by the sample mean of interior gray values. A textured region with noise is characterized by a set of gray value sample means (texture vector) derived from noisy primitives. Every pixel (sample point) and its neighborhood give rise to an estimate of texture vector. Components of the estimated vector at a pixel characterize noisy primitives of a textured region grown from the pixel. Co-occurrence of noisy primitives from this grown region are calculated. Final segmentation is obtained by grouping pixels with identical estimates of texture vectors and co-occurrences, created at each pixel. Homogeneity degree δ of noisy primitives provides a basis for multiscale analysis. Computational efficiency and robustness of the proposed method are related. Experiments are reported for synthetic textures as well as real textures from Brodatz album and real gray scale and color images.*

1 Introduction

The goal of any image segmentation is to partition an image (grid of samples) into connected subsets of samples, denoted as regions, each having a uniform texture. Textures have no universal model. A variety of texture models have been derived from: (1) perceptual studies [7] (mimicking humans), (2) specific two-dimensional tasks [12], such as, automated surface inspection (textile, paint), medical image processing (semi-automated search for tissues, tumors), (3) texture gradient analyses [3] (projective distortion problem) and (4) texture imitation [5] (realism in computer graphics). Texture segmentation methods have been based on two types of models: statistical [6, 5] and structural methods [13, 11], which can be alternately viewed as pixel-based and region-based models [1]. The motivation for the work described in this paper is to develop a robust and computationally efficient hierarchical texture segmentation method.

Our work here takes a structural-statistical approach. First in Section 2, image texture segmentation is introduced using modelled textures and compared with observable (realistic) textures. A new texture segmentation method is described in Section 3. Section 4 shows segmentation performance on synthetic images (robustness and efficiency) as well as Brodatz textures and real scene textures. Concluding remarks are presented in Section. 5.

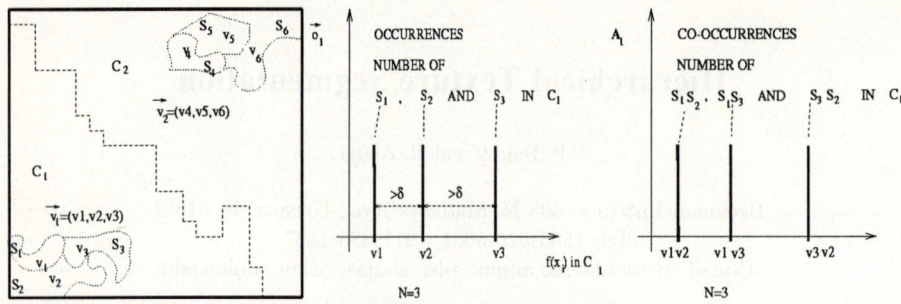

Figure 1: Textured region and its occurrence and co-occurrence of primitives.

2 Modelled and Observable Textures

The following terms are introduced next: image, texture image segmentation, texture model, texture primitive and observable texture. Input data, a two-dimensional image, is represented by a function $x_i \longrightarrow f(x_i)$, where x_i is a sample point $x_i = (x_1, x_2)$ in the domain of function f and $f(x_i)$ is the function value denoted as attribute at each location pointed to by sample point x_i. The goal of texture image segmentation is to partition the sample points x_i in an image into sets of connected sample points, denoted as regions $\{C_k\}_{k=1}^{K}$, such that each region consists of uniformly distributed texture primitives. Texture primitives are defined as piecewise constant regions $\{S_j\}_{j=1}^{J}$ (identical attributes inside S_j). Thus, each textured region C_k can be characterized by a set of attributes $f(x_i)$ describing primitives, e.g., three gray values $f_1 = v1, f_2 = v2$ and $f_3 = v3$ of the region C_1 shown in Figure 1 (left). This set of attributes for a textured region C_k is denoted by a texture vector $\vec{v}_k = (v1, v2, ...)$ containing N_k entries (a size of the vector is N_k). Two textured regions have different vectors. Each element from a texture vector $\vec{v}_k$ characterizes multiple primitives $S_j \in C_k$. The number of primitives $S_j \in C_k$ characterized by each vector component creates a component of an occurrence vector $\vec{o}_k$. The number of neighboring primitive pairs in C_k gives rise to a component of a co-occurrence matrix $\mathbf{A}_k = \{a_{i,j}\}$, e.g., a pair (S_1, S_2) occurs in the region C_1 $a_{1,2}$-times. Figure 1 (middle and right) shows occurrence and co-occurrence values of region C_1, where components on the horizontal axis are labeled with the vector values $(v1, v2, ..$ or $v1v2, ...)$. The diagonal elements of $\mathbf{A_k}$ are equal to zero; A textured region consists of uniformly distributed primitives therefore (1) components of occurrence vectors and (2) off-diagonal components of co-occurrence matrix have similar values (ideally values are identical).

A texture primitive with noise is modelled as a δ-homogeneous region and obtained by performing δ-homogeneous image segmentation [2]. δ-homogeneous image segmentation partitions the image into a set of regions $\{S_j\}_{j=1}^{J}$ with δ homogeneity and contrast α; $\alpha = \alpha_{1,2} = \| \max\{f(x_i \in S_1^{\delta})\} - \min\{f(x_i \in S_2^{\delta})\} \|$ and $\delta = \| \max\{f(x_i \in S_1^{\delta})\} - \min\{f(x_i \in S_1^{\delta})\} \|$. The noisy primitive is described by the sample mean of its interior attributes (first-order statistics). A textured region C_k with noise is characterized by a set of attribute sample means (texture vector $\vec{v}_k$), which is the set of the sample means of the noisy primitives. A distribution of primitives can be observed from the co-occurrences of noisy primitives (second-order statistics) characterized by texture vector.

Observable (realistic) textures possess properties either directly captured by the texture model or achieved by several variations of the texture model. Variations of the following ideal characteristics are considered: (1) model for a primitive, (2) distribution of primitives within a textured region, (3) random noise inside of a primitive. Two more issues related to vectors in observable textures are taken into account; a variable size of vectors and partially different vectors of neighboring textured regions - vector overlaps. Other issues related to illumination conditions (e.g., shadows, highlights, blur) are not addressed.

3 Texture Segmentation

First, the proposed method is described. It is followed by a discussion of (A) computational efficiency and (B) robustness against (1) noise in primitives, which effects vector detection, (2) nonuniform distribution of primitives tightly related to vector overlaps.

Assumptions: Let us consider that a given image contains δ-homogeneous regions S_j^δ (texture primitives) and all neighboring pairs of regions S_j^δ have contrast $\alpha > \delta$. Let us assume that a subset $sub_1\{S_j^\delta\}$ of all regions $\{S_j^\delta\}_{j=1}^J$ creates an unknown connected textured region $C_1^\delta = sub_1\{S_j^\delta\}$. The textured region C_1^δ is characterized by (1) N_1 attribute sample means of texture primitives (components of a vector $\vec{v}_1$), which are mutually more than δ apart and (2) identical values of occurrence components $\vec{o}_1$ and co-occurrence off-diagonal components $\mathbf{A_1}$ due to a uniform distribution of primitives.

Derived textured region at each sample point: Given a fixed size of vector N, a vector $\vec{v}_{x_i}$ can be built at each sample point x_i by searching in a neighborhood of x_i[1]. The first component of the vector is $v1 = f(x_i)$. The next value of the vector is any attribute, which is more than δ apart from all vector elements found before; e.g., $v2 = f(x_k)$ if $\| f(x_i) - f(x_k) \| > \delta$. From the vector $\vec{v}_{x_i}$ found at each sample point x_i, a 2δ-homogeneous textured region[2] $C_{x_i}^{2\delta}$ is created by grouping together all neighboring sample points x_k satisfying the inequality $cyc\{\| \vec{v}_{x_i} - \vec{v}_{x_k} \|\} \leq \delta$, where $cyc\{.\}$ represents subtraction of any cyclical variation of vector components. A co-occurrence matrix $\mathbf{A}_{x_i}$ of primitives S_j^δ within $C_{x_i}^{2\delta}$ is calculated.

Analysis based on size of texture vector: Three types of textured regions $C_{x_i}^{2\delta}$ exist with respect to a priori unknown textured region C_1^δ depending on a selected vector size N. If the assumed size N of a vector is identical to the actual vector size N_1 of C_1^δ, $N = N_1$, then a priori unknown region C_1^δ is identical with all

---

[1] x_i became an index of the vector $\vec{v}_{x_i}$ since every sample point creates its own vector.
[2] Notation: Derived textured region $C_{x_i}^{2\delta}$ is found at each x_i (in the subscript) and by searching for samples around x_i with attributes $\pm\delta$ apart from vector components of $\vec{v}_{x_i}$, thus spanning 2δ wide attribute range (in superscript) by each vector component.

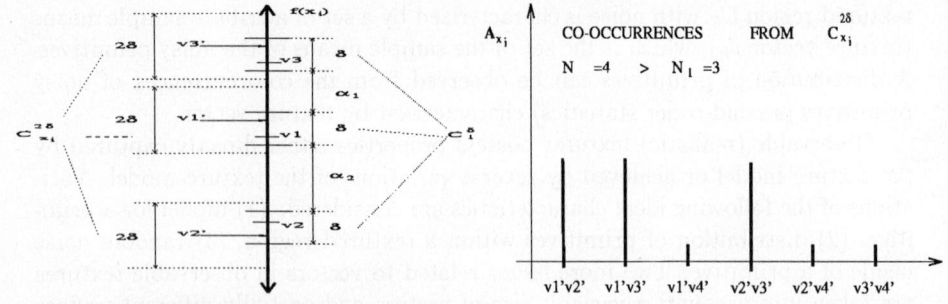

Figure 2: Vector components and co-occurrence matrix components.
Left - Vectors $\vec{v}_1 = (v1, v2, v3)$ and $\vec{v}_{x_i} = (v1', v2', v3')$ for regions C_1^δ and $C_{x_i}^{2\delta}$ in a case of $N = N_1 = 3$. Right - The co-occurrence matrix $\mathbf{A}_{x_i}$ is created from $C_{x_i}^{2\delta}$ using a vector $\vec{v}_{x_i} = (v1', v2', v3', v4')$ of the size $N = 4 > N_1 = 3$. $N = 4$ greater than the actual vector size $N_1 = 3$ of an unknown region C_1^δ with its vector $\vec{v}_1 = (v1, v2, v3)$.

regions $C_{x_i}^{2\delta}$ found, where x_i belongs to the interior of C_1^δ. In this case, vectors for a calculated region $C_{x_i}^{2\delta}$ and for a priori unknown region C_1^δ satisfy the inequality $cyc\{\| \vec{v}_1 - \vec{v}_{x_i} \|\} \leq \delta$. The vector components are shown in Figure 2 (left).

If $N > N_1$ then all N_1 components of the vector $\vec{v}_1$ (corresponding to a priori unknown region C_1^δ) are not more than δ apart from N_1 out of N components of all vectors $\vec{v}_{x_i \in C_1^\delta}$ found. The N_1 components from the vectors $\vec{v}_{x_i \in C_1^\delta}$ are identified by selecting vector components having identical values of their off-diagonal co-occurrence components from $\mathbf{A}_{x_i}$ (Figure 2 right).

If $N < N_1$ then the vector $\vec{v}_{x_i}$ as well as the co-occurrence matrix $\mathbf{A}_{x_i}$ cannot estimate global properties of a textured region.

Size of vectors: Following from the previous analysis, the goal is to select a vector size N greater or equal to the unknown true vector size N_1. Any unknown size N_1 of vector $\vec{v}_1$ is bounded by the minimum and maximum values, $N_{min} \leq N_1 \leq N_{max}$. The minimum size of the vector is two, $N_{min} = 2$, because in order to create a textured region there must be at least two primitives with distinct characteristic attributes. The maximum size of the true vector N_{max} is derived considering two neighboring textured regions with non-overlapping vectors, which span the whole attribute range $R_{total} = N_1 * (\delta_1 + \alpha_1) + N_2 * (\delta_2 + \alpha_2)$. Then the maximum size of a vector is found by setting $N_1 = N_{min} = 2$ and $N_2 = N_{max}$ (see Figure 3).

Steps in the proposed texture segmentation method:
(1) For a fixed δ, estimate the maximum size of the true vector N_{max}.
(2) Find a vector $\vec{v}_{x_i}$ of the size $N = N_{max}$ at each sample point x_i consisting of attributes from a neighborhood of x_i, which are more than δ apart.
(2) Create textured regions $C_{x_i}^{2\delta}$ at each sample point x_i consisting of connected samples having attributes not more than δ apart from at least one component of the vector $\vec{v}_{x_i}$.
(3) Calculate co-occurrences $\mathbf{A}_{x_i}$ of primitives S_j^δ described by $\vec{v}_{x_i}$ within a tex-

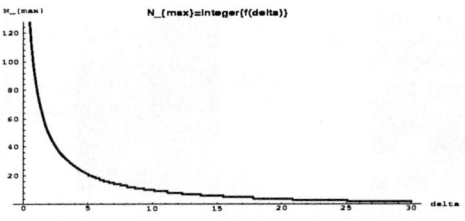

Figure 3: Maximum size of vector as a function of homogeneity δ. The graph represents $N_{max} = Integer(\frac{256}{2*\delta+1} - 2)$.

tured region $C_{x_i}^{2\delta}$.
(4) Compare vectors $\vec{v}_{x_i}$, $\vec{v}_{x_l}$ and check uniformity of off-diagonal co-occurrence components $\mathbf{A}_{x_i}$, $\mathbf{A}_{x_l}$ for all adjacent samples x_i, x_l.
(5) Assign sample points x_i, x_l into final textured regions based on (4).
Computational efficiency is achieved by (1) estimating texture vectors $\vec{v}_{x_i}$ from one-dimensional cross sections of regions, (2) lowering the dimensionality of computations to obtain derived regions $C_{x_i}^{2\delta}$ and (3) replacing co-occurrence matrices with co-occurrence vectors corresponding to regions $C_{x_i}^{2\delta}$.
Noise robustness of vector detection was analyzed for texture primitives S_j^δ having either $\alpha > \delta$ or $\delta \geq \alpha$. If $\alpha > \delta$ then there is no error in texture vector detection. If $\delta \geq \alpha$ for a primitive S_j^δ due to a random zero-mean noise then the error probability for the first vector component $v1_{x_i}$ can be expressed as $Pr(\| v1_{x_i \in S_j^\delta \subset C_1^\delta} - v1_1 \| \leq \delta) = \xi$, where $v1_1$ is the first vector component of $\vec{v}_1$ having a correct attribute sample mean based on primitives S_j^δ in C_1^δ and ξ is a confidence coefficient.
Maximal *robustness against nonuniform distribution of primitives* (denoted as RN) means that a minimal number of components from cyclically matched vectors $\vec{v}_{x_i}$ and $\vec{v}_{x_l}$ is sufficient to merge samples x_i and x_l. This means that a small vector overlap will lead to erroneous merger since overlapped components of vectors will have identical corresponding co-occurrence components. Thus, the *robustness against vector overlaps* RO is minimal. We can conclude that the following equation governs the two parameters RN and RO; $RO + RN = constant$.
Hierarchical texture segmentations are created by increasing homogeneity parameter δ and satisfying the condition $C_k^\delta \subseteq C_k^{\delta + \Delta \delta}$ for every region C_k^δ. The hierarchy is guaranteed by modifying attribute values within created regions C_k^δ at each scale δ to the sample means of created regions based on their vectors $g(x_i, \delta) = \frac{1}{M_{j,k}} \sum_{l=1}^{M_{j,k}} f(x_l \in S_j^\delta \subset C_k^\delta)$, where $M_{j,k}$ is the number of samples in all primitives $S_j^\delta \subset C_k^\delta$ described by one vector component $v1$.
Color image segmentation is a straight forward extension of gray scale image segmentation. The texture vector ($Nx1$) becomes a texture tensor (matrix $Nx3$) in the case of color images. Every component of gray scale vectors is replaced

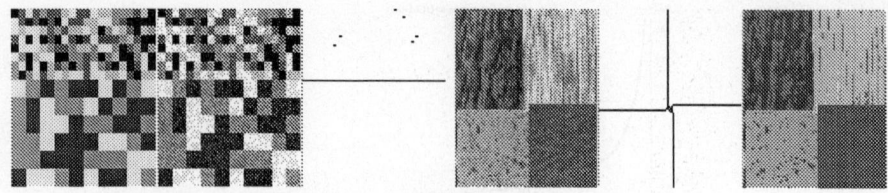

Figure 4: Segmentation of synthetic and Brodatz textures.

with a three-dimensional vector containing red, green and blue information. The rest of the algorithm is the same.

4 Performance Evaluation

Performance of the proposed texture segmentation method is judged based on (1) robustness (noise, nonuniform distribution of primitives, vector overlap), (2) computational requirements, (3) results for gray scale and color images (Brodatz textures, real scenes).

Noise robustness of vector detection: Attributes $f(x_i)$ are corrupted by an additive noise $n(x_i)$ with a known symmetric probability distribution function D_n (zero mean, standard deviation σ), $p(x_i) = f(x_i) + n(x_i)$. By introducing a uniformly distributed noise we could control two crucial parameters δ and α of primitives S_j^δ as their mutual relationship $\delta = 0 < \alpha = 50$ continuously changed to $\delta = 50 > \alpha = 0$. There is no error in vector detection for $\delta < \alpha$. It was experimentally verified that there is only a small error for $\delta > \alpha = 0$ (see Figure 4 left in a presence of vector overlaps) by observing the number of misclassified sample points (less than 2.5%).

Robustness against nonuniform distribution and vector overlap: Relationship between RO and RN ($RO + RN = const$) required to choose an optimal solution (equal weights on RO and RN). In the current implementation, the weights on importance of RO and RN are evenly distributed.

Computational requirements are proportional to the number of given sample points, the size of vector N and the number of sample points in $C_{x_i}^{2\delta}$ used for calculating co-occurrences $\vec{a}_{x_i}$. Computational time and memory are approximately linearly increasing with increasing size of vectors N. In average, one segmentation of a 2D gray scale image (100x100 samples) takes between $3-4s$ for $N=3$ and $6-7s$ for $N=5$ on Sparc 20 workstations.

Experimental results for gray scale and color images: Due to the high computational requirements for the size of vector $N = N_{max}$ at small values of δ and some empirical observations, we ran the following segmentations with a fixed size $N < N_{max}$ of a vector over a range of scales δ. The final segmentation was selected manually from the hierarchy of segmentations. Experiments with textures from *Brodatz* album [4] (see Figure 4 right) showed that as long as there were perceptually different sets of gray scale values the segmentation was performed

Figure 5: Bear and natural scene.

Figure 6: Beach scene and valley.

correctly. The proposed texture segmentation method was applied to gray scale (Figure 5) and color (Figure 6) images as a preprocessing step for object recognition. The color images were obtained from [9] and MIT database; http://www-white.media.mit.edu/vismod/imagery/VisionTexture/vistex.html. Each triple of shown images in Figures 5 and 6 contains original (left), segmentation (middle) and reconstructed image using vectors (right). Many of tested images were successfully segmented according to our perceptual criteria, but many times we could not answer the basic question needed for the evaluation, such as "What is the expected segmentation?"

5 Summary and Conclusions

We have presented a new hierarchical texture segmentation method. The approach presented has the following characteristics: (1) satisfactory segmentation performance of the statistical-structural approach to texture modeling (compared with [13]), (2) direct use of co-occurrence probabilities derived from regions of arbitrary shape, (compared with other methods using co-occurrences introduced in [6]), (3) tradeoff between computational requirements (proportional to dimensionality of computations) and noise robustness of the proposed method and (4) unsupervised hierarchical segmentation providing results indexed by the scale parameter δ. Results showing robustness against (1) noise in primitives and (2) nonuniform distribution of primitives were reported for synthetic data. Computational efficiency of the segmentation method was measured. Experiments with data from Brodatz album [4] and real scenes (gray scale and color images) were conducted.

References

[1] N. Ahuja and B. J. Schachter. Image models. *Computing Surveys*, 13:373–397, Dec. 1981.

[2] P. Bajcsy and N. Ahuja. A new framework for hierarchical segmentation using similarity analysis. In *Proceedings on the 1st Int. Conf. on Scale-Space Theory in Computer Vision*, pages 319–322, Utrecht, The Netherlands, 1997.

[3] D. Blostein and N. Ahuja. Shape from texture: Integrating texture-element extraction and surface estimation. *IEEE on PAMI*, 11(12):1233–1251, December 1989.

[4] P. Brodatz. *Textures: A photographic Album for Artists and Designers*. New York, NY, Dover, 1966.

[5] G. R. Cross and A. K. Jain. Markov random field texture models. *IEEE on PAMI*, 5(1):25–39, January 1983.

[6] R. M. Haralick. Statistical and structural approaches to texture. *Proceedings of the IEEE*, 67(5):786–804, May 1979.

[7] B. Julesz. Experiments in the visual perception of texture. *Scientific American*, 232:34–43, April 1975.

[8] J. Malik and P. Perona. Preattentive texture discrimination with early vision mechanisms. *Journal of Optical Society of America*, 7(5):923–932, May 1990.

[9] D. K. Panjwani and G. E. Healey. Markov Random Field models for unsupervised segmentation of textured color images. *IEEE Transaction on Pattern Analysis and Machine Intelligence*, 17(10):939–954, Oct. 1995.

[10] A. P. Pentland. Fractal-based description of natural scenes. *IEEE on PAMI*, 6(6):661–674, November 1984.

[11] W. H. Tsai and K. S. Fu. Image segmentation and recognition by texture discrimination: A syntactic approach. In *Proc. of the 4th Int. Joint Conf. on Pattern Recognition*, pages 560–564, Kyoto, Japan, Nov. 1978.

[12] M. Tuceryan and A. K. Jain. *The Handbook of Pattern Recognition and Computer Vision, Eds. C. H. Chen and L. F. Pau and P. S. Wang*, chapter 2.1 Texture Analysis, pages 235–276. World Scientific Company, 1992.

[13] S. W. Zucker. Toward a model of texture. *Computer Graphics and Image Processing*, 5:190–202, 1976.

Range Image Segmentation: Adaptive Grouping of Edges into Regions

Xiaoyi Jiang and Horst Bunke

Dept. of Computer Science, Univ. of Bern, CH-3012 Bern, Switzerland

Abstract. In this paper we propose an adaptive grouping algorithm to achieve a complete range image segmentation from an edge map. Its effectiveness has been extensively evaluated on three range image sets used in recent range image segmentation comparison studies. It turned out that our approach almost consistently outperforms all the region-based segmentation algorithms tested on the image sets with respect to both segmentation quality and computation time.

1 Introduction

Vision tasks based on range image analysis have been relying in most cases upon scene representations in terms of surface patches. This has led to the general agreement that the range image segmentation task is one of dividing range images into closed regions with application domain specific surface properties [2]. A complete image segmentation into regions can be achieved by both edge- and region-based approaches. Compared to region-based techniques, edge detection has some very appealing properties, in particular regular operators like convolution, simple control structure, and more precise localization of surface boundaries. As a drawback, edge detection cannot guarantee closure of boundaries for surface extraction, making a subsequent grouping and completion necessary.

In this paper we propose a simple adaptive technique for grouping edges into regions. Its effectiveness will be evaluated on three image sets used in two recent range image segmentation comparison studies [2, 7], demonstrating superior performance of a complete image segmentation to many region-based algorithms with respect to both segmentation quality and computation time. Our work is related to [4, 6]. To our knowledge, however, it is the first time that the edge-based approach is demonstrated to almost consistently outperform region-based methods on a large number of range images acquired by different types of scanners.

2 Adaptive grouping of edges into regions

We assume a binary edge map as input to our grouping algorithm. Our approach is based on the observation that any boundary gap can be closed by dilating the edge map. If the largest gap of a region has a length L, then $L/2$ dilations will successfully complete the region. However, we have no idea about the actual value

```
perform component labeling on input edge map;
RegionList = { connected regions of size > T_size };
while (RegionList != ∅) {
    delete arbitrary region R from RegionList;
    verify R;
    if (successful)
        record region R;
    else {
        perform dilation within R;
        perform component labeling within R;
        RegionList += { connected regions of size > T_size };
    }
}
postprocessing;
```

Fig. 1. Outline of the adaptive grouping algorithm.

of L before the grouping process is finished. In order not to miss any region, we potentially have to select a high value for L as maximum allowable gap length, resulting in a consistently large number of dilations applied to all regions. This is not only a unnecessary overhead in dealing with regions that are (almost) closed. But also relatively small-sized or thin regions will disappear. In the following we propose an adaptive approach that carries out the minimum number of dilations necessary for each particular region. Our method is embedded in a hypothesis-and-verification framework. It increases the number of dilations only for those regions that cannot be successfully verified. An outline of our adaptive grouping technique in C-style pseudo-code is given in Figure 1.

2.1 Hypotheses generation and verification

From the input edge map, region hypotheses can be found by a component labeling. Usually, this initial region map contains many instances of under-segmentation, i.e., multiple true regions are covered by a single region hypothesis. To recognize the correctly segmented and under-segmented regions, we perform a region test for each region R of the initial segmentation. If the region test is successful, the region R is registered. Otherwise, there still exist open boundaries within R. In this case we perform one dilation operation within R, potentially completing the boundaries. Again, a component labeling is done for R to find new region hypotheses, and these are verified in the same manner as for the initial regions. This process of hypotheses generation (component labeling) and verification (region test) is recursively repeated until the generated regions have been successfully verified or they are not further considered because of a region size smaller than a preset threshold T_{size}.

The region test starts with a plane test. The principal component analysis technique [5] is used to compute a plane function by minimizing the the sum of squared Euclidean (orthogonal) distances. The region is regarded a plane if both the RMS and average orthogonal distance of region pixels to the plane are smaller than a respective preset threshold. If this test fails, we compute a second surface approximation by means of a biquartic polynomial function

$$f(x,y) = \sum_{i+j\leq 4} a_{ij}x^i y^j.$$

This is done by a least-square method that minimizes the sum of squared fit errors

$$\sum_{k=1}^{n}(f(x_k,y_k) - z_k)^2$$

where the points $(x_k, y_k, z_k), k = 1, \ldots, n$, comprise the region under consideration. Again, the region acceptance is based on the RMS and average fit error.

Conceptually, a region test based on the biquartic surface function alone suffices for the verification purpose. Using regression theory it can be shown that such a function is an unbiased estimate of any underlying region models of order less than four including planes [4]. For two practical reasons, however, we have added the plane test. First, a quartic function approximation is computationally much more expensive than the plane test. Therefore, an initial plane test enables us to exclude the planar regions of a range image from the expensive second test. Moreover, in our experiments the quartic surface function encountered difficulties to reasonably approximate highly sloped planes, mainly due to the use of fit error in the minimization instead of the Euclidean distance. Since the desired Euclidean distance is minimized by the principal component analysis method, the plane test turns out to be more suitable for handling planar surface patches in general.

2.2 Postprocessing

We introduce three postprocessing steps to complete our adaptive grouping algorithm. Until now, the edge pixels are not considered to be part of regions. Also, the dilations necessary for the boundary closure discard pixels near region boundaries. These unlabeled pixels should be added to their corresponding regions. For this purpose we merge each unlabeled pixel to an adjacent region if the orthogonal distance (for a plane) or the fit error (for a biquartic surface patch) is tolerable.

Occasionally, it happens that the dilations link some noise edge pixels within a true region to a connected contour, and produces an over-segmentation of the region. Instances of over-segmentation can be easily corrected by merging adjacent regions. The region test described above is performed for the union of two adjacent regions, and they are merged in case of success. This operation is repeated until no more merge is possible. In our current implementation, this postprocessing step is only done for planar regions. In this case the region test is very fast. More importantly, we have a simple criterion to exclude most of the pairs of adjacent regions from an actual region test. A pair should undergo the region test only if the angle of their normals takes a small value.

Another potential problem with dilations is that small-sized or thin regions may disappear or become smaller than the minimum region size threshold T_{size}. To recover these regions, we consider all the pixels that don't belong to any region found so far. Again, region hypotheses are generated by a component labeling and verified by the region test. The successful regions complete the overall region segmentation.

3 Experimental evaluation

The adaptive grouping algorithm has been implemented in C on a Sun SparcStation. In this implementation we have utilized the edge detection method reported in [3] to generate binary edge maps. In order to make our results comparable to those of other, especially region-based, segmentation algorithms, we have chosen three image sets as test data that have been used in recent experimental range image segmentation comparison studies [2, 7]. The actual comparison was embedded in the framework proposed in [2], where objective performance metrics have been defined to compare a machine-generated segmentation with an ideal segmentation (ground truth), including the number of correctly detected, over-segmented, under-segmented, missing, and noise regions.

3.1 Segmentation into planar surface patches

As the first experimental comparison of range image segmentation algorithms, the task considered in [2] was limited to segmenting a range image into planar surface patches. Two sets of forty range images each that have manually specified ground truth served as the test basis: one acquired by an ABW structured light scanner, and the other by a Perceptron time-of-flight laser scanner. Figure 2 shows a range image (top left) of medium complexity from the ABW set, together with the ground truth (top right). One of the most complex scenes from the Perceptron set is represented in Figure 3. In [2] the forty images of each set were divided into a training set of ten images and a test set of thirty images. The training images were used to fix the parameters of a segmenter. Then, the performance metrics on the test set are the basis for comparing different segmenters. So far, four region-based algorithms have been involved in the comparison study. In the following we call them USF, WSU, UB, and UE; see [2] for a description of these algorithms. The average performance metrics per image for the four segmenters are graphed in Figure 4 against a compare tool tolerance. In essential, this tolerance value defines the degree of the overlap between a region in the ground truth and a machine-generated region to be considered as a corresponding region pair; see [2] for more details.

In our tests on the ABW and Perceptron image set, the region test was restricted to the plane test alone. We have fixed the parameter T_{size} for each image

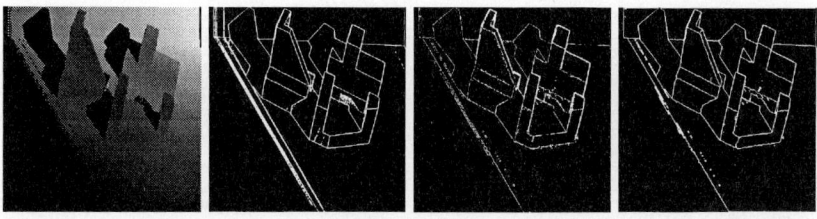

Fig. 2. An ABW image and the grouping result.

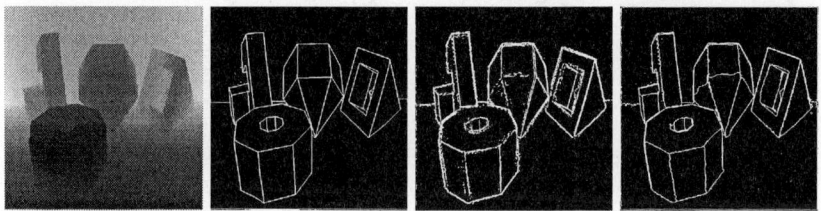

Fig. 3. A Perceptron image and the grouping result.

set after some experiments on the training data. The thresholds for the RMS and average orthogonal distance were determined by examining these quantities of all regions from the training images. The two thresholds were set such that almost all regions of the training images will successfully pass the region test. Finally, the threshold used for merging unlabeled pixels to regions was set to be a multiple of the threshold for the average orthogonal distance.

For the ABW and Perceptron image in Figures 2 and 3, the binary edge map (middle right) and the result of the adaptive grouping (right) are shown there as well. There are noise edge pixels, especially in the Perceptron images, which are noisier than the structured light data in general [2]. Many open boundaries can be observed. Partly, very long gaps exist; see, for instance, the U-shaped object in the ABW image. Our adaptive grouping technique has been successful in dealing with both problems.

The performance metrics for our method, referred to as EG, over the test sets are drawn in Figure 4. On the ABW image set, it reached the same performance as the UE algorithm, which is the best among the four region-based approaches tested so far on this set. In terms of over-segmentation, it shows even an improvement. On the Perceptron image set, our method beats all four region-based algorithms with respect to four performance metrics; the only exception is the under-segmentation metric which is located in the middle field. Table 1 presents the average results on all performance metrics for all five algorithms on both test sets at 80% compare tool tolerance, demonstrating the superior performance of our adaptive grouping algorithm.

The average processing time for the four region-based segmentation algorithms on the ABW and Perceptron test sets, per image, were 78 and 117 minutes (USF) on a Sun SparcStation 20, 6.3 and 9.1 minutes (UE) on a Sun SparcSta-

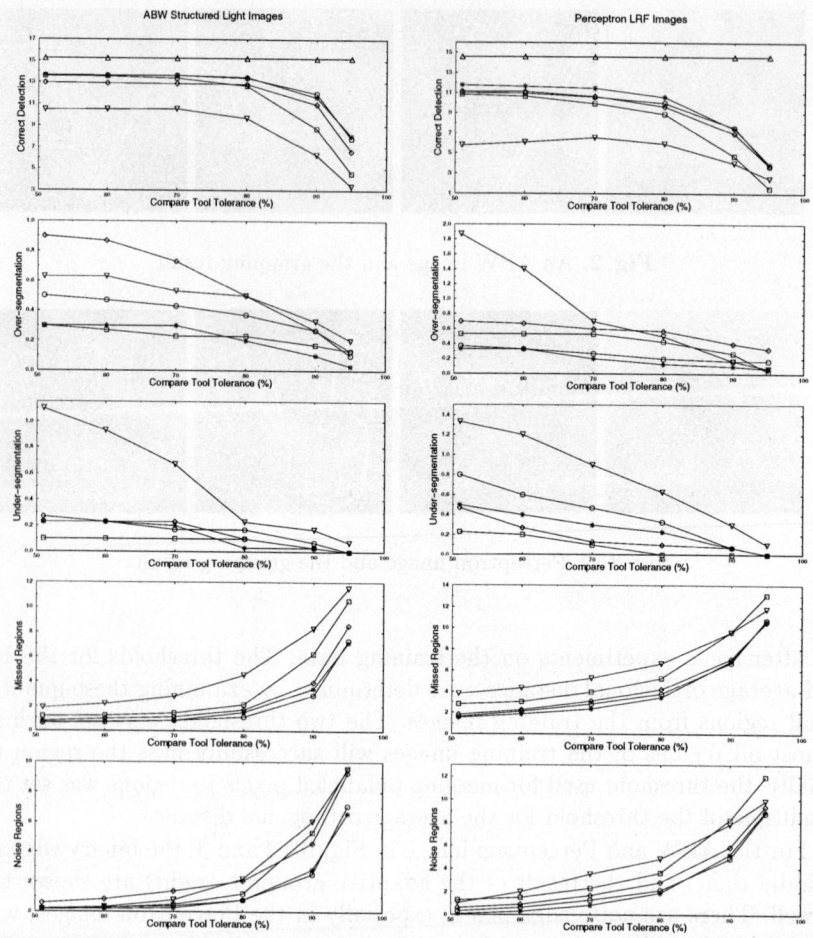

Fig. 4. Performance metrics for 30 ABW test images (left) and 30 Perceptron test images (right). Symbols: △ = average number of regions per image (maximum number of correct detections), □ = USF, ▽ = WSU, ◇ = UB, ○ = UE, ∗ = EG.

tion 5, 4.4 and 7.7 minutes (WSU) on a HP 9000/730, and 7 and 10 **seconds** (UB) on a Sun SparcStation 20. Our adaptive grouping technique requires for both test sets 15 **seconds**, per image, on a Sun SparcStation 5, including all processing steps from edge detection to postprocessing.

3.2 Segmentation into curved surface patches

Recently, the range image segmentation study has been extended to curved surfaces [7]. For this purpose an set of sixty range images were acquired by a K2T model GRF-2 structured light scanner. Similarly to the first comparison study, the set was divided into a training set of twenty images and a test set of forty

ABW 30 test images

algorithm	GT regions	correct detection	over-segmentation	under-segmentation	missed	noise
USF	15.2	12.7	0.2	0.1	2.1	1.2
WSU	15.2	9.7	0.5	0.2	4.5	2.2
UB	15.2	12.8	0.5	0.1	1.7	2.1
UE	15.2	13.4	0.4	0.2	1.1	0.8
EG	15.2	13.5	0.2	0.0	1.5	0.8

Perceptron 30 test images

algorithm	GT regions	correct detection	over-segmentation	under-segmentation	missed	noise
USF	14.6	8.9	0.4	0.0	5.3	3.6
WSU	14.6	5.9	0.5	0.6	6.7	4.8
UB	14.6	9.6	0.6	0.1	4.2	2.8
UE	14.6	10.0	0.2	0.3	3.8	2.1
EG	14.6	10.6	0.1	0.2	3.4	1.9

Table 1. Average results of all five segmenters on ABW and Perceptron test sets at 80% compare tolerance.

images. Figure 5 shows a test image of average complexity (left) and the corresponding ground truth (middle left).

Currently, two segmentation algorithms have been tested on this image set. The classical work by Besl and Jain [1], referred to as BJ, has been adapted to work on these data. We have participated in the comparison study using our adaptive grouping technique, called EG in the following. The parameters were determined in the same manner as described in Section 3.1. Figure 5 shows the edge map and the grouping result of the test image as well. Table 2 lists the performance metrics for both algorithms over the test set at comparison tolerance 80%, where each metric represents the average percentage per image; see [7] for more results. The region growing algorithm BJ has surprisingly much more difficulties on this image set than our simple grouping technique. The only exception is that our method demonstrates a higher percentage of under-segmentation. The reason lies in the nature of such an edge grouping approach. The edge detection method [3] used in our implementation is able to detect jump and crease edges but not smooth edges (discontinuities only in curvature). Therefore, two surfaces meeting at a smooth boundary will not be separated and an under-segmentation occurs. This happened in some of the test images. Since it seems that no edge detection method known from the literature can deal with smooth edges, a potential under-segmentation is not a particular weakness of our adaptive grouping technique, but an inherent problem of any edge grouping method.

For our method, the average computation time for this test set is 23 **seconds**,

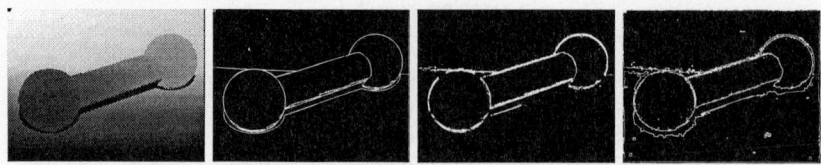

Fig. 5. A GRF-2 image and the grouping result.

algorithm	correct detection	over-segmentation	under-segmentation	missed	noise
BJ	16.1%	54.5%	3.6%	21.8%	30.8%
EG	68.2%	2.2%	12.3%	4.5%	2.6%

Table 2. Average results of the algorithms BJ and EG on GRF-2 test set.

per image, on a Sun SparcStation 5. The BJ algorithm requires typically a few hours on a Sun SparcStation 20. The difference in processing time is remarkable.

4 Conclusions

In this paper we have proposed a simple adaptive grouping algorithm to solve the boundary closure problem. Its effectiveness has been extensively evaluated on three range image sets used in recent range image segmentation comparison studies. It turned out that our approach almost consistently outperforms all the region-based segmentation algorithms tested on the image sets with respect to both segmentation quality and computation time.

References

1. P.J. Besl, R.C. Jain, Segmentation through variable-order surface fitting, IEEE Trans. on PAMI, 10(2): 167–192, 1988.
2. A. Hoover et al., An experimental comparison of range image segmentation algorithms, IEEE Trans. on PAMI, 18(7): 673–689, 1996.
3. X. Jiang, H. Bunke, Robust and fast edge detection and description in range images, Proc. of IAPR Workshop on Machine Vision Applications, Tokyo, 538–541, 1996.
4. S.-P. Liou et al., A parallel technique for signal-level perceptual organization, IEEE Trans. on PAMI, 13(4): 317–325, 1991.
5. T.S. Newman et al., Model-based classification of quadric surfaces, CVGIP: Image Understanding, 58(2): 235–249, 1993.
6. B. Parvin, G. Medioni, B-rep object description from multiple range views, Int. Journal of Computer Vision, 20(1/2): 81–11, 1996.
7. M. Powell et al., Comparing curved-surface range image segmenters, Proc. of Int. Conf. on Computer Vision, Bombay, India, 1998. (to appear)

Optimising the Complete Image Feature Extraction Chain

M. Mirmehdi[1], P. L. Palmer[2] and J. Kittler[2]

[1] Dept. of Computer Science, University of Bristol,
Bristol BS8 1UB, England
[2] Centre for Vision, Speech and Signal Processing,
Surrey University, Guildford GU2 5XH, England

Abstract. The hypothesis verification stage of the traditional image processing approach, consisting of low, medium, and high level processing, will suffer if the set of low level features extracted are of poor quality. We investigate the optimisation of the feature extraction chain by using Genetic Algorithms. The fitness function is a performance measure which reflects the quality of an extracted set of features. We will present some results and compare them with a Hill-Climbing approach.

1 Introduction

The traditional classification in most object recognition systems is that of low, medium, and high level image information processing. This transformation of signals into symbols consists of the extraction of features, the formation of a hypothesis through the grouping of multiple features, and finally, verification of the hypothesis via some form of comparison and confirmation against a pre-registered model. The medium and high level stages are highly dependent on the set of features provided via the low level feature extraction stage. They may suffer if there are too few or too many features: too few will lead to inadequate hypotheses without strong evidence, and too many will cause confusion, extra hypotheses [1] and extra computation. Figure 1 illustrates this point through over-segmentation and under-segmentation of a simple image.

There has recently been much emphasis on the failure of the primitive feature extraction techniques to provide robust features for the higher levels of processing [2–4]. Most techniques, such as corner detection, edge detection, line extraction etc., all employ thresholds that need to be varied, dependent upon the image context. The setting of the correct thresholds cannot be easily determined except through trial and error as demonstrated in Figure 1, but the chance to experiment is not always available or efficient, and even under the same context, small changes, e.g. in illumination, may require changes in the parameters of the algorithm.

In this paper, we are concerned with the extraction of robust features that the higher levels of processing can rely on. By employing a cost function which can measure the quality of the extracted features we can optimise the feature

 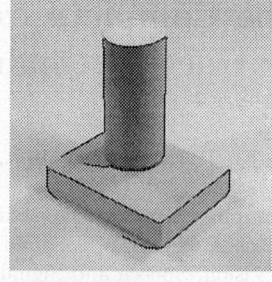 

(a) Over-segmentation   (b) Under-segmentation   (c) Better Segmentation

Fig. 1. (a) and (b) Over and under-segmentation using arbitrarily chosen parameter values in each case. (c) A more reasonable and efficient segmentation.

extraction process. We apply this to the extraction of line segments, but the same framework is equally applicable to other features such as corners, colours, and surfaces. As suggested by [3], it would be better to optimise the low level stages as a whole chain of processes, since optimality of individual processing stages does not guarantee the overall optimality of the system. Thus, they introduced a single objective function to measure the performance of a system as a whole, while the parameters of the individual stages are adjusted as necessary. This performance measure was applied by [5,6,1] within a Hill-Climbing framework to optimise the extraction of low level features for two different applications. However, Hill-Climbing approaches may get trapped in local optima. Furthermore, a parameter such as the edge detection filter width does not necessarily comply with the notion of the derivative, a notion that is essential in Hill-Climbing and calculus-based methods [7]. In this paper, the performance measure will be used as part of a fitness function in the context of a genetic optimisation process.

In Section 2 we review past uses of GAs in computer vision and present our own implementation in Section 3. Some results are discussed in Section 4, followed by conclusions in Section 5.

2 Application of GAs in Computer Vision

GAs provide a robust approach for multi-dimensional space search controlled by stochastic operators [7]. Their application in computer vision is slowly becoming widespread. [8] used simplified classification for labelling an image and applied a distributed GA to iteratively modify the labelling leading to image segmentation. [9] optimally enhanced an image through the automatic selection of an enhancement operator using GAs. [10] have implemented a GA based technique for edge detection, which in fact optimises the validity of an edge structure once a set of edge pixels is found. Our work is different in that it optimises the output of the whole line extraction stage incorporating in this parameters for the actual edge detection stage. In the same way, our work differs from that of [11]. They apply GAs to optimise the search in the Hough space for complex multi-parameter primitives. Their work complements that of [12] who used GAs to extract predefined geometric primitives from geometric data as an optimised alternative to the Hough transform. [13] have used GAs to select the optimal disparity at each pixel position within a stereo matching context.

3 GA Implementation

We aim to implement an optimisation framework for deriving a near-optimal or optimal set of parameters that will result in a useful set of line segments at the end of the low level feature extraction processing chain. This processing chain quite commonly involves edge segmentation, edge linking, and the Hough transform. We consider the corresponding parameters in conjunction with each other in our aim to achieve an overall optimisation.

We optimise N parameters, $P_i, i = 0, ..., N-1$, resulting in a N-dimensional solution space. The traditional GA chromosome representation is that of a bit string, but we employ a word scheme which eases the encoding of our continuous space parameter values. Hence, each chromosome is represented by N words or genes, where each particular gene takes on a particular label or allele. The alleles are allowed to consist of feature values in the same M label set, $L_j, j = 0, ..., M-1$. However each distinct parameter may need to take on values of different order and range. Therefore, we use a special set of weights, $W_{P_i}, i = 0, ..., N-1$, when decoding a chromosome to transform its label values to the appropriate parameter value, V_{P_i} for each P_i.

The crossover operation is applied not to all pairs of chromosomes but to a randomly selected set. When it is applied, a chromosome pair is split at some randomly chosen position along its length, for example say at the second word from the left, and the four resulting segments are recombined by joining the head of one chromosome to the tail of the other. This is a simple single point crossover. The mutation operation, which consists of selecting a random word and altering its value, is applied very rarely and at a rate of usually less than 1%.

3.1 The Optimisation Process

Figure 2 shows a skeletal view of the overall optimisation cycle. Chromosomes are randomly generated and evolved in the genospace and decoded for fitness evaluation. The fitness function evaluates a chromosome in the data space and returns the fitness value. This process continues for each chromosome over a number of generations (indicated by the dashed line in Figure 2). We do not employ any specific criteria to determine convergence and termination of the process. However, this issue is discussed in detail in [14].

3.2 Encoding, Decoding, and Constraints

For encoding, we simply define the label set L and allow the chromosomes to be generated by taking on values from the set. Once a chromosome is ready for fitness evaluation, decoding has to take place with respect to the gene values. At this stage the decoding is combined with our constraints set.

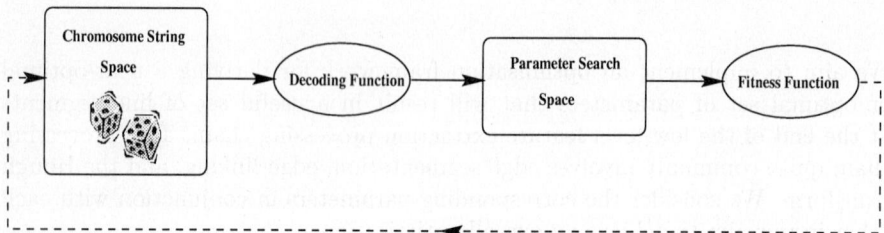

Fig. 2. Overview of the optimisation cycle

As with many other practical systems it is possible to limit and constrain the solution space. When a chromosome presents a particular solution, the labels are transformed via the weights and analysed to determine if a particular value lies outside the constrained solution space (for example an edge filter of width 1, or an upper hysteresis threshold which is smaller than the lower hysteresis threshold). If the chromosome is thus handicapped, the fitness function will not be evaluated, thus saving on computation, and an inhibiting value is returned to degrade the fitness for that particular chromosome.

Each parameter P_i has an acceptable range of values R_{P_i} where for each value $V_{P_i}, V_{P_i} \in R_{P_i} = (V_{P_{i_{MIN}}}, ..., V_{P_{i_{MAX}}})$. Thus, we require a constraint function C for each parameter which takes a particular label L_j and outputs the real-world value for that parameter, $V_{P_i} = C_{P_i}(L_j, W_{P_i})$.

3.3 The Fitness Function

The fitness function resolves whether the chromosome provides a good encoding of the solution. The fitness we use to determine the performance of the low level procedures is called the *performance quality measure*. This measure is the outcome of the whole low level stage - edge detection, linking and Hough transform and promotes a different concept to measuring performance of the individual procedures within the processing chain [15]. Hence, the fitness function consists of three major steps: generate the edge map, detect line segments, and measure the quality of lines.

The human vision system enjoys optimisation through its use of context to change parameters. A wide dynamic range is achieved by processing locally for signals amidst noise, and adapting to lighting conditions. The performance quality measure [3] acts on the hypothesised boundaries between regions and

Parameter	Functionality	Processing Stage	Range R_{P_i}
ω	mask width	edge detection	5,7,9,11,13,15
hl	hysteresis lower threshold	edge linking	0.0,...,1.0
hu	hysteresis upper threshold	edge linking	0.0,...,1.0
$k\rho$	kernel width in ρ	Hough transform	1.0,...,5.0
$k\theta$	kernel width in θ	Hough transform	3.0,...,25.0

Table 1. Table of parameters to be optimised

thus processes locally and will give quality values which reflect the statistical distribution of pixels in the area local to the hypothesised line between regions of differing grey-level value in the original image. Thus, consider a true boundary between regions of significantly different nature with mean grey-level values μ_1 and μ_2, and variances σ_1^2 and σ_2^2. The performance measure may be computed for the whole image as:

$$Q = \sum_{s=1}^{S} (P_s)^{3/2} \ln \left[\frac{(\mu_{s1} - \mu_{s2})^2}{\sigma_{s1}^2 + \sigma_{s2}^2} \right] \quad (1)$$

where the low level routines produce S segments in the image and each line segment s consist of P_s edge pixels. The measure is able to increase as more line segments are found, unless these extra segments are false alarms, in which case it decreases. It has also been shown to correlate well with measures based on ground truth for synthetic images, and has been applied successfully to real imagery [3,1]. Here, it acts as the genetic fitness function such that a measure of quality or fitness is calculated for the low level extraction processes characterised by parameters that are encapsulated in any one chromosome. The higher the quality, the more significant is the identity of the associated chromosome.

4 Examples

In the examples described below the optimisation process is limited to only a few parameters which we thought were fundamental to the quality of the results. They are the edge detection mask size, the edge linking hysteresis thresholds, and the kernel widths for ρ and θ in the Hough transform stage (Table 4). The edge detection algorithm we use is not important as long as we can optimise its parameters to give its best overall output. For our edge detector, the mask width is optimised since its size represents a tradeoff between noise sensitivity and good localisation of the edge gradients. As for hl and hu, hysteresis thresholding is a distinctive factor in the quality of the results which takes into consideration the spatial arrangement of edge strings. In optimising for $k\theta$ and $k\rho$, we establish the voting in a 2D hypothesis testing kernel which is canvassing for the orientation information of the edge pixels and their distance from the hypothesised line. Note that in the scenario presented here, the GA can only return the best of what the tools at hand can do.

In all our experiments across all the images tested, the crossover and mutation probabilities were set to 0.90 and 0.01. Our optimisation process is commenced with a different seed on each run such that the initial chromosome population generated is completely random. Initially, we ran our tests using a large population (e.g. 500) and for a large number of generations (e.g. 50). The resulting optimised parameter values were then considered as the ideal value and taken as a point of reference. We demonstrate this through the image in Figure 3 which is particularly difficult for a Hough stage as it contains a number of very close line segments. Thus the size of the θ and ρ kernels are distinctly important. Resulting line segments are shown in Figure 3(b).

Table 2 shows corresponding results for varying chromosome populations. Overall, the hysteresis thresholds were not too sensitive and could vary within certain limits before adversely affecting the quality of the picture, particularly the upper hysteresis threshold. Small variations meant the loss of only a very few prominent lines. However, if set too high, it would cause the loss of a considerable number of line segments. At the end of each generation there are a number of chromosomes representing points in the multi-dimensional space that are neighbours and of similar high quality. The quality of results for the lowest population size (10) indicates that the optimisation can be run quickly using a small population size confident that near optimal results can be reached. Figure 3(c) shows little significant difference in the line segments resulting from a population of 10 to that of 500. The optimisation process for a population of 10 over 10 generations took just over 2 minutes on an R10000 Silicon Graphics processor, while it required over 2 hours for a population of 500 over 50 generations.

We have also implemented the optimisation process using a simple Hill-Climbing approach [5]. It can be seen from Figure 3(d) that the technique suffers from the classic problem of being trapped in local optima. The corresponding optimised parameter set and quality value are shown at the bottom of Table 2.

In [14] we describe this framework in more detail. We also show other examples and discuss the issue of convergence of the results over a number of generations.

5 Conclusions

In this paper, we have shown a framework for the optimisation of the low level feature extraction chain using Genetic Algorithms. We have outlined that any number of parameters associated with the feature extraction techniques, such as the edge detector and the Hough transform, can be encoded and optimised. The process was demonstrated in detail for five different parameters for a difficult image with a number of close line segments. Our implementation determines values for a combined parameter set across the processing chain rather than for individual parameters for each single processing step.

The role of optimisation is very important for arriving at confident interpretations of signals into symbols and it has an inherent place in vision systems [1,4].

Computationally, genetic algorithms may still not be completely feasible within a real-time image processing system, however, much research is going on in utilising the implicit parallelism within them and they are becoming increasingly popular in image processing, e.g. see [7,10-12]. We conclude from our results that a quick solution can be reached by optimising with a small population and for a small number of generations, since GAs are so robust and efficient in locating a global maxima in the parameter search space. The current framework can be used as part of a *training stage* to find near-optimal parameter values for situations where the imaging environment is constrained. It also removes the need for trial and error parameter value determination for everyday use.

Acknowledgement - This work was supported by the DRA, Farnborough, UK. The authors also wish to thank J.C. Clarke for the use of the *pens* image.

References

1. M. Mirmehdi, P.L. Palmer, J. Kittler, and H. Dabis. Complex feedback strategies for object recognition. *Submitted to IEEE Transcations in Image Processing*, 1996.
2. M. Mirmehdi, P.L. Palmer, J. Kittler, and H. Dabis. Complex feedback strategies for hypothesis generation and verification. In *Proceedings of the 7th British Machine Vision Conference*, pages 123-132. BMVA Press, 1996.
3. P.L. Palmer, H. Dabis, and J. Kittler. A performance measure for boundary detection algorithms. *CVGIP: Image Understanding*, 63(3):476-494, 1996.
4. B. Draper, A. Hanson, and E. Riseman. Knowledge-directed vision: Control, learning and integration. *Proceedings of the IEEE*, 84(11):1625-1637, 1996.
5. M. Mirmehdi, P.L. Palmer, and J. Kittler. Framework for control of parameters of low-level processes in early vision. *Submitted to ECIS Workshop on Computational Vision*, 1995.
6. M. Mirmehdi, P.L. Palmer, J. Kittler, and H. Dabis. Hypothesis generation and verification using complex feedback strategies for object recognition. Technical report, University of Surrey VSSP-TR-2/96, 1996.
7. D. Goldberg. *Genetic algorithms in search, optimization & machine learning*. Addison-Wesley, 1989.
8. P. Andrey and P. Tarroux. Unsupervised image segmentation using a distributed genetic algorithm. *Pattern Recognition*, 27(5):659-673, 1994.
9. S.K. Pal, D. Bhandari, and M.K. Kundu. Genetic algorithms for optimal image enhancement. *Pattern Recognition Letters*, 15(3):261-271, 1994.
10. S. Bhandarkar, Y. Zhang, and W. Potter. An edge-detection technique using genetic algorithm-based optimization. *Pattern Recognition*, 27(9):1159-1180, 1994.
11. E. Lutton and P. Martinez. A genetic algorithm for the detection of 2d geometric primitives in images. In *Proceedings of International Conference on Pattern Recognition*, pages 526-528, 1994.
12. G. Roth and M. D. Levine. Geometric primitive extraction using a genetic algorithm. *IEEE PAMI*, 16(9):901-905, 1994.
13. H. Saito and M. Mori. Application of genetic algorithms to stereo matching of images. *Pattern Recognition Letters*, 16(8):815-821, 1995.
14. M. Mirmehdi, P.L. Palmer, and J. Kittler. Genetic optimisation of the image feature extraction process. *Pattern Recognition Letters*, 18(4):355-365, 1997.
15. W. K. Pratt. *Digital Image Processing*. Wiley and Sons, 1978.

Population	Fitness	ω	hl	hu	$k\rho$	$k\theta$
10	3197.8	7	0.248	0.75	1.896	8.045
100	3313.2	7	0.289	0.668	2.241	6.321
500	3318.7	7	0.227	0.75	1.724	7.471
Hill-Climbing	2750.9	7	0.533	0.91	3.086	5.266

Table 2. Table of fitness values for different populations

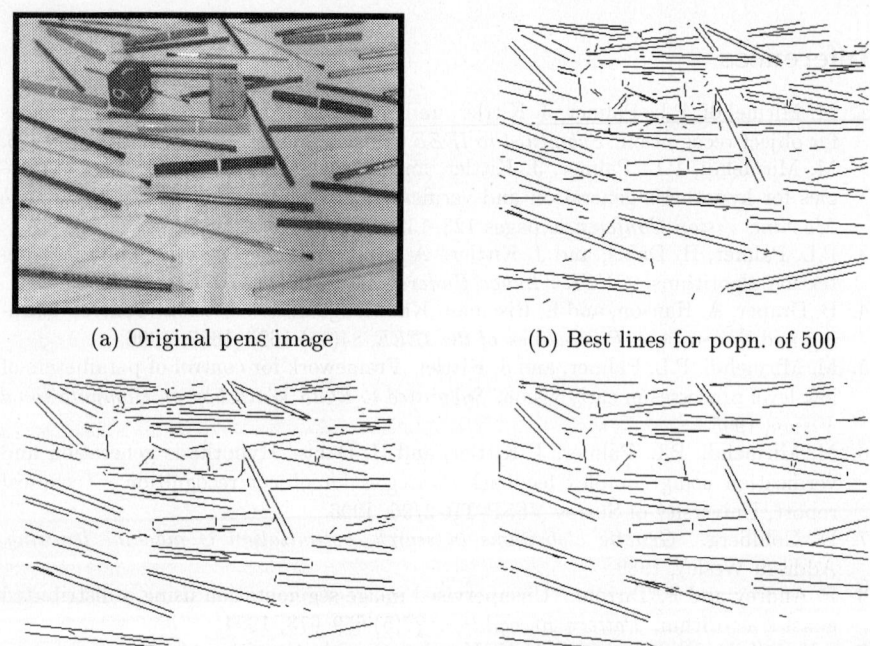

(a) Original pens image   (b) Best lines for popn. of 500

(c) Best lines for popn. of 10   (d) Best lines by Hill-Climbing

Fig. 3. (a) Original pen image with best optimised line sets for (b) population of 500 over 50 generations, (c) population of 10 over 10 generations, and (d) Hill-Climbing optimisation.

A Unified Framework for Salient Curves, Regions, and Junctions Inference*

Mi-Suen Lee and Gérard Medioni

Institute for Robotics and Intelligent Systems
University of Southern California, Los Angeles, CA 90230, USA
email: misuen, medioni@iris.usc.edu

Abstract. We present a unified computational framework to generate descriptions in terms of regions, curves, and labelled junctions, from sparse, noisy, binary data in 2-D. Each input site can be a point, a point with an associated tangent direction, a point with an associated tangent vector, or any combination of the above. The methodology is grounded on two elements: tensor calculus for representation, and non-linear voting for communication. Each input site communicates its information (a tensor) to its neighborhood through a predefined (tensor) field, and therefore casts a (tensor) vote. Each site collects all the votes cast at its location and encodes them into a new tensor. A local, parallel routine then simultaneously detects junctions, curves and region boundaries. The proposed approach is non-iterative, and the only free parameter is the size of the neighborhood, related to the scale. We illustrate the approach with results on a variety of images, then outline further applications.

1 Introduction

The general goal of image understanding is to make high level inference such as shape and/or motion description for an environment captured by one or more images. Our research goal is to develop a unified computational framework for the inference of viewer-centered descriptions in terms of curves, regions and junctions in the scene from noisy oriented or non-oriented binary data, without invoking specific object or scene model. By introducing the use of a computational technique called *tensor voting* as communication framework, a robust, non-iterative, and threshold-free method is devised to achieve this goal.

The inference of curves and regions from sparse noisy data is traditionally studied in the field of perceptual grouping. Based on the type of input, we can classify previous works into 3 areas, namely, dot clustering [1][13], line grouping [2][10], and illusory contour inference [11][12]. Most previous works take either the fit-and-split or the fit-and-merge approach, resulting in iterative algorithms which require parameter tuning.

Recently a number of non-iterative methods [7][11][12] have been developed for 2-D curve completion that involve the computation of a measurement for every image point which signifies the presence of perceptually significant structures. Among them, Guy and Medioni[4] have introduced a salient structure estimation technique called vector voting that is able to handle the tasks of interpolation, discontinuity

---

* This research was supported by NSF Grant under award No. IRI-9024369.

detection, and outlier identification simultaneously. They have obtained impressive results for the inference of multiple undirected curves and illusory contours. While other methods[7][12] give equally good results, it is the potential for generalization that make Guy and Medioni's approach promising. Applying the same methodology to surface inference, they have obtained exciting results [5] in 3-D.

Extending Guy and Medioni's methodology, our tensorial framework of salient structure inference broadens the original method significantly, for it augments the non-linear voting approach with a mathematical foundation which accounts for many heuristic measures in the original work, unifies the representation of different features and the inference of various structures, and allows us to incorporate *polarity* for the inference of regions.

We present our unifying tensorial framework for salient structure inference in section 2. We then illustrate the simultaneous inference of junctions, curves and regions using our salient structure inference engine with results on point and line segment data in section 3. And finally, we conclude this paper with a discussion and outline further application in section 4.

In the this paper, scalars are denoted by italic letters, e.g. l, tensors are denoted by bold capital letters, e.g. T, vectors are denoted by bold lower-case letters, e.g. e. The unit length vector for e is denoted as $\hat{e}$.

2 Salient Structure Inference Engine

The building block of our computational framework for inferring salient structure is the procedure that simultaneously interpolates curves, identifies discontinuities, and detects outliers, which we called the *salient structure inference engine*. The input to this procedure are local estimates of orientation associated with a directed or undirected tangent. Multiple orientation estimations are allowed for each location. The output of this procedure is a description for the input that compose of labeled junctions and directed or undirected curves.

2.1 Overview of the inference engine

Our salient structure inference engine is a non-iterative procedure that makes use of tensor voting to infer salient geometric structures from local features. Figure 1 illustrates the mechanisms of this engine. In practice, the domain space is digitized into discrete cells. We use a construct call a saliency tensor, to be defined in section 2.2, to encode the saliency of various features such as points and curve elements. Local feature detectors provide an initial local measure of surface features. The engine then encodes the input into a sparse saliency tensor field. Saliency inference is achieved by allowing the elements in the tensor field to communicate through voting. A voting function specifies the voter's estimation of local orientation/discontinuity and its saliency regarding the collecting site, and is represented by tensor fields. The vote accumulator uses the information collected to update the saliency tensor at locations with features and establish those at previously empty locations. From the densified saliency tensor field, the vote interpreter produces saliency maps for various features from which salient structures can be extracted. Once the generic saliency inference engine is defined, it can be used to perform many tasks, simply by changing the voting function.

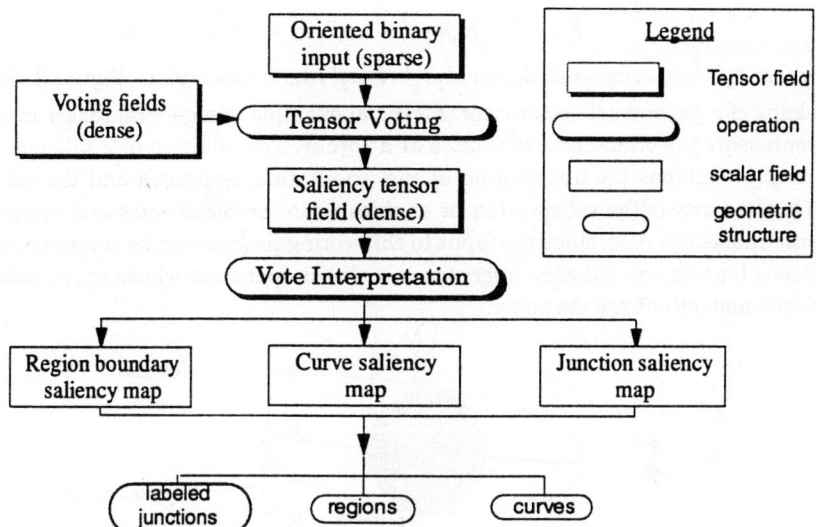

Figure 1 Flowchart of the Salient Structure Inference Engine

2.2 Saliency Tensor

The design of the saliency tensor is based on the observation that discontinuity is indicated by the variation of orientation when estimating curve locally. Intuitively, the variation of orientation can be captured by an ellipse in 2-D. Note that it is only the *shape* of the ellipse that describe the orientation variation, we can therefore encode saliency as the *size* of the ellipse. One way to represent an ellipse is to use the distribution of points on the ellipse as described by the covariance matrix. By decomposing a covariance matrix S into its eigenvalues λ_1, λ_2 and eigenvectors $\hat{e}_1, \hat{e}_2$, we can rewrite S as:

$$S = \begin{bmatrix} \hat{e}_1 & \hat{e}_2 \end{bmatrix} \begin{bmatrix} \lambda_1 & 0 \\ 0 & \lambda_2 \end{bmatrix} \begin{bmatrix} \hat{e}_1^T \\ \hat{e}_2^T \end{bmatrix} \quad (1)$$

Thus, $S = \lambda_1 \hat{e}_1 \hat{e}_1^T + \lambda_2 \hat{e}_2 \hat{e}_2^T$ where $\lambda_1 \geq \lambda_2$ and $\hat{e}_1$ and $\hat{e}_2$ are the eigenvectors correspond to λ_1 and λ_2 respectively. The eigenvectors correspond to the principal directions of the ellipse and the eigenvalues encode the size and shape of the ellipse. S is a *linear* combination of outer product tensors and, therefore a tensor.

Every saliency tensor thus has 3 parameters, λ_1, λ_2, and θ, where θ is the orientation of the major axis. This saliency tensor is sufficient to describe feature saliency for undirected curves. When dealing with boundaries of regions, we need to associate a polarity to the feature orientation to distinguish the inside from the outside. Polarity saliency, which relates the polarity of the feature and its significance, will be dealt with in section 2.4.

According to the spectrum theorem [3], a general saliency tensor S can be expressed as a *linear* combination of a stick tensor, one describes a perfect orientation estimation, and a plate tensor, one describes total orientation uncertainty, as:

$$S = (\lambda_1 - \lambda_2)\hat{e}_1\hat{e}_1^T + \lambda_2(\hat{e}_1\hat{e}_1^T + \hat{e}_2\hat{e}_2^T) \qquad (2)$$

where $\hat{e}_1\hat{e}_1^T$ describes a stick and $(\hat{e}_1\hat{e}_1^T + \hat{e}_2\hat{e}_2^T)$ describes a plate. Figure 2 shows the shape of a general saliency tensor. A typical example of a general tensor is one that represents the orientation estimates at a corner. The addition of 2 saliency tensors simply combines the distribution of the orientation estimates and the saliencies. This linearity of the saliency tensor enables us to combines votes and represent result efficiently. Also, since the input to this voting process can be represented by saliency tensors, our saliency inference is a closed operation whose input, processing token and output are the same.

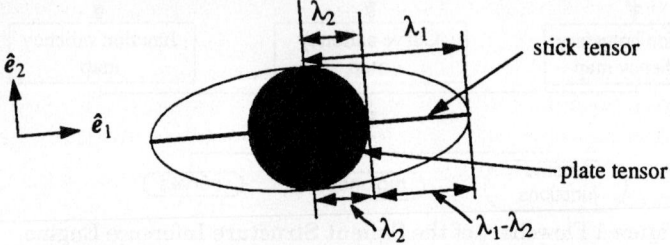

Figure 2 Decomposition of a general saliency tensor

2.3 Tensor Voting

Given a sparse feature saliency tensor field as input, each active site has to generate a tensor value at all locations that describe the voter's estimate of orientation at the location, and the saliency of the estimation. The value of the tensor vote cast at a given location is given by the voting function which characterize the structure to be detected. Due to the general properties of the function (detailed in [8]), this vote generation process can be implemented as a convolution with two saliency tensor fields, one encodes the votes due to the stick component of the voter, and one encodes the votes due to the plate component of the voter. Moreover, the plate voting field is generated from the stick voting field, which unifies the process for handling various orientation information.

After all the input site saliency tensors have cast their votes, the vote interpreter builds feature saliency maps from the now densified saliency tensor field by computing the eigensystem of the saliency tensor at each site. According to equation (2), each saliency tensor can be broken down into two components, $(\lambda_1-\lambda_2)\hat{e}_1\hat{e}_1^T$ corresponds to an orientation, and $\lambda_2(\hat{e}_1\hat{e}_1^T + \hat{e}_2\hat{e}_2^T)$ corresponds to an isotropic junction. The curve saliency therefore is measured by $(\lambda_1-\lambda_2)$, with tangent estimated as $\hat{e}_1$, and the junction saliency is measured by λ_2. Notice that the feature saliency values do not depend only on the saliency of the feature estimation, but also are determined by the distribution of orientation estimates.

Since the voting function is smooth, sites near salient structures will have high feature saliency values. The closer the site to the salient structure, the higher the feature saliency value the site will have. Structures are thus located at the *local maxima* of the corresponding feature saliency map. The vote interpreter hence can extract features by non-maxima suppression on the feature saliency map. To extract structures represented by oriented salient features, we use the methods developed in [6]. By expressing a maximal curve as a zero-crossing curve, a local iso-contour marching process [9] can be used for the extraction process

2.4 Incorporating polarity

So far, we have only considered the inference of undirected curves from undirected features. However, in real applications, we often deal with bounded regions. The simplest encoding for region is to associate a polarity, which denotes the inside of the object, with each curve element along the region boundary. It is precisely this polarity information that is being expressed in the output of an edge detector. We therefore need to augment our representation scheme to express this knowledge, and to use it to perform directed curve and region grouping.

In the spirit of our methodology, we attempt to establish at every site in the domain space a measure we called polarity saliency which relates the possibility of having a directed curve passing through the site. In particular, a site close to two parallel features with opposite polarities should have low polarity saliency, although the undirected curve saliency will be high.

Figure 3 depicts the situation where the voter has the polarity information. Since polarity is only defined for directed features, a scalar value from -1 to 1 encodes polarity saliency. The sign of this value indicates which side of the stick component of the tensor is the intended direction. The size of this value measures the degree of polarization at a site. Initially, all sites with features have polarity saliency of -1, 0, or 1 only.

Figure 3 Encoding Polarity Saliency

To propagate this polarity information, we assign either -1 or 1 to each tensor vote as polarity saliency, which is determined by the polarity of the voter and the estimated connection between the voter and the site.

Inferring directed feature saliency

Once votes are collected, directed curve saliency map can be computed by combining polarity saliency with undirected curve saliency. Since polarity is associated with an orientation, we need to take the orientation into account when we infer polarity saliency. An intuitive way to determine polarity is to compute the vector sum $u(x, y) = \sum_i v_i(x, y)$ of these directed votes $v_i(x,y)$'s at every site (x,y). The length of the resulting vector gives the degree of polarization and the resulting direction relates the indented polarity. We thus assign to every site a polarity saliency $PS(x,y)$ as:

$$PS(x, y) = \text{sgn}(u(x, y) \bullet \hat{e}_1(x, y))|u(x, y)| \qquad (3)$$

where $\hat{e}_1(x,y)$ is the major direction of the saliency tensor obtained at site (x,y).

Directed curve elements are those which have both high undirected curve saliency and polarity saliency. The natural way to measure directed curve saliency hence is to take the product of the undirected curve saliency and the polarity saliency. Once the directed curve saliency map is computed, region boundaries can be extracted by the method described in section 2.3.

3 Salient Structure Inference

Having defined the salient structure inference engine, we proceed to describe how to apply it to infer salient curves, regions and junctions in 2-D.

3.1 Curve inference

While the salient structure inference engine defined in above performs structure inference efficiently, it is the definition of the voting function that determines the effectiveness of the inference. For curve inference, we use circles as the connecting paths for orientation estimation and the Gaussian function to model the decay DF of voter's influence with path length s and curvature ρ as:

$$DF(s,\rho) = e^{-\left(\frac{s^2+c\rho^2}{\sigma^2}\right)}$$

where c is the constant that reflects the relative weight of path length and curvature and σ is the scale factor that determines the rate of attenuation. The arguments for choosing this voting function are detailed in [8].

Applying the voting function defined in above and depicted in Figure 4, we have obtained results on noisy data, one of which is shown in Figure 5.

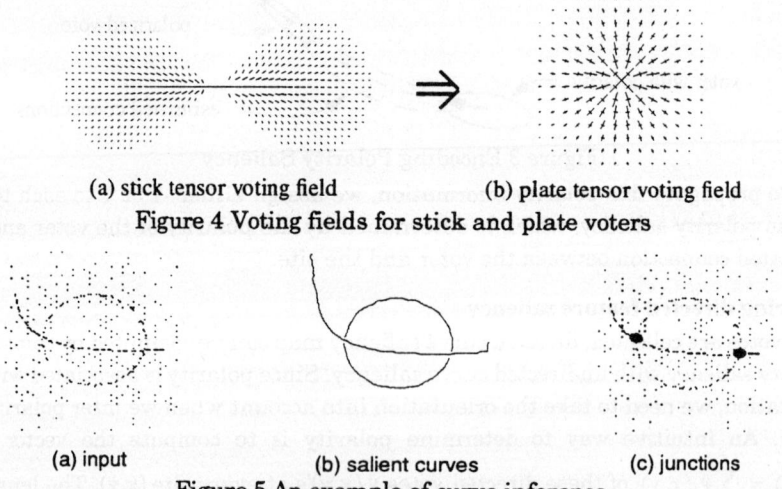

(a) stick tensor voting field   (b) plate tensor voting field
Figure 4 Voting fields for stick and plate voters

(a) input   (b) salient curves   (c) junctions
Figure 5 An example of curve inference

3.2 Region inference

While the incorporation of polarity allows us to infer regions properly, this polarity information is not always available. When data are sparse and irregular, local boundary detection becomes hard. A typical example is shown in Figure 6(a). Since boundary detection is about locating discontinuities, it is possible to use the salient structure inference engine to detect points on the boundary. In the spirit of our methodology, we again seek to compute at every site in the domain space a measure we call boundary saliency which relates the possibility of having the site being on the boundary of a region. A boundary point has the property that most of its neighbors are on one side. We therefore can identify boundary points by computing the directional distribution of neighbors at every data point. This local discontinuity estimation is similar to the orientation estimation for salient structure inference, except that accurate orientation estimate is irrelevant to discontinuity estimation and thus does not require the use of tensor. Therefore the voting function for boundary inference can be characterized as a radiant pattern with strength decays with distance from the center. Since a point on the boundary will only receive votes from one side while a point inside the region will receive votes from all directions, the size of the vector sum of the polarized votes should indicate the "boundariness" of a point, that is, the boundary saliency. On the other hand, the direction of the resulting vector relates the polarity information at the site. Figure 6(b) presents the result of this boundary inference on the 2-D data set depicted in Figure 6(a). Observe that points are not labeled in absolute terms as borders or non-borders, but are presented with reservations as indicated by the boundary saliencies.

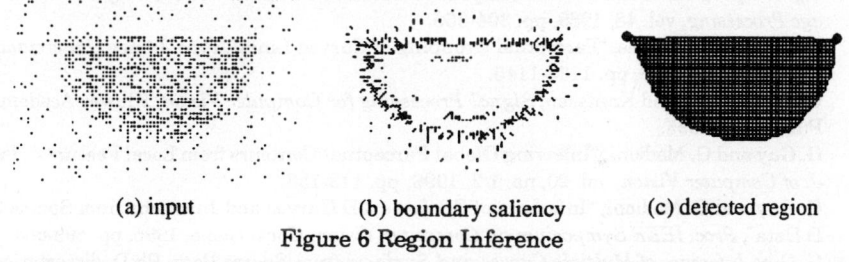

(a) input   (b) boundary saliency   (c) detected region
Figure 6 Region Inference

Having obtained the polarity information and identified the boundary points, the inference of region can proceed by using the boundary saliencies as the initial point saliency. Figure 6(c) depicts the resulting region inferred from the boundary points shown in Figure 6(b). Note that we not only find the region, but also accurately find the corners. We depict in Figure 7 another result of applying our inference engine to dot clusters.

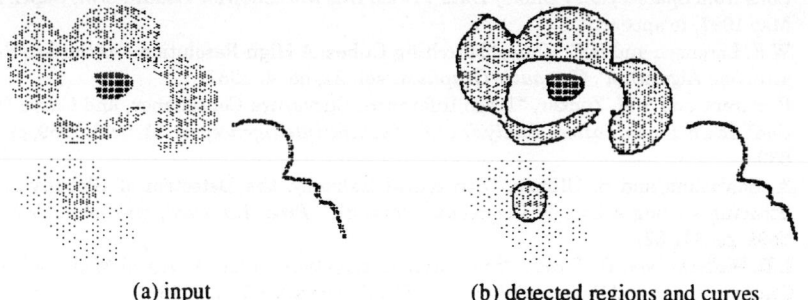

(a) input   (b) detected regions and curves
Figure 7 Sparsely distributed point set

4 Conclusion and Discussion

We have presented a unifying computational framework for the inference of multiple salient junctions, curves and regions from sparse noisy data. The methodology make use of a technique called tensor voting, which combines the efficiency of non-linear voting and the strength of tensor representation, to allow simultaneous detection of discontinuities and outliers while interpolating curves and region boundaries. The proposed method is non-iterative, requires no initial guess and thresholding, and the only free parameter is scale.

Notice that the problem of inferring salient structures from noisy data in fact is common among many computer vision tasks, we believe our approach can be used to solve a number of mid-level vision problems. We have extended our method to 3-D and applied it to compute shape descriptions from local disparity measurements and have obtained some promising results. Other potential applications includes image segmentation using edge and polarity information obtained by edge detectors, and shape from motion using local estimation of optical flow in image sequences.

References

[1] N. Ahuja and M. Tuceryan, "Extraction of early perceptual structure in dot patterns: integrating region, boundary, and component Gestalt", *Computer Vision, Graphics and Image Processing*, vol. 48, 1989, pp. 304-356.

[2] J. Dolan and R. Weiss, "Perceptual Grouping of Curved Lines", *Proc. Image Understanding Workshop*, 1989, pp. 1135-1145.

[3] G.H. Granlund and Knutsson, *Signal Processing for Computer Vision*, Kluwer Academic Publishers, 1995.

[4] G. Guy and G. Medioni, "Inferring Global Perceptual Contours from Local Features", *Int. J. of Computer Vision*, vol. 20, no. 1/2, 1996, pp. 113-133.

[5] G. Guy and G. Medioni, "Inference of Surfaces, 3D Curves and Junctions from Sparse 3-D Data", *Proc. IEEE Symposium on Computer Vision*, Coral Gable, 1995, pp. 599-604.

[6] G. Guy, *Inference of Multiple Curves and Surfaces from Sparse Data*, Ph.D. dissertation, Technical Report IRIS-96-345, Institute for Robotics and Intelligent Systems, University of Southern California.

[7] F. Heitger and R. von der Heydt, "A Computational Model of Neural Contour Processing: Figure-Ground Segregation and Illusory Contours", *Proc. Int. Conf. Computer Vision*, 1993, pp. 32-40.

[8] M.S. Lee and G. Medioni, "Inferred Descriptions in terms of Curves, Regions, and Junctions from Sparse, Noisy Binary Data", *Proc. Int. Workshop on Visual Form*, Capri, Italy, May 1997, to appear.

[9] W.E. Lorensen and H.E. Cline, "Marching Cubes: A High Resolution 3-D Surface Reconstruction Algorithm", *Computer Graphics*, vol. 21, no. 4, 1987.

[10] P. Parent and S.W. Zucker, "Trace Inference, Curvature Consistency, and Curve Detection", *IEEE Trans. Pattern Analysis and Machine Intelligence*, vol. 11, no.8, 1989, pp. 823-839.

[11] A. Sha'ashua and S. Ullman, "Structural Saliency: the Detection of Globally Salient Structures using a Locally Connected Network", *Proc. Int. Conf. on Computer Vision*, 1988, pp.321-327.

[12] L.R. Williams and D. Jacobs, "Stochastic Completion Fields: A Neural Model of Illusory Contour Shape and Salience", *Proc. Int. Conf. Computer Vision*, 1995, pp. 408-415.

[13] S.W. Zucker and R.A. Hummel, "Toward a Low-Level Description of Dot Clusters: Labeling Edge, Interior, and Noise Points", *Computer Vision, Graphics, and Image Processing*, vol. 9, no.3, 1979, pp.213-234

Learning Multiscale Image Models Of 2D Object Classes

Benoit Perrin, Narendra Ahuja and Narayan Srinivasa
The Beckman Institute for Advanced Science and Technology
University of Illinois at Urbana-Champaign
405 N. Mathews Avenue, Urbana, IL 61801

Abstract

This paper is concerned with learning the canonical gray scale structure of the images of a class of objects. Structure is defined in terms of the geometry and layout of salient image regions that characterize the given views of the objects. The use of such structure based learning of object appearance is motivated by the relative stability of image structure over intensity values. A multiscale segmentation tree description is automatically extracted for all sample images which are then matched to construct a single canonical representative which serves as the model of the class. Different images are selected as prototypes, and each prototype tree is refined to best match the rest of the class. The model tree for the class is that tree which is best supported over all the initializations with different prototypes. Matching is formulated as a problem of finding the best mapping from regions of example images to those of the model tree, and implemented as a problem in incremental refinement of the model tree using a learning approach. Experiments are reported on a face image database. The results demonstrate that a reasonable model of facial geometry and topology is learnt which includes prominent facial features.

1 Introduction

This paper is concerned with learning the canonical gray scale structure of the images of a class of objects. Structure is defined in terms of the geometry, layout and photometric relationships among salient image regions that are common across given object images. The use of such structure based learning of object appearance is motivated by the relative stability of image structure over intensity values across multiple views. Since region boundaries in image often correspond to scene discontinuities in illumination, albedo, objects, etc., these boundaries are more invariant to changes in illumination level and viewpoint than the intensity values themselves. Thus, any descriptions of objects inferred in terms of the region structures are expected to be fairly stable representations of object images.

The region structure of an image may be represented by a tree that captures the different geometric and photometric scales, and geometric and topological interrelationships characterizing the image regions. Large regions are said to have a coarse spatial scale while smaller sizes are said to be associated with finer spatial scales. Given a tree that represents such a multiscale segmentation of an image, the goal is to extract a tree that serves as the common denominator of the trees corresponding to all object images. We use an algorithm that automatically extracts the tree representation of an image without any *a priori* knowledge of the scales present. The trees derived from different object images are matched to construct a canonical representation or model that provides maximal degree of match to all the trees. The model tree for the class is derived as that tree which is best supported over all the initializations. The matching of each prototype tree to the rest of the class is formulated as a problem of finding the best matches from regions of example images to those of the model tree, and implemented as a problem in incremental learning. The learning is realized using a modified Fuzzy Adaptive Resonance Thoery (Fuzzy ART) architecture that incrementally refines a prototype tree based on the features of the matched regions of example images.

2 Previous Work

There have been previous efforts at finding canonical aspects of the images from a single class. However, most of these are limited to relatively unstable features such as the responses of an edge detector or, edge segments. Shams [15] presents an approach to applying canonical graph representation for pattern recognition. He uses local features of orientation and intensity edges for recognizing objects such a plane or a tank. Von der Marlsburg et al. [17] uses an architecture for pattern recognition based on graph matching to recognize faces [8] and to achieve invariance with respect to some location and orientation variabilities [7]. To improve upon the unreliability of the isolated local features such as used in these methods, [5] uses simple homogeneity criteria to split and merge tiles defined by an *a priori* tessellation method to obtain more stable segmentations, and thus more meaningful regions. However, *a priori* chosen, low level criteria do not yield meaningful candidate regions for canonical class descriptions from real images which is the objective of this paper. Methods have been developed that integrate *a priori* domain knowledge into the segmentation which improves segmentation at the expense of making the methods domain specific [19, 16, 18, 12, 3, 6].

Moghaddam and Pentland [9] present a probabilistic method for 2D object detection. This method requires the *a priori* identification of important object features. It has been applied to faces and hands images, gives interesting results even with few eigenspaces, and seems robust with respect to noise and data variations. Murase and Nayar in [11] describe a 3D method based on eigenspace generation. Instead of using intrinsic shape information, which is often difficult to extract due to lighting variations, they match appearance. As in [9], a suitable number of eigenspaces for each object must be found. Hornegger and Niemann in [4] propose a statistical framework for learning, localizations and identifications of objects. One of the most prominent connectionist methods is based on the radial-basis function (RBF) network introduced by Poggio and Girosi [14] to learn a mapping from an input space to an output space. It was applied by Poggio and Edelman [13] to recognize three-dimensional stick figures from two-dimensional images. Mukherjee and Nayar in [10] use a learning algorithm to obtain the network parameters using basis functions generated from wavelet decomposition of different training images of objects. The work described in this paper is aimed at obtaining a canonical representation or model of the object images such that the model is explicitly defined in terms of relatively low level, and thus domain independent, features, which are also relatively independent of imaging conditions.

3 Overview of Learning Approach

Each image is converted into a segmentation tree [1] in which the levels are indexed by a photometric scale parameter. Initially, the prototype is stored in the form of the segmentation tree of one of the sample object images. The segmentation trees of several other samples are then presented to the learning system one image at a time. For every sample image presented, each of its regions at the coarsest scale is a candidate for matching with the prototype regions at the same scale. The set of all single region candidates constitutes a *unary* hypothesis set. In addition to single region candidates, additional regions are also generated as candidates by merging two and three neighboring regions in the region adjacency graph (RAG) of the sample image segmentation tree giving rise to *binary* and *ternary* hypothesis sets, respectively. The union of unary, binary and ternary hypothesis sets of a sample image constitutes the hypothesis set for that sample image. A coarsest-scale hypothesis set consisting of unary, binary and ternary hypotheses is analogously generated for the prototype. These sample and prototype hypotheses sets are paired to provide the best matches of region features such as area, centroid and shape (measured as the eccentricity of the best fitting ellipse). The matching is performed using a learning algorithm. At the finer scales, the hypothesis sets are generated, for both the prototype and the sample image, from regions that are the children and/or neighbors of matched regions at the next coarser level. While these hypothesis sets also contain unary, binary or ternary intrascale regions, the generation of these regions is guided by the matching process at a coarser scale. This prevents the creation of irrelevant hypothesis sets at finer scales.

The prototype hypothesis sets that find matches with each sample image hypothesis set, at each scale, are modified in order to reflect the features in the sample image that are different from the prototype. In addition to the modifications to the prototype features, the learning algorithm also stores the frequency with which each region of the prototype hypothesis set is matched with the regions belonging to the sample image hypothesis set of the training samples. Matched prototype regions with the highest frequencies indicate the regions that are most salient. The final result of the learning process is the set of regions that is most salient based on all training samples starting with a given prototype. While this approach is stable with respect to different prototype initializations, we use the commonly extracted regions across several initializations as the canonical representation for the object class. This removes any bias due to specific prototype initializations.

4 Choice of Features for Region Matching

Several features are used to determine the quality of match between a pair of regions. The work in this paper focusses on learning canonical models of facial image structure. These images contain a face in the center and there is little background area. All the sample images are normalized to the prototype image such that each face image appears in the center and has about the same size. Region centroids represent their relative positions, and are used as one type of feature. Region area is used as another feature. The eccentricity of the region is used as its shape feature, and is approximated by that of its best-fitting ellipse.

In order to derive these three normalized features, the two-dimensional moment of order $(p+q)$, for an NXM discretized image $g(x,y)$ is defined as $m_{pq} = \sum_{y=0}^{M-1} \sum_{x=0}^{N-1} x^p y^q g(x,y)$. The centroid (c_x, c_y) of each region is defined as $c_x = m_{10}/m_{00}$ and $c_y = m_{01}/m_{00}$ where m_{00} is the area of the region. This area is computed as the number of pixels within each region. To compute the eccentricity, we first compute the central moments μ_{pq} of the region by replacing x and y in the expression for m_{pq} by expressions $(x - c_x)$ and $(x - c_y)$. These moments are then normalized with respect to scale as $\nu_{pq} = \mu_{pq}/m_{00}^{\frac{x+q}{2}}$. The normalized eccentricity η, which gives a shape measure invariant to translation and scale, is computed as $\eta = \sqrt{(\nu_{20} - \nu_{02})^2 + 4\nu_{11}^2}/m_{00}$.

At the coarsest scale only, the area and centroid features were used. The eccentricity information was not used at this scale because, the regions at the coarsest scale are normally large. These large regions are not very stable in it shape feature. By traversing to finer scales from the coarse level region, its features take a more definite shape. For example, a face region at the coarsest scale, extracted as a single region or by merging two or three regions, is usually not well defined. However, at finer scales, the regions that are found within it can correspond to the eye or mouth and these are far more well defined in their shape. Thus, at the coarse scale, only the area and centroid were used as features for the matching process. At finer scales, the area, centroid and its eccentricity were used.

5 Criteria for Merging Regions

The merging of regions is an important and necessary step in our approach. This is because lighting effects lead to smooth shading in images. During the multiscale segmentation, this may further lead to splitting of shaded regions with subregions at arbitrary locations, and thus to the creation of spurious regions. To alleviate this problem, methods need to be developed to merge erroneous subregions into the original parent region. Since there is no sharp change in gray-level value across the border between spurious regions, the gray-level gradient across the border will be low. This property is used to detect mergable regions. If the gradient at most of the border pixels has a shallow slope, then the two regions can be merged. The exact condition for merging depends upon the definition of the terms *most of the border pixels* and the *degree of shallow slope*. For this purpose, we use two thresholds: T_{per} (*most of the border pixels* means more than $T_{per}\%$) and T_{gr} (*degree of shallow slope* means slope smaller than T_{gr}). To compute these thresholds, the histogram of gradient values at the border pixels is plotted. An example of the histogram for the image of a car is shown in Figure 1. Experiments suggest that T_{gr} is best if located at the tail end of the steep part of curve, just before it flattens. The threshold T_{per} for the number of border points is selected such that T_{gr} is greater than T_{per}. In this paper, T_{per} was set typically between 80 and 90%. Thus, two regions R_1 and R_2 can be merged if $|B| * 100/|N| > T_{per}$ where B is a border point. A point on the R_1-R_2 region boundary is a border point if $G(B) < T_{gr}$ where $G(B)$ is gradient at the border point. N is the total number of border points between regions R_1 and R_2. The term $|x|$ here refers to the sum of the number of elements in x. An important consequence of reducing these spurious regions is the reduction in the number of regions that have to be merged to

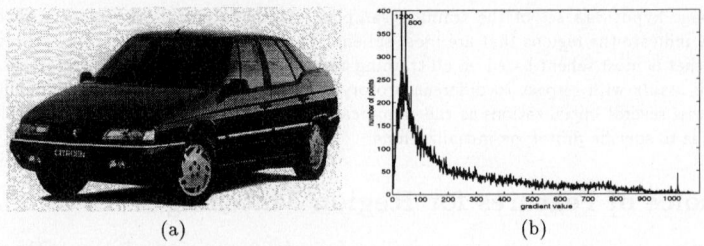

(a) (b)

Figure 1: (a) Original picture of the car on which experiments were based. (b) Histogram of the gradient values on border points for the car.

generate the hypothesis sets. This increases the efficiency of the matching and learning process. The three features that define each region have to be re-computed after merging of two or more regions occurs. They are computed from the parent regions unless the regions have already been merged in other training samples and their values are available.

6 Coarse-To-Fine Generation of Hypothesis Sets

At the finest scale, the segmented image typically contains several small regions. Finding the canonical representation at this scale would require the generation of several mergings. Therefore, it is desirable to use the knowledge that comes from a matching at a coarser scale to guide the matching at a finer scale. Using the segmented tree, the algorithm searches for matching regions among the children of the regions that have been matched at the previous coarser scale level. This confines the search to interesting areas, thus limiting the number of hypotheses. Because the coarsest scale has fewer regions than the other scales, it can be used to initialize the process without expensive computations.

At each scale, the hypothesis sets are generated for the sample and prototype images in two steps. First, two sets of unary regions or *unary hypotheses* are extracted from the segmented images at the current scale. One is extracted from the prototype RAG, and another from the current sample RAG. To select the unary hypotheses at the coarsest scale level, all the regions of RAG are selected for matching. Below this level, hypotheses are selected in two ways. Let us assume that a match has been found between prototype and sample RAG. Then for the finer scales, children of these matched regions are selected to compose a pair of unary hypothesis set. Neighbors of these children are also added to the sets. This process is repeated for every matched region of the preceeding coarse level.

For the regions in the prototype that are not children of matched regions, a centroid-based hypothesis generation strategy is used, where the region that has a nearby centroid in the sample image is selected. Neighbors of these two regions are also selected to form two sets of unary hypotheses. The centroid-based method generates small but numerous sets of hypotheses. However, experiments have shown that, in most cases, the number of matches obtained from using the first strategy is more important than from the centroid-based method. Therefore, the first strategy can still be used for objects or scenes where the positioning of regions is not precise by possibly expanding the neighborhood. Once the pair of unary hypothesis sets is formed, unary hypotheses are merged to generate binary and ternary hypotheses. The output of the hypothesis generation process is, therefore, two sets of regions which are candidates for matching.

7 Learning Algorithm and Architecture

In this approach, a learning architecture adapted from Fuzzy ART algorithm [2] was chosen to perform matching because it is self-organizing, robust to noise, and massively parallel, which makes it useful for on-line pattern recognition applications. One of the most important properties of the architecture used is that the majority of processing involves simple compare and add operations, resulting in an efficient learning algorithm.

7.1 Fuzzy ART Representation and Notation

In developing an algorithm to perform the above computation, we have used fuzzy set theory [20] to represent classes as well as perform computations [2]. For concreteness, we will explain the notation for the 2-dimensional space; it generalizes to other spaces in a straight forward manner.

Class: A class is represented by specifying two diagonally opposite vertices of its rectangle. This is done by a vector consisting of the coordinates of one vertex followed by the complement (with respect to 1) of the coordinates of the diagonally opposite vertex. Thus, for example, the output class represented by the rectangle defined by vertices (x_1, y_1) and (x_2, y_2) is represented by the vector $(x_1, y_1, 1 - x_1, 1 - y_1)$. For a (class consisting of) a single point (vertex) (x_1, y_1), the representation is the 4-tuple $(x_1, y_1, 1 - x_1, 1 - y_1)$, denoted by its *weight* vector $\mathbf{W}$.

Norm: The norm $|V|$ of a vector (class) is the sum of the city block distances of the class from the points $(0, 0)$ and $(1, 1)$.

Distance: The distance between a sample point and a class rectangle is denoted by the city block distance to the nearest point in the class.

AND Operation: The fuzzy AND (or $\wedge$) between two classes is the vector whose elements are obtained by taking pairwise MIN of the corresponding elements of the operand vectors. AND of two classes (points) denotes the result of adding one to the other, possibly requiring expansion.

Choice Function: The choice function is used to determine the class defined by its weight $\mathbf{W}$ that is closest to a given point. Given a new point $\mathbf{I}$ and a class $\mathbf{W}$, the choice function is defined as $\frac{|\mathbf{I} \wedge \mathbf{W}|}{|\mathbf{W}|}$, which assumes highest value for that class which is at shortest distance from $\mathbf{I}$. If the choice function is one for a given class, then the class is a *fuzzy subset choice* for input $\mathbf{I}$. This means that the input $\mathbf{I}$ is completely contained within the class $\mathbf{W}$. If more than one category is a fuzzy subset choice, then a small but positive parameter α is added to the denominator to break the tie such that the class that maximizes $|\mathbf{W}|$ among the fuzzy subset choices is chosen.

Vigilance function: The vigilance function is used to enforce the restriction on class size. For example, given a sample $\mathbf{I}$ and a class $\mathbf{W}$, $\mathbf{W}$ is allowed to (expand and) include $\mathbf{I}$ if the value of the vigilance function, defined as $\frac{|\mathbf{I} \wedge \mathbf{W}|}{|\mathbf{I}|}$, is no smaller than a certain *a priori* (user specified) threshold ρ called the vigilance parameter.

7.2 The Modified Fuzzy ART Algorithm

The modified Fuzzy ART algorithm consists of two layers as shown in Figure 2. The F_1 layer is called the *input* layer while the F_2 layer is called the *prototype* layer. The F_1 layer receives each region m belonging to the sample image hypothesis set as an input vector $\mathbf{S_m}$. This vector is defined as: $\mathbf{S_m} = (S_{m1}, S_{m2}, \cdots, S_{mM})$ where the parameters S_{mi} (for $i = 1, \cdots M$), in general, represent the characteristic features of each region such as area, centroid, color, etc.

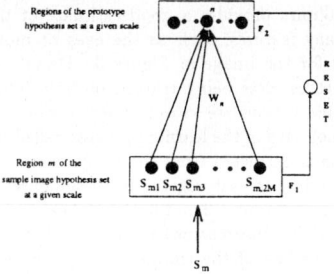

Figure 2: The modified fuzzy ART network.

The learning algorithm is outlined now. First, initialize the components of the weight $\mathbf{W}_n$ corresponding to each region n of the prototype hypothesis set to the M-dimensional feature value vector $\mathbf{P}_n = (P_{n1}, P_{n2}, \cdots, P_{nM})$. Select a region m from the sample image hypothesis set as input. Then, present the feature vector $\mathbf{S}_m = (S_{1m}, S_{2m}, \cdots, S_{Mm})$ to F_1. This input and its complement are stored as a single vector $\mathbf{I}$ in F_1. Compute the class $n \in \mathbf{P}_n$ that is closest to the input $\mathbf{I}$ using the choice function T_n as $T_n = \frac{|\mathbf{I} \wedge \mathbf{W}_n|}{|\mathbf{W}_n|}$. Form the hypothesis that the selected class n is the appropriate classification for the given input. Then, test the hypothesis using the vigilance criterion as $\rho_n = \frac{|\mathbf{I} \wedge \mathbf{W}_n|}{|\mathbf{I}|} \geq \rho$ where ρ is the vigilance parameter set by the user.

If class n satisfies the vigilance criterion, then the input $\mathbf{I}$ is added to the list L_{mn}. If there are more regions in the sample image hypothesis set, then repeat the above steps for these additional regions. When all the regions are processed, determine the maximum value in the list L_{mn} for each region n of the prototype hypothesis set. For each such region, update the weights W_n using the features of the matched region m as $\mathbf{W}_n = \beta |\mathbf{I} \wedge \mathbf{W}_n| + (1-\beta)\mathbf{W}_n$ where β controls the rate at which the features of the matched input region m are allowed to refine the weight $\mathbf{W}_n$. Typically, $\beta \leq 0.2$ during training so that only the regions that have extremely good matches are allowed to refine the weights $\mathbf{W}_n$. The bound on the size of the hyperrectangle, $|D_n|$, for each class n can be defined as $|D_n| \leq M(1-\rho)$ where M is the number of features in the input. Thus, if the vigilance parameter ρ is small, the size of the hyperrectangles are bigger and vice versa. The training process continues until the input feature space is covered with hyperrectangles. The frequency of winning for all winning nodes in F_2 is incremented by 1. If there are more training samples, then repeat the above steps for the new samples. It should be noted that there is a separate Fuzzy ART network for each scale. At the end of the learning process, the regions with the highest frequency of winning are the regions that correspond to the 2D model for the given training samples.

8 Experiments and Results

The proposed algorithm was tested for 20 different initializations of face images (from the O.R.L. face database) with all initializations being frontal views in a neutral position. For each initialization, 400 training images of 92x112 each were used to train the network. These images were presented in different orders. The segmentation transform used four different scales. The scales are ranked from level 0 to 3 with the coarsest scale corresponding to level 0 and the finest scale to level 3. The feature vector for matching consisted of area and centroid information for all scales. The canonical representations have been found to be most stable with the following range of fuzzy ART parameters, from scale 0 to 3 where $\alpha = (0.85, 0.9, 0.9, 0.9)$ and $\beta = (0.1, 0.2, 0.1, 0.2)$. This range of parameters allows the algorithm to converge slowly to canonical features.

8.1 Results for Different Initializations

To illustrate the effect of different initializations, consider the original gray level images of two different faces as shown in Figure 3(e) and Figure 4(e). These images are segmented into quite different initial regions by the multiscale transform as shown in columns (a) and (d) of these figures. The coarse scale segmentation on top of the column provides a good estimate of the quality of initialization: while the image in Figure 4 has many features, such as the eyes or mouth in the right place, it is more difficult to find these features for the image in Figure 3. Despite these different initializations, the algorithm extracts similar features after being trained on 400 different face images. In column (b) and (f) the most frequently found regions are shown. These regions are approximated with ellipses and superimposed on a face image not used in the learning process called the *neutral face*. The area, centroid and eccentricity for each region is extracted from the weight vector as the mean of the weight vector and of its complement corresponding to each region. In these figures, the 8 best regions of the coarsest scale level (level 1), 10 best regions for level 2 and 3 and 15 best regions for the two last pictures (scale level 4 without and with eccentricity information) are shown. This selection was based on whether the regions were matched at least for half of the training examples presented. Furthermore, it has been observed that the extracted features are similar among all initializations for this choice of number of regions at each scale. Columns (c) and (g) show the number of times each pixel of the image has been matched by the 25 most frequently found regions. Here, the darker the pixel, the more frequently it has been matched. Because the silhouette of the face is often found at the coarsest level as can be seen in columns (c), the images for the coarsest level are much darker i.e., more regions are matched. The most frequently matched regions show that the algorithm concentrates on the area with the obvious features: eyes, mouth.

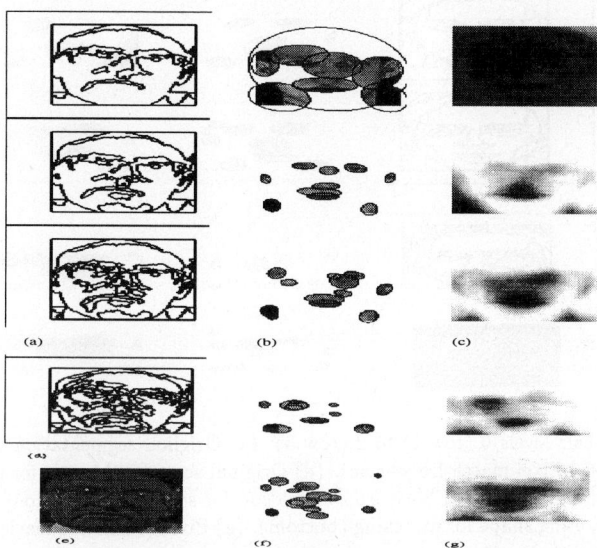

Figure 3: For scale levels 0 (row 1) to 2 (row 3): (a) Original segmentation tree. (b) Ellipse approximation. (c) Pixel match frequencies. (d) Original segmented image for scale level 3. (e) Original gray-scale image. (f) Ellipse approximation for scale level 3 without using shape for matching (top), using shape for matching (bottom). (g) Pixel match frequencies without shape (top), with shape (bottom).

8.2 Results after Postprocessing

As mentioned in Section 3, the results obtained from different initializations are used to select the most commonly found regions across all these initializations. Results of this postprocessing is illustrated using Figure 5. The regions represents the final canonical representation at each scale level. This representation was computed using the most frequently occurring regions (ranging from 8 for scale 0 to 15 for scale 3) across all initializations. The Figures (b) are the elliptic approximations of the most frequently matched regions. Match frequencies from the postprocessing show a clear difference between the matched and unmatched regions. This enables us to extract the most frequent regions. For each of these stable regions, the pixels that have been most frequently matched are represented in columns (c). The darker pixels correspond to the more frequently matched regions. At finer scale levels, there exists a clear threshold to separate the most frequent regions from others. The effect of thresholding (60 % percent of the maximum frequency) the stable regions in columns (c) is shown in columns (d). By superimposing the thresholded regions in columns (d) into a single image, we obtain the result shown in columns (a). Despite the variabilities in the database and with a limited range of possible matching regions (only up to ternary hypothesis sets), these results show that the algorithm has been able to efficiently extract the main features of faces.

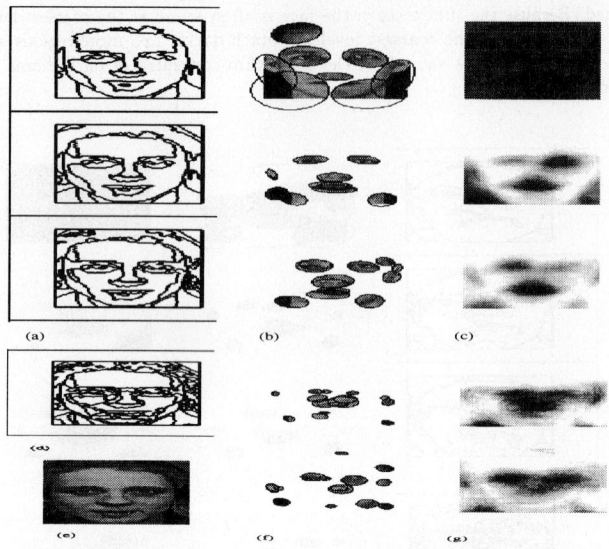

Figure 4: For scale levels 0 (row 1) to 2 (row 3): (a) Original segmentation tree. (b) Ellipse approximation.(c) Pixel match frequencies. (d) Original segmented image for scale level 3. (e) Original gray-scale image. (f) Ellipse approximation for scale level 3 without using shape for matching (top), using shape for matching (bottom). (g) Pixel match frequencies without shape (top), with shape (bottom).

References

[1] N. Ahuja. A transform for multiscale image segmentation of integrated edge and region detection. pages 1211–1235, 1996.

[2] G. A. Carpenter, S. Grossberg, and D. B. Rosen. Fuzzy ART: Fast stable learning and categorization of analog patterns by an adaptive resonance system. *Neural Networks*, 4:759–771, 1991.

[3] J. R. Beveridge et al. Segmenting images using localized histograms and region merging. *International Journal of Computer Vision*, 2(3):311–347, 1989.

[4] J. Hornegger and H. Niemann. Statistical learning, localization and identification of objects. In *Proc. IEEE Conf. on Computer Vision and Pattern Recognition*, pages 914–919, 1995.

[5] S. L. Horowitz and T. Pavlidis. Picture segmentation by a directed split-and-merge procedure. In *Proc. International Conference on Pattern Recognition*, pages 424–433, 1974.

[6] I. Y. kim and H. S. Yang. A systematic way for region-based image segmentation based on markov random field model. *Pattern Recognition Letters*, (15):969–976, 1994.

[7] W. K. Konen, T. Maurer, and C. von der Malsburg. A fast dynamic link matching algorithm for invariant pattern recognition. *Neural Networks*, 7(6):1019–1030, 1994.

[8] M. Lades, J. C. Vorbrüggen, J. Buhmann, J. Lange, C. von der Malsburg, R. P. Würtz, and W. Konen. Distortion invariant object recognition in the dynamic link architecture. *IEEE Transactions on Computers*, 42(3):300–310, 1993.

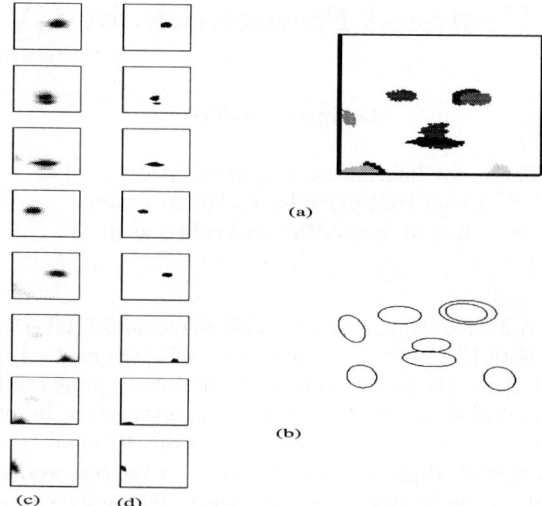

Figure 5: Scale level 1: (a) Shape of the stable regions. (b) Ellipse approximation of the stable regions.(c) Pixel match frequencies for each detected feature.(d) Result of the thresholding (c).

[9] B. Mogghaddam and A. Pentland. Probabilistic visual learning for object detection. In *Proc. IEEE International Conference on Computer Vision*, pages 786–793, 1995.

[10] S. Mukherjee and S. K. Nayar. Automatic generation of grbf networks fo visual learning. In *Proc. IEEE Conf. on Computer Vision and Pattern Recognition*, pages 794–800, 1995.

[11] H. Murase and S. K. Nayar. Visual learning and recognition of 3-d objects from appearance. *International Journal of Computer Vision*, (14):5–24, 1995.

[12] A. M. Nazif and M. D. Levine. Low level image segmentation:an expert system. *IEEE Transactions on Pattern Analysis and Machine Intelligence*, 6(5):555–577, 1984.

[13] T. Poggio and S. Edelman. A network that learns to recognize three-dimensional objects. *Nature*, 343:236–266, 1990.

[14] T. Poggio and F. Girosi. Networks for approximation and learning. *Proceedings of the IEEE*, 78:1481–1497, 1990.

[15] S. Shams. Multiple elastic modules for visual pattern recognition. *Neural Networks*, 8(9):1439–1456, 1995.

[16] J.M. Tenenbaum and H. G. Barrow. Experiments in interpretation-guided segmentation. *Artificial Intelligence*, 8:241–274, 1977.

[17] C. von der Malsburg and E. Bienenstock. A neural network for the retrieval of superimposed connection patterns. *Europhysics Letters*, 3(11):1243–1249, 1987.

[18] D. I. Waltz. Generating semantic descriptions from drawings of scenes with shadows. Technical Report A. I. Memo 1271, M. I. T. Artificial Intelligence Laboratory, 1972.

[19] Y. Yakimovsky and J. A. Feldman. A semantics-based decision theory region analysis. In *Proc. International Joint Conference on Artificial Intelligence*, pages 580–588, 1973.

[20] L. Zadeh. Fuzzy sets. *Information Control*, 8:338–353, 1965.

3D Model Centered Framework for CV and VR

Michihiko MINOH

Center for Information and Multi Media Studies
Kyoto University, Kyoto 606-01 Japan
E-mail: minoh@media.kyoto-u.ac.jp

Abstract. A 3D model, particularly a 3D shape model, takes an important role in both CV and CG in the context of image media application such as VR. Here, 3D model centered framework is proposed in which the importance of 3D model processing is emphasized on. In our framework, images are considered as constraint of the 3D model processing, which gives clear strategy of taking an image in the real world. The research based on the framework are described as examples to show the advantages of our framework.

1 Introduction

Research in Computer Vision(CV) mainly focus on how to extract information from images, particularly 3D shape information of objects in an image. Since the problem of reconstructing the 3D shape of the object from the image is ill-posed, constraint has to be introduced to have the problem solved. As the constraint, geometric, statistical and smooth constraint are often used because these are easy to be formulated. And sometimes, the same reason makes almost all the objects which are handled in research rigid.

From the viewpoint of the technology, the most important thing is what we are going to do with the information extracted by the CV methods. In this sense, CV is not a purpose but a tool. The tool is used for the specific purpose. The CV technology has to have an application context.

One well known context is robotics, in which CV serves as visual sensor to make a robot understand the environment. In this context, real time constraint is important because the robot has to interact in real world.

Another important context is media application. In this context, CV takes a role to input information to the computer and Computer Graphics(CG) takes a role to output information from the computer. The computer works as an image media and exchange information with a user. If the computer presents information in three dimensional form, it becomes Virtual Reality(VR) environment.

In this paper, the role of the 3D models in the context of image media is emphasized on. The 3D model has to be put in the center of the processing both in CV and CG. In other words, the main object to be processed should not be an image but a 3D model. The image is only referred to in order that the 3D model is processed.

2 3D Model Centered Framework

Considering the context of image media, particularly VR applications, 3D models take an important role. Here, a new framework which puts 3D model at the center of the process is proposed. The key point of this framework is the object for processing, which is not an image but a 3D model.

Usually, most research of Computer Vision process images intensively and try to reconstruct objects in the image with a priori constraint. Since images are just appearances of the 3D real world and what we expect the system to do is to obtain 3D information, why is it necessary to process the images intensively?

This question leads us to the 3D model centered framework in the context of image media. A typical 3D model is given in advance to the system, and the system processes the 3D model referring to the images. This imposes strong constraint on how to take the images, in other words, how to control a camera.

The main process for 3D models is "deformation." Since the 3D models are given in advance, it is not necessary to generate 3D models. Instead, it becomes important to deform the given 3D models referring to the images.

This framework gives us several advantages:

- Computer Vision methodology is easy to cooperate with Computer Graphics and Virtual Reality techniques.
- Since the main processing is deformation of 3D models, it could be processed in real time.
- Considering 3D models as the processing object and images as constraint, constraint becomes explicit and the criterion to control the camera becomes clear.

These advantages are discussed from the viewpoint of CV and VR in the following subsections.

2.1 CV Problems

The problem setting of Computer Vision i.e. 3D reconstruction from an image, is intrinsically ill-posed. To make the problem solvable, additional constraint has to be necessary as is shown in Fig.1. However, we do not know how much constraint is necessary to turn the problem solvable. If the problem is simple enough to be represented in mathematical form, it is clear how many additional constraint is necessary. But it is not a practical case.

One of the main stream of CV research is the model based techniques, which represent the necessary constraint as the form of "model". With the model, an image is tried to be interpreted. Even if the model is represented in the form of 3D shape, the image is mainly processed by referring to the model.

In our framework, the 3D models are the object for processing, and they are not the representation of the constraint. Instead, the image works as the constraint of deforming the 3D model(see Fig.1). The problem in this case is not ill-posed, because the problem is to find a mapping between the 3D real world

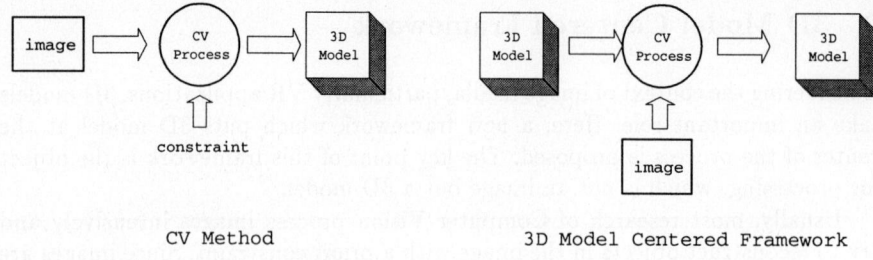

Fig. 1. Input, output and constraint in CV.

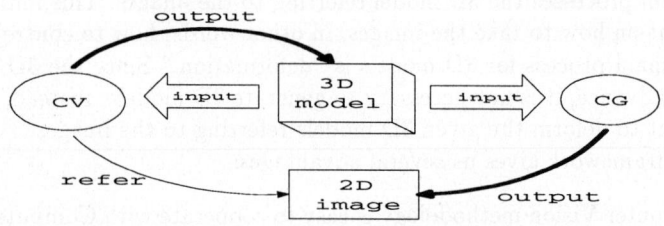

Fig. 2. The role of 3D model in relation with CV and CG.

and the 3D virtual world in the computer. The problem turns out to be a search problem, although the search space is too huge.

Therefore, we still need constraint to restrict the search space. The given image takes the role. In the CV method, the constraint is introduced by the assumption the researchers impose on, which often makes the method impractical. On the other hand, in our framework, the constraint is additive, i.e. if the process lacks constraint, the system could iteratively take advantage of another image until the constraint becomes enough for the processing. If the system has the environment to control a camera and to obtain the image derived by the processing, it is not necessary to give the system a priori constraint.

2.2 VR Related Problems

In the VR related applications, CV method serves as an input, and CG method does as an output. Generally, the input of CV is an image and the output is a 3D model. On the contrary, the input of CG is a 3D model and the output is an image. Since the image is used for both input and output to the VR system, the key point to bridge both CV and CG is the 3D model.

To complete the link from CV to CG, CV has to generate 3D models from the images. There are research to generate 3D models by CV methods[1, 2], but

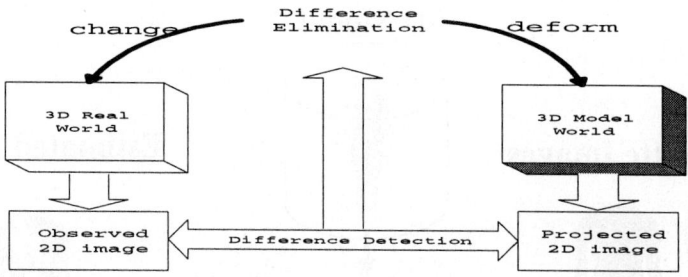

Fig. 3. 3D model processing by the difference detection.

it is very difficult for CV to generate a complete 3D shape model in practical situation. On the other hand, most research in CG consider the 3D model as input data and focus on how to generate more realistic image. In this way, CV and CG cannot be linked for ever in this framework.

Alternative way is the proposed 3D model centered framework in which 3D models are given in advance(see Fig.2). The 3D model in the computer is deformed by referring to the image obtained from the real world observation. Since the 3D model world exists from the beginning of the processing, CG can generate images even if the CV process has not finished. In other words, CV and CG processes are working in parallel. This is one of the most important characteristics in VR applications, because a human interacting with the virtual world acts in real time.

3 Human Shape and Motion Estimation

In the 3D model centered framework, the system mainly processes the 3D model referring to the image[3, 4]. Analysis of human shape and motion, gesture and face becomes important particularly in the context of image media. A computer system has to interact with a human user, so the input of the system is an image and voice of him/her. In this section, three kinds of research concerning to the human shape, pose and motion are discussed.

One is concerning to measure individual human shapes. In this research, the 3D shape model of a typical human is given in advance. Referring to the images, the system deforms the human model the silhouettes of which are fit to the images. As a result, the 3D shape model of the human in the images is obtained.

The other research are concerning to measure the pose and motion of a human. The point of this research is to use a precise human model of a specific individual, who appears in the image. By changing the pose of the 3D shape model, the pose of the real human is estimated. If the input is an image sequence, the motion of the human body is estimated.

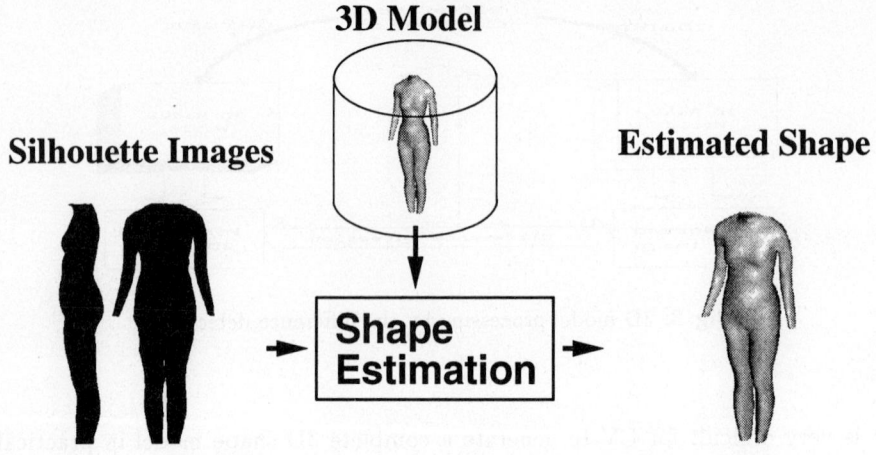

Fig. 4. Shape estimation schema and result.

The point of our research is, as shown in Fig.3, the difference detection between the image obtained in the real world and the image generated based on the 3D models in the system. If the difference is detected, there are two possible ways to eliminate the difference. One is to deform the 3D model in the system and the other is to change the real world. If the system has a tool to change the real world like a robot, the latter choice can be taken. Here, such a thing is not assumed, so the former way is considered.

The human body model given in advance is generated by the measured data of a real human. The measurement has done in the human measurement device made by NKK. By approximating the depth data by the surface patches, the 3D human shape model is generated[9].

3.1 3D Model Generation by Deformation

The shape difference between two human bodies does not come from the structure of the human body but from the shape of each part of it. Since it is difficult and expensive to measure an individual human shape with a specific 3D measurement device, it is convenient to generate an individual 3D human shape by deforming a typical 3D model already measured. Here, we use two silhouette images, one is the front view and the other is side view of the target human body, which are taken with a calibrated camera.

The schematic diagram of the process is shown in Fig.4. By detecting the feature points of the contour of the silhouettes both in the given images and projected images, the correspondence of the feature points is determined. Based on the correspondence, Free Form Deformation[7, 8] is applied to the 3D human shape mode, which also deforms the contour of the projected silhouette images.

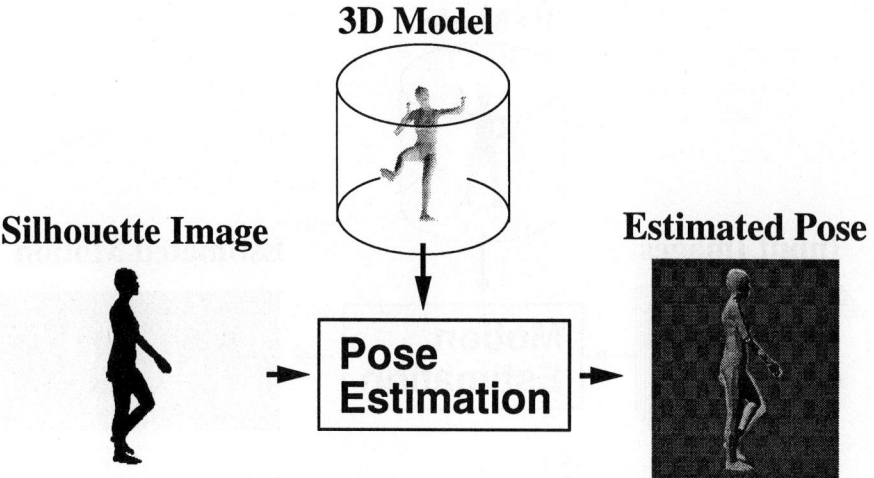

Fig. 5. Pose estimation schema and result.

Next, the 3D shape model is again deformed under the constraint of the local similarity by the energy minimization method. Since the information available to the deformation is obtained only by the silhouette images, the other information is augmented by the 3D human body model itself as the constraint functions.

An experimental result is shown in Fig.4. To evaluated preciseness of the result, three measured human shape models are used in the experiment. The average error between the original 3D shape and the deformed 3D shape is about 5mm. For the purpose of VR applications, it is considered to be small enough. However, for the other applications, more precision would be necessary.

3.2 3D Pose Estimation

Suppose that we have an image of a human and that the human is measured in a 3D shape measurement device to generate a 3D human shape model. The camera parameters is assumed to be known. In this case, the difference of the shape between the 3D human model and the human in the real world comes from the difference of the pose. By changing the pose of the 3D human shape model, and by referring to the image, the pose of the human in the image is estimated as is shown in Fig.5[10].

The 3D human model is represented by the patches and the tree data structure is employed for the model. By changing the joint angle of the part in the order of the depth first search in the tree structure, the pose of the human body is estimated step by step. In each step, the silhouette of the part is evaluated by comparing the silhouette in the image, and several candidate values of the joint angle are calculated. Then, each candidate value is evaluated based on how much the projection of the part with the specified angle value is covered the

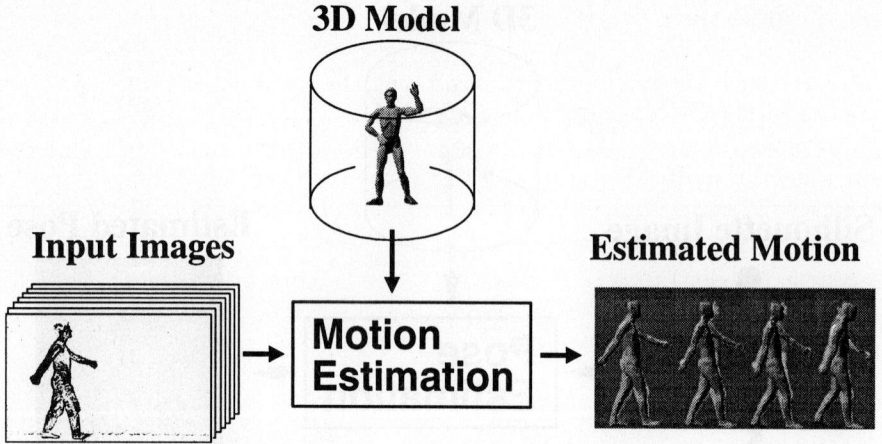

Fig. 6. Motion estimation schema and result.

given silhouettes in the image. The best candidate value is selected and the pose is estimated.

An experimental result is shown in Fig.5. The problem to estimate the pose of the human model is intrinsically a search problem. However, the search space of the problem is too huge to find the best solution. If the search goes in the wrong direction due to the limit number of the candidates, the process fails with the wrong pose whose silhouette does not cover the silhouette of the image. It turns out to be a trade-off problem between the processing speed and the preciseness of the results. Since no domain specific constraint is introduced, the method is applicable to any articulated object.

3.3 3D Motion Estimation

Instead of an image, an image sequence is used to estimate the motion of the human body. The assumptions are the same as the estimation of the pose of the human body. The schema is shown in Fig.6.

The motion estimation process distinguishes the parts of the human body into 3 kinds of status; occluded, moving and stationary. The moving region is detected by the frame difference images. The region of the difference image is also divided into 3 parts; generated moving region, continuous moving region and vanishing moving region. Based on the node mode and the moving region characteristics, the motion of the human body is estimated[11, 12].

One of the results is shown in Fig.6. Since the body motion has to be described together with the body shape, the motion itself can not be described alone. Our method does not separate the body motion from the body shape. Though the estimation is not so correct compared with the data obtained by a motion capturing method, it could be more natural when the body motion is represented in a virtual space.

4 Conclusion

The 3D model centered framework in the context of image media is proposed. In the framework, the 3D model is mainly processed by taking images as constraint of the process. In the framework, CV and CG can cooperate with each other and the process could be in real time. Showing several examples in our laboratory, the usefulness and effectiveness of the framework is discussed.

But, this is just the beginning. Many interesting problems such as how to represent and describe 3D models, how to process in real time and how to control a camera etc. are left for the future.

References

1. Kanade, T., Rander, P.: Virtualized Reality: Constructing Virtual Worlds from Real Scenes. IEEE MultiMedia, **4-1** (1997) 34–47
2. Katkere, A., Moezzi, S., Kuramura, D. Y., Kelly, P., Jain, R.: Toward Video-based Immersive Environments. Multimedia Systems, **5-2** (1997) 69–85
3. Rehg, J. M., Kanade, T.: Digital Eye: Vision -Based Hand Tracking for Human-Computer Interaction. Proc. of the Workshop on Motion of Non-Rigid and Articulated Objects, (1994) 16–22
4. Attwood, C. I., Sullivan, G. D., Baker, K. D.: Model-based Recognition of Human Posture Using Single Synthetic Images, Proc. of the Fifth Alvey Vision Conference (1989) 25–30
5. Bharatkumar, A. G., Daigle, K. E., Pandy, M. G., Qin, C., Aggarwal, J. K.: Lower Limb Kinematics of Human Walking with the Medial Axis. Proc. of the Workshop on Motion of Non-Rigid and Articulated Objects, (1994) 116–123
6. Rohr, K.: Incremental Recognition Of Pedestrians from Image Sequences. Proc. of CVPR (1993) 8–13
7. Celniker, G., Gossard, D.: Deformable Curve adn Surface Finite-Elements for Free Form Shape Design. Computer Graphics, **25-4** (1991) 257–268
8. Ruprecht, D. H.: Free Form Deformation with Scattered Data Interpolation Methods. Geometric Modeling, Springer-Verlag Wien NewYork (1993) 267–281
9. Sakaguchi, Y., Minoh, M., Ikeda, K.: PARTY: Grid Generation for Human Body and Paper Pattern by the Geometrical Constraint Method. Trans. of IEICE Japan **J77-DII** (1994) 2210–2219 (In Japanese)
10. Kameda, Y., Minoh, M., Ikeda, K.: Three Dimensional Motion Estimation of a Human Body Using a Difference Image Sequence. Asian Conference on Computer Vision **2** (1995) 181–185
11. Kameda, Y., Minoh, M., Ikeda, K.: Three Dimensional Pose Estimation of an Articulated Object from its Silhouette Image, Asian Conference on Computer Vision (1993) 612–615
12. Kameda, Y., Minoh, M., Ikeda, K.: A Human Mothin Estimation Method Using 3-Successive Video Frames Proc. of VSMM'96 (1996) 135–140

Image-Based Geometrically-Correct Photorealistic Scene/Object Modeling (IBPhM): A Review

Zhengyou Zhang[1,2]

[1] INRIA, 2004 route des Lucioles, BP 93, F-06902 Sophia-Antipolis Cedex, France
[2] ATR HIP, 2-2 Hikaridai, Seika-cho Soraku-gun, Kyoto 619-02 Japan
e-mail: zzhang@sophia.inria.fr, zzhang@hip.atr.co.jp

Abstract. There are emerging interests from both computer vision and computer graphics communities in obtaining photorealistic modeling of a scene or an object from real images. This paper presents a tentative review of the computer vision techniques used in such modeling which guarantee the generated views to be geometrically correct. The topics covered include mosaicking for building environment maps, CAD-like modeling for building 3D geometric models together with texture maps extracted from real images, image-based rendering for synthesizing new views from uncalibrated images, and techniques for modeling the appearance variation of a scene or an object under different illumination conditions. Major issues and difficulties are addressed.

Keywords: photorealistic modeling, image-based rendering, multiple-view geometry, photometric models, CAD, camera calibration, 3D reconstruction, uncalibrated images, domain knowledge, illumination variation.

1 Introduction

Considerable effort in computer graphics has been devoted, on the one hand, to the development of complex computer aided design (CAD) systems which aim at modeling the geometric and material attributes of the objects in the environment, and on the other hand, to the development of systems which try to reproduce the light propagation under physical laws in order to generate photorealistic renderings. Unfortunately, besides it is a labor-intensive process, this traditional approach has difficulty in creating realistic photographs because the geometry of objects found in the real world is very complex and the subtle light effect is difficult, if impossible, to model.

Computer vision, although it may promise much more, can be considered as an inverse process of computer graphics, namely the process of recovering 3D structure from 2D images. Furthermore, real images directly capture the material properties of objects under real world illumination. Therefore, the combination of computer vision and computer graphics allows us to create, directly from real images, realistic photographs of the environment from viewpoints different from the original ones. Indeed, there are emerging interests in applying the computer vision techniques to the interdisciplinary field of virtual reality, such as human-computer interface, 3D scene/object reconstruction and image-based rendering.

Both vision and graphics technologies have difficulties in producing complex geometric models in great details, but by appropriately using the visual details contained in the original images, we can achieve photorealistic modeling without replying on a complex model. Therefore, compared with the traditional computer graphics modeling, image-based photorealistic modeling (IBPhM) has a number of advantages including

- a weaker dependency of scene complexity because many details need not to be explicitly modeled,
- a simpler geometric model because of the previous reason, which also implies the ease of model acquisition,
- no need of physical simulation because the realism is in the original images,
- less computational requirement because of all previous reasons.

There are two approaches to IBPhM:

CAD-like modeling: A scene is represented by a 3D model together with texture maps extracted from real images.

Image-based rendering (IBR): A scene is represented as a collection of images. New images are generated from the original images.

It seems that several researchers prefer IBR to CAD-like modeling, but I argue that it depends on applications and on the available input information. Among the techniques reviewed below, those *based on camera geometry* are usually more appropriate to use 3D models rather than a collection of images. This is for three reasons:

- CAD-like modeling is less memory demanding than IBR, because the data redundancy in the original images are used in building 3D models and are later discarded.
- The conventional rendering pipeline can be used once a CAD-like model is available, while it is not designed for IBR. One can of course expect that special IBR-dedicated hardware will be developed in the near future.
- Both approaches require the establishment of feature correspondences between images. IBR is not easier in this task.

However, IBR does have an advantage over CAD-like modeling at the current stage of development. That is, uncalibrated images can be used in IBR to generate new views. Use of uncalibrated images implies that only a 3D *projective* model can be built. Lack of metric measures in such models makes it difficult to use the conventional rendering pipeline.

The well-known Apple QuickTimeVR system [1] creates a series of environment maps at key locations in the scene. An environment map records the light arriving from all directions at a point. The user is then able to look around a scene from these *fixed* locations. The fixed location constraint can be relaxed in four ways. The first tries to model the apparent motion of pixels (i.e., optical flow) from one camera location (viewpoint) to another, which allows a smooth view interpolation [2, 3]. The second goes further by trying to capture the complete flow of light in a region of the environment [4–6]. In free space, the light field is a 4D function, and an image is a 2D slice of the 4D light field. The success of this technique depends on having a high sample rate, which implies to acquire and save (after proper compression) a large number of images. The third approach relies on the depth value (or disparity) at each pixel [7, 8]. The depth values are usually obtained with user interaction. Given the depth value, a point can be reprojected from vantage points. The fourth approach is called image transfer, which uses point correspondences between images to synthesize a new one without explicitly reconstructing the structure [9, 10]. The last two are however mathematically equivalent. Except the second approach, all the other three require the establishment of correspondence of pixels or feature points between images, which can be a very difficult task and is usually done manually or semi-manually. The second approach has, however, its own limit. It requires the acquisition and storage of a large number of images from known camera viewpoints. It is mainly studied in the computer graphics community, and will not be reviewed in this paper. In Sect. 2, mosaicking techniques to build environment maps are reviewed. It is a 3D representation of a 3D scene. In Sect. 3, we describe the multiple-camera geometry, and how a CAD-like model is built. In Sect. 4, techniques to synthesize new views from original images are reviewed. In reviewing these techniques, pixel or feature point correspondences are assumed to be given. Appendix A will review the commonly used techniques for solving the correspondence problem.

When creating a *dynamic* virtual environment, illumination variation must be considered because it has an important effect on the person navigating in it. This has many applications including product advertisement on the Web where a user wants to examine the appearance of the product under various illumination conditions from different viewpoints. Section 5 reviews techniques which model illumination variations from real images.

2 Mosaicking

When the scene is a planar surface or when the images are taken from the same point of view (i.e., the camera undergoes a pure rotation around the optical center), images are related by a linear projective transformation called a collineation or a homography. More precisely, if $\mathbf{m} = [u, v]^T$ and $\mathbf{m}' = [u', v']^T$ are respectively the projected point of the same space point in the first and second image, then they are related by

$$\rho \begin{bmatrix} u' \\ v' \\ 1 \end{bmatrix} = \begin{bmatrix} H_{11} & H_{12} & H_{13} \\ H_{21} & H_{22} & H_{23} \\ H_{31} & H_{32} & H_{33} \end{bmatrix} \begin{bmatrix} u \\ v \\ 1 \end{bmatrix} \quad \text{or more compactly,} \quad \rho \widetilde{\mathbf{m}}' = \mathbf{H} \widetilde{\mathbf{m}} \qquad (1)$$

where ρ is an arbitrary scale factor, and $\mathbf{H}$ is a non-singular matrix defined up to a scale factor. Here, we have used the following notation: for any vector $\mathbf{x} = [x_1, x_2, \ldots]^T$, $\widetilde{\mathbf{x}} = [x_1, x_2, \ldots, 1]^T$ (i.e., 1 is added as the last element).

Because of relation (1), images can be stitched or mosaicked into a larger and/or higher resolution image. A full-view panorama is possible if images cover the whole viewing space. This is important because the full-view panorama allows us to create an environment map, which can be used to quickly generate new views within the environment. A nice survey of the mosaicking techniques is already available [11]. Several automatic techniques have recently been developed [12–14].

If one of the conditions mentioned at the beginning of this section is not satisfied, there will be misregistration caused by motion parallax. Shum and Szeliski [15] have developed a technique called *deghosting*, which divides each image into small patches, estimates patch alignment, and finally warps each image locally.

Another possibility to obtain a panorama is to use special hardware, such as a camera system with conic mirror [16], hyperboloidal mirror [17], or paraboloidal mirror [18]. However, such a panoramic image is limited by the resolution.

3 CAD-like modeling

Building a CAD-like model from an image sequence consists of the following steps:

Calibration: Recover the external (position and orientation) and internal parameters of the camera for each image in the sequence.

Shape modeling: Build, usually with manual interaction, a 3D geometric model of the scene/object, which is usually a polyhedral approximation.

Texture modeling: Build a texture map for each face of the polyhedral model from original images.

Different calibration techniques will be reviewed shortly. Automatic shape modeling is difficult, and at the current stage of development, one is usually contented with manual initialization followed by automatic refinement by computer vision techniques. Techniques for texture modeling follow much the same way as for image mosaicking, as described in Sect. 2. A face of the polyhedral model is planar, which is seen in several images. These image patches are related by a homography. Possibly different parts of each face are seen from different viewpoints. Mosaicking techniques can therefore be used to build a complete representation of the appearance of the face. Sometimes, image affine transformation, as an approximation to homography, can be used.

Regarding the camera calibration problem, there are essentially three paradigms:

Photogrammetry. The calibration of each camera is performed by observing a calibration object whose geometry in 3-D space is known with a very good precision, and can be done very efficiently [19]. In the stereo setup, the calibration object should be seen simultaneously from all cameras, which is quite problematic when we need to capture a complete view of a scene or an object using many cameras. The solution to this is to follow the structure-from-motion path, either using only one camera [20–22] or using a stereo rig [23–25]. The idea is to recover the rigid displacements of the cameras from visual information by establishing correspondences of features (points, line segments, lines, curves, etc.).

Self-calibration. Techniques in this category do not use any calibration object. Just by moving a camera in a static scene, the rigidity provides in general two constraints [26, 27] on the cameras' internal parameters from one camera displacement by using image information alone (image point or line correspondences). Therefore, if images are taken by the same camera with constant internal parameters, point correspondences between three images are sufficient to recover both the internal and external parameters which allow us to reconstruct 3-D structure up to a similarity [28, 29]. Self-calibration can also be done for uncalibrated stereo rig, where the internal parameters and the relative orientation of the cameras and the motion of the stereo rig are all unknown [30–32]. More knowledge about the camera internal parameters and camera motion will simplify the computation, and more precise and robust results can be obtained.

Domain knowledge. If there is no constraint at all on either the internal parameters or the external parameters of different cameras, we can only achieve a projective reconstruction of the scene [33–35]. This is not enough for many applications which require 3D

Euclidean modeling. However, we usually have domain knowledge of the imaged scene. Such knowledge includes Euclidean location of a point, parallelism, distances between two points, angles between two lines, ratios of distances, etc. This allows us to compute the projective transformation which brings the projective structure back to a Euclidean coordinate system [36–38].

4 Image-based rendering

As mentioned in the introduction, I do not consider here techniques based on interpolation from dense image sequences. Furthermore, I do not consider the case where cameras' internal parameters are known, because I believe that CAD-like modeling is usually more appropriate than IBR. We consider here only the case of uncalibrated images without using explicit 3D models. The main steps in image-based rendering with uncalibrated images are listed below:

1. Establish point correspondences between images
2. Estimate the epipolar geometry between images
3. Build a representation of the scene using matched points
4. Specify the desired position of the new image
5. Transfer the scene representation into the new image
6. Map textures (colors) from the original images to the new images

The most crucial and difficult part is Step 1 and Step 2. Step 1 will be reviewed in Appendix A. The second step consists in estimating the fundamental matrix between two images [39], the trifocal tensor between three images [40], or the **P**-matrices (camera projection matrices) between N images [41]. A complete review of techniques for estimating the fundamental matrix and projective reconstruction is done by Zhang [42]. A good technique for estimating the trifocal tensor is developed in [43]. The PhD thesis of Laveau [44] addresses the issues of estimating **P**-matrices between N images.

The representation of the scene depends on the number of matched points. If full pixel correspondences are available, the scene is probably better to be represented by the original images, and the later *texture mapping* step can use linear or bilinear interpolation to find the color of a point in the new image. If only a sparse set of point matches are available, we can divide each image into a set of triangular patches using points as vertices. Texture mapping can then be realized through affine transformation or even better through plane projective transformation (homography).

Given a set of point/pixel correspondences and the epipolar geometry, we have *implicitly* a 3D projective description of the scene with respect to a projective coordinate system. Step 4 is then to specify the desired position of the new image (i.e., its **P**-matrix with respect to the projective coordinate system). There are 15 degrees of freedom, and one can imagine how difficult the task is. This really limits the usefulness of uncalibrated images. There are two possibilities to get around this difficulty: use a reference image if it is available for the desired position; use domain knowledge to obtain a quasi-Euclidean structure [38].

Image transfer is simply a shortcut of 3D reconstruction followed by projection onto the new image according to the **P**-matrix. There are however a number of problems in practice, the occlusion problem in particular. There are probably several points which are transferred to the same location of the new image. One should decide which point is visible. If the structure is Euclidean, z-buffer technique can be used. This is however not useful for uncalibrated images, because a 3D point is projective and defined up to a scale factor. The solution is to use *oriented* projective geometry [44].

The transfered point in the new image is in general a nonlinear function of the points in the original images. It is, however, linear if affine camera model (including orthographic, weak-perspective and paraperspective projections) is used. Another case where the transfer is linear is the following: if the original views have parallel optical axis which is orthogonal to the line joining the optical centers, then a new view on the same line with parallel optical axis is a linear combination of the original views. The view morphing technique described in [10] is based on this observation. It first rectifies the original images to have aligned scan lines, then produces an intermediate rectified image through linear interpolation, and finally postwarps it to obtain the desired image. With this technique, one is able to generate an image sequence corresponding to a camera moving along a line joining the optical centers of the two original images. However, if postwarping is not chosen appropriately, there will be a significant projective distortion in the generated views.

5 Modeling the illumination variation

All the techniques described up to now assume a static scene with *fixed illumination*. Illumination variation must be considered when creating a dynamic virtual environment because it has an important effect on the person navigating in it. Traditionally, techniques based on some reflectance models such as ray tracing are used. The most widely used model is the Lambertian where surface radiance depends only on the irradiance of the surface and not on the observer's viewpoint. One usually observes a significant deviation of this model from reality. Recently, there is an important effort on the estimation of surface reflectance of natural materials using more general functions including the BRDF (Bidirectional Reflection Distribution Function) [45–47]. The state-of-the-art, however, does not yet allow us to estimate the BRDF reliably enough to be used in practice. Within computer vision, researchers follow a different approach to model the illumination variation for object recognition. The idea is to capture illumination and object reflectance directly from real images. For Lambertian surfaces of arbitrary texture without self-occlusion and self-shadows, three images taken with non-collinear light sources are enough to completely determine the structure of the illumination manifold [48, 49]. In other cases, more (but usually only a few) images are required [50–52]. The Karhunen-Loéve transform or Singular-Value Decomposition (SVD) [53] is typically used to compute the basis images which capture the essential information relevant to the reflectance and illumination variations. Finally, we observe that, except Shashua's work [48], all other works do not consider 3D geometric information. Shashua represents 3D geometry of an object as a linear combination of two images because an approximate (affine) camera model is used, and uses only three images taken under different illumination conditions to represent the photometric property. However, three images are usually not sufficient to represent the photometric property of a complex object. Recently, Zhang [54] developed a system which uses full perspective camera model and as many images under different illumination conditions as possible. The scene/object is represented by a 3D geometric model together with a set of basis images which capture the essential photometric property of the scene/object under different illumination condition. He is then able to generate photorealistic views simulating both changes in point of view and in illumination condition.

Instead of recovering shape from 2D images, Sato et al. [55] use a light-stripe range finder and a color CCD camera to acquire a sequence of range images and a sequence of color images of an object. Range images are merged to reconstruct the surface shape of the object. The reconstructed shape and the color images are then used to determine the surface reflectance properties of the object. Finally, they are able to synthesize images with realistic shading effects under arbitrary illumination conditions.

6 Concluding remarks

There are emerging interests from both computer vision and computer graphics communities in obtaining photorealistic modeling of a scene or an object from real images. This paper has presented a tentative review of the computer vision techniques used in such modeling which guarantee the generated views to be geometrically correct. The topics covered include mosaicking for building environment maps, CAD-like modeling for building 3D geometric models together with texture maps extracted from real images, image-based rendering for synthesizing new views from uncalibrated images, and techniques for modeling the appearance variation of a scene or an object under different illumination conditions.

There are a number of open issues:

– What is the difference in rendering mechanism between CAD-like modeling and IBR? Which one is more efficient? There is probably no universal answer, and it is likely task-dependent (processing time, hardware accelerators, memory requirement, field of view of the scene to be modeled, etc).
– How to obtain a coherent representation of a large scale scene from a large number of images? Geometric reasoning should play a fundamental role in this regard.
– The intensity/color is almost always different from one image to another because of dynamic change in camera gain, even if images are perfectly aligned. How to fuse different observations in order to avoid intensity/color discontinuity in the final texture maps? This is known as *image blending*.
– How to compute the reflectance properties of a complex scene from real images?
– Matching, the eternal problem in computer vision, although alleviated in IBPhM through user interaction, still needs considerable work. Before having a perfect image matching, we need to answer the following question: How to tolerate false matches in generating photorealistic views?

And there are many more.

A   Image matching techniques

When the epipolar geometry is unknown, there are mainly two categories of techniques. The first is optical-flow based. The representative work is the pyramidal hierarchical motion estimation framework proposed by Bergen et al. [56]. The second is feature-based. The representative work is the robust image matching technique through the recovery of the unknown epipolar geometry proposed by Zhang et al. [57].

When the epipolar geometry is known or recovered, there are many stereo matching techniques available, such as correlation, relaxation and dynamic programming. See, e.g., textbook [19] for a general discussion.

Acknowledgment. Because the short time available for the preparation of this paper, many important references could be overlooked. Please email me.

References

1. S. Chen, "QuickTime VR - an image-based approach to virtual environment navigation," in *Computer Graphics, Annual Conference Series*, pp. 29–38, ACM SIGGRAPH, 1995.
2. S. Chen and L. Williams, "View interpolation for image synthesis," in *Computer Graphics, Annual Conference Series*, pp. 279–288, ACM SIGGRAPH, 1993.
3. T. Werner, R. Hersch, and V. Hlavac, "Rendering real-world objects using view intepolation," in *Proc. Fifth International Conference on Computer Vision*, (Cambridge, Massachusetts), pp. 957–962, June 1995.
4. A. Katayama, K. Tanaka, T. Oshino, and H. Tamura, "A viewpoint independent stereoscopic display using interpolation of multi-viewpoint images," in *Stereoscopic displays and virtual reality systems II* (S. Fisher, J. Merritt, and B. Bolas, eds.), vol. 2409 of *Proc. SPIE*, pp. 11–20, 1995.
5. S. Gortler, R. Grzeszczuk, R. Szeliski, and M. Cohen, "The Lumigraph," in *Computer Graphics, Annual Conference Series*, pp. 43–54, ACM SIGGRAPH, 1996.
6. M. Levoy and P. Hanraham, "Light field rendering," in *Computer Graphics, Annual Conference Series*, pp. 31–42, ACM SIGGRAPH, 1996.
7. L. McMillan and G. Bishop, "Plenoptic modeling: An image-based rendering system," in *Computer Graphics, Annual Conference Series*, pp. 39–46, ACM SIGGRAPH, 1995.
8. P. Debevec, C. Taylor, and J. Malik, "Modeling and rendering architecture from photographs: A hybrid geometry- and image-based approach," in *Computer Graphics, Annual Conference Series*, pp. 11–20, ACM SIGGRAPH, 1996.
9. O. Faugeras and S. Laveau, "Representing three-dimensional data as a collection of images and fundamental matrices for image synthesis," in *Proc. International Conference on Pattern Recognition*, (Jerusalem, Israel), pp. 689–691, Computer Society Press, Oct. 1994.
10. S. Seitz and C. Dyer, "View morphing," in *Computer Graphics, Annual Conference Series*, pp. 21–30, ACM SIGGRAPH, 1996.
11. S. Kang, "A survey of image-based rendering techniques," Tech. Rep. CRL 97/4, Digital Equipment Corporation, Cambridge Research Lab, Aug. 1997.
12. R. Szeliski and H.-Y. Shum, "Creating full view panoramic image mosaics and environment maps," in *Computer Graphics, Annual Conference Series*, pp. 251–258, ACM SIGGRAPH, 1997.
13. I. Zoghlami, O. Faugeras, and R. Deriche, "Using geometric corners to build a 2d mosaic from a set of images," in *Proc. IEEE Conference on Computer Vision and Pattern Recognition*, (San Juan, Puerto Rico), pp. 420–425, IEEE Computer Society, June 1997.
14. S. Peleg and J. Herman, "Panoramic mosaics by manifold projection," in *Proc. IEEE Conference on Computer Vision and Pattern Recognition*, (San Juan, Puerto Rico), pp. 338–343, IEEE Computer Society, June 1997.
15. H.-Y. Shum and R. Szeliski, "Construction and refinement of panoramic mosaics with global and local alignment," in *Proc. 6th International Conference on Computer Vision*, (Bombay, India), IEEE Computer Society Press, Jan. 1998.
16. Y. Yagi and S. Kawato, "Panorama scene analysis with conic projection," in *Proc. IEEE International Workshop on Intelligent Robots and Systems*, pp. 181–187, July 1990.
17. K. Yamazawa, Y. Yagi, and S. Kawato, "Omnidirectional imaging with hyperboloidal projection," in *Proc. IEEE/RSJ International Conference on Intelligent Robots and Systems*, pp. 1029–1034, July 1993.
18. S. Nayar, "Catadioptric omnidirectional camera," in *Proc. IEEE Conference on Computer Vision and Pattern Recognition* (G. Medioni, R. Nevatia, D. Huttenlocher, and J. Ponce, eds.), (San Juan, Puerto Rico), pp. 482–488, IEEE Computer Society, June 1997.

19. O. Faugeras, *Three-Dimensional Computer Vision: a Geometric Viewpoint*. MIT Press, 1993.
20. J. Aggarwal and N. Nandhakumar, "On the computation of motion from sequences of images — a review," *Proc. IEEE*, vol. 76, pp. 917–935, Aug. 1988.
21. T. Huang and A. Netravali, "Motion and structure from feature correspondences: A review," *Proc. IEEE*, vol. 82, pp. 252–268, Feb. 1994.
22. Z. Zhang, "Motion and structure from two perspective views: From essential parameters to euclidean motion via fundamental matrix," *Journal of the Optical Society of America A*, vol. 14, no. 11, 1997. In Press.
23. Z. Zhang and O. Faugeras, *3D Dynamic Scene Analysis: A Stereo Based Approach*. Springer-Verlag, Berlin, New York, 1992.
24. Z. Zhang, "Iterative point matching for registration of free-form curves and surfaces," *The International Journal of Computer Vision*, vol. 13, no. 2, pp. 119–152, 1994. also Research Report No.1658, INRIA Sophia-Antipolis, 1992.
25. Z. Zhang, "Motion of a stereo rig: Strong weak and self calibration," in *Recent Developments in Computer Vision* (S. Li, D. Mital, E. Teoh, and H. Wang, eds.), vol. 1035 of *Lecture Notes in Computer Science*, pp. 241–254, Springer-Verlag, Berlin, 1996.
26. S. J. Maybank and O. D. Faugeras, "A theory of self-calibration of a moving camera," *The International Journal of Computer Vision*, vol. 8, pp. 123–152, Aug. 1992.
27. Q.-T. Luong, *Matrice Fondamentale et Calibration Visuelle sur l'Environnement-Vers une plus grande autonomie des systèmes robotiques*. PhD thesis, Université de Paris-Sud, Centre d'Orsay, Dec. 1992.
28. Q.-T. Luong and O. Faugeras, "Self-calibration of a moving camera from point correspondences and fundamental matrices," *The International Journal of Computer Vision*, vol. 22, no. 3, pp. 261–289, 1997.
29. R. Hartley, "An algorithm for self calibration from several views," in *Proc. IEEE Conference on Computer Vision and Pattern Recognition*, (Seattle, WA), pp.908–912, 1994.
30. Z. Zhang, Q.-T. Luong, and O. Faugeras, "Motion of an uncalibrated stereo rig: self-calibration and metric reconstruction," *IEEE Transactions on Robotics and Automation*, vol. 12, pp. 103–113, Feb. 1996. Short version appeared in the *Proc. International Conference on Pattern Recognition*, volume I, pages 695–697, Jerusalem, Israel, Oct. 1994.
31. A. Zisserman, P. A. Beardsley, and I. D. Reid, "Metric calibration of a stereo rig," in *Proc. Workshop on Visual Scene Representation*, (Boston, MA), June 1995.
32. F. Devernay and O. Faugeras, "From projective to euclidean reconstruction," in *Proc. IEEE Conference on Computer Vision and Pattern Recognition*, (San Francisco, CA), pp. 264–269, IEEE, June 1996.
33. O. Faugeras, "What can be seen in three dimensions with an uncalibrated stereo rig," in *Proc. 2nd European Conference on Computer Vision* (G. Sandini, ed.), vol. 588 of *Lecture Notes in Computer Science*, (Santa Margherita Ligure, Italy), pp. 563–578, Springer-Verlag, May 1992.
34. R. Hartley, "Projective reconstruction and invariants from multiple images," *IEEE Transactions on Pattern Analysis and Machine Intelligence*, vol. 16, no. 10, pp. 1036–1040, 1994.
35. G. Xu and Z. Zhang, *Epipolar Geometry in Stereo, Motion and Object Recognition*. Kluwer Academic Publishers, 1996.
36. R. Hartley, R. Gupta, and T. Chang, "Stereo from uncalibrated cameras," in *Proc. IEEE Conference on Computer Vision and Pattern Recognition*, (Urbana Champaign, IL), pp. 761–764, IEEE, June 1992.
37. R. Mohr, B. Boufama, and P. Brand, "Understanding positioning from multiple images," *Artificial Intelligence*, vol. 78, pp. 213–238, 1995.
38. Z. Zhang, K. Isono, and S. Akamatsu, "Euclidean structure from uncalibrated images using fuzzy domain knowledge: Application to facial images synthesis," in *Proc. 6th International Conference on Computer Vision*, (Bombay, India), IEEE Computer Society Press, Jan. 1998.

39. Q.-T. Luong and O. D. Faugeras, "The fundamental matrix: Theory, algorithms and stability analysis," *The International Journal of Computer Vision*, vol. 17, pp. 43–76, Jan. 1996.
40. A. Shashua, "Algebraic functions for recognition," *IEEE Transactions on Pattern Analysis and Machine Intelligence*, vol. 17, no. 8, pp. 779–789, 1995.
41. Q.-T. Luong and T. Viéville, "Canonical representations for the geometries of multiple projective views," *Computer Vision and Image Understanding*, vol. 64, pp. 193–229, Sept. 1996.
42. Z. Zhang, "Determining the epipolar geometry and its uncertainty: A review," *The International Journal of Computer Vision*, 1997. In Press. Updated version of INRIA Research Report No.2927, 1996.
43. P. Torr and A. Zisserman, "Robust parameterization and computation of the trifocal tensor," *Image and Vision Computing*, vol. 15, pp. 591–605, 1997.
44. S. Laveau, *Géométrie d'un système de N caméras. Théorie, estimation et applications*. PhD thesis, École Polytechnique, May 1996.
45. M. Oren and S. Nayar, "Generalization of the lambertian model and implications for machine vision," *The International Journal of Computer Vision*, vol. 14, pp. 227–251, Apr. 1995.
46. L. Wolff, "Generalizing Lambert's law for smooth surfaces," in *Proc. 4th European Conference on Computer Vision* (B. Buxton, ed.), vol. II, (Cambridge, UK), pp. 40–53, Apr. 1996.
47. J. Koenderink, A. van Doorn, and M. Stavridi, "Bidirectional reflection distribution function expressed in terms of surface scattering modes," Research Report UU-PAhp-046, Utrecht State University, 1995.
48. A. Shashua, *Geometry and Photometry in 3D Visual Recognition*. PhD thesis, Massachusetts Institute of Technology, 1992.
49. S. Nayar and H. Murase, "Dimensionality of illumination in appearance matching," in *Proc. IEEE International Conference on Robotics and Automation*, (Minneapolis, Minnesota), pp. 1326–1332, Apr. 1996.
50. R. Epstein, P. Hallinan, and A. Yuille, "5±2 eigenimages suffice: An empirical investigation of low-dimensional lighting models," in *Proc. IEEE Workshop on Physics Based Modeling in Computer Vision*, (Cambridge, Massachusett), pp. 108–116, June 1995.
51. P. Belhumeur and D. Kriegman, "What is the set of images of an object under all possible lighting conditions?," in *Proc. IEEE Conference on Computer Vision and Pattern Recognition*, pp. 270–277, June 1996.
52. G. Hager and P. Belhumeur, "Real-time tracking of image regions with changes in geometry and illumination," in *Proc. IEEE Conference on Computer Vision and Pattern Recognition*, June 1996.
53. G. Golub and C. van Loan, *Matrix Computations*. The John Hopkins University Press, 1989.
54. Z. Zhang, "Modeling geometric structure and illumination variation of a scene from real images," in *Proc. 6th International Conference on Computer Vision*, (Bombay, India), IEEE Computer Society Press, Jan. 1998.
55. Y. Sato, M. Wheeler, and K. Ikeuchi, "Object shape and reflectance modeling from observation," in *Computer Graphics, Annual Conference Series*, pp. 379–387, ACM SIGGRAPH, 1997.
56. J. Bergen, P. Anandan, K. Hanna, and R. Hingorani, "Hierarchical model-based motion estimation," in *Proc. 2nd European Conference on Computer Vision* (G. Sandini, ed.), vol. 588 of *Lecture Notes in Computer Science*, (Santa Margherita Ligure, Italy), pp. 237–252, Springer-Verlag, May 1992.
57. Z. Zhang, R. Deriche, O. Faugeras, and Q.-T. Luong, "A robust technique for matching two uncalibrated images through the recovery of the unknown epipolar geometry," *Artificial Intelligence Journal*, vol. 78, pp. 87–119, Oct. 1995.

Measuring Object Surface Shape and Reflectance Properties

Yoichi Sato[1], Mark D. Wheeler[2], and Katsushi Ikeuchi[1]

[1] Department of Electrical Engineering and Electronics
Institute of Industrial Science, The University of Tokyo, Tokyo 106, Japan
[2] Apple Computer Inc., 1 Infinite Loop, MS:301-3M, Cupertino, CA 95014, USA

Abstract. An object model for computer graphics applications should contain two aspects of information: shape and reflectance properties of the object. A number of techniques have been developed for modeling object shapes by observing real objects. In contrast, attempts to model reflectance properties of real objects have been rather limited. In most cases, modeled reflectance properties are too simple or too complicated to be used for synthesizing realistic images of the object.
In this paper, we propose a new method for modeling object reflectance properties, as well as object shapes, by observing real objects. First, an object surface shape is reconstructed by merging multiple range images of the object. By using the reconstructed object shape and a sequence of color images of the object, parameters of a reflection model are estimated in a robust manner. The recovered object shape and reflectance properties are then used for synthesizing object images with realistic shading effects under arbitrary illumination conditions.

1 Introduction

As a result of significant advancement of graphics hardware and image rendering algorithms, the 3D computer graphics capability has become available even on low-end computers. In addition, the rapid spread of the internet technology has caused a significant increase in the demand for 3D computer graphics.

However, it is often the case that 3D object models are created manually by users. That input process is normally time-consuming and can be a bottleneck for realistic image synthesis. Therefore, techniques to obtain object model data automatically by observing real objects could have great significance in practical applications.

An object model for computer graphics applications should contain two aspects of information: shape and reflectance properties of the object. A number of techniques have been developed for modeling object shapes by observing real objects. In contrast, attempts to model reflectance properties of real objects have been rather limited. In most cases, modeled reflectance properties are too simple or too complicated to be used for synthesizing realistic images of the object. For example, if only observed color texture or diffuse texture of a real object surface is used (e.g., texture mapping), correct shading effects such as highlights cannot be reproduced correctly in synthesized images. On the other hand, object reflectance properties can be represented accurately by a bidirectional reflectance distribution function (BRDF). If a BRDF is available for the object surface, shading effects can be, in principle, reproduced correctly in synthesized images. However, the use of BRDF is not practical because measurement of BRDF is usually very expensive and time-consuming [13]. In practice, we cannot obtain a BRDF for real objects with various reflectance properties.

In this paper, we propose a new method for modeling object reflectance properties, as well as object shapes, from multiple range and color images of real objects. Unlike previously proposed methods [1, 6, 7, 10], our method can create complete object models, i.e., not partial object shape, with non-uniform reflectance properties. First, the object surface shape is reconstructed by merging multiple range images of the object. By using the reconstructed object shape and a sequence of color images of the object, parameters of a reflection model are estimated in a robust manner. The key point of the proposed method is that, first, the diffuse reflection components and the specular reflection component are separated from the color image sequence, and then, reflectance parameters of each reflection component are estimated separately. Unlike previously reported methods, this approach enables reliable estimation of surface reflectance properties which are not uniform over the object surface, and which include specularity as well as diffusely reflected lights. We demonstrate the capability of our object modeling technique by synthesizing object images with realistic shading effects under arbitrary illumination conditions.

This paper is organized as follows: Sect. 2 describes our image acquisition system for obtaining a sequence of range and color images of the object. Section 3 explains reconstruction of the object surface shape from the range image sequence. Section 4 describes our method for estimating reflectance properties of the object using the reconstructed object shape and the color image sequence. Object images synthesized using the recovered object shape and reflectance properties are shown in Sect. 5. Concluding remarks are presented in Sect. 6.

2 Image Acquisition System

The object whose shape and reflectance information is to be recovered is mounted on the end of a robotic arm. The object used in our experiment is a ceramic mug whose height is about $100mm$. Using our image acquisition system, a sequence of range and color images of the object is obtained as the object is rotated at a fixed angle step. Twelve range images and 120 color images were used in our experiment shown in this paper.

A range image is obtained using a light-stripe range finder with a liquid crystal shutter and a color CCD video camera. 3D locations of points in the scene are computed at each image pixel using optical triangulation. Each range-image pixel represents an (X, Y, Z) location of a corresponding point on an object surface. The same color camera is used for acquiring range images and color images. Therefore, pixels of the range images and the color images directly correspond.

The range finder is calibrated to produce a 3×4 projection matrix Π which represents the projection transformation between the world coordinate system and the image coordinate system. The location of the robotic arm with respect to the world coordinate system is also found via calibration. Therefore, the object location is given as a 4×4 transformation matrix T for each digitized image.

A single incandescent lamp is used as a point light source. In our experiments, the light source direction and the light source color are measured by calibration. The gain and offset of outputs from the video camera are adjusted so that the light source color becomes $(R, G. B) = (1, 1, 1)$.

3 Surface Shape Modeling

A sequence of range images of the object is used to construct the object shape as a triangular mesh. For reconstructing object shapes as a triangular mesh model from multiple range images, we used the volumetric method developed by Wheeler, Sato, and Ikeuchi [14]. The method consists of the following four steps, each of which is briefly described in this section.

1. Surface acquisition from each range image

The range finder in our image acquisition system measures 3D points on an object surface. Therefore, we need to convert the measured 3D points into a triangular mesh. This is done by connecting two neighboring range image pixels based on the assumption that those points are connected by a locally smooth surface. If those two points are closer in a 3D distance than some threshold, then we consider them to be connected on the object surface. In Fig. 1 (a), 4 out of 12 input range images of the mug are shown as triangular meshes.

2. Alignment of all range images

All of the range images are measured in the coordinate system fixed with respect to the range finder system, and they are not aligned to each other initially. Therefore, after we obtain the triangular surface meshes from the range images, we need to transform all of the meshes into a unique object coordinate system.

To align the range images, we use a transformation matrix T which represents the object location for each range image (Sect. 2). Suppose we select one of the range images as a key range image to which all other range images are aligned. We refer to the transformation matrix for the key range image as T_{merge}. Then, all other range images can be transformed into the key range images coordinate system by transforming all 3D points $P = (X, Y, Z, 1)$ as $P' = T_{merge} T_f^{-1} P$ where $f = 1 \ldots n$ is a range image frame number.

3. Merging based on a volumetric representation

After all of the range images are converted into triangular patches and aligned to a unique coordinate system, we merge them using a volumetric representation. First, we consider imaginary 3D volume grids around the aligned triangular patches. Then, in each voxel, we store the value, $f(x)$, of the signed distance from the center point of the voxel, x, to the closest point on the object surface. The sign indicates whether the point is outside, $f(x) > 0$, or inside, $f(x) < 0$, the object surface, while $f(x) = 0$ indicates that lies on the surface of the object.

This technique has been applied to surface extraction by several researchers [4, 2, 3]. The novel part of our technique is the robust computation of the signed distance. Our technique computes the signed distance by using a new algorithm called *the consensus surface algorithm* [14]. In the consensus surface algorithm, a quorum of consensus of locally coherent observation of the object surface is used to compute the signed distance correctly, which eliminates many of the troublesome effects of noise and extraneous surface observations in the input range images, for which previously developed methods are susceptible.

Figure 1 (c) shows two cross sections of the volumetric data constructed from the input range images of the mug. A darker color represents a shorter distance to the object surface, and a brighter color represents a longer distance.

4.Isosurface extraction from volumetric grid

The volumetric data is then used to construct the object surface as a triangular mesh. The marching cubes algorithm [8] constructs a triangular mesh by traversing zero crossings of the implicit surface, $f(x) = 0$, in the volume grid. Here, the marching cube algorithm was modified so that it handles holes and missing data correctly [14]. Figure 1 (d) shows the result of triangular mesh reconstruction. In this example, 3782 triangles were generated from the volumetric data.

The marching cube algorithm generally produces a large number of triangles whose sizes vary significantly. Thus, it is desirable to simplify the reconstructed object surface shape by reducing the number of triangles. We used the mesh simplification method developed by Hoppe et al. [5] for this purpose. In our experiment, the total number of triangles was reduced from 3782 to 488.

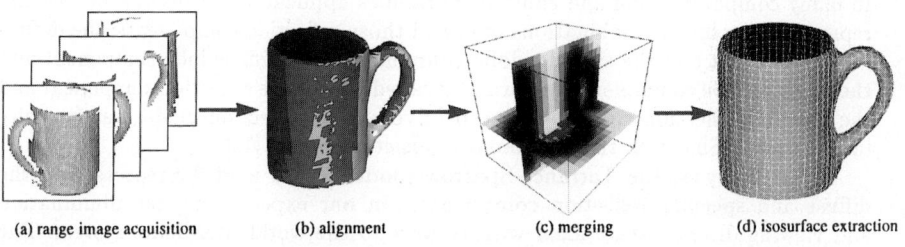

Fig. 1. Shape reconstruction by merging range images: (a) Input surface patches (4 out of 12 patches are shown), (b) Result of alignment, (c) Obtained volumetric data (two cross sections are shown), (d) Generated triangular mesh of the object shape (3782 triangles)

Polygonal normals computed from a triangular surface mesh model can approximate real surface normals fairly well when the object surface is relatively smooth and does not have high curvature points. However, accuracy of polygonal normals becomes poor when the object surface has high curvature points and the resolution of the triangular surface mesh model is low, i.e., a smaller number of triangles to represent the object shape.

This becomes a problem especially for the task of reflectance parameter estimation. For estimating reflectance parameters at a surface point, we need to know three directions at the surface point: the viewing direction, the light source direction, and the surface normal. As a result, with incorrectly estimated surface normals, small highlights observed within each triangle cannot be analyzed accurately, and therefore they cannot be reproduced in synthesized images. For this reason, we compute surface normals at regular grid points (20 × 20 points in our experiment) within each triangle using the 3D points from the range images.

The surface normal at a grid point P_g is determined from a least squares best fitting plane to all neighboring 3D points whose distances to the point P_g are shorter than some threshold. This surface normal estimation method has been used by other researchers for other applications [4].

The surface normals computed at regular grid points within each triangle are used both for reflectance parameter estimation and for rendering color images of the object.

4 Surface Reflectance Modeling

After the object shape is reconstructed, we measure reflectance properties of the object surface using the reconstructed shape and the input color images. First, the two fundamental reflection components (i.e., the diffuse and specular reflection components) are separated from the input color images. Then, the parameters for the two reflection components are estimated separately. Separation of the two reflection components enables us to obtain a reliable estimation of the specular reflection parameters. Also, the specular reflection component (i.e., highlight) in the color images does not affect estimated diffuse reflection parameters of the object surface.

4.1 Reflection model

In many computer vision and computer graphics applications, reflection models are represented by linear combinations of two of those reflection components: the diffuse lobe component and the specular lobe component. The diffuse lobe component and the specular lobe component are normally called the diffuse reflection component and the specular reflection component, respectively. This reflection model was formally introduced by Shafer as the dichromatic reflection model [11].

In our analysis, the Torrance-Sparrow model [12] is used for representing the diffuse and specular reflection components. In our experiments, the illumination and viewing directions are fixed with respect to the world coordinate system. The reflection model used in our analysis is given as

$$I_m = K_{D,m} \cos\theta_i + K_{S,m} \frac{1}{\cos\theta_r} \exp\frac{-\alpha^2}{2\sigma^2} \qquad m = R, G, B \qquad (1)$$

where θ_i is the angle between the surface normal and the light source direction, θ_r is the angle between the surface normal and the viewing direction, α is the angle between the surface normal and the bisector of the light source direction and the viewing direction, $K_{D,m}$ and $K_{S,m}$ are constants for the diffuse and specular reflection components, and σ is the standard deviation of a facet slope of the Torrance-Sparrow model.

In this paper, we refer to $K_{D,R}$, $K_{D,G}$, and $K_{D,B}$ as the diffuse reflection parameters, and $K_{S,R}$, $K_{S,G}$, $K_{S,B}$, and σ as the specular reflection parameters.

4.2 Mapping color images onto object surface shape

For separating the diffuse and specular reflection components and for estimating parameters of each reflection component, we need to know a sequence of observed colors at each point on the object surface as the object is rotated.

We represent world coordinates and image coordinates using homogeneous coordinates. A point on the object surface with Euclidean coordinates (X, Y, Z) is expressed by a column vector $P = [X, Y, Z, 1]^T$. An image pixel location (x, y) is represented by $p = [x, y, 1]^T$. As described in Sect. 2, the camera projection transformation is represented by a 3×4 matrix Π, and the object location is given by a 4×4 object transformation matrix T. We denote the object transformation matrix for the input color image frame f by T_f ($f = 1 \ldots n$). Thus, using the projection matrix Π and the transformation matrix T_{merge} for the key range image (Sect. 3), the projection of a 3D point on the object surface in the color image frame f is given as

$$p_f = \Pi T_f T_{merge}^{-1} P \qquad (f = 1 \ldots n) \qquad (2)$$

where the last component of p_f has to be normalized to give the projected image location (x, y).

The observed color of the 3D point in the color image frame f is given as the (R, G, B) color intensity at the pixel location (x, y). By repeating this procedure for all frames of the input color image sequence, we get an observed color sequence for the 3D point on the object surface

4.3 Reflection component separation from color image sequence

We now describe the algorithm for separating the two reflection components. This separation algorithm was originally introduced by Sato and Ikeuchi [9].

Using three color bands, red, green, and blue, the coefficients $K_{D,m}$ and $K_{S,m}$, in (1), generalize to two linearly independent vectors,

$$\mathbf{K_D} = [K_{D,R} \ K_{D,G} \ K_{D,B}]^T \qquad \mathbf{K_S} = [K_{S,R} \ K_{S,G} \ K_{S,B}]^T \qquad (3)$$

unless the colors of the two reflection components are accidentally the same.

First, the color intensities in the R, G, and B channels from input images of the object are measured for each point on the object surface as described in Sect. 4.2. The three sequences of intensity values are stored in the columns of an $n \times 3$ matrix M. Considering the reflectance model (1) and two color vectors in (3), the intensity values in the R, G, and B channels can be represented as

$$\begin{aligned}
M &= [\mathbf{M_R} \ \mathbf{M_G} \ \mathbf{M_B}] \\
&= \begin{bmatrix} \cos\theta_{i1} & E(\theta_{r1}, \alpha_1) \\ \cos\theta_{i2} & E(\theta_{r2}, \alpha_2) \\ \vdots & \vdots \\ \cos\theta_{in} & E(\theta_{rn}, \alpha_n) \end{bmatrix} \begin{bmatrix} K_{D,R} & K_{D,G} & K_{D,B} \\ K_{S,R} & K_{S,G} & K_{S,B} \end{bmatrix} \\
&= [\mathbf{G_D} \ \mathbf{G_S}] \begin{bmatrix} \mathbf{K_D}^T \\ \mathbf{K_S}^T \end{bmatrix} \\
&\equiv GK
\end{aligned} \qquad (4)$$

where $E(\theta_r, \alpha) = (\exp(-\alpha^2/2\sigma^2))/\cos\theta_r$, and the two vectors $\mathbf{G_D}$ and $\mathbf{G_S}$ represent the intensity values of the diffuse and specular reflection components with respect to the illuminating/viewing directions θ_i, θ_r, and α. The vectors $\mathbf{K_D}$ and $\mathbf{K_S}$ represent the diffuse and the specular reflection color vectors, respectively.

From (1), it can be seen that the distribution of the specular reflection component is limited to a fixed angle, depending on σ. Thus, if the angle α is sufficiently large at a point on the object surface, an observed color at the point should represent the color of the diffuse reflection component which gives $\mathbf{K_D}$. The angles α, θ_i, and θ_r are computed using the object transformation matrix T_f ($f = 1 \ldots n$) and the camera projection matrix Π as follows. The light source location is acquired via calibration, and the camera projection center can be computed from the projection matrix Π. Also, the surface normal at the surface point of the object model for the color image frame f can be computed by rotating the surface normal at the surface point by the object transformation matrix T_f. Using the light source direction, the viewing direction and the surface normal, α, θ_i, and θ_r are computed.

Also, in our experiments, the specular reflection color vector $\mathbf{K_S}$ is directly measured as the light source color by a calibration procedure.

Then, the two reflection components represented by the matrix G are obtained by projecting the observed reflection stored in M onto the two color vectors $\mathbf{K_D}$ and $\mathbf{K_S}$ as

$$G = MK^+ \tag{5}$$

where K^+ is the 3×2 pseudoinverse matrix of the color matrix K.

Once we get the matrix K, the matrix G can be calculated from (5). Each of the diffuse and specular reflection components is given as

$$M_D = \mathbf{G_D K_D}^T \qquad M_S = \mathbf{G_S K_S}^T. \tag{6}$$

4.4 Diffuse reflection parameter estimation

Using the diffuse reflection component separated from the observed color sequence, we now can estimate the diffuse reflection parameters ($K_{D,R}$, $K_{D,G}$, and $K_{D,B}$) without undesirable effects from the specular reflection component (i.e., highlights). Using the angle θ_i computed as stated in the previous section, the diffuse reflection parameters are estimated by fitting the reflection model (the first term of (1)) to the separated diffuse reflection component. Hence, the estimated diffuse reflection parameters are not affected by the particular shadings in the observed images, e.g., the effect of the light source can be factored out.

The diffuse reflection parameters are estimated at regular grid points within each triangle just as the surface normals in Sect. 3 are estimated. The resolution of the grid of points is 80×80 in our experiment, while it is 20×20 for the surface normal estimation. The higher resolution is necessary to capture details of the diffuse reflection texture on the object surface. Figure 2 (a) shows the result of the diffuse reflection parameter estimation where the estimated parameters are visualized as surface texture on the mug.

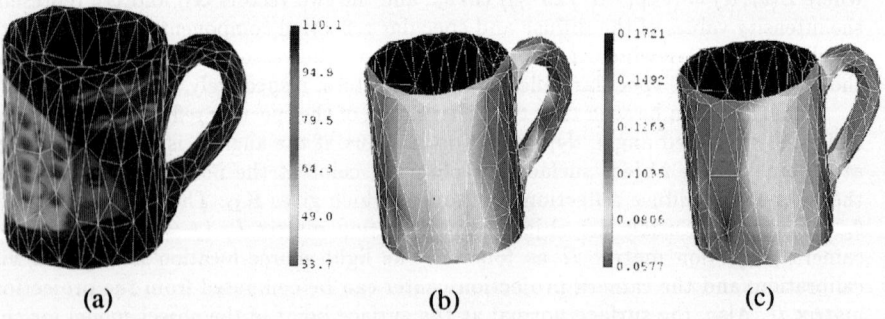

Fig. 2. Estimated diffuse and specular reflection parameters

4.5 Specular reflection parameter estimation

As in the diffuse reflection parameter estimation, the specular reflection parameters ($K_{S,R}$, $K_{S,G}$, $K_{S,B}$, and σ) are also computed using the angle θ_r and the angle α. However, there is a significant difference between estimation of the diffuse and specular reflection parameters. Unlike the diffuse reflection component, the specular reflection component is usually observed only from a limited range of viewing directions. For a finite set of views, the specular reflection component will only be observed over a small portion of the object surface in the input color image sequence. For much of the object surface, we cannot estimate the specular reflection parameters. Even if the specular reflection component is observed, the parameter estimation can become unreliable if the specular reflection component is not observed strongly, or if the separation of the two reflection components is not performed well.

For the above reasons, we decided to use a slightly different strategy for estimating the specular reflection parameters. Since the specular reflection parameters may only be estimated sparsely over the object surface, we use interpolation to infer the specular reflection parameters over the entire surface.

Selection of Surface Points for Parameter Estimation. For the specular refection parameters to be estimated reliably, the following three conditions are necessary at a point on the object surface. All of the three conditions contribute to reliable separation of the diffuse and specular reflection components.

1. The two reflection components must be reliably separated. Because the diffuse and specular reflection components are separated using the difference of the colors of the two components (Sect. 4.3), these color vectors should differ as much as possible. This can be examined by saturation of the diffuse color. Since the light source color is generally close to white (saturation = 0), if the diffuse color has a high saturation value, the diffuse and specular reflection colors will be different.

2. The magnitude of the specular reflection component is as large as possible.

3. The magnitude of the diffuse reflection component is as large as possible. Although this condition might seem to be unnecessary, we empirically found that the specular reflection parameters can be obtained more reliably if this condition is satisfied.

An evaluation measure: $v =$ *diffuse saturation* $\times$ *max specular intensity* $\times$ *max diffuse intensity* is used to represent how well these three conditions are satisfied. In our experiments, 100 vertices with the largest values were chosen according to our evaluation measurement v.

Interpolation of Estimated Specular Parameters. In our experiment, the camera output is calibrated so that the specular reflection color (i.e., the light source color) has the same value from the three color channels (Sect. 2). Therefore, only one color band was used to estimate the specular reflection parameters (K_S and σ) in our experiment.

After the specular reflection parameters K_S and σ were estimated at the 100 selected vertices, the estimated values were linearly interpolated based on a distance on the object surface, so that the specular reflection parameters were obtained at regular grid points within each triangle of the object surface mesh. The resolution of the grid points was 20×20 in our experiment, while the resolution was 80×80 for the diffuse reflection parameter estimation. In general, specular reflectance does not change so rapidly as diffuse reflectance, i.e., diffuse texture on the object surface. Therefore, the resolution of 20 × 20 was enough to capture the specular reflectance of the mug.

Interpolated values of the two specular reflection parameters are shown in Fig. 2 (b) and (c). The obtained specular reflection parameters were then stored in two specular reflection parameter images (a K_S image and a σ image) just as estimated surface normals were stored in the surface normal image.

5 Image Synthesis

Using the measured object surface shape and reflectance parameters, we can synthesize color object images under arbitrary illumination/viewing conditions.

Figure 3 shows synthesized images of the object with two point light sources. Note that the images represent highlights on the object surface naturally. For comparing synthesized images with the input color images of the object, the object model was rendered using the same illumination and viewing directions as some of the input color images. Figure 4 shows two frames of the input color image sequence as well as two synthesized images that were generated using the same illuminating/viewing condition as the input color images. It can be seen that the synthesized images closely resemble the corresponding real images. In particular, highlights, which generally are a very important cue of surface material, appear on the side and the handle of the mug naturally in the synthesized images.

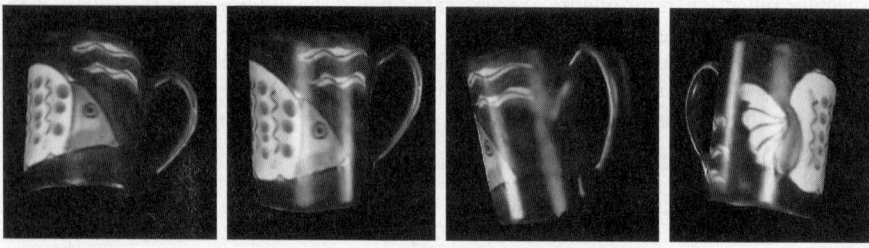

Fig. 3. Synthesized object images

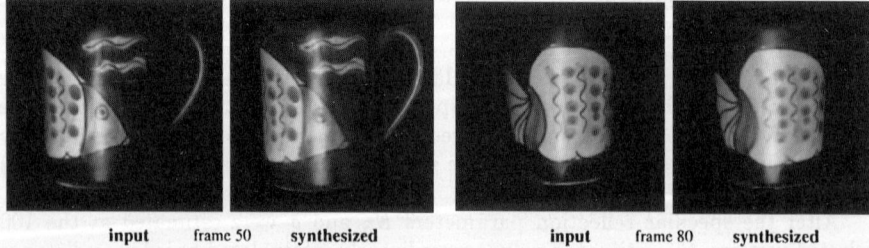

Fig. 4. Comparison of input color images and synthesized images

6 Conclusion

We have explored automatic generation of photorealistic object models from observation. Achieving photorealism in synthesized object images requires accurate modeling of shape and reflectance properties of the object. In this paper, we have presented a new paradigm for acquiring object shape and reflectance parameters from range and color images.

The object surface shape is reconstructed by merging multiple range images of the object. By using the reconstructed object shape and multiple color images of the object, parameters of the Torrance-Sparrow reflection model are estimated. For estimating reflectance parameters of the object robustly, our method is based on separation of the diffuse and specular reflection components from a color image sequence. Using separated reflection components, reflection model parameters for each of the two components were estimated separately. In particular, the specular reflection parameters were successfully obtained by identifying suitable surface points for estimation and by interpolating estimated parameters over the object surface.

Our experiments have shown that our object modeling method can be effectively used for synthesizing realistic object images under arbitrary illumination and viewing conditions.

References

1. R. Baribeau, M. Rioux, and G. Godin, "Color reflectance modeling using a polychromatic laser sensor," IEEE Trans. on Pattern Analysis and Machine Intelligence, vol. 14, no. 2, pp. 263-269, 1992.
2. B. Curless and M. Levoy, "A volumetric method for building complex models from range images," Computer Graphics (SIGGRAPH '96 Proceedings), pp. 303-312, 1996.
3. A. Hilton, J. Stoddart, J. Illingworth, and T. Windeatt, "Reliable surface reconstruction from multiple range images," Proceedings of European Conference on Computer Vision '96, pp. 117-126, 1996.
4. H. Hoppe, T. DeRose, T. Duchamp, J. McDonald, and W. Stuetzle, "Surface reconstruction from unorganized points," Computer Graphics (SIGGRAPH '92 Proceedings), pp. 71-78, 1992.
5. H. Hoppe, T. DeRose, T. Duchamp, J. McDonald, and W. Stuetzle, "Mesh Optimization," Computer Graphics (SIGGRAPH '93 Proceedings), pp. 19-26, 1993.
6. K. Ikeuchi and K. Sato, "Determining reflectance properties of an object using range and brightness images," IEEE Trans. on Pattern Analysis and Machine Intelligence, vol. 13, no. 11, pp. 1139-1153, 1991.
7. G. Kay and T. Caelli, "Inverting an illumination model from range and intensity maps," CVGIP: Image Understanding, vol. 59, pp. 183-201, 1994.
8. W. E. Lorensen and H. E. Cline, "Marching cubes: a high resolution 3D surface construction algorithm," Computer Graphics (SIGGRAPH '87 Proceedings), vol. 21, no. 4, pp. 163-169, 1987.
9. Y. Sato and K. Ikeuchi, "Temporal-color space analysis of reflection," Journal of Optical Society of America A, vol. 11, no. 11, pp. 2990-3002, November 1994.
10. Y. Sato and K. Ikeuchi, "Reflectance analysis for 3D computer graphics model generation," Graphical Models and Image Processing, vol. 58, no. 5, pp. 437-451, September 1996.
11. S. Shafer, "Using color to separate reflection components," COLOR Research and Application, vol. 10, no. 4, pp. 210-218, 1985.
12. K. E. Torrance and E. M. Sparrow, "Theory for off-specular reflection from roughened surface," Journal of Optical Society of America, vol. 57, pp. 1105-1114, 1967.
13. G. J. Ward, "Measuring and modeling anisotropic reflection," Computer Graphics (SIGGRAPH 92 Proceedings), vol. 26, no. 2, pp. 265-272, 1992.
14. M. D. Wheeler, Y. Sato, and K. Ikeuchi, "Consensus surfaces for modeling 3D objects from multiple range images," DARPA Image Understanding Workshop, 1997.

Robust Image Composition Algorithms for Augmented Reality

Marie-Odile Berger and Gilles Simon

INRIA Lorraine/ CRIN-CNRS, BP 101, 615 rue du Jardin Botanique, 54602
Vandœuvre les Nancy cedex, France

Abstract. We present our augmented reality system for image composition. We have worked with a view to avoiding strong and tedious interactions with the user. In this paper, we especially stress on the robust temporal registration method we have devised. An original method for resolving occlusions is also presented.

1 Introduction

In the past few years, virtual reality has attracted a great deal of media attention. The idea is to immerse a user into a completely computer-generated virtual world. Unfortunately, these environments often lack realism and the user is cut off from any view of the real world outside. Moreover, the numerical simulation of virtual environments is most of the time cost expensive.

On the contrary, augmented reality (AR) allows the user to interact with the real world in a natural way. Augmented reality systems aim at enhancing the user's vision with computer generated imagery but does not attempt to replace the real world. This explain why interest in AR has substantially increased in the past few years and medical, manufacturing or urban planning applications have been developed [11,5,3].

We focus in this paper on the problem of image composition for video sequences, which is one of the key point for numerous AR applications. Indeed, to make AR systems more effective, the computer generated objects must be blended convincingly with the real images.

Requirements for a realistic composition

The first challenge to be solved is to correctly retain up-to-date the scene-camera pose relationships over relative motion. This temporal registration allows the image of the computer generated objects to be computed for each frame of the video sequence. The registration task must be achieved with special care because the human visual system is very good at detecting even small misregistrations.

Unfortunately, ensuring temporal registration is not sufficient to perform realistic composition. Other significant visual cues to the human perceptual system must be considered: for instance, proper occlusion resolution between real and virtual objects is highly desirable in composition systems. Other photometric interactions between real and virtual objects (continuity of lighting, shadowing...) should also be considered.

Instrumenting the scene?
Augmented reality problems can often be solved by using either algorithmic solutions or sensor based solutions. For instance, the registration problem can be solved using position sensors (as Polhemus sensors). Easily detectable landmarks placed in the scene can also be used to make the registration process easier [5]. However, instrumenting the real world world is not always possible, especially for vast or outdoor environments. Thus, vision based object registration is an interesting and cheaper approach that leaves the environment unmodified. Hence, a wide variety of applications can be considered with such methods.

We focus in this paper on image composition methods which do not involve neither landmarks nor sensors. We only assume that the 3D model of some parts of the scene is known; it will be used for object based registration. This hypothesis is generally not restrictive for practical applications because the main structures of the scene are often known (ground, main objects in the scene ...).

We describe in the next section an overview of our augmented reality system. We then present the robust solutions we have devised for resolving the temporal registration problem (section 3) and the occlusion problem (section 4).

2 Overview of our Augmented Reality System

Before giving the overview of our system, we discuss the methods able to solve the temporal registration problem and the occlusion problem.

Camera calibration: If a large number of 2D/3D point or line correspondences are available, the registration reduces to a classical calibration process which allows the intrinsic parameters as well as the pose to be computed. Otherwise, a straightforward process is to calibrate the camera before shooting the video sequence. The underlying assumption is that the intrinsic parameters remain constant as the camera moves; we then have only to compute the camera pose for each frame. We use this latter solution because in practice, a small number of points can be extracted with sufficient accuracy, especially for outdoor scenes.

Viewpoint computation: Pose recovery has been extensively studied in the past few years. Two broad classes of methods can be distinguished: the most classical one uses object based registration; this means that 3D knowledge is needed to compute the pose from image/model correspondences. The other alternative is basically 2D: if the projection of a sufficient number of points are observed from different positions, the camera pose can be recovered up to a scale factor [10]. Unfortunately, these approaches turn out to be sensitive to inaccuracies in 2D feature measurements. For sake of efficiency, we therefore use object based registration.

Matching: Object based registration is a matching process between models and images. For video sequences, a reasonable assumption is that the user can locate objects in the first image frame. The matching process in the subsequent frames is then often achieved by using template matching (correlation). Since a single outlier can have a large effect on the resulting pose, special care is often taken to reduce possible false matches. For instance [11] uses geometric invariants to check and select only successfully tracked points. Other methods [6] use a velocity model and a Kalman filter to better predict the position of the image feature. Unfortunately the use of a velocity model imposes regularity constraints on the camera motion; this can be inappropriate for augmented reality applications for which the scene is often shot by a moving observer.

We therefore advocate a less constraining approach. Instead of attempting to refine the matching process, we prefer to use a robust statistical method to compute the pose from the matching induced by the tracking process. Unlike most existing systems, registration is achieved from points, lines or free form curves. Another original aspect of our system lies in its ability to handle occlusions between the real scene and the computer generated objects.

The system is initialized with known camera parameters and a user specified set of four 3D-2D corresponding points pointed out by the user. This allows the initial pose to be computed (with the method of Dementhon and Davis). Then, the 2D features corresponding to the visible model features are automatically determined. Once initialized, the system follows a three step loop:

- **Tracking**: The set of features is tracked in the current image using a curve-based tracker that we have previously developed [2]. Among the set of tracked curves, a small number may be misdetected or completely erroneous (outlier). See for instance the primitives 4 and 5 in Fig. 2.d.
- **Robust temporal registration**: Correspondences are generally maintained during tracking. Unfortunately, even a single tracking error can have a large effect on the resulting pose. For point features, robust approaches allow the point to be categorized as outlier or not [7]. When curved features are considered, the problem is not so simple. We have then devised a robust algorithm capable of extracting the parts of the features that match the 3D model and to compute the pose in a robust manner (see section 3).
- **Resolving occlusion and image composition**: We propose in section 4 a contour based method that allows the occlusions to be solved without 3D reconstruction of the scene (see section 4).

3 Robust Statistical Methods for Temporal Registration

3.1 Robust Estimation

Pose recovery amounts to compute the rotation R and the translation t which map the world coordinate system on the camera coordinate system. $[R, t]$ is represented by 6 parameters $p = [p_1...p_6]$. For 2D/3D point correspondences, a classical way to compute the pose is to minimize the reprojection error

$$min_p \sum r_i^2 = min_p \sum d(m_i, proj(RM_i + t))$$

It is well known that the least square estimation is not robust to noise because the larger the residual r_i is, the larger is its influence on the estimate. To tackle this issue, statisticians have suggested many robust estimators. Among them, the two most popular are the M estimator and the Least Median Square method (LMS) [8]. The LMS method consists in minimizing the median of the squared residuals $\min_\mathbf{p} \mathrm{med}_i r_i^2$. This method is able to handle data sets which contain less than 50% outliers but is not very accurate. But the main drawback is that the minimum has to be searched in the space of possible estimates, that can be very large!

Since the rate of outliers is generally less than 50% for practical applications, we prefer to use M-estimators which can be reduced to a weighted least square problem. The M estimators try to reduce the effect of outliers by minimizing a function of the residuals

$$\min_\mathbf{P} \sum_{i=1}^{n} \rho(r_i), \qquad (1)$$

where ρ is a continuous, symmetric function with minimum value at zero. Such estimators prove to be well suited to cases where the rate of outliers is approximately less than 20%. Table 1 lists three commonly used ρ functions and their derivative. Among these estimators, some are more restrictive than others: when Tukey's influence function is null for residuals larger than a threshold c, Cauchy's influence remains larger than zero while decreasing, whereas Huber's influence remains constant.

Table 1. Three commonly used M-estimators.

Type	$\rho(x)$	$\psi(x)$
Huber $\begin{cases} \text{if } \|x\| \le c \\ \text{if } \|x\| > c \end{cases}$	$\begin{cases} x^2/2 \\ c(\|x\| - c/2) \end{cases}$	$\begin{cases} x \\ c * \mathrm{sgn}(x) \end{cases}$
Cauchy	$\frac{c^2}{2} \log\left(1 + \left(\frac{x}{c}\right)^2\right)$	$\frac{x}{1+\left(\frac{x}{c}\right)^2}$
Tukey $\begin{cases} \text{if } \|x\| \le c \\ \text{if } \|x\| > c \end{cases}$	$\begin{cases} \frac{c^2}{6}\left[1 - \left(1 - \left(\frac{x}{c}\right)^2\right)^3\right] \\ c^2/6 \end{cases}$	$\begin{cases} x\left(1 - \left(\frac{x}{c}\right)^2\right)^2 \\ 0 \end{cases}$

3.2 A Robust Two Stage Statistical Method for Pose Computation

In our system, features of various types are considered: point, lines and curves. Defining outliers for curved features is not so simple, as some parts of the 2D curves can perfectly match the 3D model whereas other parts can be erroneously matched. Let us define

- C_i be a 3D curve, described by the chain of 3D points $\{M_{i,j}\}_{1 \le j \le l_i}$
- c_i be the projection of C_i in the image plane, described by the chain of 2D points $\{m_{i,j}\}_{1 \le j \le l_i}$, where $m_{i,j} = Proj(\mathbf{R}M_{i,j} + \mathbf{t})$

- c'_i be the detected curve (tracked curve) corresponding to C_i, described by the chain of 2D points $\{m'_{i,j}\}_{1 \leq j \leq l'_i}$.

A simple solution would be to perform a one stage minimization

$$\min \sum_{i,j} \rho(d_{i,j}) \qquad (2)$$

where $d_{i,j} = Dist(m'_{i,j}, c_i)$ is the distance between $m'_{i,j}$ and the curve c_i.

Unfortunately, this method is unsatisfactory because it merges all the features into a set of points, and makes no distinction between local errors (when a feature is only partially well localized), and gross errors (when the position of a feature is completely erroneous). However, these two kinds of errors are not identical, and not treating them separately induces a great loss of robustness and accuracy.

By contrast, we propose to perform a robust estimation in a two-stage process: a *local stage*, which computes a robust residual for each feature, and a *global stage* which minimizes a robust function of these residuals. The local stage reduces the influence of erroneous sections of the contours (features 1 and 4 on Figure 2.d), whereas the global stage discards the *feature outliers*, *i.e.* contours which are completely erroneous, or which contain too large a portion of erroneous points (feature 5 on Figure 2.d).

The local stage
The aim of this stage is to reduce the influence of erroneous sections of the features: to perform this task, the residual error r_i of curve C_i is computed by a robust function of the distances $\{d_{i,j}\}_{1 \leq j \leq l'_i}$. We use the M-estimation technique by taking $r_i^2 = \frac{1}{l'_i} \sum_{j=1}^{l'_i} \rho(d_{i,j})$. Since this estimate must not be too restrictive, we have hence chosen Huber's function for the local stage, which has proved to be a good choice in our experiments [9].

The global stage
This stage fits the tracked 2D features to the projection of the 3D features, by minimizing the residuals r_i which are computed for each couple of 3D/2D features. We use Tukey's function, which is restrictive enough to suppress the influence of outliers, but which takes all the data into consideration.

3.3 Results

We present in this section an application of our method to an augmented reality application: the illumination of the bridges of Paris [3]. The aim was to test several candidate illumination projects for a number of bridges of the Seine. We therefore want to replace the bridge in the sequence with its lighting simulation. A 300-image panoramic sequence of the Pont Neuf was shot at dusk time from another bridge. Because the images are dark and noisy, only 6-8 curves can be tracked in each frame (Figure 2.b). The solid lines correspond to the tracked 2D features, whereas the dashed lines correspond to the projection of the corresponding model features (black is used for the features which are not - yet - used). The result of the tracking in the 12th image is shown in Figure 2.c. The reader may notice that the tracking process fails for feature 5. Figure 2.d shows the re-projection of the model features after the robust pose computation. Despite the bad accuracy of the model, the result is visually convincing. In order the reader to be aware of the parts of the curve which are less taken into account in the computation, we have drawn in black the points for which the residual is greater than c (c is defined in Table 1. Roughly speaking, these points are the ones for which the weight in the computation is decreased because their residual is too large. It must also be noticed that feature 5 is considered as an outlier and is removed from the set of tracked features (discarded features are drawn in black).

Since new features may appear while old ones disappear, the set of model features that are tracked in the sequence must be dynamically updated. The method we use to achieve this task is described in [9].

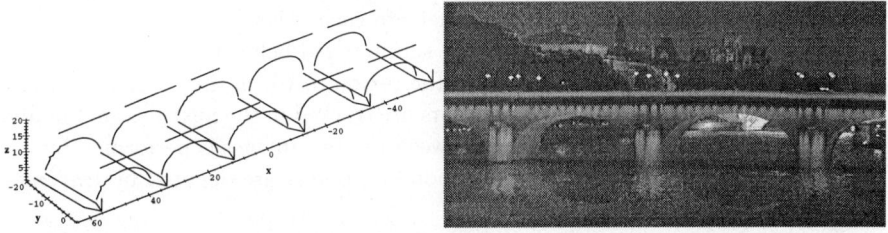

Fig. 1. (a)The complete wire-frame model of the bridge; (b)Final composition.

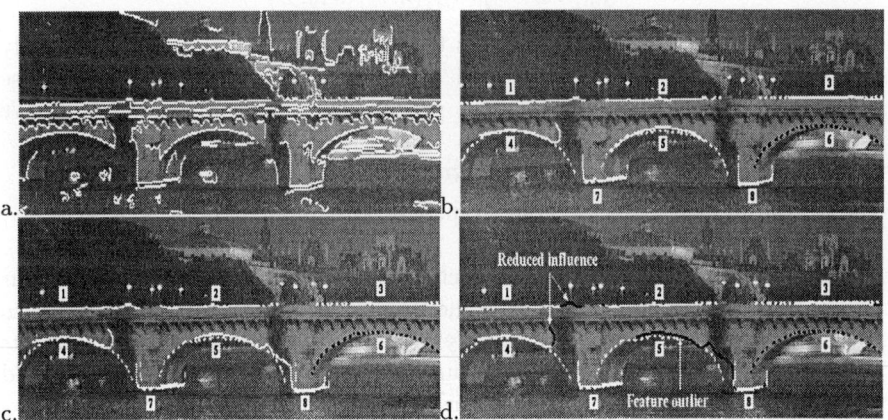

Fig. 2. Temporal registration for the 12th image. (a) Edge map. (b) 2D features before tracking (image 10). (c) Tracking in image 12 (projections of the 3D features are those of image 10). (d) Robust pose computation.

4 Resolving Occlusions

Most of the time, augmented reality systems simply overlay computer generated images and only attempt to minimize object registration errors. However, such methods are effective only when there are no occlusions between the real and the computer generated objects.If the model of the 3D scene is known, as in [4], the problem can easily be solved. Otherwise, resolving occlusion could theoretically be achieved by inferring a dense map from two consecutive images. Unfortunately, despite new advances in 3D reconstruction, the depth map lacks accuracy and cannot be used as is. Instead of performing 3D reconstruction, we propose to use a contour based approach. Our aim is to find, among the contours in the scene, those belonging to the boundary of the occlusion mask. Our approach stems from the fact that for a real scene containing only rigid objects, the boundary of the mask is only composed of contours present in the image and of occluding contours of the virtual object. This is obviously not true anymore if the objects are deformable or can penetrate each other [1].

The main steps of our algorithm are summarized below [1].

The contour chains are extracted in the region to which the computer generated object V corresponds. These contours are tracked in the next image. Finally, the matching of the contours points between the two images is performed by using the epipolar constraint. Two corresponding points are denoted by (m_1, m_2) in the sequel.

The heart of our system is the labeling stage which allows each contour point m_1 to be labeled with *in front of* or *behind* depending on the relative position of the corresponding point of the scene and of the computer generated object. To this aim, let us define f_{m_1} (Fig. 3):

$$f_{m_1} : Z \to proj_{I_2}(m1_x, m1_y, Z)$$

where $m_1 = (m1_x, m1_y)$ and $proj_{I_2}$ is the projection in image I_2. It can easily be seen that $f_{m_1}(Z)$ can be expressed as an homographic function of Z whose coefficients depend on the calibration process and on the image point m_1. We have

$$m_2 = f_{m_1}(Z_{real}) \quad and \quad m_{obj} = f_{m_1}(Z_{obj})$$

Due to the monotony of the homography, it is therefore easy to compare Z_{real} and Z_{obj}.

The last step is to recover the occluding mask from the set $\mathcal{H}$ of contour points labeled *in front of*. Because some labeling errors may occur, we resort to a regularization approach. The underlying idea is to add regularity constraints which will produce the most regular curve resting on $\mathcal{H}$. Starting from a closed curve outside $\mathcal{H}$, we use active contour models to obtain such a result. An example is shown in Fig 3.b.

5 Conclusion

This paper has presented an image composition system capable of ensuring temporal registration in a robust and autonomous way. Significant results on video

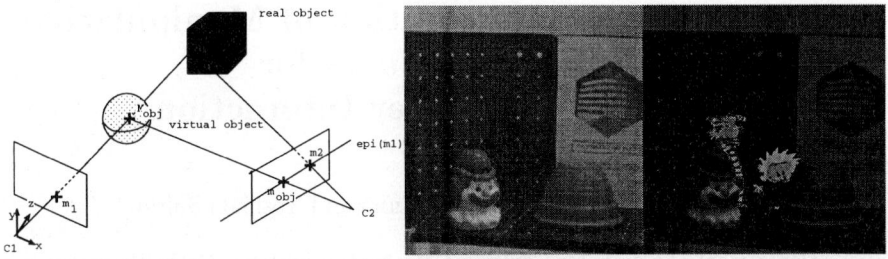

Fig. 3. (a)The relative positions of the real and the virtual objects; (b-c) An example of image composition.

image sequences can be seen at URL http://www.loria.fr/isa. For architectural applications, the size of the database which describes the objects is sometimes huge. Thus, further investigations concern the way to infer the more pertinent 3D features to be tracked.

References

1. M.-O. Berger. Resolving occlusion in augmented reality : a contour based approach without 3D reconstruction. In *CVPR 97, Puerto Rico (USA)*, pages 91–96, 1997.
2. M.-O. Berger. How to Track Efficiently Piecewise Curved Contours with a View to Reconstructing 3D Objects. In *ICPR 94, Jerusalem (Israel)*, volume 1, pages 32–36, 1994.
3. M.-O. Berger, C. Chevrier, and G. Simon. Compositing Computer and Video Image Sequences: Robust Algorithms for the Reconstruction of the Camera Parameters. In *Eurographics'96, Poitiers, France*, volume 15, pages 23–32, August 1996.
4. D. Breen, R. Whitaker, E. Rose, and M. Tuceryan. Interactive Occlusion and Automatic Object Placement for Augmented Reality. In *Eurographics'96, Poitiers, France*, 1996.
5. G. Ertl, H. Müller-Seelich, and B. Tabatabai. MOVE-X: A System for Combining Video Films and Computer Animation. In *Eurographics*, pages 305–313, 1991.
6. D. Koller, K. Daniilidis, and H. H. Nagel. Model-Based Object Tracking in Traffic Scenes. In *ECCV 92, Santa Margherita Ligure (Italy)*, pages 437–452, 1992.
7. S. Ravela, B. Draper, J. Lim, and R. Weiss. Tracking Object Motion Across Aspect Changes for Augmented Reality . In *ARPA Image Understanding Worshop, Palm Spring (USA)*, August 1996.
8. P. Rousseeuw and A. Leroy. *Robust Regression and Outlier Detection*. Wiley Series in Probability and Mathematical Statistics. Wiley, 1987.
9. G. Simon and M.-O. Berger. A two-stage robust statistical method for temporal registration from features of various type. In *ICCV 98, Bombay (India)*, 1998.
10. C. Tomasi and T. Kanade. Shape and Motion from Image Streams under Orthography: A Factorization Method. *IJCV*, 9(2):137–154, 1992.
11. M. Uenohara and T. Kanade. Vision based object registration for real time image overlay. *Journal of Computers in Biology and Medecine*, 1996.

Context-Based Recognition of Manipulative Hand Gestures for Human Computer Interaction

Kang-Hyun Jo[*], Yoshinori Kuno and Yoshiaki Shirai

Department of Computer-Controlled Mechanical Systems, Osaka University
Email: jkh@seri.re.kr

Abstract. This paper presents a system recognizing manipulative hand gestures like grasping, moving, holding an object(s) with both hands, and extending or shortening of the object(s) in the virtual world using contextual information. Contextual information is represented by a state transition diagram, each state of which indicates possible gestures at the next moment. Image features obtained from extracted hand regions are used to judge state transition. When we use a gesture recognition system, we sometimes move our hands unintentionally. To solve this problem, our system has a rest state in the state transition diagram. All unintentional actions are considered as taking a rest and ignored. In addition, the system can recognize collaborative gestures with both hands. They are expressed in a single state so that the complexity in combination of gestures of each hand can be avoided. We have realized an experimental human interface system. Operational experiments show promising results.

1 Introduction

Vision-based interfaces with which we can give orders to computers by hand gestures have attracted much interest [1], [2] [3] [4]. However, most conventional systems can deal only with pointing gestures [5], [6], [7], [8]. We can point at an object in a 3-D virtual world and move it by hand gestures with these systems. However, if we want to handle object parts more freely in the virtual world to design an object by combining these parts, we need various other manipulative operations, such as grasping, moving, holding an object with both hands, extending, shortening, and putting down on the desk. An interface system should recognize manipulative gestures corresponding to these operations. However, some of these gestures are similar if we look at each separately, thus difficult of recognition. We propose a use of contextual information to solve this problem. We use the virtual reality system for a certain purpose. Thus, we can reduce the number of possible next operations following a particular operation to a small number. We do not take an operation that does not lead the status closer to the goal. We represent this knowledge by a state transition diagram.

---

[*] He is currently with System Engineering Research Institute, Taejon, Korea.

Human manipulative gestures are often conducted by the two hands. Thus, vision-based interfaces should have the capability of recognizing collaborative gestures with both hands. In this system, each collaborative gesture is expressed in a single state in the state transition diagram so that the complexity in combination of gestures of each hand can be avoided.

Conventional human interface systems consider only meaningful gestures. However, we sometimes move our hands unintentionally. Human interface systems must discern such human's unintentional actions from intentional manipulative gestures and respond only to the latter. Our human interface system has a rest state in the state transition diagram. If the system considers that the current human action might be unintentional because the extracted features are out of expected ranges, the system takes a rest in the rest state until it can make a more certain decision.

This paper presents our context-based manipulative hand gesture recognition method. It also describes an experimental human interface system using the proposed method.

2 Context-based approach to recognize human gestures

In this paper, we present a human interface system that enables a human user to manipulate objects in the virtual 3D world. In the virtual space, he may point at an object, grasp it, move it, bring it into the work space, and may change its size or put it down on another object. These actions will not happen independently. We can choose possible actions (hand gestures) following each action. We represent these relationships by a state transition diagram. Since this diagram limits the number of possible recognition classes, the system can recognize the next gesture using simple image features.

2.1 Feature extraction

In order to obtain hand features, hand regions need to be extracted. Each hand region can be extracted by thresholding in hue information of color images. In the current implementation, we have to wear a green glove and a red one on each hand to realize fast reliable feature extraction. Fig.1 shows an example of hand region extraction.

The feature vector, $O(t)$, is calculated from this segmented hand regions. The hand feature vector consists of the following elements (see Fig. 2):

1. $area$: the region area,
2. cx : x coordinate of the centroid of the region,
3. cy : y coordinate of the centroid,
4. $fmax$: the finger tip length from the centroid, which is the maximum of $f(\theta)$ indicating the distance between the centroid and the hand region contour in direction θ,
5. fx : x coordinate of farthest point from the centroid,

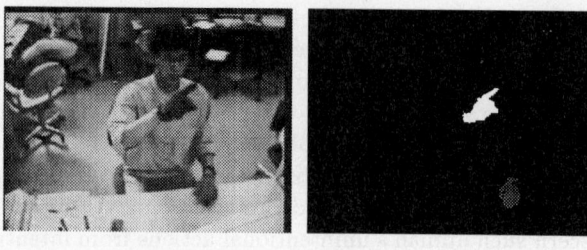

(a) Original image. (b) Extracted hand regions.

Fig. 1. Example of hand region extraction.

6. fy : y coordinate of farthest point,
7. $fmean$: the mean of $f(\theta)$ for all θ.

We use two cameras placed upper front of the user and above. The feature vectors are calculated for each hand in each image.

Except certain pointing gestures, we cannot recognize gestures from a frame of an instance. Thus, the system observes gestures for an appropriate length of interval determined for each state.

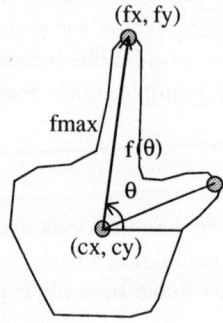

Fig. 2. Computed features.

2.2 Gesture recognition

Fig. 3 shows the state transition diagram used in the system. Table 1 describes the conditions for state transition. Except pointing gestures such as Direct and

Move, the system uses feature vectors for several frames to recognize gestures. To cope with the variation of motion speed, the system examines in most cases whether particular feature values are increasing or decreasing as shown in Fig.4 rather than checks their exact values.

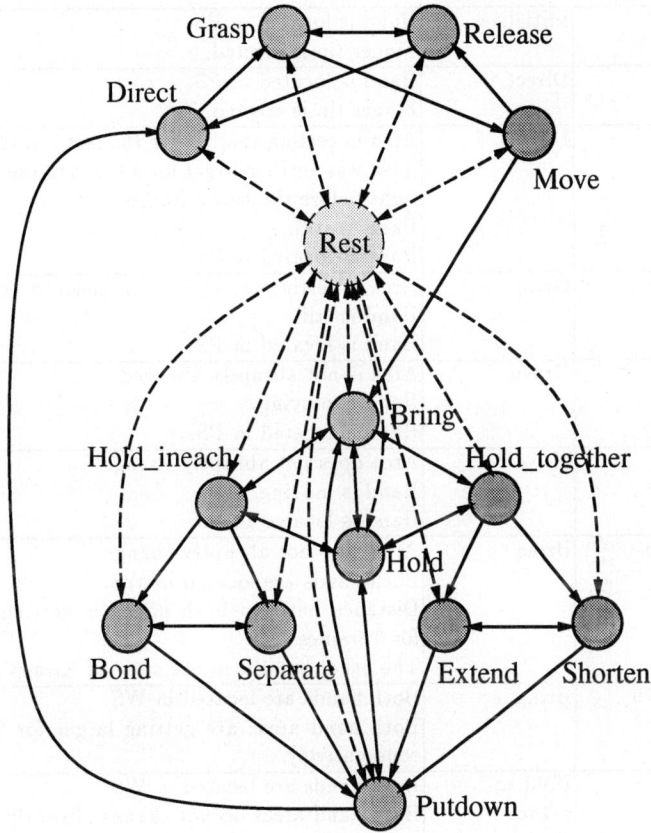

Fig. 3. State transition diagram(self loops are omitted).

In Fig. 3, transition from and to 'Rest' is depicted by dotted lines. If a new observation does not match any possible state transition conditions from the current state, the system changes it to the Rest state. While in the Rest state, if the system observes a feature vector that satisfies one of the possible transition conditions from the previous state, the state is changed according to the condition.

Table 1. Conditions for state transition

PS : Palette space where object components are located.
WS : Work space where target objects are manipulated.
The finger tip is erected : fmax > fmean + threshold.
Area is getting larger: area > prev_area + thresh_area, where prev_area is the area in the previous frame and thresh_area is a threshold with respect to area.

New state	Previous state	Conditions
Direct	Initialize	Hand is located in PS. Finger tip is erected.
Direct	Direct	Hand is located in PS. Finger tip is erected.
Grasp	Direct	Area is getting smaller for the last 2 frames. Area was getting larger for 3 times in the 8 frames right before the last 2 frames. Hand is still. Hand is located in PS.
Release	Grasp	Area is getting larger for 3 frames consecutively. Hand is still. Hand is located in PS.
Move	Grasp	Area is not abruptly changed. Hand is moving. Hand is located in PS.
Bring	Move	Area does not abruptly change. Hand is moving. Hand is located in PS.
Hold_together	Bring	Area does not abruptly change. Both hands are located in WS. Distance between both hands is getting smaller for 3 frames. The other hand is in the state of 'Grasp'.
Putdown	Bring	Both hands are located in WS. Both hand areas are getting larger for 3 frames consecutively.
Extend	Hold_together	Both hands are located in WS. Both hand areas do not change abruptly. Distance between both hands is getting larger.

2.3 Collaborative gestures with both hands

Human manipulative gestures are often conducted by the two hands. Thus, vision-based interfaces should have the capability of recognizing collaborative gestures with both hands. In this system, we allot a state for each manipulative gesture with both hands. This solves the combinatorial problem that arises when a state is given for each hand and manipulative gestures are represented by the

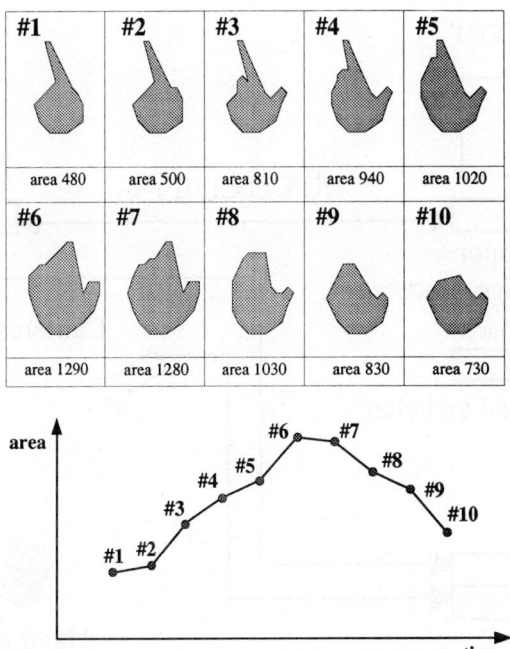

Fig. 4. Example of recognition of 'Grasp'.

combination of both hand states. As shown in Fig. 3, Hold_together is assigned to the manipulative gesture for holding an object with both hands. If this has been recognized, the object will be extended, shortened or released after then.

3 System configuration and operation experiments

We have developed an experimental human interface system using the proposed recognition method. Fig. 5 shows the system configuration. The current system uses two Unix machines(Sun's SS5 and SS20), each for feature extraction and graphics rendering, respectively. The latter displays two views for the user as shown in Fig. 6. One is a usual perspective view of the virtual world. The other is a top view of the virtual world to help the user to understand positional relations of objects.

We consider the following scenario for operational experiments. There is a palette space (PS) in the virtual world where object components are floating. We Direct (point at) a desired object in PS and Grasp it by hand gestures. Then, we Move it and Bring it to the work space (WS), which is a space on the desk in the virtual world. We hold the object by both hands (Hold_together) and Extend it or Shorten it to change the object size to desired one. Then, we bring another

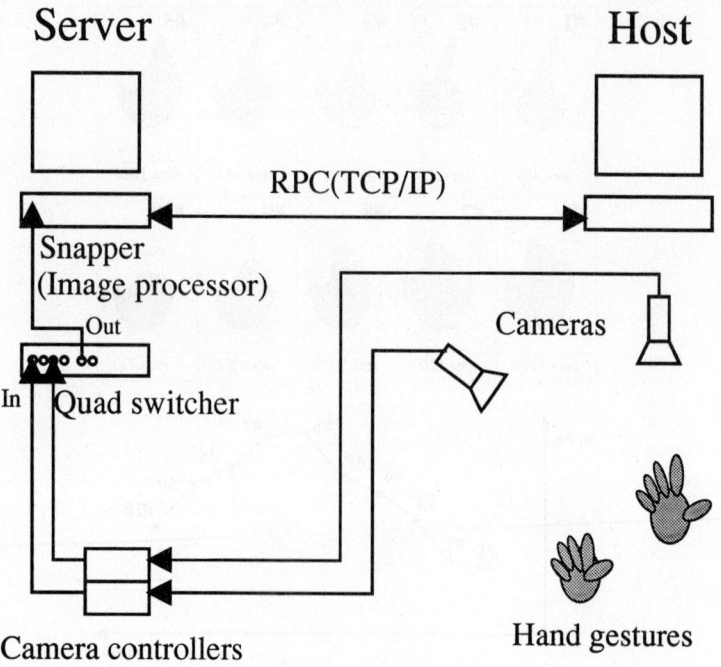

Fig. 5. System configuration.

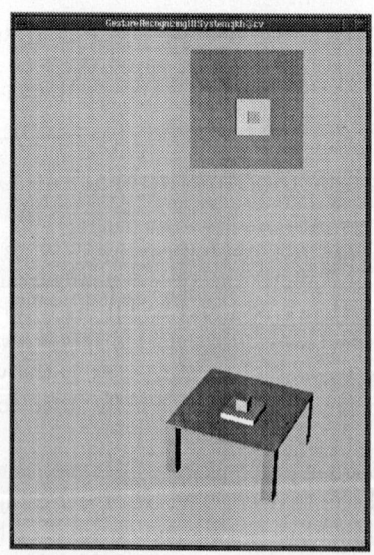

Fig. 6. Display example.

object component from PS and pu it down on the first object (Putdown).

We carried out operational experiments according to the above scenario. We were able to build such an object as shown in Fig. 6 by hand gestures even if we added any unintentional movements of the hands during operation.

4 Conclusion

We have proposed a method of recognizing various manipulative gestures using contextual information. The method can ignore unintentional movements of the hands to recognize only meaningful hand gestures. It can also recognize collaborative gestures with both hands. We have developed a human interface system using the proposed method. Operational experiments show promising results.

This work was supported in part by the Ministry of Education, Science, Sports and Culture under the Grand-in-Aid for Scientific Research (09555080, 09650471), by the Kurata Foundation under the Kurata Research Grant, and Kayamori Foundation of Informational Science Advancement.

References

1. T.S. Huang and V.I. Pavlović, "Hand gesture modeling, analysis, and synthesis", Pooc. Int. Workshop on Automatic Face- and Gesture-Recognition, pp.73-79, 1995.
2. P. Maes, T. Darrell, B. Blumberg, and A. Pentland, "The ALIVE system: wireless, full-body interaction with autonomous agents", MIT Media Lab. Perceptual Computing Technical Report No. 257, 1995.
3. R. Kjeldsen and J. Kender, "Visual hand gesture recognition for window system control", Pooc. Int. Workshop on Automatic Face- and Gesture-Recognition, pp.184-188, 1995.
4. Francis K.H. Quek, "Unencumbered gestural interaction", IEEE Multimedia, Vol.4, No 3, pp.36-47, 1996.
5. M. Fukumoto, K. Mase, and Y. Suenaga, "Real-time detection of pointing actions for a glove-free interface", Proc. IAPR Workshop on Machine Vision Applications '92, pp.473-476, 1992.
6. R. Cipolla, P.A. Hadfield, and N.J. Hollinghurst, "Uncalibrated stereo vision with pointing for a man-machine interface", Proc. IAPR Workshop on Machine Vision Applications '94, pp.163-166, 1994.
7. K.-H. Jo, Y. Kuno, and Y. Shirai, "Invariance based human interface system using realtime tracker", Proc. Second Asian Conf. on Computer Vision '95, Vol.2, pp.22-27, 1995.
8. K.-H. Jo, K. Hayashi, Y. Kuno and Y. Shirai, "Vision-based human interface system with world-fixed and human-centered frames using multiple view invariance", Trans. IEICE Information and Systems, Vol.E79-D, No.6, pp.219-228, 1996.

An Algorithm for Recursive Structure and Motion Recovery under Affine Projection

Miroslav Trajković and Mark Hedley
Department of Electrical Engineering, Sydney University, NSW 2006, Australia
E-mail: {miroslav, hedley}@ee.usyd.edu.au

Abstract. In this paper we present an algorithm for structure and motion (SM) recovery under affine projection from video sequences that is suitable for real time applications. The algorithm tracks the motion of a single structure, be it an object or the entire scene itself, allowing for any type of camera motion. This could be used for example to track the motion of a vehicle in a warehouse (single object, static camera) or for visual navigation from a moving platform (track scene from moving camera). The algorithm requires a set of features to be detected in each frame, and that at least four features are correctly matched between each three consecutive frames. Compared to previous algorithms, this novel algorithm has a lower computational cost, dynamically detects outliers and allows for previously lost features to reappear in the sequence. The algorithm has been tested on real image sequences, and compared to other algorithms we have found that our algorithm has both a smaller error and a lower computational time.

1. Introduction

In any machine vision task the objects of most interest, and the most demanding to handle, are those that are moving. The reason for this is that once we have information about static objects, by virtue of their being static, we do not need to update information about them at the video rate. Moving objects on the other hand should be tracked frame by frame, and these are the objects that the machine vision system will usually need to respond to (for example collision avoidance). A fundamental machine vision task then is to identify and track moving objects in a video sequence, and to determine their structure, which is assumed to be rigid. In this paper we will approach the problem of determining the motion and structure of a single rigid object moving in the field of view of a static camera. The same algorithm can be used if there is a moving camera in a static environment to determine the motion of the camera and the structure of the environment.

One of the most widely used approaches to structure and motion (SM) recovery is the factorization algorithm proposed by Tomasi and Kanade [8]. Image feature points are tracked over several frames and the SM parameters are computed using a singular value decomposition (SVD). This algorithm cannot be used for real time applications as this is a batch method and can only compute the parameters once all measurements have been made. The algorithm is also computationally very expensive (due to the use of the SVD). Morita and Kanade [5] developed a sequential version that has almost the same accuracy but has a much lower computational burden. The issues not addressed in these papers were how to prevent outliers incorrectly biasing the result, and how to handle a set of feature points that changes dynamically.

Held [1] partially extended the Morita and Kanade algorithm to include new features. The video sequence was divided into subsequence, and the features tracked over the subsequence. To track between the subsequences there needed to be common tracked features between consecutive subsequences. The algorithm was verified using synthetic measurements only, with added Gaussian noise. McLauchlan et al. [2,3] used the variable state-dimension filter (VSDF) to handle the problem of missing and new features. They posed the structure from motion problem as a parameter estimation problem and solved it by using the Extended Kalman Filter (EKF). They updated the structure and *only the last* motion estimate at each time step and therefore achieved a low computational cost $O(k)$ compared to when the complete motion matrix has been updated. The authors mentioned outlier rejection, but they gave neither a method nor a reference to handle this problem. Another limitation of their algorithm is that features that have been obscured and reappear are treated as new feature points, and are not recognised as a previously tracked feature.

In this paper we present an algorithm that iteratively computes the SM parameters at each time step. It has the same computational cost $O(n+k)$ as [2] but it will update both the structure and complete motion parameters, so it converges to the true solution. It is made possible by noting that structure should remain constant over time (rigid body assumption) and by updating it at each time step. Motivated by work of Reid and Murray [6] we have extended their algorithm so that it can easily include new features and recover the complete structure and motion providing that for each three consecutive frames there are at least four correctly matched feature points. Furthermore, our algorithm detects outliers at each time step. Finally we developed a procedure that checks whether newly appeared features are genuinely new, or old features that have reappeared. The correct classification of these new features both decreases the computational cost of the algorithm and improves its accuracy.

2. Background

In this paper we assume an affine camera model which has the form:

$$p = MS + t \qquad (1)$$

where S is 3×1 is the world coordinate of a feature point, p is 2×1 image projection of S, M is 2×3 projection matrix and t is 2×1 translation vector. This model is a generalisation of the orthographic camera model [5,8] and is a good approximation of the perspective projection when the change in depth is small compared to the average distance from the optical center. This condition is almost always satisfied for independently moving objects. If the object of interest undergoes rigid motion the image projections will not change if we fix the world coordinates of the object of interest and change camera parameters M and t accordingly. Hence, without loss of generality, we may assume that scene is static and that the camera is moving.

Given k projections (images) of n scene points we want to recover the camera motion parameters $M(j)$, $t(j)$, $j = 1,2,\cdots,k$ and scene structure parameters $S_i, i = 1,2,\cdots,n$ (world vectors of tracked scene points). These parameters are related by the measurement equation:

$$p_i(j) = M(j)S_i + t(j) \qquad (2)$$

where $p_i(j)$ is the projection of ith point onto the jth image, and i and j vary from 1...n and 1...k respectively. To compute the the SM parameters using the measurement equations, we must minimise the cost function:

$$C(k) = \sum_{j=1}^{k}\sum_{i=1}^{n} \|p_i(j) - M(j)S_i - t(j)\|^2 \qquad (3)$$

For real time implementation the parameters must be computed following the acquisition of each frame. It may be shown that the translational component $t(j)$ at each frame is given by the centroid of the feature locations $t(j) = \overline{p_i(j)}$ and equation (2) can now be rewritten as $w_i(j) = M(j)S_i$ where $w_i(j) = p_i(j) - t(j)$, or in matrix form:

$$W = MS \qquad (4)$$

where $W = [w_i(j)]_{k \times n}$, $M = [M(1)^T \cdots M(k)^T]^T$ and $S = [S_1 \cdots S_n]$, and this is the equation obtained (although in different way) by Tomasi and Kanade [8]. To solve it, they employed singular value decomposition of measurement matrix and showed that $M = U_3 \Sigma_3^{1/2}$ and $S = \Sigma_3^{1/2} V_3^T$, where U_3, Σ_3 and V_3 are submatrices of U, Σ and V corresponding to the three largest singular values of W. However the structure and motion parameters can be determined without computing the full SVD, since

$$W^T W = V \Sigma^2 V^T \text{ and } WW^T = U \Sigma^2 U^T \qquad (5)$$

hence the structure and motion parameters are given by the eigenvector decomposition of $W^T W$ and WW^T respectively.

Note that equation (4) does not have unique solution. In fact, if A is an arbitrary invertible 3×3 matrix the matrices MA and $A^{-1}S$ are valid solutions [8]. To ensure a unique solution, the first two rows of M are fixed, and now in equation (3) j will range from 2...k [3].

3. Algorithm Details

3.1 Structure and Motion Recovery

Unlike Morita and Kanade who have computed motion and structure by iteratively updating equation (4) and McLauchlan who applied EKF to the measurement equation (2), we employ direct minimisation of the cost function (3). For some initial number of frames (usually 3) we initialise the motion and structure matrices by solving equation (4). Once we have the matched features at time $k+1$, and have computed values M, t and S at time k, we wish to find the updated parameters that minimise

$$C(k+1) = C = \sum_{j=2}^{k+1}\sum_{i=1}^{n} w_{ij} \|p_i(j) - M(j)S_i - t(j)\|^2 = \sum_{j=2}^{k+1}\sum_{i=1}^{n} w_{ij} \|v_i(j)\|^2 \qquad (6)$$

where w_{ij} is a binary weight which is unity if feature i is present in frame j, and zero otherwise.

It is convenient to group the motion parameters M and t into a single vector defined as $m = [M_{11} M_{12} M_{13} t_1 M_{21} M_{22} M_{23} t_2]^T$. To determine the structure and motion parameters that minimise the cost function we must solve:

$$\frac{\partial C}{\partial m(j)} = \sum_{i=1}^{n} w_{ij} D_i^T v_i(j) = 0, \ j = 2, 3, \ldots, k+1$$

$$\frac{\partial C}{\partial S_i} = \sum_{j=2}^{k+1} w_{ij} E^T(j) v_i(j) = 0, \ i = 1, 2, \ldots, n$$
(7)

where $D_i = -\partial v_i/\partial m$ is a simple function of S_i, and $E = -\partial v_i/\partial S_i = M$. For simplicity the time parameter j has been dropped. Equation (7) forms a non-linear set of equations to be solved. There are three possible approaches to doing this:

1. Use the previously determined values of M and S_i to calculate D_i and E, then linearise the equations using

$$v_i(j) = \hat{v}_{ij} + \frac{\partial v_i}{\partial m} \Delta m(j) + \frac{\partial v_i}{\partial S_i} \Delta S_i = \hat{v}_{ij} - D_i \Delta m(j) - E^T(j) \Delta S_i$$
(8)

From this we can rewrite the system as a linear set of equations to be solved for $\Delta m(j)$ and ΔS_i. This is the set of equations obtained in [3], and the algorithm has linear convergence.

2. Linearise all terms in equation (7), rather than assuming that the derivatives are constant. From this a different set of linear equations is determined that are similar to those in the previous technique, but this has quadratic convergence.

3. The simplest method is to solve for $m(j)$ using the previous values for the derivatives and S_i. Then use the updated parameters to solve for S_i. This is alternated until convergence. This method has linear convergence, but the lowest computational cost.

We used the second method because it has quadratic converge, and usually found three to five iterations sufficient. For the initial estimates of the parameters at each stage of the algorithm we used the parameters determined in the previous stage.

3.2 Matching and Outlier Rejection

Before we can recover the structure and motion parameters we need to have matched feature points (here corners) between consecutive images. This usually involves matching two sets of corners using a cross correlation of a window about each corner [9]. This procedure will often give some small number of incorrect matches (outliers), which can significantly bias the structure and motion estimates. These outliers should be detected so they can be ignored. For this we developed a technique based on robust statistics, using the Least Median of Squares (LMedS) estimator [4], which is described in this section.

Matching between two images.

Having matched the corners between two consecutive images, we randomly select four matched corners to obtain a sample matrix $W_S = [\mathbf{x}_1 \quad \mathbf{y}_1 \quad \mathbf{x}_2 \quad \mathbf{y}_2]^T$ where $\mathbf{x}_i$ is a vector containing the x coordinates of the four selected corners in image i, and $\mathbf{y}_i$ is likewise defined. We can decompose this 4×4 matrix using the SVD:

$$W_S = M_S S_S + t_S.$$

where M_S and t_S can be found using equation (5), then we find the structure vector S_R for the remaining set of matched corner points C_R by solving the equation

$$W_R = M_S S_R + t_S,$$

where W_R is the measurement matrix corresponding to C_R. The position error from the model (M_S, t_S) is given by

$$D = W_R - (M_S S_R + t_S)$$

The standard variation of the error vector $\varepsilon = [|d_i|]$ is computed using the equation [4]

$$\sigma^0 = 1.4826\left(1 + \frac{5}{n-p}\right)\operatorname*{median}_{i}(\varepsilon_i)$$

where n is number of points in C_R and p is the number of sample points (4 in our case). The number of samples of four sets of matched corners which have to be taken to assure that at least one consists only of inliers with probability higher than γ is given by

$$N_S = \frac{\log(1-\gamma)}{\log(1-(1-\pi)^p)}$$

where π is expected percentage of outliers in the sample. We find the sample with the lowest standard variance, and use the corresponding vector ε.

The next step is outlier rejection. For each point we determine the initial weight w_i.

$$w_i = \begin{cases} 1 & \text{if } \varepsilon_i \leq 2.5\sigma^0 \\ 0 & \text{if } \varepsilon_i > 2.5\sigma^0 \end{cases}$$

Then we compute the *robust* standard deviation estimate as

$$\sigma = \sqrt{\sum_i w_i \varepsilon_i^2 \Big/ \left(\sum_i w_i - p\right)}$$

Finally, outliers are found as points whose residual error is outside the confidence interval 2.5σ.

Matching between three images

By matching feature points between three images we can make use of the rigidity constraint. For each set of three frames, the corners are matched between the consecutive frames and the outliers rejected. Then, given the position of a feature in any two images, and the motion parameters M and t, its position in the third image is uniquely determined. Hence, by using this procedure, we can detect those outliers which are "correctly" matched over each pair of consecutive images (satisfy the epipolar constraint) but do not satisfy the rigidity constraint over three frames.

3.3   New Feature Points

New feature points will be included in the measurement matrix if they appear and are matched in three consecutive frames. Consider frames n-2, n-1, n and n+1, and let C_1 denote a set of features which are matched over frames n-2 to n, while C_2 denotes the set of features which have been matched *only* over frames n-1 to n+1. Let P be an arbitrary feature from C_2 and let $p^{(n)}$ denote its image projection in the nth frame. Since the motion (and structure) parameters are estimated over the first n frames, the structure parameter S of this point is found by solving the set of equations,

$$\begin{bmatrix} p^{(n-1)} \\ p^{(n)} \end{bmatrix} = \begin{bmatrix} M^{(n-1)} \\ M^{(n)} \end{bmatrix} S + \begin{bmatrix} t^{(n-1)} \\ t^{(n)} \end{bmatrix}.$$

Hence the initial parameters for the algorithm in Section 3.1 have been found.

Before we add the point from C_2 to the measurement matrix, and perform an update, we first check whether this point is genuinely new or whether it is an old feature that has reappeared. Let P_i be a previous feature that had disappeared and P_l be an arbitrary feature from C_2. We can say that P_i and P_l represent the same feature if the world distance between them is small enough, *i.e.* if

$$d_{il} = \|\Delta S\| = \|S_i - S_l\| < \alpha \tag{9}$$

where α is an unknown threshold to be determined.

While we do not know what is *small* in the structure space (especially because it is only unique up to an affine transformation), we can say what is small in the image space (typically three to five pixels). If the features are close in structure space then the distance between their projections in the image plane has to be small but the reverse need not be true. We use the criteria that the two points are the same one if the distance between their image projections is less than threshold β for all camera locations. Mathematically, this condition can be expressed as

$$\max_{\Delta S, j} \left(\|p_i(j) - p_l(j)\| \right) \leq \beta . \tag{10}$$

The difference between image projections is given by $\Delta p = p_i - p_l = M \Delta S$ where the index j has been dropped for clarity. By applying the Euclidean norm to both sides we obtain $\|\Delta p\|_E \leq \|M\|_S \|\Delta S\|_E$ where the index S denotes the spectral norm, which is defined as a square root of the maximum eigenvalue of the matrix $M^T M$, i.e. $\|M\|_S = \sqrt{\max \lambda(M^T M)}$. Since the spectral norm is subordinate to Euclidean norm there exists at least one ΔS such that equality holds, and from equation (10) we get $\|\Delta p\|_E = \|M\|_S \|\Delta S\|_E \leq \beta$. If $\lambda_j = \|M(j)\|_S$, then by combining (9) and (10) the threshold $\alpha = \beta / \lambda$ where λ is the largest value of λ_j.

4. Experimental Results

4.1 POV image sequence

This artificially generated sequence consists of 90 frames of an object rotating around the fixed axes (see Fig. 1 a–c). The rotation is a constant four degrees per frame. The corners are detected and matched using an algorithm similar to [9]. This is a difficult sequence, because no single corner appears in all the frames, and because the number of points is small – it varies from 6 to 13 per frame. For this reason, both the [8] and [5] are not applicable to this sequence as the number of features is not constant. The VSDF algorithm may be used, but in its original form (without the reappearance of features) this will have a growing number of features as old features reappear, this will lead to slower computation and greater errors in the calculated parameters.

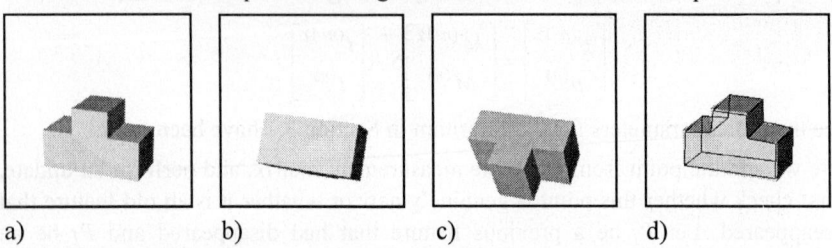

a)   b)   c)   d)

Fig. 1. POV sequence: 1st (a), 30th (b) and 60th (c) image in the sequence of 90 images. d) consistent features and recovered structure (the lines are shown for visualisation only).

Using our algorithm we obtained a total of 22 features, and by disregarding all of them which appeared in less than 20 frames we were left with total of 17 features which is quite close to the 16 corners on the object. Corners that have been effectively tracked over all the frames are shown on Fig 1d. As we can see they are all close to the true locations, including those which are currently occluded or not detected. Some distortion is present due to imperfection of the affine model, but the structure has been correctly recovered.

4.2 Hotel Sequence

This sequence consists of 197 points tracked over 181 frames. Corners were detected in the first image and their location in consequent images were determined using an optical flow technique [7] (see Fig. 2 a,b).The SM parameters were calculated using the SVD algorithm and our algorithm, which produced the same results (without outlier rejection), but the former took about an order of magnitude longer to compute. As in [3], we have found that updating motion parameters over all previous frames in each step is unnecessary. A very important step here is the outlier detection. Since we have matches over all frames, the rigidity constraint was checked not over last three frames only, but over the whole sequence. The comparison of mean squared error with, and without outlier removal is given in Fig. 2c.

5. Conclusion

This paper addresses the problem of structure and motion recovery. The first problem is speed and we presented a fast recursive algorithm that updates the *complete* motion and structure after each time step and have shown that it gives the same results as in

[8]. Furthermore, we have developed an outlier detection technique that decreased the error even further. The paper is also one of the first that deals the problem of feature dropout and reappearance. We have developed a procedure that that allows previously lost features to reappear in the sequence and demonstrated that it works in practice.

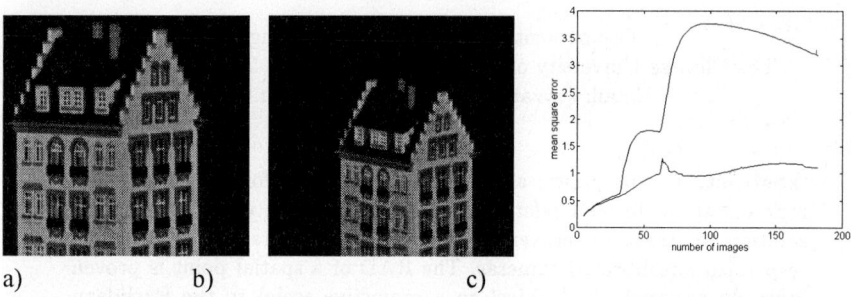

a)   b)   c)

Fig.2 1st (a) and 181st (b) image from the hotel image sequence; c) Plots of the mean square error (cost function) with (lower curve) and without (upper curve) outlier rejection

Acknowledgments

Financial support for this project was provided by the Australian Research Council. The hotel image sequence was provided by the "Modeling by Videotaping group" in the Robotics Institute, Carnegie Mellon University.

REFERENCES

[1] Held A., "Piecewise Shape Reconstruction by incremental factorization", *Proc. 7th British Machine Vision Conference, vol I*, UK, 1996, pp. 333-342.
[2] McLauchlan P. et al. "Recursive affine structure and motion from image sequences" *Proc. 3rd ECCV, vol. I*, Stockholm, pp. 217-224, 1994.
[3] McLauchlan P. and Murray D.. "A unifying framework for structure and motion recovery from image sequences" *Proc. 5th ICCV*, MIT, pp. 314-320, 1995.
[4] Meer P. et al. "Robust regression methods for computer vision: a review", *Intl. Journal Computer Vision* vol. 6., pp. 59-70, 1991.
[5] Morita T. and Kanade T., "A sequential factorization method for recovering shape and motion from image streams", *Proc. 1994 ARPA Image Understanding Workshop, vol. II*, pp. 1177-1188, Monterey, CA1994.
[6] Reid I. D. and Murray D. W., "Active Tracking of Foveated Feature Clusters Using Affine Structure", *Intl. J. Computer Vision*, vol. 18, pp. 41-60, 1996.
[7] Shi J. and Tomasi C., "Good Features to Track". *IEEE Conference on Computer Vision and Pattern Recognition*, June 1994, pp. 593-600.
[8] Tomasi C. and Kanade T., "Shape and motion from image streams under orthography: A factorization approach", *Intl. J. Comp. Vision*, vol. 9, pp. 137-154, 1991.
[9] Trajković M. and Hedley M., "Fast feature detection and matching for machine vision", *Proc. 7th British Machine Vision Conference, vol. I*, Edinburgh, UK, 1996, pp. 91-100.

Relative Affine Depth: Structure from Motion by an Uncalibrated Camera

Zhong-Ying Zhang and Hung-Tat Tsui

Department of Electronic Engineering,
The Chinese University of Hong Kong, Shatin, N. T., Hong Kong
Email: {zyzahng, httsui}@ee.cuhk.edu.hk

Abstract. In this paper, a new projective model for 3D information representation, termed *relative affine depth* (RAD), is derived for the solution to structure recovery of an object at arbitrary positions with respect to uncalibrated cameras. The RAD of a spatial point is proven inversely proportional (subject to a projective scale) to the Euclidean depth of a spatial point in a camera-centered coordinate frame, which is sequentially obtained in a straitforward way. The point matches are accomplished using a feature point tracking algorithm developed by us. Experiments with real objects show good results for shape recovery.

1 Introduction

In recent years, projective and affine reconstruction have significantly influenced the researches on structure from motion (SFM) [1]. In the cases where metric measurements for object shapes are not essential, people attempt to use projective or affine framework to formulate the camera projection and image transformation. Under these frameworks, difficulties such as camera calibration encountered in the metric SFM methods can be avoided without losing much fidelity of the structure recovery from multiple images [1, 6, 11, 12, 13]. Remarkably, the inherent linear mathematicasis of the projective and affine models permits the applications of compact matrix manipulation schemes to the interpretations of the mechanisms of scene's imaging and image matching over the multiple views. Even more importantly, the projective and affine methodologies make possible the application of various geometrical and algebraic invariants inherently existing in the projective transformations to different vision tasks including the difficult SFM problems [9]. For these reasons, investigating SFM methods modulo a projective or affine ambiguity of the structure recovery has been attracting more and more attentions [7, 2, 8, 3, 4, 11, 12].

Amnon Shashua and Nassir Navab have investigated the problem of shape recovery of 3D objects under the projective and affine frameworks [11, 12]. They formulated the transformation between two views in projective space by deriving an intermediate parameter termed *relative affine structure*, from which a relative 3D structure of an object is calculated up to affine ambiguity.

In this paper, we present our work on the development of a simple algorithm for SFM subject to projective assumptions. The work is an extension of Amnon

Shashua and Nassir Navab's but is different from their work in a major aspect: we start our work from a newly-defined intermediate affine structure parameter termed *relative affine depth (RAD)* so that our algorithm applies to an object at arbitrary positions and in arbitrary poses with respect to the camera. Whereas in Shashua's method, a non-general assumption on the viewing geometry is needed and the camera intrinsic parameters are assumed known.

Throughout the paper, the world is considered as a 3D projective space $\mathcal{P}^3$ and images a 2D projective space $\mathcal{P}^2$. An object is represented by its distinct structural points in such a paradigm. $\Psi, \Psi' \subset \mathcal{P}^2$ are two arbitrary views with their projection centers being O, O' $\subset \mathcal{P}^3$, respectively. The symbol $\cong$ denotes equality up to a scale, which is commonly encountered in GL_n, a group of $n \times n$ matrix transformations. PGL_n is accordingly defined as the projective group.

2 Principles and Algorithms

2.1 Relative Affine Depth

In Figure 1, π is a plane determined by three arbitrary non-colinear points $\boldsymbol{P}_i \in \mathcal{P}^3, i = 1, 2, 3$. $\boldsymbol{p}_i \in \Psi$ and $\boldsymbol{p}'_i \in \Psi'$ are $\boldsymbol{P}_i$'s projections on the two views, respectively. $\tilde{V}$ is the intersection of the line $\overline{OO'}$ with π, and v and v' are those with Ψ and Ψ', respectively. Let A be a homography of $\mathcal{P}^2$ determined by the equations $A\boldsymbol{p}_i \cong \boldsymbol{p}'_i, i = 1, 2, 3$, and $A\boldsymbol{v} \cong \boldsymbol{v}'$.

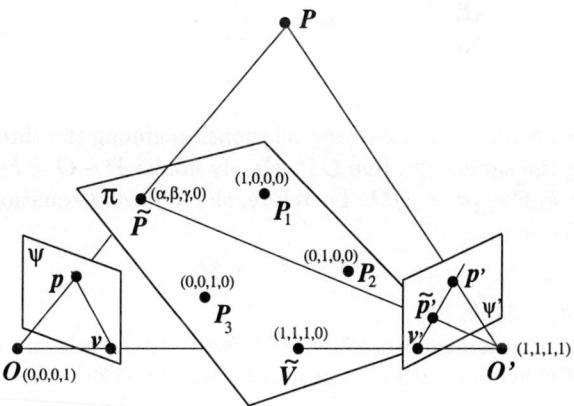

Fig. 1. The general projective geometry of perspective stereo views.

By assigning the concerned points with the canonical projective coordinates as shown in Figure 1, Shashua showed that, for any point $P \in \mathcal{P}^3$ with its images on Ψ and Ψ' being p and p', respectively, there exists an equation [12],

$$p' \cong Ap + kv' = A_c P_h, \qquad (1)$$

where k is a scalar which is independent of Ψ', $A_c = [A \mid v']$ is a camera transformation matrix between the two views, and P_h is therefore regarded as an instance of P in homogeneous coordinates in $\mathcal{P}^3$. Since k is a parameter derived with respect to the first view and is invariant to any other subsequent views, it was therefore termed the *relative affine structure* by Shashua [12].

In [12], a fourth point $P_0 \notin \pi$ is chosen to scale A or v' in Equation (1) such that $p'_0 \cong A p_0 + k_0 v'$. Then the scaled k becomes a parallax (we denote it by ρ) between the two views, similar to that in the results obtained by Koenderink and Van Doorn in [7]. ρ is given by the explicit form as,

$$\rho = \frac{k}{k_0} = \frac{z_0}{z} \frac{d}{d_0}. \tag{2}$$

where d and d_0 are the distances of P and P_0 from the plane π, z, and z_0 are the depths of these two points in a camera-centered coordinates frame, respectively.

The parallax term in Equation (2), however, should be of a different form if deeper insights are explored into the geometry of our paradigm, and an Euclidean depth can be represented in terms of new intermediate structural parameters. In the sequel we discuss the problem and give our results with the following theorems.

Theorem 1. *In Figure 1, let $\tilde{P}$ denote the intersection of the line $\overline{OP}$ with π, k_1 the ratio of $|\overline{OP}|$ to $|\overline{O\tilde{P}}|$, where $|\cdot|$ is the dimension of $(\cdot)$. Then k_1 is related to k in Equation (1) as follows,*

$$k_1 = \frac{1}{1-k}. \tag{3}$$

Proof. In Euclidean space, following relationship among the three points, $O, \tilde{P}$, and P, along the same sight line $\overline{OP}$, clearly holds: $P - O = k_1(\tilde{P} - O)$. Thus we have $P = k_1\tilde{P} - (k_1 - 1)O$. Therefore, the following equation holds up to a scale of $1/k_1$,

$$P \cong \tilde{P} - k'O, \tag{4}$$

where $k' = (k_1 - 1)/k_1$.

By assigning canonical projective coordinates to the relevant points as shown in Figure 1, following equation is proven to hold by [12]:

$$P \cong \begin{pmatrix} \alpha \\ \beta \\ \gamma \\ 0 \end{pmatrix} - k \begin{pmatrix} 0 \\ 0 \\ 0 \\ 1 \end{pmatrix} = \mu \begin{pmatrix} \alpha' \\ \beta' \\ \gamma' \\ 0 \end{pmatrix} + k \begin{pmatrix} 1 \\ 1 \\ 1 \\ 1 \end{pmatrix}, \tag{5}$$

where $\tilde{P}' = (\alpha', \beta', \gamma', 0)^\top$ is the intersection of the sight line $\overline{O'P}$ with π (not depicted in Figure 1). Please note that, $\tilde{P}, \tilde{V}$, and $\tilde{P}'$ are colinear on the intersecting line of the epipolar plane $\widehat{POO'}$ with π, and $\tilde{P}'$ lies at the opposite end

of $\overline{O'P}$. From above observations, an equation follows (please refer to [12]) as below,
$$\mu p' = Ap + kv', \tag{6}$$
where A is the homography between the two views of $\mathcal{P}^2$, determined by the equations $Ap_i \cong p'_i, i = 1, 2, 3$, and $Av \cong v'$. It can be seen that Equation (6) is identical with Equation (1) with a scale of μ. By comparing Equation (4) and Equation (5), we see that it is reasonable to put $k' = k$. The theorem is therefore proved.

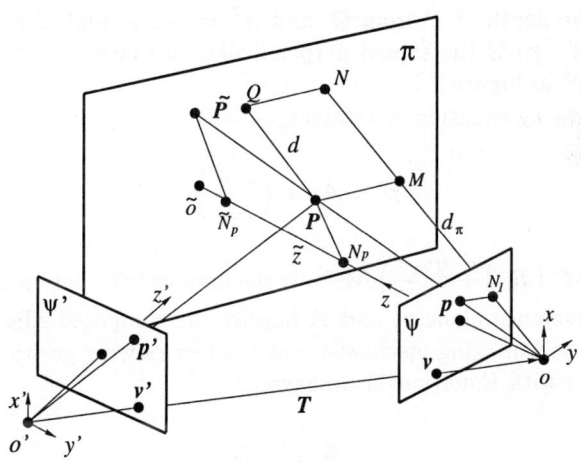

Fig. 2. Geometric illustration for the proofs of Theorem 2 and Theorem 3.

The explicit form of k in Equations (3) is given in Theorem 2.

Theorem 2. *In Figure 2, let π denote a reference plane, O and O' two view centers. Let p and p' denote the projections of $P \notin \pi$ in the two views, d and d_π be the distances of P and O from π, respectively. And let $T = (x_t, y_t, z_t)^\top$ be the translation vector from O' to O, and z and z' the depths of P from O and O', respectively. Then the relative affine parameter k in Equation (1) is of the form,*
$$k = \frac{z_t}{z} \frac{d}{d_\pi}. \tag{7}$$

Proof. If M_i and M'_i denote the intrinsic parameter matrices of the camera at the two viewing positions, then $zM_i^{-1}p$ and $z'M_i'^{-1}p'$ are the Euclidean coordinates of P with respect to the first and the second camera frames, respectively.

Let $R \in \mathcal{O}_3^+$ denote the orthogonal rotation matrix from O' to O, then the following equation holds:
$$z'M_i'^{-1}p' = zR^{-1}M_i^{-1}p + T. \tag{8}$$

From Equation (8), following equation is obtained directly:

$$p' \cong M'_i R^{-1} M_i^{-1} p + \frac{1}{z} M'_i T. \qquad (9)$$

Obviously, the two sides of the above equation are equal to each other up to a scale of z'/z.

From Figure 2, it is easy to see that $\overline{ON}$, the distance of O from π, denoted by d_π, is,

$$d_\pi = n^\top (\tilde{z} M_i^{-1} p), \qquad (10)$$

where $\tilde{z}$ is the depth of $\tilde{P}$ from O, and $n^\top$ is the normal of π. Similarly, $d = d_\pi - n^\top (z M_i^{-1} p)$ is the signed perpendicular distance of P from π, i.e., the length of $\overline{MN}$ in Figure 2.

Thus following equation is obtained,

$$p' \cong Ap + \left(\frac{z_t d}{z d_\pi}\right) v', \qquad (11)$$

where $A = M'_i \left(R^{-1} + \frac{Tn^\top}{d_\pi}\right) M_i^{-1}$, is the homography between the two views due to the reference plane π, and A implies the composed effects of the camera's motion and imaging mechanisms at the two viewing positions. Comparing Equation (11) with Equation(1) we have,

$$k = \frac{z_t}{z} \frac{d}{d_\pi}, \qquad (12)$$

which finishes the proof of the theorem.

In Theorem 1 and Theorem 2, k is the so-called *relative affine structure* in [12] and it exists with respect to no reference point. It is in fact relevant to the both views (due to z_t). In the following theorem, we show that k is independent of the second view if a fixed point is chosen to scale it, and therefore, this scaled k is proved an invariant to the camera's subsequent movements. The theorem further derives a relative Euclidean depth of a space point from the relative affine structural parameters. We call the relative Euclidean depth the *relative affine depth*.

Theorem 3. *Let $P_0 \notin \pi$ be an arbitrarily selected space point of $\mathcal{P}^3$, k_0 its corresponding relative affine structure and z_0 its Euclidean depth from O. Then k of any other point of $\mathcal{P}^3$ scaled by k_0 is independent of the second view and the ratio of z to z_0 is,*

$$\lambda = \frac{z}{z_0} = \frac{k_0 - 1}{k - 1}. \qquad (13)$$

where λ is defined as the relative affine depth (RAD).

Proof. From Equation (12), we have,

$$k_0 = \frac{z_t}{z_0}\frac{d_0}{d_\pi}.$$

Then the same result as that shown in Equation (2) is obtained if k in Equation (12) is scaled by k_0, i.e.,

$$\rho = \frac{k}{k_0} = \frac{z_0}{z}\frac{d}{d_0}. \tag{14}$$

ρ is obviously independent of the second view. Also from Equation (12), we have,

$$k = \frac{z_t}{z}\frac{d_\pi - \boldsymbol{n}^\top(z\boldsymbol{M}_i^{-1}\boldsymbol{p})}{d_\pi} = \frac{z_t}{z}\left[1 - \frac{\boldsymbol{n}^\top(z\boldsymbol{M}_i^{-1}\boldsymbol{p})}{d_\pi}\right] = \frac{z_t}{z}(1-k_1).$$

By substituting above equation with Equation (3), we have,

$$\frac{z_t}{z} = k - 1. \tag{15}$$

For $\boldsymbol{P}_0$, similar equation exists

$$\frac{z_t}{z_0} = k_0 - 1.$$

Equation (13) thus obviously holds.

2.2 The Algorithm for SFM

In the following algorithm, we use three non-colinear matched points and the epipoles over the two images to linearly recover the homography $\boldsymbol{A}$. In this case, a minimum of *eleven* general matched points are necessary for the recoveries of the homography $\boldsymbol{A}$ and the epipoles $\boldsymbol{v}$ and $\boldsymbol{v}'$.

Algorithm 1. *Algorithm for Affine SFM*

1. For eight point matches across two views, $\boldsymbol{p}_i \sim \boldsymbol{p}'_i$, $i = 1, 2, \cdots, 8$, determine the fundamental matrix $\boldsymbol{F}$ with the equation $\boldsymbol{p}'_i{}^\top \boldsymbol{F}\boldsymbol{p}_i = 0$, and then solve for $\boldsymbol{v}$ and $\boldsymbol{v}'$ with $\boldsymbol{F}\boldsymbol{v} = 0$ and $\boldsymbol{F}^\top \boldsymbol{v}' = 0$ [1, 2].
2. Determine the homography $\boldsymbol{A}$ by any three non-colinear point matches and the epipoles satisfying $\boldsymbol{A}\boldsymbol{p}_i = \boldsymbol{p}'_i$, $i = 9, 10, 11$ and $\boldsymbol{A}\boldsymbol{v} = \boldsymbol{v}'$, respectively.
3. For an arbitrary point $\boldsymbol{P}_0$ that is not coplanar with the three control points for determining $\boldsymbol{A}$, compute its corresponding relative affine structure k_0 with following equation:

$$k = \frac{(\boldsymbol{p}' \times \boldsymbol{v}')^\top \cdot (\boldsymbol{A}\boldsymbol{p} \times \boldsymbol{p}')}{\|\boldsymbol{p}' \times \boldsymbol{v}'\|^2}. \tag{16}$$

4. Similarly, for any other space point $\boldsymbol{P}$, compute its k with Equation (16), and then determine the relative affine depth λ by Equation (13).

3 Experiments

Figure 3 (a) and (b) show the two views of a calibration box. Each of the box's three visible surfaces perpendicular to each other are printed with an array of 9×9 white circles on a dark background. The crosses printed on the papers pasted on the wall serve as the control points for solving for the affine epipolar constraint condition and the homography in the SFM problem. Correspondences of the white circles on the two vertical sides over the two images are determined by applying the affine tracking algorithm developed by us [13].

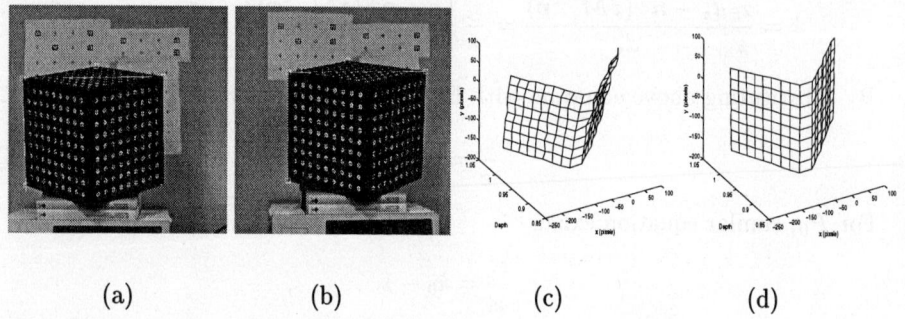

(a) (b) (c) (d)

Fig. 3. Two views of the calibration box and the recovered shapes of the two front surfaces of the calibration box: (a) and (b): the stereo views of the box. The crosses on the wall, indicated by the black squares are chosen as the control points for solving for the fundamental matrix as well as the homography between the two views. (c) and (d): the recovered shapes of the two front surfaces of the calibration box. (c) The wall is used as the reference plane. (d) The smoothed version of (c).

The lower-left vertex of the calibration box is chosen as the reference point for the computation of the relative affine depth λ. The centroids of all the 162 circles distributing on the two vertical surfaces of the box are used to construct two perpendicular meshes. Figure 3 (c) and (d) show the recovered shapes of the two surfaces with the wall serving as the reference plane.

4 Conclusion

In this paper, we have shown the performance of a novel relative affine depth (RAD) method on shape recovery. The relative affine structure is used in this paper as an intermediate parameter for the derivation of the more useful structural parameter, i.e., the RAD. We associate the RAD with a ratio between Euclidean distances of space object points. Since this relative depth is valid in the relative affine sense, we refer to it as the relative affine depth.

The major advantages of SFM by means of the relative affine depth are: firstly, the relative depth of an object visible to a camera from two arbitrary viewing positions can be simply computed from the projectively determined parameters. This virtue leads to the second advantage: the elaborate task of camera calibration can be omitted when using this technique to achieve the shape recovery. Although it is inherently projective and the recovered structures have the projective skews, this method is of great potential to many visual tasks, such as mobile robot navigation, robot arm manipulation, active gaze, etc., where metric shape measurements may not be regarded as essential. The experiments show that the SFM method presented in this paper can provide useful recovered shapes of 3D objects.

Acknowledgment

The authors are grateful to the support from the RGC Earmarked Research Grant, UPGC Ref No. CUHK 507/95E.

References

1. O. Faugeras, "Stratification of three-dimensional vision: Projective, affine, and metric representations," *J. Opt. Soc. Am. A,* Vol.12, No.3, 1995.
2. O. Faugeras, "What can two images tell us about a third one?" *Int'l J. of Computer Vision,* Vol.18, 1996.
3. R. Hartley, "Projective reconstruction and invariants from multiple images," *IEEE Trans. PAMI,* Vol.16, No.10, 1994.
4. R. Hartley, "A linear method for reconstruction using multiple uncalibrated images," *Proc. Int'l Conf. Computer Vision,* 1995.
5. R. Hartley, "In defense of the eight-point algorithm," *IEEE Trans. PAMI,* Vol.19, No.6, 1997.
6. K. Kanatani, "Computational projective geometry," *CVGIP: Image Understanding,* Vol. 54, No. 3, 1991.
7. J. Koenderink and A. Van Doorn, "Affine structure from motion," *J. Opt. Soc. Am. A,* Vol.8, No.2, 1991.
8. R. Mohr, et al., "Relative 3D reconstruction using multiple uncalibrated images," *Proc. IEEE Conf. CVPR,* New York, June 1993.
9. J. L. Mundy and A. P. Zisserman, *Geometric Invariance in Computer Vision,* MIT Press, Cambridge, MA, 1992.
10. L.S. Shapiro, A. Zisserman, and M. Drady, "3D motion recovery via affine epipolar geometry," *Int'l. J. Computer Vision,* Vol. 16, 1995.
11. A. Shashua, "Projective structure from uncalibrated images: Structure from motion and recognition," *IEEE Trans. PAMI,* Vol.16, No.8, 1994.
12. A. Shashua, "Relative affine structure: Canonical model for 3D from 2D geometry and application," *IEEE Trans. PAMI,* Vol.18, No.9, 1996.
13. H. T. Tsui, Z. Y. Zhang, and S. H. Kong, "Feature tracking from an image sequence using geometric invariants," *Proc. IEEE Computer Society Conf. on Computer Vision and Pattern Recognition,* San Juan, USA, June, 1997.

The Eigenspace Method for Rigid Motion Recovery from less than Eight Point Correspondences

Miroslav Trajković and Mark Hedley
Department of Electrical Engineering, Sydney University, NSW 2006, Australia
E-mail: {miroslav, hedley}@ee.usyd.edu.au

Abstract

In this paper we propose a new method for the estimation of rigid motion from two monocular images when less than eight point correspondences are available. The motion parameters are found using the essential matrix. By employing previously unused constraints on essential matrix, we show that it can be estimated through the minimisation of a two-dimensional cost function defined over the space of all possible directions of translation. The new formulation is easier to understand and implement than previously proposed approaches, and has a low computational cost. The algorithm has been evaluated on synthetic data. Our experiments show that that the new method is capable of finding all solutions and that choice of initial state is not critical.

1 Introduction

Motion is a powerful clue for an observer to explore and interact with the environment. It is well known that given two images of the static environment taken by a moving camera, it is possible to recover the motion of the camera (rotation and direction of translation) between two acquired images, and the relative depths of the scene points in a reference frame.

At least five pairs of matched feature points between two images are needed to determine the motion between images. Furthermore, if eight or more matched points are available then a linear algorithm may be used to recover the motion [5]. While the case of eight or more matched points has been widely studied [1, 5], considerably less research has been done for the case where there are between five and seven matched points. The case of five point matches has been investigated by ([2], [9]), and they have shown that up to ten solutions may exist in general. Netravali *et al* have also given a generalisation of their algorithm for the case of six and seven point matches. When six or more point matches are available there is, in general, a unique solution for camera orientation [9], and in the case of seven point matches a closed form for the motion may be found (which requires solving a cubic equation).

In this paper, we present a simple algorithm for recovering 3D motion (rotational and translational) and relative depths when less than eight point matches are available. We formulate the problem in a new way that leads to a simpler algorithm for recovering the motion (*cf.* [2, 9]). As a consequence our algorithm is easier to implement and understand, and it appears to be more stable.

The paper is organised as follows. In section two we give the necessary background to the projective transformation and introduce the essential matrix and some of its properties. In section three we give a brief review of the related work. Section four presents methods for recovering rigid motion from five, six and seven point matches. Experimental results are presented in section five. A discussion of results and possible extensions are given in section six and concluding remarks are given in the last section.

2 Background

2.1 Notation

In this paper we assume a pinhole camera model. The coordinate system is fixed on the camera with the origin coinciding with the projection centre of the camera and the z-axes coinciding with the optical axes and pointing toward the scene. The focal length is assumed known, and without loss of generality set to unity. Therefore, the equation of the image plane is $z=1$.

Let us consider an arbitrary point before and after the motion. The following notation is introduced and will be used throughout the paper. $\mathbf{P} = (X,Y,Z)^T$ and $\mathbf{P'} = (X',Y',Z')^T$ denote spatial vectors of the point before and after the the motion, while $\mathbf{p} = (x,y,1)^T = \mathbf{P}/Z$ and $\mathbf{p'} = (x',y',1)^T = \mathbf{P'}/Z'$ denote corresponding image vectors.

A function svd that maps a 3×3 matrix to a 3-dimensional vector is defined as

$$\text{svd}(A) = (\lambda_1, \lambda_2, \lambda_3)^T$$

where λ_1, λ_2 and λ_3 are singular values of $\mathbf{A}$ in decreasing order.

2.2 Motion Model

The motion model that is assumed in this paper is a rigid motion model which can be decomposed into a rotation and a translation and is described by the following equation:

$$\mathbf{P'} = R \cdot \mathbf{P} + \mathbf{T} \tag{1}$$

and after some rearrangement (see [11]) the following equations is obtained:

$$\mathbf{p'}^T E \mathbf{p} = 0 \tag{2}$$

where $\mathbf{T}_0$ denotes direction of translation and $E = [\mathbf{T}_0]_\times R$. The matrix E is called the essential matrix and has been found (along with the fundamental matrix which is corresponding matrix for an uncalibrated camera) to be very useful in Computer Vision for motion recovery ([5]) and motion segmentation. The essential matrix has the following properties that we will use throughout the paper.

Lemma 1. $\mathbf{T}_0$ is a unit eigenvector of E^T corresponding to the zero eigenvalue.

Lemma 2. $\text{svd}(E) = (1\ 1\ 0)^T$ or equivalently, eigenvalues of EE^T are 1, 1 and 0.

For proofs see [4, 11].

Given n point correspondences, equation (2) can be rewritten as linear equations in the elements of E and we obtain

$$A\mathbf{E} = \mathbf{0}, \tag{3}$$

where A is a $n \times 9$ measurement matrix easily determined from (2) and $\mathbf{E} = [e_{11}\ e_{21}\ e_{31}\ e_{12}\ e_{22}\ e_{32}\ e_{13}\ e_{23}\ e_{33}]^T$, for $E = [e_{ij}]_{3\times 3}$.

If eight or more matches are available then (except for some degenerate cases, see [6]) $\mathbf{E} = \sqrt{2}\mathbf{h}$ where $\mathbf{h}$ is the unit eigenvector associated with the smallest eigenvalue of $A^T A$, and R and $\mathbf{T}$ can be computed using method described in [11]. The advantage of this method is that it is linear and simple to compute, but it can not be used if the number of points is between five and seven, in which case a solution still can be found.

3 Prior Research

The motion recovery problem has been extensively studied by the computer vision community for the last fifteen years. There are two types of commonly used motion recovery algorithms – linear (described in the previous section) and non-linear. Non-linear algorithms may be broadly divided into two classes: 1) General methods based on non-linear estimation which simultaneously solve for camera orientation and relative depths through the minimisation of a non-linear cost function; and 2) Subspace methods (*e.g.* [3]), which subdivide the problem and solve separately for rotation, direction of translation and relative depths. The second class of methods is more efficient, as it usually first involves the minimisation in a two or three dimensional space to find the direction of translation or rotation, then the computation of other parameters by solving linear equations.

While both linear and non-linear algorithms have been proposed for the situation where eight or more point matches exist, the case for less than eight point matches has been investigated only in two papers ([2] and [9]). In both a complicated polynomial system needed to be solved to recover the motion.

Faugeras and Maybank [2] used two different approaches in their paper. The first one was based on the properties of essential matrix, which was introduced by [5]. They employed the fact that the set of essential matrices can be described by polynomial equations and investigated the multiplicity of solutions using algebraic geometry. A similar approach was taken by Natravaly *et al* [9], a set of three cubic and five linear equations need to be solved. This is even more computationally demanding to implement. The second approach of [2] was based on projective. Using this approach a high order polynomial needed to be solved. The problems with this approach are that it is difficult to understand and it is difficult to implement, as it requires extensive symbolic algebra.

Heeger and Jepson [3] assumed small rotation between two frames and have shown that in this case motion equation is bilinear what enabled them to find direction of translation through the optimisation of the cost function in the two-dimensional space (unit sphere). After that, they found rotation and relative depths by solving linear system. The main disadvantage of this method is that it assumes a small rotation, so it can not be applied in general.

In this paper, we propose a subspace method, which is specially designed for the case when less than eight motion correspondences are available and is based on evaluation of the essential matrix E given the constraints described in the previous section. The advantages of this method are: 1) it is simple; and 2) it requires only a minimisation in the compact two-dimensional space.

4 Finding motion from less than eight points

In this section we describe how to compute motion parameters when fewer than eight point matches are available. We will separately consider three cases: five, six and seven point matches. We will show that when seven point matches are available, the unique solution can be found by solving a cubic equation, and that in case of five or six point matches the camera motion can be recovered through the minimisation of two-dimensional functions defined on the unit sphere.

4.1 Computation of motion from five point matches

Our method is based on solving equation (3) with the constraints that E satisfies the properties of an essential matrix. For clarity, we will rewrite this here:

$$A\mathbf{h} = 0, \quad \text{subject to} \quad \text{svd}(E) = (1\ 1\ 0)^T. \tag{4}$$

As A has rank 5 (A is 5×9 matrix), the solution space of equation $A\mathbf{h} = 0$ will be a hyperplane spaned by eigenvectors $\mathbf{h}_1, \mathbf{h}_2, \mathbf{h}_3$ and $\mathbf{h}_4$ of the matrix $A^T A$ associated with its four zero eigenvalues. Hence, the solution of equation (4) may be written as

$$\mathbf{h} = a_1\mathbf{h}_1 + a_2\mathbf{h}_2 + a_3\mathbf{h}_3 + a_4\mathbf{h}_4, \quad \|\mathbf{a}\| = 1, \tag{5}$$

where a_1, a_2, a_3, a_4 are the coefficients to be found. The essential matrix corresponding to $\mathbf{h}$ is given by

$$E = a_1 E_1 + a_2 E_2 + a_3 E_3 + a_4 E_4 \tag{6}$$

where $E_1, \cdots, E_4$ are matrices associated with eigenvectors $\mathbf{h}_1, \cdots, \mathbf{h}_4$, and our task is to find a_1, a_2, a_3, a_4 which are solution to equation

$$\text{svd}(a_1 E_1 + a_2 E_2 + a_3 E_3 + a_4 E_4) = (1\ 1\ 0)^T.$$

This may also be formulated as a minimisation problem

$$\min_{\mathbf{a}} \left\| \text{svd}(a_1 E_1 + a_2 E_2 + a_3 E_3 + a_4 E_4) - (1\ 1\ 0)^T \right\|.$$

The space of solutions for $\mathbf{a}$ may be further reduced in the following way. If E is an essential matrix, then according to Lemma 1 for the direction of translation $\mathbf{T}$ (unit vector) we have:

$$E^T\mathbf{T} = (a_1 E_1 + a_2 E_2 + a_3 E_3 + a_4 E_4)^T \mathbf{T} = a_1 E_1^T\mathbf{T} + a_2 E_2^T\mathbf{T} + a_3 E_3^T\mathbf{T} + a_4 E_4^T\mathbf{T} =$$
$$[E_1^T\mathbf{T}\ \ E_2^T\mathbf{T}\ \ E_3^T\mathbf{T}\ \ E_4^T\mathbf{T}][a_1\ \ a_2\ \ a_3\ \ a_4]^T = B\mathbf{a} = \mathbf{0}$$

i.e. the search space for $\mathbf{a}$ can be reduced by noting that $\mathbf{a}$ must lie in the null space of B. The rank of B is 3 (it is a 3×4 matrix), so $B^T B$ has a single eigenvector with zero eigenvalue.

The cost function with the reduced search space can be defined as a function of $\mathbf{T}$ (hence defined over a two dimensional space) as follows:

- Pick up an arbitrary unit vector $\mathbf{T}$.
- Form matrix $B = [E_1^T\mathbf{T}\ \ E_2^T\mathbf{T}\ \ E_3^T\mathbf{T}\ \ E_4^T\mathbf{T}]$.
- Compute $\mathbf{a}$ as the eigenvector of $B^T B$ associated with its zero eigenvalue.
- Compute $\mathbf{l} = (l_1\ \ l_2\ \ l_3)^T = \text{svd}(a_1 E_1 + a_2 E_2 + a_3 E_3 + a_4 E_4)$; and

- Compute the cost function as $F_c(\mathbf{T}) = (l_1 - 1)^2 + (l_2 - 1)^2$

This cost function is always positive and will have minimum $F_c(\mathbf{T}) = 0$ for those values of $\mathbf{T}$ for which $(l_1 \; l_2 \; l_3)^T = (1 \; 1 \; 0)^T$. The search space is small (the unit hemisphere which is compact two-dimensional space) and the computational complexity of the evaluation of the cost function is low. In practice, to find all solutions, we performed the minimisation several times (the Simplex method, implemented in MATLAB, was used) with randomly chosen initial values. Usually all solutions were found. An example of typical cost function is shown in Figure 1.

4.2 Six point matches

When six point matches are available, matrix A has rank six and $A^T A$ has three zero eigenvalues. The solution of equation (4) can be written as

$$\mathbf{h} = a_1 \mathbf{h}_1 + a_2 \mathbf{h}_2 + a_3 \mathbf{h}_3, \quad \|\mathbf{a}\| = 1,$$

where a_1, a_2, a_3 are coefficients to be found and $\mathbf{h}_1, \cdots, \mathbf{h}_3$ are eigenvectors of $A^T A$ associated with its zero eigenvalues. As with 5 point matches, the essential matrix cor-responding to $\mathbf{h}$ is given by $E = a_1 E_1 + a_2 E_2 + a_3 E_3$ where E_1, E_2, E_3 are matrices associated to eigenvectors $\mathbf{h}_1, \cdots, \mathbf{h}_3$. Again, using Lemma 1 we obtain

$$E^T \mathbf{T} = [E_1^T \mathbf{T} \; E_2^T \mathbf{T} \; E_3^T \mathbf{T}][a_1 \; a_2 \; a_3]^T = B\mathbf{a} = 0$$

The necessary condition for above system to have a non-zero solution is $\mathrm{rank}(B) = 2$, and we can employ another constraint on $\mathbf{T}$, e.g. $E_1^T \mathbf{T} \cdot ((E_1^T \mathbf{T}) \times (E_1^T \mathbf{T})) = 0$. This equation may be written after some rearrangement as a homogenous third order polynomial in $\mathbf{T}$, but is very difficult to use in practice. Therefore, for this case, we employ the same minimisation procedure as for five points. The only difference is that vector $\mathbf{a}$ is three-dimensional and that it is computed as the eigenvalue of B associated with its smallest eigenvalue.

4.3 Seven point matches

With seven point matches, matrix A has rank seven and $A^T A$ has two zero eigenvalues. The solution of equation (4) can be written as

$$\mathbf{h} = a_1 \mathbf{h}_1 + a_2 \mathbf{h}_2, \quad \|\mathbf{a}\| = 1,$$

The essential matrix corresponding to $\mathbf{h}$ is given by $E = a_1 E_1 + a_2 E_2$ where E_1 and E_2 are matrices associated to eigenvectors $\mathbf{h}_1$ and $\mathbf{h}_2$. The equation to be solved is:

$$\mathrm{svd}(a_1 E_1 + a_2 E_2) = (1 \; 1 \; 0)^T.$$

Using the restriction that $\det(E) = 0$, coefficients a_1 and a_2 may be found by solving a cubic polynomial. As this polynomial can have three solutions, it is necessary to check above equation and choose the solution that satisfies it. This method has been previously described and was used for estimation of the fundamental matrix [7].

5 Experimental Results

In this section we present results of our algorithm applied to synthetic motion fields consisting of five or six point matches. Synthetic motion fields were generated as in [3], using a random depth map. The focal length was set to unity and the image was 9×9 pixels. The pixel width was set to 0.1 and all the points in first frame have integer (in unit of pixels) positions. Experiments were performed for various values for translation and rotation, and the motion fields used here were generated by rotation around axis (80°, 10°) (in spheric coordinates) for the rotation angle of 16° and various translation vectors.

5.1 Five point matches

When five point matches are available the number of solutions corresponding to each set of point matches is up to 10. A solution is feasible if it yields a 3D reconstruction in which all the points are on the same side of the camera for both camera positions (if all the depths are negative, than we simply change the sign of translation vector and the sign of depths).

We give two examples of results using our algorithm. In the first, we applied rotation [80° 10° 16°] and translation [1 1 2] and obtained the set of five point matches .

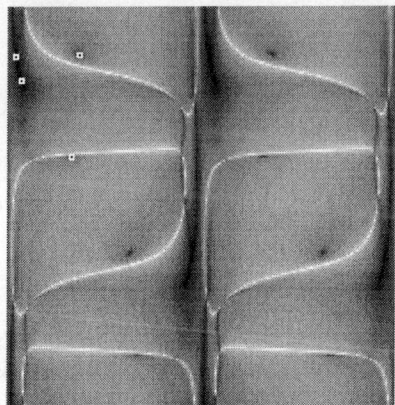

Figure 1. Cost function corresponding to the above set of point matches. Solutions are marked with white squares with a black point in the middle. The solutions are shown in one quadrant only. Note that each solution has three counterparts in other quadrants but they all yield the same motion and structure parameters (up to a sign). The axes the are polar coordinates of the direction of translation (T_0).

For this configuration we found all four solutions, with all of them being feasible. The results are not shown for the lack of space, for more details see [10]. The cost function is shown in Figure 1. Light areas indicate a high value for the cost function, while dark areas indicate low cost. Solutions (minima) are marked as white squares with a black dot in the centre.

In most of our experiments we found four, or less frequently six, solutions. To check how the algorithm works when all ten solutions exist, we tested it using data for which we know that ten solutions exist. Let us consider a set of image correspondences **a**↔**a**′, **b**↔**b**′, **c**↔**c**′, **d**↔**d**′, **e**↔**e**′. It is shown in [2] that if **a, b, c** and **a′**, **b′**, **c′** are two triads of mutually orthogonal vectors and **d**=**e′**, **e**=**d′**, these image correspondences yield ten solutions. If this set is exposed to a small perturbation it should still yield ten real solutions. One such set of data was generated, and all ten solutions were found. For further details see [10].

5.2 Six Point Matches

With six point matches, a unique solution exists in general, and in all our experiments, we were able to find this unique solution.

The cost function associated with an arbitrary set of six points is shown in Figure 2. As can be seen (and as obtained by our algorithm), the cost function has four minima. However, when the second criteria ($A\mathbf{h}=0$) is applied, only one of these solutions remains valid.

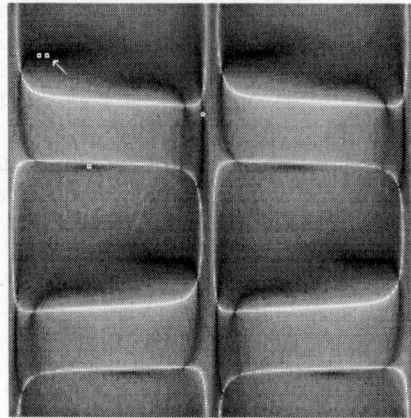

Figure 2. Cost function associated to the six point correspondences. The white arrow point to the solution for which $\|A\mathbf{h}\|=0$.

6 Discussion

The main characteristic of our algorithm is that it employs the fact that the essential matrix corresponding to the five point matches must lie on a known four-dimensional hyperplane. Thus, if the direction of translation is known, the essential matrix (and accordingly rotation and relative depths) can be recovered using a linear algorithm. Consequently, the problem of finding camera orientation is solved through the optimisation of the cost function in the space of all possible directions (on the unit sphere).

If this fact were not employed, then even if the direction of translation were known, finding the rotation vector would be still complicated, and not possible in closed form. Therefore, optimisation in a higher dimensional space would have to be employed.

One immediate application of our algorithm is rigidity checking. In [12], rigidity checking was performed using six point matches by non-linear optimisation in an 11-dimensional parameter space (motion parameters and relative depths of the scene points in the reference frame). It seems that our algorithm would produce a much simpler solution to this problem. Namely, it is enough to find minima of the cost function described in Section 3.2 and check if minimum is below some threshold. This can be done using the two dimensional optimisation. Moreover, it can be readily visualised (*e.g.* see Fig. 2).

7 Conclusion

In this paper we have presented a new approach to the problem of determining the motion between two images where the number of matched points is less than eight. The main advantages of our algorithm are *simplicity* and *low computational cost*. Not only that, but the formulation of the problem is considerably simpler than previous presentations.

The algorithm was implemented and tested using simulated data. When five point correspondences were given, we usually found four solutions which is in line with observations made in [2]. For those configurations that yield ten decomposable solutions, the algorithm found all solutions. For six and seven point matches, the unique solution was always found.

References

[1] Aggarwal J.K. and Nandhakumar N., "On the computation of Motion from Sequences of Images – A review", *Proceedings of the IEEE*, vol. 76, pp. 917-935, 1988.

[2] Faugeras O. and Maybank S., "Motion from Point Matches: Multiplicity of Solutions", *Intl. Journal Computer Vision*, vol. 4, pp. 225-246, 1990.

[3] Heeger D. and Jepson A., "Subspace Methods for Recovering Rigid Motion I: Algorithm and Implementation", *Intl. Journal Computer Vision*, vol. 7, pp. 95-117, 1992.

[4] Huang T. and Faugeras O., "Some Properties of the E Matrix in Two-View Motion Estimation", *IEEE PAMI*, vol. 11, 1310-1312, December 1989.

[5] Longuet-Higgins, H.C., "A computer algorithm for reconstructing a scene from two projections", *Nature* 293, pp.133-135, 1981.

[6] Longuet-Higgins, H.C., The reconstruction of a scene from two projections – configurations that defeat the 8-point algorithm",*Proc. IEEE 1st Conf. On Artif. Intell. Applications*, pp. 395-397, 1984.

[7] Luong Q.T. et al, "On determining the fundamental matrix: analysis of different methods and experimental results", *INRIA TR-1894*, INRIA-Sophia Antipolos.

[8] McReynolds D. and Lowe D., "Rigidity Checking of 3D Point Correspondences Under Perspective Projection", *IEEE Trans. PAMI*, vol. 18, pp. 1174-1185, 1996.

[9] Netravali A. et. al. "Algebraic Methods in 3-D Motion estimation from Two-View Point Correspondences", *Intl. Journal of Imaging Systems and Technology*, vol. 1., pp. 78-99, 1989.

[10] Trajković M. and Hedley M. "The eigenspace method for motion recovery from less than eight point correspondences", (manuscript in preparation).

[11] Weng J., Huang. T and Ahuja N., "Motion and Structure from Image Sequences", Springer-Verlag, 1993.

3D Shape and Motion Analysis from Image Blur and Smear: A Unified Approach

Y. F. Wang
Department of Computer Science
University of California
Santa Barbara, CA 93106

Ping Liang
College of Engineering
University of California
Riverside, CA 92521-0425

Abstract

This paper addresses 3D shape recovery and motion estimation using a realistic camera model with an aperture and a shutter. The **spatial blur** *and* **temporal smear** *effects induced by the camera's finite aperture and shutter speed are used for inferring both the* **shape** *and* **motion** *of the imaged objects.*

1 Introduction

In this paper, we address the problem of 3D shape recovery and motion estimation using a realistic camera model. Even though the pin-hole camera model has been prevalent in computer vision and graphics for its simplicity and mathematical tractability, this model hides certain shape cues that can be very powerful in 3D analysis. For example, if a more realistic camera model is used—which takes into consideration the aperture and shutter speed settings of the camera—the depth of field and motion blur effects become significant in the image formation. Depth of field induces *spatial blur*, where the amount of image blur is effected by the imaged object's apparent depth. Motion blur induces *temporal smear*, the effect of which is affected by the speed of the imaged object.

Traditional shape and motion analysis utilizing a pin-hole camera model does not consider the image blur and smear effects induced by a finite aperture and/or shutter speed. However, these effects often times cannot be ignored. Unless the imaged object is at a known distance to be sharply focused or the sensor is truly a pin-hole camera, *spatial blur* due to a finite aperture and depth of field will occur. Furthermore, unless the shutter speed is such that during the time the shutter opens the scene object's motion is negligible, *temporal smear* due to object motion need be considered. These degradation effects do exist in practical applications. For examples, often times in navigation an obstacle's location (depth) may not be known in advance (hence the object may not be focused correctly). A camera may not be equipped with a motorized lens with computer-controlled zoom or focus rings, hence, adaptation to objects at different depths might not be possible. The illumination level might be low, which requires extended exposure. These effects should be considered in designing robust depth inference and motion estimation algorithms.

Algorithms that utilize image blur for image analysis have recently been proposed. Most notable are the "shape-from-focus" and "shape-from-defocus" approaches [2, 4, 5, 7] for 3D shape recovery. In "shape-from-defocus," a single camera, or a pair of cameras with a specially-designed beam-splitter placed in front of them [5] used to produce a pair of identical images except for the aperture size and therefore the depth of field. The amount of image blur in the two images can be shown to be a simple function of one variable: the distance between the viewer and the imaged point. Hence, to estimate the depth of an object, we need only compare the corresponding points in the two images and measure the change in the image blur.

Image motion estimation can be roughly classified into gradient-based, feature-matching, and spatial-temporal filtering approaches. Recently, a "motion-from-smear" framework [1] was proposed as another alternative. [1] studies motion estimation from the *smear effect in an image sequence*, but does not consider possible image blur due to defocusing. The proposed algorithm unifies these two approaches and takes into account both *finite-aperture spatial blurring* and *finite-shutter-speed temporal smear* effects for shape and motion computation.

It is nontrivial to extend shape and motion analysis to allow both spatial blur and temporal smear. As shown later in Sec. 2, *indistinguishable visual effects can be produced non-incidentally by an object which is out of focus but otherwise stationary, or by a moving object which is in focus, or by objects with different combinations of the degree of out-of-focusness and speed of motion in between.* Hence, image interpretation in the presence of spatial blur and temporal smear can be highly ambiguous. This ambiguity in a sense is similar to the "aperture problem" well known in the optical flow analysis [3]. However, the problem is much more severe in a general shape and motion analysis allowing blur and smear, as *it may not be possible to uniquely identify the source of the image blur even at image locations with a multitude of gradient directions (e.g., a corner).* Hence, it is important to isolate the image blur induced by a finite aperture and shutter speed to enable shape and motion analysis, respectively.

2 Technical Description

Fig. 1 depicts the image configuration where blur due to finite aperture size and smear due to non-zero exposure time are shown. D_0 and d_0 denote the object and image distances that achieve a perfect focus, assuming that all lens aberration effects are ignored. For objects located at a distance D other than D_0, image blur is induced by a spatial "spillage" of pixel values into adjacent pixels, and the amount of image blur is affected by the lens aperture and the object's distance. Furthermore, we assume that an imaged object can undergo a general motion, denoted by V in Fig. 1. Object motion performs a temporal "integration" of pixel values along the trajectory when the shutter opens and induces image smear as a result.

The problem we address in this paper is this: Given that the camera is modeled as a finite aperture sensor instead of a pin-hole model, and that the shutter may be open long enough for the imaged object's motion to be registered (e.g., in a low light scenario), the recorded images are subject to both spatial blur and temporal smear degradation. How can we reliably estimate the shape and motion of the imaged object from both image blur and smear? In the next section, we will first review the "shape-from-focus" approach for shape recovery using static, finite-aperture images. Then we will discuss the difficulty in analyzing both blurred and smeared images, and present our formulation for deducing shape and motion from such images.

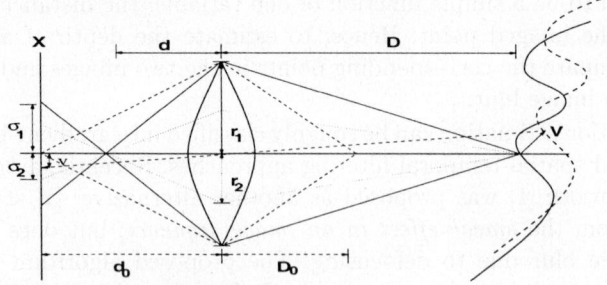

Fig. 1. Image configuration with spatial blur and temporal smear

2.1 Shape-from-Defocus using Static, Finite-Aperture Images

To simplify the analysis, we assume that the point spread function of the optical system is approximated by a Gaussian function of standard deviation σ.[1] As shown in Fig. 1, two images, taken of the same object but with different aperture opening r_1 and r_2 will show different degrees of blurring denoted by σ_1 and σ_2. It can be shown that the amount of image blur is related to the object's perceived distance D by [5]

$$D = \frac{Fd_0}{d_0 - F - k\sigma f}, \quad (1)$$

where F is the focal length and f the f-number of the lens. k is a proportionality constant which can be determined through camera calibration. The blurred image, $g(x,y)$, can be represented as the convolution of an un-blurred image, $f(x,y)$ (taken using a pin-hole camera), and a Gaussian kernel of variance σ:

$$g(x,y) = \int_{x'}\int_{y'} f(x-x', y-y') \frac{1}{\sqrt{2\pi}\sigma} e^{-\frac{x'^2+y'^2}{2\sigma^2}} dx'dy'.$$

Fourier transform of the above equation is

$$G(\omega_x, \omega_y) = \frac{1}{\sqrt{2\pi}} F(\omega_x, \omega_y) e^{-\frac{\sigma^2(\omega_x^2+\omega_y^2)}{2}}.$$

If two images are taken of the same scene using different aperture settings of σ_1 and σ_2, then the ratio of their Fourier coefficients reveals the relative blurring:

$$\frac{G_1(\omega_x, \omega_y)}{G_2(\omega_x, \omega_y)} = \frac{e^{-\frac{\sigma_1^2(\omega_x^2+\omega_y^2)}{2}}}{e^{-\frac{\sigma_2^2(\omega_x^2+\omega_y^2)}{2}}}, \text{ or}$$

$$ln\frac{G_1(\omega_x, \omega_y)}{G_2(\omega_x, \omega_y)} = \frac{1}{2}(\sigma_2^2 - \sigma_1^2)(\omega_x^2 + \omega_y^2), \quad (2)$$

which provides one constraint on σ_1 and σ_2. Furthermore, observe that the two aperture settings are constrained by the perceived depth as shown in Eq. 1, hence, we have:

$$ln\frac{G_1(\omega_x, \omega_y)}{G_2(\omega_x, \omega_y)} = \frac{1}{2k^2}((\frac{Fd_0}{f_2 D} - \frac{d_0-F}{f_2})^2 - (\frac{Fd_0}{f_1 D} - \frac{d_0-F}{f_1})^2)(\omega_x^2 + \omega_y^2).$$

If the camera parameters are known, the only unknown in the above equation is D. After D is solved for, we can recover σ_1 and σ_2 using Eq. 1. In some implementations [5], one of the cameras is assumed to be a pin-hole camera and produces images with a negligible blur (or $\sigma = 0$). Eq. 2 can then be used to derive the other σ.

[1] Researchers have used different approximations to the PSF, such as Gaussian [5] and pillbox functions [7].

2.2 Shape and Motion using Blurred and Smeared Images

A more complicated scenario is when the shutter is open long enough for object motion to be registered. The images thus produced suffer from both *spatial blur* and *temporal smear* degradation. We establish the following fact:

Proposition: *For a single blurred and smeared image, the same visual image blur effects can be non-incidentally produced by a multitude of combinations of finite-aperture blur and motion smear, with different degrees of out-of-focusness and object motions. This is true for image areas with a single gradient direction (e.g., an edge) and multiple gradient directions (e.g., a corner).*

The proposition leads to the following observations:
- It is not beneficial to analyze a single blurred and smeared image by itself. Multiple images, or multiple sequences of images will be needed.
- The aperture problem in optical flow [3] is exacerbated by the presence of blur and smear. As the presence of multiple gradient directions (e.g., a corner) in a small neighborhood theoretically allows the recovery of local motion using the local flow analysis, which is not the case for blurred and smeared images.
- Mechanisms probably will be needed to isolate the image blur effects induced by a finite aperture and shutter speed. This will lead to a more robust estimation of shape (based on the finite-aperture blur effect) and motion (based on the nonzero-exposure-time smear effect).

To illustrate, we will assume that the effect of object motion, if any, over a small image neighborhood and a short time period can be characterized by a constant, nonzero flow velocity denoted by (u, v). Now consider two extreme scenarios: degradation in one comes entirely from finite aperture blur (i.e., no object motion) and in the other entirely from nonzero exposure time smear (i.e., a pin-hole camera is used with infinite depth-of-field). Denote the images in these two cases as g_b (blur only) and g_s (smear only). Then these two images can be related to the un-blurred and un-smeared image (f) by:

$$g_b(x,y) = \int_x' \int_y' f(x-x', y-y') \mathcal{B}(x', y') dx' dy' \text{ and} \quad (3)$$

$$g_s(x,y) = \int_t f(x-ut, y-vt) \mathcal{S}(t) dt$$

where $\mathcal{B}$ and $\mathcal{S}$ denote the blurring and smear mechanisms. The proposition above suggests that g_b and g_s can be indistinguishable even at places with a high information content (e.g., edges and corners). We will illustrate first for the case where the the image neighborhood contains a single step edge.

Without loss of generality, we assume the edge aligns with the y-axis, or

$$f(x,y) = \begin{cases} a & x < 0 \\ b & x \geq 0 \end{cases}.$$

Then we can simplify the expressions of g_s and g_b into

$$g_b(x) = \int_x f(x-x') \bar{\mathcal{B}}(x') dx', \text{ and}$$

$$g_s(x) = \int_t f(x-ut) \mathcal{S}(t) dt \text{ where}$$

$$\bar{\mathcal{B}}(x') = \int \mathcal{B}(x', y') dy'.$$

We consider the following $\mathcal{B}$ and $\mathcal{S}$ functions: $\mathcal{B}$ can be either a 2D Gaussian (e.g., in [5]) or a pillbox function (e.g., in [7]), and $\mathcal{S}$ can either a box (a fast shutter with negligible open and close times) or a trapezoid function (a slow shutter with nonzero rising and trailing edges). Their expressions are

	B or S	$\bar{B}$
Gaussian	$\frac{1}{\sqrt{2\pi}\sigma}e^{-\frac{x^2+y^2}{2\sigma^2}}$	$e^{-\frac{x^2}{2\sigma^2}}$
pillbox	$\frac{1}{\pi r^2}\Pi(\frac{\sqrt{x^2+y^2}}{r})$	$\frac{2}{\pi r^2}\sqrt{r^2-x^2}$
box	$\Pi(\frac{t}{T})$	
trapezoid	$\Lambda(\frac{t}{R}) + \Pi(\frac{t-R}{T-R-F}) + \Lambda(-\frac{t-T}{F})$	

where σ denotes the variance of the Gaussian function, r denotes the radius of the pillbox function, T is the length of time the shutter opens, and R and F are the rise and fall times in a trapezoid function, respectively. Π is the unit-length rectangular function (1 between 0 and 1 and 0 otherwise) and Λ is the unit-length ramp function ($\Lambda = \int \Pi dt$).

Sample 1D Gaussian, pillbox, box, and trapezoid functions are plotted in Fig. 2(a) and edge profiles result from Gaussian or pillbox blurring, and from box or trapezoid smear are shown in Fig. 2(b). As can been seen that the four edge profiles are extremely alike. The reason for the similarity is that both out-of-focus blur and motion smear induce an "accumulation" of neighboring pixel values. The accumulation results from either a spatial "spillage" by blurring or a temporal "integration" by smear. The degradation filters have very similar shapes as shown in Fig 2(a) and produce similar accumulation effects.

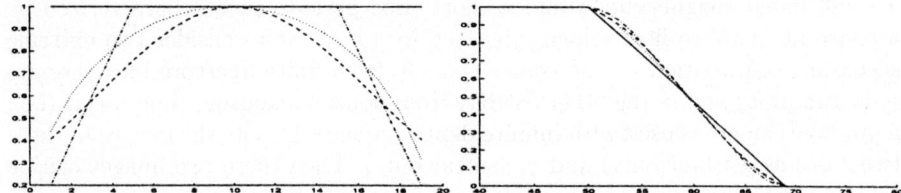

Fig. 2. (a) Different image blur and smear filters, and (b) blurred and smeared 1D edge profiles. Solid: box, dashdot: trapezoid, dot: pillbox, and dashed: Gaussian.

Even in a neighborhood with multiple gradients, it can still be impossible to make such an assertion. The reason is that consider, for example, a 90° corner with edges aligned with the coordinate axes. A 2D motion in a 45° direction in between the x and y axes will induce the same smear effect along both the x and y edges, which in turn can be attributed to out-of-focus blur. Furthermore, many combinations of out-of-focus blur and motion smear along the bisecting direction of the corner can explain the phenomenon equally well. Hence, unlike in an optical flow analysis, a reliable estimate of motion is not possible even in the presence of multiple gradient directions in a neighborhood.

To enable a robust shape and motion estimation in the presence of both blur and smear, it is important that the degradation effects induced by out-of-focus blur and motion smear be identified and isolated. We present an approach below.

2.3 Mathematical Formulation

Combining the two expressions in Eq. 3 taken into consideration both blur and smear, a blurred and smeared image (g) is related to the un-blurred and un-smeared one (f) by:

$$g(x,y) = \int_T^{T+\Delta t} \int_{x'} \int_{y'} f(x-ut-x', y-vt-y') \frac{1}{\sqrt{2\pi}\sigma} e^{-\frac{x'^2+y'^2}{2\sigma^2}} dx'dy'dt,$$

where (u,v) denote an object pixel's image velocity and Δt denote the length of time the shutter opens. Fourier transform of the above equation is

$G(\omega_x, \omega_y)$

$$= \int_x \int_y \{\int_T^{T+\Delta t} \int_{x'} \int_{y'} f(x-ut-x', y-vt-y') \frac{1}{\sqrt{2\pi}\sigma} e^{-\frac{x'^2+y'^2}{2\sigma^2}} dx'dy'dt\} e^{-i(\omega_x x+\omega_y y)} dxdy$$

$$= \int_T^{T+\Delta t} \int_{x'} \int_{y'} \int_x \int_y f(x-ut-x', y-vt-y') e^{-i(\omega_x x+\omega_y y)} dxdy \frac{1}{\sqrt{2\pi}\sigma} e^{-\frac{x'^2+y'^2}{2\sigma^2}} dx'dy'dt$$

$$= F(\omega_x, \omega_y) \int_T^{T+\Delta t} \int_{x'} \int_{y'} \frac{1}{\sqrt{2\pi}\sigma} e^{-\frac{x'^2+y'^2}{2\sigma^2}} e^{-i(\omega_x x'+\omega_y y')} dx'dy' e^{-i(\omega_x u+\omega_y v)t} dt \quad (4)$$

$$= F(\omega_x, \omega_y) \frac{1}{\sqrt{2\pi}} e^{-\frac{\sigma^2(\omega_x^2+\omega_y^2)}{2}} \int_T^{T+\Delta t} e^{-i(\omega_x u+\omega_y v)t} dt$$

$$= F(\omega_x, \omega_y) \frac{1}{\sqrt{2\pi}} e^{-\frac{\sigma^2(\omega_x^2+\omega_y^2)}{2}} e^{-i(\omega_x u+\omega_y v)(T+\frac{\Delta t}{2})} sinc((\omega_x u+\omega_y v)\frac{\Delta t}{2}) \Delta t .$$

As shown in the above equation, two degradation mechanisms due to both spatial blur and temporal smear are present. Suppose that two image sequences are recorded of the same scene using two different aperture settings. Then we can identify four different combinations: (I) images taken over the same time interval with the same aperture setting, (II) images taken over the same time interval with different aperture settings, (III) images taken over different time intervals with the same aperture setting, and (IV) images taken over different time intervals with different aperture settings. (I) produces only a single image and is not very interesting. (II) has been shown to facilitate 3D shape inference. It will be shown below that (III) and (IV) can be used to recover object motion from blurred and smeared images.

Taking the ratio of the Fourier transform of two images taken with (possibly) different apertures over (possibly) different times using Eq. 4, we have

$$\frac{G_1(\omega_x,\omega_y)}{G_2(\omega_x,\omega_y)} = \frac{e^{-\frac{\sigma_1^2(\omega_x^2+\omega_y^2)}{2}} e^{-i(\omega_x u+\omega_y v)(T_1+\frac{\Delta t_1}{2})} sinc((\omega_x u+\omega_y v)\frac{\Delta t_1}{2})\Delta t_1}{e^{-\frac{\sigma_2^2(\omega_x^2+\omega_y^2)}{2}} e^{-i(\omega_x u+\omega_y v)(T_2+\frac{\Delta t_2}{2})} sinc((\omega_x u+\omega_y v)\frac{\Delta t_2}{2})\Delta t_2} . \quad (5)$$

For scenario (iii) above, if two images are taken with the same aperture setting but over different time periods then the ratio is simplified to

$$\frac{G_1(\omega_x,\omega_y)}{G_2(\omega_x,\omega_y)} = \frac{e^{-i(\omega_x u+\omega_y v)(T_1+\frac{\Delta t_1}{2})} sinc((\omega_x u+\omega_y v)\frac{\Delta t_1}{2})}{e^{-i(\omega_x u+\omega_y v)(T_2+\frac{\Delta t_2}{2})} sinc((\omega_x u+\omega_y v)\frac{\Delta t_2}{2})} .$$

Assume that the shutter speed is the same and the image velocity stays constant over adjacent image frames, we have:

$$ln\frac{G_1(\omega_x,\omega_y)}{G_2(\omega_x,\omega_y)} = i(\omega_x u+\omega_y v)(T_2-T_1) .$$

The blurring effect is canceled out and (u,v) can be recovered.

For the scenario (iv) above, observe from Eq. 4 that blurring affects only the magnitude of the system impulse response, while smear affects both the magnitude and the phase. Hence, if only the phase component is considered, again with the same shutter speed Δt, the ratio in Eq. 5 is simplified and different image blur σ is eliminated, or

$$phase\{\frac{G_1(\omega_x,\omega_y)}{G_2(\omega_x,\omega_y)}\} = (\omega_x u + \omega_y v)(T_2 - T_1).$$

In either case, (u, v) can be isolated and estimated from the image sequence. A theoretical analysis of the accuracy of the formulation can be found in [6].

3 Experimental procedures

Since "shape-from-blurring" formulation has been extensively studied in the literature, our experiments instead concentrated on the "motion-from-smear" formulation. To verify the correctness of the proposed motion estimation algorithms, experiments using both synthetic and real images were conducted. Due to the page limitation, we will present here only the results on synthetic and real blurred and smeared images.

Two superimposed sinusoids, both moving with speeds of 1 pixel/frame at $0°$ and $90°$ degrees orientation for a perceived velocity of (1,1) pixels/frame, were used. A motion-smeared image was generated by adding three adjacent images together. A second smeared image was generated by shifting the first smeared image by (2, -1) pixels. Various amounts of white noise, with a maximum amplitude of 1, 5, 10, and 20 intensity levels (out of a maximum of 255) were added to these motion-smeared images to simulate real-world sensor and environmental noise. The estimated motion vectors and their histograms are shown in Fig. 3 for a noise level of 20 intensity levels. As can be seen from Table 1 that the algorithm performed almost perfectly for low noise levels. Even at a high noise level (10/255), the results were fairly accurate.

Table 1. Estimated mean motion (ground truth (2,-1)) and the standard deviation in the estimation for the synthetic data

Noise magnitude	Mean velocity horizontal	vertical	Standard deviation horizontal	vertical
2/255	1.9999	-0.9999	0.0007	0.0016
5/255	2.0004	-1.0003	0.0019	0.0044
10/255	2.0100	-0.9716	0.0999	0.3002
20/255	1.9007	-0.9105	0.7819	0.5875

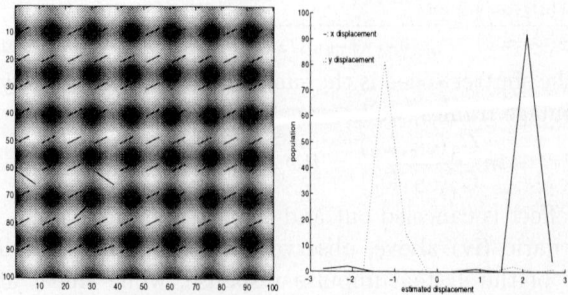

Fig. 3. 2D motion simulation results for the noise level of 20.

Our real experimental setup comprised a stationary camera viewing objects on a mobile platform. The objects were shifted to the right and downward slightly at each position, and two images with two different aperture stops ($f/1.8$ and $f/2.8$) were taken at each position. Three adjacent images from a single aperture stop were added to simulate the smearing effect.

Four types of combinations in motion computation are possible: Using the pair of images from the $f/1.8$ sequence, using the pair of images from the $f/2.8$ sequence, using the first image from the $f/1.8$ sequence and the second image from the $f/2.8$ sequence, and using the first image from the $f/2.8$ sequence and the second image from the $f/1.8$ sequence. Because no ground truth was available, we instead computed the histogram and verified the consistency. The combined results from the four analysis are displayed in Fig. 4.

As seen from these figures, motion estimation in the regions occupied by the 3M box and the cylinder was fairly consistent and correctly predicted the right and downward motion. The motion estimation in the upper right corner, which imaged a far-away wall, was less consistent. The inconsistency was probably caused by that the wall lacked any discernible features and was at a large distance that its motion relative to the camera was small. Motion estimation could easily be fooled by background noise and varying lighting condition.

4 The Concluding Remarks

We broaden the scope of "shape-from-blurring" to allow the camera to have not only a finite aperture but also a non-zero exposure time. Hence both spatial blur and temporal smear were used for inferring the shape and motion of objects.

References

1. W. G. Chen, N. Nandhakumar, and W. Martin. Image Motion Estimation from Motion Smear — A New Computational Approach. *IEEE Trans. Pattern Analy. Machine Intell.*, 18(4):412–425, 1996.
2. B. K. P. Horn. Focusing. Technical Report 160, MIT, 1968.
3. B. K. P. Horn. *Robot Vision*. The MIT Press, Cambridge, MA, 1986.
4. S. K. Nayar and Y. Nakagawa. Shape from Focus. *IEEE Trans. Pattern Analy. Machine Intell.*, 16(8):824–831, 1994.
5. A. Pentland. A New Sense for Depth of Field. *IEEE Trans. Pattern Analy. Machine Intell.*, 9(4):523–531, 1987.
6. Y. F. Wang and P. Liang. 3D Shape and Motion Analysis from Image Blur and Smear: A Unified Approach. In *Proceedings of International Conference on Computer Vision*, Bombay, India, 1997. to appear, (also accepted for oral presentation at Asian Conference on Computer Vision, Hong Kong, 1998).
7. M. Watanabe and S. K. Nayar. Telecentric Optics for Computational Vision. In *Proc. European Conf. Comput. Vision*, Apr. 1996.

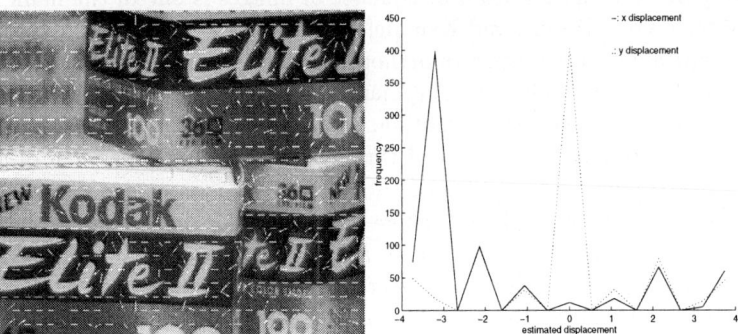

Fig. 4. Final results by combining those from four different estimation procedures

3D Line's Extraction from 2D Spatio-temporal Image Created by Sine Slit

Pingtao Wang, Katsushi Ikeuchi, Masao Sakauchi

Institute of Industrial Science, University of Tokyo, Tokyo 106, Japan

Abstract. This paper describes a new method for extracting 3D line's parameters from a three-dimensional(3D) Spatio-Temporal Images(STIs). This method assumes that images are taken densely so that its temporal continuity from one frame to next one is approximately equal to the spatial continuity in an individual image. The method utilizes knowledge of the camera motion to cut a 3D STI with a sine slit, thereby forming a two-dimensional(2D) STI which contains much less data than that of the original 3D STI. The significant 3D lines in the scene can be reconstructed from only one 2D STI without the corresponding process, which is necessary in general methods based on 3D STI. For straight line camera motions, a 2D STI has a linear structure on the 3D line parameters that makes them easier to analyze. The process computes the position and direction vectors of the 3D lines and determines all their apexes. In this paper, we first describe the creation of a 2D STI from a 3D STI and the loci of 3D lines in the 2D STI, then give the algorithm for calculating 3D line parameters from the 2D STI, and finally examine 3D line extracting accuracy depending on the parameters of the sine slit. Some experimental results are presented to verify the effectiveness of the method.

1 Introduction

Recovering 3D information from a sequence of images is one of the main issues in computer vision. Tomasi and Kanade[1] developed a factorization method to recover shape and motion using an orthographic projection model, and obtained robust and accurate results. Azarbayejani and Pentland[2] presented a formulation for recursive recovery of motion, pointwise structure, and focal length from feature correspondences tracked through an image sequence. For other approaches, see Huang's review paper[5] .

Because there is no single 2D image which contains enough information to enable reconstruction of the 3D scene, a computer vision approach is usually based on multiple images, including stereo pairs and image sequences acquired by a moving camera. It is theoretically possible to elicit a 3D scene by using more than one image. A 3D spatio-temporal image(STI) is a typical multiple image, or a volume of 2D images, in which temporal continuity from one image to next image is approximately equal to the spatial continuity in an individual image. A 3D STI can be used to reconstruct the 3D scene, but the methods come at the cost of much more data to process, and the added complexity of additional viewpoints which, if unknown, must usually be determined from the images.

In order to avoid the corresponding procedure, Bolles and Baker developed a technique which cuts EPIs(Epipolar-Plane Image) in a 3D STI[3] [4] which gives both traces of 3D points and spatio-temporal events such as the occlusion of an object by another and scene structure from motion. A simple EPI is a slice of a 3D STI along a scan line, shown as in figure 1(a). The depth of the points in a 3D scene can be determined by analyzing its locus, which is usually a slanted line in an EPI. All points of the 3D scene can theoretically be obtained by processing a set of EPIs. But, with an EPI method, it is impossible to extract 3D lines which are more useful for visual sensation.

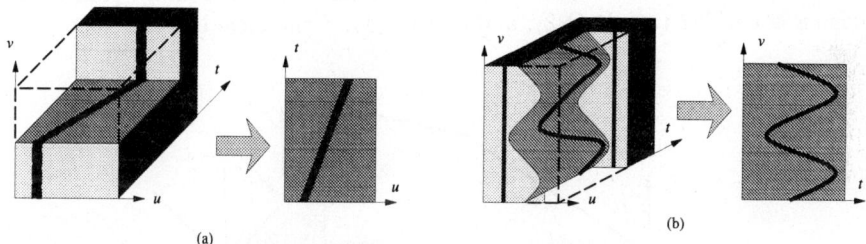

Fig. 1. (a) Epipolar-Plane Image, (b) 2D Spatio-Temporal Image with Sine Slit

We shall be describing a new technique for reconstructing 3D line segments in a static scene from an extended sequence of images with a straight line camera motion. There are also many other methods which estimate 3D line segments by matching 2D line segments between successive images [6] [7]. Matching is one of the most difficult steps in structure estimation. Our technique is similar to the Bolles and Baker method in that we take the images in close succession to omit the line segment matching, and cut the 3D STI to form a 2D STI which is shown in figure 1(b). In contrast with an EPI method, a sine slit, which crosses all scan lines or epipolar planes rather than only one and scans all features flowing into frames, is used here. To simplify the process, we assume that the camera motion parameters are known. Thus, instead of having to estimate both the camera motion parameters and the object shape from the data, as is commonly done in motion analysis methods, we estimate only object shapes. This assumption is same as the case of EPI methods. On the other hand, the estimated features here are the parameters (position and direction vectors) of 3D line segments, not the depth of the 3D scene points which are estimated in an EPI method.

This paper is organized as follows. Section 2 describes the creation of a 2D STI from a 3D STI. Section 3 addresses the proposed algorithm with which the all parameters of a 3D line is calculated from one 2D STI. Section 4 examines the effect of the parameters of the sine slit on the accuracy of the estimation. Section 5 gives some experiment results. Section 6 concludes the paper with a discussion.

2 From 3D STI to 2D STI

In this section, we describe how a 2D STI is created from a 3D STI and how a 3D straight line appears in a created 2D STI.

2.1 Coordinate System and 3D Line's Projection

The camera model and coordinate systems are shown in figure 2. $OXYZ$ is the world coordinate system. $O_t X_t Y_t Z_t$ is the camera coordinate system, which has a translation vector $S = ss$ and rotation matrix R to the world coordinate system, where $s = (s_X, s_Y, s_Z)^T$ is a unit vector and s is the distance between O and O_t. Camera's direction is $O_t X_t$-axis. Plane cuv is the image plane. Here, axis cu is parallel to axis $O_t Z_t$, and cv to $O_t Y_t$. f the focal length.

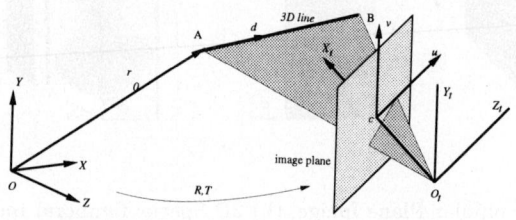

Fig. 2. Coordinate system and the projection of a 3D line

In the world coordinate system, a 3D straight line is determined by its direction vector $d = (d_X, d_Y, d_Z)^T$ and its position vector $r_0 = (X_0, Y_0, Z_0)^T$, and can be mathematically expressed as

$$r \times d = r_0 \times d \ . \tag{1}$$

Here, r is a point on the line.

For the convenience, we define two notions here,

Projection plane: the plane which crosses both the 3D line and O_t.
Normal vector of the projection plane: the normal vector of a 3D line's projection plane. It is defined mathematically as

$$Q = r_0 \times d \ . \tag{2}$$

In the camera coordinate system $O_t X_t Y_t Z_t$, the 3D line's parameters become to

$$\begin{aligned} r_{0_t} &= R(r_0 - S) \\ d_t &= Rd \\ Q_t &= r_{0_t} \times d_t = RQ - R(S \times d) \ . \end{aligned} \tag{3}$$

2.2 A 3D Line in a 2D STI

A 3D Line's Parameters and Its Projection. As to a point on a 3D line, we assume that its projection on the image plane is (u_t, v_t). Because the vector $r_t = (f, v_t, u_t)^T$ is in the projection plane, we get

$$r_t \cdot Q_t = 0 \ . \tag{4}$$

From (3) and (4), we obtain

$$(R^{-1} r_t) \cdot (S \times d) = (R^{-1} r_t) \cdot Q \ . \tag{5}$$

If we define $r = R^{-1} r_t$, which is the expression of the projection vector r_t in the world coordinate system, we get the relation among r, 3D line's parameters d and Q, and camera's translation vector S,

$$r \cdot (S \times d) = r \cdot Q \ . \tag{6}$$

The above relation is considered in the world coordinate system. The camera's rotation has been compensated by considering the projection vectors at the world system. Below, the camera's rotation will not be considered except specially mentioned.

A 3D Line in 3D STI. If we consider that $S = ss$, (6) becomes

$$s = \frac{r \cdot Q}{r \cdot (s \times d)} \ . \tag{7}$$

Without loss of generality, we let camera move in plane OXZ. This means $s = (s_X, 0, s_Z)^T$. Furthermore, a 3D line, which is parallel to the camera's moving direction, cannot be reconstructed from the image sequences. That is to say, the direction vector of the 3D line, which can be reconstructed here, can be expressed as $d = (p, 1, q)^T$. Assuming $r = (f, v, u)^T$ and $Q = (Q_X, Q_Y, Q_Z)^T$, we can write (7) as

$$s = \frac{fQ_X + vQ_Y + uQ_Z}{(us_X - fs_Z) - vQ_d} \tag{8}$$

with $Q_d = qs_X - ps_Z$. (8) shows that a 3D line should take shape of curved surface in the 3D STI with $s, u,$ and v as its three axes.

Types of Slit and 3D Line in a 2D STI. In order to estimate a 3D line's parameters, we can rewrite (8) to a linear form

$$fQ_X + vQ_Y + uQ_Z + svQ_d = s(us_X - fs_Z) \ . \tag{9}$$

It is obvious that Q and Q_d can be obtained from (9) if $f, v, u,$ and sv are not linearly related. One may think of some quadratic curves, such as parabolic or hyperbolic curves, which can satisfy the nonlinear restraints. We might rather say that a sine slit is an desirable one, because a sine slit both scane almost all features flowing into and out of the image plane, and satisfy the nonlinear restraints. Another reason for selecting a sine slit here is that the process of estimating the parameters from a 2D STI will be simplified by using a sine slit. This will be discussed in section 3. A sine slit can be mathematically denoted by

$$u = a\sin(\frac{2\pi v}{K}) \tag{10}$$

here, a is the amplitude, and K is the period of the sine slit. With a sine slit, (9) becomes

$$fQ_X + vQ_Y + a\sin(\frac{2\pi v}{K})Q_Z + svQ_d = s(a\sin(\frac{2\pi v}{K})s_X - fs_Z) \ . \tag{11}$$

This means that a 3D line becomes a curve in a 2D STI which uses v and s as its two axes.

3 Estimation of A 3D Line

From (11), estimating the parameters of a 3D line from a 2D STI is equivalent to extracting four parameters of a curved line in a 2D image. There has been some research in the extraction of more than two parameters' curves such as circles and ellipses in a 2D image using the Hough Transformation (HT)[8]. Those methods can also be used here to estimate the parameters of a 3D line from a 2D STI.

3.1 Hough Transformation

In general, a four-dimensional HT is necessary to estimate four parameters. It is a time and memory consuming process. As to that a sine slit is used and the specifications of our problem, the four-dimensional HT can be simplified to two steps of two-dimensional HT.

Let (s_1, v_1) and (s_2, v_2) be two points of one curve in a 2D STI, and there is the relation $v_2 = v_1 \pm kK, k = 1, 2, \cdots$ between v_1 and v_2. Then, we have

$$\sin(2\pi\frac{v_1}{K}) = \sin(2\pi\frac{v_2}{K}) \ . \tag{12}$$

We can also get

$$(v_1-v_2)Q_Y+(s_1v_1-s_2v_2)Q_d = s_1(a\sin(\frac{2\pi v_1}{K})s_X-fs_Z)-s_2(a\sin(\frac{2\pi v_2}{K})s_X-fs_Z) \ . \tag{13}$$

Here, we refer to the two points with the relation (12) as *dual-points*. In the same way as with Random Hough Transformation(RHT) [9], parameters Q_Y and Q_d can be obtained from a set of dual-points. Then, parameters Q_X and Q_Z can also be calculated by dealing with Q_Y and Q_d as known parameters in (11).

3.2 Calculation of the Parameters of 3D Lines

Given Q_X, Q_Y, Q_Z and Q_d by RHT, from (2) and constraint $\boldsymbol{Q} \cdot \boldsymbol{d} = 0$, the parameters of a 3D line can be calculated with following equation.

$$\begin{cases} \boldsymbol{d} = (p,1,q)^T = \left(-\frac{Q_d Q_z + s_X Q_Y}{Q_X s_X + Q_Z s_Z}, 1, \frac{Q_d Q_X - s_Z Q_Y}{Q_X s_X + Q_Z s_Z}\right)^T \\ \boldsymbol{r}_0 = (X_0, Y_0, Z_0)^T = \left(Q_z + pY_0, \frac{qQ_X - pQ_z}{p^2 + 1 + q^2}, qY_0 - Q_X\right)^T \end{cases} \quad (14)$$

4 Error Analysis

For difference between an estimated 3D line segment and its original one, we define the following three types of errors:

Direction Error is the angle between a estimated estimated 3D line and its original one.

Position Error is defined as $\frac{|\boldsymbol{r}_0 - \boldsymbol{r}_0'|}{|\boldsymbol{r}_0|}$ with $\boldsymbol{r}_0'$ is the estimated position vector.

Length Error is defined as $|1 - \frac{len'}{len}|$ with len' and len are the estimated length and the true length of the segment respectively.

Obviously, the value of the three errors above would be zero, if the extracted 3D line segment is the same as the original one.

Some simulations are made to analyze how all errors vary with the amplitude and period of a sine slit. The results are shown in figure 3. Here, the horizontal axes are the amplitude of sine slit, and the vertical axes are errors. The relation between errors and period of sine slit are shown by different curves in figures. From the results, we can conclude that errors will decrease with the increase of sine slit amplitude, but will change very little with sine slit's period. The conclusion can also be explained as follows: as the amplitude of sine slit increases, the object can be captured from additional different angles and positions, and the precision of the estimation increases undoubtedly.

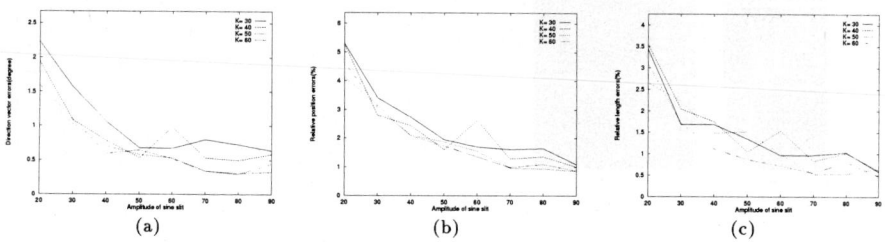

Fig. 3. (a) Direction errors, (b) position errors, (c) length errors

5 Experiments

Some frames of a house is shown in figure 4(a). With 250 image sequences, the created 2D STI is shown in figure 4(b), and its edge image and the extracted segments are shown in figure 4(c) and 4(d) respectively. After determining the apexes of all segments, the three views of the house are shown in figure 4(e).

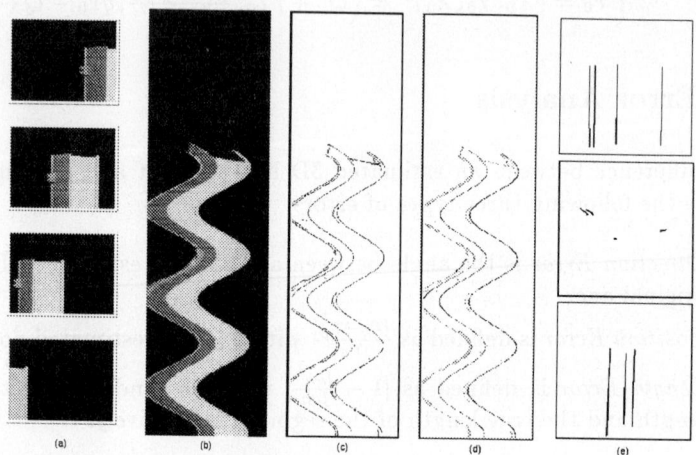

Fig. 4. (a) Image sequences, (b) 2D STI, (c) edge image, (d) segment image, (e) Results: front view(upper), top view(middle), side view(bottom)

Another experiment was conducted with images of some bars. The original images and the results are shown in figure 5(a) — (e).

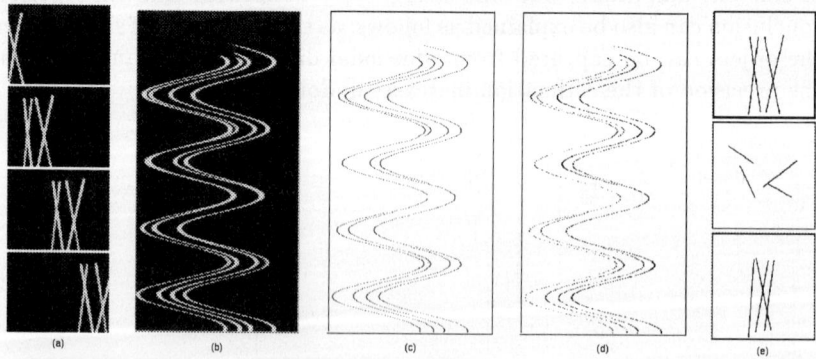

Fig. 5. (a) Image sequences, (b) 2D STI, (c) edge image, (d) segment image, (e) Results: front view(upper), top view(middle), side view(bottom)

6 Conclusion

In this paper, we presented a general analysis of how a 2D STI is created from a 3D STI, and the estimation of a 3D line segment from the created 2D STI is accomplished. The selection of the slit is also discussed. We showed that a sine slit is a desirable one for a straight line camera motion. From a created 2D STI with a sine slit, 3D line's parameters form a linear structure and can be estimated without directly processing the 3D STI. The error analysis was done using simulations. The effectiveness of the proposed method was demonstrated by using real images.

References

1. Tomasi C., Kanade T.: Shape and Motion from Image Streams under Orthography: a Factorization Method. Intern. Journal of Comput. Vision. **9**(1992) 137-154.
2. Azarbayejani A., Pentland P.: Recursive Estimation of Motion, Structure, and Focal Length. IEEE Trans. PAMI, **17**(1995) 562-575.
3. Bolles R., Baker H., Marimont D.: Epipolar-Plane Image Analysis: An Approach to Determining Structure from Motion. Intern. Journal of Comput. Vision. **1** (1987) 7-55.
4. Baker H., Bolles R.: Generalizing Epipolar-Plane Image Analysis on the Spatiotemporal Surface. Intern. Journal of Comput. Vision. **3** (1989) 33-49.
5. Huang T. S., Netravali A. N.: Motion and structure from feature correspondences: A Review. Proceedings of the IEEE, **82**(1994) 252-268.
6. Weng J. Y., Huang T. S., Ahuja N.: Motion and structure from line correspondences: closed-form solution, uniqueness, and optimization. IEEE Trans. PAMI. **14**(1992) 318-336.
7. Zhang Z. Y.: Estimating motion and structure from correspondences of line segments between two perspective images. IEEE Trans. PAMI. **17** (1995) 1129-1139.
8. McLaughlin R. A.: Randomized Hough Transform: Improved Ellipse Detection with Comparison. Technical Report, University of Western Australia(http://ciips.ee.uwa.edu.au/Reports/).
9. Xu L., Oja E.: Randomized Hough Transform(RHT): Basic mechanisns, algorithms, and computational complexities. CVGIP: Image Understanding **57**(1993) 131-154.
10. Zheng J. Y., Tsuji S.: Panoramic Representation for Route Recognition by a Mobile Robot. Intern. Journal of Comput. Vision. **9**(1992) 55-76.

Toward Non-intrusive Motion Capture

A. Bottino, A. Laurentini, P. Zuccone
Dipartimento di Automatica ed Informatica, Politecnico di Torino, Italy
bottino@polito.it, laurentini@polito.it

Abstract. Shape-from-silhouettes, a simple approach to 3D shape understanding, can also be used for recovering posture and motion of human bodies. Silhouettes are easy to obtain from intensity images, do not require foreign objects attached to the subject, can directly provide a 3-D reconstruction of the body, or drive model-based motion capture. The purpose of this work is to demonstrate the effectiveness of this approach in a virtual environment, and to investigate number and position of stationary cameras suitable for precise reconstruction of the 3-D posture and motion of the human body.

1. Introduction

Understanding the human body posture and motion is a much debated and challenging computer vision problem, since the human body is a highly complex and flexible object. In addition, many postures produce self-occlusion, which makes this task more difficult. Many techniques have been reported, different for the data used (areas and contours in intensity images, optical flow, laser range data, etc.) and for the approach to motion recovery. References to many of them can be found in the papers of Rohr [1] and Leung and Yang [2].

Many approaches are model-based. The human body is represented with some kind of model whose 3-D posture and motion is matched with the physical data. Stick articulated models (as in [2]) idealize the human skeleton. Cylinders and generalized cylinders [1], [3], [4], deformed superquadrics [5], geons [6], parametric solids and finite elements [7], [8], have been used to build models which mimic more or less closely the human body. Motion estimation has been improved by predictive Kalman filtering [1], [7], [8].

Capturing the motion of the human body is an important practical issue. Realistic animation of 3-D characters for movies, TV and 3-D games are driven by motion data obtained from human performers. Many other application areas exist, as telerobotics, ergonomics, crash simulation with dummies, biomechanics and sport performance analysis. Several commercial "motion capture" (MC) equipment exist at present. They are based on the idea, developed some years ago(see for instance [9], [10], [11]), of tracking key points, usually joints, of the subject. This can be done with either of two technologies: magnetic and optical tracking. Both techniques require more or less bulky objects to be attached to the body. These objects disturb the subject and more or less affect his gestures. For some application, as analyzing sport performances, this could be a serious drawback.

The purpose of the authors is to develop an alternative approach, on one hand capable of overcoming the problems highlighted for commercial devices in some application areas, and on the other hand sufficiently simple and robust as to allow the implementation of practical equipment. Our approach is based on multiple 2-D silhouettes of the body extracted from 2-D images. Other non-intrusive approaches, as in [2], are based on outlines extracted from one image. However in these cases posture recovering can be underconstrained, and no direct 3-D reconstruction can be performed. The main features of our approach are the following.

Silhouettes are easy to obtain from intensity images. From silhouettes: i) a direct reconstruction of the 3-D shape of the body can be computed with a technique known as *volume intersection* (VI), and ii) the 3-D posture and motion of models of the human body can be obtained by fitting a model to the reconstructed volume.

Number and position of the cameras, subject to practical constrains, strongly affect the accuracy of the reconstruction of the human body, and of the estimation of posture and motion of a model. Studying this problem is a necessary step toward effective motion capture from multiple silhouettes. The purpose of this paper is twofold: 1) demonstrating in a virtual environment a system based on this approach; 2) investigating the precision of both 3-D direct reconstruction and model-based posture and motion identification for several postures of the body, different arrangements of cameras, and various resolutions.

The content of this paper is as follows. In Section 2 we discuss the reconstruction of 3-D shapes and the determination of the posture of 3-D models from silhouettes. In Section 3 we describe the virtual environment and the algorithms for determining posture and motion. In Section 4 we present and discuss the experimental results obtained.

2. The Multiple Silhouette Approach to Motion Capture

Reconstructing 3-D shapes from 2-D silhouettes is a popular approach in computer vision. The VI technique (Fig. 1) recovers a volumetric description R of the object O by intersecting the solid regions of space C_i within which each silhouette S_i constrains the object to lie ([12], [13], [14], [15], [16], [17]). R is a bounding volume which more or

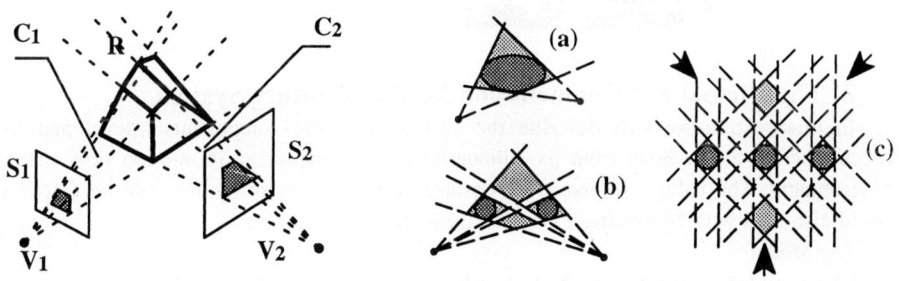

Fig. 1   Fig.2

less closely approximates O, depending on the viewpoints and the object itself.

Despite the simplicity of the basic idea, the VI approach raises a number of theoretical questions, such as: which objects are exactly reconstructable; which is the closest approximation that can be obtained for non reconstructable objects; what can be inferred about the unknown object from the reconstructed object. The recently introduced geometric concept of visual hull provides answers to these questions [17], [18], [19].

Using VI for reconstructing the human body requires to face some difficulties. Bad placement and insufficient number of cameras produce bulges which affect the correct placement of the model (see Fig. 2(a) for a 2-D example). In addition, because of the complex shape of the human body, this technique can produce "phantom" volumes, that is protrusions not corresponding to real limbs. In the 2-D example of Fig.2(b) two silhouettes are generated by two objects but VI reconstructs two more objects and, without further information, we could be unable to tell the "phantom" objects from the real ones. In this case, a third viewpoint easily overcomes the problem. However, three viewpoints could be not sufficient to avoid "phantoms". For instance, in Fig. 3, three evenly spaced viewing directions produce two "phantom" objects. Also in this case an additional viewpoint, or a better choice of the viewing directions, overcomes the problem. Thus choosing the number and position of the cameras is crucial for reconstructing 3-D bodies, and investigating this point is among the goals of this paper.

Model-based posture understanding based on VI consists in fitting a model of the human body to the volume reconstructed by VI. In principle this is exactly the same than fitting in 2-D the projections of the model to the various silhouettes (at least for objects of unknown shape), since the information provided by the set of silhouettes or by the reconstructed volume is the same. However, fitting in 3-D allows to easily visualize the bulges or "phantom" volumes which affect the recovered posture. For this reason, in our study we fit the model in 3-D, at the expense of some more computation.

In model-based motion capture the "phantom" problem is by far less severe, since exploiting continuity is a powerful tool for fitting the "true" volumes in ambiguous cases.

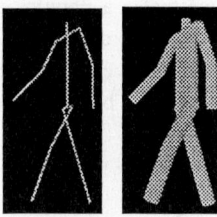

Fig 3 - the skeleton (a) and the dummy (b)

3. The Virtual Environment and Motion Capture System

In this section we will describe the various software components developed for reconstructing a 3-D body from its silhouettes and for capturing its motion in a virtual environment. The volume intersection, posture determination an motion capture software could also deal with silhouettes extracted from real world images.

The Model

The skeleton of the human body has been modeled as a tree of 14 rigid segments connected by ball joints (see Fig.3(a)). The lengths of the segments agree with the average measures of the male population [21]. The body is modeled by cylinders of various width centered about the segments (see Fig.3(b)), except for the trunk, whose section is a rectangle with smoothed angles. Specifying a posture requires to specify a vector $\wp$ with 31 parameters. A convenient program can drive the dummy in a walk sufficiently realistic for our purposes (see Fig. 4).

Cameras and Silhouettes

Any number of stationary cameras can be located anywhere in the virtual 3-D environment (see for instance Fig. 5). Each virtual camera is defined according to the Tsai's camera model [19]. Its position and orientation are specified with a view center, an optical axis, and an up vector. The camera provides in the image plane a frame of 512·512 two level pixels. The silhouettes have been obtained using Open GL.

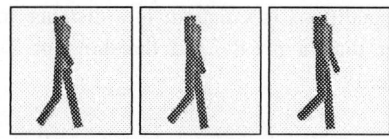 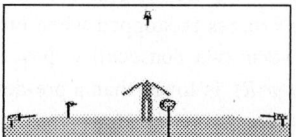

Fig.4 - The walking dummy   Fig. 5 - The virtual environment

The Volume Intersection Algorithm

Provision has been also made for images produced by a real camera, and the back-projection of the silhouettes can be performed using Tsai's camera model and calibration data [20].

The VI algorithm, which works at various resolutions, outputs the boundary voxels of the reconstructed volume R. Its running time is dependent on the number of boundary voxels, and thus approximately on the square of the linear resolution.

The outline of the algorithm is as follows. A 3-D point P is an internal point, belonging to R, if each projection of P (according to the camera model) in an image plane belongs to the corresponding silhouette. A voxel is a *boundary voxel* if some, but not all of its vertices belong to R. After finding with a simple heuristic one boundary voxel, the algorithms checks the six adjacent voxels and selects as boundary voxel those which share with the first voxel a face whose vertices are not all interior or all exterior. By recursively applying this rule, all the boundary voxels are found. In Fig. 6 we show some outputs of the VI algorithm at various resolutions.

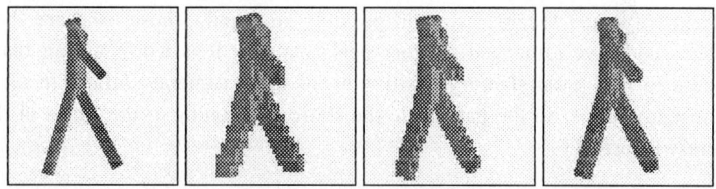

Fig. 6 - The original posture and the volumes reconstructed at increasing resolutions (voxel sizes of 49, 35 and 21 mm)

Determining the Posture of the Model

Fitting the model to the volume R obtained by VI requires to minimize some measure of distance between the dummy and the volume reconstructed. To this purpose, we consider the squared distances between the centers C_i of the boundary voxels and the skeleton of the dummy. More precisely, we assign to each center the smallest of the distances from all the segments of the skeleton. The distance $d_j(C_i)$ between the point C_i and the segment j is the smallest distance between the point and a point of the segment. Let S_j, j = 1, ..., 14, be the set of centers assigned to each segment. We define the distance function as:

$$D(\wp, R) = \sum_{j=1}^{14} w_j \cdot \sum_{\forall C_i \in S_j} d_j^2(C_i)$$

The purpose of the weights w_i is to enhance the contribution of the smallest parts of the model in order to obtain similar posture errors for trunk and limbs. For minimizing $D(\wp,R)$ we use the gradient method in the 31-dimension space of the position parameters. The process is stopped when either of two conditions is satisfied: 1) a distance measure between two consecutive postures is lower than a pre-defined threshold or 2) when $\Delta D(\wp,R)$ is lower than a pre-defined threshold.

Recovering the Motion of the Model

In order to recover the motion of the dummy, the above procedure is applied several times to multiple sequences of frames. Exception made for the first time, the starting position of the model is that obtained from the previous set of silhouettes. Since each time the dummy is very close to its final position, the computation of the new posture requires relatively few steps. In addition, some sort of implicit filtering takes place, since possible local minima of the distance function due to "phantom" volumes are avoided.

4. Experimental results

In this section we present and discuss some of the most significant results relative to the analysis of motion and posture of the walking dummy. The test has been performed in the following condition:

- we consider a full gait cycle (two steps of about 1 m each), recorded in 42 frames
- 3 different voxel sizes for the VI algorithm are used (49, 35, and 21 mm)
- two different camera arrangements with 4 and 5 cameras have been set. In the second case the additional camera has a vertical optical axis shooting the walking dummy from above (see Fig. 7). The optical axes of the first 4 cameras all converge into the origin and all lie in an horizontal plane at a distance of 1 m from the floor. The optical center of the fifth camera is located at 4 meters from ground level. Although better arrangements of 4 and 5 cameras could possibly be found, those used are the result of several tests and give rather satisfactory performances. It is worth noting that we have been unable to find a satisfactory positioning of three cameras. Infact, in all the cases tested, for many frames in the gate cycle the static positioning of the limbs of the dummy was severely incorrect.

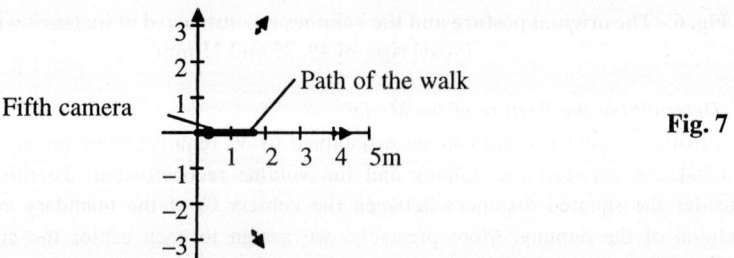

Fig. 7

4.1 Volume Intersection Accuracy

The accuracy of the reconstruction obtained by VI algorithms is defined as the ratio between the reconstructed volume and the volume of the original object [12], [13], [14].

In order to minimize the consequences of the quantization, the reconstructed volume is computed with the formula:

$$rec.\,volume = \sum_{inner\,voxels} s^3 + \sum_{boundary\,voxels} \frac{voxel\,vertices \in R}{8} s^3, \quad s = voxel\,size$$

The reconstruction accuracy obtained with 4 and 5 cameras is plotted for all frames of the gait cycle and three resolutions in diagrams 1 and 2.

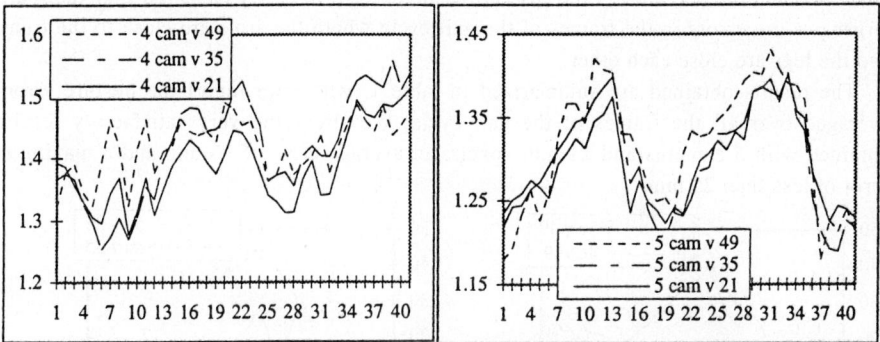

Diag. 1,2 - **Volume intersection accuracy for 4 and 5 cameras**

Even with 5 cameras, these data show a rather coarse reconstruction for many postures. Since decreasing the voxel size provides perceptible but not substantial improvements, low accuracy is essentially due to the low number of cameras in relation with the complex shape of the body. However, as we will show in the next subsection, these rather rough volumes allow to determine with good precision the posture of the dummy.

4.2 Posture Determination in the Gait Cycle

In this subsection we present the experimental results concerning the accuracy of the postures of the model determined in the gait cycle. For summarizing the overall difference between the true posture and the posture determined with our algorithm, we compute an average posture error for each frame as follows:

$$\sum_{i=1}^{14} L_i \cdot d_i \Big/ \sum_{i=1}^{14} L_i$$

where L_i is the length of each segment and d_i is the average distance between corresponding points of each real segment and the corresponding reconstructed segment (see Fig. 9):

$$d_i = \int_0^1 D_i(t)dt$$

$$D_i(t) = \|P_{i1}(t), P_{i2}(t)\| = \|S_{i1} + (E_{i1} - S_{i1})t, S_{i2} + (E_{i2} - S_{i2})t\|$$

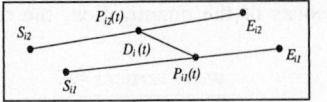

Fig. 9

The average posture errors for each frame of the gait cycle, expressed in mm, are reported in diagrams 3, 4, and 5 for decreasing voxel size. In each diagram we plot the errors for both the 4 and 5 cameras arrangements.

Substantial error reduction is provided by the fifth camera, since it strongly reduces phantoms and bulges, and by the resolution increase. It is easily observed that there are larger posture errors in the frames of the gait cycle where the arms are close to the trunk and the legs are close each other.

The results obtained are summarized in table 1, where we report the posture errors averaged over all the frames of the gait cycle. We stress the very satisfactory results obtained with 5 cameras and 21 mm voxels: an average error of 16 mm and a maximum error of less than 25 mm.

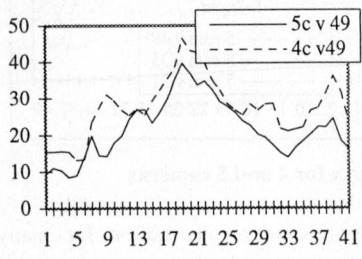

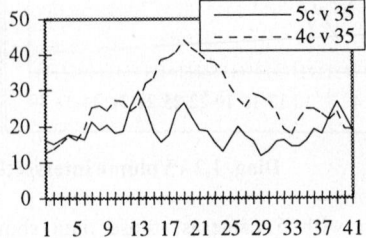

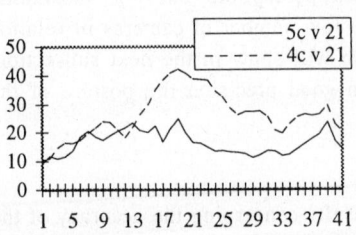

Diag. 3, 4, 5 - Average posture error in millimeters vs. camera number (voxel size 49, 35, 21 mm)

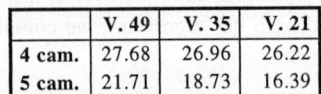

	V. 49	V. 35	V. 21
4 cam.	27.68	26.96	26.22
5 cam.	21.71	18.73	16.39

Table 1 - Posture errors in millimeters averaged over all frames

5. Concluding Remarks and Future Work

We have demonstrated in a virtual environment a technique for determining posture and motion of the human body. This technique is based on multiple silhouettes obtained with a set of TV cameras, and does not require foreign objects attached to the subject. This could be of paramount importance in application areas such as sport performance analysis. The experimental results obtained relative to a walking dummy show that precise posture determination and motion capture can be obtained with four or five cameras in a sufficiently large workspace. Another interesting result is that the precision of the motion captured is relatively unaffected by reconstructing the 3-D dummy at low resolution. Most of the software developed could be also applied to silhouettes extracted from real world images.

Further work both in virtual and real world is needed for transforming the idea into an effective practical tool. In virtual world, we plan more extensive tests to explore the precision achievable with our technique. In particular, we plan to extend our motion capture experiments beyond the walking dummy, and to study other kinds of motions, as running and jumping. We are also evaluating the use a more realistic dummy, based on a dynamic model, and some filtering methods for improving precision. Next step will be applying our algorithms to real image sequences, and comparing the results obtained with those of the conventional motion capture algorithms.

REFERENCES

[1] K. Rohr, "Toward model-based recognition of human movements in image sequences," *CVGIP: Image Understanding,* vol. 59, no.1, pp.94-115, 1994

[2] M. K. Leung and Y. H. Yang, "First sight: a human body outline labeling system," *IEEE Trans. on PAMI,* vol. 17, no. 4, pp.359-377, 1995

[3] D. Marr and H. K. Nashihara, "Representation and recognition of the spatial organization of three-dimensional shapes," *Proc. R. Soc. London,* vol.B 200, pp.269-294,1978

[4] R. Mohan and R. Nevatia, "Using perceptual organization to extract 3D structures," *IEEE Trans. on PAMI,* vol. 11, no. 11, pp.1121-1139, 1989

[5] F. Solina and R. Bajcsy, "Recovery of parametric models from range images from range images: The case for superquadrics with global deformation," *IEEE Trans. PAMI,* vol. 12, no. 2, pp.131-147, 1990

[6] I. Biederman, "Recognition-by-components: A theory of human image understanding," *Psychological Rev.,* vol.94, no. 2, pp.115-147, 1987

[7] A. Pentland and B. Horowitz, "Recovery of nonrigid motion and structure," *IEEE Trans. on PAMI,* vol. 13, no. 7, pp.730-742, 1991

[8] A. Pentland and S. Sclaroff, "Closed-form solutions for physically based shape modeling and recognition," *IEEE Trans. on PAMI,* vol. 13, no. 7, pp.715-729, 1991

[9] R. Rashid, "Toward a system for the interpretation of moving light displays," *IEEE Trans. on PAMI,* vol. 2, no. 6, pp.574-581, 1980

[10] J. Webb and J. Aggarwal, "Structure from motion of rigid and jointed objects," *Artif. Intell.* vol.19, pp. 107-130, 1982

[11] H. Lee and Z. Chen, "Determination of 3D human body posture from a single view," *Comput. Vision, Graphics and Image Processing,* vol.30, pp. 148-168, 1985

[12] N. Ahuja and J. Veenstra, "Generating octrees from object silhouettes in orthographic views," *IEEE Trans. on PAMI,* Vol.11, pp.137-149, 1989

[13] Noborio et al., "Construction of the octree approximating three-dimensional objects by using multiple views", *IEEE Trans. on PAMI,* Vol.10, pp.769-782, 1988

[14] M. Potemesil, "Generating octree models of 3D objects from their silhouettes in a sequence of images", *Comput.Vision Graphics Image Process.,* Vol.40, pp.1-29, 1987

[15] C. H. Chian and J. K. Aggarwal, "Model reconstruction and shape recognition from occluding contours", *IEEE Trans.on PAMI,* Vol.11, pp.372-389, 1989

[16] W. N. Martin and J. K. Aggarwal, "Volumetric description of objects from multiple views", *IEEE Trans. on PAMI,* Vol.5, pp.150-158, 1983

[17] P. Srinivasan et al., "Computational geometric methods in volumetric intersection for 3D reconstruction", *Pattern Recognition,* Vol.23, pp.843-857, 1990

[18] A. Laurentini, "The visual hull concept for silhouette-based image understanding," *IEEE Trans. Patt. Anal. Machine Intell.,* vol. 16, no.2, pp. 150-162, 1994

[18] A. Laurentini, " How far 3-D shapes can be understood from 2-D silhouettes," *IEEE Trans. Pattern Anal. Machine Intell.,* vol.17, no.2, pp.188-195, 1995

[19] A. Laurentini, " How many 2d Silhouettes does it take to reconstruct a 3d object," *Computer Vision and Image Understanding,* vol.67, no.1, July, pp.81-87, 1997

[20] R. Y. Tsai, "A versatile camera calibration technique for high-accuracy 3D machine vision metrology using off-the-shelf TV cameras and lenses," *IEEE Journal of Robotics and Automation,* Vol.RA-3, no.3, pp.323-344, 1987

[21] S. Fantin, M. Dias, "3D Virtual Mannequin Database" ENEA, document E65334/ENEA/WP2-03.00, 1/12/1994.

Appearance Based Visual Learning and Object Recognition with Illumination Invariance

Kohtaro Ohba[1], Yoichi Sato[2] and Katsusi Ikeuchi[2]

[1] Mechanical Engineering Laboratory, MITI, Tsukuba, 305, JAPAN
[2] Institute of Industrial Science, The University of Tokyo, Tokyo, 106, JAPAN

Abstract. This paper describes a method for recognizing partially occluded objects to realize a bin-picking task under different levels of illumination brightness by using the eigen-space analysis. In the proposed method, a measured color in the RGB color space is transformed into the HSV color space. Then, the hue of the measured color, which is invariant to change in illumination brightness and direction, is used for recognizing multiple objects under different levels of illumination conditions. The proposed method was applied to real images of multiple objects under different illumination conditions, and the objects were recognized and localized successfully.

1 Introduction

Recently, visual learning methods based on the eigen-space analysis have shown a potential to realize the robust object recognition system, [1] and [2]. And also, it can solve some of these difficulties, such as requirement for real-time processing, difficulty in segmentation, and difficulty in obtaining appropriate object models. In the eigen-space analysis, object models are *learned* from a series of images taken in the same environment as in the recognition mode.

Although promising, the current eigen-space analysis is based on the assumption that objects are not occluded in images. Therefore, to apply the eigen-space analysis for partially occluded objects, we proposed to divide appearances into small windows, referred to as *eigen-windows* [4] and to apply eigen-space analysis to each eigen-window.

One drawback of the eigen-window method is that only a limited number of images can be used for learning object models, and therefore, all possible illumination directions cannot be taken into account.

In this paper, to overcome that drawback, we propose to use the color measurement *hue*, which is illumination invariant, in the eigen-window method. To demonstrate the effectiveness of the proposed method, we applied the method to real images taken under different illumination directions and brightness.

2 Eigen-Window Method

2.1 Eigen-Space Technique

Let M be the number of the images $z_1, z_2, \cdots, z_M$ in a training set related to each rotation of view points θ_1 and θ_2. Each image z_i, with dimensions $N \times N$, has been converted into a column vector of length N^2.

By subtracting the average brightness c of all the images, we obtain the training matrix of the size of N^2 by M,

$$Z = [z_1 - c, z_2 - c, \cdots, z_M - c]. \quad (1)$$

This covariance matrix $Q = ZZ^T$ provides a series of eigenvalues λ_i and eigenvectors $e_i (i = 1, \cdots, N^2)$, where each corresponding eigenvalue and eigenvector pair satisfies:

$$\lambda_i e_i = Q e_i. \quad (2)$$

For reducing memory requirement, we ignore eigenvectors corresponding to small eigenvalues $e_i (i > l)$. These eigenvectors do not affect object recognition results significantly. Once we obtain the remaining eigenvectors, we can construct the eigenvector matrix $E = [e_1, e_2, \cdots, e_l]$ which projects an image z_i (dimension N^2) into the eigen-space as an *eigen-point* g_i (dimension l).

$$g_i = E^T (z_i - c). \quad (3)$$

The eigen-space analysis can drastically reduce the dimension of the images (N^2) to the eigen-space dimension (l) while preserving enough dominant features to reconstruct the original images.

2.2 Eigen-Window Technique

To reduce the disturbance effects such as image shift and occlusion, we propose to select small windows in the original images. Each of the selected small windows is then analyzed by using the eigen-space analysis as described in the previous section. We call this method *the eigen-window method*. Figure 1 shows the overview of the method.

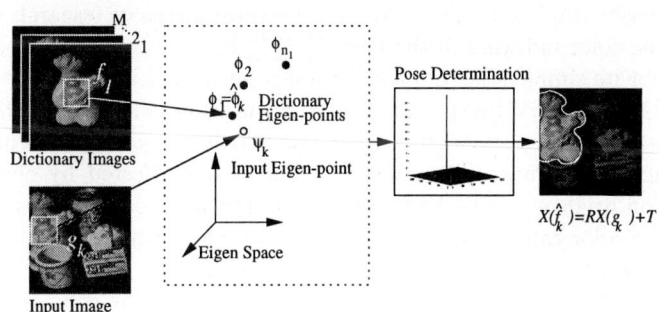

Fig. 1. Eigen-Window Technique.

The training set of eigen-windows is given as:

$$F = [F^1, F^2, \cdots, F^M], \tag{4}$$

where F^i denotes the collection of eigen-windows from the ith training image. Each F^i has the form $[f_1 - c, f_2 - c, \cdots, f_{n_i} - c]$, where f_j denotes the jth eigen window in the ith training image; n_i denotes the number of eigen-windows in the ith image; and c is the average intensity value across all eigen-windows in the whole training set. In Figure 1, the white square denotes one of the training eigen-windows.

Please see more details about matching operation, voting operation, pose determination and selection of effective eigen-windows in [4].

3 Illumination Invariance

In our previous work [4], we have shown that the eigen-window method can successfully recognize and localize an object in b/w input images which contain multiple objects with specularity, even if the input images contain significant amount of noise, occlusion, image shifting and scaling change.

However, the method was based on an assumption that the location and brightness of a light source are fixed. Therefore, the method did not take into accounts shading variation such as highlights on object surfaces. For instance, if an object exhibits specularity, the object appearance can change drastically with different illumination directions, which confuses recognition and localization of the object.

To overcome this limitation, we propose to use an illumination invariant measure for the eigen-window method. By using the illumination invariant measure, the eigen-window method can be used successfully for recognizing and localizing multiple objects under different illumination conditions.

3.1 Illumination Invariance: Hue

Instead of black-and-white intensity images, we use RGB color images in the modified eigen-window method. Actually, several pieces of research works were done on the color indexing in the past [5] - [7], but we would like to use the *hue* criterion for its simpleness, and a color image measured in the RGB color space is converted to a HSV image (H: Hue, S: Saturation, V: Value). In these three parameters, the hue parameter is the value which represents color information, e.g., without brightness. Therefore, the hue is not affected by change of the illumination brightness and direction if the following two conditions hold: 1) the light source color can be expected to be almost white, and 2) a saturation value of object color is sufficiently large.

In Figure 2, object color is represented by three color components S_1, S_2, and S_3. In the RGB color space, those three color components are Red, Green and Blue. Then, the light source color I is given as $I = (1, 1, 1)$. To define hue, saturation and intensity, one pair from three components, Red, Green, and Blue, have to be assigned to S_1' and S_2'. Usually, Red and Green are assigned as $S_1' = R$ and $S_2' = G$.

We conducted a simple experiment using a color test chart to see how hue is affected under different levels of illumination brightnesses. The result is shown in Figure 2 (a). In Figure 2 (b), we can see that hue remains almost constant over a wide range of illumination brightness for many color blocks.

However, for some color blocks, the value of hue does change with different levels of illumination brightnesses. For instance, the black-white color blocks in the last row of the color chart (color blocks #30-#35), red (color block #12) and magenta (color block #24).

That is because the saturation of color blocks #30-#35 is not sufficiently large, i.e., they are very close to gray. Also, hue has a discontinuity at 0 and 2π. That is the reason for unstable hue of the color blocks #12 and #24.

To obtain the value of hue reliably, we propose to use three criteria: *intensity value*, *saturation*, and *phase*.

[Intensity Value]
$$if\ V < V_t\ then\ H = 0, \tag{5}$$

[Saturation]
$$if\ S < S_t\ then\ H = 0, \tag{6}$$

[Phase]
$$if\ H < \Delta P_t\ or\ \|H - 2\pi\| < \Delta P_t\ then\ H = 0. \tag{7}$$

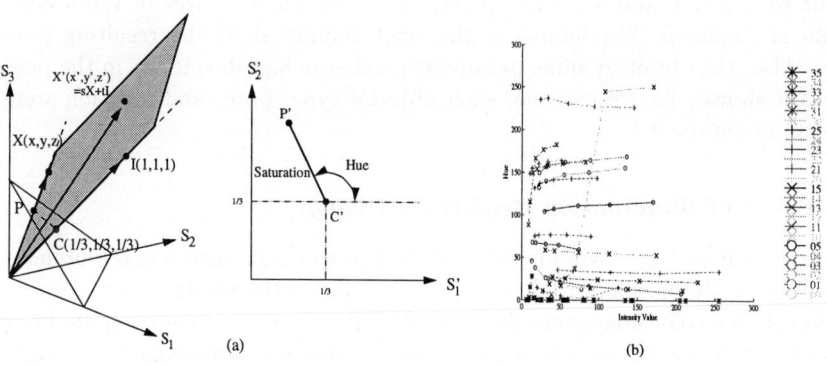

Fig. 2. HSV. (a)HSV space; (b) Hue-Intensity space.

4 Experimental Results

The proposed method was used for recognition and localization of objects in three test cases. In the first case, the same illumination condition was used both for training and for input images. In the second case, input images were taken under different levels of illumination brightnesses. In the last case, input images were taken with different light source locations.

4.1 Object Recognition and Localization with Hue Image

First, a set of training eigen-windows was obtained as described in Section 2. The training images were taken at $\theta_1 = [-20, 0, 20]$ and $\theta_2 = [0, 10, 20, \cdots, 350]$ for three different objects, *mug*, *bird*, and *tylenol*. We refer to the original images as $type(\theta_1, \theta_2)$. For example, the image $mug(-20, 60)$ denotes the image for the mug taken at the position $\theta_1 = -20 deg$ and $\theta_2 = 60 deg$. One hundred eight images were taken for each of the objects by using the experimental setup.

Then, eigen-windows were selected in each training image by using the detectability, similarity and reliability measurements. The number of eigen-windows for each of the objects was initially more than 8,000. After the three measurements were applied, less than 2,000 of the training eigen-windows were finally obtained. Then, these eigen-windows were projected to produce eigen-points according to the equation (3).

One input image containing multiple objects was taken as shown in left hand side of Figure 3. In the input image, there are 7 objects, *duck*, *mug*, *barney*, *bird*, *stop-sign*, *tylenol*, and *tylenol-cold*. First, eigen-windows were selected in the input image by using the detectability measure. Then, we established correspondences between the input eigen-windows and the training eigen-windows by using the similarity between their eigen-points.

The recognition and localization results are shown in figures in the middle column in Figure 3. The figures in the right column show the resulting pose spaces. Also, the obtained affine parameters and standard deviations in the pose space are shown. As we can see, each object's type, pose, and location were successfully obtained.

4.2 Effect of Illumination Brightness Change

The same training eigen-window set was applied to input images taken under a wide range of illumination brightness. Figure 4 shows the result.

The original color images are shown in the left column, and the computed hue images are shown in the middle column. The localization and recognition results are shown in the right column. The affine parameters and standard deviations of the pose space are also given in the figure.

The hue images did not change significantly with different levels of illumination brightness. The main difference between the hue image for the brightest illumination and that for the darkest illumination is that hue values were not computed over a large portion of the object surfaces. This is because intensity

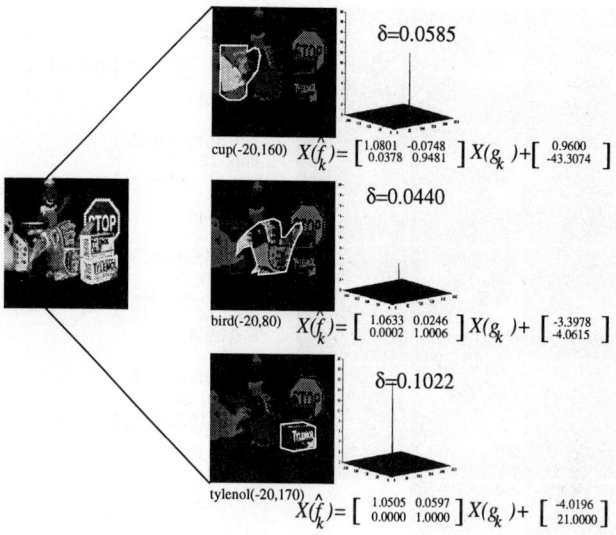

Fig. 3. Recognition Result

values were so small that hue values were set to zero as the background value according to the equation (6).

The experimental results show that the proposed method works even when input images are taken with different levels of illumination brightness. The object was recognized and localized successfully.

4.3 Effect of Different Light Source Positions

The proposed method was also applied to input images taken with different light source locations. As the light source position changes, the appearance of objects in input images changes drastically. Therefore, changing light source position makes recognition and localization of objects even harder than changing illumination brightness.

In this experiment, four different light source positions were used as shown in Figure 5(a). The left column images of Figure 5(b) show the input images taken with each of the four light source positions. The middle column images of Figure 5(b) show the obtained hue images. The right column images present the recognition and localization results. The affine parameters and standard deviations of pose-space are also shown in the figure.

Note that, in this experiment, there was no ambient illumination. Hence, the appearance of the objects change significantly with different light source positions. Nevertheless, the mug was correctly recognized and localized except in the input image for the light position 1. In this case, hue values were not

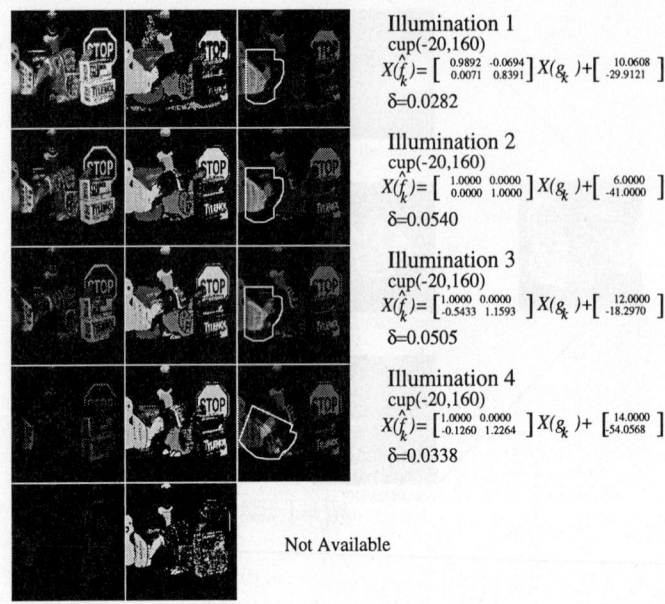

Fig. 4. Object Recognition Results with Illumination Change

obtained over a large portion of the object surface because of shadow casting on the surface.

5 Conclusion

In this paper, by using hue, which is an illumination invariant measure, the eigenwindow method was extended further for recognition and localization of objects in images taken under changing illumination conditions. To use hue information of input images reliably, we introduced three criteria for computing hue values: intensity value, saturation, and phase.

The proposed method was applied to real images, and the method recognized and localized objects successfully even in images taken under significantly different illumination conditions.

References

1. M. A. Turk and A. P. Pentland, "Face Recognition Using Eigenfaces," Proc. CVPR 1991, pp.586-591, 1991.
2. H. Murase and S. K. Nayar, "Visual Learning and Recognition of 3-D Objects from Appearance," *International Journal of Computer Vision*, Vol.14, No.1, pp.5-24, 1995.

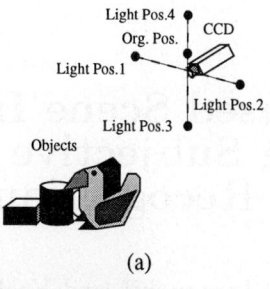

(a)

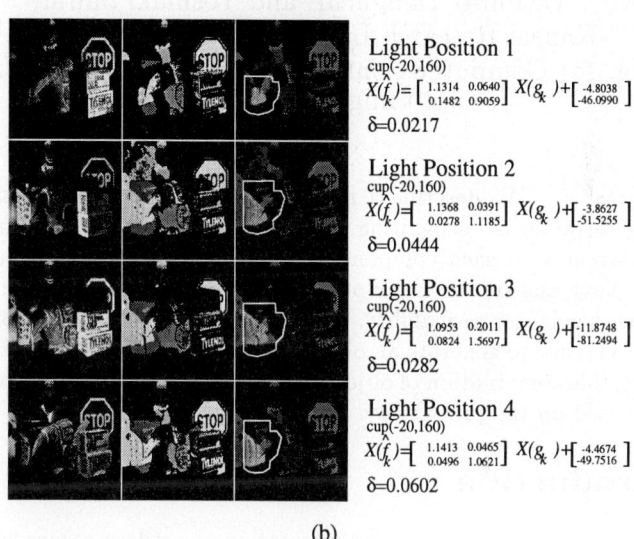

Light Position 1
cup(-20,160)
$$X(\hat{f}_k) = \begin{bmatrix} 1.1314 & 0.0640 \\ 0.1482 & 0.9059 \end{bmatrix} X(g_k) + \begin{bmatrix} -4.8038 \\ -46.0990 \end{bmatrix}$$
$\delta = 0.0217$

Light Position 2
cup(-20,160)
$$X(\hat{f}_k) = \begin{bmatrix} 1.1368 & 0.0391 \\ 0.0278 & 1.1185 \end{bmatrix} X(g_k) + \begin{bmatrix} -3.8627 \\ -51.5255 \end{bmatrix}$$
$\delta = 0.0444$

Light Position 3
cup(-20,160)
$$X(\hat{f}_k) = \begin{bmatrix} 1.0953 & 0.2011 \\ 0.0824 & 1.5697 \end{bmatrix} X(g_k) + \begin{bmatrix} -11.8748 \\ -81.2494 \end{bmatrix}$$
$\delta = 0.0282$

Light Position 4
cup(-20,160)
$$X(\hat{f}_k) = \begin{bmatrix} 1.1413 & 0.0465 \\ 0.0496 & 1.0621 \end{bmatrix} X(g_k) + \begin{bmatrix} -4.4674 \\ -49.7516 \end{bmatrix}$$
$\delta = 0.0602$

(b)

Fig. 5. Effect of Light Source Direction. (a)Light Source Position; (b)Recognition Result.

3. C. Tomasi and T. Kanade, "Shape and Motion without depth," *Proc. of the Third International Conference in Computer Vision*, Osaka, Japan, December 1990.
4. K. Ohba and K. Ikeuchi, "Recognition of the Multi Specularity Objects using the Eigen-Window," *Proceeding of International Conference on Pattern Recognition*, August 1996.
5. G. Healey and D. Slater, "Global color constancy: recognition of objects by use of illumination-invariant properties of color distribution," *Journal of Optical Society of America*, Vol.11, No.11, pp.3003-3010, Nov. 1994.
6. M. J. Swain, "Color Indexing," *International Journal of Computer Vision*, Vol.7, No. 1, pp.11-32, 1991.
7. B. V. Funt and G. D. Finlayson, "Color Constant Color Indexing," *IEEE Trans. on Pattern Analysis and Machine Intelligence*, Vol.17, No.5, pp.522-529, May 1995.

Evidence-Based Scene Interpretation Considering Subjective Certainty of Recognition

Yasuhiro Taniguchi[†] and Yoshiaki Shirai[††]
Kansai Research Laboratories, Toshiba Corp.[†]
Mech. Eng. for Computer-Controlled Machinery Osaka University,[††]
E-mail taniguti@krl.thosiba.co.jp

abstract : We describe a method of interpretation of outdoor scenes considering certainty of recognition. First, probability that each uniform color region corresponds to each component of an object is calculated using region attributes. Next, the system attempts to merge uniform color regions into a new region which has larger probability. The merged region with high probability is used as an evidence to generate an object candidate. Finally, the system obtains the most reliable combination of objects among multiple consistent combinations of objects based on the probability.

1 Introduction

This paper describes a method of interpretation of outdoor scenes which include various objects occluding one another. Interpretation of such scenes involves the following problems: 1) a scene includes objects whose shape, size and color are not fixed; 2) the viewing direction is not fixed; 3) objects occlude one another. It is, therefore, not easy to straightforwardly recognize an object from local features extracted from the corresponding region of the input image. One solution to these problems is to use the certainty of recognition. A system proceeds with interpretation until the most probable interpretation is obtained.

Ohta[1] defined certainty of objects by the fuzzy measure in order to recognize outdoor scenes. The system interprets a scene based on production rules which include the certainty. However, the treatment of the certainty in production rules is heuristic. Yakimovsky[2] defined the certainty by probability. A system proceeds to an interpretation only using the ratio of the highest probability and the next highest probability. Moreover, relations among the other object regions are not considered in the early stage of interpretation. Another problem with these systems is that subparts of objects are not considered. Therefore, an object of which a large part is hidden by the other objects cannot be recognized.

We define a probability of each uniform color region corresponding to each component of an object as a certainty of recognition. A system attempts to merge uniform color regions which constitute a component so that the probability increases. Although there are many combinations of uniform regions, the system restricts the combinations using probability of a uniform region corresponding to a component. For merged regions, probability of each region being a component is also calculated. The merged region with high probability is used as an evidence to generate object candidate. An object is recognized from a set of components. Generally, many object candidates are generated by combining components. Using the probability of regions becoming components, the number of combinations can be restricted. Moreover, for each uniform color region, multiple interpretations are admitted in the recognition process. The final result is the interpretation which has the highest probability.

2 Certainty of scene interpretation

In this paper, we interpret complex outdoor scenes by combining object components which correspond to regions segmented by color and intensity properties. Here, an interpretation of a scene means determination of position and orientation of an object in the scene.

In the case that the system interprets a region set $(R_1, R_2, \cdots, R_N)$ as an object set $(O_1, O_2, \cdots, O_M)$, each of which has a fixed position and orientation in the scene, reliability of interpretation is represented by probability as follows:

$$P(O_1, \cdots O_M | R_1, \cdots, R_N) = \frac{P(R_1, \cdots, R_N | O_1, \cdots, O_M) P(O_1, \cdots, O_M)}{P(R_1, \cdots, R_N)}. \quad (1)$$

Generally, the system needs to calculate for all combinations of all positions and orientations of objects in order to interpret the scene by using equation (1). However, because the system observes the region set $(R_1, R_2, \cdots, R_N)$, the system calculates only for some combinations of objects for which the system can observe the region set $(R_1, R_2, \cdots, R_N)$. Also, since the system cannot prepare all models of objects which exist in the scene, the system interprets the region R_i as an unknown component if the system cannot find out an appropriate model from the model set.

2.1 Condition of probability calculation

Generally, a region does not belong to two or more objects except in the case of a transparent object. Here, we assume that there is no dependency between objects in the scene and between attributes of the region to make probability calculation easy. If we represent a hypothesis which consists of the object set $(O_{h1}, O_{h2}, \cdots, O_{hM})$ generated from the region set $(R_1, R_2, \cdots, R_N)$ as H_h, equation (1) is represented as follows;

$$P(O_{h_1}, O_{h_2} \cdots O_{h_M} | R_1, R_2, \cdots, R_N)$$
$$= \frac{\prod_i \prod_k P(A_{i1k} \cdots A_{in_ik} | C_{i1} \cdots C_{im_i}) P(H_h)}{\sum_h \{\prod_i \prod_k P(A_{i1k} \cdots A_{in_ik} | C_{i1} \cdots C_{im_i}) P(H_h)\}}. \quad (2)$$

In fact, equation (2) is calculated only for hypotheses (combinations of objects) for which the system can observe the region set $(R_1, R_2, \cdots, R_N)$. Therefore, first, the system determines the kinds of components to which each observed region can correspond. Next, the system generates hypotheses from combinations of components which correspond to observed regions.

2.2 Object models

Models of objects are represented by a hierarchical description of the whole object and components. We deal with a component which cannot correspond to any known model of components as an unknown component.

An object is represented by a combination of components, and each component is represented by a set of attributes such as color, shape, size and so on. Here, we select attributes so that each attribute is independent of the others.

An attribute model of components is represented by a probability density function defined by authors. Since it is difficult to define the probability density function theoretically, we make it empirically by analyzing thirty kinds of experimental images. For example, a real distribution of the hue of taillight extracted from thirty kinds of images is shown in Fig.1(a). In order to represent features of distribution shown in Fig.1(a), we give a model distribution $p(A|C)$ as a combination of uniform and normal distributions as shown in Fig.1(b), where A represents a kind of attribute and C represents a kind of component. Then, the borders $A1$ and $A2$ are determined by the authors and standard deviation of normal distribution is calculated from real distribution as shown in Fig.1(a).

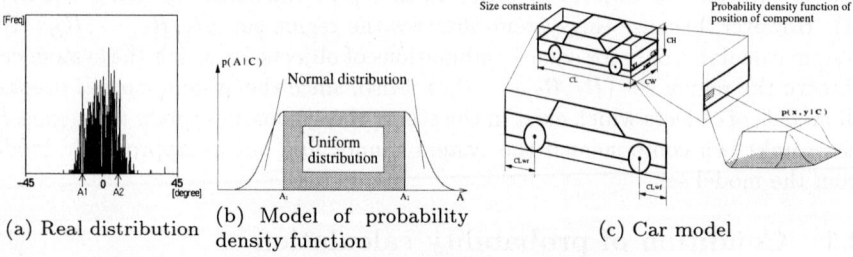

(a) Real distribution (b) Model of probability density function (c) Car model

Fig. 1: Model representation.

A model of each whole object is given by a uniform distribution which represents an object size. For example, a road is expressed by a plane whose boundaries are limited. Moreover, a car is represented by a combination of components which is fixed on a plane of a rectangular parallelepiped which surrounds a whole car. Here, a position of a component on the plane is given by a distribution represented by a combination of a uniform and normal probability density function such as an attribute of a component. A description of a car model is shown in Fig.1(c) as an example of the object model.

3 Extraction of candidate regions of components

3.1 Input information

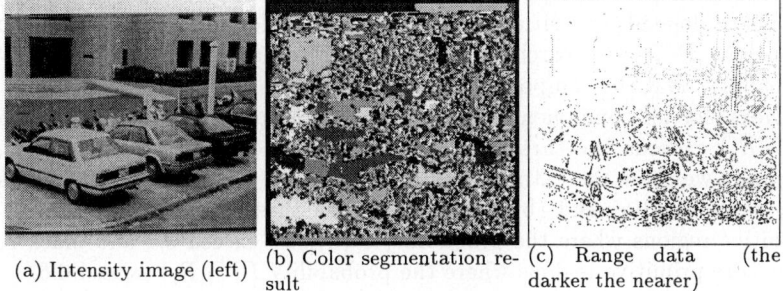

(a) Intensity image (left) (b) Color segmentation result (c) Range data (the darker the nearer)

Fig. 2: Input information.

A pair of color stereo images is used as input sensory data. A left intensity image is shown in Fig.2(a). Edges are extracted from the intensity image and range data are obtained by the feature based multi-stage stereo method[3]. Range data are shown in Fig.2(c). In this figure, the darker points are closer to the camera. The left color image is segmented into regions with homogeneous color and intensity [4]. The segmented regions are minimum units to be processed by our interpretation system. The color segmented regions are shown in Fig.2(b). For each of the unit regions, an average of disparity in the region is calculated. If a unit color region involves the range gap, the average of disparity is calculated in each subregion which is segmented by the range gap and the system uses the average in the subregion containing more range data as the average of the disparity of the unit color region.

For each of the unit regions, an area size, an average and a variance of color properties and a height from a ground plane are calculated, where color properties are intensity T, hue θ and saturation S.

3.2 Extraction of primitive region of components

First, the system examines the kinds of components to which each region segmented by image features can correspond. Next, the system selects regions with high probability (primitive regions) for each component and merges neighboring regions of primitive regions so that the probability of the region corresponding to the component increases.

A probability $P(C_l|R_j)$ that the region R_j in the image corresponds to the component C_l is represented as follows by using attribute value $(A_{j1}, A_{j2}, \cdots, A_{jm})$ of the region R_j;

$$P(C_l|R_j) = P(C_l|A_{j1}, A_{j2}, \cdots, A_{jm}). \tag{3}$$

If each attribute of the region is selected as it becomes independent, the probability $P(C_l|R_j)$ becomes as follows by using probability density function $p(A_jk|C_l)$ $(k = 1, \cdots, m)$ of region attributes which is given by the model;

$$P(C_l|R_j) = \frac{P(C_l)p(A_{j1}|C_l)\cdots p(A_{j(m)}|C_l)}{\sum_i P(C_i)p(A_{j1}|C_i)\cdots p(A_{j(m)}|C_i)}, \tag{4}$$

where the system needs to select independent attributes so that equation (4) holds. Here, in consideration of characteristics of input data, the three kinds of attributes (color, height from ground, size) are utilized.

Also, prior probability $P(C_l)$ of equation (4) is difficult to define theoretically, because it depends on a situation of a scene. Therefore, in this paper, the authors define the $P(C_l)$ by searching thirty kinds of image to determine an average probability that the component appears in scenes.

Next, for all components, the system selects regions where the probability $P(C_l|R_j)$ exceeds the threshold as primitive regions of the components. For example, we show the extracted primitive regions for license plate and asphalt regions in Fig.3(a) and (b), respectively. In these figures, gray regions show primitive regions where the probability $P(C_l|R_j)$ exceed 0.4 and black regions show core primitive regions where the probability $P(C_l|R_j)$ exceed 0.6.

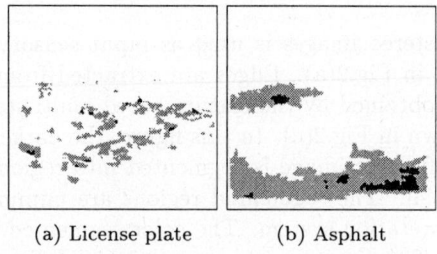

(a) License plate (b) Asphalt

Fig. 3: **Extracted primitive regions.**

3.3 Extraction of candidate regions of components by merging neighboring regions

Generally, each primitive region does not correspond to each component because it is segmented based on only image feature without considering a meaning of a region. Therefore, the system needs to extract the region corresponding to just one component by merging neighboring regions so that the probability $P(C_l|R_j)$ increases. Generally, since the region merging proceeds for all regions and for all kinds of components, computational cost is very high. However, in our method, the system has information as to the kinds of components to which each region is able to correspond. Therefore, using this information, the system restricts computational cost of merging neighboring regions.

Extracted candidate regions of components for primitive regions indicated in Fig.3 are shown in Fig.4(a) and all candidate regions corresponding to car components are shown in Fig.4(b).

4 Scene interpretation based on certainty

In this section, first, extracted candidate regions of components are transformed into a top view image of a scene. Next, the system generates hypotheses of an object set by checking a relationship between candidate regions in the top view image. Then, the most reliable interpretation is determined from among all hypotheses by using the probabilistic measure represented by equation (2).

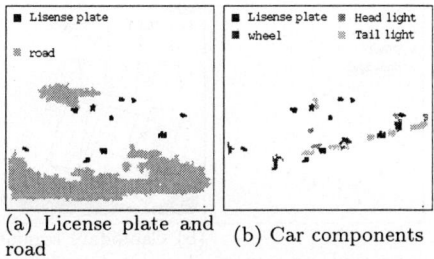

(a) License plate and road   (b) Car components

Fig. 4: Extracted candidate regions of components.

4.1 Construction of object candidates

Generally, a number of combinations of components for all objects in the scene is very large. Therefore, in our method, the system selects core regions which have high probability from candidate regions of components and searches the combination of components starting from these core regions. In experiments, the system sets the threshold of probability to 0.6 for selecting core regions. In our method, first, a center of gravity of extracted core regions of components is transformed into a top view image of a scene. The system searches the combination of components in this 2-D image. If two components which construct an object are found, the position of the other components of the object is much restricted by using an object model. For all core regions, the system searches the combination of components and generates candidates of objects whose position and orientation are fixed by the position of components.

There are cases in which a generated object candidate lies on top of another generated one. Therefore, the system selects a set of object candidates which exist in the scene without overlap. The system regards each set of object candidates as hypothetical interpretation of a region in the scene. Since each hypothesis is generated only from the core region with sufficiently high reliability, each prior probability that each hypothesis is true is given by the same value. When the probability is maximum in the case that all components in the hypothesis correspond to unknown components, the system interprets that no known object exists in this region.

4.2 Determination of Most Probable Interpretation

We present experimental results which are interpreted by using equation (2). The road candidates are indicated in Fig.5(b) as $R1$ and $R2$, which are extracted from primal candidate regions shown in Fig.4(a). There are two hypothetical interpretations of these road candidates. One is that each region is a road. The other is that each region is an unknown region. For regions $R1$ and $R2$, the probability expressed by equation (2) is calculated as being 0.61 that $R1$ is a road and 0.39 that $R1$ is an unknown region, respectively. In the same way, the probabilities are 0.72 and 0.28, respectively, for $R2$. Finally, the system interprets that both regions $R1$ and $R2$ are roads.

The center of gravity of car components shown in Fig.4(b) is transformed

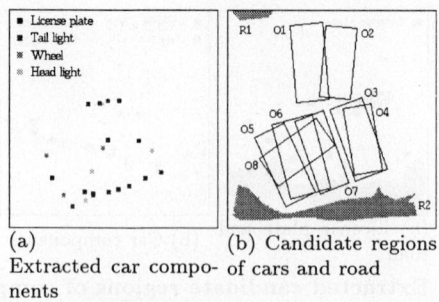

(a) Extracted car components

(b) Candidate regions of cars and road

Fig. 5: Extracted candidate regions of cars and road.

into the top view image shown in Fig.5(a). The car candidates $O1$ to $O8$ are obtained from this top view image by using constraints of components. Some regions which have overlapping candidates is extracted. For each of these regions, hypotheses without inconsistency are generated. For example, the following six hypotheses are generated for an object set $(O5, O6, O7, O8)$. The probability that each hypothesis is true is expressed as the value in parentheses shown in Fig.6. Finally, the system interprets that car candidates $O5$ and $O7$ exist in the region corresponding to this object set.

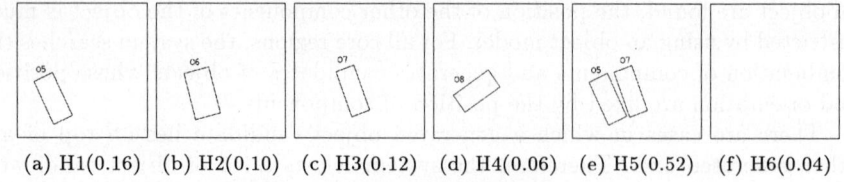

(a) H1(0.16) (b) H2(0.10) (c) H3(0.12) (d) H4(0.06) (e) H5(0.52) (f) H6(0.04)

Fig. 6: Hypotheses.

In the same way, the probability shown in Table 1 is calculated for object sets $(O1, O2)$ and $(O3, O4)$.

Table 1: Probability for object sets

$O_1 \setminus O_2$	exist	empty	$O_3 \setminus O_4$	exist	empty
exist	0	0.23	exist	0	0.58
empty	0.12	0.65	empty	0.30	0.12

Therefore, the system interprets that there are no cars in the region corresponding to the object set $(O1, O2)$ and the car candidate $O3$ exists in the region corresponding to the object set $(O3, O4)$.

A final result is obtained by merging intermediate results for all regions corresponding to all hypotheses. The interpreted car position and orientation is shown as three rectangular parallelepipeds in Fig.7.

Fig. 7: Result of interpretation.

5 Conclusion

We have proposed the method which efficiently interprets outdoor scenes by using the certainty of recognition to restrict a number of combinations between components. This certainty is defined as the probability that the region corresponds to the component of the object. Hypotheses are generated by combining components of the object. Using the components with high certainty, the system restricts the number of combinations between components in generating hypotheses. Moreover, the certainty of recognition which is defined by considering constraints between components enabled the system to recognize objects which are largely hidden by other objects.

Although, we have mainly dealt with cars as objects in the scenes, this method is useful for recognizing other artificial or natural objects.

References

[1] Y. Ohta, "Knowledge-Based interpretation of outdoor natural color scenes," *Research Notes in Artificial Intelligence 4*, Pitman Advanced Publishing Program, 1985.

[2] Y. Yakimovsky and J. Feldman, "A Semantics-Based Decision Theory Region Analyzer," *Proc 6th Int. Joint Conf. on Artificial Intelligence*, Vol.26, No.9, pp.580-588, 1993.

[3] O. Nakayama, A. Yamaguchi, Y. Shirai, and M. Asada, "A multistage stereo method giving priority to reliable matching," *Proc. of IEEE Int. Conf. on Robotics and Automation*, pp.1753-1758, 1992.

[4] A. Okamoto, Y. Shirai, and M. Asada, "Reconstruction of 3-D scene structure by integrating color and range information," *Proc. of 1st Korea-Japan Joint Conference on Computer Vision*, PP.77-80, 1991.

Robust Hypothesis Verification for Model Based Object Recognition Using Gaussian Error Model

Frederic Jurie

LASMEA - CNRS UMR 6602
Université Blaise-Pascal, F-63177 Aubière,France

Abstract. The use of hypothesis verification is recurrent in the model based recognition literature. Small sets of features forming salient groups are paired with model features. Poses can be hypothesised from this small set of feature-to-feature correspondences. The verification of the pose consists in measuring how much model features transformed by the computed pose coincide with image features. When data involved in the initial pairing are noisy the pose is inaccurate and the verification is a difficult problem.

In this paper we propose a robust hypothesis verification algorithm, assuming data error is Gaussian. We present experimental results obtained with 2D and 3D recognition proving that the proposed algorithm is fast and robust.

1 Introduction

Despite recent advances in computer vision the recognition and localisation of 3D objects from a 2D image of a cluttered scene is still a key problem. The reason for the difficulty to progress mainly lies in the combinatoric aspect of the problem.

This difficulty can be bypassed if the location of the objects in the image is known. In that case, the problem is then to compare efficiently a region of the image to a viewer-centred object database. Recent proposed solutions are, for example, based on principal component analysis (Murase and Nayar 1995, Pentland, Moghadam and Starner 1994), modal matching (Sclaroff and Pentland 1995) or template matching (Brunelli and Falavigna 1995).

But Grimson (1991) emphasises that the hard part of the recognition problem is in separating out subsets of correct data from the spurious data that arise from a single object.

Recent researches in this field are focused on the various components of the recognition problem : which features are invariant and discriminant (ter Haar Romeny, Florack, Salden and Viergever 1994), how is it possible to group features into salient parts (Jacobs 1996), how to index models (Califano and Mohan 1994), how to identify sets of data feature/model feature pairings that are consistent with an object (Olson 1993) or which similarity measures are relevant (Huttenlocher, Klanderman and Rucklidge 1993).

From these papers, one can notice that most of the proposed strategies can be described in terms of prediction - verification schemes. In a pre-processing stage model feature groups having invariant properties are stored in an hash table. Recognition

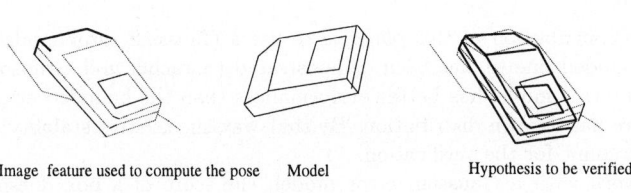

Image feature used to compute the pose   Model   Hypothesis to be verified

Fig. 1. Verification of an inaccurate pose

first consists in grouping image features into salient parts. Index keys obtained from these small groups select objects in the model base, producing sets of pairings from indexed models to data features. The transformation aligning a model with the image is usually referred as the pose of the object. Poses obtained by this method are treated as hypotheses realizing a few feature-to-feature correspondences. These hypotheses have to be verified by transforming remaining model features and by trying to match them with image features. Transformations are usually assumed to be affine.

However errors in data generally make the pose incorrect and the verification becomes more difficult (see Fig. 1 for an illustration). In that case, as pointed out by (Gandhi and Camps 1994) the noise in the data will propagate into the pose and will decrease the quality of the correspondences obtained in the verification stage.

2 Overview of the proposed approach

As explained in the introduction, object recognition systems frequently hypothesise model pose from the matching of a small number of model features. A fundamental question is how such hypotheses can be confirmed if poses are inaccurate and if data are cluttered.

The above mentioned approaches generally start from the initial set of correspondences, and try to extend it feature by feature. The pose is refined step by step, guiding for the selection of next features. This solution is not optimal since these algorithms look for only on point at a time. We experimentally observed that in many cases it leads to false solutions, specially if objects are occluded or if data are cluttered.

In this paper, the opposite approach is adopted. Assuming that the right pose belongs to a known volume of the pose space (including the initial pose) we first take into account all of the correspondences compatible with this volume. This set of poses is iteratively reduced, until it can be taken as a single pose. The computed pose (and the related correspondences) is better than when correspondences are added one by one, because influences of each possible correspondences are taken into account together. The transformation aligning models on images is supposed to be affine.

The verification stage has to be as fast as possible, because in case of complex scenes and large number of possible objects, a large number of hypotheses have to be verified. It is expensive to verify each one of these hypotheses. The proposed approach allows hierarchical verification : we compute for each possible hypothesis the volume of possible poses, and determine how many correspondences are compatible with this set of poses. This score represent the maximal number of correspondences that a unique pose of the volume may verify. Hypotheses having low score are first discarded. The remaining ones are refined. For purpose of simplicity, we assume that the sub-space is a rectangular volume of the pose space, that we call a "box".

A major contribution of this paper is to use a Gaussian error model rather than the usual bounded one (it has been demonstrated (Sarachik and Grimson 1993) that the Gaussian error model has better performances than the bounded error one). Each model feature has its own distribution. By that way the relative stability of features is taken into account for the verification.

As we work with a Gaussian error model, the score of a box doesn't consist in counting the number of correspondences, but relies on the evaluation of the maximal probability of the object to be matched knowing the set of possible poses.

3 Gaussian model error

Let f_d denotes a data feature and f_t the transformation of the feature model f by the pose P. Let $\delta = f_d - f_t$.

Data features are assumed to be normally distributed. Let $P(\delta|C)$ the probability of having δ knowing C. C means "f_t and f_d are corresponding features". In a v-dimensional feature space, the v-dimensional normal probability density function with covariance matrix Q is :

$$P(\delta|C) = (2\pi)^{-\frac{v}{2}}|Q|^{-\frac{1}{2}}exp(-\frac{1}{2}(f_t - f_d)^T Q^{-1}(f_t - f_d))$$

If features are image points, the dimensionality is 2 and $f_i = (a_1, a_2)^T$ where (a_1, a_2) is the 2d spatial position of the feature. The dimensionality is 4 for line segments. No assumption is made neither on the dimensionality, neither on the kind of parameters encoding the feature.

For purpose of simplification, in the rest of the paper we do not represent features in their original feature space. We decompose the covariance matrix Q^{-1} as : $Q^{-1} = UD^{-1}U^t$

where U is the orthogonal matrix of eigenvectors of Q and D the diagonal matrix of eigen values. The v eigen values will be denoted $\lambda_1, \cdots, \lambda_v$.

By projecting features in the eigenspace, the normal distribution for the model feature i is much simple :

$$P(\epsilon|C) = (2\pi)^{-\frac{v}{2}} \prod_{j=1}^{j \leq v} \lambda_j^{-\frac{1}{2}} exp(-\frac{1}{2} \sum_{j=1}^{j \leq v} \frac{\epsilon_j^2}{\lambda_j})$$

ϵ denotes the difference $(f_t - f_d)$ projected into the eigenspace ($\epsilon = U\delta$). There is one different covariance matrix per model feature.

4 Probability of a feature-to-feature correspondence knowing the affine transformation P

Denoting $\epsilon = U(f_t - f_d)$ where f_d is the data feature and f_t the transformation in the image reference of the feature f. The probability of having a correspondence between f_t and f_d knowing the pose P is :

$$P(C|P) = P(C|\epsilon)\frac{P(\epsilon|C)P(C)}{P(\epsilon)} = \alpha exp(-\frac{1}{2}\sum_{j=1}^{j<=v}\frac{\epsilon_j^2}{\lambda_j})$$

where : $\alpha = (2\pi)^{-\frac{v}{2}} \prod_{j=1}^{j \leq v} \lambda_j^{-\frac{1}{2}} \frac{P(C)}{P(\epsilon)}$

$P(\epsilon)$ is modelled during a training stage. This value depends on the kind of feature used. We assumed $P(C) = \frac{\mathcal{N}}{\mathcal{N} \times \mathcal{M}}$ where $\mathcal{N}$ is the number of image features and $\mathcal{M}$ the average number of model features.

The pose P is a vector of dimensionality $\mathcal{D}$, where $\mathcal{D}$ denotes the dimensionality of the pose space.

$\epsilon = U(f_t - f_d) = (\epsilon_1, \ldots, \epsilon_v)^T$ is a linear combination of U, f_d and f_t. Then we can write : $\forall j \in [1, \ldots, v] \epsilon_j = H_j \times P$ where vector H_j is a linear combination of U, f_d and f.

With these notations, we have : $P(C|P) = \alpha exp(-\frac{1}{2} \sum_{j=1}^{j<=v} \frac{(H_j \times P)^2}{\lambda_j})$

The product $H_j \times P$ can be geometrically interpreted as the distance from transformation P to the hyper plane H_j of the pose space.

The probability of the correspondence is then the weighted sum of squared distances from this transformation P to the H_j hyper planes. This geometric property is exploited in the next section.

5 Maximal probability of correspondence, assuming poses belongs a known box of the pose space

The maximal probability of correspondence, knowing that the transformation belongs to to a box (cubic volume) of the pose space, is denoted $P(C|BOX)$

The computation of this probability requires the maximisation of the quadratic function, subject to linear inequality constraints (the box of the pose space). This is a problem which can be solved in a finite number of steps with quadratic programming techniques.

One common approach uses the *Lagrange Multipliers* combined with active set methods (Fletcher 1987).

But active set methods are highly time consuming. For example, if the transformation is an orthographic scaled projection (weak perspective) the dimension of the pose space is 8, and supposing the size of the active set is 8, a 16x16 linear system is to be solved at each iteration.

Accordingly we propose a more efficient algorithm allowing to compute an approximation of this maximal value.

It consists in the 3 following steps (detailed just after the enumeration) :

1. let V the affine manifold generated by the v hyper-planes H_j given by a pair of matched features (see end of previous section for details). We first compute the position of the pose $P_0 \in V$ such that $\forall P \in V, d(c, P_0) \leq d(c, P)$ were c is the centre of the box and $d()$ the Euclidean distance.
2. if P_0 is not in the box then P_{min} is taken as the intersection of the line (c, P_0) with the convex hull of the box ($P_{min} = P_0$ if P_0 is included in the box).
3. if P_{min} is not in the box, its position is iteratively adjusted to minimise the distance $D(P)$ with $D(P) = \sum_{j=1}^{j<=v} \frac{(H_j \times P)^2}{\lambda_j}$

6 From feature correspondences to object correspondences

We assume that the probability of having a occurrence of the model M in an image subject to a pose P only depends on which model features are matched. If the model size is $\mathcal{M}$ (the number of model features), there are $2^{\mathcal{M}}$ possible configurations denoted γ :

$$P(M|P) = \sum_{\gamma \in \Gamma} P(M|P, \gamma) P(\gamma|P) \quad (1)$$

Configurations can be grouped according to their number of matches. Let $E^k, k \leq \mathcal{M}$ the set of configurations matching k model segments. Then $E^k = \bigcup_{j=1}^{j < C_\mathcal{M}^k} \gamma_j^k$, and $\Gamma = \bigcup_{i=1}^{i \leq \mathcal{M}} E^i$, the set of all possible exhaustive and mutually exclusive configurations.
Then : $P(M|P) = \sum_{\gamma \in \Gamma} P(M|P, \gamma) P(\gamma|P) = \sum_{k=1}^{k \leq \mathcal{M}} \sum_{j=1}^{j \leq C_\mathcal{M}^k} P(M|P, \gamma_j^k) P(\gamma_j^k|P)$

We can simplify this formula, as M and P are conditionally independent given γ :
$P(M|P) = \sum_{k=1}^{k \leq \mathcal{M}} \sum_{j=1}^{j \leq C_\mathcal{M}^k} P(M|\gamma_j^k) P(\gamma_j^k|P)$

The size of Γ is so large than $P(M|\gamma)$ would be difficult to learn. We simplify this expression considering that the most significant parameter for computing this probability is the number of image features matched.

That is to say : $\forall k \in \{1 \ldots \mathcal{M}\}, \forall i \in [1 \ldots C_\mathcal{M}^k] P(M|\gamma_i^k) = P(M|\bigcup_{i=1}^{i \leq C_\mathcal{M}^k} \gamma_i^k) = P(M|E^k)$

The probability $P(M|P)$ can therefore be written : $P(M|P) = \sum_{k=1}^{k \leq m} P(M|E^k) \sum_{j=1}^{j \leq C_\mathcal{M}^k} P(\gamma_j^k|P)$.

$P(M|E^k)$ is the probability of having model M knowing that k of its features are matched. It has been computed during a learning stage.

The computation of $P(E^k|P) = \sum_{j=1}^{j \leq C_\mathcal{M}^k} P(\gamma_j^k|P)$ is more tedious. The event E^k is the union of C_m^k different configurations. $P(\gamma_j^k|P)$ is the probability of that combination, given a set of correspondences. This probability can be written as a function of individual feature correspondences : $P(\gamma_j^k|P) = \prod_{i=1}^{i \leq \mathcal{M}} P(m_i \overset{b(i)}{\to})$

where $b(i)$ is a Boolean variable meaning that the model segment m_i is be (or is not) supposed to be matched in that combination ($m \overset{0}{\to} = m \not\to, m \overset{1}{\to} = m \to$)[1] If we suppose that $m_i \overset{b(i)}{\to}, i \in \{1, \ldots, \mathcal{M}\}$ are independent events, we have $P(\gamma_j^k|P) = \prod_{i=1}^{i \leq \mathcal{M}} P(m_i \overset{b(i)}{\to} |P)$.

In that case $P(E^k|P)$ is a sum of products long to be computed. We propose to use an approximation of that sum, by only taking into account its maximal terms. As each term is a product of positive values, the maximal product is obtained with maximal values. This simplification is very easy to implement : we sort probabilities $P(m_i \overset{b(i)}{\to} |P)$, and affect the k highest probabilities to the k segments that are to be matched in the configuration. Other probabilities are affected to unmatched segments.

If we suppose that I is an index function such that $\forall (k,l) \in \{0, \ldots, \mathcal{M}\}^2, P(m_{I(l)} \to |P) > P(m_{I(k)} \to |P) \Rightarrow k < l$

Then

$$P(E^k|P) = P(m_0 \overset{b(0)}{\to}, \ldots, m_i \overset{b(i)}{\to}, \ldots, m_\mathcal{M} \overset{b(\mathcal{M})}{\to} |P)$$
$$>= \prod_{j=1}^{j <= i} P(m_{I(j)} \to |P) \prod_{j=i+1}^{j <= \mathcal{M}} (1 - P(m_{I(j)} \not\to |P)) = Papprox(E^k|P)$$

All the experiments presented in that paper have been obtained with the exact values $P(E|P)$ and $Papprox(E|P)$. We always obtain exactly the same results in both case.

---

[1] The fact that a model segment m is matched is denoted $m \to$ (respectively $m \not\to$ if the segment is not matched).

7 Searching the best pose into a box of the pose space

An upper bounds of $P(M|P)$ knowing P belongs to a box of transformations (denoted $P(M|BOX)$) can be obtained by introducing values $P(C|BOX)$ instead of $P(C|P)$ in equation (1). We have $\forall P \in BOX, P(M|P) \leq P(M|BOX)$ because "a best pose" is computed individually for each model feature with no guaranty that these computed pose are equal. (It is possible to find distinct poses in the box aligning each model feature to an image feature, while there is no pose aligning correctly all the model features.)

The value $P(M|BOX)$ is however very informative and permits to make a rough selection between possibles poses, but the "real" best pose is still to be computed. We assume that when the box becomes small it can be treated as a single pose.

To reduce the size of the box around the highest values of $P(M|BOX)$, we use a recursive division of the initial box, as proposed by Breuel (1992). Recursive subdivision consists in recursively splitting the box in two parts, alternating axes. This process can be seen as a tree search. The root node corresponds to the initial box. Leaves are the smallest regions taken into account.

Breuel (1992) first proposed to explore the "best" branch (dividing on each level the box with the best evaluation) and then to backtrack the search, looking for other possible solutions. The maximal number of boxes explored and consequently the run time cannot be guaranteed. In worst cases the whole space has to be explored. That is why we recommend a N-search algorithm. N branches are explored at the same time and no backtracking is required. The maximum number of boxes evaluated is below Nh where h denotes the number of levels.

The probabilistic evaluation of sub-boxes $P(M|BOX)$ guide the search : only the best N sub-boxes are kept in the next level of division. During all our experiments N was set to 5.

The initial box is supposed to be large enough to compensate for the data errors. If we assume the error model is Gaussian the initial box should have an infinite size.

For practical reasons we adopt a more manageable definition : the size of the initial box is such that there is at least p chance it includes the correct pose. The convex volume of the feature space bounded by $P(C|\epsilon) = p$ leads to a convex volume of the pose space. The initial box is taken as the smallest box including this volume. In our experiments the box has a constant size, centred on the initial pose.

8 Application to recognition

The pose verification algorithm presented in that paper have been integrated into two different recognition applications : in the first one, a 2D recognition application, the 10 small objects (stored in a viewer-centred database) have to be recognised in cluttered noisy images. In the second application, a 3D recognition application, we try to recognise 4 polyhedral objects modelled in an object-centred reference. In both cases image and model features are line segments.

Application to 2D recognition The 10 objects are modelled by 600 2D views. A view is a collection of line segments. Recognition first consists in indexing the knowledge base with geometric invariants. In our experiments invariants are relative angles in groups of co-terminating line segments. For each image several hundred of hypothesised 2D poses (2D affine transformations) are to be verified. It takes less than 6 seconds even for

Fig. 2. Application to 2D recognition (see text for details)

images having more than 350 line segments (about 55 ms to verify one pose)[2]. Several hundred of images have been processed without observing any errors : the right pose and the correct object have always been obtained.

The Fig. 2 shows typical results. The figure has 4 columns. The two first represent successively image and line segments. Third one shown the best verification obtained. The selected model is aligned on the image by applying the best pose. The last column represents the initial pose (the pose computed from the initial set of correspondences). One can observe that the verification is performed correctly even if the initial pose is very inaccurate. This explains why the correct object and the correct pose have been chosen from the 600 views.

Application to 3D recognition Initial correspondences are produced using an alignment technique (Huttenlocher and Ullman 1990). Transformation is first a scaled orthographic projection, refined to be a perspective projection as explained by DeMenthon and Davis (1992). Perspective projection can be seen as an affine transformation if an approximation of the pose is known. This is the case for verification algorithms.

We performed *static* and *dynamic* recognition (with moving objects). Recognition time measured with 4 polyhedral models is below 2 seconds on the first image and below 400 ms on subsequent ones. The 3D initial box is centred on the previous pose. The size of the box is modified according to the quality of the previous verification. Very reliable results have been obtained, even with strong occlusions.

9 Conclusions

In this paper, we have proposed a robust solution to the pose verification problem, when the pose is inaccurate because obtained from a few feature-to-feature noisy correspondences.

It consists it refining a set of feasible transformations so that the probability of object match knowing the pose is maximal.

The proposed algorithm has been integrated in 2 different recognition applications : 2D recognition and 3D recognition with line segments. However it can be directly used with different features and different recognition strategies.

We experimentally prove that it is fast, robust to data noise and robust to occlusions. The robustness is partly due to the probabilistic framework used to describe

---

[2] times are measured on a HP-700 workstation

data-to-model correspondences. Convergence to the optimal pose would not be possible assuming bounded error model.

Application to 3D recognition shows that this pose verification algorithm can be used as a robust tracking algorithm.

References

Breuel, T.: 1992, Fast recognition using adaptive subdivisions of transformation space, *Proc. IEEE Conference on Computer Vision and Pattern Recognition*, Champain, Illinois, pp. 445–451.

Brunelli, R. and Falavigna, D.: 1995, Person identification using multiple cues, *IEEE Transactions on Pattern Analysis and Machine Intelligence* **17**(10), 955–966.

Califano, A. and Mohan, R.: 1994, Mulidimensional indexing for recognizing visual shapes, *IEEE Transactions on Pattern Analysis and Machine Intelligence* **16**(4), 373–392.

DeMenthon, D. and Davis, L.: 1992, Model-based object pose in 25 lines of code, *Proc. European Conference on Computer Vision*, Santa Margherita Ligure, Italy, pp. 19–22.

Fletcher, R.: 1987, *Practical Methods of Optimization*, wiley-interscience publications edn, John Wiley and Sons, New York.

Gandhi, T. and Camps, O.: 1994, Robust feature selection for object recognition using uncertain 2d image data, *Proc. IEEE Conference on Computer Vision and Pattern Recognition*, Seattle, Washington, pp. 281–287.

Grimson, W.: 1991, The combinatorics of heuristic search term for object recognition in cluttered environment, *IEEE Transactions on Pattern Analysis and Machine Intelligence* **13**(9), 920–935.

Huttenlocher, D. and Ullman, S.: 1990, Recognizing solid objects by alignment with an image, *International Journal of Computer Vision* **5**(2), 195–212.

Huttenlocher, D., Klanderman, G. and Rucklidge, W.: 1993, Comparing images using the hausdorff distance, *IEEE Transactions on Pattern Analysis and Machine Intelligence* **15**(9), 850–863.

Jacobs, D.: 1996, Robust and efficient detection of salient convex groups, **18**(1), 541–548.

Murase, H. and Nayar, S.: 1995, Visual learning and recognition od 3d object from appearance, *International Journal of Computer Vision* **18**(14), 5–24.

Olson, C.: 1993, Probabilistic indexing: A new method of indexing 3d model data from 2d image data, *Technical report*, University of California at Berkeley.

Pentland, A., Moghadam, B. and Starner, T.: 1994, View-based and modular eigenspaces for face recognition, *Proc. IEEE Conference on Computer Vision and Pattern Recognition*, Seattle, Washington, pp. 84–91.

Sarachik, K. and Grimson, W.: 1993, Gaussian error models for object recognition, *Proc. IEEE Conference on Computer Vision and Pattern Recognition*, New-York, pp. 400–406.

Sclaroff, S. and Pentland, A.: 1995, Modal matching for correspondence and recognition, *IEEE Transactions on Pattern Analysis and Machine Intelligence* **17**, 545–561.

ter Haar Romeny, B., Florack, L., Salden, A. and Viergever, M.: 1994, Higher order differential structure of images, *Image and Vision Computing* **12**(6), 317–325.

Shape Modeling from Multiple View Images Using GAs

Satoshi KIRIHARA and Hideo SAITO

Department of Electrical Engineering, Keio University
3-14-1 Hiyoshi Kouhoku-ku Yokohama 223, Japan
TEL +81-45-563-1141 (ext.3310), FAX +81-45-563-2773
E-mail: kiri@ozawa.elec.keio.ac.jp, saito@ozawa.elec.keio.ac.jp

Abstract

Shape modeling is a very important issue for many study, for example, object recognition for robot vision, virtual environment construction, and so on. In this paper, a new method of object modeling from multiple view images using genetic algorithms (GAs) is proposed. In this method, a similarity between model and every input image is calculated, and then the model which has the maximum similarity is found. For finding the model of maximum similarity, genetic algorithms are used as the optimization method. In the genetic algorithm, the sharing scheme is employed for efficient detection of multiple solution, because some shape may be represented by multiple shape models. Some results of modeling experiments from real multiple images demonstrate that the proposed method can robustly generate model by using the GA.

1 Introduction

Many methods have recently been studied for 3-D object modeling. These methods can be applied to object recognition for robot vision [1], construction of virtual world [2][3], and so on. Especially, virtual world is significant subject for communication using computer networks. Many previous modeling method use range data of the object shape which can generally be obtained by range finders [4] [5].However, range finders are not popular tool, and they can not obtain the surface texture data which can be used for synthesizing images presented in virtual space (image-based rendering).

Therefore, many methods for recovering range information from 2-D images taken with CCD camera have been studied [6][7]. Generally, because these methods use some multiple images, some corresponding points between each image must be detected. If the corresponding points are detected exactly, the range information can be recovered by triangulation. However, the detection of corresponding points is known as a difficult problem and thus have been studied hardly. It is especially difficult to make correspondence between multi-view images because of the occlusion and the inconsistency of background scene.

If we consider that the object problem is not estimating accurate range information but generating accurate object models, we don't have to recover the range data. For example, we can obtain the object model by the use of interactive operation system [8] in which generation of hypothesis of models and verification of the hypothesis on the multiple view images are repeated interactively. Such kind of modeling system does not require the range data of the object but verification of the generated hypothesis of models by the human.

In this study, we intend to make automatic modeling system based on the repeated operation of generation and evaluation of hypothesis. For searching the best hypothesis of model efficiently, we employ genetic algorithms (GAs) [9]. In our method, the model matching to every input images are found by applying GAs which repeat evaluation of hypotheses of the models. For the evaluation, similarity between the model and the input image at each view point is calculated, and then the model having the maximum evaluation is found by GAs.

In our previous article [10], the concept and the early results are presented, but the quality of the results are not sufficient. In this paper, we have made significant improvement for the algorithms and obtained high quality results, which are shown in the following parts of this paper.

2 Proposed method

In this study, we assumed that the object is the polyhedral such as an artificial building. Under this assumption, the object can be expressed by multiple model shapes of triangular prisms and rectangular prisms. Input images are taken by CCD camera from multiple view. The problem is estimation of shape, position, and pose of the models from the input images. **Fig. 1** shows the scheme of the shape modeling assumed in this paper.

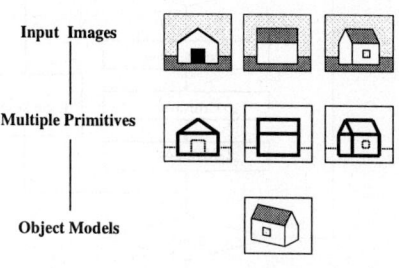

Fig. 1. Scheme of the shape modeling.

In this method, modeling is performed by maximizing an evaluation function using GAs. The evaluation function is described later.

2.1 Model parameters

As described above, we employ the model shapes of triangular prisms and rectangular prisms for representing the object surface.

We define the model parameters as the following:

1. model category (rectangles or triangles) (A),
2. length to x, y, z axis (lx, ly, lz),
3. rotation angle around y axis (bt),
4. position of center of gravity (xc, yc, zc).

2.2 Modeling by GAs

Fig. 2 shows the flow of the modeling process by GAs. First, a group of strings is produced at random. These strings represent hypothesizes of the model. All strings are evaluated, then the genetic operations are applied to them. The string which has the best evaluation is searched by repeating the genetic operation. The model parameters represented by the best strings is regarded as the final solution.

Definition of string For using GAs, the object model must be expressed as a string. As described in the previous section, the object model can be represented by some simple parameters in this method. Each parameters has 8 bits binary except for the parameter of model category. The parameter of model category has only 1 bit, because it is used for only distinguishing between rectangles and triangles. Then, the strings consist of 57 bits binary. For corresponding the hamming distance of the string to the distance of the parameter value,

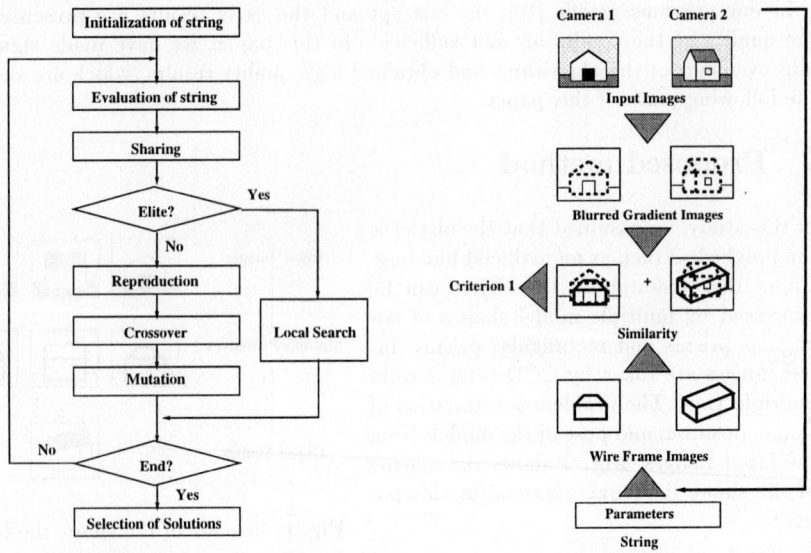

Fig. 2. The flow of the modeling process by GAs

Fig. 3. Similarity criterion

Evaluation function We define evaluation function which represents similarity between input images and the model hypothesis. The following two criterion are used in this method.

1. Similarity between the input multiple view images and the synthetic wire frame images as shown in **Fig. 3** (f_1).
2. Consistency of texture patterns on the model plane which are given by multi-view images as shown in **Fig. 4** (f_2).

First criterion (f_1) is defined as the following equation.

$$f_1 = \sum_i \frac{\sum_{x,y} B_i(x,y) S_i(x,y)}{\sqrt{\sum_{x,y} B_i^2(x,y)} \sqrt{\sum_{x,y} S_i^2(x,y)}} \quad (1)$$

where S_i represents wire frame image of the model at ith view, and $B_i(x,y)$ is blurred gradient image of the input image at ith view. $B_i(x,y)$ is calculated as

$$B_i(x,y) = G(x,y) * \sqrt{\{\frac{\partial I_i(x,y)}{\partial x}\}^2 + \{\frac{\partial I_i(x,y)}{\partial y}\}^2}, \quad (2)$$

where G(x,y) is Gaussian operator and $I_i(x,y)$ is input images at ith view.

The second criterion (f_2) is defined as the following equation.

$$f_2 = -\sum_{x,y} \sum_i \{T_i(x,y) - \frac{\sum_j T_j(x,y)}{p}\}^2 \quad (3)$$

where $T_i(x,y)$ is represents texture pattern back-projected from ith input image, and p is the number of the images back-projected on the model plane. If the model parameters are exact, every back-projected texture must be the same, then f_2 have to be 0 (maximum value).

As these two criterion, the total evaluation function f is

$$f = f_1 + \alpha f_1 f_2, \qquad (4)$$

where α is a weighting constant.

The second criterion (f_2) supports the first criterion (f_1). We previously used the following equation

$$f = f_1 + \alpha f_2, \qquad (5)$$

which sometimes gives high evaluation f even if the model does not correspond to the object input images, because f_2 can have the high evaluation value when the object surface and the background have the smooth texture. Therefore, the second criterion (f_2) support the first criterion (f_1), then the second term ($\alpha f_1 f_2$) cannot have high evaluation until f_1 has high evaluation to some degree.

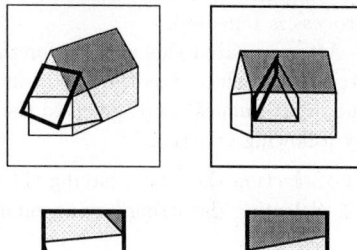

Fig. 4. Consistency of texture in criterion 2

Sharing Although GAs are generally used for finding an single solution having maximum evaluation, they can be used for finding multiple solutions simultaneously by some modifications of the algorithms. Sharing method [9] is one of the popular modification for finding multiple solutions efficiently. We employ the sharing method for finding multiple models which are included in the input images.

Sharing is an operation that decreases evaluation of strings if the strings are similar to other strings in the population of strings. By this operation, strings tend to have various values, which means that multiple solutions can be simultaneously found.

Original evaluation of a string is modified according to the number of the neighbor strings. The relationship between the modified evaluation and the original evaluation is shown in the following equation.

$$E_s(x_i) = \frac{E(x_i)}{\sum_{j=1}^{n} s(d(x_i, x_j))}, \qquad (6)$$

where $s(d) = max(1 - d/\sigma, 0)$,

x_i, x_j : ith and jth strings, $\qquad E(x_i)$: original evaluation of x_i
$E_s(x_i)$: modified evaluation of x_i, $d(x_i, x_j)$: distance between x_i and x_j
$s(d)$: sharing function of d, $\qquad \sigma$: constant determining effect of sharing
n : number of the strings in a population

In general, σ is a constant parameter. In our method, however, σ is changed in proportional to the number of generations in GAs. In the earlier generation, sharing operates effectively and many solutions can be held in the string population by the use of higher σ. In the later generation, σ has lower value, then, local optimization can be realized around the multiple solutions.

Genetic operations The flow of the modeling process by GAs is already shown in **Fig. 1**. First, the strings are initialized at random. The number of strings are 256.

Next, evaluation value is calculated for each string by the use of equation 4. The original evaluation is modified by sharing function as described in the previous section. Some elite strings are then selected according to modified evaluation. The selected elite strings are improved by a local search method. In this way, the strings having higher evaluation are selected as the elite strings.

Other strings which are not selected as elites are applied to genetic operations. Genetic operations are reproduction, crossover and mutation. First, the parent strings are selected according to modified evaluation. Then, the offspring strings are generated by one-point crossover. Some bits selected at random and reversed by mutation. This process is repeated.

After repeated this process for the number of generations defined in advance, some strings are selected as the object models. In this method, the number of the models in the input images is not given to the algorithm. Therefore the object models is selected by following criterion.

1. Selecting the string having the texture evaluation (f_2) under a threshold level.
2. Selecting the string having enough distance to the object models which are already selected.

3 Experiments

We performed some experiments for demonstrating that the proposed method is efficient for object modeling. In this experiments, some real images taken by CCD camera. The objects are a locker, a model of a roof of a house, and two tissue boxes.

These objects are located in natural environment, so the taken multi-view images have various background scene. Input images is 256 × 256, 8bit gray-scale. The camera parameters of every image are previously measured by initial experiment.

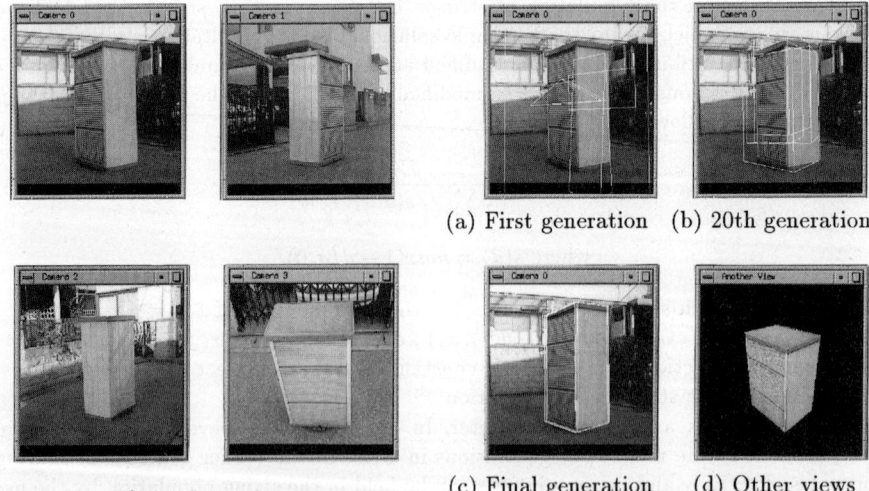

Fig. 5. Input images

(a) First generation (b) 20th generation

(c) Final generation (d) Other views

Fig. 6. Result of first experiment

First, the experimental results for the locker are shown. **Fig. 5** shows the input multi-view images. In the first generation, candidates of solution are selected at random as shown in **Fig. 6**(a). They are getting to be close in the twenties generation as shown in **Fig. 6**(b). In the final generation, the solutions are optimized completely as shown in **Fig. 6**(c). Then, we obtain the object model . Using the object model, we calculate the image form other view points (**Fig. 6**(d)).

As shown in the edge images (**Fig. 7**), it is very difficult to make correspondence of these edges or points between the multi-view images , because many background edges in one input image cannot exist in other input images. Since the proposed method does not require the correspondence between the multi-view images, the exact modeling result can be obtained.

Next, we show the experimental results for a model of a roof of a house in **Fig. 8**. In this experiment, the object is expressed by the one triangular prism model. In this case, only small texture variation is included in the background in the input images, that is different from the first experiment for the locker. As described in the section of the evaluation function, if we use the simple evaluation function of eq.(5), the strings converge some fail solutions as shown in the object in **Fig. 9**(b). However,by using the evaluation function of eq.(4), the exact model can be obtained as shown in **Fig. 8**(b) although there are little texture variation.

(a) First generation (b) Final generation

Fig. 7. Edge image

Fig. 8. Result of second experiment

 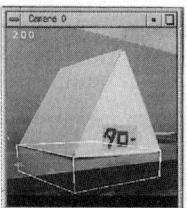 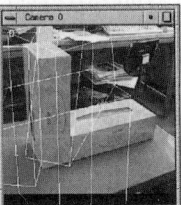 

(a) Edge image (b) Example of fail solution (a) First generation (b) Final generation

Fig. 9. Result of simple evaluation function

Fig. 10. Result of third experiment

Third, we show the experimental results for two tissue boxes. In this experiment, the object in the input images is needed to be represented two rectangle models. In the first generation, candidates of solution are selected at random as shown in **Fig. 10**(a). The result of modeling is shown in **Fig. 10**(b) .

The object consists of two tissues in this case, such as an artificial building is generally expressed by combination of some rectangle and triangle. The modeling results for two models have a larger modeling error than the modeling results for one model. This error is caused by the some difficulty in obtaining the multiple solution via GAs with sharing method. This will be solved by the future study.

In this study, GAs play the important role for obtaining the shape model efficiently because of performance of optimization of GAs.

Fig. 11 shows the distribution of the evaluation function for valuable parameters. Because the distribution for all parameters can not be displayed, we show the distribution in respect to the parameters lx and ly. It is obviously recognized that the evaluation distribution has a lot of peaks. For optimizing in such evaluation distribution, the conventional methods are not suitable because they tend to give wrong solutions of local peaks. As described in papers on GA's applications, GAs are powerful algorithm in such situations.

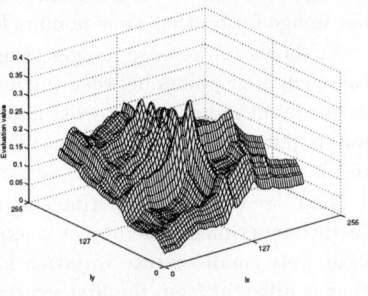

Fig. 11. The distribution of the evaluation function

4 Conclusion

In this paper, we propose a method for object modeling from multiple view images using genetic algorithms (GAs). In the proposed method, generation and verification of model hypothesis are repeated, and the hypothesis which has the best evaluation is efficiently searched by the use of GAs. Some results of object modeling experiments from real multiple images demonstrate that the proposed method can robustly generate model by using GAs.

References

1. Mark D. Wheeler and Katsushi Ikeucti : "Sensor Modeling, Probabilistic Hypothesis Generation, and Robust Localization for Object Recognition". IEEE trans. PAMI, Vol. 17, No.3, pp.252-265 (1995)
2. Carlo Tomasi and Takao Kanade : "Shape and Motion from Image Stream : a Factorization Method". Technical Report in CMU, CMU-CS-92-104 (1992)
3. Camillo J. Taylor and David J. Kriegman : "Structure and Motion from Line Segments in Multiple Image". IEEE trans. on PAMI, Vol.17, No.11, pp.1021-1032 (1995)
4. B.Parvin and G.Medioni : "B-rep from Unregistered Multiple Range Image". in Proc.of the 1992 IEEE International Conf. on Robotics and Automation, Nice, France, May, 1992, pp.1602-1607 (1992)
5. T.Masuda, K.Sakaue and N.Yokoya : "Registration and integration of multiple range images for 3-D model construction". Proceedings of the 13th ICPR, vol.1, pp.879-883 (1996)
6. S.D.Cochran, G.Medioni: "3-D Surface Description from Binocular Stereo". IEEE Trans. on PAMI, Vol.14, No.10, pp.981-994 (1992)
7. W.B.Seales, O.D.Faugeras: "Building Three-Dimensional Object Models from Image Sequence". COMPUTER VISION AND IMAGE UNDERSTANDINDING, Vol.61, No.3, pp.308-324 (1995)
8. URL: http://www.wbs.or.jp/bt/sel/skv/index.htm
9. D.Goldberg: Genetic Algorithms in search, Optimization and Machine Learning, Addison-Wesley (1989)
10. H. Saito and M. Mori: "Object Modeling from Multiple Images Using Genetic Algorithms". Proceedings of the 13th ICPR, Vol. IV, pp.669-673, (1996)

3-D Reconstruction of Multipart Self-Occluding Objects

Nebojsa Jojic[†], Jin Gu[‡], Helen C. Shen[‡] and Thomas Huang[†]

[†]Beckman Institute, University of Illinois at Urbana-Champaign
405 N. Mathews Ave, Urbana, Illinois 61801, USA
{jojic,huang}@ifp.uiuc.edu

[‡]Department of Computer Science
Hong Kong University of Science and Technology, Hong Kong
{csgjx,helens}@cs.ust.hk

Abstract. In this paper we present a method for reconstruction of multipart objects from several arbitrary views using deformable superquadrics as the models of the object's parts. Two visual cues are used: occluding contours and stereo (possibly aided by projected patterns). The object can be relatively complex and can exhibit numerous self occlusions from some or all views. Our preliminary experiments on a human body and a tailor's mannequin show that the reconstruction is more complete than in purely stereo or structured light based methods and more precise than the reconstruction from occluding contours only.

1 Introduction

In this paper we study the problem of 3D shape reconstruction of objects consisting of several parts which may partially or completely occlude each other from some views.

The usual stereo or structured light methods can not give a complete surface estimate in such cases, as either the lighting source or the cameras may not see all the parts of the object. Moreover, the correspondence problem is not a trivial task in using stereo or structured light methods.

To overcome these difficulties, we model the object with several deformable superquadrics and use occluding contours and stereo to govern part positioning, orientation and deformation. Image contours provide a crude surface estimate which guides stereo matching process. In turn the 3D points provided by the stereo cue can further refine the surface estimate and improve contour fitting.

Compared with related work [5][6], we propose a faster force assignment algorithm based on chamfer images, and we can reconstruct multiple objects and multipart self-occluding objects, which can be rigid or may not be capable to perform prescribed set of movements as required in [3]. We can use arbitrary camera configuration, unlike the case in [7] where parallel projection and coplanar viewing direction are assumed. Moreover, the whole scheme avoids the problem of merging of reconstructed surface patches from different views as in [7].

2 Deformable Superquadric Geometry and Mechanics

In this section, we describe briefly the deformable model we adopt as the 3D part model. The details of this model can be found in [6].

The deformable model is represented as a sum of a reference shape

$$\mathbf{s}(u,v) = \left[a_1 C_u^{\epsilon_1} C_v^{\epsilon_2}, a_2 C_u^{\epsilon_1} S_v^{\epsilon_2}, a_3 S_u^{\epsilon_1}\right], \tag{1}$$

and a displacement function $\mathbf{d}(u,v) = \mathbf{S}(u,v)\mathbf{q_d}$, where u and v are material coordinates; $S_w^\epsilon = sgn(\sin w) \mid \sin w \mid^\epsilon$; $C_w^\epsilon = sgn(\cos w) \mid \cos w \mid^\epsilon$; a_1, a_2, a_3, ϵ_1, ϵ_2 are global deformation parameters (stored in $\mathbf{q_s}$); $\mathbf{q_d}$ contains nodal variables (displacements at the nodes sampled over material coordinates), and $\mathbf{S}(u,v)$ is the shape matrix containing basis functions in the finite element representation of the continuous displacement function.

In addition to global and local deformations of the parts, a rigid transformation of each part is allowed. It is defined by translational and rotational degrees of freedom $\mathbf{q}_c$ and $\mathbf{q}_\theta$. Collecting all the possible degrees of freedom for a part together, we can describe the state of a deformable model in terms of a time-varying vector $\mathbf{q} = (\mathbf{q}_c^T, \mathbf{q}_\theta^T, \mathbf{q}_s^T, \mathbf{q}_d^T)^T$. Under the external forces $\mathbf{f}$ the model will move and deform according to the following equation:

$$\mathbf{C}\dot{\mathbf{q}} + \mathbf{K}\mathbf{q} = \mathbf{f_q}, \tag{2}$$

where $\mathbf{C}$ and $\mathbf{K}$ are the damping and stiffness matrices, respectively, $\mathbf{f_q}$ are the generalized external forces associated with the degrees of freedom of the model. These forces are related to the external forces derived from image data [6].

The image forces we use in this paper can be written as $\mathbf{f} = \mathbf{f}_{contour} + \mathbf{f}_{3D}$, where the first component of the force deforms the superquadric to have similar occluding contours as the imaged object and the second component governs fitting of the model to the range data provided by the stereo cue.

3 Contour Force Computation

We first find model nodes residing on the occluding contour. With respect to camera l, such nodes $\mathbf{P}_i$ should satisfy $|\mathbf{N}_i \cdot (\mathbf{P}_i - \mathbf{O}_l)| \leq \epsilon$, where $\mathbf{O}_l$ is the optical center of the camera, $\mathbf{P}_i$ is the position vector of the model node i, and $\mathbf{N}_i$ is the surface normal at this node(refer to Fig. 1). Then $\mathbf{P}_i's$ are projected onto the image plane by projection operator $\prod_l$ acquired by camera calibration, i.e. $\mathbf{p}_i^I = \prod_l \mathbf{P}_i$, where the superscript I denotes the image plane points.

For each image contour point $\mathbf{c}_k^I$, the closest model projection $\mathbf{p}_i^I$ is found. Let $\mathbf{P}_i$ be its corresponding 3D point on the model, and $\mathbf{C}_k$ be a 3D point that is projected to $\mathbf{c}_k^I$, i.e. $\mathbf{c}_k^I = \prod_l \mathbf{C}_k$. As can be seen in Fig. 1, such $\mathbf{C}_k$ should lie on the line formed by $\mathbf{O}_l$ and $\mathbf{c}_k^I$. To find a single direction for the force to bring $\mathbf{p}_i^I$ to $\mathbf{c}_k^I$ we utilize the principle of minimal action and compute $\mathbf{C}_k$ as follows:

$$\mathbf{C}_k = arg\left(min_{\mathbf{C}_k} ||\mathbf{C}_k - \mathbf{P}_i||\right), \quad given \quad \mathbf{c}_k^I = \prod_l \mathbf{C}_k. \tag{3}$$

The contour force acting on the model point $\mathbf{P}_i$ is defined as:

$$\mathbf{f}_{c_i} = k_c(\mathbf{C}_k - \mathbf{P}_i), \tag{4}$$

where k_c is a scaling constant. In the case of a pinhole camera, the solution to the Eq. 3 is simply the orthogonal projection of $\mathbf{P}_i$ onto the line $\mathbf{O}_l \mathbf{c}_k^I$. In general, $\mathbf{C}_k$ could be found independent of camera model using Eq. 3.

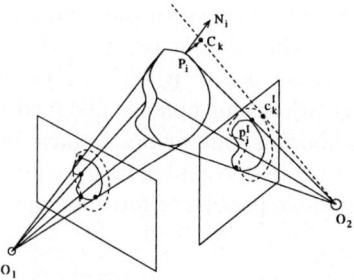

Fig. 1. Contour Forces

4 Contour Force Assignment Using Chamfer Image

To avoid a computationally expensive search for the closest model point for each contour point, we developed an algorithm for force assignment based on the chamfer image of the occluding contours with an additional index matrix containing the index of the closest contour point for each pixel in the image.

For each image, occluding contours are stored as a set of ordered lists of contour points (each list corresponds to a contour segment). The chamfer image, containing distances to the closest contour point for each image pixel, is computed using a slight modification of the well known two-pass algorithm[1]. The modification consists of adding the index matrix $\mathbf{I}$ to track the closest contour point for each pixel. When a model point $\mathbf{P}_i$ is projected to a point $\mathbf{p}_i^I = (x_i^I, y_i^I)^T$ in the image plane, its closest contour point $\mathbf{c}_k$ can be retrieved immediately using the index k stored in the index matrix $\mathbf{I}$, i.e. $k = \mathbf{I}(\lfloor x_i^I \rfloor, \lfloor y_i^I \rfloor)$. Now the force acting to bring $\mathbf{p}_i^I$ to $\mathbf{c}_k^I$ can be computed using Eq. 4.

However, this force assignment does not guarantee a good final fit, as the model points are "choosing" the data points, instead of vice versa. The problem becomes apparent when the model parts are small or relatively far away from the image contours, causing the model to be attracted only by parts of the image contours. It is also possible that all the model points are attracted to a nearly straight segment of the contour causing the superquadric to further reduce in width instead of growing. Finally, two superquadrics can be attracted to the same part of the occluding contour, even though one of them is much closer to this part than the other. We add a post-processing step to deal with these problems, by making the force assignment more similar to the usual approach to fitting, where the data points "choose" which model points they will attract.

After the first assignment based on the chamfer images we go through all the image contour points and check if they have been assigned to model points. Suppose the contour point $\mathbf{c}_k^I$ has been assigned to a model point projection $\mathbf{p}_i^I(k)$, the closest assigned contour point with a smaller index k_1 has been assigned to $\mathbf{p}_i^I(k_1)$, and the closest assigned contour point with a larger index k_2 has been assigned to $\mathbf{p}_i^I(k_2)$. Any contour point with index l satisfying $k_1 < l < k$ or

$k < l < k2$, has not yet been assigned to the model points. If any of the points $\mathbf{p}_i^I(k_1)$, $\mathbf{p}_i^I(k_2)$ is closer to the contour point $\mathbf{c}_k^I$ than its assigned point $\mathbf{p}_i^I(k)$, $\mathbf{p}_i^I(k)$ is discarded, the closer one of $\mathbf{p}_i^I(k_1)$, $\mathbf{p}_i^I(k_2)$ is assigned to $\mathbf{c}_k^I$ instead. After this, each unassigned contour point is assigned to the closer of the model points to which its neighboring contour points have been assigned. An example of the force assignment in case of complex image contours and six superquadrics is shown in Fig. 2. The whole process of force assignment and computation can

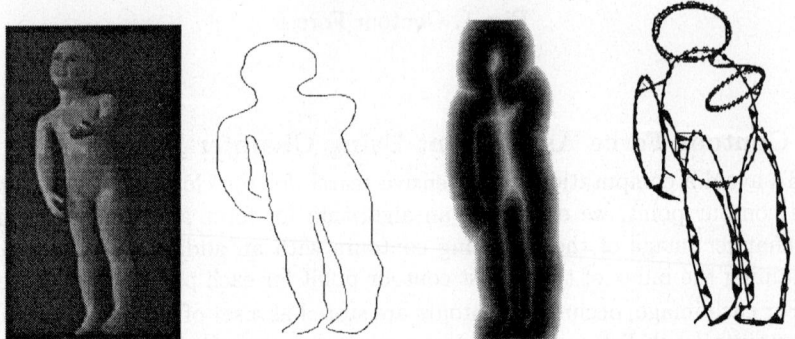

(a)Intensity Image (b) Image Contours (c)Chamfer Image (d)Force Assignment

Fig. 2. Force Assignment

be done in $O(N_m) + O(N_c)$ time, where N_m and N_c are numbers of model and contour points, while the ordinary brute force search for the closest model point for each of the contour points is done in $O(N_m \times N_c)$ time.

However, in the beginning of the fitting process, when the superquadrics are crudely positioned, oriented and sized, the described scheme can cause some errors in force assignment. To avoid that, we do the force assignment based on the brute force search in the first few time steps in solving Eq. 2. After the superquadrics global parameters have reached more reasonable values, force assignment can be based on the algorithm described above. Finally, we note that force assignment need not be done in each time step of integration of the motion equation. It can be done only every few iterations.

5 Forces Based on Stereo Aided by Structured Light

Providing that the correspondence between the features in two images are available, it is possible to reconstruct a number of 3D points on the object's surface using triangulation techniques. In case of the lack of feature points, a structured lighting source can be used to impose extra feature points on object surface.

Correspondence establishment in stereo is known to be a difficult problem. In this paper, we make use of the surface estimated by occluding contours based on the deformable superquadric models to assist the matching of feature points from two images. A match between two features is considered good only if it satisfies both the epipolar constraint and a constraint on the distance from the point, which is reconstructed by the match under consideration, to the surface estimated using occluding contours.

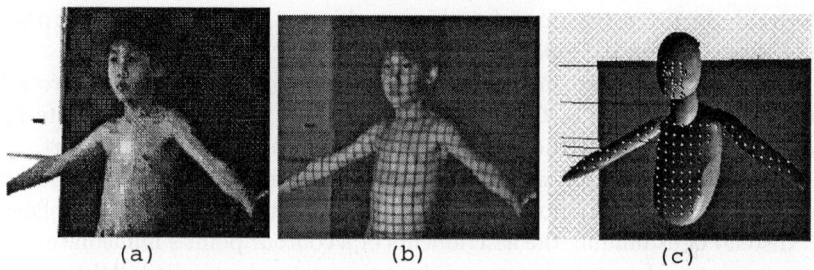

Fig. 3. Structured light provides feature points: (a)The intensity image of a human object; (b)The object illuminated by structured light; (c)The reconstructed feature points overlapped with an estimate from contours

Once the feature correspondences have been established and a number of surface points $\mathbf{R}_k$ have been reconstructed by triangulation, the part models can be further deformed to fit these points by applying forces:

$$\mathbf{f}_{3D_i} = k_{3D}(\mathbf{R}_k - \mathbf{P}_i), \qquad (5)$$

where $\mathbf{P}_i$ is the model node closest to the point $\mathbf{R}_k$ and k_{3D} is scaling constant. Brute force search is used in finding force assignment because the surface estimated from occluding contours usually provides such a good initial guess that the force assignment for fitting the 3D points $\mathbf{R}_k$ hardly varies over time.

The forces derived from stereo cue not only refine the surface reconstruction, but also may reposition superquadrics slightly so that a better contour force assignment can be achieved. Therefore, the two visual cues assist each other. They also complement each other, as occluding contours provide estimates at the parts where stereo becomes unreliable, while stereo refines the estimate at the rest of the surface.

6 Force Assignment Based on Fuzzy Clustering

We experimented with weighted force assignment based on fuzzy clustering proposed by Kakadiaris[4]. Instead of using the nearest neighbor force as described in the previous sections, each data point $\mathbf{r}_i$ should attract points on several models with forces scaled by a weight $p_{r_i,j}$ equal to the probability that this point is correctly associated with the superquadric j. Hence the force that the data point applies to the nearest point $\mathbf{x}_i^j$ on superquadric j is:

$$\mathbf{F}_{r_i}^j = p_{r_i,j} k(\mathbf{r}_i - \mathbf{x}_i^j). \qquad (6)$$

The probability $p_{r_i,j}$ is computed by fuzzy classification of data points based on their proximity to the model points:

$$p_{r_i,j} = \frac{\frac{1}{||r_i - x_i^j||}}{\sum_k \frac{1}{||r_i - x_i^k||}}. \tag{7}$$

This scheme allows a single data point to attract several superquadrics, which is desirable in reconstruction of multipart objects, as it serves the purpose of gluing the parts (for example, the arms in Fig. 4, or legs in Fig. 5).

The weighted assignment method can be directly applied to the forces $\mathbf{f}_{3D}$ and to the contour forces $\mathbf{f}_c$ described in previous sections, if the brute force search method is used in the search for the closest point on a model part.

To use fuzzy clustering with the fast force assignment algorithm proposed in Section 4, we modify the postprocessing step of that algorithm in the following way. Instead of examining the assignments of a contour point's immediate neighbors, a segment of certain length on the contour can be studied. All the model points from different parts that were originally assigned to some of the points on that contour segment, can be attracted by all the points in the observed contour segment, including the ones that were not originally assigned to any of the model points. The assignment is valid with the probability in Eq. 7, and the forces are later computed by scaling the right hand side of the Eq. 4 by this weight.

7 Experimental Results and Conclusions

We performed preliminary experiments on human-like objects with two CCD cameras and a structured lighting source in between to project a stripe pattern on the object surface. Both ordinary intensity images and images under structured lighting are taken.

In the first experiment, we imaged a real human(Fig. 5) with two cameras in the above configuration. Five deformable superquadrics (for arms, torso, neck and head) are manually positioned in the virtual 3D space so that their projections onto image plane lie relatively near the image contours (see the right figure of Fig. 4(a)). Fig. 4 illustrates several steps in integrating Eq. 2. The left column shows the model parts smoothly shaded or texture mapped and the right column shows the fitting of the model parts to the contours in one of the images.

In the second experiment, we tried to reconstruct a more difficult object - a doll with such a posture that the body parts occlude each other(see Fig. 2(a)). The object was positioned on the turning table and a total of eight camera views were used in reconstruction. Six superquadrics were initialized and in Fig. 5(b) we show the final result after fitting to both contours and the 3D points reconstructed by stereo cue. The reconstructed surface is compared to the reconstruction using only stereo correspondences obtained by manual matching to demonstrate how efficiently our scheme deals with self occlusion.

The experiments show that the contour based estimate of the surface, even by only two cameras can efficiently assist stereo matching. In the first experiment, for example, for the 80 feature points detected in the image, 74 correct matches are found purely guided by the contour based estimate of 3D surface. Usually,

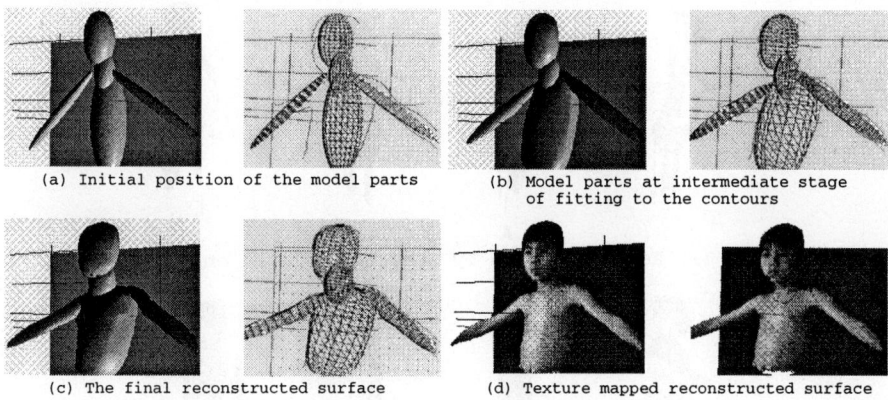

(a) Initial position of the model parts (b) Model parts at intermediate stage of fitting to the contours

(c) The final reconstructed surface (d) Texture mapped reconstructed surface

Fig. 4. Reconstruction of a human upper body

the remaining false matches are corrected as well during further refinement of the surface estimate using both cues. As can be seen from the Fig. 5(a), the stereo aided by structured light can not make estimates near the occluding contours due to absence of matched feature points there, but the combined two cues can provide a rather complete estimate of those parts that were not visible to both cameras.

The advantage of our approach to 3D reconstruction is that it does not require special camera configuration. Also, no prescribed set of movements is needed as in [3] which makes it possible to reconstruct rigid multipart object such as the doll in Fig. 2(a). This may be useful in applications such as mannequin manufacturing in the textile industry. Other applications include dressing humans in a virtual garment designed using CAD systems. For example, in [2], a virtual T-shirt is shown draped over the body reconstructed in Fig. 4(d).

In the future, improving the "gluing" of object parts together is one of the tasks we are going to work on. Additional cues, such as orientation and curvatures of the projected stripes, shading, may be added in the reconstruction algorithm. In some applications, it may be beneficial to develop an automatic part initialization algorithm based on color and texture.

Acknowledgements

This research project is supported by a grant from The Industry and Technology Development Council, (grant no. AF/122/96). The authors from University of Illinois were also supported by Army Research Laboratory under Cooperative Agreement No. DAAL01-96-2-0003. We would also like to thank Roderick Yuen, the son of Dr. Matthew Yuen, the project manager of the grant, for his help in our image acquisition.

The first author would like to thank Ioannis Kakadiaris for very useful discussions.

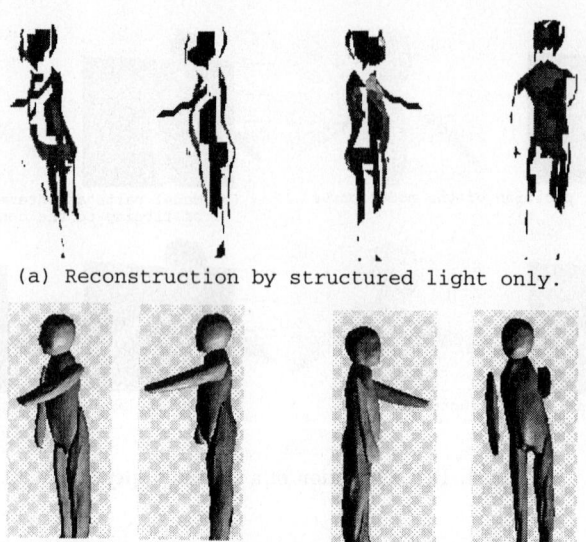

(a) Reconstruction by structured light only.

(b) Reconstruction based on integration of occluding contours and stereo aided by structured light.

Fig. 5. Reconstruction of a Mannequin

References

1. H. G. Barrow, J. M. Tenenbaum, R. C. Bollers, and H. C. Wolf. Parametric correspondence and chamfer matching: Two new techniques for image matching. *Vision-7*, pages 659–63, 1979.
2. N. Jojic and T. S. Huang. On analysis of cloth drape range data. In *these proceedings (ACCV '98)*.
3. I. Kakadiaris and D. Metaxas. Model-based estimation of 3d human motion with occlusion based on active multi-viewpoint selection. In *Proceedings 1996 IEEE Computer Society Conference on Computer Vision and Pattern Recognition*, pages 81–7, 1996.
4. I. A. Kakadiaris. *Motion-Based Part Segmentation, Shape and Motion Estimation of Complex Multi-Part Objects: Application To Human Body Tracking*. PhD thesis, University of Pennsylvania, Philadelphia, PA, 1996.
5. D. Metaxas and D. Terzopoulos. Shape and nonrigid motion estimation through physics-based synthesis. *IEEE Transaction on Pattern Analysis and Machine Intelligence*, 15(6):580–91, 1993.
6. D. Terzopoulos and D. Metaxas. Dynamic 3d models with local and global deformations: Deformable superquadrics. *IEEE Transactions on Pattern Analysis and Machine Intelligence*, 13(7):703–14, 1991.
7. Y. F. Wang and J. K. Aggarwal. Integration of active and passive sensing techniques for representing three-dimensional objects. *IEEE Transactions on Robotics and Automation*, 5(4):460–71, 1989.

On Analysis of Cloth Drape Range Data

Nebojsa Jojic and Thomas S. Huang

Beckman Institute, University of Illinois at Urbana-Champaign
405 N. Mathews Ave, Urbana, Illinois 61801, USA
Email: {jojic,huang}@ifp.uiuc.edu

Abstract. In this paper we present an algorithm for analysis of the range data of cloth drapes. The goal of the analysis is estimation of parameters for modeling and the geometry of the underlying object. In an analysis-by-synthesis manner, the algorithm compares the drape of the model with the range data and searches for the best fit. It can be applied to any physics-based cloth model. The motivating application is fashion design using CAD systems, but the ability of the algorithm to estimate the shape of the object supporting the scanned cloth indicates the possibility of utilizing cloth models to overcome problems in human tracking algorithms, caused by clothing.

1 Introduction

Development of physics-based cloth models is very important for textile industry (3D CAD systems for clothing design), computer vision (tracking dressed humans) and virtual reality applications (model-based telepresence).

Model parameters can be derived from mechanical measurements obtained by the Kawabata Evaluation System [2], [4]. A quantitative comparison between simulated and real drape has not yet been offered, except for the comparison of "drape factors" as in [4]. However, this measure is not reliable, as different materials may have the same drape factor, but exhibit visually different draping behavior.

In this paper, we address the problem of estimating model parameters of any physics-based cloth model by comparing the model to the range data of the real drape. There are several advantages to this approach:

1) We define a measure of the model quality, the mean distance between the model and the range data, computed over the *whole* surface of the scanned cloth.

2) By minimizing this mean distance, we avoid performing mechanical measurements with expensive equipment.

3) We directly address the problem of achieving the synthetic drape as close to the real drape as the model allows. Traditional approaches concentrate only on insuring correct mechanical properties of the model.

4) We show that, by studying in an analysis-by-synthesis manner the range data of the cloth draped over an object, it is possible to estimate parts of the object's shape. This encourages research on utilizing physics-based cloth models in tracking humans.

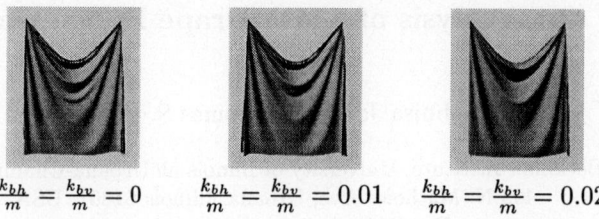

Fig. 1. 100x100 particle systems with different bending constants

2 A Cloth Model

In our experiments, we represent a rectangular piece of cloth as a particle system $\{\mathbf{P}_{i,j} : i = 1, M; j = 1, N\}$. Each particle (i, j) interacts with its neighbors and the supporting object:

1) Repelling and stretching are modeled using simple springs:

$$\mathbf{F}_{s_{ij}} = k_s \left[\sum_{k \in \{i-1,i+1\}} (u - d_{k,j})\mathbf{e}_{k,j} + \sum_{k \in \{j-1,j+1\}} (v - d_{i,k})\mathbf{e}_{i,k} + r \sum_{k \in \{i-1,i+1\}} (d - d_{k,j-1})\mathbf{e}_{k,j-1} + r \sum_{k \in \{i-1,i+1\}} (d - d_{k,j+1})\mathbf{e}_{k,j+1} \right] \quad (1)$$

where $\mathbf{P}_{i,j} = (x_{ij}, y_{ij}, z_{ij})$, $d_{k,l} = |\mathbf{P}_{i,j} - \mathbf{P}_{k,l}|$, $\mathbf{e}_{k,l} = (\mathbf{P}_{i,j} - \mathbf{P}_{k,l})/d_{k,l}$; u, v are nominal horizontal and vertical mesh spacings; $d = \sqrt{u^2 + v^2}$; k_s is the elasticity constant; r is the diagonal-to-axial strength ratio controlling shearing in the model [6].

2) Bending resistance in horizontal and vertical directions is modeled by the force:

$$\mathbf{F}_{b_{i,j}} = k_{bh}(Proj^{\mathbf{P}_{i,j}}_{\mathbf{P}_{i-1,j}\mathbf{P}_{i+1,j}} - \mathbf{P}_{i,j}) + k_{bv}(Proj^{\mathbf{P}_{i,j}}_{\mathbf{P}_{i,j-1}\mathbf{P}_{i,j+1}} - \mathbf{P}_{i,j}). \quad (2)$$

k_{bh} and k_{bv} are bending constants. $Proj^{A}_{BC}$ is the orthogonal projection of A onto the line BC.

3) The gravitational force is equal to $m\mathbf{g}$, where $\mathbf{g} = (0, -9.81, 0)[m/s^2]$ and m is the mass of the particle.

4) The external forces, $\mathbf{F}_{ext_{i,j}}$, model the interaction of the cloth with other objects.

The motion of the particles is governed by Newton's law (k_v is the damping factor):

$$\mathbf{F}_{s_{i,j}} + \mathbf{F}_{b_{i,j}} + m\mathbf{g} + \mathbf{F}_{ext_{i,j}} - k_v \frac{d\mathbf{P}_{i,j}}{dt} = m \frac{d^2\mathbf{P}_{i,j}}{dt^2}. \quad (3)$$

If only the final drape is of interest, the oscillations during the model relaxations can be avoided by setting the right hand side of this equation to zero.

Other physics-based cloth models could also be used with the algorithm explained in the following section. The model and the estimation algorithm are based on the study of the internal and external forces, but it is easy to derive these forces for energy-based deformable models, such as the models described in [2], [8], [3]. Also, the algorithm is not concerned with the discretization of the model, so it can be applied to models based on continuum approximations using finite differences or finite elements.

Interaction between the cloth model and other objects in the scene can be modeled through a collision handling routine based on the law of momentum conservation. We also designed a fast collision detection algorithm based on 3D chamfer images of the objects in the scene. Modeling a more complex cloth objects, such as garment, is also possible by cuting out parts of rectangular cloth patterns, and merging the patterns together. An example of draping a virtual T-shirt over the body of a real human reconstructed in [5] is given in Fig. 2.

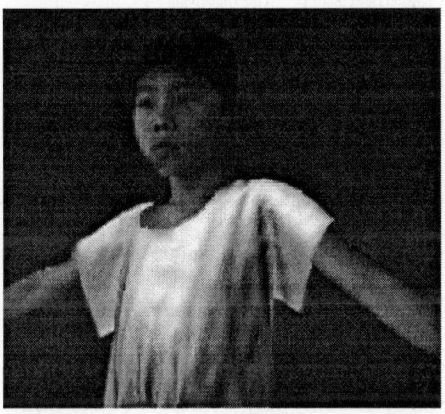

Fig. 2. A human upper body dressed in a virtual T-shirt

3 Parameter Estimation

The estimation algorithm is shown in Fig. 3. It is assumed that the cloth sample is rectangular and its size is known. M, N are chosen in advance and the spacings u, v are derived from the size of the sample. However, no assumptions are made regarding the shape of the supporting object (Fig. 3a) or cloth model parameters. Of the several model parameters in Eq. 1-3, only a few are important. k_v does not affect the final drape and m can be divided out from the Eq. 3. k_s is usually set to a high value, as cloth does not stretch visibly under its own weight. The parameters of our model that affect the final drape and should be estimated are $p1 = r$, $p2 = k_{bh}/m$, $p3 = k_{bv}/m$. For example, the effect of the bending constants is demonstrated in the Fig. 1.

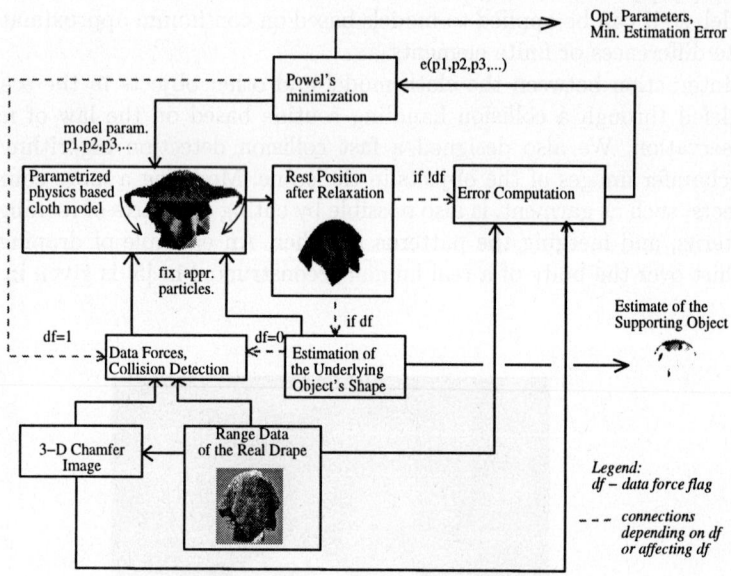

Fig. 3. Simplified block diagram of the estimation algorithm (the case of unknown draping conditions)

The range points, obtained by any available range finder, are organized into a smooth surface over which the model of the sample can be draped. The force $\mathbf{F}_{ext}$ is derived from the range data:

$$\mathbf{F}_{ext_{i,j}} = \mathbf{F}_{c_{i,j}} + \mathbf{F}_{d_{i,j}}, \quad \mathbf{F}_{d_{i,j}} = k_d(\mathbf{r}_{i,j} - \mathbf{P}_{i,j}) \qquad (4)$$

where $\mathbf{r}_{i,j}$ is the range point closest to the particle (i,j). The forces $\mathbf{F}_{d_{i,j}}$ force the model to drape similarly to the scanned cloth. With a low k_d, these forces are not sufficient to prevent the model from falling through the data surface, so it is prevented by our collision detection and handling routine, symbolically represented by force $\mathbf{F}_c$ in Eq. 4. This force is not included into Eq. 3 when it is integrated. Instead, the position and speed of a particle are directly adjusted each time step according to the law of momentum conservation [3], if collision with the range data surface is detected.

To speed up the force assignment and collision detection algorithms, we use the 3D chamfer image, the 3D matrix containing the approximate distances to the range data for any point in a certain volume [1]. In addition to this distance map, we create a 3D index matrix, containing indices of the closest range points for each entry. To create this matrix, it is necessary to keep track of the closest point during the chamfer image computation in a standard two-pass algorithm. The distance map helps to evaluate the possibility of collision of a particle with the surface of the scanned cloth, while the index matrix allows intersection tests with the right triangles on the surface and fast computation of the data forces $\mathbf{F}_{d_{i,j}}$.

In the first phase in the main loop, the model is draped over the range data and is attracted by the data at the same time (df=1 in Fig. 3) by integrating Eq. 3. Special forces are applied from the corners of the sample to the corners of the model to insure its correct positioning over the data.

After the model has assumed its rest position, the particles corresponding to the pieces of the sample that were not supported by the underlying object at the time of scanning, will satisfy the following:

$$\left|\mathbf{F}_{ext_{i,j}}\right| = \left|\mathbf{F}_{s_{i,j}} + \mathbf{F}_{b_{i,j}} + m\mathbf{g}\right| \leq \varepsilon, \tag{5}$$

where ε is a small value. For the real cloth, this equation is obviously satisfied (even with $\varepsilon = 0$) at parts not touching other objects. An example of the distribution of $|\mathbf{F}_{s_{i,j}} + \mathbf{F}_{b_{i,j}} + m\mathbf{g}|$ in the model is given in Fig. 6c. Therefore, the particles not satisfying Eq. 5 are likely to represent pieces of cloth that were supported by the underlying object, and are fixed in the next phase, simulating strong friction between the cloth and the object. The data forces are now turned off (df=0, $\mathbf{F}_{ext} = 0$), and the model is left to assume a new position under new draping conditions.

At the end of the second phase, the mean distance between the particles and the range data is computed:

$$e(p1, p2, p3, ...) = \frac{1}{MN} \sum_{i,j} \left|\mathbf{r}_{i,j} - \mathbf{P}_{i,j}\right|. \tag{6}$$

and this approximation error, as a function of model parameters $p1, p2, p3, ...$, is forwarded to a multivariate minimization routine. We use Powel's direction set minimization technique [7], as it does not require gradient information. The shape estimate is updated after each change of parameters, as the Eq. 5 is better satisfied with more accurate model parameters.

When some parts of the range data are missing, these parts are excluded from the error computation in Eq. 6. Missing data can be detected by a large distance $|\mathbf{r}_{i,j} - \mathbf{P}_{i,j}|$, which, at the end of the phase one, is supposed to be small due to the attracting data forces.

If the supporting object's shape is known a priori, the algorithm is slightly different. Collision of the cloth model with the model of the underlying object is performed and handled in each time step ($\mathbf{F}_c$ in Eq. 4), and in the second phase, there is no need to fix the positions of the particles touching the object. Instead, only the data force component of F_{ext} is turned off, but collision detection and handling is continued. The block diagram of the estimation algorithm for the general case of known draping parameters is shown in 4. This approach is useful when the test cloth sample is supported in a way that is simple to model and allows for fast collision detection, such as draping over a sphere, or suspending a sample at two or three corners.

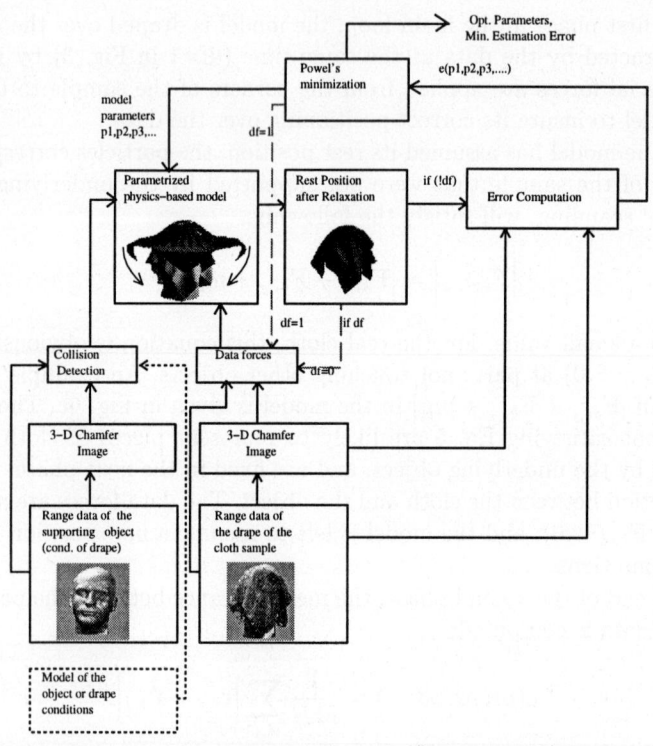

Fig. 4. Block diagram of the estimation algorithm (the case of known draping conditions)

4 Results and Conclusions

Experiments were performed on both synthetic and real range data. In the first experiment, the synthetic range data was obtained by suspending a 50x50 cloth model at two corners (Fig 5(a)). The estimation algorithm (Fig. 4) used a coarser model consisting of a 25x25 system of particles, and the conditions of drape were supposed to be known, i.e., the model was also suspended at two corners. The the mean distance between the particles of the model with estimated optimal parameters and the synthetic range data is $e=2.77$ mm. The image of the optimal 25x25 model is given in Fig. 5(b). In Fig. 5(c), the synthetic range data and the

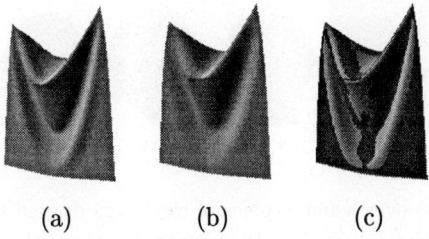

Fig. 5. Estimation algorithm results (the case of known draping conditions)

coarse model are overlapped. The "range data" is light gray and the model with optimal parameters is dark. This image shows that the position and the shape of the main fold is captured correctly, and the error is mainly due to smoothing of the coarse model.

In the second experiment, the estimation algorithm in Fig. 3 is tested. In Fig. 6a, a synthetic drape of a cloth sample 400x400mm over four spheres is given. In Fig. 6b, 6c, and 6d, the best approximation with the 25x25 model (at the end of the parameter estimation algorithm), the distribution of forces in the model and the estimated supporting shape (the contact points between the cloth and the object) are shown. The mean distance between the model and the data was 2.77mm.

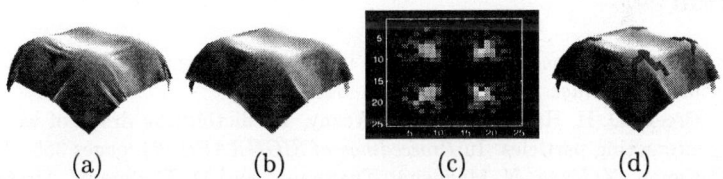

Fig. 6. Synthetic range data created with a 50x50 particle system (a); the best approximation using a 25x25 model (b); the distribution of forces in the model (c); and the estimated supporting shape overlapped with the model (d)

In the third experiment we used a Cyberware laser scanner to scan the drape of a 380x380mm cloth sample (Fig. 7b). The mean distances were 2.2mm for a 50x50 model (Fig. 7c) and 4.12mm for a 25x25 model (Fig. 7d). Note that even with a very crude 25x25 model, the estimate of the supporting object's shape (Fig. 7f) still contains a part of the nose of the underlying head (Fig. 7a), because the cloth sample was touching the nose at that point.

Our algorithm offers an alternative to using the expensive Kawabata Evaluation System for estimating cloth model parameters. Furthermore, it offers a

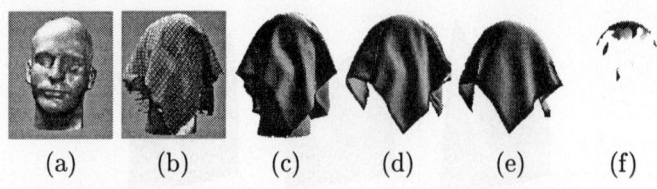

(a)　　(b)　　(c)　　(d)　　(e)　　(f)

Fig. 7. The object (a) over which a piece of cloth was draped (b), the smooth shaded surface of the range data (c), the result of approximation with a 50x50 model (d) and 25x25 model (e); and estimated shape of the supporting object, determined from the 25x25 model (f)

way to evaluate the quality of a cloth model for simulating draping behavior. The possibility of estimating an underlying object's shape encourages research on utilizing physics-based cloth models in tracking of dressed humans, as it is possible to extend the ideas presented here to studying motion of the cloth under constraints enforced by the supporting object(s).

Acknowledgments

This work was supported by Army Research Laboratory under Cooperative Agreement No. DAAL01-96-2-0003.

References

1. G. Borgefors. Distance transformations in arbitrary dimensions. *Computer Vision, Graphics, and Image Processing*, 27:321–345, 1984.
2. D. E. Breen, D. H. House, and M. J. Wozny. Predicting the drape of woven cloth using interacting particles. In *Proceedings of SIGGRAPH '94*, pages 365–372, 1994.
3. M. Carignan, Y. Yang, N. Magnenat Thalmann, and D. Thalmann. Dressing animated synthetic actors with complex deformable clothes. In *Proceedings of SIGGRAPH '92*, pages 99–104, 1992.
4. J. R. Collier, B. J. Collier, G. O'Toole, and S. M. Sargand. Drape prediction by means of finite element analysis. *Journal of the Textile Institute*, 82(1):96–107, 1991.
5. N. Jojic and T. S. Huang. 3d reconstruction of multipart, self-occluding objects. In *these proceedings (ACCV '98)*.
6. H.N. Ng, R. L. Grimsdale, and W. G. Allen. A system for modeling and visualization of cloth material. *Computer Graphics*, 19(3):423–430, 1995.
7. W. H. Press, S. A. Teukolsky, W. T. Vetterling, and B. P. Flannery. *Numerical Recipes in C: The Art of Scientific Computation*. Cambridge University Press, 1995.
8. D. Terzopoulos, J. Platt, A. Barr, and K. Fleischer. Elastically deformable models. In *Proceedings of SIGGRAPH '87*, pages 205–214, 1987.

VR Models from Epipolar Images:
An Approach to Minimize Errors in Synthesized Images

Mikio Shinya, Takafumi Saito, Takeaki Mori and Noriyoshi Osumi

NTT Human Interface Laboratories

Abstract. A new paradigm, the minimization of errors in synthesized images, is introduced to organically combine Computer Vision and Computer Graphics for Virtual Reality applications. Based on it, a powerful algorithm, called *the strip DP algorithm*, is proposed for epipolar image analysis. The algorithm reconstructs VR models from epipolar images so that the error in the images synthesized from the extracted model is minimized while geometrical consistency is maintained. The dynamic programming technique, adopted as the optimization engine, yields complete optimization at reasonable computation cost in a robust way. The strip DP algorithm is a multi-pass solution to occlusion problems, and, in each pass, it extracts connecting feature lines that are not occluded by undetermined feature lines. Experiments demonstrate its feasibility.

1 Introduction

The recent strong progress in Computer Graphics technologies has brought Computer Graphics (CG) and Virtual Reality (VR) applications into daily life. This results in increasing demands to create complex CG/VR models for fancy image generation. Unfortunately, geometric modeling technologies have advanced only in limited areas in the last decade, and most modeling still requires skilled creators performing tedious, precise, and time-consuming work. Quite naturally, people started to consider VR model generation from real images in order to by-pass the geometric modeling process and to provide easier VR model construction tools [6,5,10,8,4,1].

This paper presents a new paradigm in CG applications of CV: *minimization of errors in synthesized images*. The conventional application scheme is a simple concatenation of a CV system and a CG system, where the CV part extracts surface geometries and textures (VR models), and the CG part renders images from them. Although the CV part, namely Shape-from-X, may involve optimization in itself, the total system is an open-loop in terms of image synthesis. The error minimization paradigm can introduce a closed loop into the over-all system that tries to optimize the extraction of VR models for image synthesis. This feedback is essential to the performance and robustness of the system unifying CV and CG.

In CG applications, good models are those from which good images can be synthesized. What are good images? A simple idea is that, if the same images as the input images from which a model is derived are synthesized, they are good images. This criterion is reliable when we have a massive set of input images taken from many different view points because we can expect satisfactory image synthesis under a variety of viewing conditions similar to those of the input data.

In this paper, we particularly focus on the CG application of epipolar image analysis, and realize an error minimization paradigm. It is shown that the optimization can be achieved efficiently and robustly by the dynamic programming technique (DP), which calculates the optimum solution at $O(n^2)$ cost for n-variable optimization. The process is implemented and examined with several data sets. Preliminary results show the potential of our approach.

2 Epipolar image analysis

Epipolar analysis [3] is one of the 'shape from motion' techniques, and reconstructs 3D shape from dense image sequences. An epipolar image is a slice of an image sequence that satisfies an epipolar constraint. When the camera movement is parallel to the scan-line direction (hence perpendicular to the viewing direction), epipolar image $I_0(x,t)$ is a simple time slice of input image sequence $i(x,y;t)$, described by

$$I_0(x,t) = i(x,y_0;t),$$

where t is time, and the x-axis is in the scan-line direction. (For more general cases, see [2].) In this simple case, the trace of the image points projected from a 3D point is a straight line and its slope is proportional to the depth from the camera (Z), described as

$$dx/dt = \alpha Z,$$

where α is a constant determined by the camera parameters and the camera velocity. Therefore, extracting the traces of feature points provides their 3D depth values.

For CG applications, however, topological connections among these points are necessary to execute hidden surface removal and texture mapping in the rendering process. This reconstruction is a hard problem to solve. If the hidden parts of all the traced lines are correctly extracted, logically possible connections can be searched through symbolic operations [9]. However, it is generally difficult to detect the end/start points of edges near intersections, and small mistakes in extraction can lead to fatal errors due to the logical reasoning features.

Here, we introduce the error minimization paradigm to robustly extract geometry and textures so that the 'best' images can be synthesised in terms of errors. We do not assume that the hidden parts of feature lines are known. We also allow incorrect feature lines to be extracted. Both the selection of feature lines and the decision of occlusions are made in the optimization process. This yields robust computation by suppressing the influence of wrong results from the pre-processing stage. In short, the optimization problem addressed here can be stated in the following way: *given n extracted feature lines, select $m \leq n$ lines and determine their connections so that the difference between the synthesized images and input images is minimized.*

3 Error estimation and optimization for occlusion-less scenes

This section presents analyses on occlusion-less cases. Occlusions are then discussed in the next sections.

3.1 Errors in synthesized images

First, let us define errors when connecting two feature lines $a_1(t)$ and $a_2(t)$. By using a norm of pixel colors $||.||$, the error h can be defined by the difference between the input epipolar image $I_0(x,t)$ and the synthesized image $I_{syn}(x,t)$, as:

$$h(a_1, a2) = \int_t \int_{a_1(t)}^{a_2(t)} ||I_0(x,t) - I_{syn}(x,t)|| dx dt. \tag{1}$$

The norm can be, for example, the square sum of rgb-values,

$$||I|| = I^2 = r^2 + g^2 + b^2 \tag{2}$$

for $I = (r, g, b)$. The synthesized image I_{syn}, and hence the error function as well, depends on the rendering algorithm used. In most VR applications, simple texture mapping with the diffusive reflection model is used. In this case, I_{syn} is represented by using texture $f(s)$ defined on $0 \leq s \leq 1$, as

$$I_{syn}(x,t) = f((x - a_1(t))/(a_2(t) - a_1(t))). \tag{3}$$

Next, let us determine the texture $f(s)$ that minimizes the error h. By using Eqs. 2 and 3 and setting the deviation $\partial h/\partial f = 0$, we have the optimum texture:

$$f(s) = \int_t (a_2 - a_1) I_0((a_2 - a_1)s + a_1, t) dt / \int_t (a_2 - a_1) dt. \tag{4}$$

This equation means that the texture should be the weighted average of the input image along the flow. The error function $h(a_1, a_2)$ is then calculated as the variance of the image, represented by

$$h(a_1, a_2) = \int_0^1 [\int (a_2 - a_1) I_0^2 dt - (\int (a_2 - a_1) I_0 dt)^2] ds \tag{5}$$

3.2 Optimization with Dynamic Programming

The most important point in the previous discussion is that the error between two lines depends only on these two lines in the occlusion-less case. This feature allows us to apply the dynamic programming technique (DP), which is a powerful optimization tool, widely used in many areas, e.g., [7].

Assume that n-lines $\{l_1, ..., l_n\} = L$ are extracted as candidate feature lines. The task is to select m-lines $\{\lambda_1, ..., \lambda_m\} = \Lambda$ from L so as to minimize the total error $H(\Lambda)$. The total error H is the sum of the errors between adjacent selected lines, represented by

$$H(\lambda_1, ..., \lambda_m) = \sum_{i=0}^{m} h(\lambda_i, \lambda_{i+1}), \tag{6}$$

where $\lambda_0 = l_0$ and $\lambda_{m+1} = l_e$ are the left and right boundaries of the epipolar image. Although a naive optimization of $H(\Lambda)$ involves *n-dimensional* search, it is possible to achieve it as an $O(n^2)$ process by DP (see [7]).

4 Occlusions

Although the analysis of the occlusion-less case is straightforward, occlusion poses serious problems to DP optimization. Unlike the occlusion-less case, candidate lines may intersect each other when one occludes another. This suggests that error h associated with two lines (l_1-l_2) also depends on all crossing lines,

$$h(l_1, l_2) = h(l_1, l_2; l_3, l_4, ...).$$

Thus, DP cannot be applied in this form.

Fortunately, we found a multi-pass solution using DP that avoids NP complexity. This solution is based on the fact that the visible area of an occluding span is never affected by the occluded spans. This suggests the possibility of successively appling DP and determining the optimal connections from near to far in terms of depth.

This section presents the basic idea of dealing with occlusions using simple examples consisting of a two-layered scene (foreground and background). Although we have theoretical form for general situations, they are omitted here due to the tight space requirement.

4.1 Without self-occlusion

There are two types of occlusion to be considered: self-occlusion, in which a surface is hidden by itself or its adjacent faces, and 'circumstantial' occlusion, in which a surface is occluded by other surfaces. We first discuss the circumstantial case, and then extend the analysis to general cases. Our strategy is a multi-pass approach. In each pass, we extract spans that are occluded only by the already extracted spans. When there is no self-occlusion, this extraction can be performed in the following way.

1. Select candidate lines that cannot be occluded. A line is selected if its slope is smaller than those of all lines intersecting it. In the example in Figure 1-a, candidate lines l_4 and l_5 are selected. Let us call these lines *primal candidate lines*, or PCLs in short. Note that PCLs do not intersect each other.

2. For each pair of PCLs (p_i, p_j), evaluate the error $h_{close}(p_i, p_j)$ according to Eq. 5. The value h_{close} represents the error imposed when p_i and p_j are connected.

3. For each pair of PCLs, evaluate $h_{open}(p_i, p_j)$ that represents the minimum error imposed on the region bounded by p_i and p_j when they are assumed to be disconnected. For example, $h_{open}(l_0, l_4)$ can be given by applying DP described in Section 3.2 to lines l_0, l_1, l_2, l_3 and l_4 in the region left to l_4. More generally, h_{open} can be calculated by recursively applying Steps 1-4 to the region as in a divide-and-conquer fashion.

4. Apply DP to get

$$\min \sum h_i(\lambda_i, \lambda_{i+1}),$$

where h_i is either h_{open} or h_{close}. We set the constraint that h_i and h_{i+1} can not be h_{open} at the same time for any i. This is because, if both h_i and h_{i+1} are open, the candidate line λ_{i+1} does not really exist, in which case $h_{open}(\lambda_i, \lambda_{i+2})$ should be taken instead. Let us call this connected span a *visible span (VS)*.

5. Remove connecting PCLs and denied PCLs. Also trim VS regions from the image area to process. Repeat the process from Step 1.

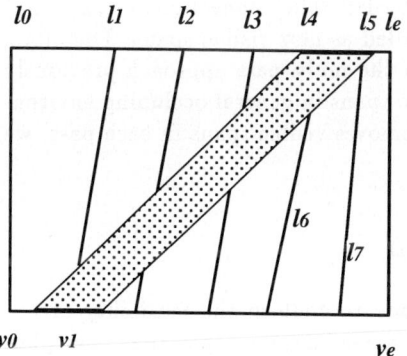
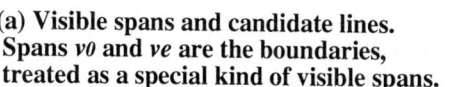
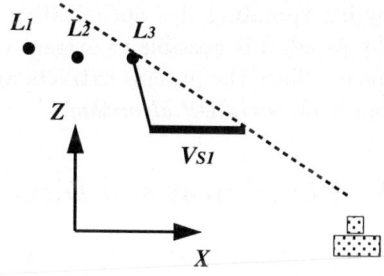

(a) Visible spans and candidate lines. Spans *v0* and *ve* are the boundaries, treated as a special kind of visible spans.

(b) The corresponding X–Z space. Span *Vs1–L3* occludes *L1* and *L2* when viewing from the right side.

Figure 1: Occlusion in epipolar images.

4.2 General occlusion

Self-occlusion makes the situation more complicated. Let us consider the situation in Figure 1, where visible span v_1 (l_4-l_5) is already extracted and all remaining lines p_i are primal candidate lines. When, for example, l_3 is connected to v_1, primal candidate lines left to l_3 are all occluded by span l_3-v_1 when viewing from the right side of v_1 (see Figure 1-b). Similarly, when l_6 connects to v_1, span v_1-l_6 occludes all PCLs right to l_6 when viewing from the left-side of v_1.

Therefore, the error $h_v(v_0, v_1)$ associated with the region between v_0 and v_1 depends only on the connectivities of the VSs and candidate lines between v_0 and v_1, represented by

$$h_v(v_0, v_1) = h_v(v_0, v_1; \mu_1, \{\lambda_i\}),$$
$$\lambda_i \in \{l_1, l_2, l_3\},$$

where μ_1 is assumed to connect to v_1. Let H_v denote the minimum value of h_v, defined as

$$H_v(v_0, v_1; \mu_1) = \min_{\lambda_i} h_v(v_1, v_2; \mu_1, \lambda_1, .., \lambda_m). \qquad (7)$$

This minimization can be performed by DP described in Section 3.2, since all l_i are primal in this example. Let us define a function H as the summation of H_v:

$$H(V_0, .., V_m; \mu_0, .., \mu_m) = \sum_i H_v(v_i, v_{i+1}) \qquad (8)$$

H represents the minimum error imposed when we choose m connections V_i-μ_i. Since Eq. 8 is in the form of DP with respect to connection (V_i, μ_i), it is possible to minimize the total error H by DP. The connections (V_i, μ_i) that provide the minimum error can be regarded as new visible spans. Therefore, by incorporating this optimization with the multi-pass approach previously discussed, it is possible to extract visible spans in general occluding environments. Since the process extracts and removes visible spans in each pass, we call it *the strip DP algorithm*.

5 Experiment and Discussion

Some preliminary experiments were made to confirm the feasibility of the strip DP algorithm. Figure 2-a shows a test data sample that was photographed by an uncalibrated motion-controlled camera. The strip DP method first extracts un-occluded visible spans, and other spans were extracted in the front-to-back order in each pass. There initially were 85 candidate feature lines(2-b), and 59 visible spans consisting of 65 feature lines were extracted in total, discarding 20 feature lines were discarded. The required computation time was 36.9 seconds in total on an SGI workstation with an R4400 at

250MHz. The image synthesized from the extracted spans is almost visually identical to the input image, and the error-power ratio was 0.018 (Figure 2-d and -e).

6 Conclusion

A new algorithm, called *the strip DP algorithm*, was proposed for epipolar image analysis. The algorithm reconstructs VR models from epipolar images so that the error in the images synthesized from the extracted model becomes minimal while geometric consistency is maintained. We adopted the dynamic programming technique as the optimization engine, because it can perform complete optimization at reasonable computation cost. The strip DP algorithm is a multi-pass solution to occlusion problems, and, in each pass, it extracts connecting feature lines that are not occluded by undetermined feature lines. Preliminary experiments demonstrated its feasibility.

References

1. Azarbayejani, A., Galyean, T., Horowitz, B., and Pentland, A "Recursive estimation for CAD model recovery", *proc. 2nd CAD-Based Vision Workshop*, (1994).
2. Baker, H. H., and Balles, R., "Generalizing epipolar-plane image analysis on the spatiotemporal surface", *International Journal of Computer Vision*, vol. 3, No.1, pp. 33-49 (1989).
3. Balles, R. C., Baker, H. H., and Marimont, D. H., "Epipolar-Plane Image Analysis: An Approach to Determining Structure from Motion", *Int. J. Computer Vision*, Vol.1, No.1, pp. 7–55 (1987).
4. Debevec, P., Taylor, C., and Malik, J., "Modeling and rendering architecture from photographs: a hybrid geometry- and image-based approach", *proc. Siggraph'96*, pp.11-20 (1996).
5. Gortler, S., Grzeszczuk, R., Szeliski, R., and Cohen, M., "The Lumigraph", *proc. Siggraph'96*, pp.43-54 (1996).
6. Levoy, M., and Hanrahan, P., "Light field rendering", *proc. Siggraph'96*, pp.31-42 (1996).
7. Ohta, Y., and Kanade, T., "Stereo by Intra- and Inter-Scanline Search Using Dynamic Programming", *IEEE PAMI*, Vol.7, No.2, pp. 139–154 (1985).
8. Seits, S., and Dyer, C., "View Morphing", *proc. Siggraph'96*, pp.21-30 (1996).
9. Yasuno, T., and Suzuki, S., "Occlusion analysis of spatiotemporal images for surface reconstruction", *proc. 4th British Machine Vision Conference* pp. 549-558 (1993).
10. Werner, T., Hersch, R, and Hlavac, V., "Rendering real-world objects using view interpolation", *proc. ICCV'95*, pp.957-962 (1995).

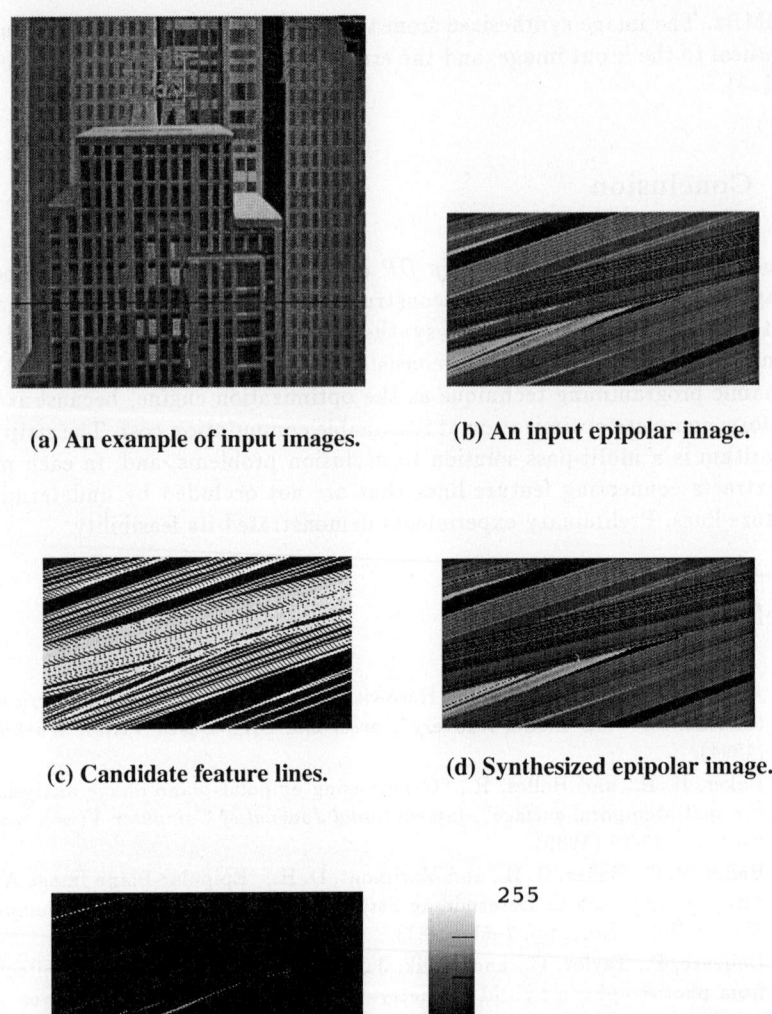

(a) An example of input images.
(b) An input epipolar image.
(c) Candidate feature lines.
(d) Synthesized epipolar image.
(e) Difference between input and synthesized images.

Figure 2: Experimental results.

Shape and Pose Parameter Estimation of 3D Multi-part Objects

Satoshi YONEMOTO, Naoyuki TSURUTA, Rin-ichiro TANIGUCHI

Department of Intelligent Systems, Kyushu University
6-1, Kasuga-koen, Kasuga, Fukuoka 816 JAPAN
yonemoto,tsuruta,rin@is.kyushu-u.ac.jp

Abstract. This paper presents an analysis-by-image-synthesis framework of shape and pose estimation of 3D multi-part objects, whose purpose is to map objects in the real world into virtual environments. In general, complex 3D multi-part objects cause serious self-occlusion and non-rigid motion. To deal with the occlusion among them, we employ both multiple calibrated cameras and time-varying sequences, since there is enough information to estimate the parameters in the sensory data. In our framework, to minimize the error between the selected measurements and the estimated model parameters, we proceed model fitting process based on proper gradient-based minimization.

1 Introduction

We have been developing a system to construct easily virtual environments by means of seamless fusion of information obtained by observing or measuring complex real world objects. As one of the actual applications, we consider 3D animation in which various kinds of natural creatures appear. Therefore, in our approach, animals are intended for the main subject, which have scarcely been considered in the past studies because of its complexity.

For designing this system, the following issue is quite important.

Required information to map the real world objects into a virtual environment consists of two kinds of information: one is a priori object model, or a priori knowledge of the objects; the other is the result of a posteriori observation, or measured object parameters. To construct an efficient or easy-to-use mapping system, the most important point is where we establish the boundary between a priori object model and a posteriori observation.

To simplify a posteriori observation, we need a precise object model in advance, which causes the difficulty in constructing the object model. On the contrary, when we assume a simple a priori object model, a posteriori observation becomes difficult, i.e., we have to solve very difficult CV problems.

For this problem, our approach adopted here is as follows:

A priori object model: each object consists of parts represented in deformable superquadrics[6], which are connected via articulation points one another.

3.2 Object Model Description

In our method, any 3D model description can be introduced only if it is represented in parameterized-form. As a 3D parametric model here, we consider deformable superquadrics(we call DSQ). Although the various types of DSQ are designed so far[1] [5][6][13], we employ the one developed in [6], which has an advantage in the sense that it can be represented by a small number of parameters and can represent deformations such as tapering and bending.

DSQ Geometry When (η, ω) is a material coordinate system, a point on SQ **e** is[1]:

$$\mathbf{e}(\eta, \omega) = \begin{pmatrix} e_1(\eta, \omega) \\ e_2(\eta, \omega) \\ e_3(\eta, \omega) \end{pmatrix} = a \begin{pmatrix} a_1 \cdot C_\eta^{\epsilon_1} \cdot C_\omega^{\epsilon_2} \\ a_2 \cdot C_\eta^{\epsilon_1} \cdot S_\omega^{\epsilon_2} \\ a_3 \cdot S_\eta^{\epsilon_1} \end{pmatrix}, \quad (1)$$

where $-\frac{\pi}{2} \leq \eta \leq \frac{\pi}{2}$, and $-\pi \leq \omega < \pi$, and a, a_1, a_2, a_3 are scale parameters, and ϵ_1, ϵ_2 are squareness parameters, and where $C_w^\epsilon = \text{sign}(\cos w)|\cos w|^\epsilon$, and $S_w^\epsilon = \text{sign}(\sin w)|\sin w|^\epsilon$.

Using **e** on SQ, a point on DSQ **s** is expressed as [6]:

$$\mathbf{s} = \begin{pmatrix} s_1 \\ s_2 \\ s_3 \end{pmatrix} = \begin{pmatrix} (\frac{t_1 e_3}{a a_3} + 1)e_1 + b_1 \cos(\frac{e_3 + b_2}{a a_3}\pi b_3) \\ (\frac{t_2 e_3}{a a_3} + 1)e_2 \\ e_3 \end{pmatrix}, \quad (2)$$

where t_1, t_2 are tapering parameters, and b_1, b_2, b_3 are bending parameters.

From the above definitions, the shape and pose parameters of each part can be expressed as:

$$\mathbf{q} = (a, a_1, a_2, a_3, \epsilon_1, \epsilon_2, t_1, t_2, b_1, b_2, b_3, r_1, r_2, r_3, c_1, c_2, c_3)^T \quad (3)$$

where r_1, r_2, r_3 are rotation parameters, and c_1, c_2, c_3 are translation parameters for each part.

Multi-part Geometry In our method, multi-part object models consist of the 3D deformable parts(DSQ) and the structure is represented in a hierarchical tree structure(See Section 4.2).

4 Parameter Estimation of Multi-part Objects

4.1 Acquiring Initial Model

Using the modeling tool, we can build model structure and make registration of part i on the initial multi-viewpoint frames $f_v(0)$ $(v = 1, \cdots, V; i = 1, \cdots, N)$ where V is the maximum number of viewpoint and N is the maximum number of constituent parts. The registration gives:

1. Initial shape and pose parameters.
2. The correspondence between the initial model data and the initial measurements for tracking in the succeeding frames, to determine to which model sample on a part the measurement corresponds.

4.2 Estimation Strategy

In general, multi-part objects essentially include some constraints based on their structure, which are useful to reduce the search space. In a proposed method, we use a simple top-down strategy under the structural constraints to reduce the computation time. We impose the following structure constraints on objects: 1) Usual objects can be expressed as hierarchical tree structures of parts; 2) these constituent parts are connected via articulation points one another. As a result, pose parameters of a part have an influence on its descendant parts, and consequently, it leads to the quasi-optimal solution.

1. Set $i = root$.
2. Estimate the parameters of $part_i$.
3. If it has child parts, then estimate their parameters recursively(goto 2).
4. If not, stop the recursion.

4.3 Parameter Estimation of Each Part

Model Fitting Problem In our analysis-by-image-synthesis approach, the parameter estimation can be reduced into the model fitting of the measurements based on a proper numerical analysis. When the error between the model and the measurements is represented by d_j, in general, the objective function E is defined as the follows:

$$E = metric \sum_j (w_j \rho(d_j)) \qquad (4)$$

In this paper, we simply used $\rho(x) = x^2$, $w_j = 1$ and $metric$ is min. Therefore, we define the error between the model sample points and the measured points as the following equation, and minimize it:

$$E_i(s) = \sum_{v=1}^{V} \sum_{j \in {}^vP^i} \left(({}^vU_{m_j}^i(s) - {}^vU_j^i)^2 + ({}^vV_{m_j}^i(s) - {}^vV_j^i)^2 \right), \qquad (5)$$

where ${}^vP^i$ is the number of corresponding pairs on $part_i$ from viewpoint v and s is a computation time to minimize.

Details of Parameter Estimation

Step 1 Determine the correspondence between the model points and given measured points from each viewpoint frames $f_v(t)$ $(v = 1, \cdots, V)$:

(1.1) Select visible model sample points:
Since the object model consists of multi-parts, the projection of the model includes two kinds of the occlusion: 1) self-occlusion caused by a part itself, 2) occlusion caused by the other parts.
To deal with such occlusion, we have to delete model sample points do not appear on the projection.
The selection process proceeds as follows:

1. Making the projection for all parts by using a *Z-buffer* technique.
2. Collect model sample points on *part$_i$* from the projection.

In this way, points which should be measured can be selected in accordance with model-based approach(top-down processing).

(1.2) Select good measured points for tracking:

Among the measured points on *part$_i$* at previous frame, select the *trackable* measured points. In addition, to deal with model sample points newly appeared in (1.1), extract other good measured points to track in the place of lost measured points. This is performed completely in a bottom-up way.

(1.3) Determine corresponding pairs:

1. Delete the corresponding pairs for the model sample points deleted because of new occlusion in (1.1), and non-trackable measured points in (1.2).
2. Get new corresponding pairs between model sample points newly appeared because of disocclusion in (1.1) and newly found good measured points to track in (1.2). the correspondence is determined by 2D Euclidean distance similarity.

We have acquired the following information at the current frames, using the results which are obtained at the previous frame:

1) The correspondence between the model sample points and selected measured points at previous frame $f_v(t-1)$:

$$\{(^vU_{m^i_j}(t-1), ^vV_{m^i_j}(t-1))^T, (^vU^i_j(t-1), ^vV^i_j(t-1))^T\},$$

2) The correspondence between tracked measured points:

$$\{(^vU^i_j(t), ^vV^i_j(t))^T, (^vU^i_j(t-1), ^vV^i_j(t-1))^T\}$$

Using the above two consequences, we can get the initial correspondence(hypothesis) between the model sample points $(^vU_{m^i_j}(t), ^vV_{m^i_j}(t))^T$ and selected measured points $(^vU^i_j(t), ^vV^i_j(t))^T$ at current frame $f_v(t)$, which should be estimated:

$$\{(^vU_{m^i_j}(t), ^vV_{m^i_j}(t))^T, (^vU^i_j(t), ^vV^i_j(t))^T\}.$$

Step 2 Estimate the parameters iteratively:

(3.1) Tuning each parameter α_k to converge toward the orientation of gradient-descent:

$$\alpha_k(s+1) = \alpha_k(s) - \beta(s+1)\frac{\partial E_i(s)}{\partial \alpha_k(s)}, \qquad (6)$$

where $k \in \{a, \cdots, b_3, r_1, \cdots, c_3\}$ and $\beta(s)$ is the step size.

(3.2) Compute the objective function $E_i(s+1)$ for the following corresponding pairs between the model sample points(should be updated) and selected measured points(in the following formula, the flow number t is omitted. s means the iteration time):

$$\{(^vU_{m^i_j}(s+1), ^vV_{m^i_j}(s+1))^T, (^vU^i_j, ^vV^i_j)^T\},$$

where $j \in {}^vP^i$; $v = 1, \cdots, V$.

(3.3) Go to (3.1) unless the function converges.

5 Experimental Results

The proposed method is performed in the following sequence: First, to acquire a sequence of multi-viewpoint frames, real cameras are mutually calibrated, and then virtual cameras to make model views are set up in the same configuration. Then, using the interactive modeling tool, we make a model structure of a object and get initial model parameters. Finally, to extract the object, based on the initial model(and the result of registration), we estimate model parameters from frame to frame. In the experiment, to evaluate the basic performance of the proposed method, we have applied it to synthesized image data. We have created the synthetic data rendered by employing three virtual cameras orthogonally calibrated. We have used an object having serious occlusion, i.e., one consists of three parts connected by two joints, like an arm. We have used a steepest descent method as gradient-based minimization in estimating their parameters. Fig.1 shows the input model views acquired from virtual cameras and reconstruction of the estimated model views from the same viewpoint. Fig.2 shows the transition of the number of the corresponding pairs on middle part. Each bar represents the number of the corresponding pairs(the black bars are newly appeared, the grays are even lost and the whites are tracked). These results show that, in case we must depend on an unstable feature tracking method, using this model-based approach, we can select only the proper corresponding pairs between the model data and the measurement.

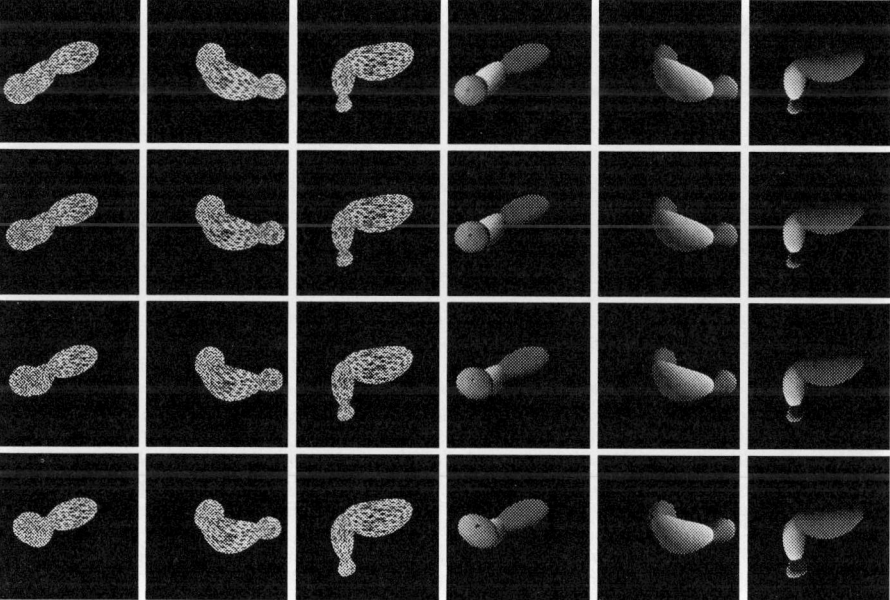

Fig. 1. (Left): Input views. (Right): Reconstruction of estimated model views from the same viewpoint. Each row indicates time-varying sequences at one time interval, and each column indicates multi-viewpoint frames(front,left,up).

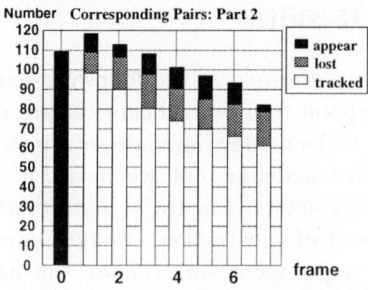

Fig. 2. the transition of the number of the corresponding pairs(on middle part).

6 Conclusions

To map efficiently the real world objects into a virtual environment, we have adopted two kinds of information: one is a priori object model, which consists of deformable parts; the other is a posteriori observation, which means recovering 3D shape and pose parameters from images. And we have explained the advantages in our approach is that non-rigid multi-part objects can be handled, and a multiple cameras are employed to deal with self-occlusion among their parts. The experiments using synthetic data have proved the basic performance of the proposed method. In future work, we plan precise experiments using real image data. Moreover, to acquire more precise model, we try to introduce local deformation into a model. To develop a real-time system is also a future work.

References

1. A.H. Barr, *Global and Local Deformations of Solid Primitives*, Computer Graphics(*Proc.*SIGGRAPH'84), Vol.18, No.3, pp21-29, 1984.
2. A. Pentland and S. Sclaroff, *Closed-Form Solutions for Physically Based Shape Modeling and recognition*, PAMI-13, No.7, pp715-729, 1991.
3. A. Pentland and B. Horowitz, *Recovery of Nonrigid Motion and Structure*, PAMI-13, No.7, pp730-741, 1991.
4. D.G. Lowe, *Fitting Parameterized Three-Dimensional Models to Images*, PAMI-13, No.5, 1991.
5. D. Terzopoulos and D. Metaxas, *Dynamic 3D Models with Local and Global Deformations: Deformable Superquadrics*, PAMI-13, No.7, pp703-714, 1991.
6. D. Metaxas and D. Terzopoulos, *Shape and Nonrigid Motion Estimation through Physics-Based Synthesis*, PAMI-15, No.6, pp580-591, 1993.
7. D.M.Gavrila, L.S.Davis, *3D model-based tracking of Humans in action: a multiview approach*, CVPR, pp73-80, 1996.
8. I.A. Kakadiaris and D. Metaxas, *3D Human Body Model Acquisition from Multiple Views*, ICCV, pp618-623, 1995.
9. I.A. Kakadiaris and D. Metaxas, *Model-Based Estimation of 3D Human Motion with Occlusion Based on Active Multi-Viewpoint Selection*, CVPR, pp81-87, 1996.
10. J.M. Rehg, T. Kanade, *Model-Based Tracking of Self-Occluding Articulated Objects*, ICCV, pp612-617, 1995.
11. J.Ohya, F.Kishino, *Human posture estimation from multiple images using genetic algorithm*, 12th ICPR, pp750-753, 1994.
12. R. Koch, *Dynamic 3-D Scene Analysis through Synthesis Feedback Control*, PAMI-15, No.6, pp556-568, 1993.
13. F. Solina and R. Bajcsy, *Recovery of Parametric Models from Range Images: The Case for Superquadics with Global Deformations*, PAMI-12, No.2, pp131-146, 1990.

Generating 3D Models of Objects Using Multiple Visual Cues in Image Sequences

Jiang Yu ZHENG, Akio MURATA, and Norihiro ABE

Kyushu Institute of Technology
680-4 Kawazu, Iizuka, Fukuoka 820, Japan
zheng@mse.kyutech.ac.jp

Abstract

This work aims at building a 3D graphics model of an object using multiple visual cues. Different materials and shapes may yield different visual cues such as corner, pattern, highlight, and contour on objects. These cues may appear as similar edges, peaks and intensities in the images. To obtain a correct shape, different shape recovery methods must be applied separately. The key problem is how to identify features for different shape recovery methods. We rotate an object and take a dense image sequence. In each Epipolar Plane Image of the rotation plane, we classify visual cues based on not only based on their image characteristics but also their motion behavior. Then we integrate shapes from different algorithms.

1. Introduction

Recent progresses on multimedia, information network, and virtual reality raise a new requirement of modeling 3D objects. Such objects can be commercial products, industrial parts, folk arts, sculptures, environment, antiques, archaeological finds, etc. New display tools, e.g., VRML, have been developed for viewing objects via network. How to input 3D models of objects conveniently and stably thus becomes a challenging problem. Although various laser range finders have been developed for measuring 3D shape, a commercial product usually limits object size. For very big or very small object, laser beam will not scan object properly because of the width of beam and the control limitation. We select video cameras (most of them have zoom function) as the input device. Receiving a standard image sequence either through network or in a video tape, we construct 3D models and return the data.

To make reconstruction process robust, we rotate an object around an axis and take an image sequence. This camera looking from the orthogonal direction to the axis is easy to be achieved. The established points must be a closed shape so as to connect them by graphics patches and displayed them as a model. Therefore, we have to use as many different visual cues as possible to obtain complete shapes. Shape recovery using the same setting has been studied on contour [1], highlight [2,3], edge [4,5], and shading [13,14] according to their motions in the corresponding epipolar plane image. This paper discusses how to combine their results for a complete model. Before quantitative data fusion for an accurate result, qualitative classification must be done to avoid a module picking up wrong points to compute.

2. Multiple Visual Cues

2.1 Physical Conditions and Visual Cues

Visual cues are related to object properties including shape, materiel, color and environment attributes including illumination, view point background, and relative motion between a camera and objects. Multiple visual cues such as silhouettes (contour), corners, patterns, texture, highlights, shadows, and shading may be visible on the objects. Unfortunately, it is impossible to determine both environment and object only from images. We hence fix the environment to achieve flexibility in modeling different objects. An observed feature can be categorized as either *fixed feature* or *moving feature* [11], depending on if it is static on the surface when the object moves. To apply shape from X methods to corresponding cues, we need to qualitatively separate image features into different cues. We analyze their image behaviors and examine their image properties such as intensity, contrast, image velocity in the consecutive images, and coherence of measured positions in the 3D space [11,15]. The relation of different cues also provides spatial constraints to each other when a feature is unable to be classified according to its own attributes.

2.2 Imaging System and EPIs

We rotate an object around an axis, A camera takes its continuous images at each rotation angle. The camera axis is orthogonal to the rotation axis as shown in Fig. 1. A vertical linear light is set at the camera side in the plane containing both rotation axis and the image origin. There is hence no shadow visible on the object.

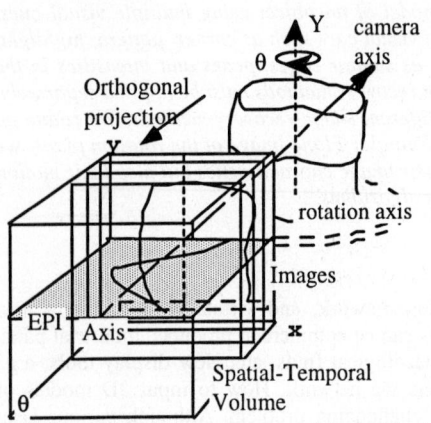 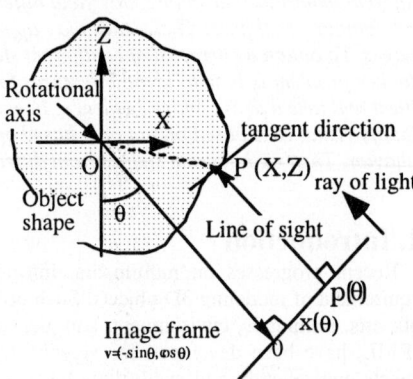

Fig. 1 Taking images of a rotating object and determining Epipolar Plane Images parallel to rotation planes.

Fig. 2 Camera and object relation in a rotation plane.

We assume orthogonal projection from the scene to the camera. We put the camera centered coordinate system C-xyz so that the x and y axes are aligned with EPI and the rotation axis in the image, respectively. The object centered coordinate system O-XYZ is located at the rotation axis and the Z and z axes in the two systems are collinear when the rotation begins. The rotation is clock-wise and the rotation angle is denoted by θ. The angle can either be controlled or can be measured in the image sequence.

For each rotation plane, we can determine its projection (a line). An Epipolar-plane image (EPI) is collected from the projection [1,12]. We will compute shape on each rotation plane using the EPI and then build the model by connecting shapes at all rotation planes.

As the object rotates, a surface point **P** moves on a circle in the rotation plane and its projection is retained in the corresponding EPI. The point P(X, Z) described in the object coordinate sub-system O-XZ is continuously projected as p(x, θ) in the EPI. According to the camera geometry in Fig. 2, the viewing direction is V(-sinθ, 0, cosθ) in the system O-XZ. The image position of the point viewed at the rotation angle θ can be written as

$$x(\theta) = \mathbf{P} \cdot \mathbf{x} = X(\theta) \cos\theta + Z(\theta) \sin\theta \qquad (1)$$

where **x** is the unit vector of the horizontal image axis. Thus, the trajectory of the point in the EPI is a sinusoidal function of the rotation angle with the period of 2π, even the point can not be distinguished as a visual cue.

3. Shape from Xs for a Rotating Object

3.1 Fixed Points and Position Estimation

Let us list conditions for using different visual cues, their motion behaviors in the EPI, algorithms for computing their positions, and accuracy of algorithms. Figure 3 gives an example of EPI showing moving traces of different visual points.

For a fixed feature such as a corner and a pattern edge on the rotational plane, its position will not change on the object during the rotation; the lines of sight looking at it from different angles always pass the same 3D point. In the EPI, a fixed point draws trajectory as a sinusoidal curve of 2π period. Multiple lines of sight through projected points on the trajectory should cross at the fixed point. In the real situation, we use the least squared error method to obtain an accurate measure of the crossing point, i.e., the estimated position.

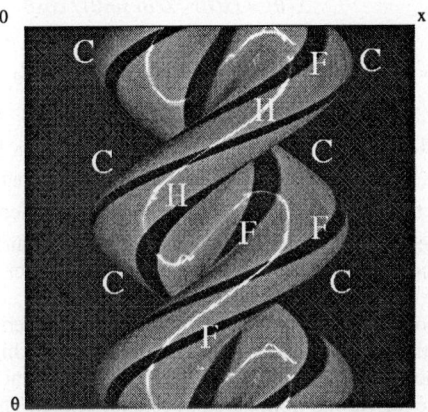

Fig. 3 An epipolar plane image in which moving features are marked. The object shape on the rotation plane is like a dumbbell which contains both convex and concave segments. H: highlight, F: fixed point, C: contour.

3.2 Contour and Surface Estimation

Moving features marked in an EPI traverse on object surface during the rotation. Under orthogonal projection, such traversing for a contour point is on the same rotation plane. At an time instance, we can still describe the relation between the contour in the image and the rim on the surface by Eq. 1 [16]. Moreover, the line of sight is the surface tangent on the rim. We can compute derivative of Eq. 1 with respect to rotation angle θ to obtain tangent of the surface (X'_θ, Z'_θ) in the rotation plane. By confining the tangent orthogonal to the image plane, we can obtain the position of the surface point as

$$\begin{bmatrix} X \\ Z \end{bmatrix} = \begin{pmatrix} \cos\theta & -\sin\theta \\ \sin\theta & \cos\theta \end{pmatrix} \begin{pmatrix} x \\ \partial x/\partial\theta \end{pmatrix} \quad (2),$$

from Eq. 1 and its derivative.

A contour trace can be extracted from a homogeneous background. In the solution, the derivative of contour position is computed in the EPI with nearby points. A degenerate case of the method is for nonconvex shape in the rotation plane. Under such situation, a contour trace in the EPI is no longer smooth and the shape unexposed as contour can not be recovered. An approach to locate such unexposed regions on objects by detecting nonsmooth points on the contour traces in the EPIs has be given by [1].

3.3 Highlight on Specular Surfaces

If an object surface has a strong specular reflection, there may exist a highlight on the surface. For objects with fewer corners and patterns, such a highlight provides information for shape recovery. When a surface point becomes a highlight, the angle of incidence is equal to the angle of reflection. Although surface points on a rotation plane may have different Y components in their surface normal, they will be highlight under the long linear light when their normal rotate to the camera and light direction.

The highlight shifts over surface points during the rotation, though these points may actually catch rays from different parts of light. If we track highlight trace in the EPI, we can compute the positions of the trace points by

$$X(\theta) = \left(\int_{\theta_i}^{\theta} \left(\frac{\partial x}{\partial \theta} \cos\theta - xctg\theta \cos\theta \right) e^{-\int_{\theta_i}^{\theta} ctg\,\theta d\theta} d\theta + X_{\theta_i} \right) e^{\int_{\theta_i}^{\theta} ctg\,\theta d\theta}$$

$$Z(\theta) = (x(\theta) - X(\theta)\cos\theta) / \sin\theta \quad (3)$$

in the θ domains [π/4, 3π/4] and [π+π/4, π+3π/4], and

$$X(\theta) = (x(\theta) - Z(\theta)\sin\theta) / \cos\theta \quad (4)$$

$$Z(\theta) = \left(\int_{\theta_i}^{\theta} \left(\frac{\partial x}{\partial \theta} \sin\theta + x\sin\theta\tan\theta \right) e^{\int_{\theta_i}^{\theta} \tan\theta d\theta} d\theta + Z_{\theta_i} \right) e^{-\int_{a}^{\theta} \tan\theta d\theta}$$

in the θ domains [-π/4, π/4] and [π-π/4, π+π/4]. In the solutions, [$X_{\theta i}$, $Z_{\theta i}$] is a known point highlighted at angle θ_i, which works as initial condition of the equation [2].

We track highlight trace from θ_i in the EPI, and estimate positions of highlight passed points by computing the integral terms in the solutions. It results a curve that is the shape in the rotation plane. This computation may include accumulated error if a highlight trace is not located precisely.

The moving trace of a highlight has some interesting characteristics. A linear shape which has zero curvature on the rotation plane will generate a horizontal highlight stripe in the EPI. At convex and concave boundaries separated by a zero curvature point or segment, highlights may appear at the same time (or viewed from the same rotation angle). They will either split or merge at the zero curvature points [2,3,10].

Table 1 gives a summary of above explained methods on their principle, result, algorithm complexity, approximation, and source of error.

4. Relations between Different Visual Cues

This section discusses the relationship between different visual cues presented in intensities, traces in EPI and estimated locations. We look at the motion of contour, highlight, and fixed point in the EPI. Figure 4 displays three shapes on the rotation plane and the traces of multi-features in the EPIs. The trace of a fixed point diverges from a trace of contour and then merges into the other trace of contour after half period of rotation. A trace of fixed point may also be cut off by a contour trace in the case of occlusion.

A trace of highlight moves over traces of surface points. If a surface shape in the rotation plane has a high curvature, a highlight keeps staying there and the highlight trace highlight is then close to sinusoidal trace. The trace of corner connects traces of highlight.

On the other hand, if a fixed point comes from an edge of pattern on the smooth surface, its trace may intersect a highlight trace in different ways, depending on the sign of curvature of the surface. For a convex surface in the rotation plane, the highlight moves slower than surface points so that its trace is more vertical than those of pattern edges in the EPI. For a concave surface, however, the highlight moves fast than surface points so that its trace is more horizontal. A crossing point of highlight and fixed point traces provides the initial position for estimating other points along the highlight trace [2].

In addition, a highlight trace may have a wave at a concave surface generated region in the EPI . It is in fact three segments of traces. A highlight splits into two at a zero curvature point. One of which moves on the concave surface and then merges with another highlight at another zero curvature point.

Visual cues	Extracted from	Image points constrain what	Recovered primitives	Computational complexity	Measured data versus output	Approximation
Fixed Points	patterns, corners	3D position	point	linear equation	redundant data single output	improved by averaging multiple output
Contour	static background	surface tangent	envolope, concave part missing	1st order derivative	single measure single output	fitting the derivative using discrete points
Highlight	specular reflection	surface normal	surface curve	1st order differential equation	output using measured data before	integral approximated by sumation

Table 1. Shape recovery methods for different visual cues and their algorithms using approximation.

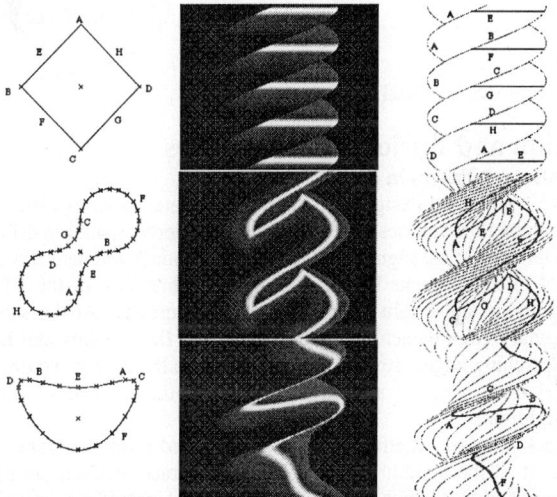

Fig. 4 Three Shapes and their corresponding traces in the EPIs. Left: shapes, Middle: EPIs, Right: traces. Motion of visual different cues are depicted. The objects rotates clockwise and the time axes (angle axes) is vertical and downward in EPIs.

As to the relation between highlight and contour, there is no intersection between their traces in the EPI, because the light is with the camera. However, if a highlight trace has no intersection with fixed points, we can draw a trace of known surface point from contour to intersect the highlight trace for acquiring the initial condition in the *shape from highlight*. Figure 5 summarizes the above discussed relations.

The basic approaches to detect fixed points, contours and highlights are filtering EPI by an edge detector and locating peaks, respectively. By using a background, a contour is easy to be located, if we search edges from both sides of the EPI towards the center line.

The contrast of a corner trace is reversed once in its sign on its trace. At the angle the corner facing the light and camera, the intensities on its both sides have no difference so that it can not be extracted as an edge. For a trace of pattern edge, however, its contrast becomes maximum when the surface normal there directs to the light and camera.

The intensity of a surface point captured by a camera is the sum of diffused and spcular components. Diffused reflectance changes at pattern edges. If a surface has a strong specular component of reflection, a highlight can be located at a sharp peak of intensity in the EPI without influenced by underlying pattern changes. However, if specular component is week,, bright pixels in the EPI contain highlights as well as bright patterns. The intensity on the highlight trace may become low when the highlight moves over a dark region on the surface. This causes difficulty in locating highlight traces.

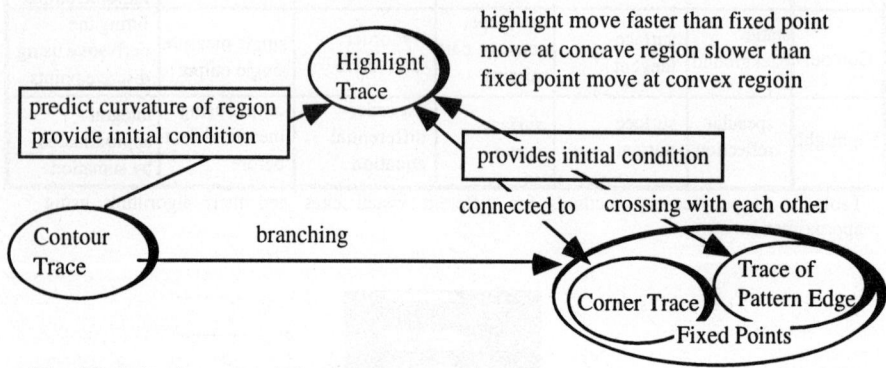

Fig. 5 Relation in traces of visual cues.

5. Classification and Fusion of Visual Cues
5.1 Separating Different Traces in EPIs

Based on above results, we design a strategy to separate traces of different visual cues for final fusion. It contains three phases based on image property, spatial distribution of measured data, and motion characteristics. Figure 6 gives detailed module and data flow.

We first extract intensity peaks higher than a given threshold in the EPI. The result may include strong highlights and bright patterns on convex surfaces. At the same time, edges are detected using ordinary edge detector. It results traces of fixed points and bounding curves at highlight traces. Searching edges from both sides of EPI to the center, we obtain contour traces. Deleting pixels belong to contour and strong highlight, the remaining edges are from fixed points and weak highlights.

The second check is to see whether the locally measured positions along a trace of edge are centralized at a point. If the condition is verified, it is a trace of fixed point; otherwise, it is a trace of moving point [11,15]. For those short traces not providing enough evidence in the examination, we leave them as candidates of fixed points since the possible computation applied to them is local and their resulted positions will not spread far away.

The third phase is to look at image velocities of highlight and fixed point. We focus on intersections of two traces in the EPI. An intersection of traces means that two visual cues are visible at the same surface point. If two conflict measures are obtained from them, only one result is true. Once their traces have an intersection, the intersecting type tells which is highlight and which is fixed point according to their relations mentioned above, if we get curvature sign of the surface from contour.

5.2. Integrating Results from Visual Cues

The integration of measured data from different algorithms is qualitative here. Fixed points have highest accuracy. The second accurate cue is contour, and then followed by highlight. We choose the result from a cue with less error, if multiple measured data are available at a surface point. At a point with one or without measured data, we just use it or connect nearby points using linear graphics patches.

In more detail cases, if we obtain data from both contour and fixed point, we choose fixed point. A long trace of fixed point can provide multiple measures of the position. By extending such a long trace to contour trace using the measured position, we can locate which contour point is overlapped with the fixed point so as to improve contour measure by measure of fixed point. On the other hand, highlight can not be measured without initial condition. Whenever a highlight trace intersects a trace of fixed point, we start estimation of highlight from the crossing point; a highlight position integrated from far away along its trace has accumulated error.

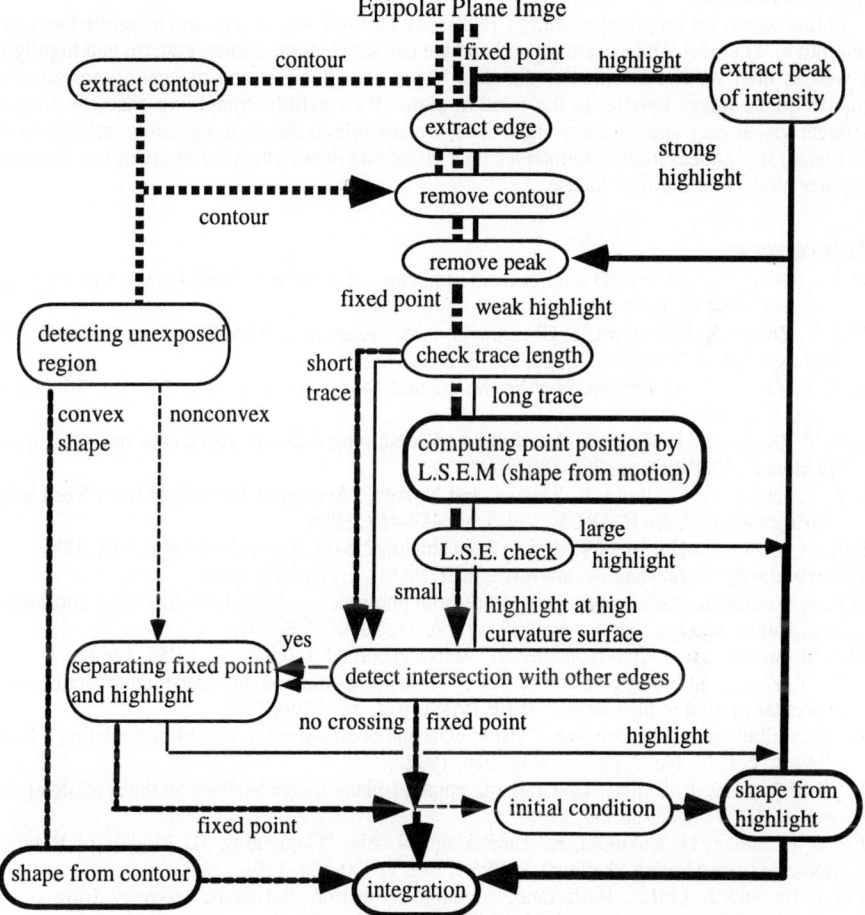

Figure 6. Modules for classification and the data flow of different visual cues.

The model construction is carried out by following contour traces. It first produces envolope of the shape in the rotation plane, and gives the unexposed areas which need further exploration[1]. Following the contour, we find fixed traces branching from it. Near coutour shapes, we pick up fixed points for integration. At concave areas or areas unexposed as contour, fixed points plays the role of yielding the shape. If highlight is involved there, we have to classify them and connect fixed points with highlight generated shape to produce a more complete surface.

6. Experiments

We have done experiments on real objects. A video camera is put 3-4m away from the objects. The object is put on a turn table. Camera lens is adjusted to long focal length to produce orthogonal projection. Due to the page limit, we give examples in the presentation of the conference. Each example will show the recovered model and shape combined from results of different *shape from X's* at EPIs. EPIs are taken at each degree of rotation. The recovered surface points are connected by triangular patches and displayed in the graphics space.

7. Conclusion

In this paper, we proposed a strategy to classify multiple visual cues and integrate measured data into a 3D model. Objects are rotated and we extract contour, corner, pattern, and highlight, as well as their motion information for data fusion. The analysis and processing are given at epipolar plane image parallel to the rotation plane. By carefully comparing characteristics of different visual cues and recovery algorithms, we are able to design a qualitative scheme to do the fusion. It produces more data than a single visual cue does, which is important in generating graphics models for multimedia use.

References

[1] J.Y. Zheng, "Acquiring 3D models from sequences of contours", IEEE PAMI, Vol.16, No.2, Feb. pp.163-178, 1994.
[2] J. Y. Zheng, Y. Fukagawa, T. Ohtsuka, N. Abe, "Acquiring 3D models from rotation and a highlight", 12th ICPR, Vol. 1, pp. 331-336, 1994.
[3] J. Y. Zheng, Y. Fukagawa, N. Abe, Shape and Model from Specular Motion, 5th ICCV, pp.92-97, 1995.
[4] J. Y. Zheng, and F. Kishino, "Verifying and combining different visual cues into a complete 3D model", CVPR92, pp. 777-780, 1992.
[5] J. Y. Zheng, H. Kakinoki, K. Tanaka, and N. Abe, "Acquiring 3D models from fixed point during rotation", 3th ICARCV, Vol. 1, pp.459-463, 1994.
[6] K. P. Horn and M. J. Brooks, "Shape from shading", MIT Press, Cambridge, MA, 1989.
[7] A. Pentland, "Local shading analysis", IEEE PAMI, 6:170-187, 1984.
[8] R. J. Woodham, "Gradient and curvature from photometric stereo including local confidence estimation", J. Opt. Soc. Amer., Vol. 11, no. 11, 3050-3068, 1994.
[9] J. Aloimonos and A. Bandyopadhyay, "Active vision", 1st ICCV, pp.35-55, 1987.
[10] J. Y. Zheng, Y. Fukagawa, and N. Abe, "3D shape estimation and model construction from specular motion in the images", IEEE PAMI, Vol. no. 5, pp. 1997.
[11] R. Vaillant and O. D. Faugeras, "Using extremal boundaries for 3D object modeling", IEEE PAMI, Vol. 14, No. 2, pp. 157-173. Feb. 1992.
[12] H. Baker, and R. Bolles, "Generalizing epipolar-plane image analysis on the spatiotemporal surface", CVPR-88, pp.2-9, 1988.
[13] J. Y. Zheng, H. Kakinoki, K. Tanaka and N. Abe "Computing 3D Models of Rotating Objects from Moving Shading", ICPR96, Vol. 1, 800-804, 1996.
[14] J. Lu and J. Little, "Reflectance function estimation and shape recovery from image sequence of a rotating object", 5th ICCV, pp. 80-86, 1995.
[15] M. Oren, S. K. Nayar, "A theory of specular surface geometry", 5th ICCV, pp. 740-747, 1995.
[16] P. Giblin, R. Weiss, "Reconstruction of surface from profiles", 1st ICCV, pp. 136-144, 1987.
[17] J. Y. Zheng, A. Murata, N. Abe, "Reconstruction of 3D Models from Specular Motion Using Circular Lights", 13th ICPR, Vol. 1, pp. 869-873, (1996).

Strategical Tracking of Polyhedral Objects by Reactive Change of Projection Pattern
- Reactive Range Finder -

Takeshi Mita[1], Shinsaku Hiura[2], Hirokazu Kato[1], Seiji Inokuchi[1]

[1] Department of Systems Engineering, Osaka University, Japan
[2] Department of Electronics and Communication, Kyoto University, Japan
email: tmita@inolab.sys.es.osaka-u.ac.jp

Abstract. This paper describes a new range finding system named Reactive Range Finder, which actively acquires the necessary range information according to each task. This system can generate a projection pattern that enables an efficient measurement adapted to the change of the scene caused by movement of the object. To confirm the efficiency of this system, we made tracking application of polyhedral object moving with 6 degrees of freedom on this system. In order to determine the position and posture of the object, it selects measurable surfaces by using the object model on-line. And it projects three spot lights to each selected surface. As the result, the image processing algorithm became easy, and reliable video-rate tracking was achieved without using particular hardware.

1 Introduction

Active range finder is widely utilized for many robotics tasks in manufacturing industries because of its high measurement accuracy. System specification such as measurement speed and measurement accuracy is different among various vision tasks. For example, highly precise measurement is necessary to get object model and high speed measurement is necessary for object tracking. Recently, high speed range finders have been developed to measure a dynamic scene. Sato[1] developed the range finder that is not only high speed but also small-sized enough to install on the hand of a manipulator. In the future, it is possible to develop both high speed (for example video-rate) and high resolution sensor. But sensor performance is not a unique factor to decide total system performance. Efficient method of range image processing is also needed to accomplish the task. The measurement, based on Active Vision[2], focusing on some areas in the scene which are important for the tasks is necessary to handle the range image sequences of dynamic scene.

On the other hand, to build practical systems in manufacturing industries, range finder and processing algorithm peculiar to each task have been developed. For example,

Inokuchi[3] proposed a vision system for depalletizer using double-slit projector. These systems are optimized by using particular and suitable hardware for each task. Therefore, the performance of each system is best for the task, but the system can not be used for more extensive field.

Our purpose is the development of range finding system that can select a measurement method and a measurement area to acquire necessary range information efficiently. We named this system "Reactive Range Finder." Reactive Range Finder generates light patterns adapted to the object shape.

Visual tracking of known objects as they move in space has been extensively studied, since it is a crucial prerequisite for solving tasks in robotics, autonomous navigation. A lot of methods using intensity image has been studied, and recently some of them achieve video rate tracking. Stark[4] proposed an approach based on active contour model, and very fast tracking was realized by using 2-D contour. But false contrast edges degrade tracking accuracy. Based on similar approach, Armstrong[5] proposed robust tracker by using a set of related primitives (high contrast edges). The robust tracker continues to track correctly, although the object is partially occluded. There still exists detection of incorrect primitives because of strong shadow. Since intensity image is vulnerable to environmental changes, range image is suitable for the task that needs reliable measurement. However there are few range finders whose measurement speed is fast enough for object tracking. Hiura[6] proposed tracking method using fast range finder with original VLSI sensor. Using graphics rendering hardware (Z-buffer) makes range image synthesis from surface model very fast. Movement of the object is calculated by comparison measured range data with synthetic images. It is possible to track a freeform, freemoved object. But the speed of the synthesis simulator has to be several times as fast as sensor speed. In tracking of the known shape object, it is not necessary to measure whole surface of the object. This can be achieved by the measurement of characteristic area of the object. Especially in case of polyhedral object, getting positions of some planes or edges make tracking possible. Thus measurement of only necessary area leads to high speed and reliable tracking.

We build tracking application of known polyhedral object on Reactive Range Finder. The system selects three visible planes by using object model, and generates a projection pattern that projects three spot lights onto each plane. As the object moves, the projection pattern is changed in real-time.

2 Reactive Range Finder

Reactive Range Finder is a range finding system based on active stereo methods and can acquire necessary information actively. This system is characterized by the ability of projection of light pattern adapted to the scene.

2.1 Measurement by Reactive Change of Projection Pattern

The development of high-speed and highly precise range finder has been made till now. It has been regarded as advantage that every pixel in image has range data. Therefore methods for processing only these dense images have been developed. It is needless to say that this range finder can be utilized in extensive fields. However it does not always have advantages in all kinds of task. In some cases, measurement speed goes down in the cause of getting needless or redundant information. From a viewpoint of measurement efficiency and cost, it is not suitable to use expensive range finder to get 1-bit information whether an object exists there.

There are some aspects for consideration of measurement strategy. For example,
(1) Task : modeling, classification, positioning, tracking, etc.
(2) Object : polyhedral object, curved object, deformable object, textured object, etc.
(3) Environment : indoor, outdoor, controlled, unknown background, etc.

Robust system for these changes is necessary. Reactive Range Finder makes efficient measurement possible by changing projection patterns according to task or object shape.

For example, a few spot lights are projected onto object surface for object tracking. Gray coded patterns[7] are projected for modeling. Pose determination of polyhedral object with three light-stripe range measurements is also possible[8]. Measurement speed and measurement accuracy are different among various methods. To achieve enough performance of the system, it is necessary to select the best method from these methods.

Reactive Range Finder is able to project arbitrary pattern, and use many measurement methods. These points are described in detail to the following.

2.2 Hardware

Reactive Range Finder needs following two functions.
(1) Project arbitrary light pattern to the scene
(2) Generate arbitrary pattern in high-speed and change them immediately

Figure 1 shows our Reactive Range Finder system. A video projector and a CCD camera are connected to a graphics workstation. The projection pattern is a graphics image drawn on the workstation screen. Typical graphics workstations have a CG rendering hardware and it is fast enough to generate projection images. And the projection pattern image is sent to the video projector as NTSC output. In this way, (1) and (2) are realized. When a spot light projection is needed, a point is drawn on the screen. When a

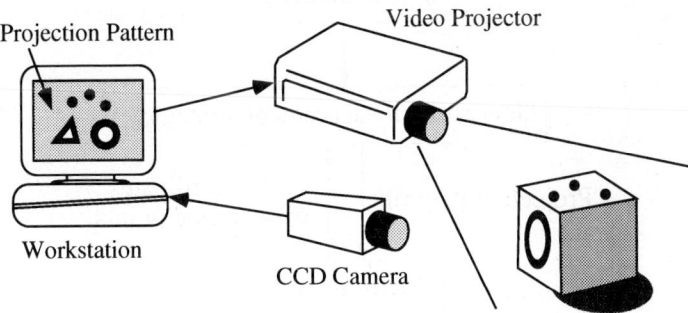

Fig. 1. Hardware component of Reactive Range Finder

slit light projection is needed, a line segment is drawn. 3-D coordinates of specified points can be acquired at the same time by using multi-spot pattern or multi-slit pattern.

2.3 Software

Reactive Range Finder is a range finding system that can use various active stereo methods such as spot light projection, slit light projection and Gray coded pattern projection, and it can change them actively according to tasks. Projection pattern, image processing and calculation of 3-D coordinate are different according to each method. So this system has each processing subroutine as software libraries. Figure 2 shows components of Reactive Range Finder. Proper selection of the library is decided automatically in "generating measurement strategy" by using knowledge about the task or 3-D object model. Ikeuchi[9] proposed a method to generate 3D-object recognition algorithms from a geometrical model for bin-picking tasks. Recoginition algorithms are generated in the form of an interpretation tree in advace. But to recognize the dynamic scene, it is necessary to generate measurement strategy on-line.

3 Tracking by Using Changeable Projection Pattern

If more than three plane equations in space are given, it is possible to track the pose of known polyhedral object. Rotation matrix is calculated from normal vectors of the planes. Translation vector is calculated from the distance between each plane and the rotation origin. Our tracking method is based on this simple principle. A plane equation is calculated from 3-D coordinates of at least three points on the plane. It is realized by projection of three spot lights onto each plane.

We built tracking system based on this method. This system generates a projection pattern automatically adapted to the object shape and its aspect by using 3-D object model. Initial position of the object is already given.

Since whether a plane is visible or not changes by movement of the object, our system selects all visible planes using object model and arranges three spot lights to each plane on-line. Invisible planes from the camera and planes with small area are excluded. So

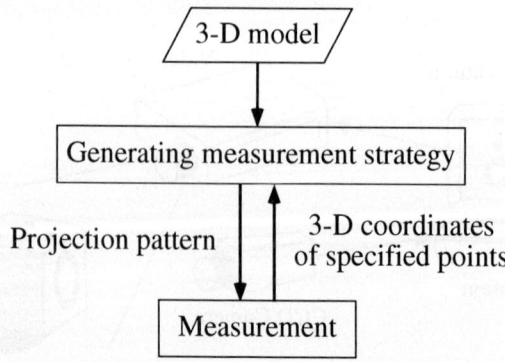

Fig. 2. Components of Reactive Range Finder

calibrated camera is needed in this method. After this process, generated light pattern is projected to the object and the image of the scene is captured. Since rough positions of projected spot lights are known, it is quite easy to detect spot areas from the image. It is needless to detect image features from the dense range image and to correspond them with the model, which needs complex calculations. Simple calculation makes out tracking system stable and real-time.

3.1 System Calibration
The perspective model based on pinhole camera is assumed. The x,y,z-axis of the 3-D world coordinate system are directed along three sides of a calibration box. The system is calibrated by projection of Gray coded patterns. Camera parameter **C** and projector parameter **P** are calculated. Both are 3x4 matrices.

If coordinate (**X,Y,Z**) in the 3-D world coordinate system is given, coordinates (**Xc,Yc**) in the camera coordinate system and coordinates (**Xp,Yp**) in the projector coordinate system are calculated as follows:

$$h_c[Xc\ Yc\ 1]^T = C[X\ Y\ Z\ 1]^T$$
$$h_p[Xp\ Yp\ 1]^T = P[X\ Y\ Z\ 1]^T$$

3.2 Polyhedral Object Model
Polyhedral object model consists of coordinates of all vertexes of the object in the 3-D local coordinate system and links between these vertexes. Each edge is represented as a link of two vertexes, and each plane is represented as a convex polygon linked with edges. And the link direction of the polygon is defined counterclockwise. The model is prepared in advance. Initial position of the object is given.

3.3 Judgment of Plane Visibility
Planes of the object utilized for tracking must be visible from both camera and projector. It is impossible to project light pattern onto the plane behind the object, therefore the selection of visible planes is necessary. First, the system projects all polygons onto camera viewing plane and projector viewing plane by using system parameter (**C** and **P**). Second, the system investigated the link direction of all projected polygons. If the direction of the link is counterclockwise, the polygon may be a visible plane. Finally, the polygon is judged whether it is occluded by other planes. Thus the system selects visible planes that have no occlusion. To track several objects at the same time, the system judges each plane visibility for all objects.

3.4 Arrangement of Spot Lights
After visible planes with no occlusion from both camera and projector are selected, the system calculates the area of each polygon on camera viewing plane and projector viewing plane. Polygons whose area is too small are rejected, because they degrade measurement accuracy. Three spot lights are projected onto the interior division points between center of gravity of the polygon and three vertexes (Figure 3).

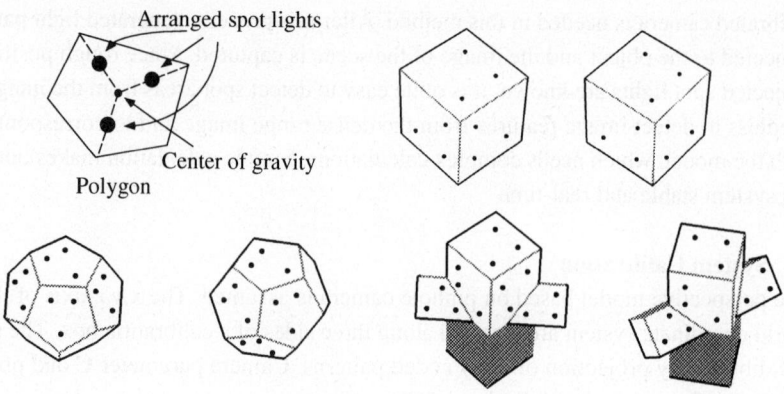

Fig. 3. Arrangement of spot lights

Thus our system generates projection pattern automatically. Next, the pattern is projected to the scene and the scene image is captured. Since it is easy to detect projected spot lights from captured image, image processing is very fast without using an image processor or parallel programming. Fast tracking is possible by using this algorithm and it is also available to track several objects at the same time.

4 Experimental Results

We built tracking prototype system based on our method. A video projector (SHARP XV-E500) and a video camera (SONY DCR-VX-1000) were connected to SGI ONYX with S-Video cable. The distance between the device and tracking object was about 1.5[m]. Figure 4 shows the scenes tracking one object moved by hand. And Figure 5 is the scenes tracking two objects at the same time. White points in each image are projected spot lights. Contours of measured object are drawn into the images as black lines. Though the object B is partially occluded by object A, the system continues to track correctly. Frame rate in Table 1 depends on the object shape and the number of visible planes. Measurement accuracy is shown in Table 2.

5 Conclusion

We have presented the idea of Reactive Range Finder and the tracking method of the polyhedral object. Our system can automatically generate the projection pattern adapted to the object shape by using 3-D object model. Even if the aspect change is occurred, the tracking operation is stable. It is also possible to track several objects at the same time. We realized real-time tracking system without particular hardware.

Tracking of freeform object is impossible with our system. However, it will be possible by using knowledge peculiar to the task such as constraints of moving direction.

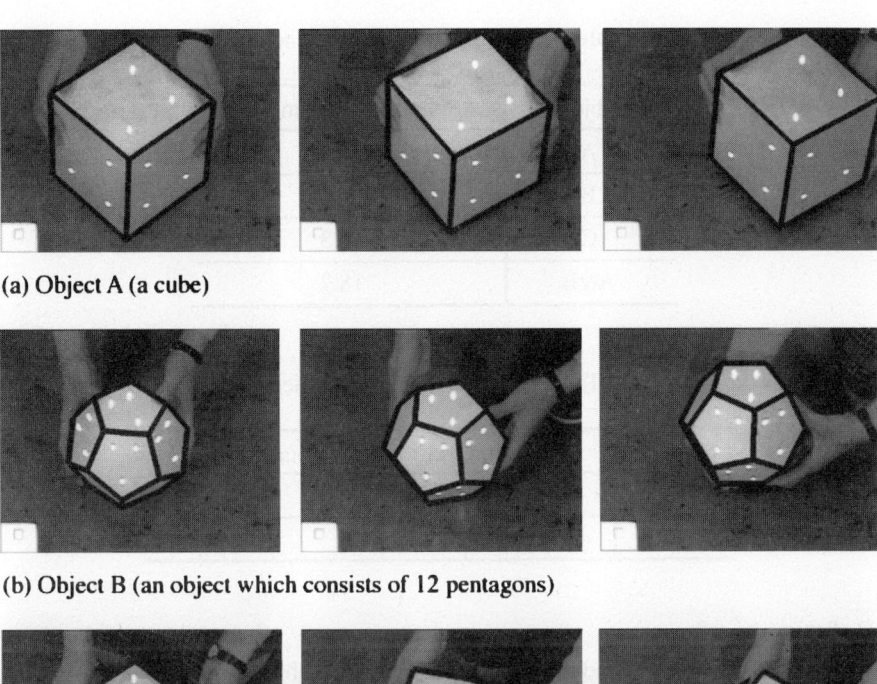

(a) Object A (a cube)

(b) Object B (an object which consists of 12 pentagons)

(c) Object C (two rectangular parallelepiped pierce with each other)

Fig. 4. Tracking one object moved by hand : White points are projected spot lights and black lines are contours of measured object. Aspect change is occured while tracking object A and object B.

Fig. 5. Tracking two objects (object A and object B) moved by hand at the same time : Object B is partially occluded by object A, but tracking operation is correctly continued.

Table 1. Average frame rate of tracking

Object	Frame Rate [frame/sec]
A	28.0
B	26.4
C	23.3
A&B	18.2

Table 2. Measument accuracies

	Translation[mm]	Rotation[deg.]
mean	-0.197	0.65
stddev	0.67	1.11

References

1. Y. Sato: Active Rangefinding and recognition with Cubicscope, Proc. of 2nd ACCV, pp.I-369-372 (1995)
2. J. Y. Aloimonos and I.Weiss : Active vision, IJCV, pp.333-356 (1988)
3. S. Inokuchi, K. Sato and Y. Ozaki : Vision system for depalletizer in automated storage, Proc. Jpara-U.S. Symposium on Flexible Automation, pp.711-714 (1986)
4. K.Stark and S.Fuchs : A method for tracking the pose of known 3-D objects based on an active contour model, Proc. of ICPR'96, pp.I-905-909 (1996)
5. M.Armstrong and A.Zisserman : Robust object tracking, Proc. of 2nd ACCV, pp.I-58-62 (1995)
6. S. Hiura, A. Yamaguchi, K. Sato and S. Inokuchi : Real-Time object tracking by rotating range sensor, Proc. of 13th ICPR, Vol.1, pp.825-829 (1995)
7. K.Sato and S.Inokuchi : Range-imaging system utilizing nematic liquid crystal mask, Proc. IEEE 1st ICCV, pp.657-661 (1987)
8. K.Kemmmotsu and T.Kanade : Uncertainty in object pose determination with three light-stripe range measurements, IEEE Trans. on Robotics & Autom, Vol.11, No.5, pp.741-747 (1995)
9. K. Ikeuchi : Generating an interpretation tree from a CAD model for 3D-object recognition in bin-picking tasks, IJCV, pp.145-165 (1987)

Autonomous Vision-Guided Robot Manipulation Control

Wey-Shiuan Hwang and John (Juyang) Weng

Department of Computer Science
Michigan State University
East Lansing, MI 48824

Abstract. This paper presents a hand-eye-head system which learns to perform temporal actions. In this system, the robot learns how to control its hand according to what is seen and the specific mission. The learning process is interactive and on-line. Several networks which store the learned information are automatically generated through learning processes. The hierarchical structure of the network allows each associative recall to be completed in $O(\log n)$ time, where n is the number of cases learned. A visual attention selection system is also implemented by using a recurrent network and an autonomous learning scheme. The recurrent network enables the learning process of the visual attention selection system to draw information not only from the current input image but also from the history. The new autonomous learning scheme combines both supervised and reinforcement learning modes so that the system gradually improves its performance through a process of trial and error.

1 Introduction

To work in an unstructured environment, robots need sensors to sense the environments and translate what they sensed to what they can relate to their actions. A basic issue is calibration. For vision-guided robotic system, the calibration is not a trivial task. Two types of approaches exist. The first type requires the human designer to explicitly model the relationships between sensors and actuators. The result is a set of equations with a number of parameters that need to be estimated. Most existing works belong to this category (e.g. [7]).

The second type of approach does not employ explicit parameterized models. But rather, a general approximator is used to map the sensor space to the actuator space. A common tool to accomplish this is the artificial neural network [7]. The first type of approach is effective when the system can be accurately modeled with a few parameters, but has problems when the system configuration is time varying or has a lot of degree of freedom. The second type does not have the latter limitation due to the generality of the neural network approximator. However, the neural networks require a huge number of iterations in learning and thus, not suited for interactive learning or task learning where each case presented must be learned immediately, on line. The system must remember each training example presented and generalize to other location based on relatively few examples learned.

For learning by watching, the work in [10] generates a sequence of robot control language as final outputs. Some efforts concentrate on learning task level [5]. In [1], it is assumed that the target will maintain a regular and repetitive movement that can be analyzed and intercepted. In [6], the three dimensional movement of a manipulator is decomposed into movements which are at most two-dimensional.

The work reported here is to study a mechanism which enables a hand-eye-head system to perform some task sequences with explicitly hand-crafting task rules into the control program. We use a self-organizing recursive partition tree (RPT) to organize the input space into a hierarchy of coarse-to-fine partitions which approximate the mapping between the sensing-task space and the temporal action space. This tree shares the common characteristic with the well known decision trees, regression trees [4] and clustering trees in the sense that uses a tree to organize information in a hierarchical manner. The major differences between our RPT and the other trees include (1) our RPT building procedure is not iterative, thus is fast. (2) our RPT building method is incremental, which allows the tree to grow according to the performance of the current tree. (3) our RPT performs inter-leaf interpolation, which results in a good approximation and generalization performance without need for time-consuming batch analysis of the entire set of data during tree construction. The method has been tested on a hand-eye-head system for vision-guided temporal tasks, such as finding a cup, grasping a cup, lifting a cup, pouring liquid from a cup into another etc.

2 Function Approximation via Learning

Based on learning, an function approximation process can be expressed as:
Given training set $S = \{(I_i, O_i) \mid i = 1 \ldots n\}$, construct $\tilde{f}$ where $O_i = \tilde{f}(I_i)$

For generalization, Bayesian method is ideal here but it is difficult to define probability distribution function from a small number of training examples. An alternative method is the *second order k-nearest-neighbor distance-based* ($KNDB - 2$) function approximator has been proposed in [9]. The $KNDB - 2$ estimates the output for I as:

$$\tilde{f}_k(I) = \sum_{i=1}^{k} \frac{w_i}{w} (O_{n_i} + (I - I_{n_i}) * (\nabla f(I_{n_i})))$$

where O_{n_i} satisfies the condition: $(I_{n_i}, O_{n_i}) \in S$, and I_{n_i} is the i-th nearest neighbor of I. $\nabla f(I_{n_i})$ is the gradient of i-th nearest neighbor of I and estimated by the Jacobian matrix. The symbol $*$ means inner product and w_i is a scalar weighting function of I and I_{n_i}. One way to choose the weighting function is to define:

$$w_i = \alpha^{-\|I - I_{n_i}\| / (\epsilon + \|I - I_{n_1}\|)}$$

Where ϵ is a small number to prevent the denominator going to zero and α is a parameter to determine how fast the weight w_i will decrease with the increasing exponent. Since the resulting approximator is based on more than one near neighbor, it will be much less sensitive to noise in S; and $\tilde{f}_k$ will be smoother.

Instead of linear search, an recursive partition tree (RPT) method was proposed in [8]. The RPT tessellates the inputs space and constructs a corresponding partition tree to store the information of the training data. The search takes only $O(\log n)$ time complexity.

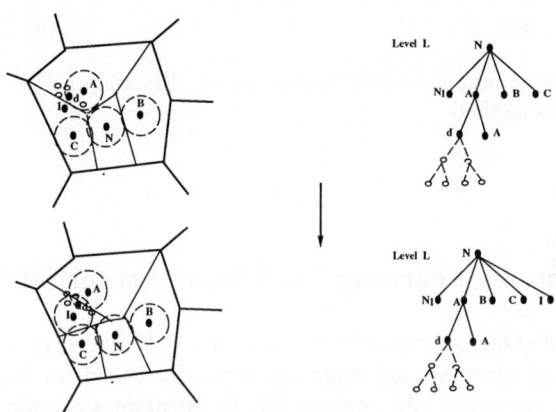

Fig. 1. An example which adoption problem will occur.

The construction of the RPT might cause the adoption problem as shown in Figure 1. Node N is a grandparent node and node A, B, C are its sons and node N_1 is its duplication. Node d is the son of node A. The empty circles mean the descendants of node d. The circles with dot lines represent the spacing at level $L+1$. Suppose we want to learn a new sample I with the existing partition tree. Although the real neighbor of I is d, node C will be picked up as its nearest node because $dis(I,C) < dis(I,A), dis(I,B), dis(I,N)$. Since the $dis(I,C)$ is less than the spacing in level $L+1$, the RPC will insert the node I at level $L+1$. The new tessellation will render node d and all of its descendants because the node I "adopt" this subspace.

One method to solve this problem is to learn S level by level in a batch mode. However, we need to reconstruct the whole partition tree if we have a new sample to learn. Carefully considering the adoption problem, we can find that the cause of adoption problem is: when a new training sample is inserted to the input space in level l, the new training sample collides with one of the siblings and thus causes the adoption problem. In Figure 2 (a), suppose P is the center of one node in level 1. N is a new training sample which will be inserted to the input space. If N collides with P, then c, which is the children of P, will be adopted by N. To solve this problem, if we insist that the tessellation should not be changed when any new node is inserted, there will be no adoption problem. In Figure 2 (b), after inserting the new training sample N, c still remains in P's region and the problem is overcome.

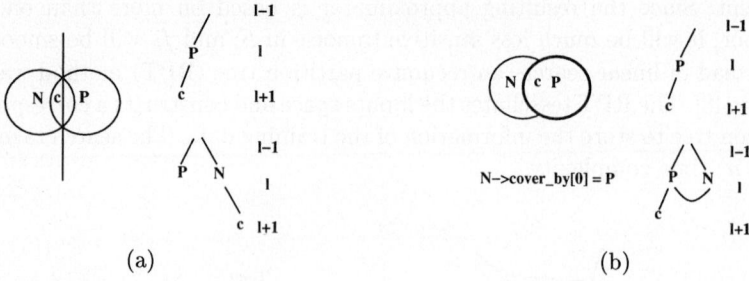

Fig. 2. (a) Adoption problem caused by tessellation. (b) Adoption problem avoided by a new tessellation method.

3 Autonomous Learning for Visual Attention Selection

The objective of visual attention is to find objects of interest in a scene. Existing studies on visual attention selection are typically based on low-level saliency measures, such as edges and texture [2]. In Birnbaum's work [3], the visual attention is based on the need to explore geometrical structure in the scene. In our case, the visual attention selection incorporates the prediction of the critical point of the learned object from the current un-centered views. Such a prediction is a result of past learning experience.

We use a coarse-to-fine approach for the visual attention selection problem. The first step is to find the approximate position of the object of interest. We use a recurrent RPT for this step. The prediction of vision attention is according to what is learned and we do not plug in any manually-modeled knowledge to the system. Once the object of interest is seen, the second step is to fine tune the visual attention to seek the more precise position of the target so that the critical point of the object is at the center of gaze. We use an autonomous learning scheme for the second step. This learning scheme combines both supervised learning and reinforcement learning modes.

The structure of a recurrent RPT is similar to the RPT as discussed in Section 2 except that:

- The input of the RPT now consists of two parts. One is from the sensor input. The second part is the context vector. The context vector specifies the current state of the system. In the case of visual attention selection, the context vector is the gaze position of the input image.
- The reset channel is designed to enable the human trainer to reset the current context vector. When a new event is going to be learned, the context vector is reset by the trainer.

In this work, a new and faster way of partitioning the space is developed. It works in the following way: As soon as a node received two samples, say

S_1 and S_2, $S_1 - S_2$ gives the normal of the hyper-plane, and the hyper-plane is placed at the midpoint between S_1 and S_2 for partitioning future samples. Computationally, if $X(S_1 - S_2) > 0.5(S_1 + S_2)(S_1 - S_2)$, X goes to the left child, otherwise X goes to the right child. Figure 3. gives a graphical explanation.

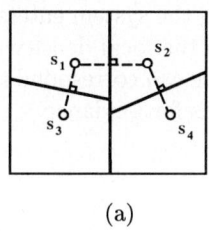
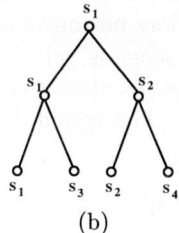

(a) (b)

Fig. 3. Fast partition and the binary RPT. Training sample sequence: S_1, S_2, S_3, S_4 (a) The partition (b) The binary RPT.

We propose a learning strategy that combines both supervised learning and reinforcement learning in an automatic fashion. The system decides whether it can try on its own depending on a signal b, associated with the current situation. The signal b is called "biological signal", simulating punishment and reward to the system. Each learned cases has an attached b number ranging from 0 to 1. The default value of b is 0.5, which is useful when the human teacher does not think punishment or reward is needed according to the system's current performance. The human teacher typically starts training the system in supervised mode, using b value 1. When the system's behavior has reached a desired level, the teacher can switch the system to autonomous mode. The mechanism is as follow:

```
Find K nearest neighbors from RPT.
If the reinforcement signal b of nearest
neighbor > 0.5
    Interpolate effector signals among
    the nearest neighbors with b >0.5
    and act.
else
    If the b of nearest neighbor > 0.25
        Explores.
    else
        Request action from teacher.
```

To explore the environment using the known examples, the output is give by:

$$O = \tilde{f}_k(I) + \delta O \otimes R$$

where δO is the difference vector between the output of the nearest neighbor and the second nearest neighbor, R is a unit random vector with a zero mean, and $\otimes$ denotes component-wise multiplication.

As can be seen, the decision is based on the biological number of the best recalled case. If the last experience gives enough confidence, the learning machine tries to interpolate (generalize) from the similar good ($b > 0.5$) cases in order to improve the performance. This interpolation is not always good, pending humans evaluation. However, since similar cases have large b values, an interpolation typical is better. If the best-matched case is such that ($0.25 < b < 0.5$), the interpolation does not make much sense. Thus, the system enters an exploration status, whose scope is automatically tuned to the local density of learned samples. Otherwise, $b < 0.25$, the best matched case can correspond to an experience of punishment. The system is "panic" and asks for guidance.

4 The Proposed System

The proposed system includes two subsystems. The vision subsystem receives stereo images as input and outputs the position of key points of a target in terms of the row and column in the right camera and the disparity (r, c, d). A coarse-to-fine approach is adopted in order to speed up the location process. The system trainer uses a graphic user interface to conduct the experiments. The visual attention is determined by an attention window in the image frame with center at (r, c). In this experiment, we assume that the size of the object does not change too much to reduce the degrees of freedom in attention control. Since the purpose of this experiment is to test our autonomous learning scheme, we have limited variation of the scene. The primary goal of this subsystem is to find the object of interest so that the robot can pick it up.

In order to perform a temporal task action—such as picking a cup up from a table, pouring water from one cup into another—the system must have available information on itself, the object, and the environment. The information needed to predict the next move of the robot arm in order to accomplish a task is:

- The type of subtask **T**: A given task can be divided into several subtasks. The representation of **T** is an integer that indicates which subtask is active.
- The current separation between robot arm and target **S**.
- The current joint angles of the robot arm **J**.

During the training phase a series of training examples are provided interactively by a system trainer through a graphic user interface. For a training example, the action can be represented as a sequence of (I, O) pairs, where $I = (\mathbf{T}, \mathbf{S}, \mathbf{J})$ and $O = (\delta \mathbf{J})$ are the input and output vectors. Training takes place according to the method described in Section 2. During an actual run of the system, each subsequent move of the manipulator in the performance of a specified task is predicted by providing the system an input vector I of the current *state of the system*. Then the knowledge base (RPT) is searched to retrieve the learned instance which best fits the current state. Finally, based on the retrieved results, an appropriate output vector O of joint increments is calculated and sent to the manipulator's controller specifying the next movement to be made.

5 Experimental Results

In the training phase of the first step, the system was trained with five different objects. Each object is presented in the different positions in the workplace. The resolution for each image was 320 by 240 pixels. The result for the first step is summarized in Table 1.

Table 1: The Results of the First Step

Average error (pixels)	Object1	Object 2	Object 3	Object 4	Object 5
Row	11.3	9.7	10.7	14.7	11.3
Column	8.7	6.7	6.3	7.7	7.3

From the approximate position obtained from Step 1, the second step used higher resolution to find the more accurate position of the objects.

In the initial supervised mode, the human trainer interactively gave the correct position of the object in the image. Eighteen images with different positions of objects were learned during this mode. Then the autonomous mode was used. First we used the same lighting and randomly placed the object within the working range for learning. Fifteen tests were then performed under the same lighting condition. Further we tested another fifteen positions under a lighting that was different from that of the training environment. The performance of the second step is shown in Table 2.

Table 2: The Results of the Second Step

Average error (pixels)	Row	Column
The same lighting	2.13	1.31
Different lighting	4.16	1.78

The result of the vision subsystem is used for the robot arm to pick up the object of interest. The accuracy in Table 3 is more than enough for this application.

For the temporal task action, we designed five subtasks: (1)approaching handle of a cup, (2)picking up a cup, (3)approaching above the cup , (4)pouring milk into another cup, and (5)putting the cup on the table. The workspace for this experiment is a $50cm \times 50cm$ horizontal plane. Thirty training sequences were used in this experiment. Each subtask was tested with the training data as well as another twenty positions in the space between the training data. The results are presented in the Table 3. There are three trials in which the pouring subtask fails to pour more than 70% of the water into the other cup. The other trials all transfer more than 70% of the water to the other cup.

Table 3: The Success Rate of Trials

	Approach handle	Pick up	Approach cup	Pour	Put on table
Resubstitution	100%	100%	100%	100%	100%
Random Test	100%	100%	100%	85%	100%

6 Discussion

Our system integrates vision and robotic manipulation in a framework that is independent of the tasks to be performed. The learning process is partially autonomous and partially supervised to take advantage of both modes. In our system, we also use visual learning techniques to improve the recognition performance.

As far as we know, this is the first attempt to build a vision-guided robot manipulator system that is of a general purpose in the sense that no task-specific decision rules are imposed, and thus, in principle, the system is potentially able to learn a wide variety of vision-action tasks. The tasks experimented with required visual searching, visual recognition, stereo-based 3D position estimation, and vision-guided manipulation for hard-to-model problem, such as pouring liquid. A future direction is to expend the method for a much larger set of tasks and a much larger number of objects.

References

1. Allen, P.K. and A. Timcenko and B. Yoshimi and P. Michelman. Automated Tracking and Grasping of a Moving Object with a Robotic Hand-Eye System. *IEEE Trans on Robotics and Automation*, 9(2):152–165, Apr 1993.
2. Martin Bichsel. *Strategies of Robust Object Recognition for the Automatic Identification of Human Faces*. Swiss Federal Institute of Technology, Zurich, Switzerland, 1991.
3. Lawernce Birnbaum, Matthew Brand, and Paul Cooper. Looking for Trouble: Using Causal Semantics to Direct Focus of Attention. In *Proc of the IEEE Int'l Conf on Computer Vision*, pages 49–56, Berlin, Germany, May 1993. IEEE Computer Press.
4. L. Breiman, J. Friedman, R. Olshen, and C. Stone, editors. *Classification and Regression Trees*. Chapman & Hall, New York, 1993.
5. Michael S. Branicky. Task-Level Learning: Experiments and Extensions. In *Proc of the IEEE Int'l Conf on Robotics and Automation*, pages 266–271, Sacramental CA, Apr 1991. IEEE.
6. Schrott, A. Feature-Based Camera-Guided Grasping by an Eye-in-Hand Robot. In *Proc of the IEEE Int'l Conf on Robotics and Automation*, pages 1832–1837, Los Alamitos CA, May 1992. IEEE Computer Society Press.
7. Walter, J.A. and K.J. Schulten. Implementation of Self-Organizing Neural Networks for Visuo-Motor Control of an Industrial Robot. *IEEE Trans on Neural Networks*, 4(1):86–95, Jan 1993.
8. J. Weng and S. Chen. Autonomous navigation through case-base learning. In *Proc of the IEEE int'l Symposium on Computer Vision*, pages 359–364, Coral Gables, FL, Nov 1995.
9. Wey-Shiuan Hwang, Sally J. Howden, and John Weng. Performaing Temporal Action with a Hand-Eye System Using the SHOSLIF Approach. In *Int'l Conf on Pattern Recognition*, volume 4, pages 35–39, Vienna, Austria, Aug 1996.
10. Yasuo Kuniyoshi, Masayuki Inaba, Hirockika Inoue. Learning by Watching: Extracting Reusable Task Knowledge from Visual Observation of Human Performance. *IEEE Trans on Robotics and Automation*, 10(6):799–822, Dec 1994.

A New Adaptive Approach on Rapid Obstacle Detection in Range Image

Zhang Qi

Post & Telecommunications Project Institute of Zhejiang Province,
8 Liu-Yuan-Qian-Dao-Lu Road, Zhaohui 2nd District, Hangzhou, 310014, P. R. China

Gu Weikang   Ye Xiuqing

Institute of Information and Intelligent System, Department of Information and Electronics,
Zhejiang University, Hangzhou, 310027, P. R. China

Abstract — New approaches for rapidly detecting the obstacles in spherical coordinate using range image sensed by Laser Imaging Range Sensor(LIRS) are presented in this paper. The algorithm that adaptively thresholds the normalized range differences is shown to have rapid and robust results for edge detection of obstacle of real outdoor range image, and is useful for obstacle detection in mobile robot. Some problems of range sensing for LIRS encountered in the implementation of the method as well as the robustness of this algorithm are discussed.

1 Introduction

LIRS are attractive to rapid obstacle detection in real-time robotic systems such as mobile robot for two main reasons: firstly, it provides range data without the computation overhead associated with passive techniques; secondly, it is largely insensitive to outside illumination conditions.

The images studied in this paper are all the real outdoor range images obtained by our LIRS which has worked successfully in actual mobile robot. This sensor is an *am-cw*(amplitude-modulated continuous-wave) LIRS that ranges by measuring the shift in phase between an emitted beam and its reflection.

The main purpose in using LIRS in mobile robot is to detect the obstacles in front of mobile robot, thus providing efficient information for path planner. Because of the discontinuities of depth or surface normal on the edges of obstacles, the information about obstacles can be obtained by detecting their edges.

Coordinate transformation that transforms the original range image with the data format of spherical coordinate into the elevation map with the data format of Cartesian coordinate is often applied in mobile robot. Although obstacles whose height is beyond a certain value can be detected by simply thresholding the map because pixel values reflect the elevation of the corresponding points in the elevation map, this method has shortcomings such as the slow processing speed and the drawback that it is unable to represent vertical planes. Another problem of this method is that the mapping between the range image and the elevation map is not one to one. Close to the sensor, several points from the range image can map to the same element in the elevation map, but farther away, the data from the range image is sparse, and there are empty elements in the map. To deal with this problem, a very time-consuming interpolation to form a continuous map must be done.

However, the algorithm presented in this paper rapidly detects obstacles directly in spherical coordinate, not needing coordinate transformation. The central idea of this is: firstly, the expected normalized range differences based on the slope being zero are obtained, then edges of obstacle are detected by applying a fixed threshold to all the pixels' quotients between actual normalized range differences and the expected normalized range differences. The algorithm which is a good trade-off between speed and efficiency is extremely fast and fairly robust as it is demonstrated on actual outdoor range images of both flat planes and uneven terrains. So it can be used for fast reliable obstacle detection by mobile robot.

2 Range Image and Its Application on Edge Detection

In this part, we describe the range image sensed by LIRS, and introduce the proposed algorithms of edge detection for obstacles using range image.

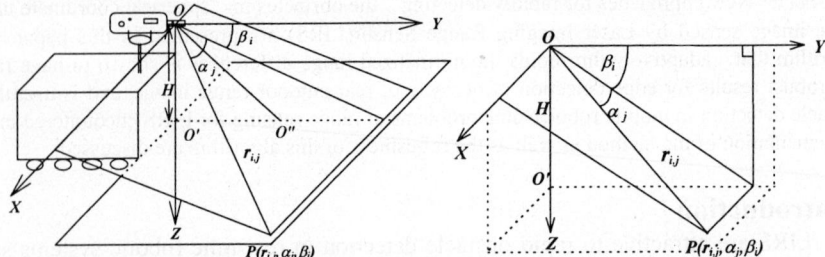

Fig. 1a Sensing frame of LIRS   Fig. 1b The coordinates of range sensing

2.1 Range Image Sensed by LIRS

Range image sensed by LIRS is naturally described by the spherical coordinate $(r_{i,j}, \alpha_j, \beta_i)$, in which $r_{i,j}$ means the range of the point *(i,j)*, α_j is the horizontal scan angle corresponding a particular column *j* in the range image and β_i is vertical scan angle corresponding a particular row *i*(Fig. 1) The Cartesian coordinate (x, y, z) of the point $(r_{i,j}, \alpha_j, \beta_i)$ can be obtained by

$$x = r_{i,j} \sin \alpha_j$$
$$y = r_{i,j} \cos \alpha_j \cos \beta_i$$
$$z = r_{i,j} \cos \alpha_j \sin \beta_i \quad (1)$$

Fig. 2 shows a real outdoor range image obtained in a basketball field in evening. Darker the pixel of range image is, nearer the corresponding point from sensor is.

Fig. 2   Outdoor range image sensed in evening

2.2 Algorithms of Edge Detection for Obstacles

There have been many proposed approaches for the edge detection of obstacle using range image. Besides the method of coordinate transformation mentioned above,

Martial Hebert etc. detected edges using zero-crossings of second derivative [1], and Minoru Asada used the gradient and curvature to detect edge [2]. Although these methods have good results, computing the second derivative, curvature and gradient are very time-consuming and it is difficult to rapid detect edges.

The fastest algorithm for edge detection in range image is maybe range differencing [3], which thresholds the difference of range, $\Delta r_{i,j}$, within a neighborhood. However, this algorithm has two shortcomings: (1) it is not robust; (2) choosing the thresholds is very difficult. For range image of perfectly horizontally flat plane, $\Delta r_{i,j}$ between two rows increase gradually from near to far(the maximum of $\Delta r_{i,j}$ is 1.78m but the minimum is only 0.039m), whereas $\Delta r_{i,j}$ between two columns increase gradually from central to bilateral(the maximum of $\Delta r_{i,j}$ is 0.111m but the minimum is 0.0m). So, large threshold levels would hide significant obstacles that are near the range sensor. Conversely, small thresholds would falsely labeled the distant flat surfaces as obstacles. In order to improve this state, someone substitute the normalized range difference $\Delta r_{i,j}/r_{i,j}$ for the $\Delta r_{i,j}$ [4]. Although choosing the thresholds in this method is easier than the above, it remains difficult(the maximum of $\Delta r_{i,j}/r_{i,j}$ between two rows is 0.073 whereas the minimum is 0.012, and the maximum of $\Delta r_{i,j}/r_{i,j}$ between two columns is 0.0046 whereas the minimum is 0.0).

If we can adaptively choose the thresholds of the normalized range differences at all pixels, it is possible to rapidly and efficiently detect the edges of obstacle. This is exactly the purpose of us. In fact, the algorithm presented in this paper automatically adjust the thresholds of actual normalized range differences point by point with the help of "the expected normalized range differences" that is calculated based on the slope being zero.

3 The Algorithm of Edge Detection in Detail

Calculating the slope of a point *(i,j)* on x-direction (or y-direction), $\partial z/\partial x$ (or $\partial z/\partial y$), is an efficient way to decide whether *(i,j)* is an edge point or not. However, computing the slope directly from range image is very time-consuming. Obviously, if there is no change of *z* on point *(i,j)*, then, $\partial z/\partial x = 0$ ($\partial z/\partial y = 0$). So, by supposing $\partial z/\partial x = 0$ ($\partial z/\partial y = 0$), simple relationship between $\Delta r_{i,j}/r_{i,j}$ and $\alpha_j, \Delta\alpha_j$ (or between $\Delta r_{i,j}/r_{i,j}$ and $\beta_i, \Delta\beta_i$) can be gotten. This $\Delta r_{i,j}/r_{i,j}$ is exactly "the expected normalized range difference" that we need.

Assume the actual normalized range differences of a point *(i,j)* on row and column direction are $\Delta_{row}^{(a)}, \Delta_{col}^{(a)}$, respectively, whereas the expected normalized range differences on row and column direction are $\Delta_{row}^{(e)}, \Delta_{col}^{(e)}$, respectively, where *a* means "*actual*" and *e* means "*expected*".

(1) Obtain the actual normalized range differences

The actual normalized range differences of point (i,j), $\Delta_{row}^{(a)}$ and $\Delta_{col}^{(a)}$, are obtained from the range image. The following normalized summations at each point (i,j) are used:

$$\Delta_{row}^{(a)} = \frac{\Delta r_{i,j}}{r_{i,j}} = \frac{(r[i][j+1] - r[i][j-1])}{r[i][j]}$$

$$\Delta_{col}^{(a)} = \frac{\Delta r_{i,j}}{r_{i,j}} = \frac{(r[i+1][j] - r[i-1][j])}{r[i][j]} \quad (2)$$

where $r[i][j]$ is the range of point (i,j) in range image.

(2) Obtain the expected normalized range difference

The expected normalized range differences are calculated based on the slope of point (i,j) being zero, namely $\Delta z/\Delta x = 0$ or $\Delta z/\Delta y = 0$.

Equation (1) gives

$$dz = \frac{\partial z}{\partial r_{i,j}} dr_{i,j} + \frac{\partial z}{\partial \alpha_j} d\alpha_j + \frac{\partial z}{\partial \beta_i} d\beta_i$$

$$dx = \frac{\partial x}{\partial r_{i,j}} dr_{i,j} + \frac{\partial x}{\partial \alpha_j} d\alpha_j$$

If β_i is held constant and substitute approximately $\Delta z, \Delta x, \Delta \alpha_j, \Delta r_{i,j}$ for $dz, dx, d\alpha_j, dr_{i,j}$ respectively, then the above equation can be written as

$$\Delta z \approx (\cos\alpha_j \sin\beta_i)\Delta r_{i,j} - (r_{i,j} \sin\alpha_j \sin\beta_i)\Delta\alpha_j$$

$$\Delta x \approx (\sin\alpha_j)\Delta r_{i,j} + (r_{i,j} \cos\alpha_j)\Delta\alpha_j$$

So, if we assume the slope on x direction being zero, i.e. $\Delta z / \Delta x = 0$, then

$$\frac{(\cos\alpha_j \sin\beta_i)\Delta r_{i,j} - (r_{i,j} \sin\alpha_j \sin\beta_i)\Delta\alpha_j}{(\sin\alpha_j)\Delta r_{i,j} + (r_{i,j} \cos\alpha_j)\Delta\alpha_j} = 0$$

namely,

$$\frac{(\frac{\Delta r_{i,j}}{r_{i,j}} \frac{1}{\Delta\alpha_j} - tg\alpha_j)\sin\beta_i}{\frac{\Delta r_{i,j}}{r_{i,j}} \frac{tg\alpha_j}{\Delta\alpha_j} + 1} = 0$$

Therefore, if β_i is held constant, the expected normalized range difference of point (i,j), $\Delta_{row}^{(e)}$, is

$$\Delta_{row}^{(e)} = \frac{\Delta r_{i,j}}{r_{i,j}} = \Delta\alpha_j \cdot tg\alpha_j \quad (3)$$

If α_j is held constant, the expected normalized range difference of (i,j), $\Delta_{col}^{(e)}$, can be evaluated by applying the similar method:

$$\Delta_{col}^{(e)} = \frac{\Delta r_{i,j}}{r_{i,j}} = -\Delta\beta_i \cdot ctg\beta_i \qquad (4)$$

(3) Edge detection of obstacles

It is very difficult to choose thresholds if only actual normalized range differences are simply used, as mentioned above. However, if the actual normalized range differences are divided by the expected normalized range differences, the quotients are good measure of the actual slope at point *(i,j)*. Large absolute values of these quotients will be formed by edges of object as well as surfaces with steep slopes. Applying a fixed threshold to the absolute values of these quotients of all pixels yields pixels that are likely to be on obstacles. In fact, the algorithm is the method that adaptively adjusts the thresholds of actual normalized range differences point by point with the help of "the expected normalized range differences".

Edges are detected row by row first, then column by column. At last, two binary image are ORed.

The quotients of normalized range differences at all pixels are defined as q_{row} and q_{col} when edges are detected row by row and column by column, respectively. The q_{row} and q_{col} are

$$q_{row} = \Delta_{row}^{(a)} / \Delta_{row}^{(e)}$$
$$q_{col} = \Delta_{col}^{(a)} / \Delta_{col}^{(e)} \qquad (5)$$

Choosing an accurate $\Delta r_{i,j}$ is the key to success. In order to ensure that the resultant $\Delta r_{i,j}$ accurately reflects the slope at the pixel *[i,j]*, we want $\Delta\alpha_j (\Delta\beta_i)$ to be as small as possible. On the other hand, decreasing the size of $\Delta\alpha_j (\Delta\beta_i)$ also decreases the magnitude of the $\Delta r_{i,j}$ which can impair the accuracy of the $\Delta r_{i,j}$ measurements. This is especially true for small $\Delta r_{i,j}$ due to the quantization of range measurements. Tests indicate that the choice of calculating $\Delta r_{i,j}$ across two columns (rows) is a good compromise between two conflicting goals; therefore, $\Delta r_{i,j}$ in (2) is calculated between the neighboring two columns(rows).

The quotients of each pixel in range image of perfectly horizontally flat plane are all approximately equal to 1.0, so a fixed threshold at all positions of the range image can be used very easily.

4 The Implementation of the Algorithm in Mobile Robot

In this part, we first discuss the "periodicity" problem of range sensing for LIRS in mobile robot which navigates in noisy and hilly environments, and present strategies for reducing the errors of obstacle detection caused by this problem. In addition, the robustness of this algorithm is also studied.

4.1 The "Periodicity" Problem of Range Sensing

For *am-cw* laser radars, the range to a target is proportional to the difference of phase. Since the phase is defined modulo 2π, it is not possible to distinguish between range r and $r+nd_a$, where $d_a = \lambda_m/2$ is called *ambiguity interval* and $n \in N$, and λ_m is the wavelength of the modulation.

Suppose two neighboring points in the same column of range image which belong to two periodicity of range are $P_1(i,j)$ and $P_2(i+1,j)$. The actual physical distance of these two points are d_1, d_2, respectively (ordinarily, $d_1 > d_2$), and the measured range are r_1, r_2, respectively.

If $\Delta d = d_1 - d_2$, $\Delta r = r_1 - r_2$, $n \in N$, then:

(1) When Δd is small, the actual range difference between these two points is small. However, in the sensed range image we can see: $r_1 \ll r_2$, because the two points belong to two different periodicity and $d_1 > d_2$. So, there is strong false edge between the two points.

(2) When Δd is large, there should be strong edge between the two points. However, there may not be edge in range image because the two points belong to two periodicities. We analyze this case as follows:

If $n_1 = [d_1/d_a]$, $n_2 = [d_2/d_a]$, $d_1' = d_1 - n_1 d_a$, $d_2' = d_2 - n_2 d_a$, where $[x]$ is the integer of x, then:

If $\Delta d = nd_a$, then $d_1' = d_2'$, $r_1 = r_2$, with result that there is no edge between P_1 and P_2;

If $\Delta d \neq nd_a$, then $d_1' \neq d_2'$, with result that there is edge between P_1 and P_2. The larger the difference between d_1' and d_2' is, the stronger the edge is. However, $\Delta r = d_1' - d_2' = \Delta d - (n_1 - n_2)d_a$, which is not equal to Δd.

In short, the periodicity problem of range sensing produce false edge or lose actual edge or make the intensity difference have an error of $(n_1 - n_2)d_a$ at the adjacent point of two periodicity of range in the resultant image of edge detection. Hence we must remove the periodicity problem in order to successfully detect edges. For usual outdoor range images, most of periodicity problem belong to the first case, in which adjacent pixels go from large values suddenly to very small values. The solution to the periodicity problem is divided into two steps:

1) Explore from near to far in each column of range image. Suppose the first point of certain column is P_0. From near to far along this column, we search N points, $P_1, P_2, ..., P_N$, where the values of range jumps from large to small suddenly. These N points belong to N different periodicities.

2) Offset all pixels in this column by applying the following rule:

$$offset(p)_0 = 0, \quad \forall p, p \in P_0 P_1$$

$$offset(p)_k = offset(p)_{k-1} + 256, \quad \forall p, p \in P_k P_{k+1}$$

where, $k = 1, 2, ..., N-1$.

Our experiments show that this method can efficiently remove the points of the first two periodicities of actual outdoor range image. However, beyond that point, measurements are usually too noisy to ensure reliable results. The image in Fig.3a has been applied a 3 × 3 median filter but the periodicity points have not been removed, whereas Fig.3b have been removed.

(a) Median-filtered but not periodicity- removed (b) Median-filtered and periodicity- removed
Fig. 3 (The original range images are on Fig. 2)

4.2 Sensitivity to Perturbations of Sensor Attitude Angles

When an LIRS is mounted on a mobile robot, its orientation, i.e. three attitude angles(the angle of roll, pitch, and yaw of the mobile robot), is very difficult to measure. So, it is useful to study the sensitivity of the obstacle detection algorithm to errors in sensor orientation and to compare this sensitivity to that of the range difference and height difference algorithms.

Table 1
Comparison of obstacle detection algorithms for sensitivity to perturbations of sensor attitude angles

perturbation of the attitude angles	height difference the max of z	range threshold the max of $r_{i,j}$	range difference the max of $\Delta r_{i,j}$	normalized range difference the max of $\Delta r_{i,j}/r_{i,j}$	this algorithm the max of q_{row}	this algorithm the max of q_{col}
$\varphi_x=\varphi_y=\varphi_z=0°$	0.00m	24.35m	1.78m	0.073	1.000046	1.005448
$\varphi_x=3°, \varphi_y=\varphi_z=0°$	-0.15m	48.57m	3.40m	0.140	1.000199	1.016800
$\varphi_y=3°, \varphi_x=\varphi_z=0°$	0.58m	42.04m	3.21m	0.136	1.000105	1.012000
$\varphi_z=3°, \varphi_x=\varphi_y=0°$	0.08m	25.51m	2.23m	0.101	1.000060	1.006400
$\varphi_x=\varphi_y=\varphi_z=3°$	0.91m	189.84m	15.98m	0.546	1.107840	1.204508

We first produce four different range images with different errors in sensor orientation, then this algorithm and the height difference, range thresholding, range difference, and normalized range difference algorithms are applied to each image. In the height difference algorithm, we threshold the height difference. In the range thresholding algorithm, we threshold simply the range, and in the range difference algorithm, we threshold the range difference. In the normalized range difference algorithm, we threshold the normalized range difference.

Table 1 summarizes the results of this experiment. These results clearly show that the algorithm presented in this paper is more robust than either the range difference or the height difference algorithms. For each case of attitude angle perturbation, the q_{row} and q_{col} only vary slightly. However, other algorithm are all sensitive to the perturbation of attitude angles, as shown on Table 1.

5 Experimental Results and Conclusion

Fig. 4 shows the results of our experiments. Fig. 4a is an outdoor range image with complicated scenes of grassland, flower terraces, trees groves and stone steps, men, chairs, etc. Fig. 4b and Fig. 4c are the threshold output of the range image row by row and column by column, respectively. Fig. 4d is the ORed binary image of Fig. 4b and Fig. 4c. The results of edge detection are satisfying, as shown in Fig. 4b, Fig. 4c and Fig. 4d in which the edges of the stone steps with the height of mere *15cm* or so are all detected reliably.

In short, this algorithm has significantly better performance than other fast edge detection methods. It works well not only for relatively flat, on-road scenes, but also for the bushed hilly terrain. All the range images are the real outdoor images obtained by our range sensor. In addition, this algorithm has rapid speed. The time for processing the range image with the size of 64×256 row by row and column by column are 0.21s and 0.24s(not including the time for filtering), respectively, using the PC486 of 66Mhz. This algorithm can be used for fast reliable obstacle detection by mobile robot, which is our final goal.

REFERENCES

1 Martial Herbert and Kanade T. Outdoor Scene Analysis Using Range Data. *in Proceedings of the 1986 IEEE Robotics and Automation Conference*, pp. 1426-1432.
2 Minoru Asada. Building A 3-D World Model For A Mobile Robot From Sensory Data. *in Proceedings of the 1988 IEEE Robotics and Automation Conference*, pp. 918-923.
3 R.T.Dunlay and D.G.Morgenthaler. Obstacle Avoidance on Roadways Using Range Data. *in SPIE Vol. 727 Mobile Robots*, Cambridge, MA, 1986, pp. 110-116.
4 M.J.Daily,J.G.Harris, and K.Reiser. Detecting Obstacles in Range Imagery. *in Image Understanding Workshop*, Los Angles, 1987.

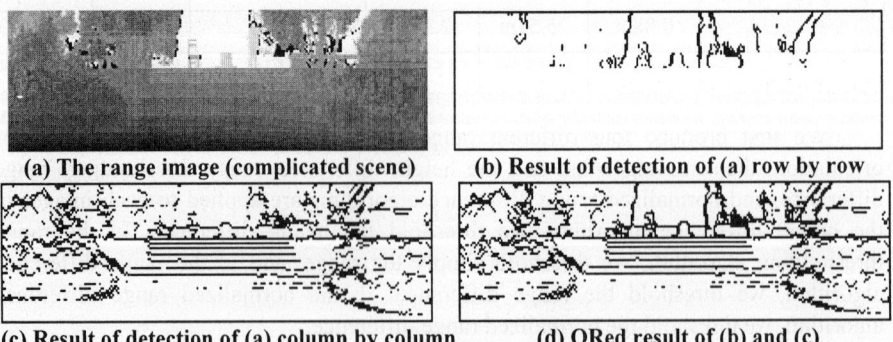

(a) The range image (complicated scene)　　(b) Result of detection of (a) row by row

(c) Result of detection of (a) column by column　　(d) ORed result of (b) and (c)

Fig. 4　　The results of edge detection of obstacles

Recognition of Shape Models for General Roads

Keiichi Uchimura and Zhencheng Hu

Department of Computer Science, Faculty of Engineering
Kumamoto University, Kumamoto, 860 Japan
E-mail: uchimura@eecs.kumamoto-u.ac.jp, ko@eecs.kumamoto-u.ac.jp

Abstract: Described here is an extraction and recognition method of road shape models from static color images of road scene. As the basic process of vision-based environment recognition, it is used extensively for the research of autonomous vehicles and ITS (Intelligent Transport Systems). The usual method is depending on the recognition result of road lane marks and generally used on expressways or highways only. In contrast, we propose a region segmentation and recognition algorithms using color and texture information, and a robust recognition method of road shape model for general roads. Experiment based on real road-scene images shows the feasibility of our approach.

1. Introduction

In recent years, with the rapidly increase of traffic flow, a lot of research on autonomous vehicle and ITS (Intelligent Transport Systems) have been done for the purpose of driving safely and smoothly [1][2]. As the first research stage of driving environment recognition system, this paper describes a road shape model's extraction method using static color images of road scene ahead.

Recognition of road region and road shape using vision-based techniques is indispensable to the research of autonomous vehicle control, obstacle detection and dangerous degree judgment [3][4]. Especially in recent years many research works focus on the extraction of road shape based on road lane marks' detection result. Ozawa's system [5][6] uses monocular images to detect the road white line and estimate the road curvature and up-down grade. The system can give driver the vehicle's position and shape of road that the driver sees forward. Ishikawa uses white line for vehicle guidance [7]. His system can select path and detect obstacles by white line detection result.

However, these kinds of algorithms are usually used on expressways or highways only. Detection of lane marks is not enough for general road recognition. First, in Japan there are still many simple roads without lane marks. Furthermore, when lane marks are rubbed by vehicle's tire, or covered by water, mud or sand, the road region is hardly to detect by the road lane marks information only. On the other hand, the road regions beside lane marks are just the high probability areas for the appearance of pedestrian, bicycles or motorbikes.

This paper presents a road shape extraction method for general roads that does not depend on the distinctness of road lane marks. We use color and texture information to segment the road image and recognize the road region, and then recognize the road shape model based on the road region's boundary as well as road lane marks.

Although this method's objects are static images, it can be also used to dynamic road image sequence, for the first several frames processing, as a bootstrap mode. It will be discussed in our future work.

2. Image Segmentation

2.1 Color Gradient Edge Detector

A color gradient operator is defined here, and it is used to extract region's boundary in a color image.

It is commonly accepted that, the same region in an image has almost the same properties (brightness, color, texture etc.), and boundaries of neighbor regions are accompanied by a discontinuity in image properties. Thus the image segmentation algorithms can be distinguished into two groups: 1) to extract the edge line as the boundaries of neighbor regions by the sharp variety of property values (edge-based methods); 2) to separate the region with almost the same property values from image directly (region-based methods).

The method we presented here belongs to edge-based methods [8]. We use variety rate of brightness and color intensity at boundaries of different regions to segment the image.

Each pixel on an inputted image has three components of the so-called red, green, and blue (R,G,B) components. The gradient of each components at pixel (i,j) in X and Y direction are defined as follows:

$$X_f(i,j) \equiv f(i,j) - f(i-1,j) \quad \text{and} \quad Y_f(i,j) \equiv f(i,j) - f(i,j-1) \tag{1}$$

where $f = R, G, B$.

And the color gradient intensity is :

$$d(i,j) = \sqrt{\sum_{f=R,G,B} (X_f^2(i,j) + Y_f^2(i,j))} \tag{2}$$

In order to suppress noise and emphasize edge points, Prewitt matrix [9] is used instead of definition (1):

$$X_f(i,j) \equiv f(i-1,j-1) + f(i-1,j) + f(i-1,j+1) - f(i+1,j-1) - f(i+1,j) - f(i+1,j+1)$$

$$Y_f(i,j) \equiv f(i-1,j-1) + f(i,j-1) + f(i+1,j-1) - f(i-1,j+1) - f(i,j+1) - f(i+1,j+1) \tag{3}$$

2.2 Binary Image and Edge Thinning Method

The color gradient image is transferred to binary image by the values of color gradient intensity, and line's width is thinned into 1 by non-maximum points restrained thinning method [10] in order to trace edge elements.

2.3 Tracing, Extending and Linking of Edge Elements

We trace the extracted edge elements, and try to link them into continuous circuit lines. The assumption used here is that boundary of an object or a part of an object is a continuous circuit line. Extending processing of edge line is started from two extremity points of each edge elements been traced. The procedure is the following:

1) Evaluate each 8-neighborhood pixels of the extremity point. The evaluating value is calculated by:

$$H(i',j') = \alpha(i',j') \times d(i',j') \qquad (4)$$

where (i', j') means 8-neighborhood pixels of point (i, j), $\alpha(i', j')$ is a weight parameter of edge direction, and $d(i', j')$ is the color gradient intensity defined by (2).

2) Extend edge line to the pixel which get the highest evaluating value and make this pixel the new extremity point.

3) Once the two extremity points are getting together or collide with other edge lines, one processing is ended.

4) Scan the whole image and end the procedure if there are no edge lines that can be extended.

2.4 Region Labeling

We use flood-filling method to label each region segmented by continuous circuit edge lines, and merge these edge points into neighbor regions respectively by the similarity of color properties.

Figure 1 shows the original image, and figure 2 shows the region segmentation result.

3. Region Recognition

To extract and recognize the characteristic region of road (pavement or road surface, white lane mark, centerline and shadow on road) from the image regions being segmented as explained above, the relative knowledge can be used. For example, recognition of road surface can use the knowledge shown below:

a) Road is level and almost on the ground surface,

b) The top of road region on the image is narrow and the bottom is wide, which can be proved by the principles of 3-D projection,

c) Road regions on the image always have low saturation and large area,

d) The neighborhood regions of white line and centerline are road regions.

and so on.

The properties of road regions (brightness, color and texture etc.) are very changeable upon the weather conditions (sunny, cloudy or rain etc.) and illuminating conditions. On this point, we present a robust recognition method for road regions.

3.1 Making the Sample of Road and the Property Model of Road Regions

A road sample window can be set at the bottom center of road scene image and a

Fig. 1. Origin image of road scene

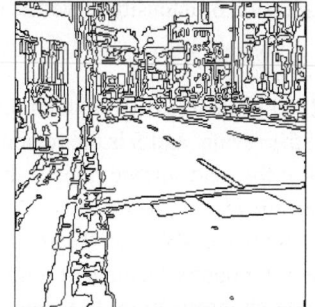
Fig. 2. Region segmentation result

built-in assumption is used that the test-bed vehicle is already on the road. We take a 50x75 pixels rectangle area in our experiment. The properties (intensity, saturation and R,G,B components) of this sample can be used to evaluate the property features of road regions (road surface, white line, centerline and shadow on road).

3.2 Extraction of Road Surface's Core Region

Now consider the regions that can meet the two conditions shown below:
1) satisfy the road surface's property features that inferred by road sample;
2) the vertical position of region's gravity center point is over a threshold.

The one with the biggest area is then chosen from these regions as the road surface's core region.

3.3 Merging of other Road Surface Regions

Neighborhood regions are evaluated while the boundary line of road surface's core region is being traced. The region with evaluation value over the threshold is merged into the road surface region. That is, region k 's evaluation value can be calculated by:

$$J_k = c_1 \times \sum_{f=R,G,B} \left| \frac{f_k - \overline{f}}{\overline{f}} \right| + c_2 \times \frac{|t_k - t|}{t} \quad (5)$$

where the first part of this formula is the dynamic color error rate and the second is the error rate of region's texture, and c_1 and c_2 are parameters for the balance of color and texture properties.

Dynamic error rate can also be called variable standard error rate. Generally on road scene image the road regions' color and texture are always changeable depending on the location. For example, the nearby road surface looks dark and coarse, the far looks bright and smooth. So it is hardly to extract all the road regions with static standard. In our approach, we change the standard value as follows: the initial color standard values are taken by R,G,B values of road surface's core region. While merging a new region, it can be renewed by:

$$\overline{f}_{new} = \frac{\overline{f}_{old} \times S_{old} + f_i \times S_i}{S_{old} + S_i} \quad (6)$$

where $f = R, G, B$, and S_{old} and S_i are areas of the road surface and the new merged region, respectively.

Dispersion of intensity served as region's texture value is calculated by:

$$t = \sum (T(i,j) - T')/S \quad (7)$$

where $T(i,j)$ is the intensity value of pixel (i,j) in the region, T' is the average intensity value of the region, and S is the region area.

All of the road surface regions then can be extracted by repeating the processing described above.

3.4 Recognition of Other Road Regions

We use property features of road regions and relative knowledge (location, shape and 3-D depth information) to recognize white line, centerline and shadow regions of road. Recognition result is shown in Fig.3.

3.5 Dealing with Small Regions

There are always some unrecognized small regions surrounding with the big road regions that have been recognized.

1) Consider the small region is surrounded by only one kind of road regions (e.g., road surface). If its area is below a threshold, it can be merged into road region. Otherwise, it may be extracted as the candidate obstacle region.

2) When the neighborhood regions is more than one kind, it can be merged into the similar neighborhood region if its area is below the threshold, and otherwise it may be extracted as the candidate obstacle region.

4. Road Shape Model

In order to grasp the general situation of the road shape ahead, we distinguish the road shape model into straight road, curve road (include circular and spiral curve) and complex road, depending on the recognition result that explained in Chapter 3.

Current vision-based road shape recognition methods only use lane marks to make the geometry shape models of road. When the road lane marks are not very distinct, or there is no lane marks at all, these methods can not give a right answer. In our approach, we also use the boundary line of road region being recognized as well as road lane marks, and so it can adapt properly to the general roads.

4.1 Reverse Projection Transformation from Image to 3-D Space

The perspective relationship of vehicle coordinate system (VCS) and the image coordinate system (ICS) is shown at Fig. 4, where the origin point of VCS is set at the camera lens' center point, and H is the height from camera lens center point to the ground. P (X,Y,Z) is a point on the VCS space, and its projection point on the image is P'(x,y). The relationship between P and P' can be explained by projection transformation as follows:

$$x = \frac{fX}{Z\cos\theta - Y\sin\theta} \quad \text{and} \quad y = \frac{f(Y\cos\theta + Z\sin\theta)}{Z\cos\theta - Y\sin\theta} \qquad (8)$$

If the camera's parameters are known and the vertical gradient of road surface is 0 (i.e., the road surface is level), points on the road surface can be reverse transformed by projection points on image. It is:

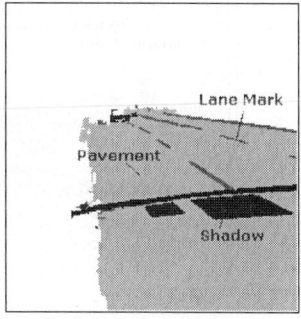

Fig.3. Recognition result

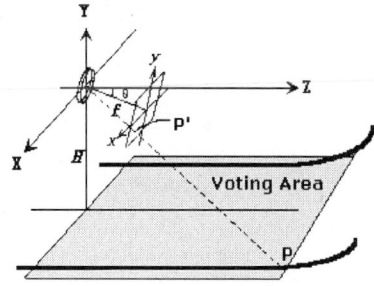

Fig.4. Relationship between camera and vehicle coordinate system

$$X_n = Y_n \cdot \frac{x_n}{y_n \cos\theta - f\sin\theta} \quad \text{and} \quad Z_n = Y_n \cdot \frac{f\cos\theta + y_n \sin\theta}{y_n \cos\theta - f\sin\theta} \tag{9}$$

where $Y_n = -H$, (x_n, y_n) is the image coordinate in ICS, and (X_n, Y_n, Z_n) is the 3-D coordinate in VCS, f is camera's focus distance and θ is the depression angle of camera.

In our approach, first of all, we thin the white line and center line regions being recognized as explained in Chapter 3. The result of thinning is called skeleton lines. When the skeleton lines is less then 2 or the number of points on the skeleton lines is below a threshold, the points on the boundary of road region can also be employed for the shape model's recognition.

These points then will be transformed into VCS space by the reverse transformation (9).

4.2 Extraction of Straight and Curve Lines

The Hough transformation is performed on the X-Z plane of VCS, for the extraction of parallel straight lines and concentric circle lines. When the distance between parallel straight lines or the radius difference of concentric circles is below a threshold, these lines can be extracted as road shape's candidate lines.

Thus, the road shape models can be recognized as straight road, curve road or complex road by the skeleton points' cover rate on these candidate shape lines.

Figure 5 shows transformation points of skeleton lines, extracted candidate lines of road shape in X-Z plane and the recognition result.

4.3 Revision of Road Region

In the case of being recognized as a straight road, the disappoint point can be determined by the projection of candidate shape lines on image. Among the lines which pass through the disappoint point on image plane, the line with the minimum least-squares error to the old road region's boundary line will be extracted as new road region's boundary, and then revise the road region.

In the case of curve road, the radius of concentric circles can be changed in a reasonable range, and projected these curve lines on image. Same to the straight one, the curves with the minimum least-squares error to the old road region's boundary line will be extracted as the new boundary, and then revise the road region.

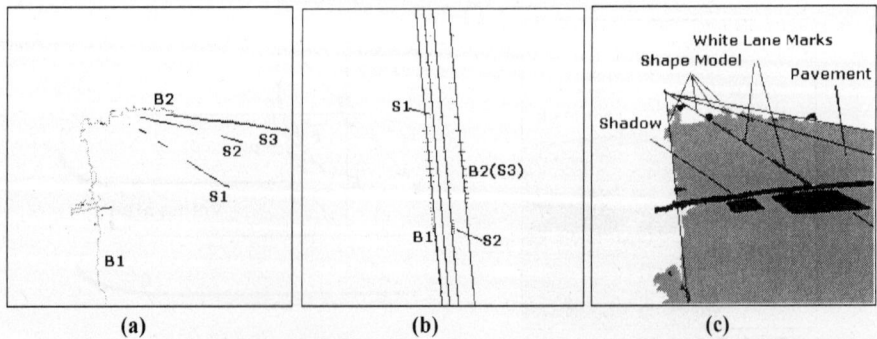

Fig. 5 Reverse projection transformation, candidate shape lines and recognition result

5. Experiments and Discussion

The original images were taken on general roads, the weather condition is fine or cloudy in the daytime. The size of images is 256x256 pixels, and each pixel being digitized to 256 levels in each color component.

Parameters of camera are: $f=50mm$, $H=1300mm$, $\theta = -5° \sim 5.6°$

And parameters of each evaluation are described in Table 1.

Figure 6 to Figure 11 show the experiment results. With these results, we can see that distinctness of road lane marks do not influence the recognition results of road by our method. It just influences the accuracy of road shape to some extent. A general road with no lane marks as shown in Fig. 6 can be extracted accurately, and road shape's recognition result is nearly correct as shown in Fig. 9.

As shown in Fig. 5(c), in a multi-lane road image with a complex background, the left shoulder of road can not be recognized, though the almost road regions are recognized correctly. The reason is that in the left shoulder region, wet areas and dry areas are mixed together. As a result, the dispersion of intensity is large, and so, error

Table 1. Parameters of each evaluation

c1	c2	α		
		$\pm 90°$	$\pm 45°$	$0°$
0.8	0.2	0.2	0.7	1.2

Fig. 6. Original image 1

Fig.7. Original image 2

Fig.8. Original image 3

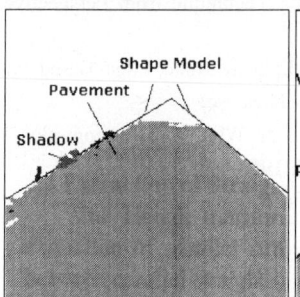
Fig. 9. Result of Fig.6

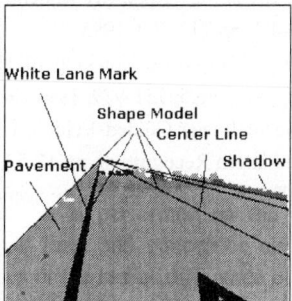
Fig.10. Result of Fig.7

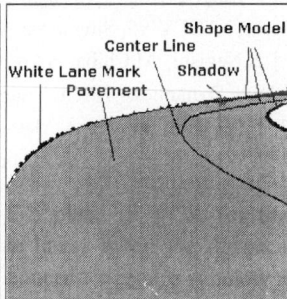
Fig.11. Result of Fig.8

rate of texture in this region is too high that this region can not be extracted in the merging processing.

A straight road like Fig.7 and a curve road as shown in Fig.8 can also be extracted and road shape model can be recognized accurately as shown in Fig.10 and Fig.11, respectively.

6. Conclusions

An image process approach for road shape's recognition is presented in this paper. The merits of the approach include a color image segmentation method based on color gradient edge detector, a robust region recognition method and the shape model's extraction. The color gradient edge detector emphasizes the variety of brightness and color on region's boundary, and it is appropriate to extract road regions from roadside's building with similar brightness or intensity. In the region recognition processing, evaluation standard is not static. It is made by property models of road surface sample that is obtained from the object image directly. And is, therefore, robust to the change of varies experiment conditions. The recognition of road shape's model also leads to the flexible general road's conditions, because it is not depending on the distinctness of road lane marks.

As the next step of our research on driving environment recognition, we will use these static image results to the processing of dynamic road image sequence as a bootstrap mode. Bad weather conditions (e.g., rain or snow) and complex road shape model will also be discussed in our future work.

References

[1] NAHSC, "Automated Highway System (AHS) System Objectives and Characteristics," NAHSC, Washington, D.C., Nov. 1995.
[2] N.Komota, "The Latest Major ITS Activities in Japan - Overview, Trends and Future Scope -," in 2nd World Congress on ITS '95, Yokohama, Japan, 1995
[3] C.Thrope, M.H.Hebert, T.Kanade, S.A.Shafer, "Vision and Navigation for the Carnegie-Mellon Navlab," IEEE Trans. on PAMI, Vol.10, No.3, pp.362-373, 1988
[4] K.Kluge, S.Lakshmanan, "Lane Boundary Detection Using Deformable Templates: Effects of Image Subsampling on Detected Lane," ACCV'95, Singapore, Vol.3, pp.141-145, 1995
[5] J.Yang, S.Guo, T.Hasegawa, S.Ozawa, "Roadway Matching Technique from Perspective Image," ACCV'95, Singapore, Vol.2, pp.747-750, 1995
[6] S.Negishi, M.Chiba, S.Ozawa, "Automatic Tracking of Highway Road Edge Based on Vehicle Dynamics," IEICE Trans. D-II, No.5, 931-939, 1994 (in Japanese)
[7] S.Ishikawa, et.al., "A Method of Image Guided Vehicle Using White Line Recognition," Proc. IEEE Computer Vision and Pattern Recognition, pp.47-53, 1986
[8] R.Nevatia, "A Color Edge Detector and Its Use in Scene Segmentation," IEEE Trans. SMC-7, pp.820-826, 1977
[9] J.M.S.Prewitt, "Object Enhancement and Extraction," in Picture Processing and Psychopictorics, B.S.Liplin and A.Rosenfeld, eds., Academic Press, 1970
[10] R.Nevatia, K.Babu, "Linear Feature Extraction and Description," Computer Graphics and Image Processing 13, pp.257-269, 1980

Visual Detection of Obstacles Assuming a Locally Planar Ground

Manolis I.A. Lourakis and Stelios C. Orphanoudakis
{lourakis,orphanou}@ics.forth.gr

Institute of Computer Science
FORTH
P.O.Box 1385, 711 10
Heraklion, Crete, Greece

AND

Department of Computer Science
University of Crete
P.O.Box 1470, 714 09
Heraklion, Crete, Greece

Abstract. Obstacle avoidance is an essential capability of an autonomous robot. This paper presents a method that enables a mobile robot to locate obstacles in its field of view using two images of its surroundings. The method is based on the assumption that the robot is moving on a locally planar ground. Using a set of point features (corners) that have been matched between the two views using normalized cross-correlation, a robust estimate of the homography of the ground is computed. Knowledge of this homography permits us to compensate for the motion of the ground and to detect obstacles as areas in the image that appear nonstationary after the motion compensation. The resulting method does not require camera calibration, is applicable either to stereo pairs or to motion sequence images, does not rely on a dense disparity/flow field and circumvents the 3D reconstruction problem. Experimental results from the application of the method on real images indicate that it is both effective and robust.

1 Introduction

The development of vehicles, capable of moving autonomously in their environment, has been one of the major efforts in the field of robotics during the last two decades. To avoid collisions, such vehicles need a means for detecting obstacles along their route. In this work, we assume a vehicle capable of acquiring images of its surroundings and propose a vision-based approach to obstacle detection.

Most approaches to visual obstacle detection exploit motion cues for locating obstacles. Furthermore, an assumption that is often made is that vehicle motion is confined to a surface that is either planar or can be approximated locally by planes [3, 2, 8, 11, 4, 14]. The existence of a planar ground gives rise to a phenomenon known as *motion parallax* in the psychophysics literature [5]: A moving observer, perceives objects extending vertically from the ground to move differently from their immediate background. Various techniques for obstacle detection based on motion parallax have been proposed. For example, Enkelmann [3] uses a calibrated camera to compute a reference flow related to the motion of the ground and then compares it with the flow estimated from images captured by a monocular observer. Inconsistencies between these two flows signal the presence of obstacles. Enkelmann assumes that the camera pursues a purely translational motion. However, such assumptions about egomotion are

---

* This work was funded in part under the VIRGO research network of the TMR Programme (EC Contract No ERBFMRX-CT96-0049).

not always valid and should be avoided when possible. Carlsson and Eklundh [2] assume a camera with unrestricted motion and predict the egomotion and the equation of the ground plane from long image sequences. Obstacles are identified in regions whose motion differs from that predicted. Jenkin and Jepson [8] apply the EM algorithm to obtain maximum likelihood estimates of the parameters of a mixture model describing the disparity field computed with phase-based techniques from a calibrated stereo pair. The probability that a point does not belong to the floor is then computed from the ownership probabilities of the mixture model. Santos-Victor and Sandini [11] employ the normal flow field estimated with an uncalibrated camera and detect obstacles lying on a planar floor by performing an inverse perspective transformation that maps the normal flow onto a horizontal (parallel to the floor) plane. Their method, however, uses an approximate parametric model of the flow generated by the ground plane, deals with outliers in an ad hoc manner and requires the camera to remain in a fixed position relative to the vehicle. Fornland [4] uses the normal flow field measured from a camera moving parallel to the ground plane to derive a linear equation relating motion parameters to the spatiotemporal derivatives of the image intensity function. Obstacles are then detected as the outliers of a robust fit estimated by RANSAC over the image points. Zhang et al present three algorithms for obstacle detection [14]. The first algorithm employs a calibrated camera to derive a linear system whose solvability implies the absence of obstacles. The second algorithm does not require camera calibration and exploits the homography of the ground plane to derive a linear system relating corresponding image coordinates in two views. Similarly to the first algorithm, inconsistency of this linear system signals the presence of obstacles. The third algorithm uses sequences of partially calibrated stereo pairs to estimate the equation of the ground plane and the height of obstacles. Due to space limitations, we do not review here approaches that do not assume a planar ground but refer the interested reader to [9].

In this work, we propose a method that uses two images to detect obstacles along the path of an autonomous vehicle. The method assumes that the ground is planar and starts by estimating the motion of the ground in the two images using a small set of matched points. Subsequently, compensation of the motion of the ground is performed by warping the second image with respect to the first according to the estimated motion. This warping registers the image of the ground in the two views, so that the obstacles are nonstationary between the two images. Finally, a change detection operation between the first and the warped second image locates the obstacles present in the scene.

The rest of this paper is organized as follows. Section 2 presents an overview of some preliminary results that are essential for the development of the proposed method. Section 3 presents the method itself. Experimental results from the application of the method on real images are presented in Section 4. The paper is concluded with a brief discussion in Section 5. A more detailed version of the current paper can be found in [9].

2 Preliminaries

2.1 Projective Geometry

In the following, projective (homogeneous) coordinates are employed to represent image points by 3×1 column vectors $\mathbf{m} = (m_1, m_2, 1)^T$.

Thus, all vectors of the form $\lambda \mathbf{m}$, where λ is an arbitrary non-zero factor, are equivalent. Regarding notation, the symbol $\simeq$ will be used to denote equality of vectors up to a scale factor. Vectors and arrays will be written in boldface.

A well-known constraint for a pair of perspective views of a rigid scene, is the *epipolar constraint*. This constraint states that for each point in one of the images, the corresponding point in the other image must lie on a straight line. Assuming that no calibration information is available, the epipolar constraint is expressed mathematically by a 3×3 singular matrix, known as the *fundamental matrix*. More specifically, assuming that $\mathbf{m}$ and $\mathbf{m}'$ are two homogeneous 3×1 vectors defining a pair of corresponding points in two images, they satisfy the following equation:

$$\mathbf{m}'^T \mathbf{F} \mathbf{m} = 0, \qquad (1)$$

where $\mathbf{F}$ is the fundamental matrix [6] and $\mathbf{m}'^T$ a row vector. $\mathbf{F}$ plays a central role in applications involving the recovery of motion and structure information from uncalibrated images.

Another important concept is the *plane homography* (also known as plane projectivity or plane collineation) $\mathbf{H}$, which relates two uncalibrated views of a plane in three dimensions. Each 3D plane Π defines a nonsingular 3×3 matrix $\mathbf{H}$ which relates the image of the plane in two views. More specifically, if $\mathbf{m}$ is the projection in one view of a point belonging to Π and $\mathbf{m}'$ is the corresponding projection in a second view, then [6]:

$$\mathbf{m}' \simeq \mathbf{H}\mathbf{m}. \qquad (2)$$

Since $\mathbf{F}$ and $\mathbf{H}$ relate vectors in homogeneous coordinates, they can be estimated only up to an unknown scale factor. $\mathbf{F}$ and $\mathbf{H}$ are related by the fact that the matrix $\mathbf{F}^T \mathbf{H}$ is skew-symmetric [6], that is

$$\mathbf{F}^T \mathbf{H} + \mathbf{H}^T \mathbf{F} = \mathbf{0}, \qquad (3)$$

where $\mathbf{F}^T$ and $\mathbf{H}^T$ are the transposes of $\mathbf{F}$ and $\mathbf{H}$ respectively.

2.2 Robust Regression

Regression analysis, i.e. the problem of fitting a model to noisy data, is a very important subfield of statistics. The traditional approach to regression analysis employs the least squares (LS) method, which is popular due to its low computational complexity. LS involves the solution of a linear minimization problem, and achieves optimal performance if the underlying noise distribution is Gaussian with zero mean. However, in cases where the noise is not Gaussian, or in the presence of *outliers*, that is observations that deviate considerably from the model representing the rest of the observations, the LS estimator becomes highly unreliable. One criterion for characterizing the tolerance of an estimator with respect to outliers is its *breakdown point*, which may be defined as the smallest amount of outlier contamination that may force the value of the estimate outside an arbitrary range. As an example, LS has a breakdown point of 0%, because a single outlier may have a substantial impact on the estimated parameters.

The *Least Median of Squares* (LMedS) estimator was originally proposed by Rousseeuw [10] and is able to handle data sets containing many outliers. LMedS involves the solution of a nonlinear minimization problem that aims at estimating a set of model parameters that best fit the *majority* of the observations. In contrast, LS tries to estimate a set of model parameters that best fit *all* the observations. Thus, LMedS has a breakdown point of 50%, a characteristic which makes it particularly attractive for the purposes of this work.

3 Obstacle Detection

The proposed method starts by extracting a set of corners from each of the two images. These corners are then matched using a similarity criterion based on normalized cross-correlation. The matching algorithm is based on that proposed in [12]. Using the matched pairs of corners, the matrix $\mathbf{F}$ defined by the two images is estimated as follows: Let $\mathbf{f}$ be the 9×1 vector defined by the 9 unknown elements of matrix $\mathbf{F}$, i.e. $\mathbf{f} = (F_{11}, F_{12}, F_{13}, F_{21}, F_{22}, F_{23}, F_{31}, F_{32}, F_{33})^T$. Then, Eq. (1) can be written as

$$(m_1 m_1', m_2 m_1', m_1', m_1 m_2', m_2 m_2', m_2', m_1, m_2, 1)\mathbf{f} = 0 \tag{4}$$

Considering N matched pairs, the N constraints given by Eq. (4) can be written more compactly as $\mathbf{Af} = \mathbf{0}$, where $\mathbf{A}$ is a $N \times 9$ matrix. The fundamental matrix is then estimated from the solution of the following minimization problem:

$$min_\mathbf{f} ||\mathbf{Af}||^2 \quad subject\ to \quad ||\mathbf{f}||^2 = 1, \tag{5}$$

where $||\ ||$ denotes the vector 2-norm. The solution to this constrained minimization problem is known to be the eigenvector of the matrix $\mathbf{A}^T \mathbf{A}$ that corresponds to the smallest eigenvalue.

As noted in [7], $\mathbf{A}^T \mathbf{A}$ is inhomogeneous in image coordinates and, therefore, ill-conditioned. To improve its condition number and to derive a more stable linear system, the coordinates of the matched corners are normalized by a pair of linear transformations $\mathbf{L}$ and $\mathbf{L}'$ as follows: $\mathbf{L}$ defines a translation of the corners in the first image, such that their centroid is brought to the origin of the coordinate system, followed by an isotropic scaling that maps the average corner coordinates to $(1, 1, 1)$. $\mathbf{L}'$ is defined similarly for corners in the second image. As shown in [7, 13], these transforms result in a more stable system, from which a fundamental matrix $\hat{\mathbf{F}}$ can be estimated. $\mathbf{F}$ is then computed from $\hat{\mathbf{F}}$ as $\mathbf{F} = \mathbf{L}'^T \hat{\mathbf{F}} \mathbf{L}$. At this point, it should be noted that there exist more accurate methods for estimating $\mathbf{F}$ [13]. However, for the purposes of the present work, the simple linear technique outlined above gives results with satisfactory precision.

Since the normalized matching pairs that are given as input to the estimation process will contain errors due to false matches and errors in the localization of corners, care must be taken so that these errors do not corrupt the computed estimate. Thus, instead of using the whole set of matched corners to estimate $\mathbf{F}$, the LMedS estimator is employed to find an estimate that is consistent with the majority of the matched corners. Using a predetermined number of iterations, LMedS picks random samples of matching pairs and computes an estimate of $\mathbf{F}$ from each of them. The estimate that yields the smallest median error is returned as the fundamental matrix which best fits the set of matched corners.

The procedure for estimating $\mathbf{H}$ is similar to that for estimating $\mathbf{F}$ above. The 9 unknown elements of matrix $\mathbf{H}$ define a 9×1 vector $\mathbf{h}$ such that $\mathbf{h} =$

$(H_{11}, H_{12}, H_{13}, H_{21}, H_{22}, H_{23}, H_{31}, H_{32}, H_{33})^T$. For each pair of corresponding points m and m', Eq. (2) yields the following pair of constraints:

$$H_{11}m_1 + H_{12}m_2 + H_{13} = H_{31}m_1m_1' + H_{32}m_2m_1' + H_{33}m_1'$$
$$H_{21}m_1 + H_{22}m_2 + H_{23} = H_{31}m_1m_2' + H_{32}m_2m_2' + H_{33}m_2' \quad (6)$$

Using N matching pairs, the $2N$ constraints given by Eq. (6), combined with the 6 constraints arising from the skew-symmetry constraint defined by Eq. (3) [2], can be written as $\mathbf{Bh} = \mathbf{0}$, where $\mathbf{B}$ is a $(2N+6) \times 9$ matrix. $\mathbf{H}$ is then estimated by solving

$$min_{\mathbf{h}} ||\mathbf{Bh}||^2 \quad subject\ to \quad ||\mathbf{h}||^2 = 1 \quad (7)$$

The solution to the above problem is the eigenvector of the matrix $\mathbf{B}^T\mathbf{B}$ that corresponds to the smallest eigenvalue. As can be clearly seen from Eq. (3) and Eq. (6), $\mathbf{B}^T\mathbf{B}$ is inhomogeneous in image coordinates. Thus, the normalization procedure that was previously employed for estimating $\mathbf{F}$ is also used for determining $\mathbf{H}$.

Assuming that at least 50% of the matched corners belong to the ground, LMedS is employed to compute a robust estimate of the ground plane homography $\hat{\mathbf{H}}$ defined by the normalized matching pairs. $\mathbf{H}$ is then computed as $\mathbf{L'}^{-1}\hat{\mathbf{H}}\mathbf{L}$. It should be noted at this point that the use of $\mathbf{F}$ in estimating $\mathbf{H}$ is not necessary. $\mathbf{H}$ has 8 degrees of freedom and, since each pair of corresponding ground corners provides 2 constraints, 4 pairs of corresponding ground corners in general position (no three corners are collinear) give rise to 8 constraints regarding the elements of $\mathbf{H}$ and, therefore, suffice to provide a solution. However, knowledge of $\mathbf{F}$ provides 5 constraints regarding $\mathbf{H}$, enabling us to estimate $\mathbf{H}$ using only 2 pairs of matching corners. Thus, the size of the random samples selected by LMedS during the computation of $\mathbf{H}$ is equal to 2.

After the homography of the ground plane has been computed, we can compensate for the motion of the ground by warping the second image with respect to the first, using bilinear interpolation and the motion defined at each image point by Eq. (2). This transformation results in the image of the ground being registered in the two views, leaving all obstacles extruding from the ground plane unregistered. Subtracting the first image from the warped one, we can declare points where the absolute value of the computed difference is above a threshold as belonging to obstacles. For more accurate results that will not be sensitive to changes in the illumination, a change detection algorithm can be employed. In some cases, change detection can produce small noisy areas that do not correspond to obstacles. A size filtering step can effectively eliminate these areas as follows. Pixels that do not belong to a connected component of a minimum size are assumed to be due to noise and can be masked out. Only regions having area greater than some predefined threshold are retained in the final obstacle map.

As mentioned in Section 1, the second algorithm for obstacle detection proposed by Zhang et al in [14] uses the homography of the ground plane, similarly to the method described in this paper. There are, however, important differences between the two methods. Zhang et al use a test based on the ratio of singular values obtained from singular value decomposition (SVD) to determine whether the linear system relating corresponding image points in two views is solvable

---

[2] Actually only 5 of these 6 constraints are linearly independent; see [6].

or not. This test requires the specification of an ad-hoc threshold, and as shown in the synthetic experiments reported in [14], is sensitive to noise. The noise sensitivity measured by Zhang et al is expected to increase when their method has to cope with real noisy data instead of simulated ones. This is due to the fact that the noise model they employ during simulation accounts only for small scale deviations from the ground plane, ignoring many other possible sources of noise. In contrast, the method proposed in this paper uses robust regression techniques to ensure that the existence of corresponding pairs of points that are contaminated by noise do not cause the obstacle detection algorithm to fail. Moreover, the algorithm by Zhang et al provides a simple yes/no answer regarding the presence of obstacles, while our method provides a map indicating the exact location of obstacles in the field of view of the observer.

4 Experimental Results

A set of experiments has been conducted in order to test the performance of the proposed method. Representative results from two of these experiments are given in this section. Both experiments were performed with the aid of stereo pairs acquired by a binocular head mounted on a mobile robot.

The first experiment refers to the stereo pair shown in Figures 1(a) and (b). The viewed scene consists of a planar floor on which lies a textured poster. A box in the middle and a flower-pot on the right side of the scene are the obstacles to be detected. White rectangles in Fig. 1(c) indicate the corners that do not conform to the estimated plane homography. Corners that agree with the estimated plane homography are marked with gray rectangles. As can be seen in Fig. 1(c), some of the corners belonging to the floor are marked as outliers after the estimation of the floor homography. These corners have been erroneously matched between the two views, forming pairs that do not satisfy Eq. (2).

Fig. 1(d) shows the right image warped according to the estimated homography of the ground plane. It is clear from Fig. 1(a) and Fig. 1(d) that image warping according to the estimated floor homography registers the image of the floor. The obstacles detected after change detection between Fig. 1(a) and Fig. 1(d) are shown in black in Fig. 1(e). No size filtering was necessary. Note that the detected obstacles correspond to the box and the flower-pot.

The second experiment is based on the stereo pair shown in Figures 2(a) and (b). A textured poster has been placed on a planar floor and a chair on the left side of the scene, along with a box on the right side, are the obstacles to be detected. Corners that do not belong to the floor are characterized as outliers by LMedS during the estimation of the ground homography, and are shown as white rectangles in Fig. 2(c). Corners belonging to the floor are marked by gray rectangles. In this particular experiment, a large number of outliers was tolerated. More specifically, LMedS concluded that 91 matched corners from a total of 201 (a percentage of about 45%) are outliers. This clearly demonstrates the robustness of the proposed method.

Fig. 2(d) shows the right image warped according to the estimated homography of the ground plane. This warping registers the image of the floor between Fig. 2(a) and Fig. 2(d). The obstacles detected after change detection between Fig. 2(a) and Fig. 2(d) are shown in black in Fig. 2(e). Again, size filtering on the output of change detection was not required. Note that the chair and the box have been successfully identified as obstacles.

Fig. 1. (a),(b) left and right view, (c) outliers detected by LMedS during the estimation of the ground homography, (d) right image warped according to ground homography, (e) detected obstacles (see text for explanation).

5 Conclusions

The capability of obstacle avoidance is crucial for a robot moving in an unknown environment. In this paper, a method for obstacle detection that has several advantages has been presented. First, it does not require any calibration information to be known. This feature is particularly attractive in the context of active vision [1], where the camera position in 3D as well as the zoom and focus are actively controlled, resulting in frequent changes in the extrinsic and intrinsic parameters of the camera. Second, the method does not require the computation of a dense set of disparities between the two views and, therefore, solving the correspondence problem for each image point is avoided. Third, there is no need for explicitly recovering the 3D structure of the viewed scene. Fourth, no restrictions on egomotion are imposed. Fifth, the method is usable either by a monocular vehicle moving in the environment or by a binocular one. Finally, the use of a robust estimator such as LMedS safeguards against errors in the input, which could otherwise have a significant effect on the accuracy of the computations.

The main disadvantage of the proposed method is that it requires at least 50% of the matched corners to be on the ground, a constraint imposed by the breakdown point of LMedS. A related shortcoming is that the method assumes that the ground is textured, in order to be able to extract corners. It should be noted, however, that most vision algorithms are expected to run into difficulties in the absence of texture.

Current research efforts are directed towards exploiting the information available in the obstacle map produced by the proposed method for deriving a complete obstacle avoidance mechanism. Such a mechanism will be used for driving a mobile robot in an unknown environment.

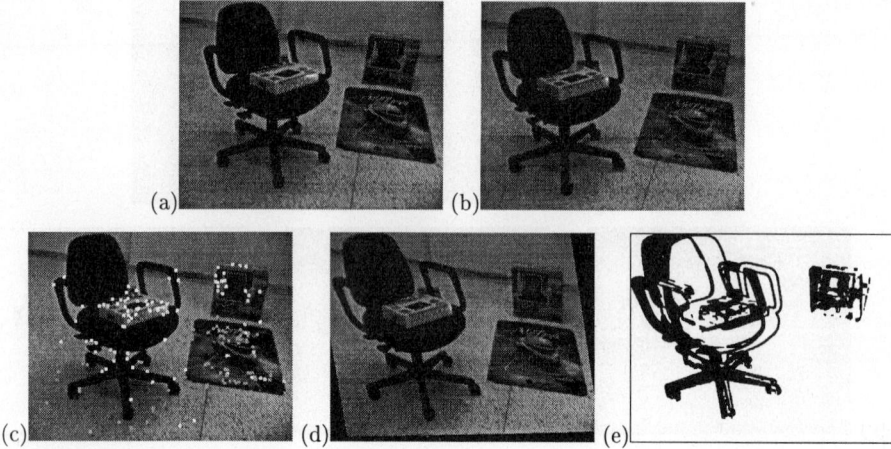

Fig. 2. (a),(b) left and right view, (c) outliers detected by LMedS during the estimation of the ground homography, (d) right image warped according to ground homography, (e) detected obstacles (see text for explanation).

References

1. J. Aloimonos. Purposive and Qualitative Active Vision. In *Proceedings of DARPA Image Understanding Workshop*, pages 816–828, 1990.
2. S. Carlsson and J.-O. Eklundh. Object Detection Using Model Based Prediction and Motion Parallax. In *Proceedings of ECCV'90, LNCS*, pages 134–138, 1990.
3. W. Enkelmann. Obstacle Detection by Evaluation of Optical Flow Fields From Image Sequences. *Image and Vision Computing*, 9(3):160–168, 1991.
4. P. Fornland. Direct Obstacle Detection and Motion from Spatio-Temporal Derivatives. In *Proceedings of CAIP'95, LNCS*, pages 874–879, Prague, September 1995.
5. J.J. Gibson. *The Perception of the Visual World*. Houghton-Mifflin, Boston, 1950.
6. R. Hartley and R. Gupta. Computing Matched-Epipolar Projections. In *Proceedings of CVPR'93*, pages 549–555, 1993.
7. R.I. Hartley. In Defense of the 8-Point Algorithm. *IEEE PAMI*, 19(6):580–593, June 1997.
8. M.R.M. Jenkin and A. Jepson. Detecting Floor Anomalies. In *Proceedings of BMVC'94*, pages 731–740, 1994.
9. M.I.A. Lourakis and S.C. Orphanoudakis. Visual Detection of Obstacles Assuming a Locally Planar Ground. Technical Report 207, ICS/FORTH, Aug. 1997.
10. P.J. Rousseeuw. Least Median of Squares Regression. *Journal of American Statistics Association*, 79:871–880, 1984.
11. J. Santos-Victor and G. Sandini. Uncalibrated Obstacle Detection Using Normal Flow. *Machine Vision and Applications*, 9(3):130–137, 1996.
12. Z. Zhang. A New and Efficient Iterative Approach to Image Matching. In *Proceedings of ICPR'94*, pages 563–565, Jerusalem, Israel, 1994.
13. Z. Zhang. Determining the Epipolar Geometry and its Uncertainty: A Review. Technical Report 2927, INRIA, July 1996.
14. Z. Zhang, R. Weiss, and A.R. Hanson. Obstacle Detection Based on Qualitative and Quantitative 3D Reconstruction. *IEEE PAMI*, 19(1):15–26, Jan. 1997.

Potential-Based Modeling of 2D Regions Using Non-uniform Source Distributions

Jen-Hui Chuang, Chi-Hao Tsai, Wei-Hsin Tsai, and Chuei-Yaw Yang

Department of Computer and Information Science
National Chiao Tung University
Hsinchu, Taiwan, R.O.C.

Abstract

One of existing approaches to path planning problems uses a potential function to represent the topological structure of the free space. [1] *Newtonian potential was used in [1] to represent object and obstacles in the 2D workspace wherein their boundaries are assumed to be uniformly charged. In this paper, the source distributions are extended to more general cases. It is shown that if the source distribution of a line segment can be uniform, linear or quadratic, the repulsion between two line segments can be evaluated analytically. Simulation results show that by properly adjusting the charge distribution along obstacle/object boundaries, path planning results can indeed be improved in terms of collision avoidance.*

1 Introduction

The goal of path planning is to determine how to move an object from its original location and orientation to the goal configuration while avoiding any collision with obstacles. The configuration space approach [2] and the critical curve approach [3] use known obstacle space as reference and plan a path barely avoiding the obstacles, e.g., moving an object while keeping in contact with the obstacles. A solution is guaranteed if there is one. However, such algorithms are more complicated than representing the free space in other simpler ways. Octree representation of the 3D free space is used in [4]. Rectangular corridors and their junctions are used in [5] to represent the free space among rectangular obstacles. Convex areas are used in [6] for path planning of a point object. Generalized cylinders and convex polygons are used in [7] to represent the free space. In [8], circular discs are used to create the generalized Voronoi diagram defined by obstacle location and shapes. The Voronoi edges are then used to derive paths for rectangular objects.

---

[1] This research is supported by R.O.C. National Science Council under NSC-85-2213-E009-126.

The above algorithms do not always choose the object configurations which best match the shape of the free space along a path. One way to obtain the best match and minimize the risk of collision is to define a repulsive potential field between the object and the obstacles. A cubic function of the distance between a point object and the obstacles is used in [9] as the potential function. An artificial repulsive potential which is the function of the shortest distance between the moving object and the obstacles is used in [10] for local planning of linked line segments. Similar local planning is done in [11] using a superquadric artificial potential function whose isopotential contours are modified n-ellipses. Boundary equations of polytopes are used in [12] to create an artificial potential function. The main advantages of the potential-based approaches include the simplicity of the representation of free space, the guidance in the object motion provided by the potential gradient, and the readiness of the extension to spaces of higher dimensions.

The Newtonian potential is used in [1] to model the free space wherein the boundaries of object and obstacles are assumed to be uniformly charged. The magnitude of the potential is unbounded near the obstacle boundary and decreases with range, which captures the basic requirement of collision avoidance. Since geometric constraints are used to guide the global motion of the object while the repulsive potential is used locally to match the shape of object to that of the free space, the local minimum of the potential function do not present a problem in path planning. In this paper, the above potential model is extended to more general cases where the polygonal boundaries are charged with source distributions which can be linear or quadratic. Simulation results show that the collision avoidance ability of the planned object path can be improved by properly adjusting the charge distribution along object/obstacle boundaries.

2 Potential due to Some Non-Uniform Source Distributions on Polygonal Region Borders

Since the potential due to a polygonal region whose border is charged can be calculated by superposing the potential due to each border segment, only the potential due to a single line segment is formulated in the following. Consider a point A at (x_0, y_0) and a finite line charge on x-axis. Suppose the charge density changes linearly or quadratically with x from x_1 to x_2. The charge density function can be expressed as $\alpha x^2 + \beta x + \gamma$ with some constants α, β, and γ. The total potential at the point A due to the charged line segment is equal to:

$$V_A = \int_{x_1}^{x_2} \frac{\alpha x^2 + \beta x + \gamma}{r} dx \tag{1}$$

with $r = \sqrt{(x - x_0)^2 + y_0^2}$, and can be evaluated analytically.

In the above discussion, we have considered the uniform, linear, and quadratic charge distributions on a line segment. In general, similar results can also be found for polygonal region boundaries. For example, the equipotential contours

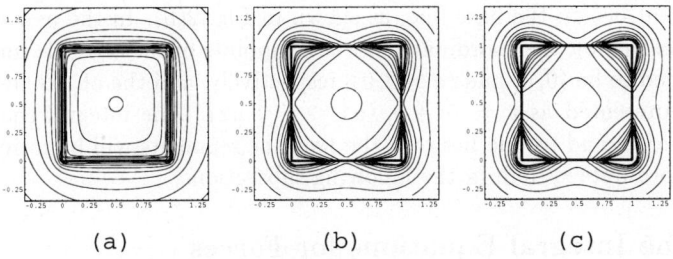

Figure 1: The potential contours due to a charged square boundary.

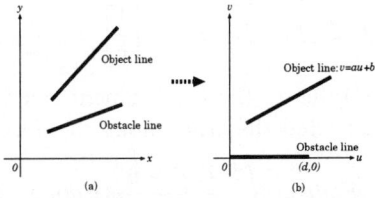

Figure 2: Coordinate transformation. (a) The original coordinate system (xy-plane). (b) The new coordinate system (uv-plane) after the transformation.

due to uniform, linear and quadratic charge distributions on a *square* are shown in Figs. 1(a), (b), and (c), respectively. The charge densities are increased linearly and quadratically from the midpoint of each edge toward the corners of the square for Figs. 1(b) and (c), respectively. Notice that the near the boundaries of the square, the equipotential contours in Fig. 1(a) are much closer to the corners of the square than that to its edges. Whereas, the equipotential contours in Figs. 1(b) and (c) are farther away from the corners, a more desirable property for collision avoidance of an object path.

3 Integral Equations of the Repulsion

Although the potential value can be used in the search for object configuration of minimal potential, more efficient search methods can be adopted if gradient of the potential, in forms of repulsive force and torque, is analytically tactable. For polygonal object and obstacles in the 2-D space, line segments of their boundaries can be used as basic elements in the calculation of repulsive force and torque needed in path planning. The repulsive force and torque between these polygonal regions can then be derived by superposing the repulsion between pairs of border segments, each contains one line segment from the the moving object and the other from one of the obstacles.

In general, each pair of the repelling line segments can have arbitrary configuration in the work space as shown in Fig. 2(a). To simplify the expressions of the repulsion between them, a coordinate system is chosen so that the obstacle line segment lies on the base line, as shown in Fig. 2(b). In the new coordinate system (uv-plane), the coordinates of the endpoints of the obstacle line segment are assumed to be $(0,0)$ and $(d>0,0)$, respectively, and the object line segment can be represented as $v = au + b$, ($u1 \leq u \leq u2$). The integral equations for repulsive force and torque between the two line segments will be formulated for such a coordinate system in the following subsections.

3.1 The Integral Equations for Forces

Consider the electric field at an object point $(u', v') = (u', au' + b)$ due to a point $(u, 0)$ on the line segment of an obstacle, we have

$$\vec{E} \triangleq (E_u, E_v) = -\nabla\left[\frac{1}{r}\right] = -\frac{1}{r^2}\hat{r}. \quad (2)$$

Thus, the force at (u', v') due to the line segment can be decomposed into two parts, one along the u-axis and the other along the v-axis,

$$F_u(u', v') = \int_{q_1}^{q_2} E_u dq = \int_0^d \frac{1}{r^2}\frac{u'-u}{r}\rho(u)du = \int_0^d \frac{u'-u}{r^3}\rho(u)du \quad (3)$$

$$F_v(u', v') = \int_{q_1}^{q_2} E_v dq = \int_0^d \frac{1}{r^2}\frac{v'}{r}\rho(u)du = \int_0^d \frac{au'+b}{r^3}\rho(u)du, \quad (4)$$

where $\rho(u)$ is the charge density along the line segment of the obstacle.

The total force between the two line segments can be calculated from

$$F_u = \int_{q_1}^{q_2} F_u(u', v') dq = \sqrt{1+a^2}\int_{u_1}^{u_2}\int_0^d \frac{u'-u}{r^3}\rho(u)\rho(u')dudu' \quad (5)$$

$$F_v = \int_{q_1}^{q_2} F_v(u', v') dq = \sqrt{1+a^2}\int_{u_1}^{u_2}\int_0^d \frac{au'+b}{r^3}\rho(u)\rho(u')dudu', \quad (6)$$

where $\rho(s)$ is the charge density along the object line, and $dq = \rho(s)ds = \sqrt{1+a^2}\rho(u')du'$.

3.2 The Integral Equations for Torques

Given any collision-free orientation of an object with non-zero torque with respect to its rotation center, the direction of the torque directly gives the direction in which the object can rotate to reach a configuration of smaller potential. Let $P = (u_0, v_0)$ be the rotation center of the object. The total torque with respect to P due to the repulsion between the two line segments is equal to

$$\tau_P = \int_{s_1}^{s_2} \vec{l}(u', v') \times \vec{F}(u', v')\sqrt{1+a^2}\rho(u')du'. \quad (7)$$

where $\vec{l}(u', v') = (u' - u_0, v' - v_0)$, and $\vec{F}(u', v') = (F_u(u', v'), F_v(u', v'))$. We will now show that with the above integral equations, the repulsive forces and torques between object and obstacles can be evaluated analytically for some simple non-uniform charge distributions along their boundaries.

3.3 Repulsion due to Simple Charge Distributions

Assume the charge density $\rho(u)$ is equal to 1, u, or u^2 for an obstacle line, and $\rho(s) = 1$, s, or s^2 for an object line, there are nine different combinations of charge distributions to be considered in evaluating the repulsion between the two line segments. For example, from (5), the repulsive force along the u-axis for these nine combinations can be obtained from

$$F_u^{mn} = F_u^{mn}(u_2) - F_u^{mn}(u_1) = (1+a^2)^{\frac{n+1}{2}} \int_{u_1}^{u_2} \int_0^d \frac{u'-u}{r^3} u^m u'^n \, du \, du' \quad (8)$$

where $m = 0, 1, 2$ is equal to the order of the charge density of the obstacle line, and $n = 0, 1, 2$ is equal to that of the object line. Similarly, the repulsive force along the v-axis and the repulsive torque can also be formulated with nine expressions according to (6) and (7), respectively. It can be shown that analytic expressions exist for all these integral equations. For example, we have

$$\begin{aligned} F_u^{01}(u') &= (1+a^2)\left(-\frac{f_1(u')}{1+a^2} + \frac{f_2(u')}{1+a^2}\right.\\ &+ \frac{a\,b\,\log(|a\,b + u' + a^2\,u' + \sqrt{1+a^2}\,f_1(u')|)}{(1+a^2)^{\frac{3}{2}}} \\ &+ \left.\frac{(-(a\,b)+d)\,\log(|a\,b - d + u' + a^2\,u' + \sqrt{1+a^2}\,f_2(u')|)}{(1+a^2)^{\frac{3}{2}}}\right) \end{aligned}$$

where $f_1(u') = \sqrt{(au'+b)^2 + u'^2}$, and $f_2(u') = \sqrt{(au'+b)^2 + (u'-d)^2}$.

In general, if $\rho(u) = \alpha_1 u^2 + \beta_1 u + \gamma_1$ and $\rho(s) = \alpha_2 s^2 + \beta_2 s + \gamma_2$, the repulsion can still be evaluated analytically through superposition.

4 Application of the Potential Model in Path Planning

4.1 A Local Planning Algorithm

For a path planning problem, the places in which the moving object is more likely to collide with obstacles are bottlenecks in the free space. In this section, a simple local planner (similar to that presented in [1]) is used to demonstrate the effectiveness of the proposed free space model in achieving different collision avoidance effects of object motion around the bottleneck regions. In the 2-D space, free space bottlenecks can be defined by the minimal distance links (MDLs) among polygonal obstacles, and object shape can be represented by its skeleton. Given an MDL, the topology of the local path is described to the local planner using the skeletal representation of the moving object.

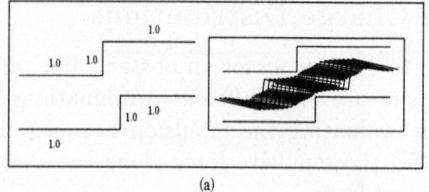 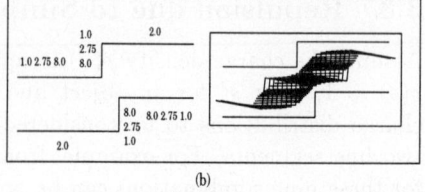

(a) (b)

Figure 3: A path planning example. The obstacle boundaries are (a) uniformly charged, (b) non-uniformly charged.

Let s_i, $1 \leq i \leq N$, denote the sequence of N selected skeleton points to cross the MDL while L denotes the line containing the MDL. The local path begins when s_1 reaches the MDL and ends when s_N leaves the MDL. The local planner performs path planning by sequentially ensuring that as each skeleton point moves onto the MDL, it stays on L while the location and orientation of the object is adjusted to minimize the Newtonian potential using repulsive torque and force discussed earlier. Additional skeleton points may be added to reduce the step size along the path, allowing for finer adjustments in the object configuration to avoid collision. To restrict the total amount of computation, a limit is put on the minimum spacing between adjacent skeleton points used in the simulation, which effectively serves as a feasibility test of the local plan.

For the simulation results presented in the next subsection, the minimal potential configurations are specified to within 1/100 of the length of the MDL in location, and within one degree in orientation.

4.2 Simulation Results

In this subsection, simulation results are presented for path planning of moving objects whose boundaries are uniformly charged for simplicity. To each obstacle segment, one, two or three numbers are attached in the illustration which give the values of the corresponding charge density at (i) any point, (ii) the two end points, or (iii) the two end points plus the midpoint of the line segment, depending on whether the distribution is of zero, first or second order, respectively.

Figs. 3 to 4 show local planning results for two object paths. In general, the minimum distance between an object path and obstacles can be reduced by the adjustment of the above charge distribution. Such an effect is achieved by readjusting the originally uniform charge distribution such that more charges are added to protruding points of obstacles. Fig. 5 shows an interesting situation where a curved object path is obtained for a straight free space passage between two parallel obstacle segments. Such an example demonstrates the flexibility of the proposed potential-based modeling of the workspace to accomondate certain path planning constraints/considerations of interest, e.g., building material, protection measure, or uncertainty in location (see [13]), which are generally non-uniform along workspace boundaries.

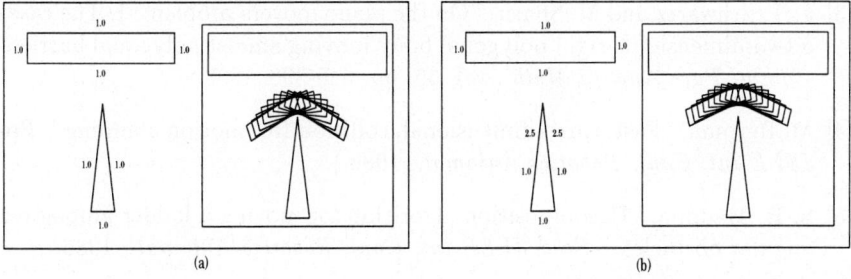

Figure 4: A path planning example. The obstacle boundaries are (a) uniformly charged, (b) non-uniformly charged.

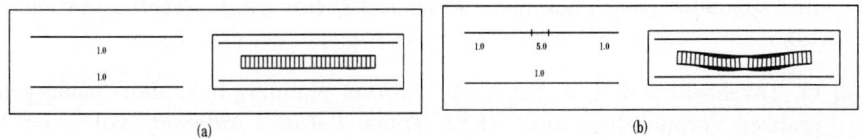

Figure 5: A path planning example. The obstacle boundaries are (a) uniformly charged, (b) non-uniformly charged.

5 Conclusions

In this paper, we propose a flexible potential field model of the workspace. The potential-based model assumes object and obstacle boundaries are charged and thus repelling one another, and special collision avoidance effects can be achieved by using some non-uniform charge distributions. The efficiency of the proposed approach is resulted from the derivation of analytic expressions of the repulsion between charged polygonal object and obstacles for more general path planning needs. A local planning procedure for deriving an object path of minimal potential based on its topological description is presented. According to the simulation results, proper selection of the charge distribution along object/obstacle boundaries can indeed improve the safety of the object path.

References

[1] J. H. Chuang and N. Ahuja, "An Analytically Tractable Potential Field Model of Free Space and Its Application in Obstacle Avoidance," to appear in *IEEE Trans. System, Man Cybern.*, vol. 28, Part B, no. 5, Oct. 1998.

[2] R. A. Brooks and T. Lozano-Perez, "A subdivision algorithm in configuration space for findpath with rotation," *Int. Joint Conf. Artificial Intelligence*, 1983.

[3] J. T. Schwartz and M. Sharir, "On the piano movers'problem: I. The case of a two-dimensional rigid polygonal body moving amidst polygonal barriers," *Comm. Pure Applied Math.*, vol. 36, pp. 345-398, 1983.

[4] M. Herman, "Fast, three-dimensional, collision-free motion planning," *Proc. IEEE Int. Conf. Robotics Automat.*, 1986.

[5] S. R. Maddila, "Decomposition algorithm for moving a ladder among rectangular obstacles," *Proc. IEEE Int. Conf. Robotics Automat.*, 1986.

[6] S. Singh and M. D. Wagh, "Robot path planning using intersecting convex shapes," *Proc. IEEE Int. Conf. Robotics Automat.*, 1986.

[7] D. T. Kuan, J. C. Zamiska, and R. A. Brooks, "Natural decomposition of free space for path planning," *Proc. IEEE Int. Conf. Robotics Automat.*, 1985.

[8] O. Takahashi and R. J. Schilling, "Motion planning in a plane using generalized Voronoi diagrams," *IEEE Trans. Robotics Automat.*, vol. 5, no. 2, pp. 143-150, 1989.

[9] C. E. Thorpe, "Path planning for a mobile robot," *Proc AAAI*, 1984.

[10] O. Khatib, "Real-time obstacle avoidance for manipulators and mobile robots," *Proc. IEEE Int. Conf. Robotics Automat.*, 1985.

[11] P. Khosla and R. Volpe, "Superquadric artificial potentials for obstacle avoidance and approach," *Proc. IEEE Int. Conf. Robotics Automat.*, 1988.

[12] Y. K. Hwang and N. Ahuja, "Potential field approach to path planning," *IEEE Trans. Robotics Automat.*, vol. 8, no. 1, pp. 23-32, 1992.

[13] J. Miura and Y. Shirai "An uncertainty model of stereo vision and its application to vision-motion planning of robot," *Proc. 13th Int. Joint Conf. Artificial Intelligence*, 1993.

A Linear Algorithm for Motion from Three Weak Perspective Images Using Euler Angles

Gang Xu and Noriko Sugimoto

Computer Vision Laboratory, Computer Science Department,
Ritsumeikan University, Kusatsu, Shiga 525, Japan

Abstract. In this paper we describe a new simple linear algorithm for motion and structure from three weak perspective projections using Euler angles. We first determine the epipolar equation between each pair of images, which determines the first and third Euler angles for the rotation between that pair of images, leaving only the second Euler angle undetermined. In the next step, combining the three rotations results in a very simple linear algorithm to determine the second Euler angles, up to a Necker reversal. Experimental results on synthetic and real images are presented. And the degenerate cases are discussed. The program can be ftped from http://www.cv.cs.ritsumei.ac.jp/noriko/motion.html

1 Introduction

It was shown by Ullman [7] that under orthographic projection 4 correspondences in 3 distinct images are needed to determine the motion and structure. Huang and Lee [2] developed a linear algorithm to solve this problem. Ostuni and Dunn [3] proposed a linear algorithm to determine motion from three weak perspective images, using the yaw-pitch-roll representation to parameterize the rotation matrix. Compared with the above two algorithms, the algorithm we propose here is more simple and all the results can be described in terms of Euler angles that can be easily understood. We will also discuss the strength and weakness compared with other motion recovery algorithms by Tomasi et al. and Shapiro et al [6, 4].

In our algorithm, the rotation matrix is parameterized by Euler angles. From each pair of images, we can determine the epipolar equation, which in turn determines the first and third of the three Euler angles. Having three distinct images enables us to determine all the second angles up to a sign, provided that the three images are not taken by coplanar cameras and the 4 points are not coplanar. Once the motion is determined, the structure can also be computed. After describing the algorithm, we will also show experimental results for both synthetic and real images.

2 Rotation Matrix Parametrized by Euler Angles

We assume that there are $n(n \geq 4)$ non-planar points in space. We also assume that there are three weak perspective images of these points and they are all matched. They are represented in the three camera coordinate systems as $X_i = [X_i, Y_i, Z_i]^T$, $X'_i = [X'_i, Y'_i, Z'_i]^T$, $X''_i = [X''_i, Y''_i, Z''_i]^T$, $i = 1, ..., n$, respectively. The relations between them can be described by

$$X_i = R_1 X'_i + t_1 , \quad X'_i = R_2 X''_i + t_2 , \quad X''_i = R_3 X_i + t_3 , \tag{1}$$

where R_1, R_2, R_3 are the rotation matrices, and t_1, t_2, t_3 are the translation vectors, respectively.

Each rotation matrix has 9 components, but the degree of freedom is only 3. There are different ways to parametrize the matrix. Ostuni and Dunn used roll-pitch-yaw, 3 rotation angles around the Z, Y, X axes [3]. Quaternions are also frequently used [1].

Let the three Euler angles for a rotation be (α, β, γ). Then the rotation matrix can be expressed as three consecutive rotations around the coordinate axes,

$$R = R(Z, \alpha) R(Y, \beta) R(Z, \gamma)$$
$$= \begin{bmatrix} \cos\alpha\cos\beta\cos\gamma - \sin\alpha\sin\gamma & -\cos\alpha\cos\beta\sin\gamma - \sin\alpha\cos\gamma & \cos\alpha\sin\beta \\ \sin\alpha\cos\beta\cos\gamma + \cos\alpha\sin\gamma & -\sin\alpha\cos\beta\sin\gamma + \cos\alpha\cos\gamma & \sin\alpha\sin\beta \\ -\sin\beta\cos\gamma & \sin\beta\sin\gamma & \cos\beta \end{bmatrix} \tag{2}$$

3 Determining the 1st and 3rd Euler Angles

For each pair of weak perspective images, e.g., the first and second images, we obtain an epipolar equation, which is in the form of,

$$f_1 x + f_2 y + f_3 x' + f_4 y' + f_5 = 0 , \tag{3}$$

where (x, y) and (x', y') are the pixel coordinates of the corresponding points. There are 4 independent parameters in this equation, and thus can be determined from mimimally 4 non-coplanar point matches. For how to optimally determine the epipolar equation from point matches, see [4, 8].

Let us assume that the horizontal and vertical image axes are mutual-orthogonal and that the horizontal and vertical spacings for each pixel is the same. Then the epipolar equation can be expressed as

$$-\sin\beta(x \sin\alpha - y \cos\alpha + sx' \sin\gamma + sy' \cos\gamma + t) = 0 , \tag{4}$$

where α and γ are the 1st and 3rd Euler angles, respectively (we omit the proof and the interested readers are referred to [4, 8]), s is the scale change between the two views due to change of distance or focal length of the camera, and t is a constant. Using (3) and (4), α, γ can be computed from f_1, f_2, f_3, f_4. Two sets of solutions are possible:

$$\alpha = atan2(f_1, -f_2) , \quad \alpha' = atan2(-f_1, f_2) ,$$
$$\gamma = atan2(f_3, f_4) , \quad \gamma' = atan2(-f_3, -f_4) . \tag{5}$$

It is easy to see that
$$\alpha - \alpha' = \pm\pi , \quad \gamma - \gamma' = \pm\pi . \tag{6}$$

The scale change s can also be determined as $s = \sqrt{\frac{f_1^2+f_2^2}{f_3^2+f_4^2}}$. It is noted that, $\mathbf{R}(Z,\alpha \pm \pi)\mathbf{R}(Y,\beta)\mathbf{R}(Z,\gamma \pm \pi) = \mathbf{R}(Z,\alpha)\mathbf{R}(Y,-\beta)\mathbf{R}(Z,\gamma)$. This means that $(\alpha \pm \pi, \beta, \gamma \pm \pi)$ is equivalent to $(\alpha, -\beta, \gamma)$. As will be clear later, both (α, β, γ) and $(\alpha, -\beta, \gamma)$ are the solutions of our linear algorithm (known as the Necker reversal). Therefore, it suffices to use any one solution of α and γ. Without loss of generality, we choose the value of α such that it is within $-\frac{\pi}{2} < \alpha < \frac{\pi}{2}$, and γ that is associated with this α.

4 Determining the 2nd Euler Angle

The 1st and 3rd Euler angles are the only two angles we can recover from two images. To determine the 2nd Euler angle, we need the 3rd image. Let the first and third Euler angles be $\alpha_1, \alpha_2, \alpha_3, \gamma_1, \gamma_2, \gamma_3$, respectively. We want to determine the 2nd Euler angles $\beta_1, \beta_2, \beta_3$. The 3 rotation matrices are not independent of each other. They are related by

$$\mathbf{R}_1\mathbf{R}_2 = \mathbf{R}_3^T, \mathbf{R}_2\mathbf{R}_3 = \mathbf{R}_1^T, \mathbf{R}_3\mathbf{R}_1 = \mathbf{R}_2^T. \qquad (7)$$

(Note that Ullman[7], Huang and Lee[2], and Ostuni and Dunn[3] all used this constraint.) The three equations are not independent. Any one is sufficient. Take the first one for example. Representing each matrix by 3 Euler angles, we have

$$\mathbf{R}(Z,\alpha_1)\mathbf{R}(Y,\beta_1)\mathbf{R}(Z,\gamma_1)\mathbf{R}(Z,\alpha_2)\mathbf{R}(Y,\beta_2)\mathbf{R}(Z,\gamma_2)$$
$$= (\mathbf{R}(Z,\alpha_3)\mathbf{R}(Y,\beta_3)\mathbf{R}(Z,\gamma_3))^T \qquad (8)$$

By defining $\omega_1 = \gamma_1 + \alpha_2, \omega_2 = \gamma_2 + \alpha_3, \omega_3 = \gamma_3 + \alpha_1$, Eq.8 can be rewritten as

$$\mathbf{R}(Y,\beta_1)\mathbf{R}(Z,\omega_1)\mathbf{R}(Y,\beta_2)\mathbf{R}(Z,\omega_2) = (\mathbf{R}(Y,\beta_3)\mathbf{R}(Z,\omega_3))^T \qquad (9)$$

For clarity, define $a_i = \cos\omega_i, b_i = \sin\omega_i, c_i = \cos\beta_i, d_i = \sin\beta_i$, for $i = 1, 2, 3$. a_i, b_i's are known, and c_i, d_i's are unknown. Expanding Eq.9 yields

$$\begin{bmatrix} a_1a_2c_1c_2 - b_1b_2c_1 - a_2d_1d_2 & -a_1b_2c_1c_2 - b_1a_2c_1 + b_2d_1d_2 & a_1c_1d_2 + d_1c_2 \\ b_1a_2c_2 + a_1b_2 & -b_1b_2c_2 + a_1a_2 & b_1d_2 \\ a_1a_2d_1c_2 + b_1b_2c_1 - a_2c_1d_2 & a_1b_2d_1c_2 + b_1a_2d_1 + b_2c_1d_2 & a_1d_1d_2 + c_1c_2 \end{bmatrix}$$
$$= \begin{bmatrix} a_3c_3 & b_2 & -a_3d_3 \\ -b_3c_3 & a_3 & b_3d_3 \\ d_3 & 0 & c_3 \end{bmatrix} \qquad (10)$$

The central components of the matrices on the left and right sides gives the simplest equation for c_2, and similarly for c_3, c_1,

$$-b_1b_2c_2 + a_1a_2 = a_3, \quad -b_2b_3c_3 + a_2a_3 = a_1, \quad -b_3b_1c_1 + a_3a_1 = a_2. \qquad (11)$$

We can solve for c_1, c_2, c_3 as

$$c_1 = \cos\beta_1 = \frac{a_3 a_1 - a_2}{b_3 b_1} = \frac{\cos\omega_3 \cos\omega_1 - \cos\omega_2}{\sin\omega_3 \sin\omega_1} , \quad (12)$$

$$c_2 = \cos\beta_2 = \frac{a_1 a_2 - a_3}{b_1 b_2} = \frac{\cos\omega_1 \cos\omega_2 - \cos\omega_3}{\sin\omega_1 \sin\omega_2} , \quad (13)$$

$$c_3 = \cos\beta_3 = \frac{a_2 a_3 - a_1}{b_2 b_3} = \frac{\cos\omega_2 \cos\omega_3 - \cos\omega_1}{\sin\omega_2 \sin\omega_3} . \quad (14)$$

Each of the β_i's has two solutions within $-\pi$ to π. The two solutions differs by a sign. c_i's solutions are not independent, since

$$b_1 d_2 = b_3 d_3 , b_2 d_3 = b_1 d_1 , b_3 d_1 = b_2 d_2 . \quad (15)$$

which are obtained from the second-row, third-column components of the rotation matrices. This means that we can have only two sets of solutions for $\beta_i, i = 1, 2, 3$, which differ by a sign. This is the well-known Necker reversal, which is confirmed by previous results of Ullman [7], Huang and Lee [2] and Ostuni and Dunn [3]. The Necker reversal can be shown by the fact that the following two equations always stand simultaneously,

$$\begin{bmatrix} X' \\ Y' \\ Z' \end{bmatrix} = R(Z,\alpha)R(Y,\beta)R(Z,\gamma) \begin{bmatrix} X \\ Y \\ Z \end{bmatrix}, \quad \begin{bmatrix} X' \\ Y' \\ -Z' \end{bmatrix} = R(Z,\alpha)R(Y,-\beta)R(Z,\gamma) \begin{bmatrix} X \\ Y \\ -Z \end{bmatrix}.$$

Our last note is about the numbers of unknowns and constraints. For any two motions, we have 6 independent rotation parameters. And all what we can recover between any two images is 2 rotation parameters, totally 6, as there are 3 image pairs for 2 motions. Thus the number of unknown motion parameters exactly equals the number of constraints. There is neither redundancy nor lack in information. Actually, there are totally 27 equations in (7), but if we substitute (14) and (15) for them, all disappear.

5 Degenerate Cases and Robustness

There are three degenerate cases in this algorithm.

(1) All points $X_i, i = 1, ..., n$ are coplanar. In this case, epipolar equation cannot be determined. For more details, see [8]. This fails any algorithm using epipolar equation.

(2) One of the second Euler angles β is zero. In this case, the rotation is only within the image plane, and images carry no information on the depth. Also the epipolar equation cannot be determined. For more details, see [8]. This fails any algorithm using epipolar equation.

(3) b_i's are zero. This corresponds to the case where the motions between the cameras are within a single plane, or the rotation is around a single axis between the 3 images. If one of b_i's is zero, then from (15), all other two become zeros, too. Therefore, other algorithms are needed to deal with this case. For a discussion, see [2].

The robustness of recovery depends on how far away the geometry is from the degenerate cases. The first two criteria are valid for every motion recovery algorithm. In comparison with algorithms that use long image sequences rather a few distinct images [5, 4], the final accuracy mainly depends on how large the second Euler angles are, but not the number of images used, while a large number of images can only improve accuracy over the error in feature point detection. As another comparison, it is of course relatively easier to match points over a few images than to match points over a long sequence.

In this sense, the algorithms using distinct images are expected to perform comparably with those using image sequences.

6 Determining the Structure

Once the rotation is recovered, the structure of the 3D points can also be determined up to a scale factor and up to a Necker reversal.

Without loss of generality, assume that the scale of the scene is the same as the first image. Since scales of the other two images are known with respect to the first image, they can be normalized to be of the same scale as the first image. Then they are equivalent to three orthographic projections.

Also assume that the 3D structure is registered in the same coordinate system as the camera taking the first image. Let the rotations be

$$\mathbf{R}_{11} = \mathbf{I}, \ \mathbf{R}_{12} = \begin{bmatrix} r_{11} & r_{12} & r_{13} \\ r_{21} & r_{22} & r_{23} \\ r_{31} & r_{32} & r_{33} \end{bmatrix}, \ \mathbf{R}_{13} = \begin{bmatrix} r'_{11} & r'_{12} & r'_{13} \\ r'_{21} & r'_{22} & r'_{23} \\ r'_{31} & r'_{32} & r'_{33} \end{bmatrix}.$$

To remove the effect of translations, we move image coordinate origins to the centroids of the matched feature points, which are the projections of the corresponding space points under weak perspective projection.

Putting the reallocated image points into a single matrix, we have

$$\begin{bmatrix} \mathbf{x}_{i1} \\ s_{12}\mathbf{x}_{i2} \\ \frac{1}{s_{31}}\mathbf{x}_{i3} \end{bmatrix} = \mathbf{A}\mathbf{X}_i, \text{ where } \mathbf{A} = \begin{bmatrix} 1 & 0 & 0 \\ 0 & 1 & 0 \\ r_{11} & r_{12} & r_{13} \\ r_{21} & r_{22} & r_{23} \\ r'_{11} & r'_{21} & r'_{31} \\ r'_{12} & r'_{22} & r'_{32} \end{bmatrix}.$$

$\mathbf{X}_i$ can be computed as $\mathbf{X}_i = (\mathbf{A}^T\mathbf{A})^{-1}\mathbf{A}^T \begin{bmatrix} \mathbf{x}_{i1} \\ s_{12}\mathbf{x}_{i2} \\ \frac{1}{s_{31}}\mathbf{x}_{i3} \end{bmatrix}$.

7 Experimental Results and Discussions

The algorithm has been tested with both synthetic and real images. Synthetic images are used for evaluation of the performance of the algorithm. A cube with each side 20 units long, is used as the object, with its 8 corners perspectively projected onto the images. For the first image, the camera coordinate system is in the same pose as the object coordinate system. The second image is synthesized after the object undergoes a rotation whose Euler angles are $(10°, 30°, 45°)$. The third image is synthesized after the object undergoes a further rotation whose Euler angles are $(10°, 30°, 45°)$. Gaussian noise of variance of 1 pixel is added to the computed image coordinates of all image points.

To see how well the computed result is close to the actual values, we list in Table 1 the recovered Euler angles for the two cases of $Z_c = 40$ and $Z_c = 200$, where Z_c indicates the distance from the camera optical center to the object. Note we only list one of the two solutions for the Euler angles.

Table 1. Recovered Euler angles (in degrees) by our algorithm ($Z_{c1} = Z_{c2} = Z_{c3}$)

	α_1	β_1	γ_1	α_2	β_2	γ_2
actual	10.000	30.000	45.000	10.000	20.000	-30.000
computed ($Z_c = 200$)	9.948	30.021	45.001	9.844	20.056	-29.799
computed ($Z_c = 40$)	10.873	24.146	43.458	6.026	18.271	-24.987

From the table, we can see that the Euler angles recovered for $Z_c = 200$ is very accurate, with errors in all angles less than $0.3°$. But when Z_c is as short as 40, the computed results deviate from the actual values quite much. To show how these values change in between the two extremes, we prepared 2 graphs, the first for the second Euler angle, and the second for the energy measuring the overall error in structure recovery. The Z_c changes from 40 to 200, by a step of 20. We can see from the graphs for the angles that the accuracy no longer improves after Z_c grows to 140. The reason is considered to be that the error due to the weak perspective projection model is no longer the main factor, while the Gaussian noise and the digitization error become dominant. As for the structure, the accuracy continues to improve till $Z_c = 200$, but at a very slow rate after $Z_c = 140$.

Finally, we show two examples using real images. The first example is a soccer ball, and the second a cube, with images shown in Fig.2. In the second example, the images are taken with the rotations being approximately around a fixed axis.

As we do not know the exact movements of the camera, we cannot compare the true rotation angles with the computed values. We only show the recovered shape of the soccer ball. Three views of the shape are given in Fig.3. These figures give a sense of how the recovered result is close to the ideal case. From them we can see the results are not only good for the soccer ball but surprisingly also for the cube.

The computed ω_i's for the soccer ball are $\omega_1 = -60.7°, \omega_2 = 30.9°, \omega_3 = 34.4°$, which are far from zero. And computed ω_i's for the cube are $\omega_1 = 1.80°, \omega_2 = -0.95°, \omega_3 = -0.97$, which are very close to zero. To look for reasons for the unexpected success, we go back to the equation (14), and take an

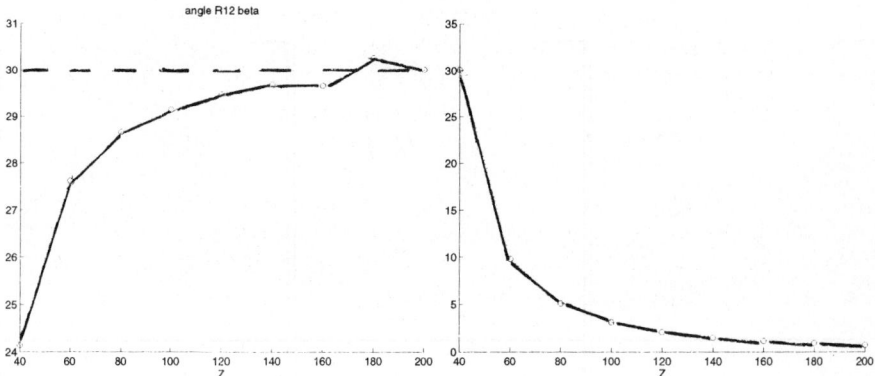

Fig. 1. The changes of the 2nd Euler angle (left) and the error in 3D positions (right) as Z_c changes from 40 to 200 units. The horizontal dotted line shows the actual value.

Fig. 2. 3 views of a soccer ball (upper row) and 3 views of a cube (lower row)

approximation as the ω_i's are very small, obtaining

$$\cos\beta_1 \approx \frac{1}{2}\left(\frac{\omega_2}{\omega_1}\frac{\omega_2}{\omega_3} - \frac{\omega_1}{\omega_3} - \frac{\omega_3}{\omega_1}\right)$$

(and two simillar expressions for the other equations) which depends on the ratio of ω_i's, but not the values themselves. As discussed earlier, if one of the ω_i's goes to zero, all others go to zero together.

8 Summary

We have in this paper presented a linear algorithm for recovering motion from 3 weak perspective images, which is more straightforward than the previous ones,

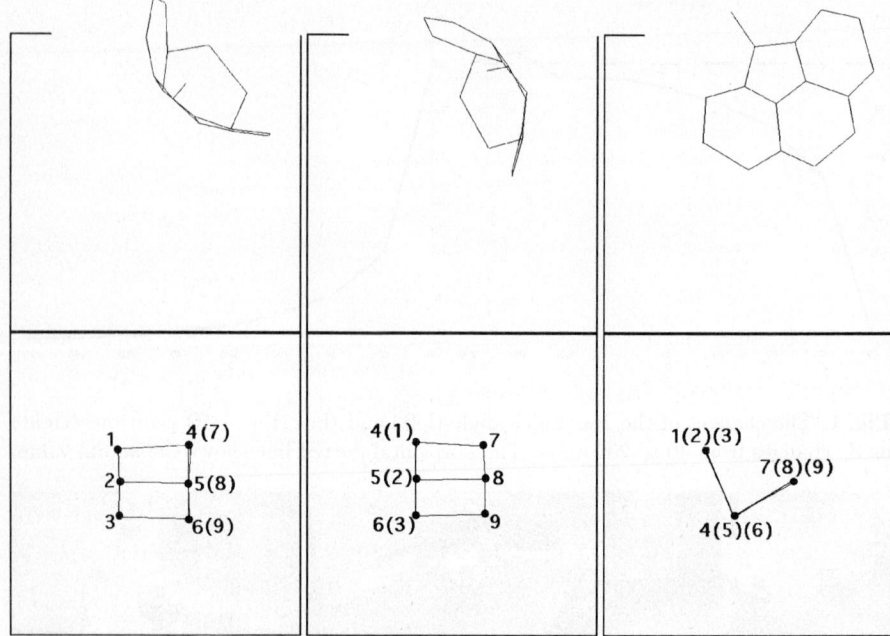

Fig. 3. Upper row from left to right: A left, right and top view of the recovered soccer ball. Lower row from left to right: A left, right and top view of the recovered cube.

and is thus easier to implement. Experimental results on synthetic and real images are presented, and discussions are given on the strength and weakness of this algorithm.

References

1. O. D. Faugeras. *Three-Dimensional Computer Vision: A Geometric Viewpoint*. MIT Press, Cambridge, MA, 1993.
2. T.S. Huang and C.H. Lee. Motion and structure from orthographic projections. *IEEE Trans. PAMI*, 11:536–540, 1989.
3. J. Ostuni and S. Dunn. Motion from three weak perspective images using image rotation. *IEEE Trans. PAMI*, 18(1):64–69, 1996.
4. L.S. Shapiro, A. Zisserman, and M. Brady. 3d motion recovery via affine epipolar geometry. *Int'l J. Comput. Vision*, 16:147–182, 1995.
5. C. Tomasi and T. Kanade. Shape and motion from image streams under orthography: a factorization method. *Int'l J. Comput. Vision*, 9(2):137–154, 1992.
6. P.H.S. Torr and D.W. Murray. Statistical detection of independent movement from a moving camera. In *Proc. British Machine Vision Conf.*, pp.79–88, 1992.
7. S. Ullman. *The Interpretation of Visual Motion*. MIT Press, Cambridge, MA, 1979.
8. G. Xu and Z. Zhang. *Epipolar Geometry in Stereo, Motion and Object Recognition: A Unified Approach*. Kluwer Academic Publishers, 1996.

On Learning Spatio-Temporal Relational Structures in Two Different Domains

Adrian R. Pearce[1], Terry Caelli[1], and Simon Goss[2]

[1] School of Computing, Curtin University, Perth WA 6845, Australia
[2] Aeronautical and Maritime Research Laboratories, DSTO, Melbourne, Australia.

Abstract. In this paper we consider the types of representations and learning procedures required to construct rules which can adequately describe relational information as it occurs in spatio-temporal sequences. A comparison of interpreting on-line hand drawings is made to the automatic generation of flight manoeuvre description based on a relational learning system we have developed, the Consolidated Learning Algorithm based on Relational Evidence Theory (CLARET). The package adapts relational learning techniques to utilise the constraints present in time series data. Our approach involves supporting queries, automatic descriptions and/or predictions from spatio-temporal action sequences.

1 Introduction

We describe a systematic approach for automatically generating human-readable descriptions (or prescriptions) from the relational structures present in spatio-temporal sequences. For example in describing flight, an approach to land manoeuvre is defined by the subsequence of different roll–pitch–yaw states of the aeroplane and different actions on the control yoke. On-line hand-drawn schematic diagrams are defined by different strokes, drawn at different times at particular orientations. Relational representation and the associated graph matching approach is a powerful architecture for interpreting visual and temporal information. In schematic hand-drawn domains it has been used for pattern recognition using graph matching and subgraph isomorphism [7]. We have developed a software package—the Consolidated Learning Algorithm (CLARET) [8]—which is based on Relational Evidence Theory [3]. The package adapts relational learning techniques [9,2] to utilise the constraints present in time series data—those of states and their continuous valued, attributed relationships (the scenery) and actions or designs (the scenario). Relational rules are generated which explicitly depict actions and relationships between states of the form,

WHILE *interpreting (or intending to achieve) goal$_i$*
  IF *this state and that state have these relationships in space* AND
    this action and that action have those relationships in time AND ...
  THEN *describe (or prescribe) sub-goal$_j$ at time t.*

In the on-line hand drawn schematic system based on CLARET [8], unconstrained schematic drawing is allowed, that is, symbols can be input with

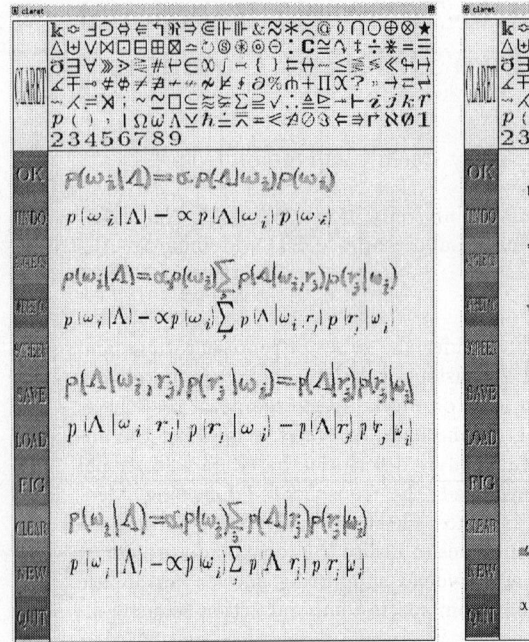 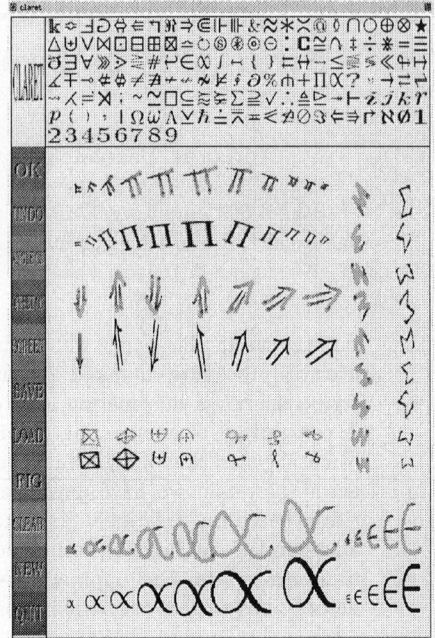

Fig. 1. Left: Scientific Symbols Application: The application window is shown comprising of 128 scientific symbols. Symbols are drawn on the central canvas and matched interpretations are back-projected by calculating orientation parameters (shown here using vertical offset). The interpretations using CLARET are shown. **Right:** Rotation, scale, and shift invariance: Symbols are shown drawn at different orientations. Matched interpretations are projected back by calculating rotation, scale, and shift orientation parameters.

no constraints on rotation, scale, or positioning. A vocabulary of symbols is provided by pasting their images across the top of the application window. Handwritten "components" are defined using an on-line digitiser which produces time-stampedposition and pressure points from pen-down to pen-up condition. Matched interpretations are projected back by calculating rotation, scale, and shift orientation parameters (see Figure 1). In the CLARET algorithm, an unknown segmented and labelled trajectory is presented to the system together with n known trajectories. First, relational descriptions are generated by extracting relationships between pattern segments Pen trajectories are first segmented into line-based descriptions using a hierarchical multi-scale polygonal approximation method [11]. Features are then extracted from individual states in continuous numerical form (see Figure 2). Second, the patterns are matched using the attribute values and labelled parts from relational descriptions. This involves the repetition of the following steps: attribute generalisation, graph matching and relational specialisation.

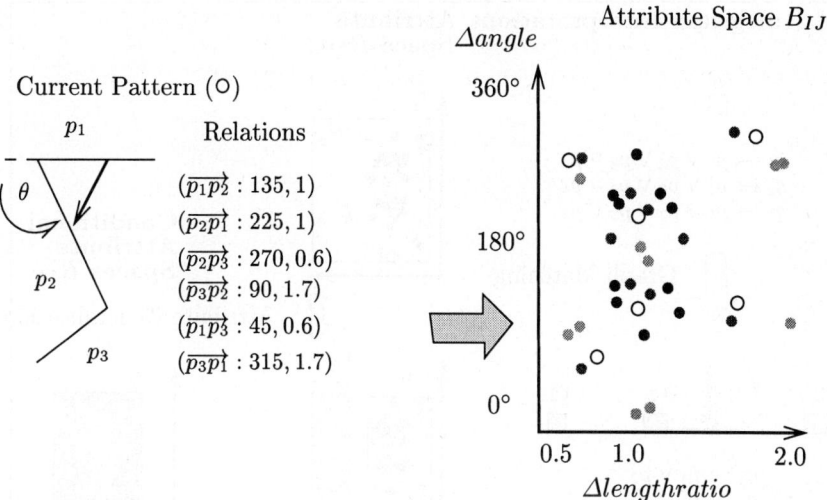

Fig. 2. Generating binary relations: For all existing (known) patterns, relations are generated which form directed, planar graph relational structures and form an attribute space B_{IJ} (closed circles). The relations $(\overrightarrow{p_Ip_J} : \Delta ANG, \Delta LEN)$ represent difference in angle (ΔANG) and length ratio (ΔLEN). Relations are also generated for the current (unknown) pattern (open circles) and are included into attribute space B_{IJ} (bottom right hand side).

The method used for generalising over attributes is based on the attribute selection and splitting technique used in decision trees, notably C4.5 [10], except that the information metric is replaced by a variance criterion. Attributes are partitioned into regions and each split results in two new regions defined by the conjunctions of attribute bounds: rules (r_i), see Figure 3 (right hand side). A least general technique is used which approximates K-means splitting [5] and minimises the attribute range of each new rule by maximising the split between rules. Each partitioned rule is replaced by two new least general rules (see Figure 3 (top middle), rule r_0 is replaced with rules r_1 and r_2). Such hashing of the observed relational attributes provides initial hypothesis about the resolution of parts required to recognise and discriminate between the known patterns.

The question can now be asked: are these rules specific enough to differentiate between the different relational structures present to correctly interpret the current pattern? Graph matching is used to solve this problem of mapping current (unknown) pattern parts to existing (known) pattern parts, by finding compatible sets of labelled rules. Inter-rule dependencies are determined for subgraphs of rules, Rulegraphs [3]. Rulegraphs are formed when labelled relations instantiate rules according to their attribute values. Two rules r_i and r_j are connected if there exists instantiations $r_i(p_Ip_J)$ and $r_j(p_Jp_K)$, such that part p_J is shared. For example, relation $\overrightarrow{p_1p_2}$ (Figure 2) instantiates rule r_1 to form $r_1(\overrightarrow{p_1p_2})$ (Figure 3). Part mappings between current and existing patterns ($q_J \mapsto p_J$) are

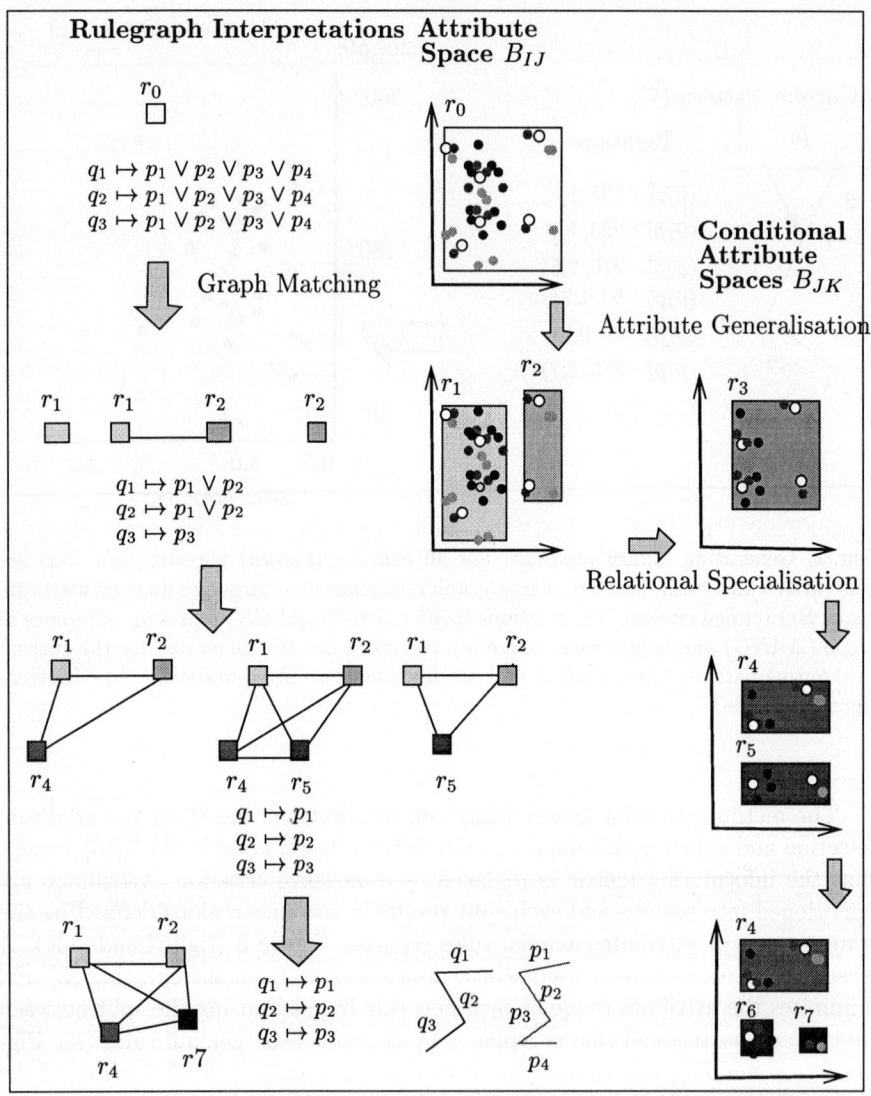

Fig. 3. The search for the interpretation of the example in Figure 2 is shown here, which involves successive applications of attribute generalisation, graph matching and relational specialisation (right hand side). A queue contains rulegraph interpretations of the current pattern in terms of all existing patterns. During the search, mappings between parts in the current pattern (open circles) are solved with respect to the existing patterns (closed circles), the shades of grey in the rules (vertices) correspond to the degree of specialisation (left hand side). A relational evidence network is used to prune the search space while guaranteeing a finite interpretation (see text for details).

solved by checking the existence of common parts in rules. This relies on checking the compatibility between rules based on the consistency of the mapping states in the current pattern with respect to the existing pattern. Rulegraph interpretations are built up by testing compatibility of new (candidate) rules r_c for possible addition to the set of already existing rules. Interpretations are made by mapping parts which instantiate rules for the current pattern with respect to each existing pattern. Initially, parts are mapped from current patterns to existing patterns for parts only in the root rule r_0. The mapping will initially be many to many, in the sense that rule r_0 is completely general (all nodes of the relational graph have the same 'colour') and has not yet been specialised, partitioned by the attribute learner.

The purpose of relational specialisation is to generate *conditional* rules by adding literals to the current clause. This creates paths through the relational structures. The method is based on Conditional Rule Generation [1] which allows for (intra-rule) dependencies between attribute states to be represented by creating paths through relational descriptions. Conditional attribute spaces B_{JK} are formed by traversing paths via relations in attribute space B_{IJ} to relations in attribute space B_{JK}. Relational specialisation forms conditional attribute spaces $B_{JK} = \{r_j(\overrightarrow{p_J p_K}) \mid r_i(\overrightarrow{p_I p_J}) \in B_{IJ}\}$ for each of the rules $r_i \in B_{IJ}$. Connected paths are traversed from instantiations $r_i(\overrightarrow{p_I p_J})$ to relations $\overrightarrow{p_J p_K}$ (via part p_J). Labelled, first order rules are generated in the form,

$$r_i(\overrightarrow{p_J p_K}, A_1, A_2, \ldots) \longleftarrow r_j(\overrightarrow{p_I p_J}, A_1, A_2, \ldots), 0.25 \geq A_1 \leq 0.95, \ldots$$

where p_I are part variables and A_i are attributes. Note that label compatibility is represented via paths through relational structures of arbitrary arity $p_I p_J \to p_J p_K \to \ldots$, which instantiate conditional rules $r_i, r_j, \ldots$ hierarchically.

To determine the best interpretation an evidential measure is required. A relational evidence network is used to impose an ordering of interpretations during the search, which maximally prunes the search space and thus determines the best interpretation in optimal time. In order to represent uncertainty in the CLARET hierarchy of rules, a model must be used which captures the dependency between specific instantiations (labelled parts) and rules (attribute states). In CLARET there is clear dependency of parts via the instantiation of conditional rules with labelled relations, and so a Bayesian network model can be formulated to include both attribute and part indexing. Each pattern ω_p is from an existing (though updatable) closed world of patterns $\omega_p \in \Omega$. Patterns give rise to relations $(\overrightarrow{p_I p_J} : A_1, \ldots, A_n) \in \Lambda$, where Λ corresponds to all relations, with unique parts p_I and p_J and attribute values, $A_1, \ldots, A_n$. Attribute values are partitioned into rules r_i which represent all possible attribute states $r_i \in \Pi$. This allows for hierarchical modelling of such processes by

$$p(\omega_p \mid \Lambda) = \sum_{i,j} \sum_{I,J,K} \prod_{I,J,K} p(r_j(\overrightarrow{p_J p_K}) \mid \overrightarrow{p_I p_J}, \omega_p) p(\overrightarrow{p_I p_J} \mid r_i(\overrightarrow{p_I p_J}), \omega_p)$$

for relations $\overrightarrow{p_I p_J}, \overrightarrow{p_J p_K} \in \Lambda$ and where rules r_i and r_j are from conditional attribute spaces $r_i \in B_{IJ}$ and $r_j \in B_{JK}$. The CLARET algorithm is designed to maximise $p(\omega_p \mid \Lambda)$ in Equation 1 while minimising the right hand side with respect to least generalisation of attributes. The formulation relates to the class of exact solutions to Bayesian Networks [6] allowing for analytical determination of probabilities during learning. The relational evidence theory techniques have been shown to compare favourably in a number of empirical comparisons to other learning methods such as Neural Networks for three dimensional object recognition [3] and schematic interpretation [8].

2 Recognising Flight Manoeuvres

Over the past few years a new class of flight simulator has been developed which has the ability to not only record the actions of pilots, instrument and simulation internal status variables, but also the time-stamped positions of objects, and dynamic entities in the three dimensional flight course relative to the pilot. In our on-going work, the "behavioural clones" approach of Sammut and co-workers is extended to include dynamic knowledge of the world [12]. In addition to the status of navigation instruments and past actions, pilots use knowledge of the world, both in own-ship (egocentric or view-dependent) and map-view (exocentric or view-independent) representations. Such information is critical in control and trajectory planning in the visual flight regime. In the agent oriented Smart Whole Air Mission Model (SWARM) decision support system, the binding of the physical process of flight (responses) with the symbolic process of tactics selection (plans) is required. This system is based on the procedural reasoning system architecture [4]. During training of the flight simulation system, the pilot selects a particular manoeuvre from an ontology of different manoeuvres and records the beginning and end of each manoeuvre and event. The aim, here, is to predict pilot actions, in a given time interval, from pilot actions, instrument settings, object co-ordinates and near-object characteristics at previous time intervals. Figure 4 shows the process of learning action in sequences of input trajectory using the CLARET learning algorithm. In recognition mode, the system dynamically binds to different manoeuvres as they occur in the input time series on-line. The system can be interactively used to either query partially enacted sequences in predictive mode or describe sequences in descriptive mode. Descriptions of manoeuvres are generated which define action sequences of the form,

WHILE *in the context of LEVEL-LEFT-TURN*
   IF *STICK-LEFT before INIT-BANK* AND
      MOUNTAIN-IN-VIEW before STICK-RIGHT AND ...
   THEN *believe intention is to TURN-TO-CROSSWIND-LEG.*

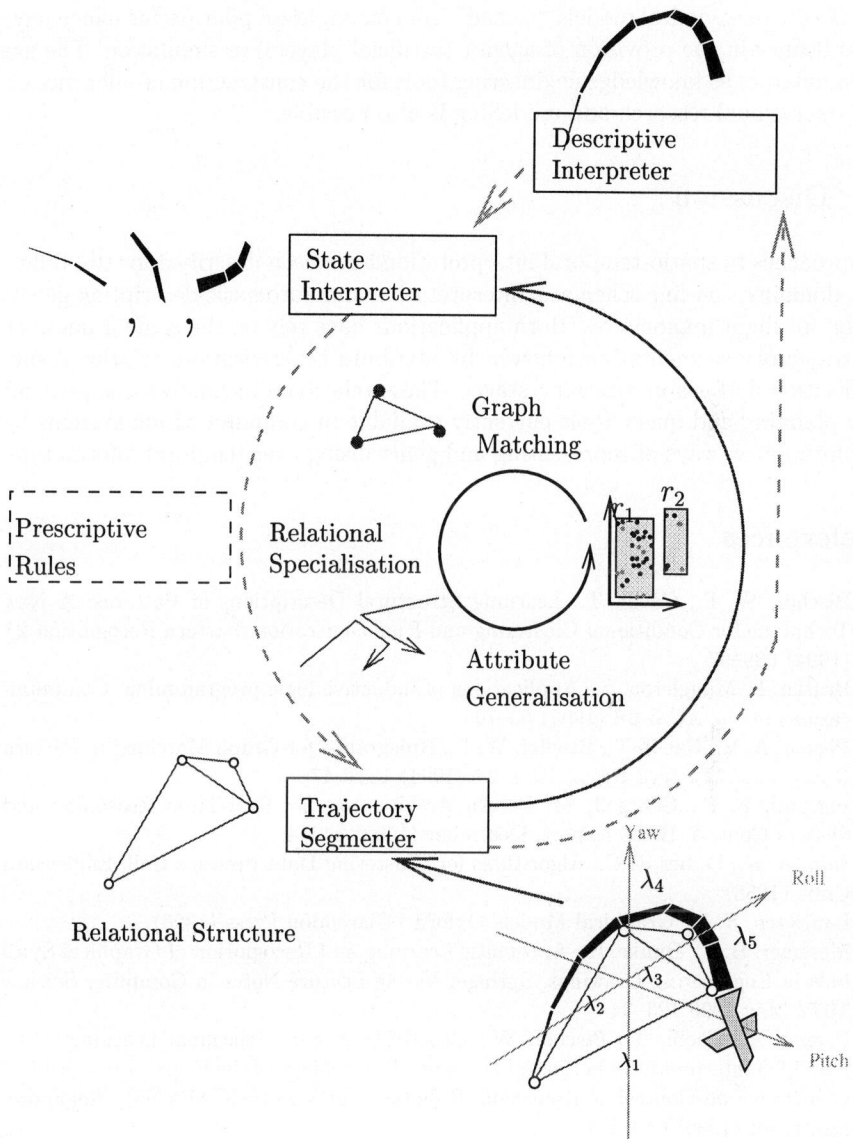

Fig. 4. Spatio-temporal Interpretation using the Consolidated Learning Algorithm (CLARET): Input flight trajectory (numeric roll-pitch-yaw and object relationships) from the flight simulator is collected and segmented into line-based descriptions. Similar segmentation and relational attribute extraction is carried out on the time-series trajectories obtained from each separate domain, e.g. controls, plane and objects. Successive applications of the matching algorithm result in both scene interpretations (states) and hierarchical scenario interpretations (actions). Hierarchical interpretation involves the successive application of attribute generalisation (numerical learning), graph matching and relational specialisation.

These behavioural models "cloned" from examples of pilot performance have significance in the provision of agency (artificial players) in simulation. The use of simulators as knowledge engineering tools for the construction of pilot models for operational research and modelling is also possible.

3 Discussion

Approaches to spatio-temporal interpretation have been described for two different domains—on-line schematic interpretation and automatic description generation for flight manoeuvres. Both applications have rely on the same framework of trajectory segmentation followed by attribute generalisation, relational specialisation and graph matching stages. These relational techniques complement the planning and query tools currently available in computer vision systems by exploring new ways of representing and generalising over temporal information.

References

1. Bischof, W. F., Caelli, T.: Learning Structural Descriptions of Patterns: A New Technique for Conditional Clustering and Rule Generation. Pattern Recognition **27** (1994) 689–97
2. Bratko, I., Muggleton, S.: Applications of inductive logic programming. Communications of the ACM **38** (1995) 65–70
3. Pearce, A. R., Caelli, T., Bischof, W. F.: Rulegraphs for Graph Matching in Pattern Recognition. Pattern Recognition **27** (1994) 1231–47
4. Ingrand, F. F., Georgeff, M. P.: An Architecture for Real-Time Reasoning and System Control. IEEE Expert, December (1992) 34–44
5. Jain, A. K., Dubes R. C.: Algorithms for Clustering Data Prentice Hall, Englewood Cliffs (1988)
6. Lauritzen, S. L.: Graphical Models Oxford : Clarendon Press (1996)
7. Messmer, B. T., Bunke, H.: Automatic Learning and Recognition of Graphical Symbols in Engineering Drawings. Springer Verlag Lecture Notes in Computer Science **1072** May 1996 123–34
8. Pearce, A., Caelli, T., Bischof, W.: CLARET: A new Relational Learning Algorithm for Interpretation in Spatial Domains. Proceedings of the Fourth International Conference on Control, Automation, Robotics and Vision (ICARV'96), Singapore, December (1996) 650–654
9. Quinlan, J. R.: Learning logical definitions from relations Machine Learning **5** (1990) 239–66
10. Quinlan, J. R.: C4.5 Programs for Machine Learning Morgan Kaufmann (1993)
11. Rocha, J., Pavlidis, T.: A shape analysis model with applications to a character recognition system IEEE PAMI **16**:4 (1994) 393–404
12. Shirazi, G. M., Sammut, C.: An Interactive Method for Learning to Control Dynamic Systems Proceedings of the Knowledge Acquisition Workshop 1996 (KA96), Sydney, Australia (1996) 60–76

An Efficient Iterative Pose Estimation Algorithm

S.H. Or[1], W.S. Luk[2], K.H. Wong[1] and I. King[1]

[1] Department of Computer Science & Engineering, The Chinese University of Hong Kong, Shatin, N.T., Hong Kong
[2] Departement Computerwetenschappen, Katholieke Universiteit Leuven, Celestijnenlaan 200A, B-3001 Heverlee, Belgium

Abstract. We propose a novel model-based algorithm which finds the 3D pose of an object from an image by breaking down the estimation process into two linear processing stages, namely the depth recovery and the pose calculation. The depth recovery stage determines the new positions of the model point set in 3D space whereas the pose calculation step is a least-square estimation of the transformation parameters between the point set formed from the previous stage and the model set. The estimates are iteratively refined until converged. The advantage of using our algorithm is that the computational cost is much reduced. We test our algorithm by applying it to both synthetic as well as real time head tracking problem with satisfactory results.

index : Pose Estimation, Real time vision, Human-computer Interface.

1 Introduction

Determining the pose (position and orientation) of a moving object from an image sequence is useful in applications such as photogrammetry, passive navigation, industry inspection and human-computer interfaces, etc. In this paper, we are interested in the model-based pose estimation algorithm. A number of work have been done by previous researchers [13, 7, 6, 4]. Some of them are optical flow based [6], which require massive computation power and are not suitable for real time applications. Others use nonlinear iterative methods such as Newton's method, which is sensitive to the initial guess supplied. Our interest is in developing an efficient algorithm which is suitable for real time applications on human motion analysis. This area is characterized by the rapid motion of the object concerned and high probability of occlusion/re-appearance of part of the object in the subsequent frames.

2 Problem formulation

We take the assumption of the pin-hole camera model with full perspective projection as shown in Fig. 1. f is the focal length of the camera and the focal plane of the camera is placed at distance f from the camera origin. The problem of pose estimation can be described as follows. Given a 3D point set $\{P_i\}, i \in 1, 2, \ldots, N$ where $P_i = (X_{P_i}, Y_{P_i}, Z_{P_i})^t, i \in 1, 2, \ldots, N$ are the coordinates of the model points at a reference instant. Assuming a rigid transformation is being applied to this point set, yielding a new set which then projected onto an image plane, producing $\{Q_i\} = (X_{Q_i}, Y_{Q_i}), i \in 1, 2, \ldots, N$, we seek R, T and $\{d_i\}, i \in 1, 2, \ldots, N$ such that

$$e^2(R, T, \{d_i\}) = \min \left(\sum_{i=1}^{N} \|d_i \hat{\nu}_i - (R \cdot P_i + T)\|^2 \right) \qquad (1)$$

e^2 is a measurement function of the current fit of rotation R, translation T, and $\{d_i\}$. d_i corresponds to the depth which determines where in 3D space should the actual point be located along the projection ray. The unit vector on this ray $\hat{\nu}_i$ is given by:

$$\hat{\nu}_i = (X_{Q'_i}^2 + Y_{Q'_i}^2 + f^2)^{-\frac{1}{2}} (X_{Q'_i}, Y_{Q'_i}, f)$$

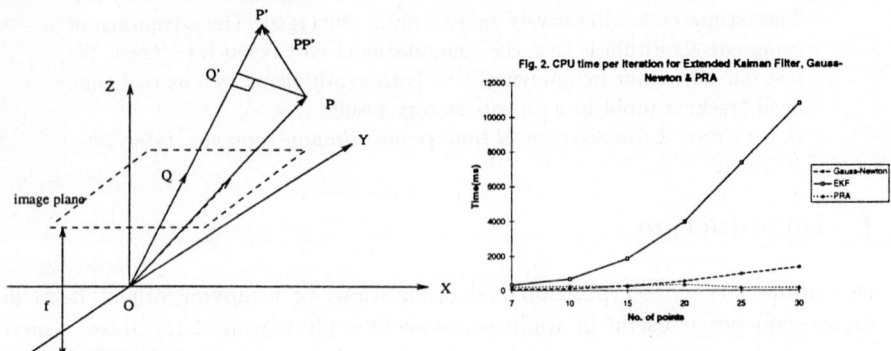

Fig. 1 Relationship between original point and projected point

Fig. 2 CPU time per iteration for Extended Kalman Filter, Gauss-Newton and PRA

In the formulation above, least square minimization methods can be applied to the $N + 6$ parameter space to find the minimum of the measurement function e^2 where N is the number of feature points of the object. This approach is adopted by [8, 11]. Since the original 3D coordinates are being transformed by a perspective projection, the objective function is thus nonlinear. To solve for the solution, iterative methods are required which is usually a time consuming process. In addition, the computational effort needed increases rapidly as the number of points increases.

3 Our Algorithm

Recently there are some work which focus on using the projection rays as a guide and perform the pose estimation in 3D space [12]. However, the approach used the iterative point matching method, which demands for much computational effort. This will not suit the requirement of a real time computer-human interface, which is what we are interested in. Inspired by the inverse projection ray approach, we propose that the minimization process be broken down into two stages: The first stage will estimate the position of all the feature points in the 3D space, i.e. $\{d_i\}$. The estimated point set will be passed to the second stage, which is a least square fitting of the model set and estimated point set. The above procedure is repeated until the result converges. By dividing the estimation process into two stages, the size of the solution space in each stage is much reduced and the cost in locating the solution decreased significantly. The resulting algorithm is very efficient. Our algorithm basically works in the following way:

1. Assume the object has no motion, determine $\{d_i\}$, $i \in 1, 2 \ldots, N$ and T.
2. Assume $\{d_i\}$, $i \in 1, 2, \ldots, N$ are determined, estimate R and T.
3. Update the state of the solution and iterate until the values of R and T less than some thresholds.

Note that we estimate both $\{d_i\}$ and T in the first stage. The point of recovering T earlier is to make our algorithm more robust since leaving T to be determined in later stage will easily lead to a result of stucking in the local minimum as indicated in some of our experiments. In the second stage, we will minimize the following objective function

$$e^2(R,T) = \sum_{i=1}^{N} \|Q'_i - (R \cdot P_i + T)\|^2 \qquad (2)$$

where Q'_i is the estimated transformed point set in the previous estimation stage, P_i is the original (model) point. Various efficient algorithms are available for the fitting of two point sets [5, 1, 9]. We choose the singular value decomposition (SVD) method [1, 9] due to its robustness in noise handling and that only a 3×3 matrix decomposition procedure is needed.

The complete algorithm is as follows:

1. Minimize the function below to estimate $\{d_i\}$, $i \in 1, 2, \ldots, N$:

$$\sum_{i=1}^{N} \|d_i \hat{\nu}_i - (P_i + T)\|^2$$

The resulting $\{d_i\}$, $i \in 1, 2, \ldots, N$ and T is given by:

$$d_i = \hat{\nu}_i^t (P_i + T) \qquad (3)$$

$$T = -\left(\sum_{i=1}^{N} A_i\right)^{-1} \left(\sum_{i=1}^{N} A_i P_i\right) \qquad (4)$$

where A_i is a 3 by 3 matrix given by

$$A_i = I - \hat{\nu}_i \hat{\nu}_i^t$$

Please refer to appendix for the derivation of the above formula.

2. Using the estimated $\{d_i\}$, $i \in 1, 2, \ldots, N$ in the previous stage, apply the SVD method to determine the rigid transformation R and T.
3. Update P_i, $i \in 1, 2, \ldots, N$ by $P_i \leftarrow R \cdot P_i + T$.

Note that A_i's are not needed to be stored explicitly. Firstly, each A_i depends only on $\hat{\nu}_i$, which is fixed before the execution of our algorithm. Therefore the value of $(\sum_{i=1}^{N} A_i)^{-1}$ in Eq. (4) can be pre-computed before the iterations, which avoids the expensive inverting operation of a matrix. Moreover, the matrix vector product $A_i P_i$ can be computed efficiently using the formula $P_i - \hat{\nu}_i(\hat{\nu}_i^t P_i)$. We also noticed the algorithm by DeMenthon and Davies [4] which requires relatively economic arithmetic operations. However, our argument for the efficiency is still valid by the following reasons. Firstly, in the approach used by DeMenthon and Davies, the algorithm requires an approximate pose estimate from the previous stage, which would inevitably contribute to the computational requirement. Moreover, their method does not guarantee the orthonormality of the resulting rotation matrix. Whereas our algorithm used the singular value decomposition, which provides an orthonormal result, hence improves the accuracy and thus justifies the increase in computation.

4 Performance comparison

We have compared the performance of two established approaches with our algorithm. The main interest here is the computational efficiency since the main concern during the motion tracking application would probably be the timing requirement as well as accuracy and stability. We chose Gauss-Newton method (by Lowe [7]) and Extended Kalman Filter (by Broida et al. [3] and Azarbayejani et al. [2]) method due to their robustness and computational efficiency. Lowe's approach, the Extended Kalman filter as well as our algorithm were implemented in C.

All the simulations are performed on a SUN Ultra 1/170 workstation. The same data set was used which involved both translation and rotation of two to four planes in 3D space (10 points per plane are used). The motion trajectory consists of increasing values in both rotation and translation for 100 frames. Monte Carlo simulation was applied to the data and Gaussian random noise with zero mean and standard deviation of one unit is added to the coordinates of each feature point. The plots of average time per frame used by various approaches are shown in Fig. 2. In the plot, our algorithm is represented using the title "PRA" which stands for *"Projection Ray Attraction"*. The time unit in our experiments is milliseconds. The computational advantage of our approach is clear from the figure. It can be seen that the time required by Lowe's approach increases linearly with the number of points whereas our approach remains roughly the same. The main reason is that Gauss-Newton method requires the solution of a $(N+6) \times (N+6)$ matrix, where N is the number of feature points measured. This matrix inverse operation is surely computationally the most demanding step. The long computation time of that of Extended Kalman filter stems from the fact that it has to solve for a $2N \times 2N$ matrix.

5 Synthetic Data Experiment

The performance of our algorithm under noisy environment is investigated. Two sets of experiments are being performed. We first describe the setup since they are common for both experiments. A number of points are randomly generated in the 3D space such that they will project on an image plane of size 2 by 2. These model points are then transformed as follows: a rotation about the axis $(1, 1, 1)$ by $6°$ followed by a translation of $(5.0, 3.0, 6.0)$. The transformed points are then projected on the image plane. The image coordinates are then digitized to a screen resolution of 512×512. The process of digitization introduces noise in this case. 100 random scenes are generated according to this setup and the digitized coordinates are used together with the original generated 3D model to determine the pose.

The errors shown in the following are all relative errors. We used quaternion to represent the rotation such that all the estimated results are in vector form. The relative error of a vector is defined by the Euclidean norm of the difference between the estimated and the true vector divided by the Euclidean norm of the true vector. The first set of experiments tests the performance of the algorithm under different number of points in the scene. The results are shown in Fig. 3a. From the plot, it is observed that the estimated rotation is relatively stable with respect to the variation of the number of points whereas the translation has the greatest improvement from 8 to 12 points. In addition, the digitization only introduces a small effect on the estimated results, as can be seen in the range of errors it is only within 3 percents.

Another set of experiments simulates the measurement noise by introducing an offset value to both digitized x and y coordinates. These offset values are normal distributed with variance of k pixels where k is the parameter to be varied in the experiments. 16 points are used in this set of experiments. Fig. 3b shows the results and our algorithm again is quite stable under the situation of increasing noise.

6 Real image testing

Pose tracking involves continuous monitoring of the pose of an object from the input image sequence. Our algorithm can be easily adapted for continuous tracking by performing estimation between successive image frames. We test our algorithm by applying it to track the head of a person in an image sequence. Our implementation is done on an SGI Indy workstation with MIPS R4400 processor. To verify the correctness of the recovered pose information, we reproduce the same motion on the workstation at the same time. Though it is difficult in this case to estimate the performance figure of the algorithm, an overall estimate of the usefulness of the algorithm can be obtained. We used the structure from motion approach to obtain the data set for the face to be tracked, following the algorithm by Szeliski [8]. In our experiment, only twelve points are used. All points are selected on the criteria that they can be tracked unambiguously throughout the sequence so as to increase the accuracy of the recovered pose. We use four points for the head boundary including ears, four points for the mouth, and four points for each of the eyebrows.

Before tracking, the user has to initialize the tracker. The selected feature points should be the same as those taken in the calibration stage. These selected feature points are then tracked by the normalized correlation method to give coordinates in successive frames. It is inevitable that error in selecting feature exists due to disagreement in the location of the feature selected and that in the calibration stage. However, according to our experience, this discrepancy is minor in that our algorithm can still maintain the tracking over a long period, say over 100 frames. The tracking information (rotation and translation) will be used to control a synthetic head model[3]. Results are shown in Fig. 4.

7 Discussion

In the experiments on head tracking, we found that our algorithm can effectively recover the motion information from the images. The main advantage of our algorithm is that it is computationally efficient. In addition, by performing the fitting of the model set in 3D object space and not on the image plane, a larger range of convergence is achieved which enables our algorithm to be more stable with respect to different initial guesses. However, our algorithm also has the following drawbacks. Firstly, due to the fact that our algorithm is formulated under the situation of two frames only, thus the error in the estimation will accumulate in the subsequent frames and may lead to failure after a large number of frames. Moreover, in the current formulation, the problem of occlusion is not being handled.

For the first problem, one solution is to bootstrap the algorithm itself after a number of frames. This is possible since our algorithm has quite a large range of convergence (approximately 40 degrees for the rotation angles from our experiments). Another solution is to cross-check the pose estimation result with two or more frames earlier so as to minimize the error itself. The second problem demands a further enhancement of our algorithm to handle the occlusion problem, which is not that difficult due to the model based nature of our algorithm. We are currently working in this direction because the task of human motion tracking including head tracking is a primary goal of our project. An exciting application of head tracking is the usage of very low bit rate transmission in video conferencing.

8 Conclusion

An efficient pose estimation algorithm is presented. By breaking down the pose estimation process into two separate linear stages, the computation requirement

---

[3] The synthetic character is a public domain implementation of the facial animation work by Keith Waters [10]

is significantly reduced. Tests on synthetic data as well as real world head tracking have demonstrated the effectiveness of our algorithm.

Appendix: Derivation of Initial Guess

We want to minimize the function below to establish the initial guess:

$$\Theta = \sum_{i=1}^{N} \|d_i \hat{\nu}_i - (P_i + T)\|^2$$

Applying partial differentiation to Θ with respect to d_i and T respectively and setting them equal to zero, we have

$$\frac{\partial \Theta}{\partial d_i} = 0 \Rightarrow d_i = \hat{\nu}_i^t (P_i + T)$$

$$\frac{\partial \Theta}{\partial T} = 0 \Rightarrow T = \frac{1}{N} \sum_{i=1}^{N} (d_i \hat{\nu}_i - P_i)$$

$$T = \frac{1}{N} \sum_{i=1}^{N} ([\hat{\nu}_i \hat{\nu}_i^t] P_i - P_i) + \frac{1}{N} \sum_{i=1}^{N} ([\hat{\nu}_i \hat{\nu}_i^t] T)$$

$$\Rightarrow \left(N \cdot I - \sum_{i=1}^{N} \hat{\nu}_i \hat{\nu}_i^t \right) T = \sum_{i=1}^{N} (\hat{\nu}_i \hat{\nu}_i^t - I) P_i$$

Writing A_i as $I - \hat{\nu}_i \hat{\nu}_i^t$, we have

$$T = - \left(\sum_{i=1}^{N} A_i \right)^{-1} \left(\sum_{i=1}^{N} A_i P_i \right)$$

References

1. K. S. Arun, T. S. Huang, and S. D. Blostein. Least-square fitting of two 3-D point sets. *IEEE Trans. Pattern Anal. Machine Intell.*, 9(5):698–700, Sept. 1987.
2. A. Azarbayejani and A. Pentland. Recursive estimation of motion, structure, and focal length. *IEEE Trans. Pattern Anal. Machine Intell.*, 17(6):562–575, June 1995.
3. T. J. Broida. Recursive 3-D motion estimation from a monocular image sequence. *IEEE Trans. Aerospace Electronic Systems*, 26(4):639–655, July 1990.
4. D. F. Dementhon and L. S. Davis. Model-based object pose in 25 lines of code. *Intl. Journal of Comput. Vision*, 15:123–141, 1995.
5. B. K. P. Horn. Closed-form solution of absolute orientation using unit quaternions. *J. Opt. Soc. Amer.*, 4:629–642, Apr. 1987.

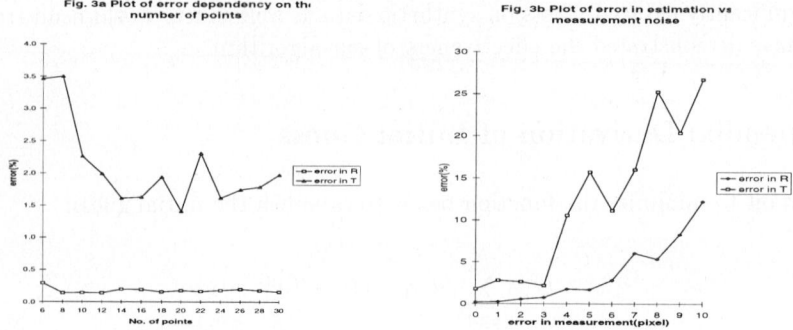

Fig. 3. Left) Plot of error dependency on the number of points, right) Plot of error in estimation vs. measurement noise

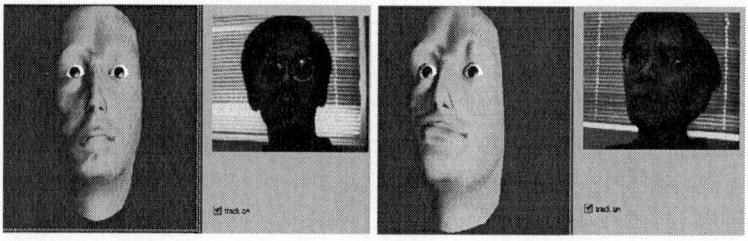

Fig. 4. Sample run of the head tracking application. Left) initialization, Right) Rotation of the head and reconstructed pose.

6. H. Li, P. Roivainen, and R. Forchheimer. 3-D motion estimation in model-based facial image coding. *IEEE Trans. Pattern Anal. Machine Intell.*, 15(6):545–555, June 1993.
7. D. G. Lowe. Fitting parameterized three dimensional models to images. *IEEE Trans. Pattern Anal. Machine Intell.*, 13(5):441–450, May 1991.
8. R. Szeliski and S. B. Kang. Recovering 3D shape and motion from image streams using non-linear least squares. *Cambridge Research Laboratory Technical Report*, Mar. 1993.
9. S. Umeyama. Least-square estimation of transformation parameters between two point pattern. *IEEE Trans. Pattern Anal. Machine Intell.*, 13(4):376–380, Apr. 1991.
10. K. Waters. A muscle model for animating three-dimensional facial expression. *Comput. Graphics*, 21(4):17–24, 1987.
11. J. Weng, N. Ahuja, and T. S. Huang. Optimal motion and structure estimation. *IEEE Trans. Pattern Anal. Machine Intell.*, 15(9):864–884, Sept. 1993.
12. P. Wunsch and G. Hirzinger. Registration of CAD-models to images by iterative inverse perspective matching. *Proc. ICPR 96*, pages 78–83, Nov. 1996.
13. Z. Zhang and O. Faugeras. *3D Dynamic Scene Analysis*. Springer-Verlag, 1992.

A New Multistage Approach to Motion and Structure Estimation by Gradually Enforcing Geometric Constraints

Zhengyou Zhang[1,2]

[1] INRIA, 2004 route des Lucioles, BP 93, F-06902 Sophia-Antipolis Cedex, France
[2] ATR HIP, 2-2 Hikaridai, Seika-cho Soraku-gun, Kyoto 619-02 Japan

Abstract. The standard 2-stage algorithm first estimates the 9 essential parameters defined up to a scale factor and then refines the motion estimation based on some statistically optimal criteria. We propose in this paper a novel approach by introducing an intermediate stage which consists in estimating a 3×3 matrix defined up to a scale factor by imposing the *rank-2 constraint* (the matrix has seven independent parameters). The idea is to *gradually* project parameters estimated in a high dimensional space onto a *slightly lower*-dimensional space, namely from 8 dimensions to 7 and finally to 5. Experiments with synthetic and real data show a considerable improvement over the 2-stage algorithm. Our conjecture from this work is that the imposition of the constraints arising from projective geometry should be used as an intermediate step in order to obtain reliable 3D Euclidean motion and structure estimation from multiple calibrated images.

Keywords: Motion and stereo, 3D reconstruction, Structure from motion, Multiple-view geometry, Gradual constraint enforcing

1 Introduction

Motion and structure from motion has been of the central interest in Computer Vision since its infancy, and is still an active domain of research. There are a large number of pieces of work reported in the literature in this domain. The reader is referred to [1–3] for a review. The problem is usually divided into two steps: (i) extract features (usually points or line) and match them between images; (ii) determine motion and structure from corresponding features. The standard approach to motion and structure estimation problem from two given sets of matched image points consists of two stages: (i) estimating the 9 essential parameters; (ii) refining the motion estimation over a five-dimensional (5D) space. Unfortunately, the results obtained using this approach are often not satisfactory, especially when the motion is small or when the observed points are close to a degenerate surface (e.g. plane). The problem is that the second stage is very sensitive to the initial guess and that it is very difficult to obtain a precise initial estimate from the first stage. This is because we perform a projection of a set of quantities which are estimated in a space of 8 dimensions, much higher than that of the real space which is 5D [4]. We propose in this paper a novel approach by introducing an intermediate stage which consists in estimating a 3×3 matrix defined up to a scale factor by imposing the *zero-determinant constraint* (the matrix has seven independent parameters, and is known as the fundamental matrix). The idea is to *gradually* project parameters estimated in a high dimensional space onto a *slightly lower*-dimensional space, namely from 8 dimensions to 7 and finally to 5. The proposed approach has been tested with synthetic and real data, and considerable improvement has been observed for the delicate situations mentioned above. Note that the constraints we use in the intermediate stage are actually *all* constraints existing between two sets of image points if the images are *uncalibrated* (the intrinsic parameters are not know). That is, we are determining projective motion and structure in this stage. Our conjecture from this work is that the imposition of the constraints arising from projective geometry should be used as an intermediate step in order to obtain reliable 3D Euclidean motion and structure estimation from multiple calibrated images. The local minimum problem is less severe in the projective framework than in the Euclidean one [5]: The optimization for the projective structure often succeeds in locating the true global minimum starting from the unreliable initial guess when the Euclidean optimization does not.

2 Notation and Problem Statement

2.1 Notation

If $\mathbf{x} = [x, y, \cdots]^T$, its augmented vector (adding 1 as its last element) is denoted by $\widetilde{\mathbf{x}}$, i.e., $\widetilde{\mathbf{x}} = [x, y, \cdots, 1]^T$. A camera is described by the widely used pinhole model. The coordinates of a 3-D point $\mathtt{M} = [x, y, z]^T$ in a world coordinate system and its retinal image coordinates $\mathbf{m} = [u, v]^T$ are related by $s\widetilde{\mathbf{m}} = \mathbb{P}\widetilde{\mathtt{M}}$, where s is an arbitrary scale, and $\mathbb{P}$ is a 3×4 matrix, called the perspective projection matrix.

The matrix $\mathbb{P}$ can be decomposed as $\mathbb{P} = \mathbf{A}\,[\mathbf{R}\ \mathbf{t}]$, where $\mathbf{A}$ is an upper-triangular matrix which maps the normalized image coordinates to the retinal/pixel image coordinates, and $(\mathbf{R}, \mathbf{t})$ is the 3D displacement (rotation and translation) from the world coordinate system to the camera coordinate system. The five parameters in $\mathbf{A}$ are known through calibration.

The first and second images are respectively denoted by I_1 and I_2. A point $\mathbf{m}$ in the image plane I_i is noted as $\mathbf{m}_i$. The second subscript, if any, will indicate the index of the point in consideration.

2.2 Epipolar Equation

We consider two perspective images of a single scene, and we want to determine the relation between the two images and the structure of the scene. Let the displacement from the first camera to the second be $(\mathbf{R}, \mathbf{t})$. Let $\mathbf{m}_1$ and $\mathbf{m}_2$ be the images of a 3-D point M on the cameras. Without loss of generality, we assume that M is expressed in the coordinate frame of the first camera. Under the pinhole model, we have the following two equations:

$$s_1\widetilde{\mathbf{m}}_1 = \mathbf{A}_1\,[\mathbf{I}\ \mathbf{0}]\,\widetilde{\mathtt{M}}, \quad \text{and} \quad s_2\widetilde{\mathbf{m}}_2 = \mathbf{A}_2\,[\mathbf{R}\ \mathbf{t}]\,\widetilde{\mathtt{M}}, \qquad (1)$$

where $\mathbf{A}_1$ and $\mathbf{A}_2$ are the intrinsic matrices of the first and second cameras, respectively. Eliminating M, s_1 and s_2 from the above equations, we obtain the following fundamental equation

$$\widetilde{\mathbf{m}}_2^T \mathbf{F} \widetilde{\mathbf{m}}_1 = 0 \quad \text{with } \mathbf{F} = \mathbf{A}_2^{-T}[\mathbf{t}]_\times \mathbf{R} \mathbf{A}_1^{-1}, \qquad (2)$$

where $[\mathbf{t}]_\times$ is an antisymmetric matrix defined by $\mathbf{t}$ such that $[\mathbf{t}]_\times \mathbf{x} = \mathbf{t} \times \mathbf{x}$ for all 3-D vector $\mathbf{x}$ ($\times$ denotes the cross product). Matrix $\mathbf{F}$ is knowns as the fundamental matrix [6, 7]. Equation (2) is a fundamental constraint underlying any two images if they are perspective projections of one and the same scene.

For convenience, we use **p** to denote a point in the *normalized* image coordinate system, i.e., $\tilde{\mathbf{p}}_1 = \mathbf{A}_1^{-1}\tilde{\mathbf{m}}_1$, and $\tilde{\mathbf{p}}_2 = \mathbf{A}_2^{-1}\tilde{\mathbf{m}}_2$. Let $\mathbf{E} = [\mathbf{t}]_\times \mathbf{R}$, which is known as the *Essential matrix*. It was introduced by Longuet-Higgins [8], and its property has been studied in the literature. Now, we can write equation (2) as

$$\tilde{\mathbf{p}}_2^T \mathbf{E} \tilde{\mathbf{p}}_1 = 0 . \tag{3}$$

Because the magnitude of **t** can never be recovered from two perspective images, we set $\|\mathbf{t}\| = 1$. The relationship between **E** and **F** is readily described by

$$\mathbf{F} = \mathbf{A}_2^{-T} \mathbf{E} \mathbf{A}_1^{-1}, \quad \text{and} \quad \mathbf{E} = \mathbf{A}_2^{T} \mathbf{F} \mathbf{A}_1 . \tag{4}$$

3 The New Multistage Motion Algorithm

In this section, we first recall the 8-point algorithm, which ignores the constraints on the essential parameters, and the nonlinear algorithm, which performs an optimization directly over the 5D motion space. Finally, we show how we can impose the zero-determinant constraint as an intermediate step in order to provide a better initial estimate for the previously mentioned nonlinear algorithm.

3.1 The Linear Criterion

Equation (3) can be rewritten as a linear and homogeneous equation in the 9 unknown coefficients of matrix **E**: $\mathbf{u}^T \boldsymbol{\epsilon} = 0$, where $\mathbf{u} = [x_1 x_2, y_1 x_2, x_2, x_1 y_2, y_1 y_2, y_2, x_1, y_1, 1]^T$, $\boldsymbol{\epsilon} = [E_{11}, E_{12}, E_{13}, E_{21}, E_{22}, E_{23}, E_{31}, E_{32}, E_{33}]^T$, $\mathbf{p}_1 = [x_1, y_1]^T$, $\mathbf{p}_2 = [x_2, y_2]^T$, and E_{ij} is the element of **E** at row i and column j. If we are given 8 or more matches and *ignore* the constraints on the essential parameters, we will be able, in general, to determine a unique solution for **E**, defined up to a scale factor. This can be done by solving, for example, the following classical least-squares problem:

$$\min_{\boldsymbol{\epsilon}} \|\mathbf{U}\boldsymbol{\epsilon}\|^2 \quad \text{subject to } \|\boldsymbol{\epsilon}\| = \sqrt{2}, \tag{5}$$

where $\mathbf{U} = [\mathbf{u}_1, \cdots, \mathbf{u}_n]^T$. The constraint on the norm of $\boldsymbol{\epsilon}$ is derived from the fact that **R** is an orthonormal matrix and $\|\mathbf{t}\| = 1$ [9]. The solution is the eigenvector of $\mathbf{U}^T \mathbf{U}$ associated with the smallest eigenvalue. This is known as the eight-point algorithm in the literature. Once we have estimated the essential matrix **E**, we can recover the motion $(\mathbf{R}, \mathbf{t})$. See [9] for more details.

The advantage of the linear criterion is that it leads to an analytic solution. However, we have found that it is quite sensitive to noise, even with a large set of data points. There are two reasons for this:

- We have omitted the constraints on the essential matrix. The elements of **E** are not independent from each other.
- The quantity we try to minimize (5) does not have much physical meaning.

3.2 Minimizing the Distances to Epipolar Lines

As was described in Sect. 2.2, $\mathbf{F}\tilde{\mathbf{m}}_1$ represents actually the epipolar line of $\mathbf{m}_1$ in the second image. If $\mathbf{m}_2$ corresponds exactly to $\mathbf{m}_1$, we would expect the distance from $\mathbf{m}_2$

to the epipolar line $\mathbf{F}\tilde{\mathbf{m}}_1$ to be zero. Thus, a natural idea is to use a nonlinear criterion by minimizing:

$$\sum \left(d^2(\tilde{\mathbf{m}}_{2i}, \mathbf{F}\tilde{\mathbf{m}}_{1i}) + d^2(\tilde{\mathbf{m}}_{1i}, \mathbf{F}^T\tilde{\mathbf{m}}_{2i}) \right), \tag{6}$$

where $d(\tilde{\mathbf{m}}_2, \mathbf{F}\tilde{\mathbf{m}}_1)$ is the Euclidean distance of point $\mathbf{m}_2$ to its epipolar line $\mathbf{F}\tilde{\mathbf{m}}_1$ in the second image. Unlike the case of the linear criterion which uses the elements of the essential matrix, we minimize the above functional over the motion parameters. Recall that we deal with calibrated cameras, i.e., $\mathbf{F}$ depends only on $\mathbf{R}$ and $\mathbf{t}$. The rotation is represented by a 3D vector, whose direction is parallel to the rotation axis and whose magnitude is equal to the rotation angle. The translation is represented by its spherical coordinates. Thus, the minimization is carried out over these five unknowns. As the minimization is nonlinear, we use the result of the analytical method as its initial guess.

3.3 3D Reconstruction

Once we know the motion $(\mathbf{R}, \mathbf{t})$, given a match $(\mathbf{m}_1, \mathbf{m}_2)$, we can estimate the 3D coordinates M by minimizing the distance between the back-projection of the 3D reconstruction and the observed image point, that is

$$\hat{\mathbf{M}} = \arg\min_{\mathbf{M}} \left(\|\mathbf{m}_1 - \mathbf{h}_1(\mathbf{a}, \mathbf{M})\|^2 + \|\mathbf{m}_2 - \mathbf{h}_2(\mathbf{a}, \mathbf{M})\|^2 \right),$$

where $\mathbf{h}_1(\mathbf{a}, \mathbf{M})$ and $\mathbf{h}_2(\mathbf{a}, \mathbf{M})$ are the camera projection functions corresponding to (1).

3.4 Maximum Likelihood Estimation

We are given n point matches $\{(\mathbf{m}_{1j}, \mathbf{m}_{2j}) | j = 1, \ldots, n\}$. Each point is assumed to be corrupted by additive independent Gaussian noise, i.e., $\mathbf{m}_{ij} = \bar{\mathbf{m}}_{ij} + \boldsymbol{\eta}_{ij}$ for $i = 1, 2$ and $j = 1, \ldots, n$, where $\bar{\mathbf{m}}_{ij}$ is the *ideal* point position if noise did not exist, and $\boldsymbol{\eta}_{ij}$ is a Gaussian random vector of mean $\mathbf{0}$ and covariance matrix $\boldsymbol{\Lambda}_{ij}$, i.e., $\boldsymbol{\eta}_{ij} \sim N(\mathbf{0}, \boldsymbol{\Lambda}_{ij})$. The noise terms in different points are independent, i.e., $E[\boldsymbol{\eta}_{ij} \boldsymbol{\eta}_{kl}^T] = \mathbf{O}$ for $j \neq l$. Let us denote the observed points by $\mathbf{x} = [\mathbf{m}_{11}^T, \mathbf{m}_{21}^T, \ldots, \mathbf{m}_{1j}^T, \mathbf{m}_{2j}^T, \ldots, \mathbf{m}_{1n}^T, \mathbf{m}_{2n}^T]^T$, which is a vector of $4n$ dimensions.

Let $\mathbf{a} = [\mathbf{r}^T, \boldsymbol{\phi}^T]^T$ be the 5-D vector composed of three parameters representing the rotation between the two images and two parameters representing the translation (see Sect. 3.2). Let $\mathbf{M}_j$ be the 3-D vector corresponding to the jth point expressed in the coordinate system associated with the first camera. The motion and structure parameters are then represented by a vector of $(5+3n)$ dimensions, denoted by $\boldsymbol{\theta} = [\mathbf{a}^T, \mathbf{M}_1^T, \ldots, \mathbf{M}_n^T]^T$. The maximum likelihood estimate, $\hat{\boldsymbol{\theta}}$, of the parameter vector $\boldsymbol{\theta}$ is given by the value of $\boldsymbol{\theta}$ which makes the observed data $\mathbf{x}$ most likely, that is, $\hat{\boldsymbol{\theta}} = \arg\max_{\boldsymbol{\theta}} \log[P(\mathbf{x}|\boldsymbol{\theta})]$, where $P(\mathbf{x}|\boldsymbol{\theta})$ is the conditional probability density of the observed data $\mathbf{x}$ given the parameter vector $\boldsymbol{\theta}$. Under the assumption that the data is corrupted by independent Gaussian noise, we have the following weighted nonlinear least-squares formulation:

$$\hat{\boldsymbol{\theta}} = \arg\min_{\boldsymbol{\theta}} \sum_{i=1}^{2} \sum_{j=1}^{n} \delta\mathbf{m}_{ij}^T \boldsymbol{\Lambda}_{ij}^{-1} \delta\mathbf{m}_{ij} \tag{7}$$

with $\delta\mathbf{m}_{ij} = \mathbf{m}_{ij} - \mathbf{h}_i(\mathbf{a}, \mathbf{M}_j)$, which is actually the sum of squared Mahalanobis distances. Due to the nonlinear nature of perspective projection, the solution to the above problem demands the use of numerical nonlinear minimization technique such as the Levenberg-Marquardt algorithm implemented in the MINPACK library [10]. An initial guess on the motion and structure is required, which can be obtained by using the techniques described previously.

3.5 Imposing the Rank-2 Constraint to Refine the Initial Motion Estimate

The two problems described in Sect. 3.2 and Sect. 3.4 are both highly nonlinear, and their solutions are very sensitive to the initial guess. The initial guess is obtained by projecting the essential parameters onto the 5D motion space. Unfortunately, the estimation of the essential parameters as described in Sect. 3.1 is very sensitive to data noise, especially when motion is small and the space points are close to a degenerate surface such as a plane. One major reason is that we have ignored the constraints which exist on the essential parameters and have used 8 parameters instead of 5 (i.e., three redundant parameters). Here, we propose to impose the zero-determinant (rank-2) constraint, and to estimate 7 parameters before projecting onto the motion space.

Indeed, there are only 7 independent parameters in a rank-2 matrix defined up to a scale factor (the scale factor and the rank-2 constraint remove two free parameters), and the fundamental matrix in the context of two uncalibrated images [6] has exactly the same properties. There are several possible parameterizations for such a matrix, e.g., one can express one row (or column) of the fundamental matrix as the linear combination of the other two rows (or columns). The following parameterization

$$\mathbf{F} = \begin{bmatrix} a & b & -ax_1 - by_1 \\ c & d & -cx_1 - dy_1 \\ -ax_2-cy_2 & -bx_2-dy_2 & (ax_1+by_1)x_2 + (cx_1+dy_1)y_2 \end{bmatrix}$$

expresses of course a matrix of rank 2, because both the third row and column are the combinations of the other two rows and columns. Furthermore, there is a nice geometric interpretation. The parameters (x_1, y_1) and (x_2, y_2) are the coordinates of the two epipoles $\mathbf{e}_1$ (projection of the optical center of the second camera in the first camera) and $\mathbf{e}_2$ (projection of the optical center of the first camera in the second camera). The remaining four parameters (a, b, c, d) define the relationship between the orientations of the two pencils of epipolar lines. To take into account the fact that the matrix is defined only up to a scale factor, the matrix is normalized by dividing the four elements (a, b, c, d) by the largest in absolute value. This parameterization breaks down if one of the two epipoles is at infinity. Our actual implementation has taken care of this problem by using several parameterizations.

The seven parameters, (x_1, y_1, x_2, y_2), and three among a, b, c, d), are estimated by minimizing the sum of distances between points and their epipolar lines. That is, we minimize the same objective function as the one (6) described in Sect. 3.2, only the minimization is conducted over the above *7D parameter space*, instead of the 5D motion space. The minimization is nonlinear, and we use the matrix estimated in (5) as the initial guess. The reader is referred to [11] for a detailed review of different techniques for estimating the fundamental matrix.

Note that this intermediate stage can be applied to normalized image coordinates as well as pixel image coordinates, because both the determinant of $\mathbf{E}$ and that of $\mathbf{F}$ should be equal to 0.

3.6 Summary of the New Multistage Algorithm

We now summarize the main steps of our multistage algorithm:

Step 1: Estimate the essential parameters with 8-point algorithm (5). The obtained matrix is denoted by $\mathbf{E}_1$.
Step 2: Estimate a rank-2 matrix, denoted by $\mathbf{E}_2$, from $\mathbf{E}_1$ by setting the smallest singular value to 0 [12, 11], and compute the seven parameters from $\mathbf{E}_2$.
Step 3: Refine the seven parameters by minimizing the sum of squared distances between points and their epipolar lines, i.e., the objective function (6). The obtained matrix is denoted by $\mathbf{E}_3$.
Step 4: Estimate the motion parameters t and $\mathbf{R}$ from $\mathbf{E}_3$ (see e.g., [9] for the details).
Step 5: Refine the motion parameters by minimizing the sum of squared distances between points and their epipolar lines, i.e., the objective function (6).
Step 6: Reconstruct the corresponding 3D points as described in Sect. 3.3.
Step 7: Refine the motion and structure estimate by using criterion (7).

If we bypass steps 2 and 3, we have a standard 2-stage algorithm. The nonlinear minimization in steps 3, 5, 6, and 7 is done with the Levenberg-Marquardt algorithm implemented in the `Minpack` library [10].

4 Experimental Results

In this section, we describe a part of our Monte-Carlo simulations to show that our new multistage algorithm yields much more reliable results than the standard one when the level of noise in data points is high or when data points are located close to a degenerate configuration. The reader is referred to [13] for more results including a set of real data with which the standard algorithm does not work while ours does.

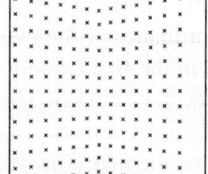
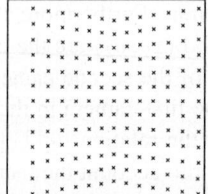
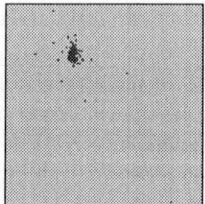
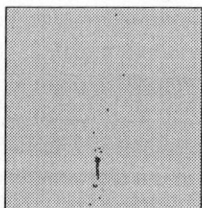

Fig. 1. Images of two planar grids hinged together with $\theta = 45°$. Gaussian noise of $\sigma = 0.5$ has been added to each grid point

Fig. 2. 3D reconstruction of the images shown in Fig. 1 with the 2-stage algorithm: Front and top views

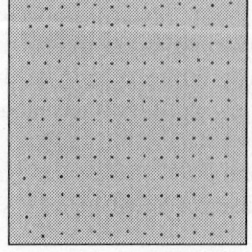
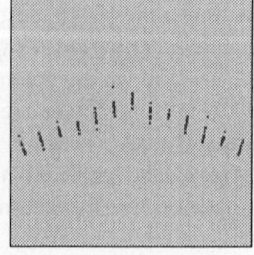

Fig. 3. 3D reconstruction of the images shown in Fig. 1 with the 3-stage algorithm: Front and top views

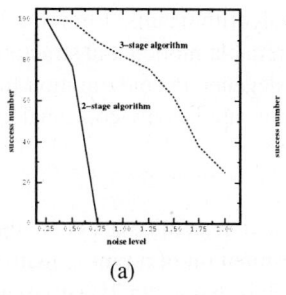
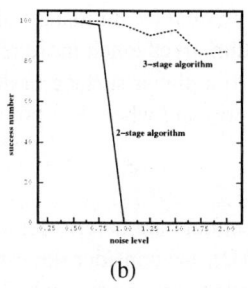
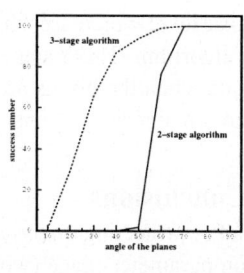

(a) (b)

Fig. 4. Comparison of the standard and new algorithms for (a) $\theta = 60°$ and (b) $\theta = 90°$ w.r.t. noise level

Fig. 5. Comparison of the standard and new algorithms for $\sigma = 0.5$ pixels with respect to the angle θ of the object

We use the same configuration as that described in [14]. The object is composed of two planar grids which are hinged together with angle $\pi - \theta$. When $\theta = 0$, the object is planar, which is a degenerate configuration for the algorithms considered in this paper. Each grid is of size 180×360 units^2. The object is placed in the scene with a distance of 530 units from the camera. The two images have the same intrinsic parameters: $\alpha_u = \alpha_v = 600$, $u_0 = v_0 = 255$, and $c = 0$. They differ by a pure lateral translation: $\mathbf{t} = [-40, 0, 0]^T$, and $\mathbf{R} = \mathbf{I}$. Small lateral motion is difficult for motion estimation because rotation and translation can be confused. The grid points are used as feature points. The x- and y-coordinates of each grid point are perturbed by independent random Gaussian noise of mean 0 and standard deviation σ pixels.

A pair of images with $\theta = 45°$ and $\sigma = 0.5$ pixels is shown in Fig. 1. The motion estimate given by the 2-stage algorithm is: $\mathbf{r} = [-7.981238e-05, -7.961793e-02, 3.779707e-04]^T$ (in radians) for rotation (which should be $[0, 0, 0]^T$), and $\mathbf{t} = [2.091024e-01 - 3.089894e-02 - 9.774055e-01]^T$ for translation (which should be $[-1, 0, 0]^T$). The corresponding 3D reconstruction is shown in Fig. 2. Clearly, the 2-stage algorithm fails. When we apply the 3-stage algorithm to the same data, we obtain $\mathbf{r} = [-6.969567e-04, -1.785441e-03, -2.330740e-04]^T$ for rotation, and $\mathbf{t} = [-9.999643e-01, -8.318301e-03, 1.468760e-03]^T$ for translation. The corresponding 3D reconstruction is shown in Fig. 3. As can be observed, quite reasonable result has been obtained with our new multistage algorithm, taking into account the fact that the object is close to a plane surface.

Now we provide more systematic and statistic results. We vary the angle θ from $10°$ to $90°$ with an interval of $10°$. We also vary the level of the Gaussian noise added to the grid points. The standard deviation σ varies from 0.25 pixels to 2.0 pixels with an interval of 0.25 pixels. For each θ and each σ, we add 100 times independent noise to the grid points. For each set of noisy data, we apply the 2-stage algorithm and our multistage algorithm. If the estimated translation vector and the true one (i.e., $[-1, 0, 0]^T$) form an angle larger than $45°$, then the algorithm is considered to have failed for this set of data. Among 100 trials for each θ and each σ, we count the number of times that the algorithm succeeds. We show in Fig. 4 the curves of the number of successes with respect to various noise levels when θ is fixed at $60°$ and $90°$, respectively. In Fig. 5, we show the curves with respect to various angles θ when the noise level σ is fixed at 0.5 pixels. A general rule is that the number of success decreases when the angle θ approaches to $0°$ and when the noise level σ increases. In all cases, our new multistage algorithm outperforms the 2-stage algorithm. The 3-stage algorithm gives much more reliable motion and structure estimate when the points are close to a planar surface (a degenerate configuration for the motion algorithms considered here) and when data points are heavily corrupted by noise.

5 Conclusions

Instead of projecting directly the essential parameters (defined in an 8D space) onto the motion parameter space (which is 5D), we consider the estimation of a rank-2 matrix defined up to a scale factor (i.e., fundamental matrix, which is defined in 7D space) as an intermediate step to determine 3D Euclidean motion and structure. The proposed approach has been tested with synthetic and real data, and considerable improvement has been observed for the delicate situations such as heavily noisy data and near-degenerate data. Our conjecture from this work is that the imposition of the constraints arising from projective geometry should be used as an intermediate step in order to obtain reliable 3D Euclidean motion and structure estimation from multiple (≥ 3) calibrated images. More details, including a robust technique which detects false matches, can be found in [13].

References

1. H. Nagel, "Image sequences - ten (octal) years- from phenomenology towards a theoretical foundation," in *Proceedings 8th ICPR, Paris, France*, pp. 1174–1185, IEEE, Oct. 1986.
2. J. Aggarwal and N. Nandhakumar, "On the computation of motion from sequences of images — a review," *Proc. IEEE*, vol. 76, pp. 917–935, Aug. 1988.
3. T. Huang and A. Netravali, "Motion and structure from feature correspondences: A review," *Proc. IEEE*, vol. 82, pp. 252–268, Feb. 1994.
4. C. Braccini, G. Gambardella, A. Grattarola, and S. Zappatore, "Motion estimation of rigid bodies: Effects of the rigidity constraints," in *Proc. EUSIPCO, Signal Processing III: Theories and Applications*, pp. 645–648, Sept. 1986.
5. J. Oliensis and V. Govindu, "Experimental evaluation of projective reconstruction in structure from motion," tech. rep., NEC Research Institute, Princeton, NJ 08540, USA, Oct. 1995.
6. Q.-T. Luong and O. D. Faugeras, "The fundamental matrix: Theory, algorithms and stability analysis," *The International Journal of Computer Vision*, vol. 17, pp. 43–76, Jan. 1996.
7. O. Faugeras, "Stratification of 3-D vision: projective, affine, and metric representations," *Journal of the Optical Society of America A*, vol. 12, pp. 465–484, Mar. 1995.
8. H. Longuet-Higgins, "A computer algorithm for reconstructing a scene from two projections," *Nature*, vol. 293, pp. 133–135, 1981.
9. O. Faugeras, *Three-Dimensional Computer Vision: a Geometric Viewpoint*. MIT Press, 1993.
10. J. More, "The levenberg-marquardt algorithm, implementation and theory," in *Numerical Analysis* (G. A. Watson, ed.), Lecture Notes in Mathematics 630, Springer-Verlag, 1977.
11. Z. Zhang, "Determining the epipolar geometry and its uncertainty: A review," *The International Journal of Computer Vision*, 1997. In Press. Updated version of INRIA Research Report No.2927, 1996.
12. R. Hartley, "In defence of the 8-point algorithm," in *Proceedings of the 5th International Conference on Computer Vision*, (Boston, MA), pp. 1064–1070, June 1995.
13. Z. Zhang, "Motion and structure from two perspective views: From essential parameters to euclidean motion via fundamental matrix," *Journal of the Optical Society of America A*, vol. 14, no. 11, 1997. In Press.
14. K. Kanatani, "Automatic singularity test for motion analysis by an information criterion," in *Proceedings of the 4th European Conference on Computer Vision* (B. Buxton, ed.), (Cambridge, UK), pp. 697–708, Apr. 1996.

Tracking a Person with Pre-recorded Image Database and a Pan, Tilt, and Zoom Camera

Yiming Ye[1], John K. Tsotsos[2], Karen Bennet[3] and Eric Harley[2]

[1] IBM T.J Watson Research Center, P.O. Box 704,Yorktown Heights, N.Y. 10598[†]
[2] Department of Computer Science, University of Toronto
[3] IBM Canada Center for Advanced Studies, North York, Ontario

Abstract. This paper proposes a novel tracking strategy that can robustly track a person or other object within a fixed environment using a pan, tilt, and zoom camera with the help of a pre-recorded image database. We define a set called the Minimum Camera Parameter Settings (MCPS) which contains just enough camera states as required to survey the environment for the target. This set of states is used to facilitate tracking and segmentation. The idea is to store a background image of the environment for every camera state in MCPS, thus creating an image database. During tracking camera movements are restricted to states in MCPS. Scanning for the target and segmentation of the target from the background are simplified as each current image can be compared with the corresponding pre-recorded background image.

Keywords: Tracking, Image database, Segmentation,

1 Introduction

The task of visually tracking objects moving in three-dimensions has received considerable attention in the computer vision community over the past few years. The task is a challenging one because it not only involves the difficulties of segmenting the target from various backgrounds, but also analysis and prediction of the target's motion. Approaches to this problem include the use of multiple cameras [1, 4], two- and three-dimensional models of the target [1, 7] and attempts to follow specific features of the moving target, such as head or hands through the use of an active camera [6]. The stability of these tracking methods is adversely affected by the complexity of the environment.

In this paper we show that some of these problems can be alleviated through the use of a pre-recorded image database and intelligent control of the camera (sensor planning). We first select a set of camera states (i.e., pan, tilt, and zoom settings) such that wherever the target may appear in the given environment, there exists at least one camera state appropriate for target recognition. The background images for these camera states are stored in an image database.

[†] yiming@vis.toronto.edu,   tsotsos@vis.toronto.edu,   bennet@vnet.ibm.com, eharley@db.toronto.edu

These same camera states are used during tracking, so that the background images form references to facilitate segmentation. We illustrate these ideas with an experiment basesd on a simple segmentation method and tracking algorithm.

2 The Minimum Set of Camera States

We first would like to choose a set a camera states such that wherever the target is in the given environment, at least one of the camera states puts the target into the field of view with good image quality. For a given recognition algorithm and fixed camera viewing angle size $\langle w, h \rangle$, the probability of successfully recognizing a target appearing in an image is high only when the distance l from the target to the camera is within a certain range. This *effective range* is such that the whole target is within the camera's field of view and the target features are represented with sufficient clarity. A set of viewing angles $\langle w_0, h_0 \rangle, \langle w_1, h_1 \rangle, \ldots, \langle w_{n_0}, h_{n_0} \rangle$ can be selected such that their effective ranges divide the space around the camera center into a layered sphere, covering the depth D of the environment. These angles can be obtained empirically or derived from geometric constraints and the requirement that the size of the target in the image remain constant from one layer to the next (see [12] for details).

Each layer of the layered sphere can be successfully scanned for the target using the corresponding angle size $\langle w, h \rangle$ by sweeping the pan and tilt parameters $\langle p, t \rangle$ of the camera. A single camera direction $\langle p, t \rangle$ produces a viewing volume which is a rectangular pyramid, the intersection of which with the spherical layer produces an effective viewing volume for camera state $\langle w, h, p, t \rangle$. A target appearing in the **effective volume** will be detected with high probability by the given recognition algorithm when the camera is in the corresponding state. To examine the entire layer for the target we need a set of camera directions, $\langle p, t \rangle$, such that the union of their effective volumes cover the whole layer with little overlap. An algorithm that generates this set given $\langle w, h \rangle$ is presented in [12]. Thus, we can produce a set of camera states $\langle w, h, p, t \rangle$ whose effective volumes cover the entire sphere around the camera to some depth D. This set becomes the Minimum Camera Parameter Settings ($MCPS$) required to track the target within the environment.

3 Segmentation

In order to detect and track a target, we must be able to segment it from the background of the image. Generally this is a very difficult task. Our strategy here is to alleviate the some of the difficulties of segmentation by using the camera states of MCPS to create a database of images, IDB_{MCPS}, of the environment without the target present, and then during tracking to use these camera states and the corresponding background images for comparison when segmenting for the target. This strategy should improve the efficiency and accuracy of segmentation. We illustrate the concept using the extremely simple segmentation strategy: *calculate the difference between the tracking image and the corresponding database image, and interpret any significant difference as target.* Presumably, more discriminating segmentation routines could also benefit from sensor planning and an image database.

Details of the difference calculation in this segmentation method are described with reference to the example in Fig. 1. Image (a) is from the image database, and image (b) is taken with the same camera state, but during tracking, after the appearance of a person. Image (c) is the color difference image (b-a) calculated as follows. The color intensity (r, g, b) of a pixel at position (x, y) in (b) is compared with the intensity (r', g', b') at (x', y') in (a), where $|x - x'| \leq n$ and $|y - y'| \leq n$. The value of constant n (typically less than 6) is chosen to compensate for errors in camera movement and depends on camera angle size. The pixel intensity in the color difference image for the position (x, y) is defined to be the triple $(|r - r'|, |g - g'|, |b - b'|)$ whose 2-norm is minimum.

Image (d) in Fig. 1 is the binary difference image obtained by converting (r, g, b) intensities first to grey intensities in the range 0 to 255, and then to black/white intensities of 0 or 255 according to a threshold (40 in this case). Some small white areas are noise, and larger white areas are target. To reduce noise, we apply standard erosion and dilation operations. Blobs are then detected as groups of connected white pixels, and blobs of size $m_i > 1000$ pixels are considered to be target. Image (e) is the same as (c), but with hash marks superimposed marking the average (x_i, y_i) pixel coordinates of target blobs. Here the algorithm found five blobs of significant size, which are assumed to represent the human. The features of the target are represented by the total mass $M = \Sigma m_i$ and the mass-averaged position of the blobs, given by $X = \Sigma m_i x_i / \Sigma x_i, Y = \Sigma m_i y_i / \Sigma y_i$, where the summation is over the blobs of sufficient size.

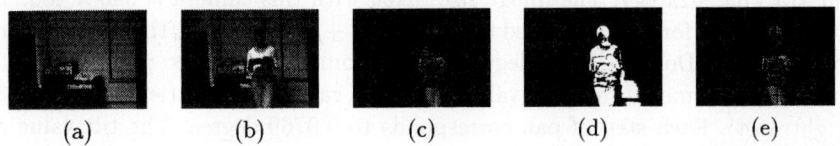

(a) (b) (c) (d) (e)

Fig. 1. Image Segmentation and Recognition Algorithm

This segmentation algorithm, although extremely simple, can successfully detect the human body, because the colors and shape of the hair, face, clothes, and other features of the human, contrast well with most backgrounds. Unfortunately the person's shadow may also be interpreted as part of the target, (cf. Fig. 1(g)), but generally this does not greatly influence the calculated mass and position of the target. In any case, a more sophisticated segmentation method can easily be substituted in this framework of tracking with an MCPS and IDB.

4 Tracking

Our tracking algorithm, using the set of camera states MCPS and the corresponding Image Database IDB_{MCPS}, continuously iterates the following four steps:

1. **If** the target is not detected then camera state is assigned by the **Where to Look Next** routine; **otherwise** the camera state is left unchanged.
2. Take an image $I^*_{\langle w,h,p,t \rangle}$ with camera state $\langle w, h, p, t \rangle$ set in Step 1.
3. Attempt to segment target from background in the image $I^*_{\langle w,h,p,t \rangle}$ with reference to the corresponding image $I_{\langle w,h,p,t \rangle}$ in IDB_{MCPS}.
4. From the results of Step 3, decide if the target is detected or not.

The **Where to Look Next** routine performs the task of selecting the next camera state $\langle w^*, h^*, p^*, t^* \rangle \in MCPS$ in an attempt to bring the target into the field of view of the camera for recognition. When there is no information regarding the whereabouts of the target, as is the case initially or later if tracking fails, then the routine simply cycles through the states of MCPS. If the target was recently in the field of view and has now moved out, then the routine uses the last known position and orientation to guess a set of next possible positions and orientations.

5 Example Experiment

In this section we describe the tracking algorithm with reference to an experiment in a fixed office environment. The camera used in our experiment is a canon VC-C1 MKII Communication Camera. The pan, tilt, and zoom of the camera are actively controlled by an SGI Indy machine through an RS-232 port The mechanical errors are relatively small, which makes this a perfect device for our tracking strategy. The image size taken with this camera is 640 × 480. The rotation angle for pan is limited to Right-Left +/- 50 degrees, the rotation angle for tilt is Up-Down +/- 20 degrees. The zoom range is 8 × power zoom. To control the camera, the pan value can take values from 0 (leftmost) to 1300 (rightmost). Each step of pan corresponds to 0.0769 degree. The tilt value can vary from 0 (lowermost) through 289 (horizontal) to 578 (uppermost). Each step of tilt corresponds to 0.0692 degree. The zoom can take values from 0 (largest camera angle) to 128 (smallest camera angle).

The tracking environment is a normal office. Figure 2(a) is a sketch of the top view of the environment. Region A is the most distant part of the office visible from the camera. Figure 2(b) gives a global view of the environment, as constructed from three camera images, with pan = 0, 525, and 1050, and constant tilt of 277 and zoom 0.

These three camera settings suffice for a complete scan of the office environment, and thus comprise the Minimum Camera Parameter Settings for our tracking task. To improve smoothness of tracking, however, we allow the pan to increment in steps of 75, from 0 to 1050, with tilt constant at 277 and zoom of 0.

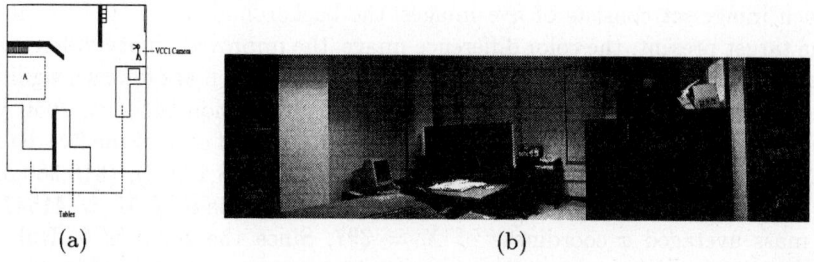

Fig. 2. (a) Top view of the tracking environment. (b) Global view of the tracking environment.

We identify these states in what follows by their pan value. One additional state, called $600'$, with pan 600, tilt 199 and zoom 55 is included to capture the more distant area A (cf. Fig. 2). We next describe the inference engine which controls the movement of the camera during tracking in the environment of Fig. 2(b):

1. Repeatedly scan the environment using camera states (pan) 0, 525, and 1050, since these comprise the Minimum Set of Camera Parameters. If a target is detected calculate its total mass M and average x-coordinate X, and Goto (2).
2. If the current zoom is 0 then select the next $< pan, tilt, zoom >$ using Method (a) below, otherwise use Method (b).
   (a) **Select pan value:** Let $p_i = i * 75$ be the current pan value, and P be the set of pan values including p_i and the next three lower and three higher pan values allowed. The set P includes all of the pan values with viewing directions that fall in the current image. The x-coordinates of the intersection of these viewing directions with the image plane are: 81, 173, 233, 320, 407, 467, and 559, from lowest to highest pan values in P. Select the next pan direction p_k from P such that the corresponding x-coordinate x_k of intersection with the image plane is closest to X.
   Select tilt and zoom values: If the next pan $p_k = 600$, and $M < 10000$, then select camera state $600'$ ($< pan, tilt, zoom > = < 600, 199, 55 >$) as the next action for tracking. (The direction and low mass imply that the person is within Region A, which being distant from the camera requires a small angle size). Else the tilt and zoom remain unchanged.
   (b) **Select pan, tilt and zoom values:** (The current zoom is 55, i.e., camera state $600'$). If $M < 31,100$ then do not change the camera state. (The direction and mass suggest that the person is still in Region A.) Else select camera state 525, ($< pan, tilt, zoom > = < 525, 277, 0 >$), as the next action. (Apparently the person has just left region A).
3. Adjust the camera to the new state, take a picture and calculate the new mass M and x-coordinate X of the target.
4. Goto Step 2 to select the next camera parameters for tracking.

The nine actions and image sets for this experiment are shown in Fig. 3.

Each image set consists of five images: the background image, the image with the target present, the color difference image, the improved binary difference image, and the color difference image overlaid with a cross mark for each significant segmented blob. The sequence in Fig. 3 begins with Action 1 in state 1050 where the human is first detected. The coordinates and mass of each of the five detected target blobs are: (x, y, m) = (309,205,16013), (332,68,13006), (318,360,5202), (422,180,5714), and (416,33,1612), yielding a total mass of M = 41547 and a mass averaged x-coordinate of X = 337. Since the zoom is 0, Rule (2a) of the inference engine applies, and the next state selected is 1050 again. In Action 2, (blobs: $(125, 170, 29670)$), the target is calculated to be at position X = 125, and according to Rule (2a) the pan must be decreased three units to 825. Action 3 (blobs: (289,115,5040),(331,212,13111),(283,35,2362)) finds the person near the center again, so the state does not change. In Action 4 (blobs: (79, 99, 4535), (50, 182, 1121), (169, 21, 5085), (109, 306, 3012), (123, 195, 1281), (175, 87, 1300)) the person is left of center, and the pan is appropriately changed for tracking to that shown in Action 5. Here the target (blobs: (279,107,8772),(221,187,1284),(291,294,2432),(299,21,3458)) is near center again, hence no camera change for Action 6. In Action 6 the target is left of center, (blobs: (210,236,1054),(227,101,4536),(260,17,2834)) suggesting a next pan value of 600, which invokes Rule (2b). This rule checks the size of the target, which being small causes an increase in zoom to that shown in Action 7. Action 7 (blobs: ((373 , 221 , 13438), (376 , 50 , 7314), (368 , 364 , 2307), (485 , 82 , 6445), (503 , 10 , 1346)) produces no change in state for the next action. In Action 8, (blobs: (137 , 204, 21174), (180 , 37, 8517), (129 , 387, 1262), (181 , 389, 1357)) the target mass increases sufficiently to reset the zoom, as shown in Action 9. Blobs found in Action 9 are: (258 , 204 , 4794), (282 , 43 , 2607), (322 , 94 , 3821), and (323 , 216 , 1031). At this point the experiment is terminated. Thus, the person was successfully tracked during a walk about the office.

6 Conclusion

This paper proposes a novel tracking strategy that can robustly track a person, or other object within an environment by a pan, tilt, and zoom camera with the help of a pre-recorded image database. We define a concept called Minimum Camera Parameter Settings (MCPS) which gives the minimum number of camera states required to detect the target anywhere within a given region. For each camera parameter setting in MCPS, we pre-record an image of the environment, and this set of camera states is used during tracking. When the target appears within an image, we segment target from background while using the corresponding background image as a reference. This can greatly simplify segmentation, and the main part of the person's body can be detected robustly. In order to guarantee smooth tracking, we can increase the number of camera states in the above process.

Since the camera is actively controlled during tracking, and segmentation is based on comparison of images taken with the same camera parameters, our

1. $<p=1050, t=277, z=0> \Longrightarrow [X=337, M=41547]$.

2. $<p=1050, t=277, z=0> \Longrightarrow [X=125, M=29670]$.

3. $<p=825, t=277, z=0> \Longrightarrow [X=315, M=20513]$.

4. $<p=825, t=277, z=0> \Longrightarrow [X=128, M=18563]$.

5. $<p=675, t=277, z=0> \Longrightarrow [X=280, M=15946]$.

6. $<p=675, t=277, z=0> \Longrightarrow [X=234, M=8424]$.

7. $<p=600, t=199, z=55> \Longrightarrow [X=402, M=30850]$.

8. $<p=600, t=199, z=55> \Longrightarrow [X=149, M=32310]$.

9. $<p=525, t=277, z=0> \Longrightarrow [X=288, M=12253]$.

Fig. 3. A tracking experiment performed in our Lab.

method requires excellent mechanical reproducibility. We tested our strategy with the Canon VCC1 Camera, and the tracking results are satisfactory. Complexity of the environment is not a problem in segmentation, however the simple segmentation algorithm which we use in this paper does depend on the constancy of the background. More sophisticated segmentation methods can also be incorporated in the same overall strategy. Our results show that through the use of a few pre-recorded background images and active control of the camera, the task of visual tracking can be simplified. This strategy may find applications in many practical situations such as human machine interaction and automated surveillance.

Acknowledgements

We would like to thank James Maclean and Gilbert Verghese for their help. This work was funded by IBM Center for Advanced Studies, Canada and the Department of Computer Science, University of Toronto.

References

1. D.M. Gavrila and L.S. Davis. 3-d model based tracking of humans in action: a multi-view approach. In *CVPR*, pages 73–79, 1996.
2. H.P. Graf, E. Cosatto, D. Gibbon, M. Kocheisen, and E. Petajan. Multi-modal system for locating heads and faces. In *International Conference on Automatic Face and Gesture Recognition*, pages 88–93, Killington, Vermont, October 1996.
3. S.X. Ju, M.J. Black, and Y. Yacoob. Cardboard people: A parameterized model of articulated image motion. In *International Conference on Automatic Face and Gesture Recognition*, pages 38–44, Killington, Vermont, October 1996.
4. I. Kakadiaris and D. Metaxas. Model-based estimation of 3d human motion with occlusion based on active multi-viewpoint selection. In *CVPR*, pages 81–87, 1996.
5. J.J. Kuch and T.S. Huang. Vision based hand modeling and tracking. In *Proceedings of International Conference on Computer VIsion*, pages 81–87, 1996.
6. B. Moghaddam, T. Darrell and A. P. Pentland. Active face tracking and pose estimation in an interactive room. In *CVPR*, pages 67–71, 1996.
7. J. Noh, D. Huttenlocher and W. Rucklidge. Tracking non-rigid objects in complex scenes. In *ICCV93*, pages 93–101, 1993.
8. N. Oliver, A. Pentland and A. Lafter. Lips and Face Real Time Tracker. In *CVPR*, 1997.
9. K. Rohr. Towards model-based recognition of human movements in image sequences. *CVGIP: Image Understanding*, 59(1):94–115, 1986.
10. C. Wren, A. Azarbayejani, T. Darrell, and A. Pentland. Pfinder: Real-time tracking of the human body. In *International Conference on Automatic Face and Gesture Recognition*, pages 51–60, Killington, Vermont, October 1996.
11. J. Yang and A. Waibel. A Real-Time Face Tracker. In *WACV*, 1996.
12. Y. Ye. Sensor Planning for Object Search. PhD Thesis, Department of Computer Science, University of Toronto, January 17, 1997.

Recovery of Motion and Structure from Optical Flow under Perspective Projection by Solving Linear Simultaneous Equations

Toshiharu Mukai[1] and Noboru Ohnishi[2]

[1] RIKEN, 2271-130 Anagahora, Shimoshidami, Moriyama-ku, Nagoya, 463 Japan
[2] Nagoya University, Furo-cho, Chikusa-ku, Nagoya, 464-01 Japan

Abstract. Determination of the three-dimensional motion and structure of an object from its optical flow is one of the most important problems in computer vision. Previous works in this field have been unsatisfactory because they require solving nonlinear simultaneous equations using iterative search. In the present paper, we propose a linear method for the recovery of motion and structure from perspectively projected optical flow of feature points which move rigidly. Furthermore, we propose a reliability measure of the recovery. Simulation results of our method are also presented.

1 Introduction

The recovery of three-dimensional motion and structure from its projected velocities on a two-dimensional screen (optical flow) is one of the most important issues in computer vision. It can be used in many fields such as three-dimensional object modeling, tracking, passive navigation, and robot vision.

This approach can be separated into two categories. One is under orthographic projection and the other is under perspective projection. Because the former is an approximation of the latter, and the approximation is true only when the object is far away from the camera, we consider only the latter in the present paper.

Some previous work on motion and structure from optical flow under perspective projection has been reported. Prazdny [7] discussed the possibility of using optical flow information to recover the relative depth of points in space. Bruss and Horn [2] proposed a method for calculating motion parameters from optical flow using the least squares technique. Adiv [1] modified Bruss and Horn's method and applied it to simulated and real images. Murray and Buxton [6] discussed many aspects of the motion from optical flow problem. However, those reports are still unsatisfactory because, in practice, those methods rely on solving nonlinear simultaneous equations using iterative search [8] (or adopting the optimal value of finite candidates as the solution [3], though that is merely an approximation of the true solution), unless very special assumptions or simplifications are made. For example, in [5, 4], points are assumed to be on the same plane and in [2], as a special case, the motion is assumed to be purely translational or purely rotational.

Solving nonlinear simultaneous equations causes two difficult problems. First, how many points are sufficient to guarantee the uniqueness of the solution? When the number of equations is equal to that of unknowns, some solution may be obtained. However, there is no guarantee that the solution is unique. Second, how can one avoid iterative search? To solve nonlinear equations, a global search is needed, unless the initial value is sufficiently near the true solution. A global search requires much computation cost and may lead to a local minimum or may not converge at all.

In the present paper, we propose a method of recovering the motion and structure from instantaneous optical flow of feature points which move rigidly under perspective projection. In order to simplify the computation and obtain clear perspectives, we adopt a spherical image screen and arrange equations in linear form. As a result, our method does not require the solution of nonlinear simultaneous equations. In addition, the uniqueness of the solution is guaranteed and improvement of recovery by increasing observed points is easily achieved. Recently, we noticed that Zhuang et al. [9] also proposed a linear method for the recovery from optical flow. The differences are 1)the view point of derivation, 2)explicit form of equations, though their fundamental principles are the same. We will also propose a reliability measure of recovery. Finally, simulation results of our method are presented.

2 Theoretical Framework

We assume that relative rigid motion of an object and an observer is detected by a monocular camera mounted on the observer. Thus we can consider, without losing generality, that the object is fixed and only the camera is moving.

Let us consider two coordinate systems. One is a camera coordinate system with its origin at the camera center, which moves with the camera. The other is a world coordinate system fixed in the world. The world coordinate system is set to coincide with the instantaneous position and orientation of the camera coordinate system at the time when optical flow is observed, without losing generality. Hereafter we use the notation such that the coordinates of a vector, e.g., q, in the world coordinate system is represented by $\tilde{q}$ in the camera coordinate system. Note that though coordinates of a vector in both coordinate systems, q and $\tilde{q}$, coincide with each other, coordinates of their time derivatives, $\dot{q}$ and $\dot{\tilde{q}}$, may not coincide because of rotation of the camera coordinate system.

The position of an observed point is expressed by a position vector r in the world coordinate system. We assume that feature points such as surface marks or surface orientation discontinuities of objects are observed. In perspective projection, the direction of a ray from the observed point to the camera center is detected by the camera. Thus we adopt a spherical pinhole camera of unit radius as a general model for imaging devices (Figure 1). In this model, the projection of the observed point r on the spherical image screen is represented by a unit vector

$$\tilde{q} = q = r/\|r\|. \tag{1}$$

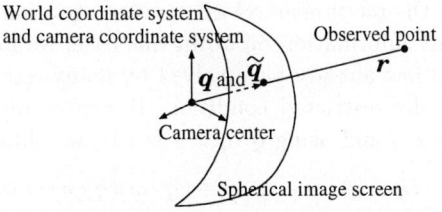

Fig. 1. Spherical image screen

The instantaneous motion of the camera can always be expressed by two components: a rectilinear (or translational) velocity and an angular (or rotational) velocity. The rectilinear velocity is denoted by the vector v in the world coordinate system and the angular velocity by the vector ω the direction of which coincides with the axis of rotation, and the norm of which is the magnitude of the angular velocity. It is easily obtained that time derivatives $\dot{q}$ and $\dot{\tilde{q}}$ of q and $\tilde{q}$, respectively, have the relationship

$$\dot{q} = \dot{\tilde{q}} + \omega \times \tilde{q}, \tag{2}$$

where $\times$ denotes an outer product.

The camera is assumed to detect $\tilde{q}$ and $\dot{\tilde{q}}$ of feature points. The purpose of the present study is the recovery of v, ω, and r of observed points from $\tilde{q}$ and $\dot{\tilde{q}}$. As easily shown, changing the scale of the entire setup does not affect the observed values, $\tilde{q}$ and $\dot{\tilde{q}}$. Therefore the scale of v cannot be obtained and it is set arbitrary when $v \neq o$. We select a unit vector as the recovered velocity when $v \neq o$. Hence the number of unknowns in motion parameters, v, ω, is five. Note that the recovery of feature point positions r also depend on the scale of v.

We use the following notation for elements of vectors and matrices: e.g., $q = [q_1\ q_2\ q_3]^T$ and $A = (a_{ij})$.

3 Recovery of Motion

We first obtain the rectilinear and angular velocities.

Consider observation of points at time t and $t+\delta t$. It is clear that $q(t), q(t+\delta t)$ and δu are on the same plane, where $q(t)$ and $q(t + \delta t)$ are projections of the observed point on the spherical image screen at time t and $t + \delta t$, respectively, and δu is the displacement of the camera center position during δt. Therefore the scalar triple product of these vectors is zero,

$$(q(t + \delta t) \times q(t)) \cdot \delta u = 0, \tag{3}$$

where $\times$ and $\cdot$ denote outer and inner products, respectively. Taking $\delta t \longrightarrow 0$ and using (2), we obtain

$$((\dot{\tilde{q}} + \omega \times \tilde{q}) \times \tilde{q}) \cdot v = 0. \tag{4}$$

Note that when the rectilinear velocity v is o, equation (4) always holds and does not include any information on ω. In this case, we must solve other equations. This problem has already been solved by many researchers (for example, see [2]) because of the restricted condition. Hereafter we assume that $v \neq o$. Expanding $(\omega \times \tilde{q}) \times \tilde{q}$ and using $\tilde{q} \cdot \tilde{q} = 1$ in (4), we obtain

$$(\dot{\tilde{q}} \times \tilde{q}) \cdot v - \omega \cdot v + (\tilde{q} \cdot \omega)(\tilde{q} \cdot v) = 0. \tag{5}$$

This is the fundamental equation of the relationship between observed values and the camera's (instantaneous) motion. Each observed point yields this equation. To calculate the five unknowns mentioned above, observation of five or more than five points is needed. By solving nonlinear simultaneous equations from these points, the camera's (instantaneous) motion may be determined. However, nonlinear simultaneous equations are usually solved by a numerical iterative method, which causes the problems mentioned in Section 1.

Therefore we proceed further to obtain equations which provide better perspectives. Regarding all vectors as column vectors and defining $b = \dot{\tilde{q}} \times \tilde{q}$ and $A = \tilde{q}\tilde{q}^T - I$ where I is the 3×3 identity matrix, (5) can be rewritten

$$b^T v + \omega^T A v = 0. \tag{6}$$

Let us define $d = [a_{11}\ a_{22}\ a_{33}\ a_{12}\ a_{23}\ a_{31}]^T$ and

$$m = [\omega_1 v_1, \omega_2 v_2, \omega_3 v_3, \omega_1 v_2 + \omega_2 v_1, \omega_2 v_3 + \omega_3 v_2, \omega_3 v_1 + \omega_1 v_3]^T.$$

When observing n points, we obtain

$$\begin{bmatrix} b^1\ b^2\ \ldots\ b^n \\ d^1\ b^2\ \ldots\ b^n \end{bmatrix}^T \begin{bmatrix} v \\ m \end{bmatrix} = o, \tag{7}$$

where we denote b from the point i as b^i, and so forth. We rewrite the above equation as $Gx = o$ for convenience. Then the size of matrix G is $n \times 9$ and that of the vector x is 9. Because the norm of x depends on the scale of v, the scale of x has no meaning. Therefore the rank of G for solving the equation is necessarily and sufficiently 8. If the rank is 9, the optimal solution in the least-squares-error sense should be obtained. The condition for making the rank of the matrix G equal to 8 is under investigation. However we expect that, on the basis of many simulations, if more than 8 points are at general positions, the rank is 8.

When $x^T = [v^T m^T]$ is obtained, ω is easily computed. From the definition of m,

$$\begin{bmatrix} v_1 & & v_2 & v_3 \\ v_2 & v_1 & v_3 & \\ & v_3 & v_2 & v_1 \end{bmatrix}^T \omega = m. \tag{8}$$

When $v \neq o$, as easily proved, the matrix is of rank 3 and ω can be obtained in the least-squares-error sense.

Note that though 8 points at general positions are needed to obtain x, it does not necessarily mean that 8 points are needed to obtain v and ω. In other words, x is a redundant expression of v and ω. However, because of this redundancy, we can obtain solutions merely by solving linear simultaneous equations.

4 Recovery of Structure

Once the velocities v, ω are determined by the above method, positions of observed points r^i ($i = 1, 2, \ldots, n$) can be estimated. As mentioned above, the scale of v cannot be obtained and point positions r^i depend on the scale. Thus we express the true rectilinear velocity as sv, where s is the unknown scale factor.

First, $\dot{q}^i$ is obtained using (2). In this section, we consider the coordinate system with its origin at the camera center, which translates with the camera but does not rotate with respect to the world coordinate system. In this coordinate system, the object points move with the rectilinear velocity $-sv$, as shown in Figure 2. From this geometry, we can obtain the following result.

$$r^i = -s(v \cdot \dot{q}^i)/\|\dot{q}^i\|^2 q^i \qquad (9)$$

Note that when $v = o$, optical flow does not include depth information. Therefore, the structure of observed points cannot be recovered.

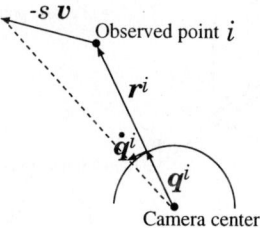

Fig. 2. Geometry of an observed point and the camera.

5 Reliability Measure of the Estimates

In almost all practical situations, optical flow cannot be obtained exactly. The errors in optical flow cause errors in the recovery of motion and structure. Hence improvement of estimates by some method (for example, a Kalman filter can be used) is important.

In most of these methods for improvement of estimation, reliability measure of estimates is needed. We consider that the reciprocal of the error of (7)

$$R_G = \|G\hat{x}\|^{-1}, \qquad (10)$$

where $\hat{x}$ is the solution of (7) and $\|\hat{x}\| = 1$, is one of candidates for reliability measure when the number of observed points n is fixed. Reliability measure when n is not fixed is under investigation.

6 Simulation Results

6.1 Errors of estimates

In the following computer simulations, we assumed the environment illustrated in Figure 3. The image screen is a part of a sphere with a 0.5 radian visual angle around the z-axis. Feature points are randomly positioned in the cube, the length of which is l, and the center of which is 50 from the camera center along the z-axis. Since the scale has no meaning, the length has no unit. From the true coordinates of a feature point and camera motion, projected position and optical flow are computed. Then they are perturbed by noise. Each $\tilde{q}$ and $\dot{\tilde{q}}$ are perturbed as

$$\tilde{q}' = \tilde{q} + n_1, \quad \dot{\tilde{q}}' = \dot{\tilde{q}} + n_2 \tag{11}$$

where n_1, n_2 are random noises which are in the tangent to the spherical image screen at the point represented by $\tilde{q}$, and $\|n_1\|, \|n_2\|$ are subjected to the Gaussian distribution with the standard deviation of μ_n [%] of an arc of the visual angle on the image screen.

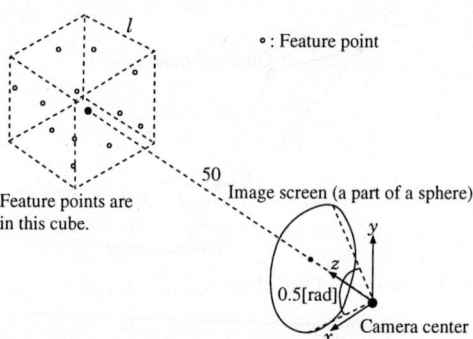

Fig. 3. Simulation environment.

Three kinds of error

$$E_{\text{vel}} = \|v/\|v\| - \hat{v}\|, \quad E_{\text{ang}} = \|\omega - \hat{\omega}\|/\|\omega\|, \quad E_{\text{pts}} = \sum_{i=1}^{n} \|r^i - \hat{r}^i\|/(nl)$$

are assumed. E_{vel} is the error of rectilinear velocity where v and $\hat{v}$ are the true and estimated rectilinear velocities, respectively. Hereafter we use similar notation. The true rectilinear velocity v is normalized because $\hat{v}$ is a unit vector. E_{ang} is the error of angular velocity and E_{pts} is the error of positions of observed points where n is the number of feature points. The estimated point position $\hat{r}^i$ is obtained from (9) and it requires the scale factor s. We used the norm of the true rectilinear velocity, $\|v\|$, as s in the following simulations. In graphs of simulation results in Figure 4, the average error of 20 trials is plotted as a point.

The following values are used as defaults in simulations and one of them is varied in each simulation: number of observed points $n = 30$, noise scale

$\mu_n = 0.2$ [%], length of a side of the cube $l = 20$, rectilinear velocity of the camera $\boldsymbol{v} = [10\ 3\ 12]^T$, and angular velocity of the camera $\boldsymbol{\omega} = [0.3\ 0.7\ 0.6]^T$ [rad]. The details of each simulation are as follows.

Simulation 1 : The number of observed points n is varied from 8 to 50. When the number is small, errors decrease with the number.

Simulation 2 : Noise scale μ_n is varied from 0[%] to 1[%]. It is shown that the method is sensitive to noise. In particular, rectilinear velocity is very sensitive.

Simulation 3 : The length of a side of the cube l is varied from 10 to 20. It is shown that, in order to obtain good estimates, observed points should be widely spread across the image screen.

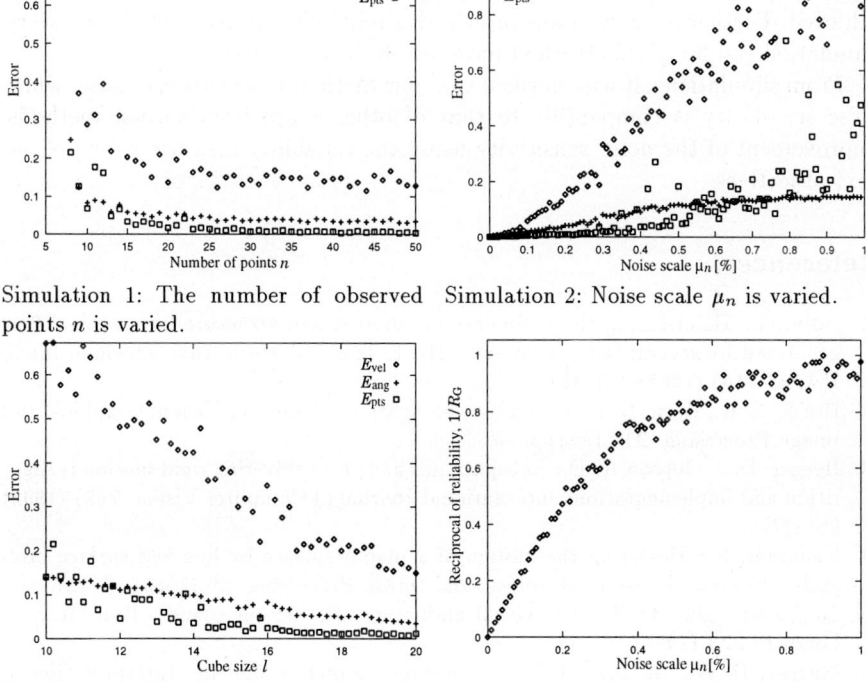

Simulation 1: The number of observed points n is varied.

Simulation 2: Noise scale μ_n is varied.

Simulation 3: The length of a side of the cube l is varied.

Simulation 4: The reciprocal of reliability measure, $1/R_G$, when noise scale μ_n is varied.

Fig. 4. Simulation results

6.2 Reliability of estimates

The reliability measure R_G is examined in Simulation 4. Simulation environment and default values are the same as those in Simulation 2. The reciprocal of the proposed reliability measure R_G when noise scale μ_n is varied from 0[%] to 1[%] is plotted, where $1/R_G$ is normalized as the maximum value is 1. Because the condition of this simulation is the same as that of Simulation 2, the errors of estimates, though they are not plotted in the figure to avoid complication, are also the same. From these graphs, it is known that $1/R_G$ increases with the errors of estimates, $E_{\text{vel}}, E_{\text{ang}}$ and E_{pts}. Therefore, when the number of observed points n is fixed, R_G can be used as a reliability measure.

7 Conclusion

We have proposed a method for the recovery of motion and structure from instantaneous optical flow of feature points which move rigidly. The tasks for solving this method are very simple. In addition, the uniqueness of the solution is guaranteed and improvement of recovery by increasing observed points is easily achieved. Furthermore, we have proposed a reliability measure of the recovery. Simulation results of this method have also been presented.

From simulations, it was clarified that our method is sensitive to noise, which noise sensitivity is comparable to that of other shape-from-motion methods. Improvement of the noise sensitivity using the reliability measure is one of our next objectives.

References

1. Adiv, G.: Determining three-dimensional motion and structure from optical flow generated by several moving objects. IEEE Trans. Pattern Anal. Machine Intell. **PAMI-7(4)** (1985) 384–401
2. Bruss, A. R., Horn, B. K. P.: Passive navigation. Computer Vision, Graphics, and Image Processing. **21** (1983) 3–20
3. Heeger, D. J., Jepson, A. D.: Subspace methods for recovering rigid motion i: Algorithm and implementation. International Journal of Computer Vision. **7(2)** (1992) 95–117
4. Kanatani, K.: Detecting the motion of a planar surface by line and surface integrals. Computer Vision, Graphics, and Image Processing. **29** (1985) 13–22
5. Longuet-Higgins, C. H.: The visual ambiguity of a moving plane. Proc. R. Soc. Lond. **B 223** (1984) 165–175
6. Murray, D. W., Buxton, B. F.: Experiments in the Machine Interpretation of Visual Motion. The MIT Press (1990)
7. Prazdny, K.: On the information in optical flows. Computer Vision, Graphics, and Image Processing. **22** (1983) 239–259
8. Weng, J., Huang, T. S., Ahuja, N.: Motion and structure from two perspective views: Algorithms, error analysis, and error estimation. IEEE Trans. Pattern Anal. Machine Intell. **11(5)** (1989) 451–476
9. Zhuang, X., Huang, T. S., Ahuja, N.: A Simplified Linear Optic Flow-Motion Algorithm. Computer Vision, Graphics, and Image Processing. **42** (1988) 334–344

Vector Coherence Mapping: A Parallelizable Approach to Image Flow Computation

Francis K. H. Quek and Robert K. Bryll

Vision Interfaces and Systems Laboratory (VISLab)
EECS Dept., The University of Illinois at Chicago
Email: quek@eecs.uic.edu

Abstract. We present a new parallel approach for the computation of an optical flow field from a video image sequence. This approach incorporates the various local smoothness, spatial and temporal coherence constraints transparently by the application of fuzzy image processing techniques. Our *Vector Coherence Mapping* VCM approach accomplishes this by a weighted voting process in "local vector space," where the weights provide high level guidance to the local voting process. Our results show that VCM is capable of extracting flow fields for video streams with global dominant fields (e.g. owing to camera pan or translation), moving camera and moving object(s), and multiple moving objects. Our results also show that VCM is able to operate under strong image noise and motion blur, and is not susceptible to boundary oversmoothing.

1 Introduction

We present a new parallel approach for the computation of an optical flow field from a video image sequence. This approach incorporates the various local smoothness, spatial and temporal coherence constraints transparently by the application of fuzzy image processing techniques. Our *Vector Coherence Mapping* VCM approach accomplishes this by a weighted voting process in "local vector space," where the weights provide high level guidance to the local voting process. Our results show that VCM is capable of extracting flow fields for video streams with global dominant fields (e.g. owing to camera pan or translation), moving camera and moving object(s), and multiple moving objects. Our results also show that VCM is able to operate under strong image noise and motion blur, and is not susceptible to boundary oversmoothing.

Barron et al [1] provides a good review of optical flow techniques and their performance. We shall adopt their philology of the field to put our work in context. They classify optical flow computation techniques into four classes. The first of these, pioneered by Horn and Schunck [2] computes optical fields using the spatial-temporal gradients in an image sequence by the application of an image flow equation. Recent examples of this approach include [3, 4]. The second class performs "region-based matching" by explicit correlation of feature points and computing coherent fields by maximizing some similarity measure. Recent examples of this approach are [5, 6]. The third class are "energy-based" methods which extract image flow fields in the frequency domain by the application of "velocity-tuned" filters. Examples of this approach include [7, 8]. The fourth class are "phase-based" methods which extract optical flow using the phase behavior of band-pass filters. Barron et al include zero-crossing approaches such as that due to Hildreth [9] under this category. Other examples of this approach include [10, 11]. Under this classification, our approach falls under the second (region-based matching) category.

The contribution of the VCM approach presented here is that it combines the correlation and constraint-based smoothing processes into a set of fuzzy image processing operations. The algorithm is parallel and obviates the iterative post process. In essence, VCM performs a voting process in vector parameter space and biases this voting by likelihood distributions that enforce the spatial and temporal constraints. Hence, VCM is similar to the Hough based approaches [12, 5]. The difference is that in VCM, the voting is distributed and the constraints enforced on each vector is local to the region of the vector. Furthermore, in VCM the correlation and constraint enforcement functions are integrated in such a way that the constraints "guide" the correlation process by the likelihood distribution. The Hough methods, on the other hand, apply a global voting space.

Our results show that VCM has good noise immunity. Unlike other approaches which use such techniques as M-estimators to enforce robustness [13, 5], the robustness of VCM lies in the fact that correlation errors owing to noise occur in image space, and have little support in the parameter space of the vectors.

2 Coherence Constraints

Spatial Coherence
Let $\mathcal{P}^t = \{p_i^t\}_{i=1}^N$ be the set of interest points detected in image I^t at time t by some suitable interest operator. Since VCM is feature-agnostic, we apply a simple image gradient operator and select high gradient points.

For a particular interest point p_i^t in image I^t, we can estimate its new position in image $I^{t+\delta t}$ by computing the correlation of the neighborhood of p_i^t in $I^{t+\delta t}$. This estimate, however, is very susceptible to image noise and chance correlations. Our approach uses the weighted aggregates of neighboring correlations to obtain a stable vector field. We define a *Normal Correlation Map* ncm for some point p_i^t to image $I^{t+\delta t}$ to be the correlation response of the region around p_i^t in I^t with an rectangular region in $I^{t+\delta t}$ centered at the coordinates p_i^t. We use *absolute difference correlation* ADC [14] to perform the correlation. Hence the ncm is:

$$\mathcal{N}(p_i^t)[m,n] = \sum_{j=x_i^t-N}^{x_i^t+N} \sum_{k=y_i^t-N}^{y_i^t+N} |I^t[j,k] - I^{t+\delta t}[m+j,n+k]| \qquad (1)$$
$$-D_x < m < D_x; \; -D_y < n < D_y$$

where $p_i^t \equiv (x_i^t, y_i^t)$, $2N+1$ is the x and y dimension of the correlation template, and D_x and D_y define the maximal expected x and y displacement of p_i at $t+\delta t$.

We define the *Vector Coherence Map vcm* at p_i^t to be:

$$vcm(p_i^t) = \frac{1}{\sum_{j=1}^{|\mathcal{P}^t|} W_i^t(p_j^t)} \sum_{j=1}^{|\mathcal{P}^t|} \mathcal{N}(p_j^t) \times W_i^t(p_j^t) \qquad (2)$$

where $0 \leq W_i^t(p_j^t) \leq 1$ is some weighting function of the contribution of the *ncm* $\mathcal{N}(p_j^t)$ of point p_j^t on the vector at p_i^t, and $\mathcal{P}^t$ is the set of interest points in I^t.

$W_i^t(p_j^t)$ permits us to enforce a variety of spatial coherence constraints on the vector field. For spatial proximity coherence, we can employ the Euclidean or Checkerboard distance respectively by applying: $W_i^t(p_j^t) = \mathcal{S}(k_1, k_2, |p_j^t - p_i^t|)$ or $W_i^t(p_j^t) = \mathcal{S}\left(k_1, k_2, max(x_j^t - x_i^t, y_j^t - y_i^t)\right)$, where $p_i^t \equiv (x_i^t, y_i^t)$ and $p_j^t \equiv (x_j^t, y_j^t)$ and $k_1 < k_2$ are weighting constants for the sigmoidal function:

$$\begin{aligned}\mathcal{S}(k_1, k_2, d) &= 1 \quad \text{for } d \leq k_1 \\ &= \frac{1 - \epsilon - \mathcal{F}(k_1, k_2, d)}{1 - 2\epsilon} \quad \text{for } k_1 < d < k_2 \\ &= 0 \quad \text{for } d \geq k_2\end{aligned} \qquad (3)$$

$0 < \epsilon \ll 1$, and $\mathcal{F}(k_1, k_2, d) = 1/\left(1 + e^{-a\left(d - \frac{k_2 + k_1}{2}\right)}\right)$ where $a = \frac{-2}{k_2 - k_1} \ln\left(\frac{\epsilon}{1-\epsilon}\right)$

Hence the *vcm* implements a voting scheme by which neighborhood point correlations affect the vector v_i^t at point p_i^t. We can convert this into a 'likelihood-map' for v_i^t by normalizing it, subject to a noise threshold T_{vcm}:

$$|vcm(p_i^t)| = \frac{vcm(p_i^t)}{peak'(vcm(p_i^t))} \qquad (4)$$

where $peak'(vcm(p_i^t))$ is the peak value of $vcm(p_i^t)$ if it is above threshold T_{vcm}, and ∞ otherwise.

$|vcm(p_i^t)|$ maps the likelihood of terminal points for vectors originating from p_i^t due to neighborhood point correlations. Computation of the dominant translation field through a video sequence is important for the segmentation of video streams [13]. *VCM* can compute such a field by setting p_i^t be some imaginary point, and using a uniform global weighting function: $W_i^t(p_j^t) = 1 \quad \forall j$. This global *vcm* corresponds to the dominant translation occurring in the frame (e.g. due to camera pan). Furthermore, a *vcm* can be computed for ANY point in image I^t whether or not it is an interest point. A *vcm* built in this way can be used to estimate optical flow at any point, so a *dense optic flow* field can be computed.

For vector tracking, however, it is undesirable to select a vector for which there is no local evidence (i.e. no appropriate correlation for p_i^t in image $I^{t+\delta t}$). This evidence may be found in the *ncm* $\mathcal{N}(p_i^t)$. To achieve this, we first normalize $\mathcal{N}(p_i^t)$ using the sigmoidal function from equation 3:

$$|\mathcal{N}(p_i^t)| = 1 - \mathcal{S}(T_W - \delta, T_W + \delta, \mathcal{N}(p_i^t)) \qquad (5)$$

where T_W is a threshold and δ controls the steepness of the sigmoidal function. We apply a fuzzy-AND operation of $|vcm(p_i^t)|$ and $|\mathcal{N}(p_i^t)|$ to obtain a 'likelihood-map' for v_i^t with both local and neighborhood support as follows:

$$\mathcal{L}_{spatial}(p_i^t) = |vcm(p_i^t)| \otimes |\mathcal{N}(p_i^t)| \qquad (6)$$

where $\otimes$ denotes pixel-wise multiplication. $\mathcal{L}_{spatial}(p_i^t)$ is the likelihood-map for v_i^t owing to spatial coherence constraints.

Temporal Coherence

To enforce a temporal coherence constraint, we employ a piecewise linear dynamic model similar to that of [15]. We introduce the expected velocity vector $\hat{v}_i^t$ for point p_i^t. We can compute $\hat{v}_i^t$ given the previous vector $v_i^{t-\delta t}$ and the previous acceleration vector $a_i^{t-\delta t}$. Assuming constant acceleration, $a_i^t = a_i^{t-\delta t}$, we can estimate: $\hat{p}_i^{t+\delta t} = p_i^t + v_i^t + \frac{1}{2}a_i^t$ and $\hat{v}_i^t = p_i^{t+\delta t} - p_i^t = v_i^t + \frac{1}{2}a_i^t$. We have $v_i^{t-\delta t}$ from the tracking history and we can estimate $a_i^{t-\delta t} = v_i^{t-\delta t} - v_i^{t-2\delta t}$.

Given an expected vector $\hat{v}_i^t$, we introduce the idea of the *scatter template*. We make the observation that the larger the expected vector, the larger would be the region of possible destination of the real velocity vector.

Hence, we apply a scatter template $\mathcal{T}_i^t$ centered at $\hat{p}_i^{t+\delta t}$. The template is applied to every point $x(k,l)$ of the *ncm* $\mathcal{N}(p_i^t)$. We can implement the scatter template $\mathcal{T}_i^t$ using sigmoid function from equation 3. For every point $x(k,l)$ belonging to $\mathcal{N}(p_i^t)$, the scatter template is:

$$\mathcal{T}_i^t(k,l) = \mathcal{S}(k_{1t}, k_{2t}, |x - \hat{p}_i^{t+\delta t}|) \qquad (7)$$

where $k_{1t} = f_1(|\hat{v}_i^t|)$ and $k_{2t} = f_2(|\hat{v}_i^t|)$ control the steepness of the sigmoid function with response to the expected vector length (function becomes steeper as $|\hat{v}_i^t|$ decreases).

This scatter template is fuzzy-ANDed with $\mathcal{N}(p_i^t)$ to obtain a new *temporal ncm* $\mathcal{N}_T(p_i^t) = \mathcal{N}(p_i^t) \otimes \mathcal{T}_i^t$ where $\otimes$ denotes pixel-wise multiplication. This applies the highest weight to the area of $\mathcal{N}(p_i^t)$ close to $\hat{p}_i^{t+\delta t}$ and suppresses the more distant values. We can compute a *temporal vcm* $vcm_T(p_i^t)$ for every point p_i^t:

$$vcm_T(p_i^t) = \frac{1}{\sum_{j=1}^{|\mathcal{P}^t|} W_i^t(p_j^t)} \sum_{j=1}^{|\mathcal{P}^t|} \mathcal{N}_T(p_j^t) \times W_i^t(p_j^t)$$

$vcm_T(p_i^t)$ can then be normalized and fuzzy-ANDed with $|\mathcal{N}(p_i^t)|$ (in the same way as equations 4 and 6) to obtain the likelihood map $\mathcal{L}_{s-t}(p_i^t)$ with both local and neighborhood *spatial and temporal* support:

$$\mathcal{L}_{s-t}(p_i^t) = |vcm_T(p_i^t)| \otimes |\mathcal{N}(p_i^t)| \qquad (8)$$

3 Boundary Conditions

We now address the question of when to begin and end a vector trace. There are 3 such boundary conditions: 1. Initialization at the first frame when no vectors exist. 2. The motion detector provides strong evidence for a moving object in a region that does not have an existing vector. 3. Equation 8 for vector tracking yield no suitable match for a point being tracked.

Case 1: First Frame Condition. Under this condition the *ncm* computation cannot be constrained by vector history. We may compute the velocities v_i^1 and v_i^2 for the first and second frames using the spatial coherence constraint alone (equation 6), and proceed with the temporal constraint equation for the third frame. Alternatively, we can compute v_i^1 and proceed using the estimate $v_i^2 = v_i^1$.

Case 2: Evidence for Vector in New Region. In this case, the motion-sensitive edge detector [14] provides strong evidence for a moving object in a region that does not have an existing vector. This is similar to case 1, except that it applies only to the region of interest, and not to the whole image.

Case 3: No Suitable Match. In the case when equation 8 yields no suitable match for a vector being tracked, three situations may be the cause: 1. Rapid acceleration/deceleration pushed the point beyond the search region; 2. The point has been occluded; or, 3. An error occurred in previous tracking.

Situation 1 is the most common cause for a loss of tracking in this case. We typically work with 30 fps data, and this sampling rate is insufficient when either the motion is too fast to compute the acceleration (or path curvature) accurately, or if there is an abrupt change in motion. To resolve the problem, we relax the temporal constraint and use $\mathcal{L}_{spatial}(p_i^t)$. If this yields a new vector, we have to decide if we continue tracking the point. To do this, we posit that there is a maximal allowable acceleration T_a. If the new vector does not violate this constraint (i.e. $|v_i^t - v_i^{t-1}| < T_a$) we proceed with the tracking. If the maximal acceleration condition is violated, we flag the point in the tracking sequence as a point of motion change and proceed as though this were a new motion.

In the current implementation, we compute both $\mathcal{L}_{spatial}(p_i^t)$ and $\mathcal{L}_{s-t}(p_i^t)$ for each interest point p_i^t (using equations 6 and 8, respectively). Currently, we do not try to recover a trace through temporary occlusion.

4 Vector Clustering

Once the vector field has been computed, we cluster the vectors using an interactive clustering algorithm presented in [14]. Vectors are clustered by vector location, direction and magnitude. The importance of each feature used during clustering can be adjusted.

5 Experimental Results

Figure 1 shows a noisy vector field obtained from the ADC response on a hand motion video sequence. The vector field computed for the same sequence produced a vector field characterized by the smooth field shown in figure 2.

Figure 4 presents the performance of VCM on a video sequence with a up-panning camera, and where the aperture problem is evident. Figure 3 shows the global *vcm* computed on a frame in the sequence. The bold line in the center of figure 4 show the correct image vector corresponding to an upward pan.

Figure 5 shows the results of the VCM algorithm for a synthetic image sequence in which two identical objects move in opposite directions at 15 pixels per frame. The correct field computed in figure 5b shows the efficacy of the temporal cohesion. Without this constraint, most of the vectors were produced by false correspondences between the two objects (figure 5). This experiment also shows that VCM is not prone to boundary oversmoothing.

Figure 7 shows the vector fields computed for a video sequence with a down panning camera and a moving hand. The sequence contains significant motion blur. The VCM and vector clustering algorithms extracted two distinct vector fields with no visible boundary oversmoothing. In the Jimmy Johnson sequence shown in figure 6, the subject is gesturing with both hands and nodding his head. Three distinct motion fields were correctly extracted by the VCM algorithm.

Figures 8 and 9 show the efficacy of the VCM algorithm on noisy data. The video sequence was corrupted with uniform random additive noise to give a S/N ratio of 21.7dB. Figure 8 shows the result of ADC correlation (i.e. using the ncm's alone). Figure 9 shows the vector field computed by the VCM algorithm. The difference in vector field quality is easily visible.

Figures 10, 11 and 12 show analysis of video sequences with various camera motions. The zoom-out sequence resulted in the anticipated convergent field (figure 10). Figure 11 shows the vector field for a combined up-panning and zooming sequence. Figure 12 shows the rotating field obtained from a camera rotating about it's optical axis.

6 Summary and Conclusions

We presented a parallelizable algorithm that computes coherent vector fields by the application of various coherence constraints. Since W in equation 2 and $\mathcal{T}$ of equation 7 may both be precomputed, the ncm is obtained by a regular convolution correlation process, and all other operations are pixelwise image multiplications, this algorithm is easily parallelizable.

The algorithm features a voting scheme in vector parameter space, making it robust to image noise. The spatial and temporal coherence constraints are applied using fuzzy-image-processing technique by which the constraints are applied to the bias the correlation process. Hence, the algorithm does not require the typical iterative second stage process of constraint application.

The experiment results presented substantiates the promise of the algorithm. VCM is capable of extracting vector fields out of image sequences with significant synthetic and real noise (e.g. motion blur). It produced good results on videos with multiple independent or composite (e.g. moving camera with moving object) motion fields. Our method performs well for aperture problems and permits the extraction of vector fields containing sharp discontinuities with no discernible oversmoothing effects.

References

1. J. L. Barron, D. J. Fleet, and S. S. Beauchemin, "Performance of optical flow techniques", *Int. Journal of Comp. Vision*, vol. 12, pp. 43–77, 1994.
2. B. K. P. Horn and B. G. Schunck, "Determining optical flow", *Art. Intel.*, vol. 17, pp. 185–204, 1981.
3. J. Weber and J. Malik, "Robust computation of optical flow in a multi-scale differential framework", *in Proc. 4th ICCV*, Berlin, Germany, May 11-14, 1993.
4. Massimo Tistarelli, "Computation of coherent optical flow by using multiple constraints", *in Proc. 5th ICCV*, MIT, Cambridge, MA, June 20-23, 1995.

5. T. Yu Tian and M. Shah, "Recovering 3D motion of multiple objects using adaptive Hough transform", *in Proc. 5th ICCV*, MIT, Cambridge, MA, June 20-23, 1995.
6. Q.X. Wu, "A correlation-relaxation-labeling framework for computing optical flow – Template matching from a new perspective", *PAMI*, vol. 17, pp. 843–853, 1995.
7. P. Anandan, "A computational framework and an algorithm for the measurement of visual motion", *Int. Jou. of Comp. Vision*, vol. 2, pp. 283–310, 1989.
8. A. Singh, *Optic Flow Computation: A Unified Perspective*, IEEE C.S.Press, 1992.
9. Ellen C. Hildreth, "Computations underlying the measurement of visual motion", *Art. Intel.*, vol. 23, pp. 309–354, 1984.
10. D. J. Fleet, *Measurement of image Velocity*, Kluwer Acad. Publ., Norwell, 1992.
11. Bernd Jahne, "Analytical studies of low-level motion estimators in space-time images using a unified filter concept", *in Proc. of the IEEE Conf. on CVPR*, Seattle, Washington, June 21-23, 1994.
12. G. Adiv, "Determining three-dimensional motion and structure from optical flow generated by several moving objects", *PAMI*, vol. 7, pp. 384–401, 1985.
13. Harpreet S. Sawhney, Serge Ayer, and Monika Gorkani, "Model-based 2D & 3D dominant motion estimation for mosaicing and video representation", *in Proc. 5th ICCV*, MIT, Cambridge, MA, June 20-23, 1995.
14. Francis Quek, "Eyes in the interface", *Int. J. of Image and Vision Comp.*, vol. 13, pp. 511–525, Aug. 1995.
15. I.K. Sethi and R. Jain, "Finding trajectories of feature points in a monocular image sequence", *PAMI*, vol. 9, pp. 56–73, Jan. 1987.

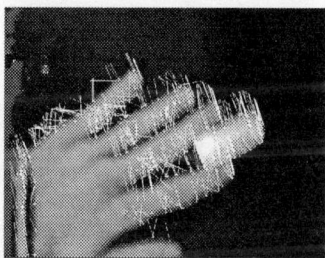

Fig. 1. Vector field obtained using ADC.

Fig. 3. Global *vcm* computed for vertical pan sequence

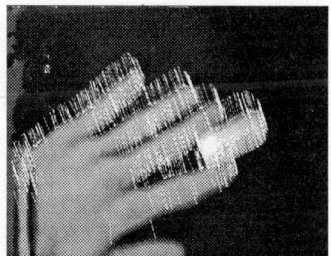

Fig. 2. Vector field obtained using VCM

Fig. 4. Vertical camera pan analyzed using VCM

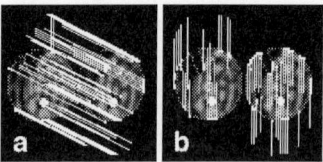

Fig. 5. Vector field obtained without and with temporal prediction respectively

Fig. 6. Hand movements detected using VCM and clustering

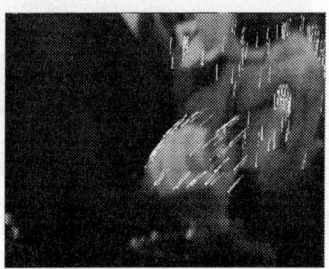

Fig. 7. Example of the vector clustering: Moving object and camera pan

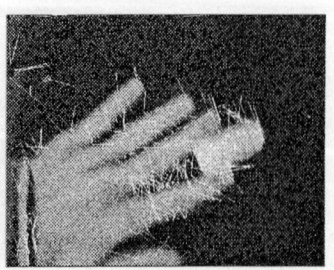

Fig. 8. Noisy frame (S/N=21.7dB) analyzed using ADC (ncm's)

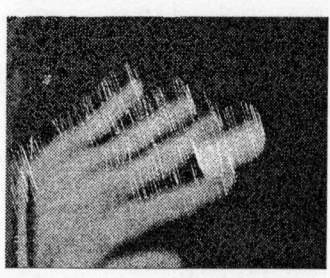

Fig. 9. Noisy frame (S/N=21.7dB) analyzed using VCM

Fig. 10. Pure zoom sequence analyzed with VCM

Fig. 11. Vector field obtained for pan and zoom sequence

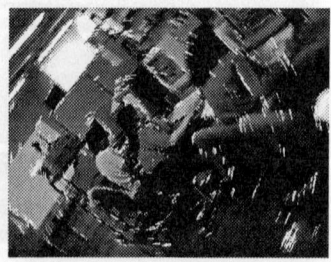

Fig. 12. Vector field obtained for camera rotating about its optical axis

Robust Motion Segmentation Using Rank Ordering Estimators

Alireza Bab-Hadiashar and David Suter
Intelligent Robotics Research Centre,
Department of Electrical & Computer Systems Engineering,
Monash University, Clayton Vic. 3168, AUSTRALIA.
E-mail: [ali, suter]@basil.eng.monash.edu.au

Abstract Robust estimators have become popular tools for solving a wide range of problems in computer vision. Despite many successes in this field, there is still a need for estimators, which are suited to specific problems such as recovering structures from multi-structural data. This paper offers an alternative approach to, and some practical insights into, the implementation of well-known rank ordering based robust estimators. The approach has been tested on synthetic and real image data for motion segmentation purposes.

1 Introduction

This paper describes an alternative approach to, and some practical insights into, the implementation of well-known rank ordering based robust estimators for motion segmentation purposes. These rank ordering based estimations are of interest because of their high break down points. High breakdown point estimators are frequently employed to solve various computer vision problems. Although some of the estimators were developed within computer vision (Stewart, 1997), almost all of them find their roots in the field of robust statistics. Indeed, the notion of fitting a parameterized model to a set of noisy data has been considered by statisticians for many years. Several robust methods have been proposed to solve many problems in different scientific fields. However, there are still problems, particularly in computer vision, for which a satisfactory method is yet to be found. This is mainly because many problems in this field are inherently complicated as they involve fitting models to multi-structural data (a set of data, which can be partitioned to different groups, each group fitting to different parametric models).

In this paper, a new approach to the general problem of fitting parametric models to multi-structural data is considered. A robust estimator named *Selective Statistical Estimator* (SSE) is developed. The SSE is based on using the least K-th Order squares (LKS) with some intuition employed to resolve the issue of what would be the best value for K.

The rest of this paper is organized as follows. The recent history of rank ordering statistics and its usage in computer vision is discussed in section 2. Section 3 offers a brief discussion on the computational cost of some of the existing robust rank ordering based estimators. The fundamental notion behind the new approach is explained in section 4. Experimental results are presented in section 5 and section 6 concludes the paper.

2 Robust Estimation

The essential idea to be developed in this paper is that too many researchers are overly concerned with retrieving the major population first. To see why this may have occurred it is perhaps instructive to look at the history of the development of robust estimators. Although the problem of fitting an a priori known model to a set of noisy data (with both biased and unbiased noise) has a long history in the statistical literature, there was not much success until the mid 80's. The break-through came in 1984, when the Least Median of Squares (LMS) was introduced (Rousseeuw, 1984). In a book published in 1987 (Rousseeuw & Leroy, 1987), the creation process of this technique is described as:

"A more complete name for the LS [least squares] method would be *least sum of squares*, but apparently few people have objected to the deletion of the word "sum"-- as if the only sensible thing to do with n positive numbers would be to add them. Perhaps as a consequence of its historical name, several people have tried to make this estimator robust by replacing the square by something else, not touching the summation sign. Why not, however, replace the sum by a median, which is very robust? This yields the *least median of squares* (LMS) estimator ... This proposal was essentially based on an idea of Hampel (1975, p. 380)".

This quote elegantly describes the context in which the LMS was born: replacing the summation (mean) by rank ordering statistics. During the first ten years of its creation, LMS has found many applications in a wide variety of engineering problems. Computer vision is no exception to this. On the other hand, different groups of researchers in computer vision have discovered limitations, which are associated with applying LMS to computer vision problems (Meer et al. 1991, Stewart 1997). The main limitation that will be considered in the remainder of the paper is the fact that the break down point of the LMS is only 50%. This means that the LMS technique needs the population recovered to have at least a majority of 50% (plus 1) of all the data, to ensure the return of the true fit. This raises a number of important questions. What would happen if the absolute majority did not exist? Does the LMS provide any means of detecting the situation where the data does not contain a population with the absolute majority?

The answer to the second question is no. Worse than that, the LMS always returns an answer. In other words, the LMS will "hallucinate" a fit to some part of data that minimizes the median of square residuals. This fit can be arbitrary far from any underlying structure embedded in the data (as minority groups). To address this problem, which is frequently encountered in computer vision tasks, a number of other estimators have been introduced by assuming some characteristics for outliers, for inliers, or for both. In a recent survey of the main robust estimators, Stewart (1997) concludes that:

"when neither $\hat{\sigma}$ [an estimate of the true scale] nor the distribution of random outliers is known, LMS should be used, although its performance degrades quickly when there are too few inliers".

The fact that the LMS performance degrades quickly when no population of data is in the absolute majority has motivated some researchers to refine the LMS to retrieve structures that do not have the absolute majority.

Lee et al. (1996) proposed a robust estimator named the Adaptive Least K-th Order Squares (ALKS). In this technique, the estimator searches for the model, which minimizes the K-th order statistics of the squared residuals where the so-called optimum value for the K is determined from the data. Following the same approach,

Miller and Stewart (1996) proposed a robust estimator named the Minimum Unbiased Scale Estimator (MUSE). In both approaches, first the proposed estimator tries to find the size of the biggest minority population (K) within the whole set of data through minimization of some scaled measure of variance. Then, the least K-th order statistics technique is used to find the embedded structure. To solve the minimization problem, both approaches rely on random sampling of the data based on the assumption that a high probability of choosing a good sample (a sample from the majority) of data can be achieved with a low number of samples. Therefore, in both approaches, the computation of scale has to be repeated for all the possible values of K, and at every stage enough sub-samples have to be gathered to ensure the reliability of the minimization results. Therefore, these types of estimation process inevitably involve expensive computation, which become infeasible for large data sets. The computation cost is discussed in the next section.

3 Computational Cost

As explained in section 2, both the ALKS and MUSE techniques rely on solving a minimization problem using the sampling technique first proposed for solving the LMS formulation (Rousseeuw & Leroy, 1987). The approximate solution, based on sampling techniques, greatly reduces the computational cost of the LMS technique and has been widely used in different applications (Meer et al., 1991 and Stewart, 1997). However, it will be shown here that the same argument may not extend to the solutions of the minimization problems arising in ALKS or MUSE techniques.

The probability of choosing at least one non-contaminated sample (with p population) out of m sub-samples from n points with ε fraction of the data being contaminated is:

$$P = 1 - (1 - (1-\varepsilon)^p)^m \tag{3.1}$$

where the population of data set n is much bigger than the population of each sample p (Rousseeuw & Leroy, 1987). The left graph in figure 3.1 shows the variation of the probability against the percentage of contamination for 30 samples. The right graph in this figure is the 3-D plot of the same function for samples up to 100.

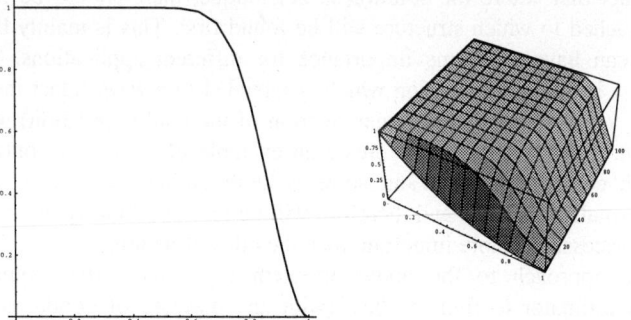

Fig 3.1: Variation of probability against the fraction of contamination for 30 samples (left) and 3-D plot of the same function for up to 100 samples (right).

This figure shows that a relatively low number of sub-samples is required while the fraction of contamination is below 0.5. As the contamination increases, the number of samples required for the same probability will increase sharply. Considering that the ALKS and MUSE both have to perform a series of sampling for all the possible values of K, the final probability of obtaining a good ensemble (correct sample for every value of K) will be the multiplication of all the probabilities achieved for each value of K. Therefore, to perform a reliable computation, a considerable number of samples must be chosen which makes the computation far more expensive that the normal LMS computation. For example, if enough samples have been gathered to achieve the probability of 0.99 for one stage, repeating the same computation for only 10 values of K brings down the probability to 0.90 and for 100 values of K, the combined probability will be only 0.37.

4 Selective Statistical Estimator

The main task considered by robust statisticians during the last three decades was to find estimators which can tolerate a large number of outliers. The fundamental assumption was that the answer to the estimation problem is unique. Uniqueness requires the structural data (as oppose to outliers) to populate more than half the data. Therefore, the best possible breakdown point for a robust estimator with a unique result is 0.5. This argument is fundamental to the LMS technique.

In computer vision problems however, the structural data may not always populate more than half the data. This happens especially in data sets containing multiple structures. Assuming multiple populations and that the fraction of the total population of the desired structure can be less than the half, the answer to the estimation process will no longer be unique. This is the dilemma that the aforementioned estimators are faced with. In order to find the best (in some sense) solution, each of these estimators propose a measure by which the desired (largest population) solution is sought (minimizing that measure). A detailed description of these estimators (and their measures) for finding the best answer is beyond the scope of this paper and an interested reader is referred to original papers for more information (Lee et al. 1996 and Miller & Stewart, 1996).

The important point, which has been apparently overlooked in the aforementioned work, is the fact that where the solution is not unique, there should be little or no importance attached to which structure will be found first. This is mainly because all the solutions can have the same importance for different applications. Therefore, going through a lengthy computation which is intended to always select the structure with a smaller variance (or with a larger fraction of the total population) may not be the best option. Indeed, it is easy to create an example of a data set containing two structures with equal population and some random outliers, in which one of the structures has smaller variance and therefore will be returned first by these estimators without being necessarily more important than the other structure.

An alternative approach to the above problem is to relax the requirement of demanding an estimator to find the best (with any measure of goodness) structure embedded in the data first. Alternatively, the estimator is asked to find an acceptable (in some sense) structure and then examine the goodness of the fit by such a structure. If the structure is indeed a solution to the estimation problem, then it can be removed from the data set and the next structure will be found using the same technique. The fundamental idea of this approach is the fact that the acceptability of any possible structure within a noisy data set is inherent in the formulation of the problem and not in the data (beauty is in the eyes of beholder!).

Having relaxed the above requirement, there exists a fundamental question, which demands a more complete answer. That is, what does *embedded structure* mean to a particular user? In other words, what condition should be substituted for the requirement of having the absolute majority? The main point emphasized in this paper is the fact that the solution to this question is not embedded in the data but it depends on the nature of the original problem, which the estimation process tries to solve. This idea is fundamental to the proposed approach and will be explained in the following.

Consider a 2-D regression problem with a noisy data set containing multiple structures. If the problem formulation regards every group of data with a population which is more than at least a quarter of the whole population (and fits nicely to a line with some measure of goodness) as a solution, the problem can have up to three solutions which can all have similar importance. In this case, there is little or no difference between the order in which any of these three solutions (with higher or lower populations, variances, etc) are found. Indeed, in a satisfactory scenario, all the structures embedded in the data set have to be recovered and tested (with some appropriate measure) and then each structure which passes the test must be considered as part of the solution for the original problem.

Also, it is important to note that if the user does not know the minimum limit on the population for any group of data to be regarded as a structure (a priori information), then every group of data with a population as small as the dimension of the problem would satisfy every measure of goodness (i.e. two points always fit perfectly to a line in the 2-D regression problem).

Considering these fundamental ideas, the SSE is proposed as a variation of the LKS where the K is proposed by the user as the lower limit of the size of populations one is interested in. After deciding what would be the minimum acceptable population for every structure (K), then one follows a similar procedure to that of the LMS. The only exception is that, instead of the sub-sample that has the smallest median of square residuals, the sub-sample that has the smallest K-th order least squares is found. This will arbitrarily return one of the structures, which has a population of at least K points. By scaling all the residuals with respect to the returned structure, using the same recipe as described by Rousseeuw and Leroy (1987), all the points belong to this structure will be found. The only difference is that the scale for SSE is found using K-th order statistics of the residuals (as opposed to the median of the residuals) with different correction factor for different K. The scale for SSE is calculated as:

$$s^0 = \Phi^{-1}(\frac{1}{2}+\frac{K}{2(N+1)})(1+\frac{5}{N-p})r_{K:N} \qquad (4.1)$$

where Φ is the unit normal cumulative distribution function, p is the dimension of the model fitted to the data, N is the population of data set and $r_{K:N}$ is the K-th ordered residual for the selected fit (see Miller and Stewart, 1996 for further discussion on adjusting the scale for the K-th order statistics). For large data sets, this can be further simplified to:

$$s^0 = \Phi^{-1}(\frac{1}{2}+\frac{1}{2k})r_{K:N} \qquad (4.2)$$

where k is the ratio N/K.

Having removed the points regarded as inliers to the first returned structure, one will apply the same procedure to the reduced data set ("outliers" with respect to the first fit) in order to recover the second structure. From now on, by adding up the number of points associated with the recovered structure and comparing it to the total number of points, it is trivial to find out when to stop the procedure.

5 Experimental Investigation

As mentioned before, the main shortcoming of the LMS approach is the fact that it will not correctly resolve the situation where the number of inliers for any structure is less than 50% of the whole data set. To show that the proposed SSE resolves situations with multiple structures, a set of examples for both synthetic and real data is included in this section.

5.1 Synthetic data

To assess the performance of the proposed estimator, a set of experiments with randomly generated data has been designed. In all of these examples, a set of 25 points is considered which are randomly generated from two linear structures (and later corrupted by 10% normal random noise) and around 30% random outliers.

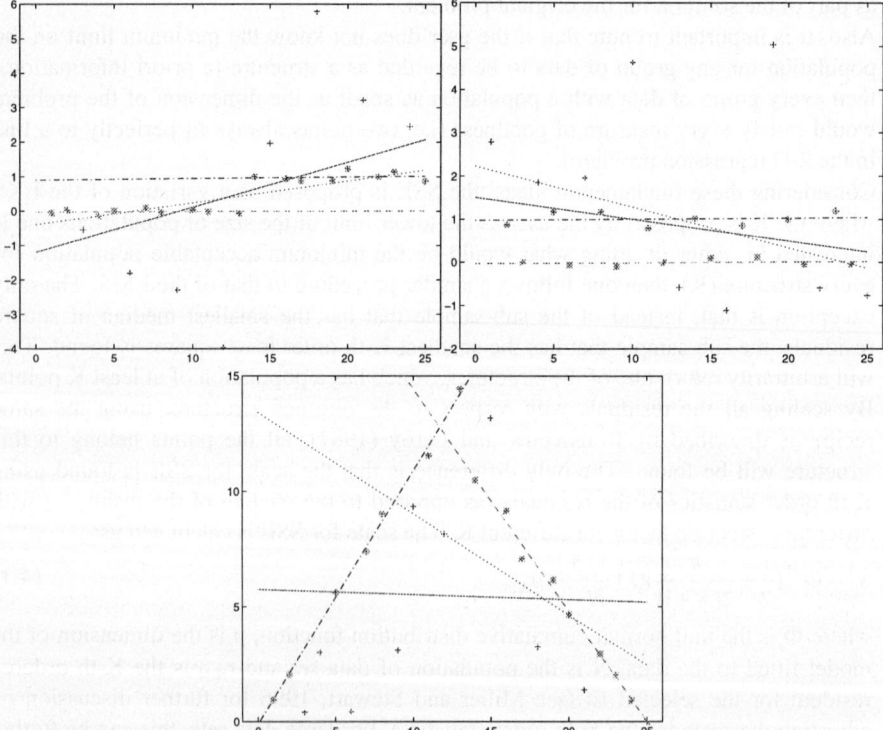

Fig 5.1: Comparison between the performance of the LS (solid), LMS (dotted) and SSE (dashed and dashdotted) for step, parallel and crease discontinuous structures over multi-structural data. The points representing the first and second SSE results are marked with cross and circle.

The populations of these two structures are 9 and 8, and there are usually 8 random outliers (sometimes a random outlier goes close to one of the structures which slightly changes the apparent populations). The SSE is implemented by looking for a structure with the population of at least 25% ($s^0 = 3.1382$) of the whole data. The results for parallel, step and crease discontinuous structures are shown in figures 5.1.

These experiments have been repeated for many different populations of points and different values for K. In all the experiments, the final result is not sensitive to the chosen value of K as long as structures exist in the data set with populations more than one K-th of the total population.

5.2 Real data

To show the performance of the proposed estimator for a real computer vision problem, motion based segmentation of a complicated scene using optic flow information is considered. By assuming that the optic flow is affine, a 2-D plane is fitted to the horizontal components of the image velocities (better results are expected where both components of flow are considered). To give an example, the Otte image sequence (Otte & Nagle, 1994) is used. This sequence is a real image sequence, recorded using a camera, which translates toward a scene. The objects in that scene are stationary, except for a Marble block, which is translating towards the left.

A snapshot taken from this sequence is shown in figure 5.2. The scene contains many sharp discontinuities in both depth and motion. The optic flow field measured for this image sequence is publicly available and it is also shown in figure 5.2. The remaining of the pictures in this figure show the performance of the SSE in six sequential stages. In these figures, white represents points where there is no data to be fitted (the optic flow data is not known at that point), black represents the unsegmented portion of the data and all the other shades (gray scale images) between these two represents a specific structure (segment) in the data set. In this experiment, it is assumed that the population of every structure must be at least 10% of the whole data set and therefore, the value for s^0 is set to 7.896 (equation 4.2).

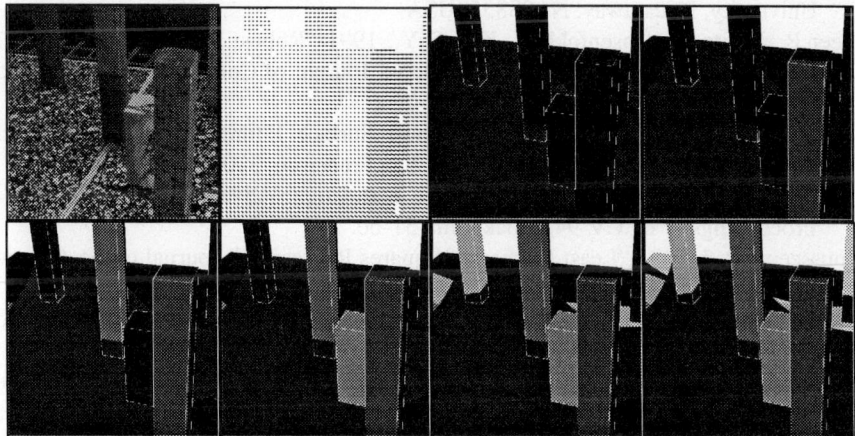

Fig 5.2: A sample image of the Otte image sequence, its measured optic flow field and its sequential segmentation in 6 stages.

It can be seen in these figures that the tabletop has been picked up at the first stage, and columns have been segmented one after the other in a sequential order. The main point which has been emphasized throughout this paper, seen clearly from these results, is the fact that there is little or no importance in the order by which the data are segmented as long as the whole data are fully investigated.

6 Conclusion

In this paper, an alternative approach to implementation of the rank-ordering estimators has been described. This approach results in a robust estimator, which can retrieve several structures out of a noisy data set. The approach is more efficient than any other competing method. A number of experiments with synthetic and real data were carried out demonstrating the effectiveness of the proposed method for motion segmentation purpose.

Although the proposed estimator is used in this paper only for motion based segmentation, it has a wide range of applications in computer vision such as segmentation of range data, intensity data, etc. In particular, the second-generation approach to the video-coding problem requires new tools for efficient segmentation of image sequences (Torres & Kunt, 1996). This is an important area in which the presented estimator will find its immediate applications (Bab-Hadiashar & Suter, 1997).

7 References

Bab-Hadiashar A., Suter D., 1997 "Motion Segmentation Using Robust Statistics and Spatial Continuity" Proceedings of International Workshop on Image Analysis and Information Fusion IAIF'97, Adelaide, Australia, to appear.

Hampel F.R., 1975 "Beyond Location Parameters: Robust Concepts and Methods" Bulletin of International Statistical Institute, 46, 375-382.

Lee K., Meer P., Park R., 1996 "Robust Adaptive Segmentation of Range Images", Technical Report, Department of Electrical & Computer Engineering, Rutgers University, Piscataway, NJ 08855, USA.

Meer P., Mintz D., Rosenfeld A., Kim D.Y., 1991 "Robust Regression Methods for Computer Vision: A review" International Journal of Computer Vision, 6(1): 59-70.

Miller J., Stewart C. 1996 "MUSE: Robust Surface Fitting using Unbiased Scale Estimates", Proceedings of CVPR'96, San Francisco, 300-306.

Otte M., Nagel H. H. 1994 "Optical Flow Estimation: Advances and Comparisons" Proceedings of ECCV'94, Stockholm, 51-60.

Rousseeuw P. J. 1984 "Least Median of Squares Regression" Journal of the American Statistical Association, 79, 871-880.

Rousseeuw P.J., Leroy A.M., 1987 "Robust Regression and Outlier Detection", John Wiely, New York.

Stewart C., 1997 "Bias in Robust Estimation Caused by Discontinuities and Multiple Structures", IEEE Transactions in Pattern Analysis and Machine Intelligence, to appear.

Torres L., Kunt M., 1996 "Video Coding the Second Generation Approach", Kluwer Academic Publishers, The Netherlands.

Optical Flow in the Scale Space

Qing Yang and Song De Ma

National Laboratory of Pattern Recognition
Institute of Automation, Chinese Academy of Sciences
P. O. Box 2728, Beijing 100080, P. R. China
Email : qyang@prlsun6.ia.ac.cn, masd@prlsun2.ia.ac.cn

Abstract

There exists optical flow in the multiscale representation of an image if this representation is viewed as an image sequence in the "time" domain. The ill-posed tracking problem in the scale-space can be robustly solved by this method. The perceptual effect of any multiscale representation can be greatly improved by the so-called "pull-back" technique.

1 Introduction

The scale-space theory is of crucial importance in computer vision and it has been intensively studied [15] [9] [2] [11] [12] [16]. A critical problem in linear scale-space theory is that local features in the image may be seriously distorted at large scales. Various nonlinear diffusion equations [13][14][5] have been proposed to deal with this drawback. The limitation of nonlinear diffusion strategy is that suppressing noises (or small edges) and keeping the large scale edges are essentially contradictory in the localization scheme. In fact, in many cases the nonlinear strategy is not always encouraging for the improvement of the perceptual effect is little .

A much more thorough method is to track points in the scale-space. Although this idea has been pointed out in the literature, but we seldom perform tracking to obtain a better multiscale representation. It is often believed that this is because of computational complexity. But we argue that the main reason is that the tracking problem is ill-posed. The procedure of regularization must be introduced. This allows us to define "optical flow in the scale space" which can be viewed as a standard optical flow.

In terms of differential geometry, the multiscale representation defined by linear and nonlinear diffusion equation can be pulled back to the original image by tracing back optical flow in the scale-space. This pull-back is called the *intrinsic multiscale representation*. The distortion of this representation is very small. The concept "optical flow in the scale-space" and the pullback technique are powerful. The perceptual effect of original multiscale representation can be greatly improved by our methods.

2 Multiscale Representation and Tracking in the Scale Space

This section briefly reviews the diffusion theory and tracking strategy proposed by Lifshitz and Pizer [10].

2.1 Linear and Nonlinear Diffusion

Koenderink [9] has pointed out that the one-parameter family of images comprising scale-space can be equivalently viewed as the solution of the diffusion (or heat) equation:

$$I_t = c\Delta I \qquad (1)$$

with the initial condition $I(x, y, 0) = I(x, y)$, the original image. In the above equation c is a diffusion constant.

The diffusion equation provides a mathematical framework with which to analyze the scale-space formulation. While Koenderink restricted his analysis to the isotropic diffusion characterized by the linear heat equation, Perona and Malik [14] suggested that a nonlinear anisotropic version of the heat equation could deal with the difficulties encountered in the use of a linear scale-space. The basic idea is to modify the conductivity in a nonlinear version of the diffusion equation

$$I_t = \nabla^T c(x,t) \nabla I \qquad (2)$$

A choice of the conductivity function $c(x,t)$ is that $c(x,t) = h(|\nabla I|)$, where h is a monotonic decreasing function. Then Eq. 2 becomes

$$I_t = \nabla^T h(|\nabla I|) \nabla I \qquad (3)$$

In Eq. 2 and Eq. 3, $\sqrt{t}$ can also be viewed as the scale parameter.

2.2 Tracking Isointensity Contours in the Scale-Space: A Conventional Approach

In the multiscale representation of an image, points at different scales can be associated by some constraints. This problem is called "tracking in the scale-space". Note that superficially the inverse process of tracking is similar to image deblurring [3]. But in fact they are different problems.

The tracking strategy in [10] can be described as follows. A nonextremum at one scale is associated with some point with the same intensity at the next scale. The point moves along the steepest ascent. Local extrema are associated by the continuity constraint. Note that the extrema will disappear at some scale. All the critical points constitute an important image description for segmentation. More precisely, suppose $I(x, y)$ is an intensity image, $I(x, y, t)$ is its multiscale representation, where t is the scale parameter. The tracking paths are integral curves of the vector field $(\frac{-I_x I_t}{I_x^2 + I_y^2}, \frac{-I_y I_t}{I_x^2 + I_y^2}, 1)$. Note that the "the steepest ascent" constraint has been added here to uniquely trace the points. In order to track an extremum at some scale, we should find corresponding extremum in the neighborhood at next scale.

Under the isointensity assumption, critical points will disappear as the scale increases. Although tracking the critical points is necessary to special purposes such as image segmentation, it is not essential to other purposes. For this reason, we expect that the critical points should not disappear. Moreover, the above tracking strategy has the following limitations:

(1) Tracking extrema in the scale-space is not an easy task for it is hard to design a robust algorithm.

(2) Even for a nonextremum point, when it is close to an extremum, the tracking will become difficult. In other words, "the steepest ascent" constraint is still not enough to robustly determine the tracking path.

(3) The smoothness of linear scale representation cannot guarantee the stability of tracking. In fact it is a highly unstable problem.

In summary, tracking in the scale-space is a typical ill-posed problem which should be regularized.

3 Optical Flow in the Scale-Space: Definition and Algorithm

Tracking paths in the scale space can be viewed as *"optical flow in the scale space"*. In $I(x,y)$'s multiscale representation $I(x,y,t)$, the scale t can also be regarded as "time". This is not a new idea, for if we consider the multiscale representation as a heat-diffusion process, the two concepts "time" and "scale" can be identified. In fact there are more important reasons for us to introduce this concept:

(1) We observe that blurring an image in the scale space is similar to the following process: when a man walks toward an object along a straight line, the object becomes clearer gradually.

(2) Optical flow in the scale-space can be computed by ordinary optical-flow algorithm.

From the above analysis, when t increases from 0 to $+\infty$, $I(x,y,t)$ can be regarded as an image sequence in the time domain. Like the ordinary image sequence, *there exists optical flow in this sequence*. The perceptual effect of the blurring using linear scale-space theory is not in accordance with man's visual perception. This drawback can be overcome by using the "pull-back" technique which will be studied in the following sections.

There are many algorithms to compute the optical flow (u,v) of $I(x,y,t)$. Since the linear scale-space representation is smooth, the classical algorithm proposed by Horn and Schunck [8] is appropriate. For sake of completeness, we outline the algorithm as follows.

The intensity constraint is $I_x u + I_y v + I_t = 0$. This means the optical flow tracks the isointensity contours in $I(x,y,t)$. Adding the smoothness constraint we obtain the following error functional to be minimized

$$\int\int (I_x u + I_y v + I_t)^2 + \alpha^2(|\nabla u|^2 + |\nabla v|^2) dx dy \qquad (4)$$

where α is the weighting factor. An iterative scheme to solve the above variational problem is

$$u^{n+1} = \overline{u}^n - I_x[I_x\overline{u}^n + I_y\overline{u}^n + I_t]/(\alpha^2 + I_x^2 + I_y^2).$$

$$v^{n+1} = \overline{v}^n - I_y[I_x\overline{u}^n + I_y\overline{u}^n + I_t]/(\alpha^2 + I_x^2 + I_y^2).$$

where $\overline{u}^n$ and $\overline{v}^n$ are approximations of the local averages of u and v.

Since scale-space representations are smooth and the above iterative method is used, the computational cost is not very high. The running time of our procedure is about 4 times as long as that of the linear case. Integrating our approach and fast algorithms for scale-space such as [4] will greatly reduce the computational complexity.

4 Intrinsic Multiscale Representation

An embarrassing problem in linear scale-space theory is that image blurring using Gaussian kernel will distort features in the original image. The distortion becomes intolerable at large scales. To overcome this difficulty anisotropic diffusion theory has been intensively studied [14] [13]. Unfortunately the effect of anisotropic diffusion is discouraging in many cases. In this paper, we will use the optical flow in the scale space to deal with this problem.

The optical flow $(u(x,y,t), v(x,y,t))$ of $I(x,y,t)$ can be carried out step by step. Thus we obtain a mapping ϕ_t from the original image $I(x,y,0) := I(x,y)$ to the blurred image $I(x,y,t)$

$$(x,y) \mapsto \phi_t(x,y) := (\phi_t^X(x,y), \phi_t^Y(x,y))$$

More precisely, for fixed (x,y), $\{\phi_t(x,y)|t \geq 0\}$s is an integral curve of the vector field $(u,v,1)$. The initial point of $\phi_t(x,y)$ is $\phi_0(x,y) = (x,y)$

Let $\widehat{I}(x,y,t) = I(\phi_t^X(x,y), \phi_t^Y(x,y), t)$. Note that $\widehat{I}(x,y,t)$ is uniquely determined by $I(x,y,t)$ and the weighting factor α. For simplicity, the subscript α is omitted.

Then we obtain an image by tracing back the optical flow in the scale space

$$\widehat{I}(x,y,t) := I(\phi_t^X(x,y), \phi_t^Y(x,y), t)$$

In terms of differential geometry [6], $\widehat{I}(x,y,t)$ is the *pull-back* function of $I(x,y,t)$ by ϕ. In this paper $\widehat{I}(x,y,t)$ is called the *intrinsic multiscale representation* of $I(x,y)$ at scale t.

$\widehat{I}(x,y,t)$ and $I(x,y,t)$ are related by a coordinate transformation. In this sense they are almost the same. But their perceptual effect are quite different. $\widehat{I}(x,y,t)$ is much more natural than $I(x,y,t)$. Since regularization is introduced to compute the optical flow, the tracking is well-defined and robust. The distortion of the edge is very small. As a matter of fact, the distortion of isointensity contours that do not merge with others [10] tends to zero when $\alpha \to 0$ ($\alpha = 0.1$ is good enough in practical situations).

5 Experimental Results

Let us see the experimental results of our method on real images.

Fig. 1 shows a squirrel sitting on the grass.

Fig. 1. A squirrel sitting on the grass (adapted from Zhu (1996). This image will be used as the original image in Fig. 2.

In Fig. 2, the left column ((a), (b), (c), (d)) shows the multiscale representation of the image in Fig. 1 at 4 scale-levels from a conventional linear scale-space. The right column ((a)', (b)', (c)', (d)') shows the corresponding intrinsic representations. The two images in the same row have the same scale. That is, (a)', (b)', (c)', (d)' are the pullbacks of (a), (b), (c), (d) respectively. We can see that the effect of their pullbacks are much better. For example, Fig. 2(d) shows the linear scale-space representation at some scale, it is very vague, but if we pull it back to the original image (Fig. 1), it becomes clear (Fig. 2(d)').

The perceptual effect of linear scale representation is not natural. When one observes an object at some scale (or distance), small details may be missed, large structures may be vague. But we cannot imagine the squirrel in Fig. 1 will become a monster in Fig. 2(d). On the contrary, Fig. 2(d)' is acceptable.

We can see that both of the linear and nonlinear diffusion strategies are not appropriate for dealing with the squirrel image. However their effect can be greatly improved by the optical flow and pullback techniques proposed in this paper. As a matter of fact the linear and nonlinear diffusion strategies have very small difference in our paper. For this reason, we prefer the linear diffusion.

Fig. 3 is another example which is similar to Fig. 2, see the caption.

6 Conclusion

We have developed the scale-space theory and proposed two new concepts: optical flow in the scale-space and intrinsic multiscale representation. The main

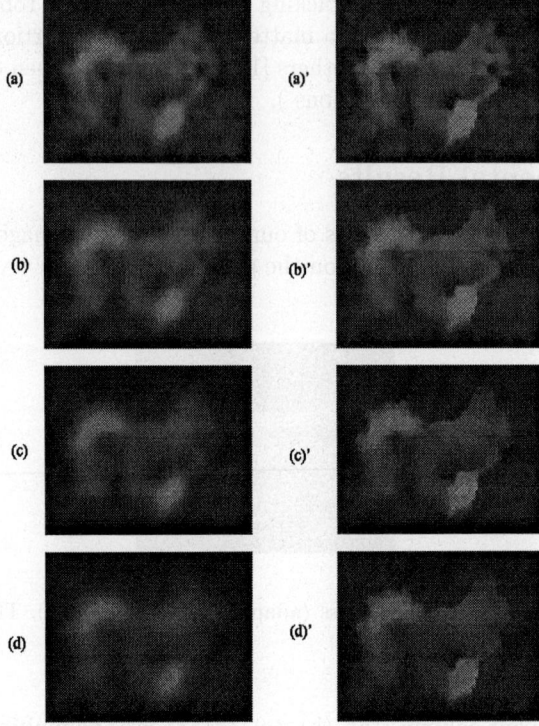

Fig. 2. Left column ((a), (b), (c), (d)): the conventional linear scale-space representations of the image shown in Fig. 1 at five scale-levels. Right column ((a)', (b)', (c)', (d)'): the corresponding intrinsic representation of the left column. The two images in the same row have the same scale parameter. That is , (a)', (b)', (c)', (d)' are the pull-backs of (a), (b), (c), (d) respectively.

novelty of the approach is that regularization is introduced to solve the tracking problem. The linear and nonlinear multiscale representation can be pulled back to the original image. This pull-back is the so-called "intrinsic multiscale representation". The advantages of intrinsic representation are: (1) It is a general method that can greatly improve the effect of any multiscale representation; (2) The distortion of features is very small. When the scale increases, global edges are preserved while noises and small edges are smoothed.

References

1. R.Adams and L. Bischof, "Seeded region growing," *IEEE Transactions on Pattern Analysis and Machine Intelligence*, vol. 16, no. 6, June 1994.
2. J. Babaud, A.P. Witkin, M. Baudin, and R.O. Duda, "Uniqueness of the gaussian kernel for scale-space filtering," *IEEE Trans. PAMI*, vol. 8, no. 1, pp. 26-23,1986.

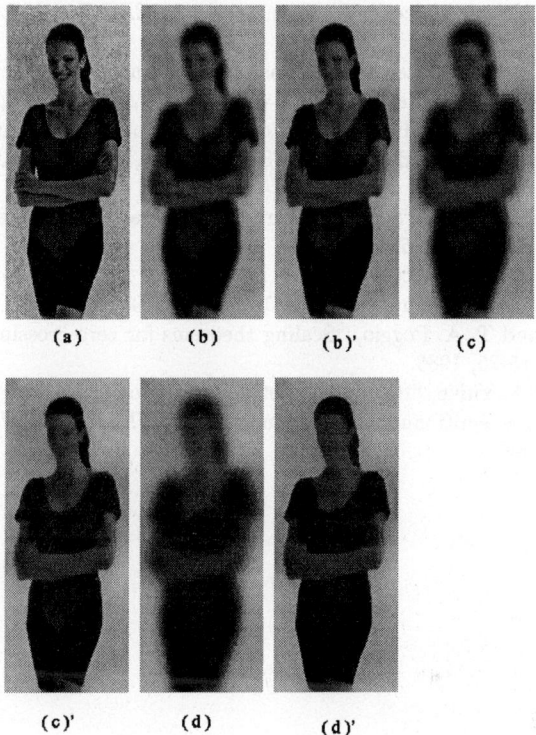

Fig. 3. (a) Image of a woman (adapted from [17]) . (b),(c),(d): the linear scale-space representation at three scale-levels. (b)',(c)', (d)' are the corresponding intrinsic representations of (b),(c), (d) respectively.

3. J. Biemond, R.L. Lagendijk, R.M. Mersereau, "Iterative methods for image deblurring," *Proc. IEEE*, vol. 78. no. 5, pp. 856-883,may 1990.
4. G. Cong and S.D. Ma, "Dyadic scale space," *Proc. of ICPR'96*, pp. 399-402, 1996.
5. L.M.J. Florack, A.H. Salden, B.M. ter Harr Romeny, J.J. Koenderink, and M.A. Viergever, "Nonlinear scale-space," *Image and Vision Computing*, vol. 13, no.4, May 1995.
6. M. Golubitsky and V. Guillemin, *Stable Mappings and Their Singularities*, Springer-Verlag, 1973.
7. R.D. Henkel, "Segmentation in scale space", *Proc. of CVIP'95*, Prague, 1995.
8. B.K.P Horn and B.G. Schunck, "Determining optical flow," *Artificial Intelligence*, vol. 17, pp. 185-204,1981.
9. J.J. Koenderink, "The structure of images," *Biol. Cybern.*, vol. 50, pp. 363-370,1984.
10. L.M. Lifshitz and S.M. Pizer, "A multiresolution hierarchical approach to image segmentation based on intensity extrema." *IEEE TPAMI*, vol. 12, no. 6,pp. 529-540, 1990.

11. T. Lindeberg, "Scale-space for discrete signals," *IEEE Trans. PAMI*, vol. 12, pp. 234-254, 1990.
12. T. Lindeberg, "Scale-space theory:A basic tool for analysing structures at different scales," *Journal of Applied Statistics*, vol. 21, no. 2, pp. 225-270, 1994.
13. K.N. Nordstrom, "Biased anisotropic diffusion–A unified regularization and diffusion approach to edge detection," CS-Tech. Rep., Univ. of California, Berkley, 1989.
14. P.Perona and J. Malik, "Scale space and edge detection using anisotropic diffusion," *IEEE TPAMI* 12: 629-639,1990.
15. A. P. Witkin, "Scale-space filtering," in *Proc. 8th Int. Joint Conf. Art. Intell.*, (Karlsrube, West Germany), pp. 1019-1022, Aug, 1983.
16. A. L. Yuille and T. A. Poggio, "Scaling theorems for zero crossings," *PAMI*, vol. 8, No. 1, pp. 15-25, 1986.
17. S.C. Zhu and A. Yuille, "Region competition: unifying snakes, region growing, and Bayes/MDL for multiband image segmentation" *IEEE TPAMI*, vol. 18, No. 9, September, 1996.

Motion Detection in Temporal Clutter

Phillip M. Ngan

Industrial Research Limited
P. O. Box 2225, Auckland
New Zealand
email: p.ngan@irl.cri.nz

Abstract. A motion detection technique is presented which exhibits robust properties suitable for general operation in a surveillance environment. The technique is sensitive to transient perturbations present in an image sequence while being insensitive to temporal clutter. Its performance is robust to scale and offset variations in the input sequence. The system is implemented as a pixel based temporal recursive filter, which requires modest computational resources to process each new input frame. Analysis of filter properties and experimental results for a real image sequence are reported.

1 Introduction

A filter that distinguishes between *persistent* and *transient* intensity variation at a pixel site is presented in this paper. A pixel site in an image sequence is said to exhibit persistent variation when a finite set of intensity values reoccurs at that site over a fixed period of time. The degenerate case of persistent variation is that of a static background, which simply comprises a sequence of constant valued pixels. A pixel site whose intensity oscillates periodically between values is also said to exhibit persistent variation. Another name for this is *temporal clutter*. By contrast, a transient event is considered to occur at a pixel site when intensities that have not occurred in the recent past appear momentarily and then disappear. In a surveillance application, a person walking through the field of view is an important example of a transient event. The filter presented in this paper exhibits high sensitivity to transient events and relative insensitivity to persistent events. Such a filter reduces the false alarm rate in the presence of temporal clutter compared to the standard background subtraction technique.

Processing to reduce sensitivity to temporal clutter can be achieved at different levels. The filter presented in this paper operates at the pixel level, and so serves as an effective preprocessing step for algorithms that reject temporal clutter at higher levels of processing.

The work presented here is hedged by practical considerations. First, the algorithm is designed to run using modest amounts of computing resource. Despite the limited resources, the algorithm is required to operate at or near video frame rate (25 frames per second). Second, the system must be able to adapt to slow changing lighting conditions, such as those related to the time of day. Rapidly varying lighting conditions, for example caused by changing cloud cover or the turning on and off of internal lights, are not considered.

This paper is an abridged edition of a Technical Report [Ngan 1997].

2 Background

Karmann and associates performed work in motion detection in the early 1990's for the purpose of traffic flow monitoring [Karmann et al. 1990]. A motion mask is generated by comparing the difference between an input image with a background image on a pixel by pixel basis. The background image is recursively updated with a dynamic equation within a Kalman filter framework. The background image is a linear combination of the current input image and previous background image. At each pixel, the weight used in the combination assumed one of two possible values, depending on whether the pixel represents the static background or moving object. Motion at each pixel site is detected by applying a threshold to the difference between the current image and background image. The threshold value varies as a function of the noise power present in the reference image.

Karmann's approach has since been extended in several ways. Koller, Weber, and Malik [Koller et al. 1994] updated not just a single background image but a vector of filtered images to improve the robustness of the detection process. Ridder, Munkelt, and Kirchner [Ridder et al. 1995] addressed problems associated with low velocity motion and stop-start motions by incorporating non-linear elements in the Kalman gain to slow the incorrect adaption of slow moving foreground objects into the background estimate. Makarov [Makarov 1996] updated the filter not on intensity images but binary edge images to overcome global intensity offsets associated with rapid changes in scene illumination. Our work differs from previous systems in that particular attention has been given to the suppression of temporal clutter, while maintaining sensitivity to salient foreground motions.

The issue of insensitivity to temporal clutter has been addressed at levels higher than pixel level processing. For example, Letang, Rebuffel, and Bouthemy [Letang et al. 1993] specifically addressed the problem of distinguishing between what they call, *static background*, *moving object*, and *parasitical motion*, by the use of a Bayesian framework. Although this approach systematically exploits prior information to assist the motion detection process, its iterative implementation demands significant computing resources for frame-rate operation.

3 Filter Description

An image that comprises j rows and i columns has $k = i \times j$ pixels. The proposed filter operates on each pixel individually and is therefore replicated k times, once for each pixel in the image. The aim of the filter is to generate a binary value M_n to indicate the presence of motion by the value 1, or lack of motion by the value 0. The subscript n denotes the temporal sequence index. The sampling interval is assumed constant and is typically dictated by the frame rate of the input video source. Let x_n be the input intensity value for the n-th sample at a given pixel location and let b_n be the corresponding background pixel from the reference image. Processing begins at sample $n = 0$ and so the first input pixel value is given by x_0. It is convenient to assume $\{x_n\}$ is a causal sequence, that is $x_n = 0$ for $n < 0$ [Oppenheim and Schafer 1989].[1] Motion detection algorithms that operate by comparing the current image against a reference image have the following general form.

$$M_n = \begin{cases} 1 \text{ if } |f(x_n, b_n)| > t_n \\ 0 \text{ otherwise} \end{cases} \quad (1)$$

The value given by the function f is usually proportional to a distance measure between the input x_n and the reference background image b_n. Motion is said to occur when the distance is greater than a threshold value t_n. Conversely, if the distance is less than the threshold, the pixel is assumed part of the background.

In this paper, we regard the output of the function $f(x_n, b_n)$ to be the output of the motion detection filter y_n, which is given by the equation,

$$f(x_n, m_n) = y_n = \frac{x_n - m_n}{l_n + \epsilon} \quad (2)$$

where m_n and l_n are defined below. The term ϵ is a small scalar constant to avoid the indeterminant situation which occurs when dividing by a zero norm.

The background reference pixel b_n is served by an estimate of the temporal mean m_n of the input sequence. We have chosen to implement this as the fading mean of the input pixel values because it can be rapidly calculated by the following recursive equation.

$$m_n = \alpha x_n + (1-\alpha)m_{n-1}, \quad 0 < \alpha < 1 \quad (3)$$

The value l_n is given by the fading L_1 norm of the difference between the current image and the reference image. It also has the same recursive form.

$$l_n = \alpha|x_n - m_n| + (1-\alpha)l_{n-1}, \quad 0 < \alpha < 1 \quad (4)$$

The L_1 norm in the denominator of (2) acts as a normalisation factor for the difference $x_n - m_n$. A normalised output allows the threshold to be defined by a constant value, t. We conjecture that a constant threshold t, chosen to work in a certain setting, will be effective in that setting whether or not temporal variations occur in the background of the scene. If the background contains a steady-state variation, the value of both the numerator and denominator of (2) change by the same proportion, and thus cancel the contribution of the variation in the filter output. However, the transient response of the denominator is α times slower than the denominator and therefore the filter responds readily to transient events. It is the low sensitivity to steady-state variations coupled with

---

[1] In the filter implementation, initial values are chosen to minimise the startup phase instead of using causal values.

high sensitivity to transients that gives rise to the motion detection properties of this filter. The following section examines the properties of the filter in detail.

Note that the value of α is restricted between zero and one on an open bounded region. On the one hand, the parameter α cannot assume the value zero because it would always ignore the current input and thus cannot be considered a filter. On the other hand, α cannot be unity since the mean would perfectly track the input, and the norm would always be zero.

4 Filter Properties

4.1 Invariance to input offset and scale

The long-term output of the filter does not change when offsets and scale factors are applied to the filter input values. Transient values do occur at the filter output in response to the initial application of the offset and scale changes. But this is followed by an exponential decay back to the original value; the rate of decay is determined by the weighting factor α. These invariant properties allow the filter to be operated robustly in the presence of real-world changes such as variable lighting conditions, provided the initial transient responses can be tolerated.

4.2 Impulse response

Although the system is non-linear, an examination of its impulse response reveals insights into the transient response of the system. The input used is a delta sequence which is zero everywhere except at time q when it has the value of one. That is,

$$\delta_{n-q} = \begin{cases} 1 \text{ if } n = q \\ 0 \text{ otherwise} \end{cases} \quad (5)$$

The impulse response is given as follows.

$$y_n = \begin{cases} 0 & \text{if } n < q \\ \dfrac{1}{\alpha} & \text{if } n = q \\ \dfrac{-1}{\alpha(n-q-1)+1} & \text{if } n > q \end{cases} \quad (6)$$

The filter response is zero before the impulse occurs which is expected of an unbiased filter. The instant at which the impulse occurs, the filter has an output of $\frac{1}{\alpha}$, which significantly is independent of the magnitude of the delta. Moreover, since the filter is dependent on only current and past values, $\frac{1}{\alpha}$ will be the value of the first non-zero output term for any arbitrary input sequence, not just an impulse sequence. After the occurrence of the impulse, the terms become negative and their magnitudes monotonically converge to zero. It can be seen that α affects the output magnitude and the rate of convergence towards zero.

5 Initialisation

The filter undergoes a startup phase when the values for the mean and norm transition from initial values to steady-state values. The duration of the startup phase largely depends on the value of α. The smaller the value of α, the longer the startup phase. The duration of the startup phase can be significantly reduced by setting the initial value of the mean m_0 to the value of the first term in the input sequence x_0, and the initial value of the norm l_0 to some estimate for the quiescent norm. A value of 3 was found to be satisfactory.

6 Example results

An image sequence was captured at 25 frames per second of a person walking from right to left across the field of view and towards the camera in an indoor environment. Two bushes present in the background of the scene fluttered in the draft of an electric fan. Figure 1 shows the input sequence sub-sampled at 1 second intervals in the left column. The corresponding filtered image sequence and the resultant motion mask are shown in the centre and right columns respectively. The output clearly highlights the moving person, especially the leading edge of his outline against the background. Also highlighted are the leading edges of the shirt and tie, and the person's shadow. In these images, the filter response to the moving bushes is imperceptible in the output sequence except at frame 650 in the left-centre of the image (although this may be lost in the image reproduction). Another feature in the output sequence is a swath of low output values that trails behind the person. The intensity disturbance caused by the person at a pixel raises its norm value and consequently suppresses the filter output value. The duration over which the suppressed output values appear is inversely proportional to the value of α.

A motion mask representing the region of the moving object can be obtained by thresholding the output image using a constant valued threshold.

A closer examination of the input and output pixel values is given in Figure 2, which shows the temporal sequence of intensity values of the pixel located at $(92, 123)$. Oscillations at a frequency of approximately 1.5 Hz present in these sequences are caused by a leaf on a fluttering bush periodically obscuring the wall behind the bush. Samples 400 to 450 display a high amplitude of oscillation but subsequent samples have a low amplitude. This low frequency envelope of the background amplitude is caused by the side-to-side movement of the fan unit as it moves alternately over the bushes. Evidence of the person walking past this pixel is given by the intensity spike at sample 670. This transient event in the sequence is manifested in the output by a corresponding spike. Clearly a simple constant threshold is sufficient to extract the presence of the foreground transient event, and yet at the same time exclude any response due to the variable background.

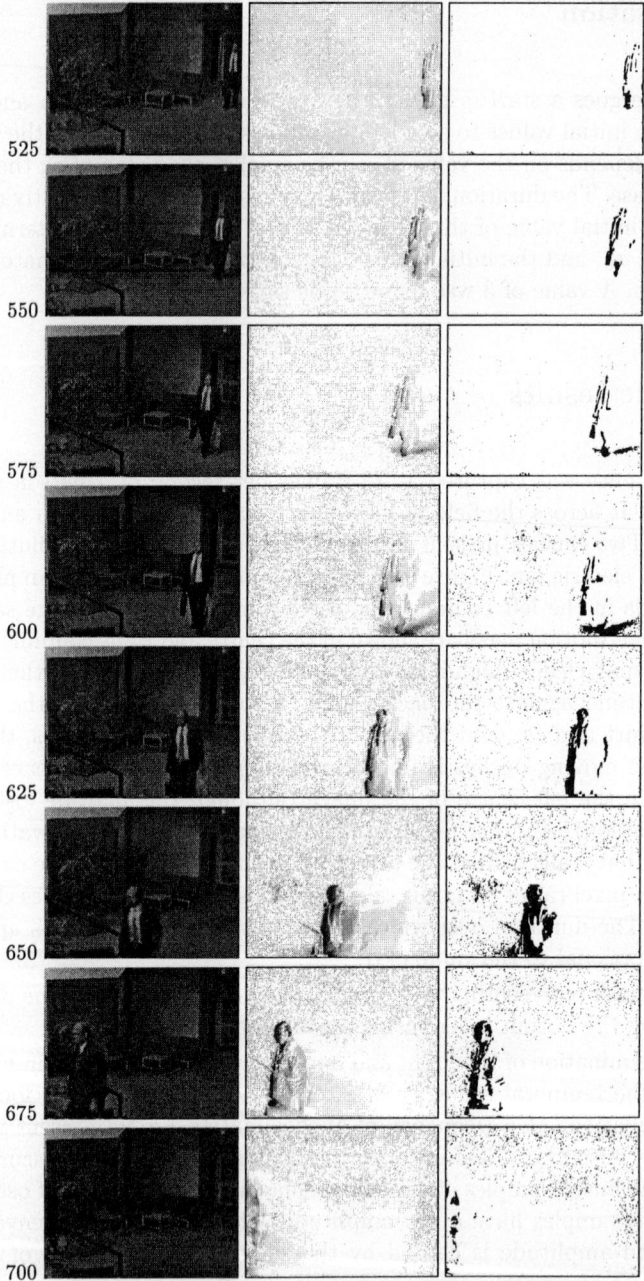

Fig. 1. The input sequence in the left column of images comprises a static background, two bushes that move in a draft, and a person walking through the scene. The filter output, shown in the central column, exhibits a high response corresponding to the person, but effectively ignores the motion of the bushes. The right column illustrates the extraction of a binary motion mask using a constant valued threshold, $t = 3.0$. The bush motion is perceptible in the left-centre of frame 650 in the motion mask sequence.

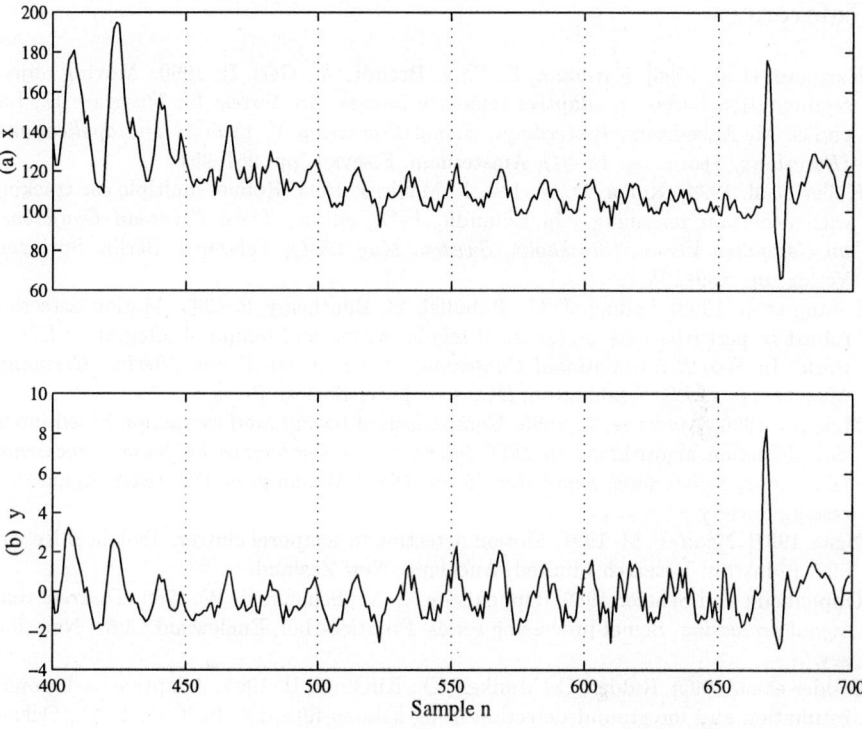

Fig. 2. (a) Intensity values for samples 400 to 699 of the input sequence at pixel $x(92, 123)$, and (b) corresponding output sequence.

7 Conclusions and future work

We have presented a signal processing technique to detect motion in the presence of temporal clutter. The filter presented has high sensitivity to transient events (such as when a person walks through a scene) and low sensitivity to persistent oscillatory events (such as moving bushes in the background of the scene). The filter exhibits desirable properties that enhances its robustness when operating in natural situations, namely invariance to intensity offset and scale.

Extensions to the work presented in this paper include the development of a gating function to preclude updating the background in regions where moving objects are considered to exist c.f. [Koller et al. 1994]. Since the moving object is not part of the background, it should not contribute to the updating of the background. Such a facility would reduce the severity of the leading edge of the moving object and create a mask that more uniformly covers the whole extent of the object. A second extension is to use an optimal value of α. Such a value can be derived using a parameter tuning procedure that minimises the variance between the expected input values and the background.

References

[Karmann et al. 1990] Karmann, K.-P.; v. Brandt, A.; Gerl, R. 1990. Moving object segmentation based on adaptive reference images. In Torres, L.; Masgrau, E.; Lagunas, M. A., editors, *Proceedings, Signal Processing V: theories and applications, (Barcelona, Spain, sep 18-21)*, Amsterdam. Elsevier, pp. 951-954.

[Koller et al. 1994] Koller, D.; Weber, J.; Malik, J. 1994. Robust multiple car tracking with occlusion reasoning. In Eklundh, J.-O., editor, *Third European Conference on Computer Vision, (Stockholm, Sweden, May 1994), Volume 1*, Berlin. Springer-Verlag, pp. 189-196.

[Letang et al. 1993] Letang, J. M.; Rebuffel, V.; Bouthemy, P. 1993. Motion detection robust to perturbations: a statistical regularization and temporal integration framework. In *Fourth International Conference on Computer Vision (Berlin, Germany, May 11-14, 1993)*, Washington, DC,. Computer Society Press.

[Makarov 1996] Makarov, A. 1996. Comparison of background extraction based intrusion detection algorithms. In *1996 International Conference on Image Processing, (Lausanne, Switzerland, September 16-19, 1996)*, Washington, DC. IEEE Signal Processing Society.

[Ngan 1997] Ngan, P. M. 1997. Motion detection in temporal clutter. Technical Report 693, Industrial Research Limited, Auckland, New Zealand.

[Oppenheim and Schafer 1989] Oppenheim, A. V.; Schafer, R. W. 1989. *Discrete-time signal processing*. Signal processing series. Prentice-Hall, Englewood Cliffs, New Jersey.

[Ridder et al. 1995] Ridder, C.; Munkelt, O.; Kirchner, H. 1995. Adaptive background estimation and foreground detection using kalman-filtering. In Kaynak, O.; Özkan, M.; Bekiroğlu, N.; Tunay, İ., editors, *Proceedings of International Conference on recent Advances in Mechatronics, ICRAM'95*, Boğaziçi University, 80815 Bebek, Istanbul, TURKEY. UNESCO Chair on Mechatronics, pp. 193-199.

A Novel Fast Three-Step Search Algorithm for Block-Matching Motion Estimation

William Booth, James M. Noras and Donglai Xu
Department of Electronic and Electrical Engineering
University of Bradford, Bradford, BD7 1DP, UK

Abstract

A fast three-step search (FTSS) block matching algorithm for motion estimation is described. The method is based on the real world image sequence's characteristic of centre-biased motion vector distribution and uses centre-biased checking point patterns and a small number of search locations to perform fast block matching. Computational complexity is reduced by employing an 11 × 11 search window rather than the traditional 15 × 15 window. Simulation results are presented which show that the FTSS algorithm provides competitive performance at faster speed when compared with similar techniques such as full search (FS), three-step search (TSS), new three-step search (NTSS) and four-step search (4SS). The advantages of the novel algorithm for hardware implementation are discussed with particular relevance to applying it to video coding.

Keywords

Compression; block matching; video coding; three step search; motion vector estimation; CIF video sequences.

1. Introduction

Motion estimation plays an important role in motion-compensated video compression because of its ability to exploit high temporal correlation between successive frames of an image sequence. Although many types of motion estimation algorithms have been developed, the simplicity of the block matching technique has made it the natural choice for most video compression standards, including MPEG [1], [2], H.261 [3], and H.263 [4]. The approach adopted in block-matching algorithms is first to divide each frame into blocks, typically 16 × 16 pixels. A motion vector is then calculated for each block in the current frame by searching for the best matching block within a limited search area in the reference/previous frame. Compression is achieved by using this best matched block pointed to by the motion vector as the predictor for the current block.

Of the block matching techniques reported in the literature, the full search (FS) method provides the optimal solution by exhaustively evaluating all the possible candidate blocks within the search range in the reference frame. However, massive computation is required in the implementation of FS. In order to speed up the process by reducing the number of search locations, many fast algorithms such as the 2-D logarithmic search (LOGS) [5], the conjugate directional search [6], the three-step search (TSS) [7], the new three-step search (NTSS) [8], and the four-step search (4SS) [9], have been developed. Chen et al [10] have demonstrated that the first three methods can easily be trapped into a local minimum, thereby degrading performance.

The NTSS and 4SS take into account the fact that the distribution of the global minimum of the block-matching distortion in real world video sequences is centred at zero. They produce smaller motion compensation errors and are faster on average than TSS. However, for some image sequences containing a lot of fast-moving events, the computational requirement of NTSS and 4SS may be higher than TSS. Furthermore, for the real-time or VLSI implementation of motion estimation, the worst case computational requirement should be considered instead of the average computation.

This paper describes a novel fast three-step search (FTSS) algorithm based on the centre-biased motion vector distribution characteristic. The major objective of the work was to follow the basic ideas embodied in the NTSS and 4SS techniques but to attempt to improve the speed of operation without compromising performance by using a smaller number of search points. Because of the intention to follow up this work with a hardware realisation, simplicity of construction was an important consideration in the development of the algorithm.

2. The Fast Three-Step Search (FTSS) Algorithm

As with earlier block matching techniques a maximum search range of ± 7 pixels in both horizontal and vertical directions is utilised by FTSS and the mean absolute error (MAE) is used as an appropriate estimate of the block distortion measure (BDM). For a given (x,y), the MAE between $block(m,n)$ of the current frame and $block(m+x, n+y)$ of the previous (reference) frame is defined as:

$$MAE_{(m,n)}(x,y) = \frac{1}{256} \sum_{i=-7}^{7} \sum_{j=-7}^{7} |f_k(m+i, n+j) - f_{k-1}(m+x+i, n+y+j)|$$

where $f_k(i,j)$ and $f_{k-1}(i,j)$ are the pixel intensities at position (i,j) of the current frame k and the previous frame $k-1$ respectively, and the $block(m,n)$ is the block with its upper left corner at position (m,n) of a frame. The first step of the algorithm employs exactly the same procedure as is used in the 4SS, namely a centre-biased search pattern with nine checking points on a 5×5 window. But after the first step is completed, the FTSS differs markedly from 4SS. The details of the FTSS algorithm are summarised below:

Step 1: The minimum BDM point is found from a nine-checking-points pattern on a 5×5 window located at the centre of the 15×15 searching area as shown in Figure 1(a). If the minimum BDM point is found to coincide with the centre of the search window, then go to step 3 else go to step 2.

Step 2: The search window size is maintained at 5×5 with a search pattern chosen after considering the two alternatives:

a) If the previous minimum BDM point is located at one of the corners of the previous search window, then five additional checking points are considered. An example is shown in Figure 1(b) where black circles and grey circles represent additional and previously evaluated pixels respectively.

b) If the previous minimum BDM point is located at the middle of any horizontal or vertical edge of the previous search window, then three additional checking points are considered. An example is shown in Figure 1(c) where black circles and grey circles represent additional and previously evaluated pixels respectively.

Step 3: The search window is reduced to 3 × 3 around the minimum BDM point found in step 2 as shown in Figure 1(d) and the direction of the overall motion vector is taken to be the minimum BDM point among these nine search points, of which eight are new.

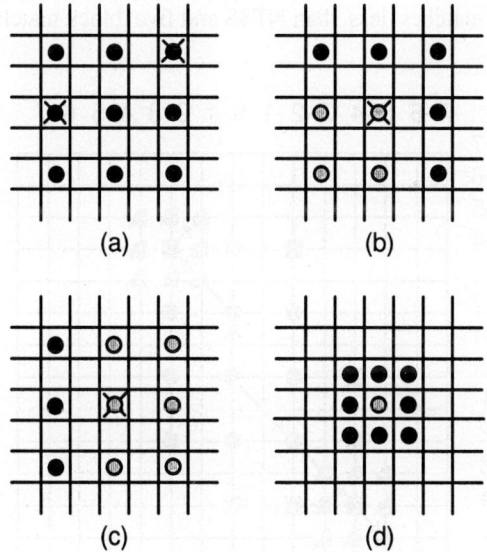

Figure 1. Search patterns of the FTSS
(a) First step centred on centre pixel
(b) Second step centred on a corner pixel
(c) Second step centred on a middle pixel
(d) Third step

In contrast to the use by previous block matching algorithms [7 - 9] of a 15 × 15 displacement window to cover the 15 × 15 search range, this three-step search pattern works with an 11 × 11 displacement window to cover the same search range in anticipation of significantly improving the speed of operation. However, reducing the window size carries the attendant risk of increasing distortion errors, particularly for sequences containing fast motion or large displacements. The justification for adopting the above approach is based on the fact that generally the block motion field of real-world image sequences varies so smoothly and slowly that high centre-biasing of the global minimum motion vector distribution will apply even for sequences containing fast motion and camera zooming. Thus the probability of creating distortion errors is likely to be extremely small for most video sequences.

Because there are some overlapping check points on the 5 × 5 search window in the second step of FTSS, the total number of checking points will vary from a minimum of (9 + 8)=17, when step 2 can be skipped, to (9 + 5 + 8)=22 in the worst case. Pictorial demonstrations of two search paths are provided by the examples shown in Figure 2. In the upper search path, a total of 20 checking points is needed to estimate that the motion vector is (1,-5). For the worst case example shown in the lower search path, 22 checking points are required when the estimated motion vector is (-5,5). For this latter case, the computational complexity of FTSS is three block matches less than TSS, eleven block matches less than NTSS and five block matches less than 4SS.

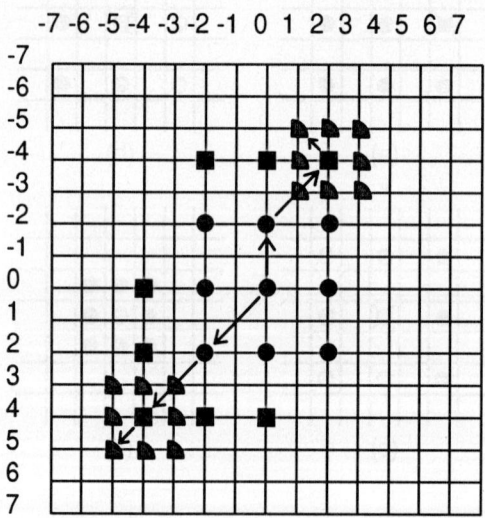

Figure 2. Two examples of FTSS search paths

3. Simulation Results

In order to test the speed and accuracy of FTSS and to measure its relative performance against other block matching techniques such as NTSS, 4SS and FS, simulations were carried out using three well known CIF video sequences, namely "Bike," "Football," and "Tennis". In each case the block size was fixed at 16 × 16 and the maximum motion in the search area was ±7 pixels in both the horizontal and vertical directions. The first 85 frames of these video sequences, each of which contains large displacements and fast motion, were used for test purposes. In the "Tennis" sequence, camera zooming and panning were also involved.

The performance of FTSS, 4SS and FS in terms of mean square error (MSE) between the estimated frames and the original frames are compared pictorially in Figures 3(a), 3(b) and 3(c).

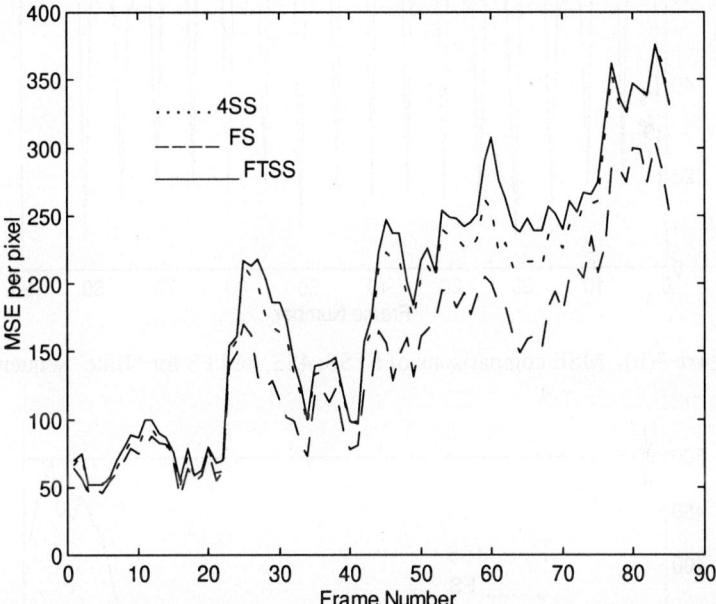

Figure 3(a). MSE comparisons of FTSS, 4SS, and FS for "Tennis" sequence.

The earlier three step search techniques TSS and NTSS have not been considered here, partly because they have been demonstrated to be inferior to 4SS but chiefly because they would overcomplicate these figures had they been included. The figures show that FS is always the most accurate technique as expected and that FTSS and 4SS have similar distortions for the vast majority of the frames. However, FTSS is slightly less accurate than 4SS for small fractions of the total time of the "Tennis" and "Football" sequences when very fast motion is involved. A full comparison of the relative accuracy of all the relevant block matching techniques as measured by the average MSE over the first 85 frames is presented in Table 1. From this table, it is clear that the distortions of FTSS are 1.8% worse on average than those of 4SS but 1.7% and 10.9% better on average than those of NTSS and TSS, respectively. The degradation in distortion performance of FTSS compared with 4SS would be less than 1% if the more appropriate statistical measure of root mean square error were employed.

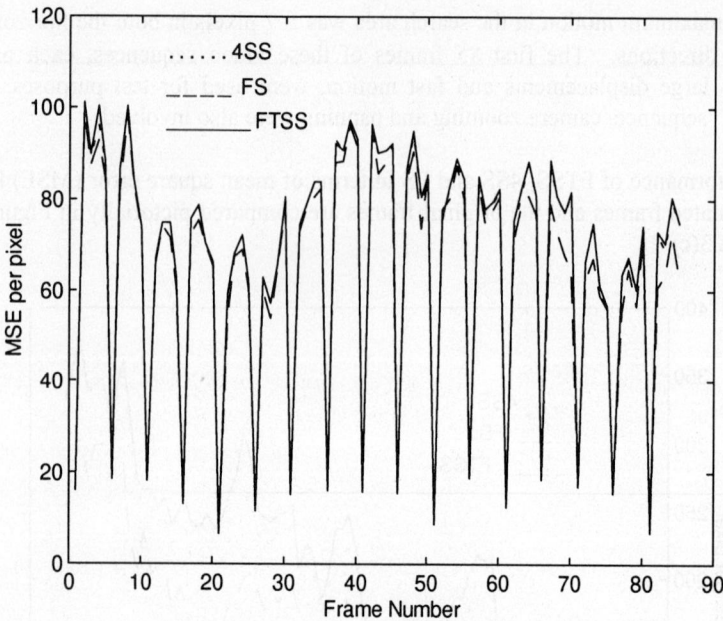

Figure 3(b). MSE comparisons of FTSS, 4SS, and FS for "Bike" sequence.

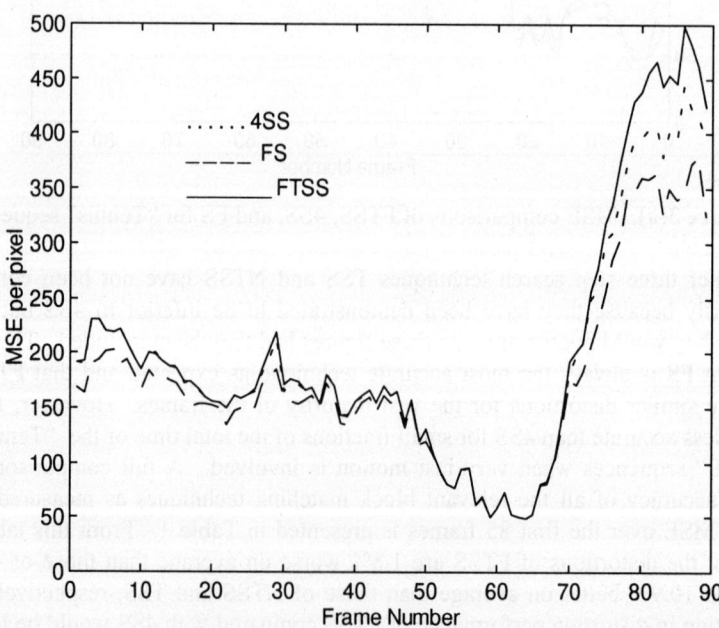

Figure 3(c). MSE comparisons of FTSS, 4SS, and FS for "Football" sequence.

TABLE 1: Average MSE of the first 85 frames of three CIF sequences

Searching Algorithm	Tennis	Bike	Football
FS	143.93	56.87	175.71
TSS	221.89	71.19	219.21
NTSS	203.02	60.11	206.69
4SS	189.28	58.35	205.90
FTSS	193.77	59.12	208.95

Speed of operation is a vitally important parameter for comparing the performance of fast block matching algorithms and a convenient measure of this is provided by the average number of search points required for estimating a motion vector. Table 2 contains relevant data on the average number of search points over the first 85 frames using the five algorithms of interest and shows that FTSS is the fastest technique for all the three CIF sequences considered.

TABLE 2: Average number of search points per motion vector estimation for the first 85 frames of three CIF sequences

Searching Algorithm	Tennis	Bike	Football
FS	225	225	225
TSS	25	25	25
NTSS	22.68	19.58	19.78
4SS	19.88	18.11	18.29
FTSS	17.93	17.52	17.59

Averaged over these sequences, the speed of FTSS is 6.1%, 17% and 41.4% greater than 4SS, NTSS and TSS, respectively. Table 2 also demonstrates that the speed-up factor is most pronounced for frame sequences containing fast movements or large displacements. In addition, it is noteworthy that the average number of search points using FTSS is usually close to its minimum value of 17. This provides a strong argument in favour of the decision to use a reduced search window area. Overall, in terms of trading off accuracy and speed, FTSS represents a significant improvement over all previously reported fast block matching algorithms.

4. Conclusions

Based on the centre-biased global minimum motion vector distribution characteristic of a real-world image sequence, a novel FTSS algorithm for fast block-based motion estimation has been developed. Experimental results show that this FTSS algorithm is significantly faster, as measured by the average number of search points, than other well-known three step search methods such as TSS and NTSS and also outscores them in terms of average mean square error. Although FTSS exhibits marginally higher distortion errors than 4SS for frames containing fast motion or large displacements, this is more than compensated for by the faster speed of operation. From the point of view of hardware implementation FTSS possesses the advantageous features of a regular search pattern structure and a relatively small number of search points. Work is currently in progress to create an associated hardware architecture.

References

1. ISO/IEC JTC1/SC29/WG11, ISO/IEC CD 11172: Information Technology, MPEG-1 Committee Draft (1991).
2. ISO/IEC JTC1/SC29/WG11, ISO/IEC CD 13818: Information Technology, MPEG-2 Committee Draft (1993).
3. International Telecommunication Union, Video Codec for Audio-visual Services at p*64 kbits, ITU-T Recommendation H.261 (1993).
4. International Telecommunication Union, Video Coding for Low Bitrate Communication, ITU-T Draft H.263 (1995).
5. Jain J. and Jain A. Displacement measurement and its application in interframe image coding, IEEE Transactions on Communication **COM-29**, 1799-1808 (1981).
6. Srinivasan R. and Rao K. R. Motion compensated interframe image prediction, IEEE Transactions on Communication **COM-33**, 1011-1015 (1985).
7. Koga T., Iinuma K., Hirano A. and Ishiguro T. Motion-compensated interframe coding for video conferencing, Proceedings of National Telecommunication Conference, pp. G5.3.1-G5.3.5, New Orleans, LA, USA (November 1981).
8. Li R., Zeng B. and Liou M.L. A new three-step search algorithm for block motion estimation, IEEE Transactions on Circuits and Systems for Video Technology **4**, 438-442 (1994).
9. Po L. M. and Ma W. C. A new four-step search algorithm for fast block motion estimation, IEEE Transactions on Circuits and Systems for Video Technology **6**, 313-317 (1996).
10. Chen L., Chen W., Jehng Y. and Chiueh T. An efficient parallel motion estimation algorithm for digital image processing, IEEE Transactions on Circuits and Systems for Video Technology **1**, 438-442 (1991).

Moving Vehicle Detection and Tracking in Image Sequences

Yi Lu, Jason Miller and Tie Qi Chen
Department of Electrical and Computer Engineering
The University of Michigan-Dearborn
Dearborn Michigan 48128-1491, U.S.A.
Voice: 313-593-5028, Fax: 313-593-9967
yilu@umich.edu

Abstract

This paper presents an algorithm that tracks moving vehicles in an outdoor environment. The algorithm was developed based on four heuristics derived from the properties of moving vehicles, maximum velocity, small velocity changes, coherent motion, and continuous motion. The algorithm was developed based on four heuristics on moving vehicles, namely maximum velocity, small velocity change, coherent motion and continuous motion. The algorithm has three major computational components, compute pixel motion vectors using a search window and a matching window, eliminate irrelevant moving pixels using a proximity filter and directional filters, and track the moving vehicles through multiple image frames using a proximity filter. The algorithm has been implemented and tested on image sequences taking in outdoor environments. The test results are presented in the paper.

1. Introduction

Tracking vehicles in outdoor scenes has a number of important applications such as video surveillance, intelligent-highway systems, etc. Most of the vehicle tracking systems available involve single objects moving under indoor or controlled lighting conditions. The most significant vehicle tracking systems which have been tested under complex outdoor conditions are described in [DTN92, LiH93]. Tracking moving vehicles in video images can be considered as a problem in two-dimensional motion detection, specifically, translation detection. Like many other motion analysis problems, the very first step, also a very challenging step in vehicle tracking is to establish image correspondence. In order to determine motion from time-varying image sequences, it is necessary to establish, either explicitly or implicitly certain type of correspondences between images. There are two issues involved in correspondence problem, the matching element and matching methods. The matching elements, or tokens, can vary significantly from one approach to another. Some approaches may use intensity-based matching others may use feature-based matching[WHA93] with features such as edges and corners. There are three major approaches to estimating two-dimensional translation: Fourier method, matching and method of difference measurement[HuT81]. The Fourier method works only if we have an isolated moving object with a uniform zero background. The most common matching method is to divide an image into small segments. For each segment in image I_i is matched with the image segments in I_{i+1}. A cost function is defined to find the best match. The size of the segments plays a critical roll in the final results. In general, using larger the segments leads to better noise immunity through averaging and can also disambiguate potential matches in areas of weak texture or potential aperture problems. However, larger windows fail where they straddle motion discontinuities, or in general where the motion varies significantly within the window. The method of difference measure attempts to guide the motion analysis with information about image differences. It computes the dissimilarity measure between the overlapping areas of regions in the two images and then analyze the difference measure for motion analysis. In particular in locating moving objets in an image sequence, one should ignore the stationary background, and the difference between two images provides information on the area of changing image and therefore actually reflect the motion of the object. This area of changing image can be tracked easily using this method.

In the problem of moving vehicle tracking, we need to analyze the results from the image correspondence problem to identify moving vehicles from its environment and other moving objects, and track the vehicles' motion. In particular in an outdoor scene, images contain cluttered backgrounds, and many other objects such as grass, clouds, etc. move simultaneously, which made the problem of tracking moving vehicles more difficult. Gardner and Lawton described a model-based vehicle tracking approach in [GaL96]. This model-based technique uses gravity and vehicle models, along with a vehicle motion model to constrain subsequent processing. Rosenfeld reported a geometric model based method to detect and track vehicles. The method uses a geometric model to search for silhouettes. The geometric model was obtained from Computer Aided Design(CAD) models. The matching of images to silhouettes is performed using probes, which are simple mathematical functions which operate locally on pixel values. An empirical probability density function of the probe values is obtained from a local region of the image, and is used to estimate the probability that the silhouette corresponds to a discontinuity in the image.

In an outdoor environment, moving vehicle tracking is difficult because there are many objects can move either randomly or with certain directions. For example, clouds in the sky, wheat or grass moving by the wind, tree leaves, telephone lines or electrical lines, etc. Pure data driven approaches will not be sufficient solve this problem. We need to develop heuristic knowledge that can identify moving vehicles from other moving objects.

In this paper, we describe an algorithm that was developed based on heuristic knowledge derived from attributes of vehicles, its trajectory, and outdoor environment. The algorithm has three major components, estimate motion vectors, eliminate non-vehicle pixels, and track through multiple image frames.

2. Tracking Moving Vehicles in an Outdoor Environment

The general vehicle tracking problem can be formulated as follows. We are given a sequence of images $I_t(x, y)$ containing moving vehicles, where $t = 0, 1, \ldots$. Without losing generality, we assume $I_t(x, y)$ were formed by locally displacing a reference image $I_0(x, y)$ with horizontal and vertical displacement fields $u_t(x,y)$, $v_t(x,y)$, such that

$$I_t(x+ u_t, y + v_t) = I(x, y)$$

where moving vehicles in $I_t(x+ u_t, y + v_t)$ are regions of points sharing common motion vectors. Moving vehicle tracking problem is to find out the regions with common motion vectors (u_t, v_t). It is a nontrivial problem when the environment is an outdoor scene which often has cluttered background and moving objects other than the target vehicles. In this application, we are considering the stationary observer, namely the camera is mounted on a stationary device such as a building, or stationary vehicle, etc.

The vehicle tracking algorithm we developed has three computational components, estimate motion vectors using a search window and matching window, identifying motion vectors of moving vehicles, and tracking moving vehicles through multiple image frames. The entire algorithm was developed based on the following heuristics derived from vehicle engineering.

• Maximum velocity. In general, we can estimate the maximum velocity of a moving vehicle in the scene, which limits the moving distance of each pixel. This maximum velocity can be used to define a search window. Figure 1. illustrates this concept. The bigger square represents the search window, which can be derived by the maximum moving velocity of vehicles in the image sequence.

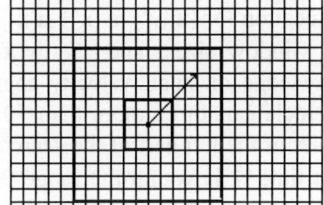

Figure 1. Illustration of search window and matching window.

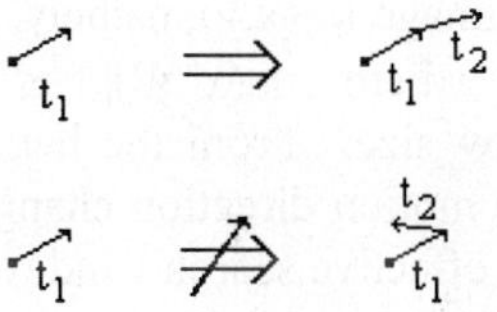

Figure 2. Possible (top) and impossible (below) velocity change between two image frames.

• Small velocity changes. Since vehicles have finite mass, vehicles are constrained to travel at small angle changes. Therefore it is reasonable to assume that the velocity change of a vehicle is always within the ranges of $-90° < \theta < 90°$, or $90° < \theta < 270°$ if it is moving in horizontal direction, $0° < \theta < 180°$, or $180° < \theta < 360°$ if it is moving towards or away from the camera. This heuristic knowledge is illustrated in Figure 2. According vehicle engineering, when the time interval between t1 and t2 is small, the motion vector illustrated at the top of Figure 2 is possible, but the one illustrated at the bottom of Figure 2 is impossible because the large angle change in motion.

• Coherent motion. Vehicles are spatially coherent objects, therefore, a vehicle appear in successive images as regions of points sharing a "common motion." Figure 3 illustrates this concept.

• Continuous motion. Moving vehicles have similar motion vectors across several image frames. Therefore we can track vehicles based on the motion vectors obtained through the previous frames.

The above heuristic knowledge has been used in the development of the three major computational components in our vehicle tracking algorithm.

2.1 Motion Estimation

Motion estimation has been well studied in the computer vision field. One fundamental approach is to match one portion of image I_t with one portion in I_{t+1} using image features such as edges, contour or intensity. In general, the outcome of matching based on image features very much depends on the accuracy of the image feature extraction algorithms, for examples, edge detection, contour extraction, etc. In this project, we use intensity feature. One problem in using intensity feature is that it does not give sufficient information to detect moving vehicles. Considering a window of pixels of non-zero motion vectors, we could be looking at randomly moving grass, tree leaves, clouds in the sky, etc. Therefore, heuristics are applied to the moving objects to identify vehicles.

In estimating motion vectors, we use the heuristics of maximum moving speed and small velocity change to optimize our search. We use two windows (see Figure 1) in our motion estimation component. The smaller window in Figure 1 is called matching window, it is used to estimate the similarity between two portions of images I_t and I_{t+1}. The larger window in Figure 1 is called search window, it is used to limit the search for the possible location of a particular pixel in the image frame I_{t+1}. According to the maximum velocity heuristic, a pixel (x', y') in $I_t(x,y)$ has a maximum translation in image $I_{t+1}(x,y)$, namely, the new location for (x',y') is (x'+ u(x', y'), y' + v(x', y')), where $|u(x',y')| \leq$ search_window_size and $|v(x',y')| \leq$ search_window_size. From the heuristic knowledge of small velocity changes, we know that the motion direction change of a vehicle cannot be more than 90 degrees. Therefore the effective search window is a subset of the left half of the search window when the orientation of the motion vector θ at time t1 is within the range 90° ≤ θ ≤ 270°, and the effective search window is the a subset of the right half of the search window (see Figure 4) when the orientation of the motion vector θ at time t1 is within the range 0° ≤ θ ≤ 90°, or 270° ≤ θ ≤ 360°.

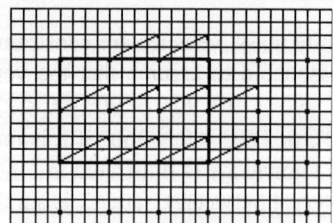

Figure 3. Pixels belonging to the same vehicle have similar motion.

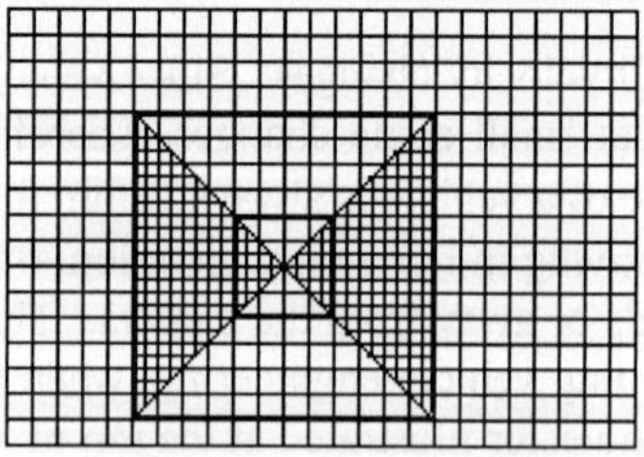

Figure 4. Effective search window derived from heuristics.

Within the effective search window, the new location of (x', y') is found by using the maximum likelihood, which consists of minimizing the squared error over the effective search window within I_{t+1} (x,y). Mathematically, we use the following likelihood function:

$$\Phi(u_i, v_j) = \sum_q \sum_p (I_t(x_i+p, y_j + q) - I_{t+1}(x_i + u_i + p, y_j + v_j + q))^2$$

where p and q should be within the matching window. The new location of (x_i, y_j) within image I_{t+1} (x,y) is at (x_i', y_j') where x' = x + u, y' = y + v, $-p \leq u \leq p$ and $-q \leq u \leq q$ and $\Phi(u, v) < \Phi(ui, vi)$, where (ui, vi) is within the search window. The motion vector for (x_i', y_j') is (dx, dy), where dx = $x_i'- x_i$, dy = $y_j' - y_j$. The concept of effective window does not only reduce the computational time but also provide more accurate search results since the search window is defined more accurately, the probability of mismatching is reduced.

Much research has been done in the optimal matching between the two frames I_t and I_{t+1}. The size of the matching window plays an important role in motion estimation. In general larger windows lead to better noise immunity through averaging and can also disambiguate potential matches in areas of weak texture. However, larger windows fail where the motion varies significantly within the window. Okutomi and Kanade described an adaptive window matching and multiple frames[OkK93]. Szeliski and Shum used quadtree splines to estimate motion[SzS96]. Our study showed that these methods require more computational time, and in many case, they do not give as much better results. Instead, we use a window of 9x9 in the computation of maximum likelihood.

2.2 Moving vehicle detection

From motion estimation, we obtained motion vectors at every image pixel. Pixels have nonzero motion vectors are candidates for moving objects. As we indicated earlier that in an outdoor environment, many objects other than vehicles can be moving. However, since vehicle motion is a rigid body motion, which means every point on the vehicle must have the same motion vector, many other moving objects do not have this property, for example, grass, telephone or electrical wires. Therefore we have applied the coherent motion heuristics to eliminate the non-vehicle objects. The following rules are used to implement this heuristics.

• applying a proximity filter to the base pixel and its neighbors.

Two motion vectors are in the proximity orientation if they have the same sign at the dx orientation, which implies both vectors are moving towards right (dx > 0) or left(dx < 0) or dy orientation, which implies both motion vectors are either moving upward or downward. The proximity filter states that if base pixel (x,y) in I_{t+1} has α number of neighbors that have the same **sign** in the motion vector (dx, dy), then keep the point, otherwise eliminate the point. The parameter α can be set by the user based on either a priori knowledge on the direction of the moving vehicle or on the moving vehicle size. If we know the moving direction of the vehicle, then we know the base point and its α neighbors must have the same sign in either dx or dy direction. The most conservative way is to set $\alpha =1$, which requires at least one of the eight neighbors of the base pixel to have the similar motion direction. If two pixels have the same sign in their dx direction, say dx > 0, it implies the two pixels both move East, if both pixels have dy < 0, it means the two pixels both move South, etc. If the base point has α neighbors that have the same dx or dy sign as the base point, then the base point is considered as the possible vehicle point and therefore is kept, otherwise, this base point is discarded.

• Eliminate the pixels that have negligible motion.

This rule intends to eliminate the random motion vectors. Two techniques are used. The first technique simply removes the motion vectors who magnitudes are very small. This is implemented by setting a minimum motion threshold, MIN_threshold. A point is not considered a moving vehicle pixel if the magnitude of its motion vector is less than or equal to preset threshold, MIN_threshold. This MIN_threshold can be derived from knowledge of vehicle speed. In our application, we use MIN_threshold = 1.

The second technique uses three directional filters, horizontal pass filter, vertical pass filter, and a horizontal-or-vertical pass filter. The directional filters attempt to remove all motion vectors which are not primarily horizontal or primarily vertical. The horizontal pass filter requires the displacement at the x direction to be larger than the displacement at the y direction, which means the object is moving more along the x direction. Similarly the vertical filter requires the displacement at the y direction to be larger than the displacement at the x direction, which means the object is moving more along the y direction. The horizontal filter requires

|displacement[i,j].x| - |displacement[i,j].y| ≤ XY_LIM **AND**
(|displacement[i,j].x| - |displacement[i,j].y| > 0),
the vertical filter requires
(|displacement[i,j].y| - |displacement[i,j].x|) ≤ XY_LIM) **AND**
(|displacement[i,j].y| - |displacement[i,j].x|) > 0),
and, the horizontal-or-vertical filter requires
| |displacement[i,j].x| - |displacement[i,j].y| | ≤ XY_LIM),
where displacement[i,j].x and displacement[i,j].y represent the displacement of the current pixel (i,j) at direction x and y respectively, and XY_LIM is a threshold which can be estimated based on the knowledge of vehicle velocity.

2.3 Moving vehicle tracking in multiple frames

After the first two frames, I_0 and I_1, we obtain a set of moving regions using the methods described in the above two subsections. For subsequent image frames, we only track the points in these regions. In addition, we compare the last computed motion vectors with the current motion vectors to further eliminate the non-vehicle objects in the successive frames I_2, I_3, ... We obtain the motion vectors at the current frame using the method described in the previous subsection. The motion vectors are then processed by applying the proximity filter to the current motion vector and the past motion vectors of its neighbors. This is implemented as follows. Assume the current image frame is I_t. t > 1. A pixel (x, y) in I_t. whose current motion vector is (dx, dy) is considered as part of a moving vehicle if either dx or dy has the same **sign** as its or at least one of its eight neighbors' previous motion vector in the x direction or in the y direction respectively. Figure 5 illustrates this concept. Figure 5 (a) shows motion vectors of a vehicle calculated from image frame I_t. Figure 5 (b) shows the likely motion vector of a vehicle pixel in image frame I_{t+1}. The motion vector shown in Figure 5 (c) is not likely to occur for a corresponding vehicle pixel in image frame I_{t+1}. This constraint basically requires the pixel's motion not only to be in the same general direction as the surrounding points between each frames in the **current** image frame, but also in the same general direction as the surrounding points from **past frames**. This technique is effective in eliminating random motion such as grass.

After we obtain individual motion vectors and eliminate the non-vehicle pixels, we group the pixels that have nonzero motion vectors and are connected in the image space together to form a region which is considered as a possible moving vehicles. A threshold can be set to filter out small regions that are impossible to be considered as the target vehicle. The grouping is implemented by generating a binary image based on the motion vectors. If an pixel has a motion vector equal to 0, then this pixel takes the value 0 in the binary image, otherwise takes the value 1 in the binary image. Then we apply a connected components algorithm described in to the binary image obtain the regions, which are subsequently mapped back to the original images to obtain the moving objects.

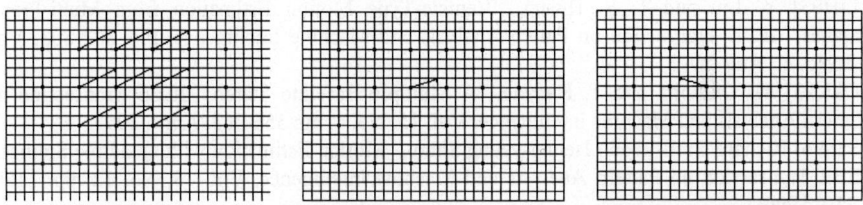

(a) Motions vectors of a vehicle in image frame I_{t1}.

(b) A likely motion vector for a corresponding vehicle pixel in image frame I_{t1+1}.

(c) An unlikely motion vector for a corresponding vehicle pixel in image frame I_{t1+1}.

Figure 5. Likely and unlikely motion vectors of a moving vehicle in two images frames.

3. Experiments and Conclusion

We have described a moving vehicle detection and tracking algorithm. The algorithm was developed based on four heuristics on moving vehicles, namely maximum velocity, small velocity change, coherent motion and continuous motion. The algorithm has three major computational components, compute pixel motion vectors using a search window and a matching window, eliminate irrelevant moving pixels using a proximity filter and directional filters, and track the moving vehicles through multiple image frames using a. We have implemented the algorithm and have tested on a number of image sequences. Figure 6 proximity filter shows the result of the algorithm applied to one image sequence. The image sequence was captured by a video camera mounted on a tripod in a highway scene. The image sequence contains moving objects such as grass, moving clouds and two moving vehicles. Figure 6 (a) and (b) shows the result obtained from image frame 1 and 2. For illustration, all results were presented by superimposing bounding boxes of connected motion vectors to the original image frames. Figure 6 (a) shows the motion vectors obtained after motion estimation step. Figure 6 (b) shows the result after the elimination of non-vehicle pixels. Figure 6 (c), (d), (e) and (f) show the result of the tracking of the two vehicles across frame 3, 4, 5 and 6 respectively. This examples shows that the two vehicles are tracked through out the entire image sequence, and noise and other moving objects are eliminated.

4. Acknowledgment

The authors would like to thank Dr. Thomas Meitzler at the U. S. Army TACOM for providing the video image data.

5. Reference

[GaL96] Warren F. Gardner and D. T. Lawton, "Interactive Model-Based Vehicle Tracking," IEEE Transaction on Patern Analysis and Machine Intelligence, Vol. 18, No. 11, PP. 1115-1121, Nov. 1996

[HuT81] T. S. Hung and R. Y. Tsai, "Image Sequence Analysis: Motion Estimation," *Image Sequence Analysis*, Edited T. S. Huang, Springer-Verlag, 1981

[LiH93] Y. Liu and T. S. Huang, "Vehicle-Type Motion Estimation from Multi-Frame Images," IEEE Transaction on Patern Analysis and Machine Intelligence, PP. 802-808, Vol. 15, No. 8, Aug. 1993

[OkK93] M. Okutomi and T. Kanade, "A Multiple Baseline Stereo," IEEE Transaction on Pattern Analysis and Machine intelligence, Vol. 15, No. 4, pp. 353-363, April, 1993

[SzS96] Richard Szeliski and Heung-yeung Shum, "Motion Estimation with Quadtree Splines," IEEE Transaction on Pattern Analysis and Machine intelligence, Vol. 18, No. 12, pp. 1199-1210, 1996

[WHA93] Juyang Weng, Thomas S. Huang, Narendra Ahuja, Motion and Structure from Image Sequences, Springer-Verlag, 1993.

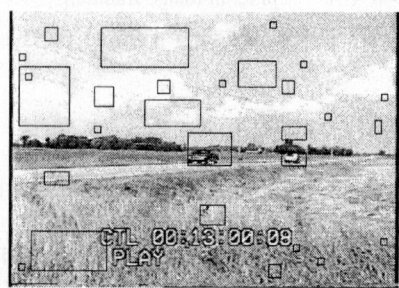 

(a) Moving objects detected based on motion vectors

(b) After eliminating the non-vehicle pixels.

(c) After pixel grouping and small object elimination.

(d) Moving objects are tracked in the third frame.

Figure 6. Test result.

Gesture Recognition from Image Motion Based on Subspace Method and HMM

Yoshio IWAI, Tadashi HATA, and Masahiko YACHIDA[1]

Department of Systems and Human Science, Osaka University
Toyonaka Osaka 560, Japan
E-mail: iwai@sys.es.osaka-u.ac.jp

Abstract. This paper proposes a method to recognize human gestures from an image sequence based on Hidden Markov Model and Karhunen-Loève Transform. As our method uses the motion vector field of the scene for recognition, it is robust for variety in the background of the scene and it doesn't require the users to wear a sensor or a marker. The motion vector field of the scene is projected to an eigen-subspace for data compression and is used as the input symbols for the HMM. View-based methods generally fail when the user translates. Our method is robust for recognition to the user's translation because the recognition window automatically fits the user by tracking the user's face.

1 Introduction

Hand gesture is one of the typical methods of nonverbal communication for human beings and we naturally use various gestures to express our own intentions in everyday life. Sensing of human expression is very important for human-computer interactive applications such as virtual reality, gesture recognition, and communication. In recent years, a large number of studies have been made on machine recognition of human expression.

Most methods of previous works require magnetic sensors, markers or constraints such as the background color, the light conditions and the user's position. It is, however, better for human beings that these constraints or sensors are not required. A vision system is suitable for human-computer interactions since this involves passive sensing and the human gestures of the hand, body, and face can be recognized without any discomfort for the user.

A lot of methods for the recognition of gestures from images have been proposed. Takahasi et al.[1] used spatio-temporal edges for the recognition of gestures based on a DP matching method. Yamato et al.[2] used matrices sampled sparsely from images and Hidden Markov Models for the recognition. The sparse matrices are quantized by the Categorized VQ method and are used as input for a discrete HMM. Campbell et al.[3] used the 3-D positions and 3-D motion of face, left hand, and right hand extracted from stereo images and HMM for the recognition. Watanabe et al.[4] used an eigen-subspace for a real-time gesture recognition system. In this method, the trajectories of the gesture in the eigen-subspace are recognized by comparing them with the trajectories of the model, which were generated in advance. Nagaya et al.[5] used the vector representation of the raw image sequence and the trajectories in the vector space for gesture recognition. The assumptions of this method fail, if the background is changed.

In this paper, we propose a gesture recognition system, based on a Hidden Markov Model and Karhunen-Loève Transform, which can recognize gestures in the various backgrounds and which doesn't require any sensors or markers.

2 Outline of System

The processes of our system are shown in Figure 1. Our system consists of two parts: data compression part and gesture recognition part. Our method uses the motion vector field of the scene for recognition. The motion vectors are calculated by a template matching method and are projected to the eigen-subspace for data compression. The projected motion vectors are decomposed into principal components by KL transform. The decomposed principal components are quantized and then they are symbolized to use the input data of discrete HMMs. Each HMM models a different gesture in advance.

First, we describe the data compression by KL transform, and then we describe the gesture recognition by HMM.

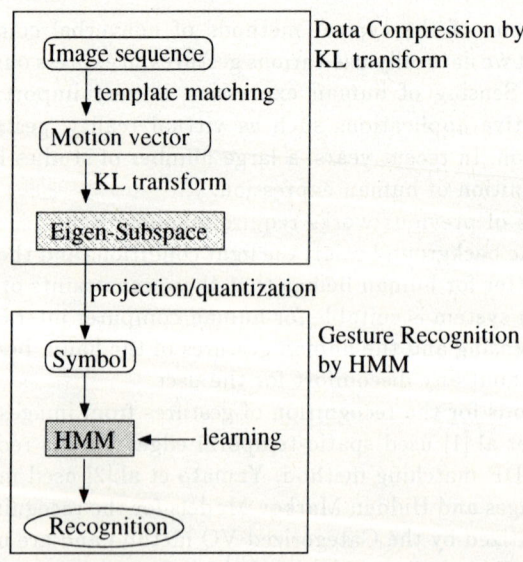

Fig. 1. The process of our system

3 Eigen-subspace for Gesture Recognition

3.1 Extraction of optical flows from an image

We use optical flows as data for gesture recognition. The advantage of using optical flows is insensibility of the static background, so our method can be used in any background. The disadvantage of using optical flows is the instability near the occluding edges, but we use the only important flows extracted by principal component analysis. We extract optical flows by a template matching method. A sample of the motion vector is shown in Figure2.

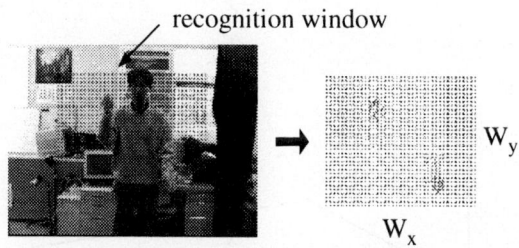

Fig. 2. motion flow extracted by template matching

The problems of template matching are the large computation cost and the sensitivity to noise such as the change of the intensity, the rotation of the template. We address the former problem by using a image processor and the latter problem by using HMM for recognition because HMM recognition method uses many images, symbolizes the input data containing random noise, and determines a gesture from an average of the input symbol.

The estimated motion vectors $V(x,y) = (Vx(x,y), Vy(x,y))$ are vectorized in raster scan order as follows:

$$V = \{V(0,0), V(1,0), \cdots, V(Wx, Wy)\}, \qquad (1)$$

where (Wx, Wy) is the size of the recognition window shown in Figure 2. The motion flow field, which is used as the input data of HMM, are sampled within the recognition window.

Detecting and fitting the recognition window: The problem which we must be concerned with is the segmentation problem where a person makes gestures. When the recognition window is put on the input image at a fixed point, the input vectors of HMMs will be changed by the position of a person and HMMs will fail to recognize gestures. We must, therefore, fit the size and the position of the recognition window according to the position of the user. The recognition window is automatically fitted to a person based on the size and position of his face detected by our previously proposed method[6].

3.2 Subspace made by KL transform

Many applications of KL transform directly use pixel values of an image as multivariate data, but we use optical flows in the equation of many models (image sequences) of gestures to make eigen-subspace for applications in various backgrounds. From the stand point of data compression, it is better to use fewer components for representation of gestures. It is well known that i-th eigen vector is not needed if the accumulated contribution ratio c_i^{acc} is greater than 80 %. In our experiment, 30 eigen vectors are needed for recognition of five gestures and the accumulated ratio is 58.8 %. We show a sample trajectory of a gesture projected into the eigen-subspace in Figure 3.

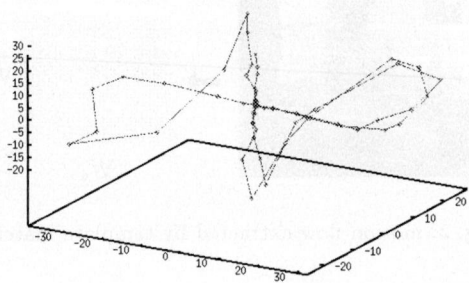

Fig. 3. A trajectory of a gesture(drum) in eigen-subspace

4 Gesture Recognition by HMM

4.1 Symbolization of Coefficient of Principal Components

We must symbolize the coefficients p_{ij} of the principal components derived from the previous section, before we use a discrete HMM for recognition. The coefficients $p_{ij}(j = 1, 2, \cdots, k)$ are quantized (annotated by $Q(.)$) by the nearest-neighbor clustering and put into a sequence of symbols O_t.

4.2 Gesture Recognition by HMM

It is a good method for gesture recognition to use HMMs because HMMs are robust for the temporal effects of the variance of the observable symbols.

The raw input data X_n is already converted into a sequence of discrete observation symbols $O_t = \{Q(p_{i1}), Q(p_{i2}), \cdots, Q(p_{ik})\}$ at each frame by the method described above. Each HMM H_i models a different gesture in advance.

It is determined which of HMMs is most likely to have the generated sequence $O = \{O_0, O_1, \cdots, O_t\}$ by the following equation:

$$I = \max_i {}^{-1}(P_j(O|H_i)), \qquad (2)$$

where $P_j(O|H_i)$ is the probability of HMM H_i that the observation symbols O can be gesture I.

4.3 Learning gesture models by HMM

Learning of a HMM is equal to estimating the parameters $\{\pi, A, B\}$ of the HMM. The algorithm which we use for estimation of the HMM parameters is the Baum-Welch Algorithm. The system iteratively calculates the above parameters until these paramters are converged. Therefore, it takes many times and a lot of training images.

5 Experimental Results

We have made several experiments to evaluate our proposed method. The size of input images was 240×320 pixels and the size of the recognition window was 165×195 pixels. The size of the template image was 9×9 pixels and the range of the search region was 27×27 pixels. The number of the dimension of the motion flow field was $33 \times 39 \times 2 = 2574$.

5.1 Experiment A

We used five gestures for these experiments which are similar to playing musical instruments such as the drum, guitar, piano, violin, and castanets shown in Fig 4. We used 60 images per gesture to make the eigen-subspace and totally used 300 images to make it.

The number of states of HMM was 5 and the number of symbols was 27. All gestures were classified within 40 frames.

The following table is the recognition rate of the gesture of the other 4 persons whose gestures are not learned by HMMs. We prepare 3 image sequences in various backgrounds for each gesture and each person, so $3 \times 5 \times 4 = 60$ image sequences are used.

gesture	drum	guitar	piano	violin	castanets
recognition rate (%)	73.3	100.0	66.7	100.0	93.3

Table 1. The recognition rate

Fig. 4. Sample images of models of gestures.

5.2 Experiment B

In experiment A, the user must face the camera. It is not better to recognize gestures from only one view point. We varied the view point of the camera horizontally from -30 degrees to +30 degrees stepped by 10 degrees as shown in Figure 5. The recognition results are shown in Figure 6. It is successful to recognize the gesture from +20 degrees to -10 degrees. The recognition rates and the convergence time of HMMs gradually down with the magnitude of the user's rotation.

All gesture can be classified by HMMs within 20 degrees of user rotation, but the gesture (castanets) is slowly converged since the output symbol of gesture (castanets) partially contains the symbol of other gesture (drum). we needed more symbols for the distinction of gestures in order to recognize these gestures

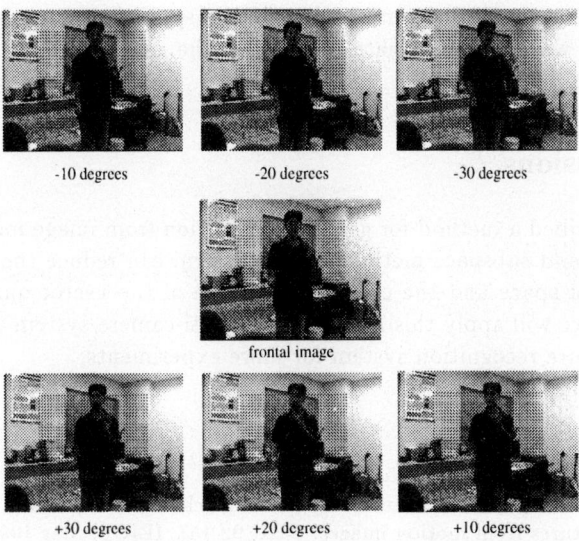

Fig. 5. The views of the camera

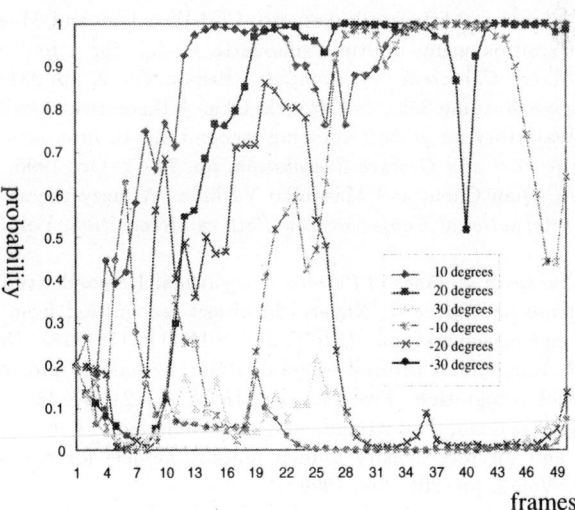

Fig. 6. The recognition results

accurately. In this case, the computation time of symbolization is important and our approach will be appropriate because of the reduction of the input data space.

6 Conclusions

We have described a method for gesture recognition from image motion by combining HMM and subspace method. KL transform can reduce the dimension of the input data space and the computation time of the vector quantization. In future work, we will apply this method to a multi-camera system and develop a real-time gesture recognition system for more experiments.

References

1. Katsuhiko Takahashi, Susumu Seki, and Ryuichi Oka. Spotting recognition of human gestures from motion images. PRU 92-157, IEICE, Mar 1993.
2. Junji Yamato, Shoji Kurakake, Akira Tomono, and Kenichiro Ishii. Human action recognition using HMM with category-separated vector quantization. *IEICE*, J77-D-II(7):1311–1318, Jul 1994.
3. Lee W. Campbell, David A. Becker, Ali Azarbayejani, Aaron F. Bobick, and Alex Pentland. Invariant features for 3D gesture recognition. In *International Conference on Automatic Face and Gesture Recognition*, pp. 157–162, Killington, Oct 1996.
4. Takahiro Watanabe, Akitoshi Tsukamoto, Chil-Woo Lee, and Masahiko Yachida. Gesture recognition using multiple silhouette models for a real-time interactive system. In *Asian Conference on Computer Vision*, Vol. 2, pp. 235–239, 1995.
5. Shigeki Nagaya, Susumu Seki, and Ryuichi Oka. A theoretical consideration of pattern space trajectory for gesture spotting recognition. In *International Conference on Automatic Face and Gesture Recognition*, pp. 72–77, Oct 1996.
6. Haiyuan Wu, Qian Chen, and Masahiko Yachida. A fuzzy theory based face detector. In *International Conference on Pattern Recognition*, Vol. 3, pp. 406–410, 1996.
7. Errki Oja. *Subspace Methods of Pattern Recognition*. Research Studies Press, 1983.
8. Hiroshi Murase and Shree K. Nayar. 3d object recognition from appearance — parametric eigenspace method. *IEICE*, J77-D-II(11):2179–2187, Nov 1994.
9. Lawrence R. Rabiner. A tutorial on hidden markov models and selected applications in speech recognition. *Proceedings of IEEE.*, 77(2):257–285, 1989.
10. G. Rigoll, A. Kosmala, J. Rottland, and C. Neukirchen. A comparison between continuous and discrete density hidden markov models for cursive handwriting recognition. Vol. 2, pp. 205–209, 1996.
11. Hiroshi Murase and Shree K. Nayar. Illumination planning for object recognition in structure environments. pp. 31–38, 1994.

Identifying Faces under Varying Pose Using a Single Example View

Dadet Pramadihanto   Yoshio Iwai   Masahiko Yachida   Haiyuan Wu

Department of Systems and Human Science, OSAKA UNIVERSITY
1-3 Machikaneyama, Toyonaka, Osaka, Japan 560

Abstract. Identifiying human faces with only a single face model per person is a difficult task, because the input face image may vary with position, expression and pose. This paper describes a flexible face identification system based on Gabor wavelet representation and flexible neural network matching. The face is represented by a hexagonal graph of neurons where each node contains local feature information extracted by Gabor wavelet transform at the corresponding position. An innovative flexible neural network matching is employed to finding out the exact correspondence of local features between the model and the input image based on the local feature similarity and neighborhood grouping neurons. The matching process is evaluated by competition rule based on the correlation of neuron activation in the model layers and input layer. Experiments with face images that include the variations of rotation-in-plane, rotation-in-depth, and the change of facial expression are also presented.

1 Introduction

The problem of face recognition has attracted researchers because not only faces represent a challenging class of naturally textured 3D objects, but also because of many applications for automatic face identification. Identification of a face is difficult because faces form a class of fairly similar objects, where all faces consist of the same facial features in roughly the same geometrical configuration [2]. The detailed 3D structure of the face makes the 2D image change if the face undergoes rotation-in-depth or if the light source changes direction. The non-rigidity of faces which may be caused by changes in facial expression [2]. The appearance of particular faces changes as time goes on, such as age, medical treatment, weight fluctuation, styles, etc.

There has been some recent work in this direction, such as pose-invariant recognition using multiple view-based [1,9], but in a real application we can not expect to have all views required to perform identification. Another approach is discussed in [10] using synthesized virtual views from one example for multiple view face identification. They need face images with known pose to teach the network to be able to synthesize the virtual views. In the flexible matching approach discussed in [7], the input image is deformed in 2-D to match the example view. The deformation, which is like a local 2-D warp of the image, allows the matching of input and example view even though they may differ in expression or out-of-plane rotation. In [8], a technique for transforming Gabor jet features across rotation-in-depth has been investigated.

This paper deal with face identification when only single pose of a face is available. Are we still able to perform face identification if the input image differ in imaging condition or pose from the model ? If so, until what limit is the system able to perform identification ? Our approach uses the hexagonal grid face features representation extracted using Gabor wavelet transformation and a flexible neural network matching.

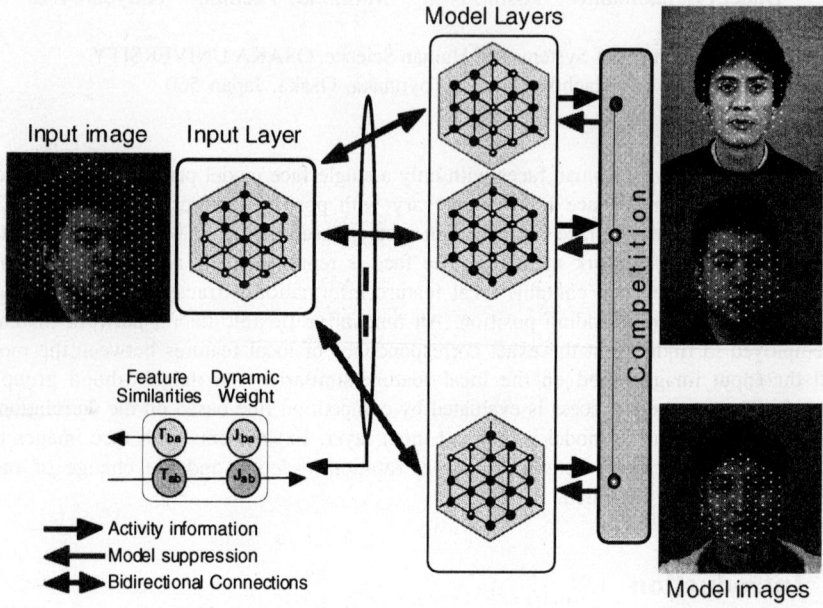

Figure 1. Face identification system.

2 Face Representation

2.1 Gabor Wavelet

Features can be generated from the image by filtering operations. The strength of features appears in an image at a certain position is generated by a filter operation with the corresponding filter kernel. Gabor wavelet transformation of an image is defined as the convolution process of image with a family of Gabor kernels:

$$\psi_{k,\theta}(x,y) = \frac{k^2}{\sigma^2}\exp\left(-\frac{k^2(x^2+y^2)}{2\sigma^2}\right)\left\{\exp(ik(x\cos\theta+y\sin\theta))-\exp\left(-\frac{\sigma^2}{2}\right)\right\}, \quad (1)$$

$$\theta = \frac{\kappa\pi}{N_\theta} \quad \text{with} \quad \kappa = 0,...,N_\theta-1 \quad \text{and} \quad \psi_{\mu,k,\theta} = \frac{1}{2^\mu}\psi_{k,\theta}\left(\frac{x}{2^{\mu/2}},\frac{y}{2^{\mu/2}}\right). \quad (2)$$

The first factor in (2) is the normalized Gaussian window. Its width is controlled by the parameter, $\sigma=2\pi$. The second factor is the sinusoidal component of the wavelet. Its (x,y) term is projected in the θ direction to determine the complex exponential.

The kernels are corrected for their DC value by the second term in the bracket of equation (1). These Correction makes the integral of the kernel vanish and provide robustness against varying brightness in the image [11]. Sensitivity against orientation depends on the orientation direction, θ, and multiscale representation of Gabor transformation can be achieve by scaling the Gabor kernel (2). Here, N_θ is the number of orientations and N_μ is the number of scales and µ is the scale index.

The convolution of an image with the Gabor kernel produces complex features for every center location, orientation and scale. The magnitude of this complex feature varies slowly with position. Conversely, the phase of these complex features changes drastically with translation or varies with wavelet frequency. We ignored the phase information and only the magnitude information is taken into account.

2.2 Hexagonal Graph Representing Faces

A face is represented as a sparse hexagonal grid. The use of the hexagonal grid representation is motivated by its structure which incorporates a uniform connection property, greater angular resolution, and a higher degree of symmetry [3, 4].

Individual face models are represented as hexagonal grids originating at the center of face. The grid only covers the image area that contains the face, as indicated by the white dots in figure 2(b). A hexagonal graph Y representing a face consists of Q number of "node"s at grid positions, $\mathbf{b}_q$, q=1, ..., Q. Here, $\mathbf{b}$ is the hexagonal coordinates. Each node has information about the distances to other nodes and a Gabor feature vector as results of Gabor wavelet transformation centered at the grid point. The Gabor feature vector is composed of $N = N_\theta \times N_\mu$ elements, which are the magnitude response of the transformation at different orientations and scales. The individual face model becomes an array of Gabor features vector, $\mathbf{M} = \{\mathbf{m}_1 ... \mathbf{m}_Q\}$, where $\mathbf{m}=\{m_1 ... m_N\}$ is the Gabor feature vector and a matrix $\mathbf{D}=\{d_{11} ... d_{qq}\}$ that described the distances among nodes. If there are N model faces. These model faces can be represented as an array of individual face models, that is $\{\mathbf{M}_1 ... \mathbf{M}_N\}$.

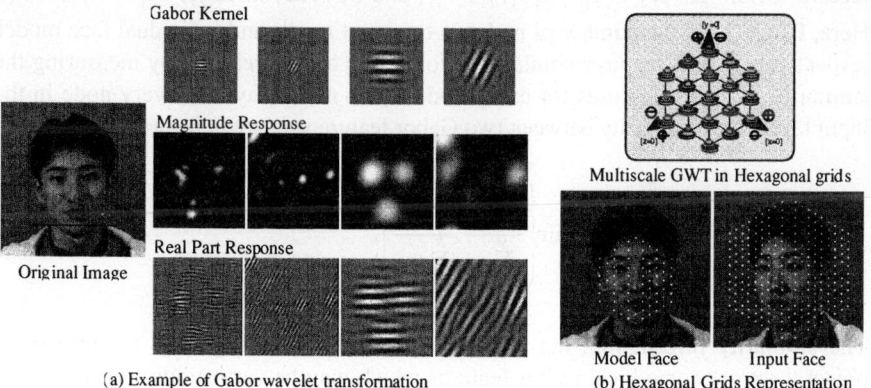

(a) Example of Gabor wavelet transformation (b) Hexagonal Grids Representation

Figure 2. Face representation using Gabor wavelet and hexagonal grid structure. (a) Examples of Gabor wavelet transformation. (b) Hexagonal grid representation.

In contrast to the model face, the hexagonal grid in the input face cover the hole image area (fig 2b). Similar to the model images, the input image is also defined as a sparse hexagonal grid of Gabor wavelet transformations. A hexagonal graph X representing an input face consists of P "node"s on the grid at position a_p, p=1, ..., P. Here, **a** is the hexagonal coordinates. The input image becomes an array of Gabor feature vectors, $\mathbf{R}=\{\mathbf{r}_1 ... \mathbf{r}_P\}$, here $\mathbf{r}=\{r_1 ... r_N\}$ is the Gabor feature vectors and a matrix $\mathbf{E} = \{e_{11} ... e_{pp}\}$ that described the distances among nodes or grid points.

3 The Flexible Neural Network Matching

The flexible neural network matching is constructed by making use of the information of local features similarity and neighborhood grouping of neurons. Flexible neural network matching is similar to the dynamic link architecture discussed in [5, 6]. The differences between these two matching model can be stated as follows. First, in our approach, the connection between input and model layer are in bidirectional fashion instead of unidirectional. These connection provides reevaluation mapping from the model layer to input layer and provides matching signal as to which model the input receives stronger connections. Secondly, these bidirectional connections incoorporated with the delayed self-inhibition of each neuron produce continuously moving neuronal group activation in the layers. These features provide fully self-organized matching process. Third is the arrangement of neurons in the input and model layers, in our model the neurons are placed according to the hexagonal grids instead of rectangular grids. This arrangement provides a uniform and symmetrical connections between neurons in a layer, which considerably a useful structure for invariant matching. Finaly, the best match between an input face and the model faces is decided by a competition mechanism. A model layer which have strong correlation of neurons activation to the input layer will suppress the other models.

3.1 Local Features Similarity

An input image and individual face models are represented as array of Gabor feature vectors $\mathbf{R}=\{\mathbf{r}_1 ... \mathbf{r}_P\}$, $\mathbf{r}_i=\{r_1 ... r_N\}$ and $\mathbf{M}=\{\mathbf{m}_1 ... \mathbf{m}_Q\}$, $\mathbf{m}_j=\{m_1 ... m_N\}$. Here, P and Q are the number of nodes in the input image and individual face model, respectively. Local features similarity information can be realized by measuring the similarity of Gabor features for every node in the model layer to every node in the input layer. This similarity between two Gabor feature vector can be defined as follow

$$T(\mathbf{r}_i, \mathbf{m}_j) = \frac{\sum_{k=1}^{N} m_k r_k}{\sum_{k=1}^{N} m_k \sum_{k=1}^{N} r_k} \min\left(\frac{\sum_{k=1}^{N} m_k}{\sum_{k=1}^{N} r_k}, \frac{\sum_{k=1}^{N} r_k}{\sum_{k=1}^{N} m_k}\right). \tag{3}$$

The similarity information may contain several local ambiguities, because in an object there are some local Gabor features which may be very similar to each other. Furthermore, these ambiguities become more complicated if we try to match an input image to all face models simultaneously. This ambiguity problem can be reduced by incorporating neighborhood grouping of local features in the matching process.

3.2 Input and Model Layers

The neurons in the input and model layers are fully connected. Their dynamics are described in equation (4a) and (4b). Between an input layer and model layers, there are all to all interlayer connections where their strength are defined by local features similarity between neurons T and dynamic weights J.

$$\frac{dx_a}{dt} = -\alpha x_a + \sum_a G_{aa'} S(x_{a'}) - \beta \sum_{a'} S(x_{a'}) - \varepsilon h_a + \eta \max_b J_{ab}^m T_{ab}^m S(y_b^m), \quad (4a)$$

$$\frac{dy_b^m}{dt} = -\alpha y_b^m + \sum_{b'} G_{bb'} S(y_{b'}^m) - \beta \sum_{b'} S(y_{b'}^m) - \varepsilon h_b^m + \eta \max_a J_{ba}^m T_{ba}^m S(x_a). \quad (4b)$$

$$S(q) = \frac{1}{1 - e^{-\lambda(q-q_0)}} \quad ; \quad G_{aa'} = \gamma \exp\left(-\frac{(\|a-a'\|)^2}{2\sigma^2}\right) \quad (5)$$

$$dist_{hexa} = (\|a-a'\|) = \sqrt{\tfrac{1}{2}\left[(i_1-i_2)^2 + (j_1-j_2)^2 + (k_1-k_2)^2\right]} \quad (6)$$

$$\frac{dh_a}{dt} = \begin{cases} \zeta_{high}(S(x_a)-h) & if(S(x_a)-h) > 0 \\ \zeta_{low}(S(x_a)-h) & if(S(x_a)-h) \le 0 \end{cases}. \quad (7)$$

The right-hand-side of equation (4a) and (4b) is composed of five terms. The second and third terms realize the idea of neighborhood neurons grouping in the input and model layers by generation of blob activity. These factors have effects as Gaussian convolved excitation, which performs local excitation by generating clusters of activity, and global inhibition, which makes the generated clusters compete against each other. The strongest cluster will finally become an equilibrium state of the layer dynamics. This equilibrium state is called as a blob. The size of this equilibrium state is determined by the strengths of excitation and inhibition.

The activity of neurons in the input and model layers are determined from the present state of x by the sigmoidal nonlinearity function S. The neurons are arranged in a symmetrical hexagonal coordinate frame and are excitatorily connected through the Gaussian interaction kernel G. The indice "a" and "b" are the hexagonal coordinates {i,j,k}. The euclidean distances between two points in the symmetrical hexagonal coordinate frame (6) is discussed in [3,4].

Blobs tend to be generated at the center of a layer. In order to perform the matching mechanism, the blob should be generated at different locations during the matching process. The delayed self-inhibition (the fourth terms in eq.4a and 4b) are added in order to achieve blob generation at different locations in each iteration cycle. These delayed self-inhibition are defined as a leaky integrator in (7) with decay constant ζ. Their effect are to memorize the current position of the blob and use this information in the next cycle to push the blob to different location.

The other parameters can be described as follows: α is the self-inhibition strength, β is the global inhibitory strength, ε is the delay self-inhibition strength, γ is the interaction kernel weight, and σ is the width of the gaussian window.

3.3 Dynamic Weight Updating Procedure

The dynamic weights are updated at every iteration cycle after the input and model layers achieve their equilibrium solutions. The updating procedure consists of two rules, which are a growing rule and a normalization rule. The growing rule lets dynamic weight grow according to the correlation between connected neurons. The normalization rule prevents the dynamic weights from infinite growth and induces competition where only one weight survive per neuron and suppresses all others.

$$J_{ba}^m = \frac{J_{ba}^m + \varepsilon J_{ba}^m T_{ba}^m S(y_b) S(x_a)}{\sum_{a'} \left(J_{ba'}^m + \varepsilon J_{ba'}^m T_{ba'}^m S(y_b) S(x_{a'}) \right)} \qquad (8)$$

3.4 Matching Evaluation

In order to make the face models compete against each other, a competition mechanism is necessary. We model this mechanism as follows

$$\frac{dc^m}{dt} = \varepsilon_c c^m \left(\sum_{q=1}^{Q} S(y_q) - \max\left[c^m \sum_{q=1}^{Q} S(y_q) \right] \right) \qquad (9)$$

The value of c is in the range from 0 to 1. If the value of c^m of the m-th model is below the threshold, the activity is suppressed. At the end of the matching process, only one model will survive.

4. Experimental Results

Our databases are composed of grey scale images with the size of 256 x 256 pixels. Each contains images of 13 different people. The first is the model database that contains frontal views of each person with a natural expression. The second is a test database that contains several images with different poses and appearance, that is rotation-in-depth, rotation-in-plane, and change in facial expression.

Figure 3 shows an example recognition process using a test face rotated-in-depth about 30 degrees from the frontal view. The matching is evaluated using competition mechanism described in section 3.4. The evolution of competition dynamics during the matching step is shown in the middle row. In this example one can see that the models compete against each other, and finaly only one model is survived and become the winner of the competition. All results reported in this paper, the best match model is decided after 400 matching steps. The reason is to make sure that the neural group activations in both layers already synchronize to each other and the dynamic weights have already achieved a stable configuration.

Rotation-in-depth. Face identification system with pose variation involving the rotation in depth is a very difficult task. It can be understood that the face features are change against this variation, and that some information may be lost or occluded by other features. The experimental results with rotation angle less than 20 degrees, about 30 degrees and about 45 degrees are shown in table 1.

Facial expressions. If there a local change in the image, it will only effect the local/few grid feature vectors not the whole feature vectors. Our matching system is performed based-on the dynamic weights connections strength, and these connection are defined by local information from the image. As a result, some connections to one model layer may be weak, but in general it is still strong enough in comparison to the connections to other model layers.

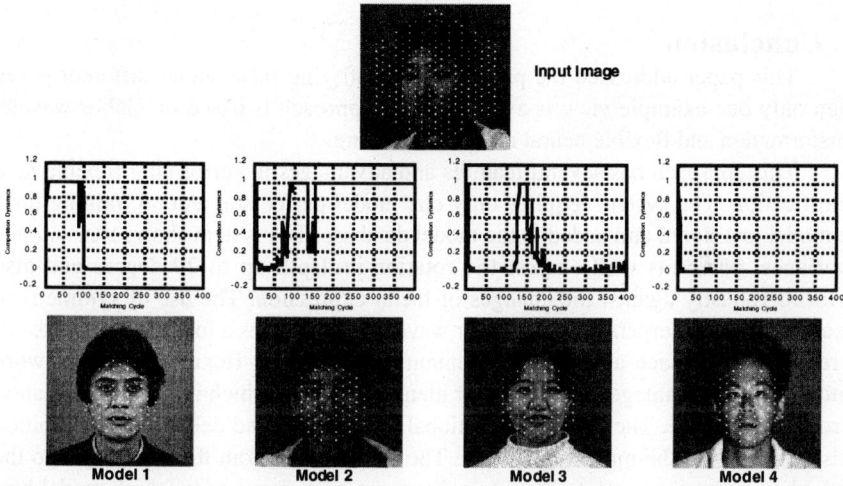

Figure 3. An Example of face identification mechanism. On the top is the input image and the bottom row shows the model images. The middle row shows the competition evolution during the matching process.

Table 1. The Identification rate (against 13 frontal view models)

Test Images	Correct	Rate (%)
52 rotated in depth (10,20 degrees)	46	88.5
26 rotated in depth (30 degree)	14	53.8
26 rotated in depth (45 degree)	5	19.2
26 frontal view with different facial expression	22	84.6
52 rotated in plane (10,20,30,40 degrees)	47	90.4

Rotation-in-plane. In performing recognition against the variation of rotation-in-plane, an addition treatment on deciding the similarity between Gabor features is necessary. Gabor wavelet at a certain orientation is only sensitive to detail of this orientation. This wavelets are not invariant against rotation in plane. With the use of 6 different orientations, we can search the maximum global similarity among them, and decided under which Gabor features sequence the similarity among them should be calculated. By taking the advantage of using the hexagonal grid structure, which have symmetrical distance between neighboring neurons, the identification of face against the rotation-in-plane can be achieved with a fairly good results.

Scalings. Basically the system is not reliable in recognizing faces under the variations of scale. The reason is that the grid points for extracting local features using Gabor wavelets is fixed. We have done an experiment with only a few images. This small experiment suggested that the face identification system still able to recognize face with ±10% scaling variations from the model faces.

5. Conclusion

This paper addressed the problem of identifying faces under different poses when only one example view is available. Our approach is based-on Gabor wavelet transformation and flexible neural network matching.

This approach has several features and advantages in performing flexible face identification. The system requires only one model per person and no training time, the model just stored and added to the model database. The system shows translational invariance, performs well against the rotation-in-depth up to 20 degree and also performs robustly against the changes of facial expression. The use of symmetrical hexagonal grids incorperated with Gabor wavelet transform as a local features detector in representing a face and the arrangement of neurons in flexible neural network matching have advantages especially for identifying faces which include the variation of rotation-in-plane. The use of bidirectional connections and delayed self-inhibition adds flexibility to the matching system. The connections from the input layer to the model layers are necessary to provide a discrimination signal as to which model best fits the input layer. The delayed self-inhibition provides a mechanism to automatic positioning the neuron group activity at different location for each matching step.

References

[01] D.J. Beymer. Face recognition under varying pose. In Proc. IEEE conf. on Comp. Vision and Patt. Recog., 756-761, 1994.

[02] D.J.Beymer. Face recognition from one example view. A.I. Memo No.1536, Artificial Intelligence Laboratory, Massachusetts Institute of Technology, 1995

[02] I. Her. A symmetrical coordinate frame on the hexagonal grid for computer graphics and vision. Trans. of the ASME, Jour. of Mechanical Design, 155:447-449, 1993.

[03] I. Her. Geometric transformation on the hexagonal grid. IEEE Trans. on Image Processing, 4(9):1213-1222, 1995.

[04] W. Konen and C. von der Malsburg. Learning to generalize from single examples in the dynamic link architecture. Neural Computation, 5:719-735, 1993.

[05] W. Konen, T. Maurer and C. von der Malsburg. A fast dynamic link matching algorithm for invariant pattern recognition. Neural Network, 7:1019-1030, 1994.

[06] M. Lades et.al. Distortion invariant object recognition in the dynamic link architecture. IEEE Trans. on Computers, 42(3):300-311, 1993.

[07] T. Maurer and von der Malsburg. Single-view based recognition of faces rotated in depth. In Proc. of the Intl. Wrkshp on Auto. Face- and Gest. Recog., 248-253, 1995.

[08] A. Pentland, B. Moghaddam, T. Starner. View-based and modular eigenspaces for face recognition. In Proc. of the IEEE Conf. on Comp. Vis. and Patt. Recog., 84-91, 1994.

[09] T. Poggio and D. Beymer. Learning networks for face analysis and synthesis. In Proc. of the Intl. Wrkshp on Auto. Face- and Gesture Recog., 160-165, 1995.

[10] L. Wiskott, J.M. Fellous, N. Kruger, von der Malsburg. Face Recognition and Gender Determination. In Proc. Intl. Wrkshp on Auto. Face and Gest. Recog., 92-97, 1995.

Multiple Camera Based Human Motion Estimation

Akira Utsumi[1], Hiroki Mori[2], Jun Ohya[1] and Masahiko Yachida[2]

[1] ATR Media Integration & Communications Research Laboratories,
2-2 Seikacho Sorakugun Kyoto 619-02, Japan
[2] Faculty of Engineering Science, Osaka University,
1-3 Machikaneyama-cho Toyonaka-shi Osaka 560, Japan

Abstract. We propose a human motion detection method using multiple-viewpoint images. We employ a simple elliptic model and a small number of reliable image features detected in multiple-viewpoint images to estimate the pose (position and normal axis) of a human body, where feature extraction is employed based on distance transformation. The COG (center of gravity) position and its distance value are extracted in the process. These features are robust against changes in human shapes caused by hand/leg bending and produce stable pose estimation results. After a pose estimation, a "best view" is selected based on the estimation result and further processing is performed including body-side detection and gesture recognition (in a 2D image of the selected view). This viewpoint selection approach can overcome the problem of self-occlusions. We confirmed the stability of the system through experiments.

1 Introduction

Human motion detection using vision techniques is becoming a popular research field. In a vision based system, the information required for reconstruction can vary depending on the application. Various models have been proposed and various image features have been utilized corresponding to the different target tasks.

To track human positions, most systems extract robust feature points in an image and perform matching of these feature points between frames. For instance, Segen et al. developed a single camera system which can track multiple persons[1]. The system extracts feature points from an input image, creates paths and clusters them. Each cluster corresponds to one person. Cai's system extracts human regions from input images by background subtraction, and gets the correspondence between frames based on position, velocity and intensity information (Cai et al. [2, 3]). These approaches concentrate on detecting and tracking human positions stably, and there is no link to posture and gesture detection.

Gesture recognition research has a different point of view. Its purpose is to detect and to categorize motion into several gestures. Azarbayejani et al. proposed a system that extracts human head and hand positions from input image sequences by using blob features and reconstructs a 2D model by detecting correspondences of the positions between frames [4, 5]. Johnson et al. generate programs to extract human hand positions from a front view image of a human by using a genetic programming approach[6]. Their programs perform well for images similar to those used for the learning. These approaches deal with a lot more information on human posture than position tracking research, but they are not concerned about changes in the human position or orientation.

The above systems are basically 2D feature based and most of them utilize a restricted number of viewpoints. When the position and orientation of a person change drastically, the detection of feature points becomes difficult because of the change of appearance. This is called the self-occlusion problem. (Cai et al. mentioned the use of multiple cameras in [3]. However, they switch their cameras manually and do not integrate multiple-camera information.)

On the other hand, some researchers are investigating more general approaches. Some employ a very high DOF 3D articulated model to handle a full human body [7, 8, 9, 10]. The idea of using a 3D model with the same DOF as the target is very straightforward; information on the target object which is built into the model can improve the stability. This approach, however, is still not completely free from the self-occlusion problem if the viewpoints are restricted (Gavrila et al. also employed many viewpoints to reduce this problem).

As for the object description point of view, a 3D model is certainly flexible and adaptable for a variety of applications. However, the search space becomes quite large and its reconstruction requires a huge computational cost. We consider that most applications do not require full 3D descriptions and that the direct usage of 2D features detected from an appropriate viewpoint is practical enough.

In this paper, we present a viewpoint selection approach. We have applied this technique to hand posture estimation and have achieved good results [11]. In our implementation, the hand posture was estimated using a small number of reliable features detected in multi-viewpoint images, and further processing (detection of finger bendings) was performed by selecting "best view" images using the pose estimation results. Here, we show that the same approach is also valid in human motion detection.

2 System Overview

The diagram of our method is outlined in Figure 1. First, the image input from each camera is segmented based on intensity information. Then, distance transformation is applied to each binary image and a COG position and its DT (distance transformation) value are determined based on the transformed image. We can estimate the human pose (position and normal axis) using these features detected in multiple viewpoints. Finally, the process of viewpoint selection is performed based on the pose estimation result and further processing (to the 2D features including body-side detection and gesture recognition) is done within the selected view image.

In the next section, we describe the pose estimation method using multiple-viewpoint information. In section 4, we will focus on the body-side detection based on viewpoint selection.

3 Multi-camera based 3D Reconstruction of Human Model

3.1 Algorithm

Human Model

Figure 2 shows a 3D human model that we use in our system. In this system, we model a human as an elliptic pillar. We assume the center axis of the pillar to be equal to the rotation axis and it is perpendicular to the ground.

COG Detection

We first perform region segmentation on the input images and separate them into two parts (human region and background region). To simplify the process,

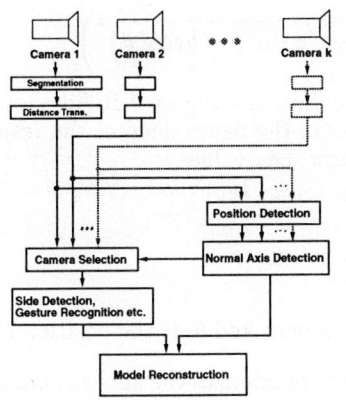
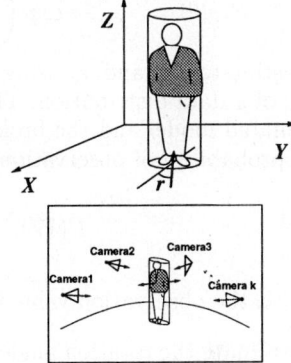

Fig. 1. System Configuration   **Fig. 2.** Human Model

the subjects wear white shirts, the background has black walls and segmentation is performed based on intensity information (Figure 3 Left).

We apply DT to the segmentation result. As a result, each pixel is a certain distance from its nearest boundary pixel (Figure 3 Right). We extract the pixel having the maximum value in the transformed image as a COG (center of gravity) point.

Position Estimation

We consider that each detected COG position in a frame corresponds to the same position on the human body, and determine the human position by matching the corresponding points between images.

Normal Axis Estimation

From the top view, a human can be described as an ellipsoid (Figure 4 Left).

The normal axis of a human body (the rotation angle) is determined based on COG skeleton values detected by multiple cameras. The rotation angle is estimated from the derived COG skeleton values with a human observation model. As we mentioned above, we employ an elliptic model. By assuming a weak perspective projection, the width s detected by the camera located at $\theta = 0$ (which corresponds to a skeleton value, i.e., the pixel value at a COG point in a distance transformed image) can be described as follows.

$$s = \frac{1}{L}\sqrt{a\sin^2\theta + b\cos^2\theta} \qquad (1)$$

Here, L is the human-camera distance and a, b are constants. θ is the rotation angle of the human body.

Allowing for Gaussian error in the observation, the probability $P(s|\theta)$ that the body width (a skeleton value) s is observed is

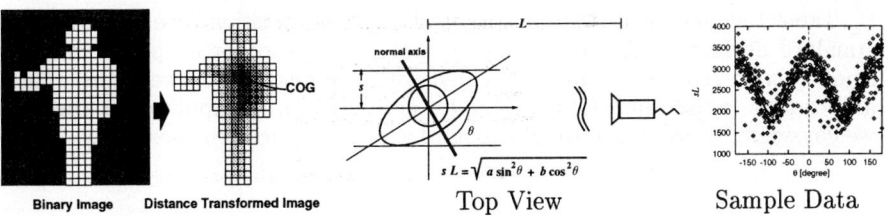

Fig. 3. COG Detection   **Fig. 4.** Elliptic Model

$$P(s|\theta) = \frac{1}{\sqrt{2\pi}\sigma_s} exp\left(-\frac{1}{2\sigma_s^2}\left(s - \sqrt{a\sin^2\theta + b\cos^2\theta}\right)^2\right) \quad (2)$$

We estimate a, b and σ_s using the sample data set. Figure 4 Right shows an example of a data distribution. The solid line in the figure denotes the result of the estimated model and the broken lines show $\pm\sigma_s$ values.

The probability of observation $\boldsymbol{W}(s_1, s_2, \cdots, s_k)$ is

$$P(\boldsymbol{W}|r) = \prod_{i=1}^{k} P(s_i|r - \theta_{c_i}). \quad (3)$$

Here, s_i is the observation value for the ith camera and θ_i is the position of the camera.

We estimate the rotation angle at the r that maximizes the above conditional probability $P(\boldsymbol{W}|r)$.

3.2 Experiments

We performed our experiments with three circularly located cameras. They are positioned at regular intervals (every 30 degrees). Subjects are placed in the center of the camera circle. All cameras are calibrated in advance. In these experiments, we recorded images using three video disc systems (three camera images). Consequently, the state of our system can be characterized as accelerated. We placed a subject near the center of the cameras and took images while changing the axis at a rate of 10 degrees per frame. The video disc systems recorded 36 × 3 frame images. The distance between each camera and the subject was about three meters.

The experiments were performed for five patterns ('stand' pose for three subjects, and the 'arms up' pose and 'sit down' pose for one subject each). The COG position and its distance transformed value were calculated for each image. An example image of each pose is appeared in figure 7. The model parameters a, b and σ described in the previous section were determined with the observed data. In this experiment, though the observed data had small differences against changes in the subjects and poses, it appeared stable.

In the following experiments, we used values calculated with all of the five pattern data as model parameters. Here, the horizontal axis denotes θ (the angle between the camera axis and human normal axis) and the vertical axis denotes the product of s the COG DT value) and L the human-camera distance). The solid line denotes our model and the distance between the solid line and the broken line denotes the standard deviation σ. Using these parameters, the position and normal axis of a human were estimated by the method described in the previous sections. The input images were identical to those used for the parameter learning.

a) **Position Estimation**

Table 1 shows the estimation result of a human position (average error and standard deviation). As shown in this table, the errors for all poses are quite small (about 7cm maximum) compared with the distance between the human and camera (about 3m). Of course, a lot depended on the application area, but we consider the errors are small enough for most application uses.

b) Normal Axis Estimation

Table 1. Position Estimation Results (cm)

	'stand'		'arms up'		'sit down'	
	x	y	x	y	x	y
av. err.	2.49	0.16	3.95	0.70	1.85	0.65
SD	3.19	1.98	5.75	3.14	7.01	4.89

The next measurement addressed rotation angle detection. For the measurement, images where a subject rotated at a position central to all of the cameras were used. Figure 5 shows the rotation angle detected by the system. The points denote the detected angle and each solid line gives the set-up condition (the actual subject's axis). The subject maintained one pose during one sequence of rotation. As can be seen, this detection was stable and robust for the different of subjects and human postures. These results can be improved by increasing the number of cameras. Later, we plan to address estimation stability issues including the relationship between the stability and the number of cameras.

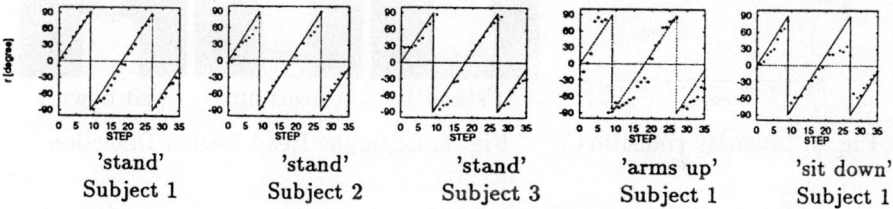

'stand'   'stand'   'stand'   'arms up'   'sit down'
Subject 1   Subject 2   Subject 3   Subject 1   Subject 1

Fig. 5. Normal Axis Estimation

4 2D Image Processing based on Best-View Selection: Body Side Detection

4.1 Best-View Selection

In the previous section, we described a method of estimating an object's posture by integrating multiple-camera information. After the target posture is determined, we can select a "best view" (camera) to retrieve further information. The best view can be altered depending on the information or features to be retrieved. In any case, however, we consider that it can be determined theoretically or statistically. In the following subsections, we focus on human body-side detection. In the previous section, only the axis was determined and no determination was made on which side was the front and which was the back.

For the detection, we utilize the difference in head region intensity for the body sides (front (face) and back (hair)). What is the best view to retrieve the feature? Figure 6 shows an example of an intensity transition based on changes in the angle between the human normal axis and camera axis. As can be seen, the intensity gives a peak when the angle is zero. In this case, it should be reasonable to select the viewpoint that has the most perpendicular axis to the body, to determine the body side from its image. Of course, a more statistical analysis is necessary, but this justifies our choice regarding the 'best' camera. The camera whose axis is closest to the human normal axis is selected, and further processing is done in its image.

By selecting a "best view", further 2D processing can be done to achieve robustness and reduce the cost.

4.2 Body Side Detection

Head Region Determination

Here, we detect the side of the human body and finally determine the full orientation. For the side detection, we utilize the difference of intensity in a human head region (we assume that the human face and (back) hair are different in their brightness).

We detect the human face region from the selected distance transformed image. In doing so, we utilize the fact that the neck points have smaller values in the resulting distance transformed image and determine the face position in the 2D image. As we already know the 3D position (and also the camera position) we can magnify the size of the head region properly.

Figure 7 shows an example of head region extraction. In each figure, the rectangle denotes the detected region. We calculate the average intensity of this region and utilize it for further processing.

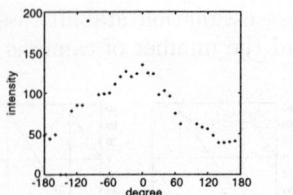

Fig. 6. Intensity transition   **Fig. 7.** Example: Head Region Detection

Body Side Detection

We calculate the average intensity of the detected region and determine the body side from the information. As mentioned above, our system determines the human normal axis by using multiple skeleton images; however, the side of the human body has not been determined yet.

Now, let us consider two states. One is a state where the front of a human appears in a selected camera (front, F), and the other is a state where the back of a human appears (back, B).

We call the face seen from the selected camera in a previous frame $t-1$ s_{t-1}, and call another face $\bar{s}_{t-1}$. We denote the probability that s_{t-1} in the last frame is the front of a human body with $P(s_{t-1} = F)$, the probability that it is the back of a human body with $P(s_{t-1} = B)$. Though the current state of the human body is unknown, the face which is seen from the camera selected in the current frame t can be one of $s_t = s_{t-1}$ (i.e. $\bar{s}_t = \bar{s}_{t-1}$) and $s_t = \bar{s}_{t-1}$ (i.e. $\bar{s}_t = s_{t-1}$). We call the probability of the former case P_a and the probability of the latter case P_b (here, $P_b = 1 - P_a$).

In the current implementation, these probabilities are,

When the assumption of $s_t = s_{t-1}$ makes the orientation difference of the neighboring frames smaller than of $\bar{s}_t = s_{t-1}$,

$$P_a = k, P_b = 1 - k \tag{4}$$

otherwise,

$$P_a = 1 - k, P_b = k. \tag{5}$$

(Here, k is a constant value close to 1. ($k < 1$))

By using these equations, the probabilities of the current state $P'(s_t = F)$, $P'(s_t = B)$ can be calculated as follows.

$$\begin{pmatrix} P'(s_t = F) \\ P'(s_t = B) \end{pmatrix} = \begin{pmatrix} P_a & P_b \\ P_b & P_a \end{pmatrix} \begin{pmatrix} P(s_{t-1} = F) \\ P(s_{t-1} = B) \end{pmatrix} \quad (6)$$

To simplify the problem, we denote the probability density of the observation of intensity ν when the state is 'front' (F) with $P(\nu|F)$ and the probability density of the observation of intensity ν when the state is 'back' (B) with $P(\nu|B)$. We assume that these two density functions can be presented by using Gaussian distributions.

$$P(\nu|F) = \frac{1}{\sqrt{2\pi}\sigma_f} \exp\left(\frac{-(\nu - u_f)^2}{2\sigma_f^2}\right) \quad (7)$$

$$P(\nu|B) = \frac{1}{\sqrt{2\pi}\sigma_b} \exp\left(\frac{-(\nu - u_b)^2}{2\sigma_b^2}\right) \quad (8)$$

Here, σ_f and σ_b denote the standard deviation and u_f and u_b denote the average intensity.

For the observation ν, the probability that the front of the human body faces the selected camera, $P(s_t = F)$, can be calculated as follows.

$$P(s_t = F) = P(F|\nu) = \frac{P(\nu|F)P'(s_t = F)}{P(\nu|F)P(F) + P(\nu|B)P(B)} \quad (9)$$

$$P(s_t = B) = P(B|\nu) = \frac{P(\nu|B)P'(s_t = B)}{P(\nu|F)P(F) + P(\nu|B)P(B)} \quad (10)$$

Here, $P(F|\nu) + P(B|\nu) = 1$.

The system compares $P(F|\nu)$ and $P(B|\nu)$, and it determines whether the face currently appearing is the front (if former is larger) or the back (if latter is larger). $P(s_t = F)$ and $P(s_t = B)$ are used for the calculation in the next frame.

4.3 Experiments

We applied the body-side detection method to the image sequence used in the experiment described in 3.2b. Figure 8 shows the result (for subject 1 ('stand')). In the top figure, the final estimated angle which includes body-side information is shown. Each point denotes a detected angle and the solid line gives the set-up condition. Number of a selected camera is show in the figure below. As can be seen, the human orientation including body-side information is detected very stably by using a selected camera image.

Furthermore, we applied our system to a real-time image sequence, where a person is walking around. Figure 9 shows a detected trajictory and a snapshot of the input sequence. Our system tracked motion of the person including its orientation stably and correctly. In the current implementation, the processing speed was about three frames per second.

5 Conclusions

We have proposed a method of detecting human motion using multiple camera images. The position and normal axis of a human body are reconstructed from a small number of reliable image features detected in multiple-viewpoint images. After that, a "best view" is selected based on the estimation result and further processing including body-side detection and gesture recognition is carried out. We confirmed the stability of the results through experiments using real images.

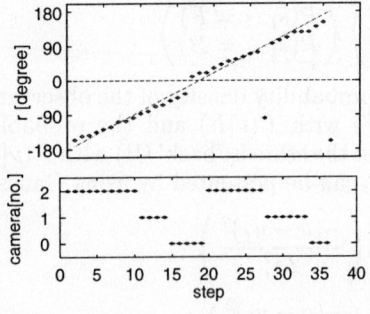 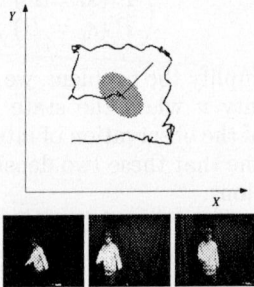

Fig. 8. Body-Side Detection for Subject 1 ('stand') **Fig. 9.** Motion Tracking for Subject 1 ('stand')

Our approach solves the self-occlusion problem without requiring a detailed 3D modeling and reduces the computational costs for both the model construction and its reconstruction.

Future work includes extending of our posture estimation technique to deal with multiple persons. Verification of its availability as a gesture based man-machine interface system will also be addressed.

References

1. Jakub Segen and Sarma Pingali. A camera-based system for tracking people in real time. In *Proc. of 13th International Conference on Pattern Recognition*, pages 63–67, 1996.
2. Q. Cai, A. Mitiche, and J. K. Aggarwal. Tracking human motion in an indoor environment. In *Proc. of 2nd International Conference on Image Processing*, pages 215–218, 1995.
3. Q. Cai and J. K. Aggarwal. Tracking human motion using multiple cameras. In *Proc. of 13th International Conference on Pattern Recognition*, pages 68–72, 1996.
4. A. Azarbayejani and A. Pentland. Real-time self-calibrating stereo person tracking using 3-d shape estimation from blob features. In *Proc. of 13th International Conference on Pattern Recognition*, pages 627–632, 1996.
5. C. Wren, A. Azarbayejani, T. Darrell, and A. Pentland. Pfinder: Real-time tracking of the human body. In *SPIE proceeding vol.2615*, pages 89–98, 1996.
6. M. Patrick Johnson, P. Maes, and T. Darrell. Evolving visual routines. In *Proc. of Artificial Life IV*, pages 198–209, 1994.
7. D. M. Gavrila and L. S. Davis. 3-d model-based tracking of humans in action: a multi-view approach. In *Proc. of CVPR*, pages 73–80, 1996.
8. K. Rohr. Towards model-based recognition of human movements in image sequences. *CVGIP: Image Understanding*, 59(1):94–115, 1994.
9. M. Yamamoto and K. Koshikawa. Human motion analysis based on a robot arm model. In *Proc. of CVPR*, pages 664–665, 1991.
10. J. O'rourke and N. J. Badler. Model-based image analysis of human motion using constraint propagation. *IEEE Pattern Anal. Machine Intell.*, 2(6):522–536, 1980.
11. Akira Utsumi, Tsutomu Miyasato, Fumio Kishino, and Ryohei Nakatsu. Hand gesture recognition system using multiple cameras. In *Proc. of 13th International Conference on Pattern Recognition*, pages 219–224, 1996.

An Autonomous Facial Caricaturing Based on a Model of Visual Illusion
— Experimental Modeling of Visual Illusion —

Kazuhito Murakami, Mikiko Takai and Hiroyasu Koshimizu

School of Computer and Cognitive Sciences, Chukyo University
101 Tokodate, Kaizu-cho, Toyota, 470-03 JAPAN
Tel : +81-565-45-0971   Fax : +81-565-46-1299
E-mail : murakami@sccs.chukyo-u.ac.jp

Abstract

In the PICASSO system, a facial caricaturing system, some visual illusions such as the Wundt-Fick illusion and the Ponzo illusion for example, are applied to evaluate the shapes of the facial parts such as eyebrows, nose, mouth and face contour, in the deformation process of caricature generation. In many cases, as well-deformed caricatures are evaluated to be successful, it is confirmed that the utilization of the visual illusion is effective to evaluate the results of caricatures. In this paper, the models of mathematical expressions of visual illusions are presented together with the results of human experiments. And, a method to apply visual illusions and to control the exaggeration process is proposed.

1. Introduction

Many researches have been reported to generate a facial caricature by computer. Many of them discuss on the method to generate a caricature, but there are few researches which refer to autonomous evaluation of the results or the works drawn by the system. Computer would answer that it is a good caricature if the caricature matches in shape or gray-level to the original image. On the contrary, human vision are impressed as a good caricature if it is properly deformed or exaggerated. This fact shows that it is reasonable to introduce the mechaninsm of human vision to the evaluation process of facial caricaturing.

From this point of view, first we led out the mathematical models of some famous visual illusion figures experimentally, then applied this model to the evaluation process of caricature generation in order to realize a KANSEI evaluation basis for caricature generation. And we realized a method to generate a caricature which gives us more accelerated impression of visual illusion. In the section 2, the basic strategy to evaluate a caricature and to control the deformation process is introduced. In the section 3, visual illusions and their experimental and mathematical modelings are summarized. In the section 4, a practical method to generate and evaluate a caricature is described, and the experimental results are presented concurrently in the following sections.

2. Strategy for Caricature Evaluation by Visual Illusion

Since the facial caricature is usually described by line-drawings, the evaluation should be executed basically by comparing two figures defined by line-drawings. What kind of facial expression should be utilized to judge the caricature good or no-good? In general, the evaluation is realized by comparing a photograph or a TV image with the caricature. This evaluation is based on a comparison between the "undeformed face" and the "deformed face". So that, the above comparative evaluation can be formalized generally by the following Eq.(1), where b is the exaggeration weight, $g(b)$ is a generalized measure, and θ_0 is an evaluation threshold.

$$\frac{g(b)}{g(0)} < or = or > \theta_0 \qquad (1)$$

In Eq.(1), if the value $g(b)$ which varies according to the deformation process becomes equal to or larger than $g(0) \times \theta_0$, let the caricature be deformed in the suitable state and let the deformation process stop. And, if $g(b)$ is smaller than $g(0) \times \theta_0$, it means that the caricature is not yet sufficiently deformed and let the deformation process proceed.

It is possible, of course, to apply the visual illusion measure such as the Wundt-Fick illusion and the Ponzo illusion to implement the generalized measure $g(b)$. And Eq.(1) can also be used as the evaluation equation to judge whether the caricature is well-deformed or not. It is simple and easy to apply visual illusions to the shape of the figure which can be physically obtained by using image processing or image measurement technique. But in the actual human vision world, the shapes of the figure should be evaluated by their appearances activated by the sensory factors of human vision, KANSEI. So, it is required to prepare the perceptual visual illusion measure $g'(b)$ in which the physical visual illusion measure $g(b)$ is modified by a coefficient (or a function) α as follows.

$$g'(b) = g(b) \times (1 + \alpha) \qquad (2)$$

3. Modeling of Visual Illusion

3.1 Visual Illusion

Figure 1(a) shows an example of the Wundt-Fick figure illusion. Although the length of the horizontal line is physically equal to the length of the vertical line, the divided horizontal line segment looks perceptually shorter than the vertical one.

Figure 1(b) shows an example of the Müller-Lyer visual illusion. Two horizontal lines are physically of equal length, however, two wings at the edge of the segment give an impression that the left horizontal line segment looks shorter than the right one. The effect of the illusion varies according to the direction and length of the wings. By corresponding these figure primitives to the facial parts, it is expected that this kind of visual illusions have a positive effect on the recognition and exaggeration of faces.

Figure 1(c) shows an example of the Ponzo figure illusion. Although two horizontal lines are physically the same length, the upper horizontal line segment adjacent to the top corner looks perceptually longer than the lower horizontal line segment. The effect of the illusion varies according to the length of lines and their crossing angle.

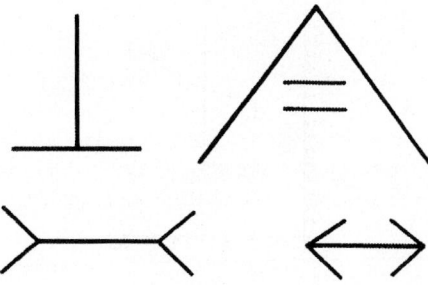

Fig. 1. Examples of visual illusions; (a) Wundt-Fick(left), (b) Müller-Lyer(bottom) and (c) Ponzo(right).

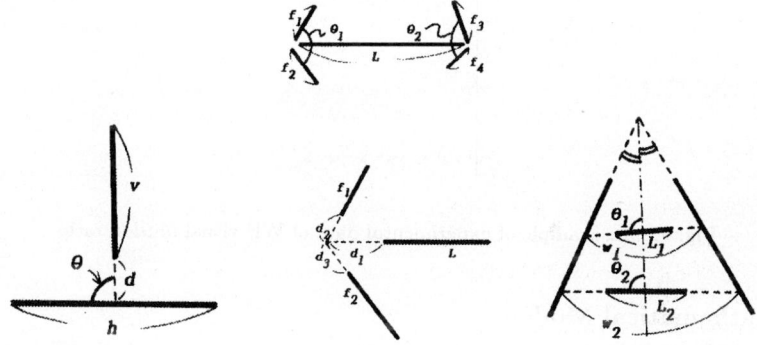

Fig. 2. Definitions of variables; (a) Wundt-Fick(left), (b) Müller-Lyer(center) (c) and Ponzo(right).

In our system, three types of visual illusions above mentioned are utilized to control the deformation process.

3.2 Experiments

The models of these three visual illusions, that is, the practical functions and values of Eq.(2) are estimated by following perception experiments using 30 students testees. Firstly, let variables which represent the length, angle, distance, and so on, be defined on the compound figures as shown in Fig.2. Secondly, the ratio of visual illusion is examined by (a) showing a figure that one parameter is changed and (b) answering how much the object figure is longer or shorter than the reference figure. Figure 3 shows an experimental data of Wundt-Fick illusion. The length of the horizontal line segment is fixed and the vertical one is varied. The x-axis and y-axis represent v and the perceived ratio v/h, respectively. Figure 4 shows another data which the angle θ between two line segments (h, v) is varied.

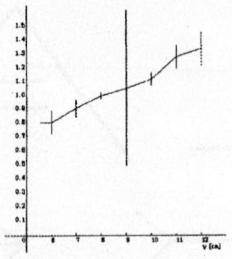

Fig. 3. An example of experimental data of WF visual illusion ratio.

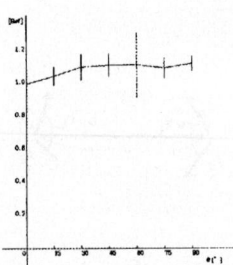

Fig. 4. An example of experimental data of WF visual illusion ratio.

3.3 Mathematical Models

From these experimental data, statistical values are estimated and some mathematical models are obtained. For the Wundt-Fick visual illusion, the ratio of visual illusion is approximately represented by

$$g_{WF}(b) = \frac{v(b)}{h(v)} \times (1 + \alpha) \qquad (3)$$

$$\alpha = 0.15 \times e^{1-(\frac{v}{2h} + \frac{h}{2v})} \cdot \sin\theta \cdot e^{-\frac{d}{v}}. \qquad (4)$$

Here, h and v are the horizontal and vertical lengths, and d and θ are the gap and their crossing angle, respectively. The correction coefficient α is, of course, a function of the exaggeration weight b, but it is omitted in the equation.

In the same way, the effect of Müller-Lyer visual illusion varies according to the length $L, f_1 \sim f_4$ of lines, gaps $d_1 \sim d_3$ and their crossing angles θ_1 and θ_2 as shown in Fig.2. From the experimental results, Müller-Lyer visual illusion g_{ML} is given by the following equations.

$$g_{ML}(b) = 1 + \beta \qquad (5)$$

$$\beta = 0.20 \times \beta_1 \times \beta_2 \times \beta_3 \qquad (6)$$

$$\beta_1 = \frac{16 \cdot f_1 \cdot f_2 \cdot f_3 \cdot f_4}{L^4} \times e^{-4(1-\frac{\sqrt{f_1 \cdot f_2} + \sqrt{f_3 \cdot f_4}}{L})} \qquad (7)$$

$$\beta_2 = \frac{1}{\sqrt{2}}\{\sin\theta_1 \cdot e^{(\pi/4-\theta_1)} + \sin\theta_2 \cdot e^{(\pi/4-\theta_2)}\} \tag{8}$$

$$\beta_3 = \frac{1}{2}\{e^{-(\frac{d_1}{L}+\frac{d_2}{f_1}+\frac{d_3}{f_2})} + e^{-(\frac{d_4}{L}+\frac{d_5}{f_3}+\frac{d_6}{f_4})}\} \tag{9}$$

Of course, while β is a function of the exaggeration weight b practically, b is omitted in this expression.

Ponzo visual illusion g_{PZ} also varies its effect according to the length L_1 and L_2 of lines and their crossing angles θ_1 and θ_2 as shown in Fig.2. From the experimental results, Ponzo visual illusion g_{PZ} is given by the following equations.

$$g_{PZ}(b) = \frac{L_1}{L_2} \cdot (1+\gamma) \tag{10}$$

$$\gamma = 0.10 \times \frac{L_1}{L_2} \cdot \frac{w_2}{w_1} \cdot \cos\theta_1 \cdot \cos\theta_2 \cdot e^{1-(\frac{L_1}{w_1}+\frac{w_1}{L_2})/2} \tag{11}$$

4. Caricature Generation and Evaluation

4.1 Deformation by In-betweening Method

In the PICASSO, the representation of faces is restricted only to line drawings in order to put the focus exclusively on the shape information of the faces. In this paper, let the facial caricaturing be described as a coordinate transformation from a primal sketch of a face **P** to its caricature **Q**. The principle of facial caricaturing in the PICASSO can be summarized as the extraction of individuality features obtained by comparing an averaged neutral face (let it be "mean face") and the exaggeration of the characteristic individuality features. These procedure can be expressed qualitatively as

$$\mathbf{Q} = \mathbf{P} + b \cdot (\mathbf{P} - \mathbf{S}) \tag{12}$$

where, **P** is a primal input face, **S** is a reference face, **Q** is a deformed caricature and b is a parameter of exaggeration weight.

Concretely by using x and y coordinates of each corresponding points of **P**, **S** and **Q**, it becomes as follows.

$$\begin{aligned} x_i^{(Q)} &= x_i^{(P)} + b \cdot (x_i^{(P)} - x_i^{(S)}) \\ y_i^{(Q)} &= y_i^{(P)} + b \cdot (y_i^{(P)} - y_i^{(S)}) \\ i &= 1, 2, ..., N(=455) \end{aligned} \tag{13}$$

The mean face is defined by

$$x_i^{(S)} = \sum_{j=1}^{M} \frac{x_i^{(P_j)}}{M}, \quad y_i^{(S)} = \sum_{j=1}^{M} \frac{y_i^{(P_j)}}{M} \tag{14}$$

where $x_i^{(P_j)}$ and $y_i^{(P_j)}$ are the x and y coordinates for the i-th point of the j-th normalized data. It is obvious that the mean face seems to be very ordinal from

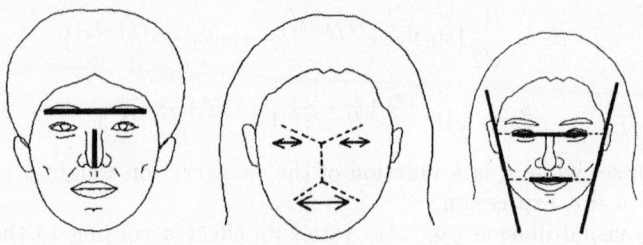

Fig. 5. Applied examples of visual illusions; (a) Wundt-Fick(lrft), (b) Müller-Lyer(center) (c) and Ponzo(right).

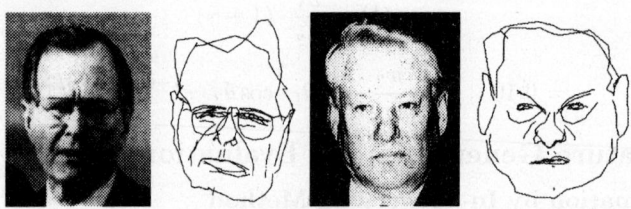

Fig. 6. The best caricatures controlled by Wundt-Fick visual illusion measure.

the viewpoint of the facial individualism. In other words, it is intended in this paper that the individuality features of the face should be implicitly extracted simultaneously by introducing the mean face. Therefore, it becomes possible to explicitly extract the individuality features of the given face **P** by comparing with this mean face **S**.

4.2 Evaluation by Visual Illusion

The Wundt-Fick illusion can be applicable, for example as shown in Fig.5, to the line $h(b)$ connected between eyebrows (or irises) as a horizontal stroke and the line $v(b)$ defined by nose as a vertical stroke. The measures used in PICASSO to evaluate whether the Wundt-Fick illusion would occur or not were defined by Eq.'s(3), (4) and following (15) and (16).

$$h(b) = \max\{x_i; i \in (eyebrow_{right})\} - \min\{x_i; i \in (eyebrow_{left})\} \qquad (15)$$

$$v(b) = \max\{y_i; i \in (nose)\} - \min\{y_i; i \in (nose)\} \qquad (16)$$

In the same way, the Müller-Lyer figure would be popped out on the combination of facial parts, for example, eyes, wrinkles of mouth, eyelids and nose and wrinkles of mouth, and so on.

Lastly, Ponzo illusion can be applied to the face contour and its internal facial parts such as eyes, nose, mouth, and so on.

5. Control of Deformation Process

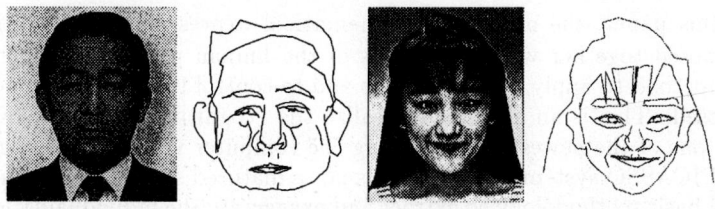

Fig. 7. The best caricatures controlled by Ponzo visual illusion measure.

5.1 Experimental Results by Wundt-Fick Visual Illusion

Figure 6 shows several experimental results controlled by the Wundt-Fick visual illusion. The figures are the best caricatures generated by PICASSO in accordance with the rule that the achievement of the caricature becomes best when the $g_{WF}(b)$ value reaches or exceeds $g_{WF}(0) \times 1.10(110\%)$. 1.10 (110%) is a value which is obtained as one of the most typical value from the experimental psychology fields. In the almost all cases of the experiments, the exaggeration process was controlled by g_{WF} value.

5.2 Experimental Results by Ponzo Visual Illusion

It is easily supposed that the appearances of the face will vary in accordance with the relative relation (size, shape and location) between the face contour and the facial parts included in it such as eyes, nose, mouth, and so on. In the PICASSO, firstly it is judged and evaluated whether the shape of the face contour is roundish, triangle or invert triangle by measuring $f_t(b)$, $f_m(b)$ and $f_b(b)$, which represent the facial width at the top, middle and bottom of the face, respectively. Figure 7 shows several experimental results controlled by the Ponzo visual illusion. The figures are the best caricatures generated by PICASSO in accordance with the rule that the achievement of the caricature becomes best when the $g_{PZ}(b)$ value reaches or exceeds $g_{PZ}(0) \times 1.06(106\%)$. 1.06 (106%) is a well known value as a result obtained in the experimental psychology.

5.3 Autonomous Deformation in Caricaturing

A few faces among the experimental results tend not to increase the $g_{WF}(b)/g_{WF}(0)$ value according to the increase of the exaggeration weight b. In such kind of case, the exaggeration process should be controlled by $f_t(b)$, $f_m(b)$ and $f_b(b)$ values and the Ponzo illusion succeedingly to the Wundt-Fick illusion.

It is expected to realize a meta-mechanism which evaluates these multiple visual illusions autonomously in order to search out the best achievement of the caricature.

As the PICASSO has a function to discover and exaggerate the facial individuality features and furthermore to evaluate the achievement, it can be said that PICASSO offers an interesting theme from the viewpoint of the "cognitive vision" or "KANSEI vision" to imitate the human visual behavior by computer.

6. Conclusions

In this paper, the models of mathematical expressions of visual illusions are proposed together with the results of the human perception experiments. And, a method to apply visual illusions and to control the exaggeration process is proposed. This is an attempt to realize the "cognitive vision" or "KANSEI vision", one of the newest theme among the computer vision, by computer. As for our PICASSO system, of course, it is not a matured system and it can barely realize a basic methodology to extract and exaggerate the individuality features. On the other hand, it is noteworthy from the viewpoint of the cognitive vision that the human visual illusions could provide the possibility to evaluate the achievement of the caricatures by PICASSO itself.

Concretely speaking, a method to control the exaggeration process autonomously basing on the Wundt-Fick visual illusion is proposed, and an autonomous strategic mechanism to evaluate and control the extraction and exaggeration of the individuality features is realized. As this is, of course, only a clue to realize a true deformation mechanism, a multiple examination of the utilization of the effective visual illusions such as Müller-Lyer illusion should become important hereafter, and this is the primal one of the coming subjects to facilitate the growth of PICASSO.

Acknowledgments

This paper was partially supported by 1996-7 Grant-in-Aid for General Scientific Research, 1996-7 IMS Research Promotion, 1996-7 the Promotion of Advanced Software Enrichment (IPA) and the Hori Information Science Promotion Foundation. We would like to express thanks to these supports.

REFERENCES

[1] Brennan,S.E.: "Caricature generation", Degree of Master of Science in Vision Studies at MIT (Sep.1982)
[2] Dewdney,A.K.: "Computer recreation :computer caricature", Science, pp.160-165 (Dec.1986) (in Japanese)
[3] Sakai,T and Kanade,T.: "Image analysis of human faces by computer", Trans. IECE, Vol.56-D, No.4, pp.683-721 (Apr.1973) (in Japanese)
[4] Shiono,M., Takeda,T. and Murayama,T.: "System for caricatured portrait drawing from facial photographs", The Journal of the Institute of Television Engineers of Japan, Vol.42, No.12, pp.1380-1386 (Dec.1988) (in Japanese)
[5] Akimoto,T. and Suenaga,Y.: "Facial image synthesis from front/side view and 3D base model", Technical Report of IEICE, PRU88-47 (1988) (in Japanese)
[6] Aizawa,K., Yamada,Y., Harashima,H. and Saito,T.: "Modeling a person's face and synthesis of facial expressions for use in model based synthesis image coding system", Technical Report of IEICE, IE87-2 (1987) (in Japanese)
[7] Imai,S.: "Figures of Optical Illusions", Science Pub. (1984) (in Japanese)
[8] Murakami,K., Koshimizu,H., Nakayama,A. and Fukumura,T.: "Facial caricaturing based on visual illusion —a mechanism to evaluate caricature in PICASSO system—", IEICE Trans. of Information and Systems, Vol.E76-D, No.4, pp.470-478 (1993)
[9] Murakami,K. and Koshimizu,H.: "Generation of Expressive Emotional Caricatures", Journal of Inst. of Television Eng. of JAPAN, Vol.50, No.10, pp.1515-1521 (1996) (in Japanese)

3D Estimation of Facial Muscle Parameter from the 2D Marker Movement using Neural Network

Takahiro ISHIKAWA [†], Hajime SERA [†], Shigeo MORISHIMA [†],
Demetri TERZOPOULOS [††]

Faculty of Engineering, Seikei University [†]
3-3-1 Kichijoji-kitamachi, musashino,
Tokyo. 180, Japan
Phone: +81-422-37-3726
E-mail: takahiro@ee.seikei.ac.jp

University of Toronto [††]
6 King's Colledge Road Toronto
Ontario Canada
Phone: +1-416-978-7777
E-mail: dt@vis.toronto.edu

Abstract

Muscle based face image synthesis is one of the most realistic approach to realize life-like agent in computer. Facial muscle model is composed of facial tissue elements and muscles. In this model, forces are calculated effecting facial tisssue element by contraction of each muscle strength, so the combination of each muscle parameter decide a specific facial expression. Now each muscle parameter is decided on trial and error procedure comparing the sample photograph and generated image using our Muscle-Editor to generate a specific face image.

In this paper, we propose the strategy of automatic estimation of facial muscle parameters from 2D marker movements using neural network. This corresponds to the non-realtime 3D facial motion tracking from 2D image under the physics based condition.

1 Introduction

Recently, research into creating friendly human interfaces has flourished remarkably. Such interfaces smooth communication between a computer and a human. One style is to have a virtual human appearing on the computer terminal who can understand and express not only linguistic information but also non-verbal infomation. This is similar to human-to-human communication with a face-to-face style and is sometimes called a Life-like Communication Agent[3][6][7]. In the human-human communication system, facial expression is the essential means of transmitting non-verbal information and promoting friendliness between the participants. We have already developed a facial muscle model as a method of synthesizing realistic facial animation.

The facial muscle model is composed of facial tissue elements and muscle strings. In this model, forces effecting each facial tissue are calculated by contraction of each muscle.

So a combination of muscle strengths decides a specific facial image. Currently, however, we have to manually determine each muscle parameter by trial and error, comparing the synthesized image to a photograph. This paper proposes a method of automatic estimation of facial muscle parameters from 2D marker movements in a face-frontal image. At first, small colored circle makers are attached to a subject's face to measure the quantity of transformation of the face when an expression appears. Then we can find out the difference between any specific expression and a neutral expression.

A neural network which has learned several patterns of facial expressions can convert the marker movements into muscle parameters. This neural network can realized an inverse mapping of the image synthesis process from the 3D muscle contraction to the 2D point movement in the display. So this is also 3D motion estimation from 2D point tracking in captured image under restriction of physics based face model. The facial image is then re-synthesized from the facial muscle model to estimate the difference from the original image both subjectively and objectively. We also tried to generate animation using the captured data from the image sequence. As a result, we can get and synthesize images which give an impression close to the original.

2 Facial muscle model[1][2]

2-1 Layered Dynamic Tissue Model

The human skull is covered by deformable tissue which has five distinct layers. Four layers (epidermis, dermis, subcutaneous connective tissue, and fascia) comprise the skin, and the fifth layer comprises the muscles of facial expression. In accordance with the structure of real skin, we employ a synthetic tissue model constructed from the elements illustrated in Figure 1, consisting of nodes interconnected by deformable springs (the lines in the figure)[2]. The epidermal surface is defined by nodes 1, 2, and 3, which are connected by epidermal springs. The epidermal nodes are also connected by dermal-fatty layer springs to nodes 4, 5, and 6, which define the fascia surface.

Fascia nodes are interconnected by fascia springs. They are also connected by muscle layer springs to skull surface nodes 7, 8, 9. The facial tissue model is implemented as a

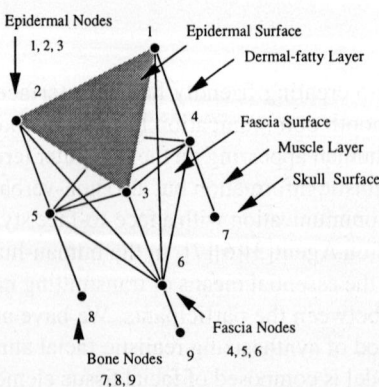

Fig. 1. Triangle shin tissue prism element

collection of node and spring data structures. The node data structure includes variables to represent the nodal mass, position, velocity, acceleration, and net force.

The spring data structure comprises the spring stiffness, the natural length of the spring, and pointers to the data structures of the two nodes that are interconnected by the spring. Newton's laws of motion govern the response of the tissue model to force [2][5]. This leads to a system of coupled, second-order ordinary differential equations that relate the node positions, velocities, and accelerations to the nodal forces. The equation for a generic node i is as follows:

$$m_i \frac{d^2 \mathbf{x}_i}{dt^2} + \gamma_i \frac{d\mathbf{x}_i}{dt} + \tilde{\mathbf{g}}_i + \tilde{\mathbf{q}}_i + \tilde{\mathbf{s}}_i + \tilde{\mathbf{h}}_i = \tilde{\mathbf{f}}_i$$

m_i is the nodal mass at node i
γ_i is the damping coefficient
$\tilde{\mathbf{g}}_i$ is the total spring force at node i
$\tilde{\mathbf{q}}_i$ is the total volume preservation force at node i
$\tilde{\mathbf{s}}_i$ is the total skull penetration force at node i
$\tilde{\mathbf{h}}_i$ is the total nodal restoration force at node i
$\tilde{\mathbf{f}}_i$ is the total applied muscle force at node i

2-2 Modified facial muscle model

The facial muscle model had two kinds of muscle models at forehead, namely, Frontaris and Corrugater. Frontaris pulls up the eyebrows and makes wrinkles in the forehead. Corrugater pulls the eyebrow together and makes wrinkles between left and right eyebrows. But those muscles can't pull down the eyebrows and make the eyes thin. So a new "Orbicularis oculi" muscle model is appended[4]. According to the anatomical chart shown in Fig. 2, the Orbicularis oculi is separated into an inside part and an outside part.

The inside part makes the eye close softly and the outside part makes eye close firmly. To make muscle control simple around the eye area, the Orbicularis oculi is modeled with a single function in our model. Normally, muscles are located between a born node and a fascia node. But the Orbicularis oculi has an irregular style, whereby it is attached between fascia nodes in a ring configuration; it has 8 linear muscles which approximate

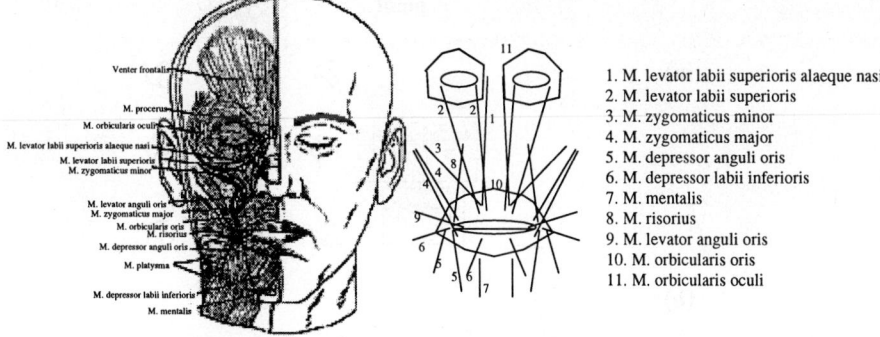

1. M. levator labii superioris alaeque nasi
2. M. levator labii superioris
3. M. zygomaticus minor
4. M. zygomaticus major
5. M. depressor anguli oris
6. M. depressor labii inferioris
7. M. mentalis
8. M. risorius
9. M. levator anguli oris
10. M. orbicularis oris
11. M. orbicularis oculi

Fig. 2. (a) Mm. faciales (b) muscle model
(quote from Sincher and Tandler : Anatomie Zahnarte)

a ring muscle. Contraction of the ring muscle makes the eye thin. The final facial muscle model has 12 muscles in the forehead area and 27 muscles in the mouth area.

3 Muscle Parameters

3-1 Feature Points

A marker is attached on each feature point of the subject's face to measure and model facial expression. A feature point is chosen for each muscle, from the grid point in face model which gives the biggest movement when contracting the muscle. If the feature point has already been chosen by another muscle, the point which gives second biggest movement is chosen as the feature point. Some feature points are appended and modified manually to make each feature point move more independently. We defined 16 feature points in the forehead area and 26 feature points in the mouth area as shown in Fig. 3.

3-2 Face Area Division

A simpler neural network structure can help speed the convergence process in learning and reduce the calculation cost in mapping, so the face area is divided into the three sub-areas indicated in Fig. 4. They are the mouth area, left-forehead area, and right-forehead area, which each give independent skin motion. Three independent networks are prepared for these three areas.

3-3 Neural Network Structure

A layered neural network finds a mapping from feature point movements to muscle parameters. A four-layer structure is chosen to effectively model the non-linear perfor-

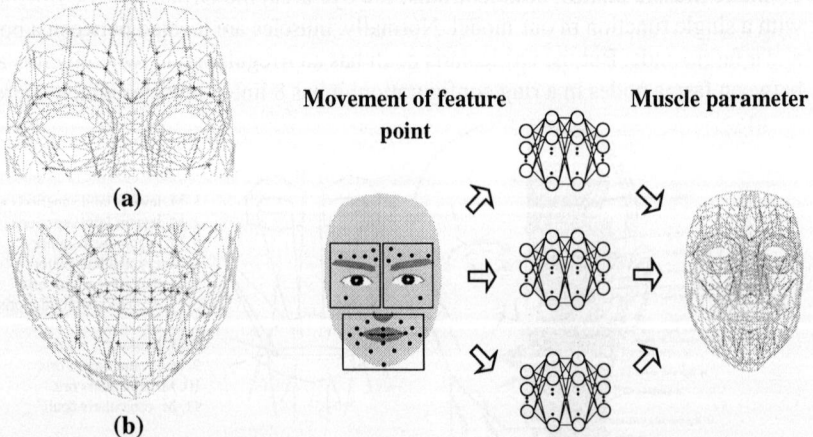

Fig. 3. (a) Feature points of forehead area (b) Feature points of mouth area

Fig. 4. Decison of facial muscle parameter by 3 Neural Network

mance. The first layer is the input layer, which corresponds to 2D marker movement.

The second and third layers are hidden layers. Units of the second layer have a linear function and those of the third layer have a sigmoid function. The fourth layer is the output layer, corresponding to muscle parameters, and it has linear units.

Linear functions in the I/O layers are introduced to maintain the range of I/O values. Feature point movements have 2 dimensions, so the number of input-layer units is double the number of feature points. The number of output-layer units is the number of muscles in each sub-area. The number of units in the hidden layer is decided heuristically. For 2 forehead sub-area, neural network consists of 16 units in input layer, 20 units in hidden (Second, Third) layer and 8 units in output layer. For the mouth sub-area, the neural network consists of 52 units in the input layer, 60 units in the hidden (Second, Third) layer, and 28 (27 muscles + a parameter for jaw rotation) units in the output layer.

3-4 Learning Patterns

Learning patterns are composed of the individual contraction of each muscle and their combination. In the case of individual motion, contraction of each muscle between maximum strength and neutral is quantized into 11 steps. In the combination case, we create 6 basic facial expressions consisting of anger, disgust, fear, happiness, sadness and surprise, and quantize the difference between neutral and each of these also into 11 steps. The number of learning patterns in the individual muscle case is 77 (7 muscles x 11 steps) for each forehead sub-area, and that in the combination case is 66 (6 expressions x 11 steps). So the total number of learning patterns is 143 in each forehead sub-area. In the mouth area, each muscle contraction does not happen individually.

So all learning patterns are composed of combinations. Learning patterns have basic mouth shapes for vowels "a", "i", "u", "e" and "o", and a closed mouth shape for nasal consonant "n". Also 6 basic expressions are appended as in the forehead area, and jaw rotation is specially introduced. Thus a total of 13 actions are selected for training, and they are also quantized into 11 steps. The number of learning patterns is 143 for the mouth sub-area. Each pattern is composed of a data pair: muscle parameter vector and feature point movement vector. Neural networks were trained using Back Propagation. The learning pattern was increased gradually according to strength of muscle parameter. (Fig. 5)

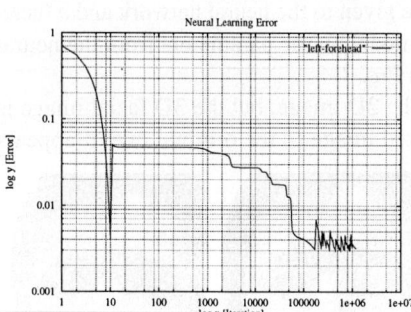

Fig. 5. Learning error at left-forehead area

3-5 Normalization

In order to absorb individual variations in human faces, each feature point movement is normalized by a standard length to decide the facial geometrical feature. The forehead area and mouth area each have local axes normalized by the local standard length. In the forehead area, Dot A is inside the left eye. Dot B is inside the right eye. Dot O is halfway between Dots A and B. The x-axis goes horizontally through Dot O, and the y-axis goes vertically through Dot O. Dot E is positioned at the hair-line, on the y-axis.

The standard lengths for the forehead area are distance AB for the x-axis and height OE for the y-axis. In the mouth area, Dot A is left edge of lip. Dot B is right edge of lip. Dot O is halfway between Dots A and B. The x-axis goes horizontally through Dot O and the y-axis goes vertically through Dot O. Dot E is the top of the nose, on the y-axis. Dots C and D are where the x-axis intersects the edge of face. The standard length for the mouth area is distance CD for the x-axis and height OE for the y-axis.

4 Evaluation

4-1 Closed Test

To confirm whether the learning process of the neural network is successfully completed or not, we input the same data used in learning into the input layer of neural network and resynthesize the facial image using muscle parameters from the output layer of the neural network. Example results are shown in Figs. 6. There are slight differences in fine detail in the re-generated image, but subjectively speaking, the overall facial features and expressions are almost the same as the original image. So the mapping rules work well for the training data.

4-2 Open Test

We attach markers on a real human's face and get movements of the markers from 2D images captured by a camera when any arbitrary expression is appearing. After normalization, the movement values of the markers are given to the neural network and a facial image is re-generated using the facial muscle model on the parameters from the neural network output. Example images are shown in Fig. 7.

Muscle parameters are decided only from the 2D image, but the 3D facial image is well regenerated. By comparing the regenerated image to the original one, it appears

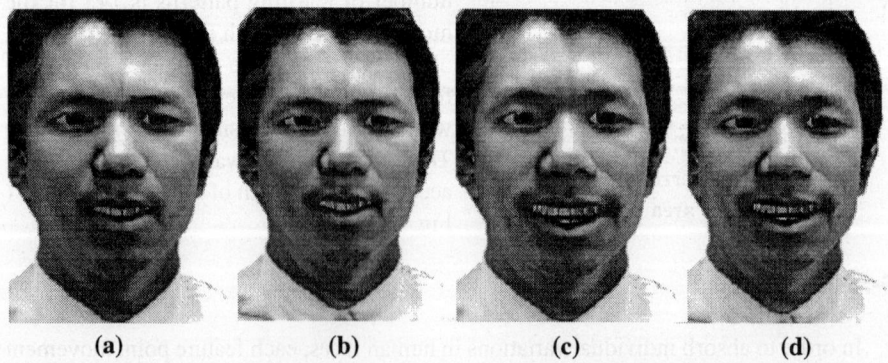

Fig. 6. (a) Disgust (original) (b) Disgust (resynthesis)
(c) Happiness (original) (d) Happiness (resynthesis)

that the regenerated image gives a weaker impression than the original one in expression production. But the facial features are well reproducted. Also, some exceptions occur when a facial feature which cannot be generated by any muscle combination in our model is given as the test sample. In the example image in Fig. 7(a)(b), the upper and lower lips are being pulled up at the same time. In our current muscle model, the lower lip does not move up beyond the standard position because the muscle action is only contraction. It is necessary to improve the muscle model by including expansion and relocation of muscles to solve this problem. Also strict rules are needed for the correspondence between the coordinates of markers on the real face and on those on the model face. This is under examination.

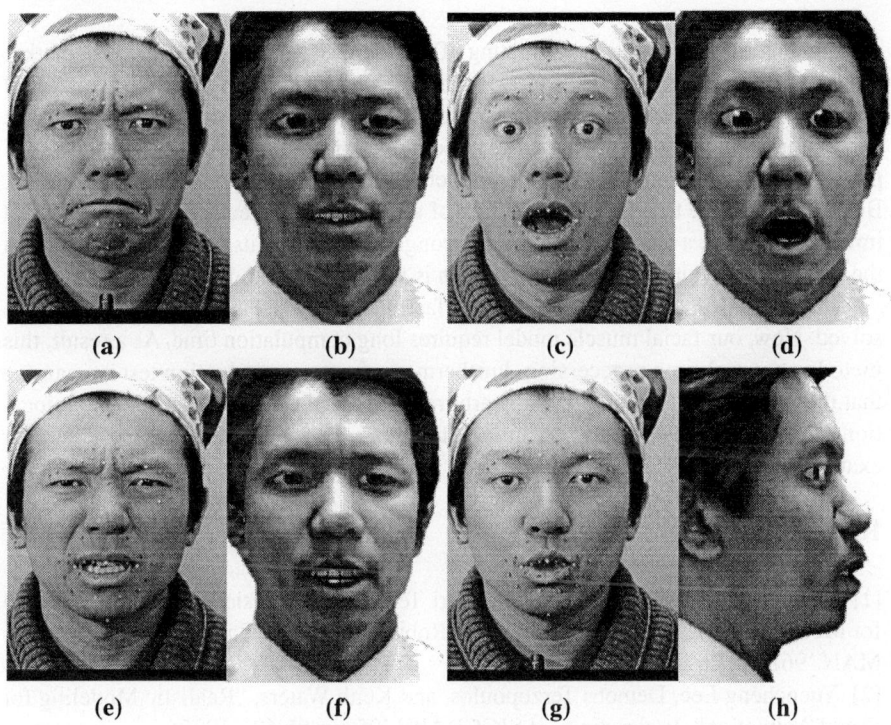

Fig. 7. (a) Anger (original) (b) Anger (resynthesis)
(c) Surprise (original) (d) Surprise (resynthesis)
(e) Disgust (original) (f) Disgust (resynthesis)
(g) Vowel-'u' (original) (h) Vowel-'u' side (resynthesis)

4-3 Animation

Feature points are given to a neural network frame by frame and then a facial animation is generated from the output muscle parameter sequence. An animation scene is the transition from one basic expression to another for all pairs of 6 basic expressions. All of the 2D feature point movements from one expression to another are linearly interpolated and then a set of feature point movements for each transition frame is decided. After that, these 2D feature point movements are converted to muscle parameters frame by frame using a neural network, and a facial image sequence is completed by synthesizing the resulting 3D facial expression for each frame. This is the test of the effect of interpolation on our parameter conversion method based on the generalization of the neural network. Training data includes only the transitions from neutral to any of the 6 basic expressions, so the transition from one of the 6 expression to another is not learned in the training process. After our subjective evaluation, slight differences from the original one are found in texture and fine features. But basically the overall facial features, impression and motion were very close to the original.

5 Conclusion and Discussion

A method of automatically estimating 3D facial muscle parameters from 2D marker movements is presented in this paper. Parameter conversion from 2D to 3D works well when the model is fitted to the target person's face precisely in the 2D image and the expression variation is within the range of combinations of basic expressions and vowel pronounciations. We currently fit the model to the facial image by manual operation. But it's impossible to decide the location of all grid points precisely in the real facial image. Of course, a marker's movement strongly depends on its initial location and on the target person, and parameter conversion is very sensitive to its effects.

Thus the correspondence between a real face and the model is the next problem to be solved. Now, our facial muscle model requires long computation time. As a result, this metod is not real-time proccessing. Furthermore, from our evaluation test we can see that there are limits to generating any arbitrary expressions with our model, so relocation of the muscles and a new definition of the physics for the new muscles are under examination.

Reference

[1] Hajime Sera, Shigeo Morishima, Demetri Terzopolus, "Physics-based Muscle Model for Mouth Shape Control", Proceedings of Robot and Human Communication '96(ROMAN '96), pp. 207-212(1996)
[2] Yuencheng Lee, Demetri Terzopoulos, and Keith Waters, "Realistic Modeling for Facial Animation", Proceedings of SIGGRAPH '95, pp.55-62, (1995)
[3] Shigeo Morishima, etc. "Life-Like, Believable Communication Agents", Course Notes #25, Siggraph (1996)
[4] Sincher and Tandler, "Anatomie Zahnarzte"
[5] Demetri Terzopoulos and Keith Waters, "Analysis and Synthesis of Facial Image Sequences Using Physical and Anatomical Models", IEEE Transactions on Pattern Analysis and Machine Intelligence, vol.15, No.6, pp.569-579(1993)
[6] S.Morishima and H.Harashima, "A Media Conversion from Speech to Facial Image for Intelligent Man-Machine Interface", IEEE JSAC, Vol.9, No.4(1991)
[7] Manabu Sudo, Shigeo morishima and Hiroshi Harashima, "A Media Conversion from English Text to Face Image for Pronunciation CAI System",Proceedings of the 1992 IEICE spring conference A-276(1992)

Appearance-Based Face Recognition under Large Head Rotations in Depth

Shaogang Gong, Eng-Jon Ong and Peter J. Loft

Department of Computer Science, Queen Mary and Westfield College, England
E-mail: {sgg,ongej,pjl}@dcs.qmw.ac.uk

Abstract. In this work we investigate appearance-based methods for face recognition on image sequences from large head rotations in depth. We describe the computational difficulties in recognising faces undergoing large pose variations and evaluate the effectiveness of different linear appearance-based methods for the problem. A framework for modelling nonlinear face pose density distributions using Gaussian mixtures was proposed for face recognition under such conditions using general and modified Hyper Basis Function networks.

1 Introduction

In this work, we address the problem of recognising faces from image sequences of head rotations in depth. Compared to the more typical scenarios of face recognition in which a single or a few isolated face images of frontal or near-frontal view are the subjects of interest, it is notoriously more difficult to recognise faces of moving people in natural scenes. This requires not only consistent detection, tracking and normalisation of faces in a given dynamic scene, but also the ability to associate, fast, face images of the same person from significantly different poses caused by continuous head rotations in depth. Essentially, head rotation in depth introduces non-linear transformations in the image plane which causes difficulties in recognition by correspondence. This problem can be addressed by either finding a view-invariant representation scheme, e.g. building a full scale 3D face model, or using a view-based representation with which a set of view-dependent linear subspaces can be modelled [4, 9]. We believe that a view-invariant single model is computationally unnecessarily complex and expensive for recognition purposes and a view-based representation [16] is computationally more efficient and psychophysically and neurobiologically more plausible [3, 10].

Face images[1] of a specific view form approximately a linear subspace with view-based representation [15]. Appearance-based face recognition schemes have been largely linear models working under the assumption that face vectors are taken from a specific narrow view, most typically the frontal view. The eigenface method [20] and its variations [15] are perhaps the most commonly used methods for appearance-based face recognition based on Principal Component Analysis

---

[1] With appearance-based approaches, we also refer to face images as face vectors in the hyper-dimensional image space.

(PCA). With PCA, a face vector can be represented as a linear combination of projections to a set of orthonormal basis vectors of a given face database. For a database of clearly linearly separable face classes, the eigenface approach can perform face recognition quite well given that reasonable alignment of the face vectors has been established. However, the eigenface approach takes no account of class labels. Projected face vectors from different classes can therefore merge together in eigen-space. In order to maximise class separation in a given data set, linear discriminants can be used based on explicitly labelling individual face vectors in the set with its class [1, 6, 19]. However, if the data set is relatively small, it is numerically not possible to perform linear discriminants on the original data set as the dimensionality is too high making the inverse of its between-class scatter matrix singular. In practice, linear discriminants may only be reasonably performed in the eigen-spaces of the data set.

2 HBF for Recognition under Head Rotations in Depth

With networks of generalised Hyper Basis Functions (HBF), a Gaussian mixture model can be used to capture the underlying density distribution of a nonlinear face-space, induced by head rotation in depth, with limited sample data [17]. With a Gaussian mixture model, a set of generalised Gaussian density functions are established in such a way that each function is tuned towards a specific subspace that is approximately linear. We consider three types of HBF network depending on the choice of the basis functions for the hidden units. We refer to them as HBF-1 (with radial Gaussians), HBF-2 (with diagonal Gaussians) and HBF-3 (with full covariance Gaussians).

An HBF network is trained in two steps. The first is an unsupervised density estimation process which aims to determine the parameters of the basis functions. An advantage of this unsupervised learning is that pose labelled face images which are difficult to obtain are not required. Instead, a large amount of unlabelled data is used and this can be made available by a real-time face detection and tracking system [11]. This also enables one to perform online re-training of the density parameters therefore adaptively re-tuning the hidden units according to those "near-miss" false recognitions. The second step in training an HBF is supervised and it determines the output weights using singular value decomposition [8].

2.1 Dimensionality Reduction and Clustering Algorithms

Due to "The Curse of Dimensionality" [2], using face images directly is likely to cause unreliable density estimation in a hyper-space given by the raw intensity resolution. Dimensionality reduction is required. Although such a reduction leads to information loss, for face recognition using HBF networks with limited training data, this is more likely to lead to an increase in performance. For an input vector of dimension d, a Gaussian mixture model of a scaled identity, diagonal and full covariance matrix requires respectively $d + 1$, $2d$ and $d(d + 3)/2$

parameters to be determined. When the available data are limited, the estimation of these parameters will be poor. An intuitive way to resolve this problem is to use face images of low resolution. However, this has its limitations since there will be a minimal image resolution one must have in order to perform any sensible recognition[2]. A more effective and principled approach for dimensionality reduction is to use PCA. This results in the face images being represented by their normalised N-dimensional pattern vectors where $N \ll d$ is the number of principal components used.

A number of techniques are available for estimating parameters of basis functions. The K-Means algorithm introduced by Moody and Darken [13] is widely used to estimate parameters of radial Gaussians. For estimating mixtures of full covariance Gaussians, an approach similar to Sung and Poggio [18] can be used. A major problem encountered while using the K-means algorithm is likely to be that each cluster can only estimate its parameters using the face vectors closest to it. Thus they are disjoint. With face pose distribution, it is more likely that certain pose clusters overlap and therefore should be able to "share points". K-means provides an unnecessary restriction on the nature of the pose clusters. An alternative approach to this form of clustering is not to explicitly restrict the data available for estimating each individual cluster. The Expectation Maximisation (EM) algorithm [5] is appropriate for such a purpose. This method is essentially an iterative method for monotonically reducing the negative log likelihood of a given data set $X = \{\mathbf{x}_1, \ldots, \mathbf{x}_i, \ldots, \mathbf{x}_N\}$

$$-\ln L = \sum_{i=1}^{N} \ln p(\mathbf{x}_i)$$

where $p(\mathbf{x})$ is the *mixture distribution* formed by a linear combination of the probability density estimates of K clusters in the mixture model and is given by $p(\mathbf{x}) = \sum_{k=1}^{K} p(x|k) P(k)$. $P(k)$ is the prior probability of the kth Gaussian.

With EM, K-means can be used to initialize the means and covariance matrices. This can speed up the convergence in the clustering process. Then, iterative approximation of the optimal clusters' parameters are performed with the following update rules: Given K clusters in a mixture model, whilst σ_i^k is the ith diagonal element of the covariance matrix of the kth cluster, x_i^n is the ith component of $\mathbf{x}_n$ and, μ_i^k is the ith component of $\boldsymbol{\mu}_k$, then

$$\mu_k^{new} = \frac{\sum_{n=1}^{N} \mathbf{x}_n P^{old}(k|\mathbf{x}_n)}{\sum_{n=1}^{N} P^{old}(k|\mathbf{x}_n)} \quad (1)$$

$$\sigma_i^{k\ new} = \frac{\sum_{n=1}^{N} (x_i^n - \mu_i^k) P^{old}(k|\mathbf{x}_n)}{\sum_{n=1}^{N} P^{old}(k|\mathbf{x}^n)} \quad (2)$$

$$P(k)^{new} = \frac{1}{N} \sum_{n=1}^{N} P^{old}(k|\mathbf{x}_n) \quad (3)$$

---

[2] Our experiments show that image resolution of at least $25 \times 25 = 625$ is required for recognition.

and the probability of a given vector **x** falling into the kth cluster is

$$P(k|\mathbf{x}) = \frac{p(\mathbf{x}|k)P(k)}{p(\mathbf{x})} \quad (4)$$

For the use of "new" parameters above, the new values are calculated in the following order: mean, covariances and prior probability.

2.2 HBF with Decoupled Inputs

With limited data, one is prevented from using both large numbers of hidden units and high input dimensionality with conventional HBF networks. This is likely to result in a suboptimal solution as insufficient information are available for capturing the face-space accurately. We derived a different network architecture, called the Decoupled Input HBF (DI-HBF), as shown in Fig. 1. With this network, one breaks the inputs into a few independent groups and dedicates a number of hidden units to each input unit group. This is computationally justified since the input dimensions are linearly independent due to the PCA-based dimensionality reduction. This forces the clustering to be performed in independently lower dimensional subspaces and therefore allow more accurate density estimation of the overall face-space. To "emerge" the estimated density distributions in the lower dimensional subspaces to a higher dimensional space, we simply superimpose the linearly independent subspaces by connecting each output unit to *all* the hidden units.

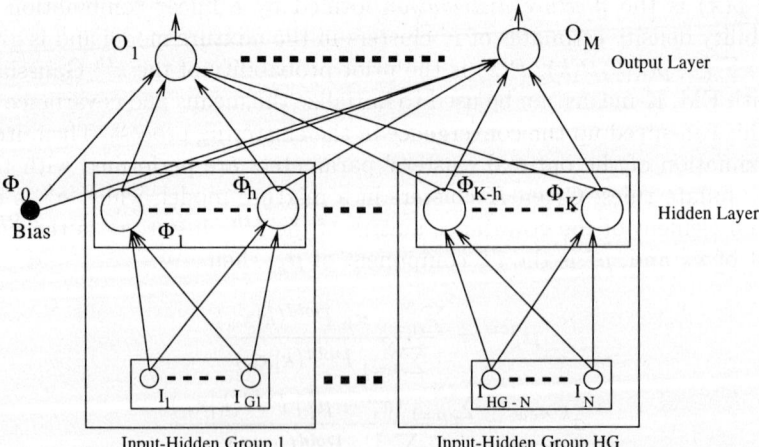

Fig. 1. A DI-HBF network has M output units, K hidden units, and N input units. However, the N inputs are sub-grouped with each group having at most $G1$ inputs and feeding *only* a corresponding group of h hidden units. In the figure shown here, there are HG groups. Each output unit is then connected with *all* the hidden units from all groups.

A DI-HBF can be trained in two stages like the HBF networks. The first stage involves estimating the parameters for the hidden unit blocks using the traditional K-means algorithm where only those input units allocated to the block are used. The second stage involves estimating the weights connecting the hidden and the output units. In this case, the hidden layer is viewed as one large group of hidden units instead of a disjoint group. The weights are then estimated using the pseudo-inverse method.

3 Experiments

The training data consist of three sets of image sequences of known faces. All face images have 56×56 = 3136 dimensions. PCA was used for dimensionality reduction. The sequences are from continuous head rotations in depth: (1) Tracked head rotation (set A with 356 images): Image sequences of head rotation from -90 to +90 were obtained [11]. With these sequences, large scale variation and translational shift (upto 15%) exist between successive frames; (2) Smooth head rotation (set B with 563 images): Sequences of head rotating from -90 to +90. The sequences were obtained under a more constrained environment with regard to lighting and background texture; (3) Labelled head rotation (set C with 95 images): Face image sequences were taken at 10 degrees head increments. Since this set was obtained at different times to the sets A and B, both the lighting and scale of the face images are different from the early sets. Examples from the three data set are shown in Fig. 2.

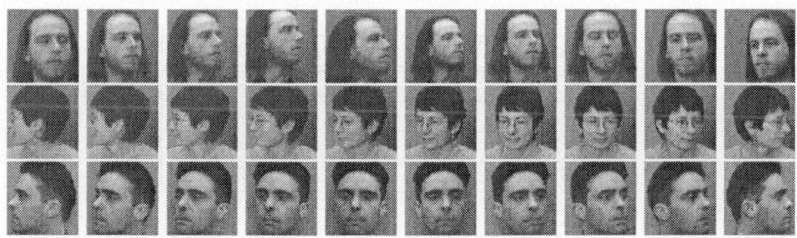

Fig. 2. Top row: Tracked head rotation, tracked face images from every tenth frame of a head rotation sequence. Middle row: Smoothed head rotation. Bottom row: Labelled head rotation, second frame of a labelled head sequence rotating in depth with 10 degrees increments.

Experiment 1 - Generalisation Ability This experiment was employed to test generalisation ability. The HBF networks were trained with the labelled sequences in set C plus synthesised data obtained through translational shifts in the (x, y) image plane for upto 15% in each direction. Two test sequences were selected from set B with face images of persons appearing quite different from those in the training sets. The results can be found in Table 1.

Method	Generalisation Ability		Scale Variation and Shift	
	On Training Data	On Testing Data	On Training Data	On Testing Data
Eigenface	16%	0%	16%	0%
Fisherface	18%	0%	18%	8%
HBF-1	84%	51%	84%	54%
HBF-2	77%	33%	77%	58%
HBF-3	64%	68%	66%	60%
DI-HBF	84%	32%	81%	50%

Table 1. Results on generalisation ability, coping with scale change and shift.

Experiment 2 - On Translational Shift & Scale Variation This experiment was designed to determine the extent to which different methods can cope with offset and scale variations. Again, the networks were trained using the labelled sequences in set C. Two test sequences were obtained from set A, both having a fair amount of scale variations and translational offsets in the images, with shifts > 10% of the image width in some cases. The results from different methods can also be found in Table 1. A further experiment was performed on the HBF and DI-HBF networks only, for which the training was performed on the smooth data from set B and testing on the tracked data from set A. The results are shown in Table 2. These show that the networks seemed to be able to cope better with scale rather than translational variations. Misclassification was almost entirely with those test images having severe translational shift. The results from *Experiment 3* further demonstrate this phenomenon.

Network	On Training Data	On Testing Data
HBF-1	95%	30%
HBF-2	96%	58%
HBF-3	99%	54%
DI-HBF	99.5%	69%

Table 2. Scale and Offset using data set B as training data.

Experiment 3 - Eliminating Shifting Effects Further experiments were undertaken with the HBF-3 network where both training and testing data were taken from set B. The testing data were formed through the extraction of different frames from the same sequences used for training. The results are shown in Table 3. The images used to form the testing data were different from those for training.

These tests show good results with over 90% correct recognition achieved. In fact the failures tended to occur when the corresponding head images were at or near profile views. This could be due to the limited data available to model the pose distribution at the profile views. The network was not able to accurately model the full pose-distribution with poor results where discontinuities exist and where data were scarce.

Forming Testing Data	On Training Data	On Testing Data
Every 5th frame	93%	93%
Every 2nd, 3rd, 7th and 8th frame	88%	90%
Every 3rd, 4th, 5th, 6th and 7th frame	91%	93%

Table 3. The Testing set was formed by extracting every n^{th} frame per ten frames from sequences in set B. The training data comprised of the remaining frames in the sequences.

4 Conclusions

With the HBF networks, Gaussian mixture models were employed to estimate the underlying face pose distribution and as a result, recognition was performed more successfully than with either of the linear models. The results from different HBF networks were comparable. However, it may not be surprising to find out that the HBF-1 network performed well in *Experiment 1* given the very limited data available for training. The DI-HBF performed better in classifying training data due to the large number of hidden units. However, for recognising images in the testing sequences, it did not fare well in generalisation when the testing images were significantly different from the training ones. One of the reasons for this was that the number of hidden units for lower dimensions was the same as those used for higher dimensional groups. This may not be the best approach as the data in higher dimensions are far more difficult to separate than those of lower dimensions. Using more hidden units for the higher dimensions and less for the lower dimensions improved its performance. Also, current DI-HBF network used only radial Gaussian clusters. A better performance hyper-ellipsoidal Gaussians produced with a full covariance matrix will be exploited. These would provide more accurate clustering of the sub-eigen-spaces. It also became apparent that the initial positioning of data vectors to basis functions also makes a big difference to final classification results. Various schemes [7, 12, 14, 21] can be applied to provide a degree of invariance to the number and centres of basis functions initially chosen.

Overall, we have shown that using HBF networks with Gaussian mixtures provides firstly an effective way to model nonlinear face pose distribution in facespace, and secondly a computationally promising approach for face recognition under large pose variations. However, for such an approach, more training data are clearly required in order to have more accurate estimation of the parameters of the Gaussian mixture models. The dimensionality of the input vector could then be extended, allowing more of the between class variation to be represented by the input space. It was also clear that translational offset has a much greater effect on the network's classification performance than the scaling in the images. Invariance to such factors as scale variation and translational shifts can be achieved by preprocessing. Gong *et. al.* [9] for example highlight how preprocessing such as Gabor wavelet transforms and Difference of Gaussian filtering can provide much invariance to changes in scale and lighting.

References

1. P. Belhumeur, J. Hespanha, and D. Kriegman. Eigenfaces vs. fisherfaces: Recognition using class specific linear projection. In *ECCV*, Cambridge, England, 1996.
2. R. Bellman. *Adaptive Control Processes: A Guided Tour*. Princeton Press, 1961.
3. H. Bülthoff, S. Edelman, and M. Tarr. How are three-dimensional objects represented in the brain? AI Memo 1479, MIT, Cambridge, Massachusetts, April 1994.
4. I. Craw. A manifold model of face and object recognition. In *Cognitive and Computational Aspects of Face Recognition*, pages 183–203. Routledge, London, 1995.
5. A. Dempster, N. Laird, and D. Rubin. Maximum likelihood from incomplete data via the em algorithm. *Journal of the Royal Statistical Society*, B-39(1):1–38, 1977.
6. K. Etemad and R. Chellappa. Discriminant analysis for recognition of human face images. In *Int. Conf. on Audio- and Video-Based Biometric Person Authentication*, pages 127–142, Crans-Montana, 1997.
7. I. Gath and A.B. Geva. Unsupervised optimal fuzzy clustering. *IEEE Transactions on Pattern Analysis and Machine Intelligence.*, 11(7):773–781, July 1989.
8. G. Golub and W. Kahan. Calculating the singular values and pseudo-inverse of a matrix. *SIAM Numerical Analysis*, 2(2):205–224, 1965.
9. S. Gong, S. McKenna, and J. Collins. An investigation into face pose distributions. In *Proc. 2nd Int. Conf. on Automatic Face and Gesture Recognition*, 1996.
10. N.K. Logothetis, J. Pauls, and T. Poggio. Spatial reference frames for object recognition: Tuning for rotations in depth. AI Memo 1533, MIT, Cambridge, 1995.
11. S. McKenna and S. Gong. Tracking faces. In *Proc. 2nd Int. Conf. on Automatic Face and Gesture Recognition*, Killington, Vermont, October 1996.
12. G.W. Milligan and M.C. Cooper. An examination of procedures for determining the number of clusters in a data set. *Psychometrika*, 50(2):159–179, June 1985.
13. J.E. Moody and C.J Darken. Fast learning in networks of locally-tuned processing units. *Neural Computation*, 1(2):281–294, 1989.
14. R.P. Nikhil, J.C. Bezdek, and C.-K. Tsao, E. Generalized clustering networks and kohenen's self-organizing scheme. *IEEE Transactions on Neural Networks*, 4(4):549–557, July 1993.
15. A. Pentland, B. Moghaddam, and T. Starner. View-based and modular eigenspaces for face recognition. In *IEEE CVPR*, Seattle, July 1994.
16. T. Poggio and S. Edelman. A network that learns to recognize three-dimensioanl objects. *Nature*, 343, January 1990.
17. T. Poggio and F. Girosi. Networks for approximation and learning. *Proceedings of The IEEE*, 78(9), September 1990.
18. K. Sung and T. Poggio. Example-based learning for view-based human face detection. Technical Report AI Memo 1512, CBCL 103, MIT, 1995.
19. D. L. Swets and J. Weng. Discriminant analysis and eigenspace partition tree for face and object recognition from views. In *Proc. 2nd Int. Conf. on Automatic Face and Gesture Recognition*, pages 192–197, 1996.
20. M. Turk and A. Pentland. Eigenfaces for recognition. *J. of Cognitive Neuroscience*, 3(1), 1991.
21. L. Xu, A. Krzyzak, and E. Oja. Rival penalized competitive learning for clustering analysis, rbf net and curve detection. *IEEE Transactions on Neural Networks*, 4(4):636–649, July 1993.

Skin-Color Modeling and Adaptation

Jie Yang, Weier Lu, Alex Waibel

School of Computer Science
Carnegie Mellon University
Pittsburgh, PA 15213, USA

Abstract. This paper studies a statistical skin-color model and its adaptation. It is revealed that (1) human skin colors cluster in a small region in a color space; (2) the variance of a skin color cluster can be reduced by intensity normalization, and (3) under a certain lighting condition, a skin-color distribution can be characterized by a multivariate normal distribution in the normalized color space. We then propose an adaptive model to characterize human skin-color distributions for tracking human faces under different lighting conditions. The parameters of the model are adapted based on the maximum likelihood criterion. The model has been successfully applied to a real-time face tracker and other applications.

1 Introduction

Human face perception is currently an active research area in the computer vision community. Locating and tracking human faces is a prerequisite for face recognition and/or facial expressions analysis, although it is often assumed that a normalized face image is available. Facial features, such as eyes, nose and mouth, are natural candidates for locating human faces. These features, however, may change from time to time. Occlusion and non-rigidity are basic problems with these features.

Color is another feature on human faces. Much research has been directed to understanding and making use of color information. Color has been long used for recognition and segmentation [1, 2] and recently has been successfully used face locating and tracking [3, 4, 5]. However, color is not a physical phenomenon. It is a perceptual phenomenon that is related to the spectral characteristics of electro-magnetic radiation in the visible wavelengths striking the retina [6]. Using color as a feature for tracking human faces has several problems. First, the color representation of a face obtained by a camera is influenced by many factors such as ambient light, object movement, etc. Second, different cameras produce significantly different color values even for the same person under the same lighting condition. Finally, human skin colors differ from person to person. In order to use color as a feature for face tracking, we have to solve these problems.

In this paper, we quantitatively investigate human skin color distributions. A common believe is that different people have different color appearances. This study shows that such a difference lies largely in intensity than color itself. By color normalization, the skin-color difference among different people can be greatly reduced. Furthermore, using goodness-of-fit techniques, we verify that under a certain lighting condition, a human skin-color distribution is a normal distribution. Based on these results, we present an adaptive parametric model to characterize human skin-color distributions for different people under different lighting conditions. Since a linear transformation of a normal distribution is still a normal distribution, the different skin-color distributions can be considered as transformed distributions from other distributions. We propose to use a linear combination of the known parameters to predict or approximate new parameters. The maximum likelihood method has been used to estimate the coefficients of the linear transformation. We investigate two cases: estimating mean vector only and estimating both mean vector and covariance matrix. We derive the maximum likelihood estimates for both cases.

2 Skin-Color Distributions

A color histogram is a distribution of colors in the color space and has long been used by the computer vision community in image understanding. For example, analysis of color histograms has been a key tool in applying physics-based models to computer vision. It has been shown that color histograms are stable object representations unaffected by occlusion and changes in view, and that they can be used to differentiate among a large number of objects [2]. In the mid-1980s, it was recognized that the color histogram for a single inhomogeneous surface with highlights will have a planar distribution in color space [7]. It has since been shown that the colors do not fall randomly in a plane, but form clusters at specific points [8]. It has been further observed that (1) human skin colors cluster in a small region in a color space; (2) human skin colors differ more in intensity than in colors, and (3) under a certain lighting condition, a skin-color distribution can be characterized by a multivariate normal distribution in the normalized color space [5]. The Figure 1 shows a face image, the skin-color occurrences in the RGB color space (256x256x256), and the skin color distribution in the normalized color space. In the following section, we justify these observations by quantitative analysis goodness-of-fit techniques.

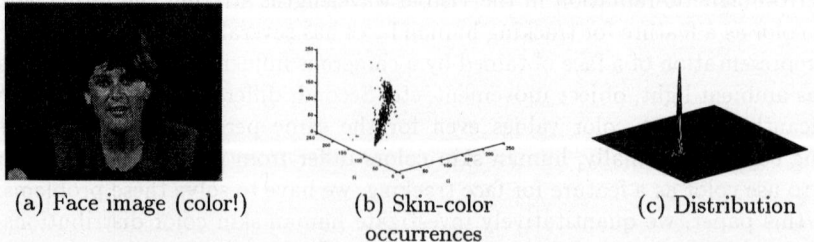

(a) Face image (color!)   (b) Skin-color occurrences   (c) Distribution

Fig. 1. An example of a human skin-color cluster and distribution

3 Quantitative Analysis and Goodness-of-Fit Test

We have built up a database which contains about 1000 face images down-loaded from the Internet and taken from our laboratory. This database covers face images of people in different races (Caucasian, African American, and Asian), genders, and the lighting conditions.

3.1 Data Analysis

Figure 2 shows in the RGB space skin color occurrences of 48 human faces randomly selected from the database. The total number of images included in such a set is limited by the memory resource of the system associated with the statistical analysis software we used, and we did not attempt to migrate the computation to a more powerful machine because through experimentation we found images beyond 20 adds little to the aggregated color pool. This attribute is further affirmed when we also analyzed several similar random sets of images and found no qualitative differences are found among them.

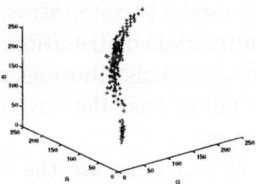

Fig. 2. Skin color cluster of 48 faces (mean values: $m_R = 188.9069$, $m_G = 142.9157$, $m_B = 115.1863$, and variances: $\sigma_R = 58.3542$, $\sigma_G = 45.3306$, $\sigma_B = 43.397$).

We further studied the effect of illumination on skin color clusters by comparing variances of skin color clusters in RGB space and the normalized color space. It has been found that variances of skin color clusters can be greatly reduced by the color normalization. Table 1 shows mean values and variances of the same color cluster in different color spaces.

3.2 Goodness-of-Fit Tests

We have observed that the skin-color distributions are Gaussian-like distributions. Unlike most of the methods used in engineering statistics assume a normal distribution of the measured data, we have tested whether the measured data of a sample do indeed have a normal distribution by goodness-of-fit techniques. Goodness-of-fit tests examine the conformity of the observed data's empirical

Table 1. Comparison of mean and variance

	RGB Space	Normalized Color Space
Mean	$m_R = 234.29$ $m_G = 185.72$ $m_B = 151.11$	$m_r = 104.22$ $m_g = 81.59$
Variance	$\sigma_R = 26.77$ $\sigma_G = 30.41$ $\sigma_B = 25.68$	$\sigma_r = 4.93$ $\sigma_g = 3.89$

distribution function with a posited theoretical distribution function. The methods of performing a test can be an analytic or a graphic approach. In the graphic approach, the most common method is *Quantile-Quantile* plot (or Q-Q plot). We will use this method to test skin-color distributions. To perform a goodness-of-fit test, we first need to formulate a null hypothesis.

NULL hypothesis:
human skin-color is normally distributed in a normalized bivariate space.

An immediate difficulty of the task is that there is no commonly agreed analytical tool available to test the normality of a bivariate distribution [9]. In order to solve the problem, we used a two step strategy: first we test the marginal distribution and then test multivariate distribution. If a multivariate distribution is a normal distribution, its marginal distributions must be normal distributions. If the marginal distributions fail to pass the normality test, there is no need to test the multivariate distribution.

The basic idea of the Q-Q plot is to use the cumulative probability of the sampling data against that of the tested distribution. A straight line indicates that we cannot reject the null hypothesis (the interested readers are referred to textbooks on the subject, e.g. [9]). When we do marginal test, we test each variable separately against the normal distribution. When we test the bivariate distribution, we test the transformed variable against χ^2 distribution [9].

We first tested marginal distributions. As the results, we could not reject the null hypothesis. Figure 3(a) and (b) are the Q-Q plots of an African American's marginal distributions. Both the normalized "r" and "g" are straight lines. We then tested bivariate distributions. Again, we cannot reject the null hypothesis. Figure 3(c) is an example of the Q-Q plot for a Caucasian skin-color distribution. The plot is a straight line except a few outliers. Therefore, we have verified that, under a certain lighting condition, the skin-color distribution of an individual can be characterized by a Gaussian distribution.

4 Maximum Likelihood Adaptation

Although under a certain lighting condition, the skin-color distribution of each individual is a multivariate normal distribution, the parameters of the distribution for different people and different lighting conditions are different. There are

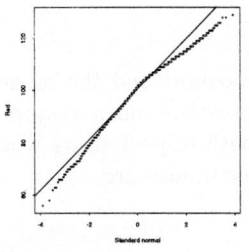 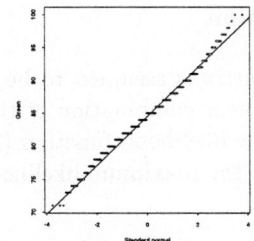 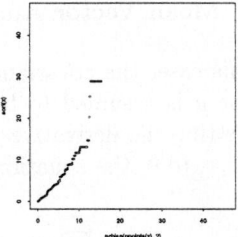

(a) Marginal test (r) of an African American (b) Marginal test (g) of an African American (c) χ^2 test of a Caucasian

Fig. 3. Examples of Q-Q plots

two schools of philosophy to handle environment changes: tolerating and adapting. Color constancy refers to the ability to identify a surface as having the same color under considerably different viewing conditions. Although human beings have such ability, the underlying mechanism is still unclear. Adaptive approach, on the other hand, is to transform the previous developed color model into the new environment. Since the Gaussian model has only a few parameters, it is possible to update them in real-time. Because the linear combination of Gaussian distributions is still a Gaussian distribution, we can consider the current Gaussian distribution is a combination of the previous distributions. One way of adaptation is to use the linear combination of known parameters to predict, or, approximate the new parameters, i.e.,

$$\hat{\mu} = \sum_{k=1}^{r} \alpha_k \mathbf{m_k}, \quad \hat{\Sigma} = \sum_{k=1}^{r} \beta_k S_k, \qquad (1)$$

where $\hat{\mu}$ is the estimated mean vector; $\hat{\Sigma}$ is the estimated covariance matrix; $\alpha_i \leq 1$ and $\beta_k \leq 1$ $k = 1, \ldots, r$, are weighting factors; $\mathbf{m_k}$ and S_k, $k = 1, \ldots, r$, are the previous mean vectors and covariance matrices.

We will use the maximum likelihood criterion to find the best set of coefficients for the prediction.

Let the sample mean and variance be

$$\bar{\mathbf{x}} = \frac{1}{N} \sum_{k=1}^{N} \mathbf{x_k}, \quad C = \frac{1}{N} \sum_{k=1}^{N} (\mathbf{x_k} - \bar{\mathbf{x}})(\mathbf{x_k} - \bar{\mathbf{x}})' \qquad (2)$$

The logarithm of the likelihood function can be written as:

$$\log L = -N \log(2\pi) - \frac{1}{2} N \log |\Sigma^{-1}| - \frac{1}{2} N tr\ \Sigma^{-1} C - \frac{1}{2} N (\bar{\mathbf{x}} - \mu)' \Sigma^{-1} (\bar{\mathbf{x}} - \mu). \quad (3)$$

Since $\log L$ is an increasing function of L, its maximum is at the same point in the space of μ as the maximum of L. We will discuss two cases: (1) adapting mean vector only; and (2) adapting both mean vector and covariance matrix.

4.1 Mean Vector Adaptation

In this case, the covariance matrix is assumed to be a constant and the mean vector μ is assumed to be a linear combination of the previous mean vectors. By setting the derivatives of the likelihood function (3) with respect to $\alpha_k, k = 1, \ldots, r$, to 0, the equations for the maximum likelihood estimates are

$$\sum_{k=1}^{r} \mathbf{m}_j' \Sigma^{-1} \mathbf{m}_k \hat{\alpha}_k = \mathbf{m}_j' \Sigma^{-1} \bar{\mathbf{x}}, \quad j = 1, \ldots, r \tag{4}$$

We can obtain α_k by solving the equation (4).

4.2 Mean Vector and Covariance Matrix Adaptation

In this case, both mean vector and covariance matrix are assumed to be a linear combination of the previous parameters. Many researchers in the field of statistics have studied this problem. In general, explicit solutions for this problem do not exist and estimates must be performed by iterative numerical techniques.

In fact, because the two sets of estimates are asymptotically independent, each set of parameters can be estimated as when the other set of parameters is known. In the following we present an EM algorithm based on the estimate procedure proposed by Anderson [10]. The basic idea of the algorithm is to iteratively estimate two sets of parameters independently. In order to iteratively estimate $\hat{\alpha}_k^{(i)}$ and $\hat{\beta}_k^{(i)}$, where the superscript (i) denotes the ith iteration.

Algorithm

1. Initialization

$$\sum_{k=1}^{r} \mathbf{m}_j' \mathbf{m}_k \hat{\alpha}_k^{(0)} = \mathbf{m}_j' \bar{\mathbf{x}}, \quad j = 1, \ldots, r,$$

$$\hat{\mu}^{(0)} = \sum_{k=1}^{r} \hat{\alpha}_k^{(0)} \mathbf{m}_k, \quad j = 1, \ldots, r,$$

$$C^{(0)} = \frac{1}{N} \sum_{k=1}^{N} (\mathbf{x_k} - \bar{\mathbf{x}})(\mathbf{x_k} - \bar{\mathbf{x}})' + (\mathbf{x_k} - \hat{\mu}^{(0)})(\mathbf{x_k} - \hat{\mu}^{(0)})'$$

$$\sum_{k=1}^{r} tr\, S_j S_k \hat{\beta}_k^{(0)} = tr\, S_j C^{(0)}, \quad j = 1, \ldots, r,$$

$$\hat{\Sigma}^{(0)} = \sum_{k=1}^{r} \hat{\beta}_k^{(0)} S_k,$$

2. Iteration

$$\sum_{k=1}^{r} \mathbf{m_j}' \Sigma^{-1} \mathbf{m_k} \hat{\alpha}_k^{(i)} = \mathbf{m_j}' \Sigma^{-1} \bar{\mathbf{x}}, \quad j=1,\ldots,r,$$

$$\hat{\mu}^{(i)} = \sum_{k=1}^{r} \hat{\alpha}_k^{(i)} \mathbf{m_k}, \quad j=1,\ldots,r,$$

$$C^{(i)} = \frac{1}{N} \sum_{k=1}^{N} (\mathbf{x_k} - \bar{\mathbf{x}})(\mathbf{x_k} - \bar{\mathbf{x}})' + (\mathbf{x_k} - \hat{\mu}^{(i)})(\mathbf{x_k} - \hat{\mu}^{(i)})'$$

$$\sum_{k=1}^{r} tr\,(\hat{\Sigma}^{(i-1)})^{-1} S_j (\hat{\Sigma}^{(i-1)})^{-1} S_k \hat{\beta}_k^{(i)} = tr\,(\hat{\Sigma}^{(i-1)})^{-1} S_j (\hat{\Sigma}^{(i-1)})^{-1} C^{(i)},$$

$$j=1,\ldots,r,$$

$$\hat{\Sigma}^{(i)} = \sum_{k=1}^{r} \hat{\beta}_k^{(i)} S_k,$$

3. If $\max(|\beta_j^{(i)} - \beta_j^{(i-1)}|, \; j=1,\ldots,r) \leq \epsilon$ for a small number $\epsilon > 0$, stop; otherwise goto step 2.

It has been shown that the solution of these estimation equations is asymptotically efficient provided that estimate of Σ is consistent [10].

4.3 Applications

Th adaptive skin-color model has been applied to many applications. The model plays a key role in the real-time face tracker [5]. The system has achieved a rate of 30+ frames/second with 305 x 229 input sequences of images on both HP and Alpha workstations. The system can track a person's face while the person walks, jumps, sits and rises. The QuickTime movies of demo sequences in different situations and on different subjects can be found in the web site http://www.is.cs.cmu.edu/. The skin-color model has been applied to other applications such as tele-conferencing [11], gaze tracking [12], and lip-reading [13].

5 Conclusions

We have proposed a statistical skin-color model for tracking human faces in real-time. We have shown that the variance of skin color clusters can be reduced by normalization by data analysis. Using goodness-of-fit techniques, we have further verified that the skin-color distribution of each individual under a certain lighting condition can be characterized by a multivariate normal distribution. Based on these results, we have proposed an adaptive skin-color model to characterize human faces different views under different lighting conditions. We have used a linear combination of the known parameters to predict or approximate new parameters. The maximum likelihood method has been used to estimate the coefficients of the linear transformation. We have investigated two cases: estimating mean vector only and estimating both mean vector and covariance matrix. In the later case, an iterative algorithm has been employed to obtain the optimal coefficients. The feasibility of the model has been demonstrated by a real-time face tracker and other applications in human computer interaction.

Acknowledgements

This research was sponsored by the Advanced Research Projects Agency under the Department of the Navy, Naval Research Office under grant number N00014-93-1-0806.

References

1. Y. Ohta, T. Kanade and T. Sakai, "Color information for region segmentation," Computer Graphics and Image Processing, Vol. 13, No. 3, pp.222-241, July 1980.
2. M.J. Swain and D.H. Ballard, "Color indexing," International Journal of Computer Vision. Vol. 7, No.1, pp. 11-32, 1991.
3. M. Hunke and A. Waibel, "Face locating and tracking for human-computer interaction," Proc. Twenty-Eight Asilomar Conference on Signals, Systems & Computers, Monterey, CA, USA, 1994.
4. T.C. Chang, T.S. Huang, and C. Novak, "Facial feature extraction from color images," Proc. the 12th IAPR International Conference on Pattern Recognition, Vol. 2, pp. 39-43, 1994.
5. J. Yang and A. Waibel, "A real-time face tracker," Proceedings of the Third IEEE Workshop on Applications of Computer Vision (Sarasota, Florida, 1996), pp. 142-147 ("Tracking human faces in real-time," Technical Report CMU-CS-95-210, CS department, CMU, 1995).
6. G. Wyszecki and W.S. Styles, "Color Science: Concepts and Methods, Quantitative Data and Formulae," Second Edition, John Wiley & Sons, New York, 1982.
7. S.A. Shafer, "Optical phenomena in computer vision," Prec. Canadian Soc. Computational Studies of Intelligence, pp. 572-577, 1984.
8. G.J. Klinker, S.A. Shafer, and T. Kanade, "Using a color reflection model to separate highlights from object color," Proc. ICCV, pp. 145-150, 1987.
9. J. D. Jobson, "Applied Multivariate Data Analysis," Vol. I: Reregression and Experimental Design, Springer-Verlag, 1991.
10. T.W. Anderson, "Asymptotically efficient estimation of covariance matrices with linear structure," The Annals of Statistics, Vol. 1, No. 1, 135-141, 1973.
11. J. Yang, L. Wu, and A. Waibel, "Focus of attention: towards low bitrate video tele-conferencing," Proceedings of 1996 IEEE International Conference on Image Processing (Lausanne, Switzerland), Vol. 2, pp. 97-100.
12. R. Stiefelhagen, J. Yang, and A. Waibel, "A model-based gaze tracking system," To appear on International Journal on Artificial Intelligence Tools.
13. U. Meier, R. Stiefelhagen, J. Yang, "Preprocessing of Visual Speech under Real World Conditions" European Tutorial & Research Workshop on Audio-Visual Speech Processing: Computational & Cognitive Science Approaches (AVSP 97).

Human Information Retrieval by Face Extraction and Recognition on TV News Images Using Subspace Method

Yasuo Ariki, Noriyuki Ishikawa and Yoshiaki Sugiyama

Ryukoku University, Seta, Otsu-shi, 520-21, JAPAN

Abstract. This paper proposes a human information retrieval system with face extraction and recognition function. A user watches the TV and if he encounters with interesting but unknown persons, he can specify them on the TV display by a mouse device. The proposed system can extract and recognize the specified persons and retrieve their information from human information database via interet and output it on a display unit as well as through voice synthesizer. We used subspace methods for a face extraction and recognition. In face extraction, the sub-image within a scanning window is projected to facial subspace and it is regarded as the facial region, if the projection amount has a local peak. In face recognition, the facial subspace of individual person is constructed at first in an observation space using respective training data. Then an input facial region is projected to the individual subspace and classified into the person with the maximum projection amount.

1 Introduction

We are getting much information everyday from TV news. They include various types of information such as politics, economics, culture, amusements, sports and so on. When we are watching TV news and if we encounter with unknown interesting persons, we have no idea to get much more information about them, because we do not know their names. This problem can be solved if we can construct a bi-directional or interactive TV system, in which we just specify the unknown human faces on a TV display.

Our purpose is to construct such kind of a bi-directional TV system which is equipped with a human face recognizer and is connected to human information database via internet. If a user encounters with interesting but unknown persons, when he is watching the TV, he can specify them on the TV display by a mouse device. The TV system can extract and recognize the specified persons and retrieve their information from human information database via interet and output it on a display unit as well as through voice synthesizer.

The system is different from conventional static hyper-media system, because TV news is live and it is difficult to make hyper-links in advance. A dynamic (or on-demand) links based on human face extraction and recognition are required on user demanding.

The human face extraction and recognition on TV videos have many problems to be solved such that the size and orientation of human faces are not limited and the face changes due to lighting and emotional expressions. In order to solve these difficult problems, we have been investigating subspace methods in a face extraction and recognition [1][2][3].

In the subspace method, most commonly used is a CLAFIC method [4]. It can present the most essential and common features of facial images in a low dimensional space, so that noise factors such as lighting, environmental noises and emotions can be removed. Moreover, the facial subspace constructed by the facial data of many persons using various orientations can represent the most typical face so that it contributes to extract facial regions with various orientations.

This subspace method is different from an eigenface method in that the eigenface method presents facial data of many persons in one feature space with low dimension [5]. On the other hand, subspace method presents facial data of an individual person in one feature subspace with low dimension. Therefore the subspaces are produced by the number of persons.

In face recognition, the facial subspace of individual person is constructed at first in an observation space using respective training data. Then an input facial image is projected to the individual subspace and classified into the person with the maximum length of the projected vector.

2 System Organization

Fig.1 shows an organization of a system we constructed for human information retrieval with face extraction and recognition function. It consists of TV video input, face extraction, face recognition, information retrieval and output. TV videos are digitized and stored on a hard disk as a digital video database. When a user is watching the digital video and if he specifies on a TV display a human face in which he is interested by a mouse device, the system can extract and recognize the specified face and retrieve his information from human information database via interet or from CD-ROM and output it on the display unit as well as through voice synthesizer. The processings at each block are described hereafter in detail.

3 TV Video Input

Our TV video database is composed of broadcasted news programs and dramas. We recorded them on 8mm tapes and converted into digital video images on the hard disk using Indigo2.

The video rate was 30 frames per second and the images were compressed by a JPEG board with 75% quality. The frame size was 240 × 320 pixels. This database was used for evaluation of face extraction, recognition and retrieval.

4 Subspace Method

The first step of the subspace method in extracting and recognizing human faces is to find orthonormal bases $V_i = \{v_{i1}, \cdots, v_{ir}\}$ from training images belonging to category ω_i as shown in Fig.2 [4]. Here r is the number of dimensions of the subspace. This can be carried out by finding axes to which the total distance from the training images is minimized. It is well known that finding orthonormal bases is equivalent with eigenvalue decomposition of the correlation matrix of the given training images.

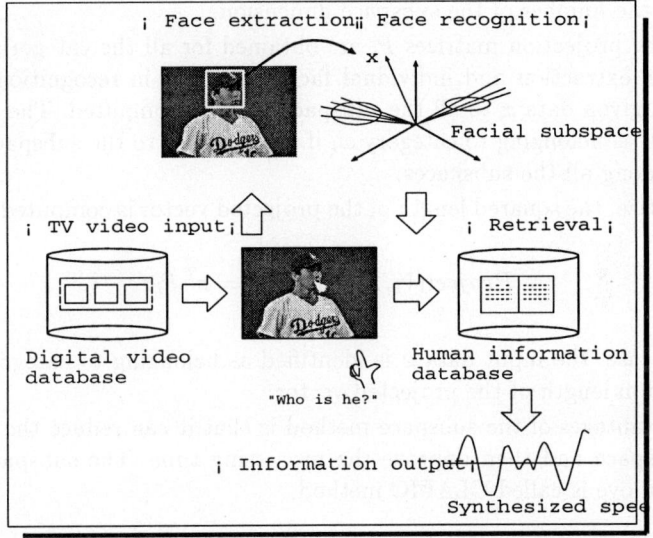

Fig. 1. System organization

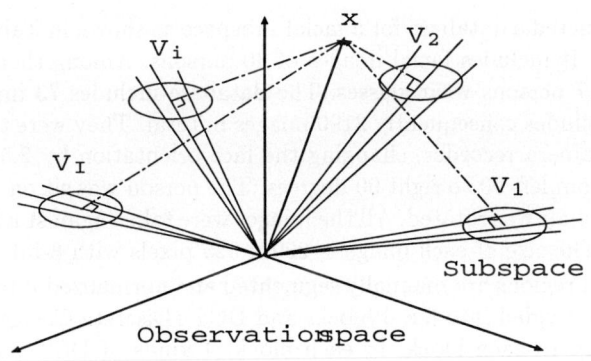

Fig. 2. Subspaces in an observation space

The second step of the subspace method is to compute a distance from an input data x to the subspace V_i. The distance is presented as follows using a projection matrix P_i by which the input data x is projected to the subspace V_i:

$$Dist(V_i, x) = \| x - P_i x \| = \| (I - P_i) x \| \qquad (1)$$

The projection matrix P_i is defined as:

$$P_i = \sum_{k=1}^{r} v_{ik} v_{ik}^T = V_i V_i^T \qquad (2)$$

where r is the number of the subspace dimension.

Once the projection matrices P_i are obtained for all the categories (a facial subspace in extraction and individual facial subspace in recognition), the distance from given data x to all the subspaces can be computed. The input data is identified as belonging to category ω_i if the distance to the subspace V_i is the shortest among all the subspaces.

In practice, the squared length of the projected vector is computed as follows:

$$Project(V_i, x) = \| P_i x \|^2 = x^T P_i x \qquad (3)$$

In this case, the input data x is identified as belonging to category ω_i with the maximum length of the projected vector.

The advantages of the subspace method is that it can reduce the dimension of feature space and then improve the processing time. The subspace method described above is called CLAFIC method.

5 Face Extraction

5.1 Database for a Facial Subspace in Extraction

We constructed a database for a facial subspace as shown in Table1 to extract facial regions. It includes facial images of 30 persons. Among them, 28 persons are male and 7 persons wear glasses. The database includes 73 images for each person and includes consequently 2190 images in total. They were taken by 8mm home video camera recorder, changing the face orientation by 2.5 degrees over 180 degrees from left 90 to right 90 degrees. The person was sit on a chair in upright and the chair was rotated. All the images were taken against a homogeneous background. The size of each image is 290 × 325 pixels with 8-bit gray values.

Then facial regions are manually segmented and normalized into 30 × 30 pixel size. They are divided into 6 × 6 blocks and DCT (Discrete Cosine Transformation) is applied to each block. In each block, 4 kinds of DCT parameters are extracted as shown in Fig.3. In the figure, The DCT parameter 1 indicates the DC component. The DCT parameter 2 and 3 corresponds with horizontal and vertical AC components computed by averaging the absolute value of horizontal and vertical AC components within the numbered regions respectively. The DCT parameter 4 indicates the higher order AC component computed by averaging the absolute value of the higher order DCT components.

Therefore each facial image is presented as a 144 (6 × 6 × 4) dimensional vector. These 144 dimensional vectors are called facial data which are free from slight changes caused by camera angle, lighting and facial expressions and are used to construct facial subspace for face extraction.

Table 1. Database for a facial subspace

Number of persons	30
Male : female	28 : 2
Glasses : non-glasses	7 : 23
Maximum orientation	±90 degrees
Orientation difference	2.5 degrees
Image size	290 × 325 pixels
Gray level	256
Feature extraction	144 DCT parameters
Number of blocks	6 × 6

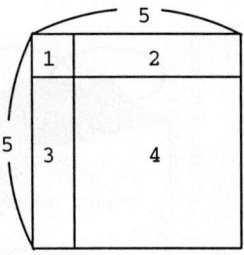

Fig. 3. DCT parameters

5.2 Extraction Method

In order to extract facial regions regardless of its orientation, one facial subspace with 180 degrees orientation range (-90 degrees to +90 degrees from the frontal position) is constructed using the face extraction database. The 13 facial data taken at every 15 degrees among 73 facial data were collected from each person to train the facial subspace. In total, 390 facial data (the 13 images × 30 persons) were used to construct one facial subspace.

In order to extract human faces with various sizes around the point specified by a user on a TV display, the input image is scanned by the window. The size ratio of the window is fixed to 10:13 and the size is changed at 19 levels from 30 × 39 to 100 × 130. At each level and at each scanning position, the window image is divided into 6 × 6 blocks and 144 DCT features are extracted. The window image (144 dimensional vector) is projected to the facial subspace and projection amount is computed. The projection amount obtained from 19 size levels are summed at each position. Fig.4 shows the bird-eye view of the projection amount summed over 19 size levels. In the figure, the bottom rectangle indicates the input image and the vertical axis to the input image indicates the projection amount. Some peaks can be seen in the figure as shown by the circles. The window images with the local peaks of the projected amount are regarded as the facial regions.

6 Extraction Result

6.1 Database for Face Extraction Evaluation

Twenty seven input images with 320 × 240 pixels of 8 bit gray levels were tested for face extraction. They were taken from TV news program and dramas. On the testing image, more than two faces with different sizes are included. Table2 shows the number of the testing images in terms of how many faces are included in the image. The back ground is complicated because the image is taken form real TV news and dramas.

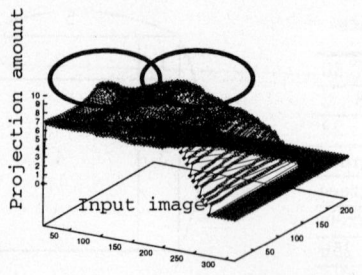

Fig. 4. Plot of the projection amount

Table 2. Content of evaluation data

One face	10 images
Two faces	13 images
Three faces	2 images
Four faces	2 images
Total number of faces	50 faces

6.2 Evaluation Method

The following three values are evaluated; extraction rate, false alarm and mis-extraction rate.

$$\text{Extraction rate} = \frac{\text{Number of correctly extracted faces}}{\text{Number of true faces}} \quad (4)$$

$$\text{Mis-extraction rate} = \frac{\text{Number of faces which were not extracted}}{\text{Number of ture faces}} \quad (5)$$

$$\text{Falise alarm} = \frac{\text{Number of falsely extracted faces}}{\text{Number of true faces}} \quad (6)$$

Here "the correctly extracted faces" are decided through comparison with true facial regions segmented by human.

6.3 Face Extraction Experiment

We carried out face extraction experiment for 27 testing images. The extraction result is shown in Table3. The highest extraction rate 92.0% was obtained by facial subspace with one dimension. The extraction rate seems to be a little low. But in the application of human information retrieval, it is not a major problem because the face position is roughly specified by a user.

7 Face Recognition

7.1 Recognition Method

Face recognition is carried out by constructing a facial subspace for each person as shown in Fig.2. Multiple facial images x_{ik} are taken for each person from TV news and dramas. Here i indicates the person and k indicates the image number. From these facial images, the correlation matrix is computed and the projection matrix P_i is finally obtained through eigenvalue decomposition. When an input image is given, the projection amount shown in Eq. (3) is computed between the input image and facial subspaces of persons. The input image is identified as the face of the person with maximum projection amount.

Table 3. Face extraction result (%)

Subspace dimension	Extraction rate	Mis-extraction rate	False alarm
1	92.0	8.0	32.0
5	84.0	16.0	40.0
10	80.0	20.0	44.0

7.2 Database for Facial Subspaces in Recognition

We constructed a database for facial subspaces as shown in Table4 to recognize facial regions. It includes facial images of 100 famous persons. Among them, 67 persons are male and 15 persons wear glasses. The database includes 8 images for each person and consequently 800 images in total. They were taken from TV news and dramas. The size of each image is 100 x 130 pixels with 8-bit gray values.

Then they are divided into 6×6 blocks and 144 DCT parameters are extracted as described in section5.1. These DCT images are called facial data which are free from slight changes caused by camera angle, lighting and facial expressions and are used for facial subspace construction.

Table 4. Database for facial subspaces

Number of persons	100
Male : female	67 : 33
Glasses : non-glasses	15 : 85
Maximum orientation	depending on person
Image size	100 × 130 pixels
Gray level	256
Feature extraction	144 DCT features
Number of blocks	6 × 6

7.3 Facial Model Construction

Using the TV video database described in Table4, we constructed the facial subspaces for 100 famous persons. The number of images used to construct the facial subspace for each person was 8 images. This is due to the limitation of the orientation range included in the TV video database.

8 Information Retrieval

8.1 Human Information Database

Human information database is available in 5 CD-ROMs. It includes about 100,000 famous persons. When the system recognizes a face specified by a user among 100 persons, it can retrieve his information from this human information database.

8.2 Retrieval Experiment for News Video Data

Table5 shows the total evaluation result in closed data from face extraction, face recognition to information retrieval for 30 persons among 100. In the table, "automatic" means that the faces are extracted automatically after position is specified by human. On the other hand, "manual" means that extraction of facial regions is completely performed by human. From the table, automatic face extraction is almost same as manual extraction. At present, 96.0% information retrieval rate is obtained.

For the open data (more than one year apart from the data used for model construction) and semi-open data (1-5 days apart from the data used for model construction), the result is shown in Table6. The information retrieval rate for open or semi-open data is lower than that for the closed data. The reason is that the face orientation is wider in the recognition evaluation of faces for open data as well as the time lapse which changes their hair style and makeup.

Table 5. Information retrieval rate for closed data(%)

Automatic	Manual
96.0	100

Table 6. Comparison of automatic information retrieval rate(%)

Closed	Semi-open	Open
96.0	60.0	10.0

9 Conclusion

In this paper, we described the system which can retrieve human information after face position was specified by the user. The basic techniques of face extraction and recognition based on the subspace method are also described. The simple experiment of the face extraction, recognition and retrieval was carried out to investigate the effectiveness of the techniques. We are now integrating the system toward more robust information retrieval. Future work will be to improve the face recognition rate.

References

1. Y.Ariki, N.Ishikawa: Integration of Face and Speaker Recognition by Subspace method. ICPR96. (1996) C456-460
2. Y.Sugiyama, Y.Ariki: Facial region tracking and Recognition by Subspace Projection. VSMM96. (1996) 225-230
3. Y.Ariki, N.Ishikawa and Y.sugiyama: Extraction and Recognition of Facial Regions by Subspace Method. ACCV95. (1995) III-733-III-742
4. E.Oja: Subspace Methods of Pattern Recognition. Research Studies Press. England. (1983)
5. M.A.Turk and A.P.Pentland: Face Recognition using Eigenfaces Proc. of IEEE Conference on Computer Vision and Pattern Recognition. (1991) 586-591

Converting Facial Expressions Using Recognition-Based Analysis of Image Sequences

Takahiro Otsuka and Jun Ohya

ATR Media Integration & Communications Research Laboratories
2-2 Hikaridai, Seika-cho, Soraku-gun, Kyoto 619-02, Japan
E-mail:otsuka@mic.atr.co.jp, ohya@mic.atr.co.jp

Abstract. A method for converting one person's facial expression into another person's is proposed. The sequence of the feature vector for each expression is modeled by using HMM with the hidden states corresponding to the different muscle conditions (relaxed, contracting, and the end of contraction). The probability of each state is evaluated for each frame and the contraction rate of each muscle is obtained from the probability of each state using a matrix representing the characteristics of other people's expressions. The experiments showed the superior realism of the expression generated by our proposed method.

1 Introduction

We are developing a new communication environment in which individual users located far apart can converse with each other by participating in a "virtual scene". In this virtual scene, the means by which the human body and face are rendered needs to be an important consideration if users are to feel that face-to-face interaction is occurring. There are two fundamental approaches to rendering of the human figure; 1) to regenerate an existing human figure as accurately as possible, and 2) to change (metamorphose) the appearance of a human figure to that of a generated character. An example of this second strategy is the Virtual Kabuki Theater[1], developed by one of the authors as a testbed. In this Kabuki theater (Kabuki is a Japanese traditional play), a user can metamorphose into a Kabuki actor in a virtual scene.

In the Virtual Kabuki Theater, however, the facial expression of a character is produced simply according to the user's facial expression. Therefore, the generated expression is constrained by the user's ability to express himself through facial expression and consequently, it can be difficult to produce such impressive expressions as an actor displays in a drama and cinema. To generate these expressions, it is not enough to be able to detect the facial motion of the user accurately. It is more important to be able to control the facial motion according to the user's emotion and intension. Consequently, it is necessary to be able to determine the user's emotion and intension from facial expressions.

Previous research on facial expression synthesis from analysis [2, 3, 4] has some relationship on this research. In these methods, the displacement of skin or facial muscle was estimated by using image processing. Then, a 3D wireframe

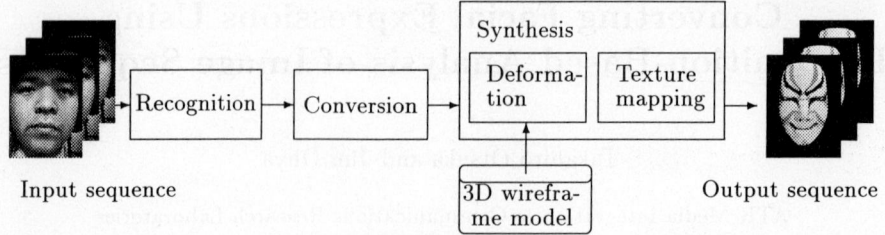

Fig. 1. Structure of the facial expression conversion algorithm

model is deformed to display facial expressions according to the degree of displacement. As the accuracy of the generated expressions with respect to that of the original face was the principal criterion, the conversion to another person's face was not considered there.

2 Facial expression conversion algorithm

In our proposed method, first, the category of facial expressions and its stage in progress are estimated using recognition-based analysis. Then, the estimated category and its stage are converted into the rate of contraction for each muscle of a different character. Finally, the 3D model of the character is deformed in proportion to the contraction rate.

The proposed method for facial expression conversion consists of the three modules shown in Fig. 1 the explanation of which is presented in the following subsections.

2.1 Recognition module

The recognition module is designed to estimate the category and its stage in progress from the input sequences and is divided into three components; image processing, feature extraction and recognition. The image processing component is designed to compute optical flows between two consecutive frames of an input sequence. The processing is applied only to the two regions surrounding the right eye and the mouth to improve efficiency because the facial muscles are concentrated around the eyes and the mouth.

The feature extraction component is concerned with obtaining low-dimensional features representing facial motion characteristic to particular expression such as eye opening, lip corner pulling, and etc. Two-dimensional Fourier transform is applied to two images in which each pixel represents the vertical and horizontal component of the velocity vector or optical flow. The lower spatial frequency coefficients are selected as the component of a feature vector because the principal component analysis of input sequences has revealed a strong contribution of those coefficients on the principal eigenvectors.

The recognition component matches the sequence of feature vectors with the models representing the pattern of each expression. For the recognition of facial expressions, various methods were proposed in computer vision community. For recognition, a simple matching operation was applied to the frame displaying maximum degree of expression. But that information is not enough for our purpose in which the category and stage in progress are also necessary. For this purpose, HMMs are selected because HMMs have a robustness on the temporal fluctuation and a learning capability from symbol (vector) sequences. In addition, HMMs provide the probability of the intermediate states from which valuable information can be obtained.

HMMs are a kind of state transition machine, but the destination of a transition and the output symbol or vector to be generated are decided from probability distributions. Therefore, an HMM is specified by the number of states N and two probability distributions. The transition probability a_{ij} is the probability of being a state S_j at the next time step from the state S_i. The output probability $b_j(V)$ is the probability of generating output vector V while being a state S_j.

HMMs can be simplified by restricting the transitions between two states. For modeling a causal temporal sequence, the left-to-right HMM shown in Fig. 2 is appropriate as both the transitions to itself and states at right-hand side are allowed. In the case of facial expression recognition, the states can be interpreted as the different conditions of the muscles, i.e., relaxed, contracting, and the end of contraction. In the following, the number of states N is assumed to be three and the interpretation of these states S_1, S_2, and S_3 are relaxed, contracting, and the end of contraction respectively. However, increasing the number of states would improve the accuracy of the model.

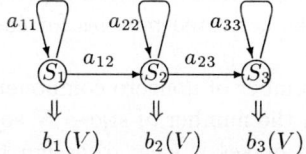

Fig. 2. Structure of left-to-right HMM ($N = 3$)

A continuous function is used to represent the output probability because the preliminary experiments[5] showed a higher recognition rate using a continuous probability density than a discrete probability distribution. The probabilities of the HMMs are estimated by applying Baum-Welch algorithm to the training sequences.

When the sequence of the feature vectors is given, the probability of a state S_i is computed in a temporal sequence using a Forward algorithm which can be expressed as:

$$p_i(t) = b_i(V_t) \sum_{j=1}^{N} p_j(t-1) a_{ji}, \quad t > 1 \qquad (1)$$

As the probability of state S_2 contains the information for the contraction of muscles, only the probability of state S_2 is send to the conversion module.

2.2 Conversion module

The conversion module generates a data set for the contraction rates of the muscles for the person to be modeled in the synthesis module. As the module is supplied only the probability of the contracting state S_2 for each expression, the module needs to evaluate the contraction rate of each muscle from the table or matrix representing the particular person's contraction rate while displaying each expression as:

$$d_i(t) = \sum_j T_{ij} p_2^{(j)}(t), \qquad (2)$$

where d_i denotes the contraction rate of the ith muscle at time t, $p_2^{(j)}(t)$ denotes the probability of state S_2 for the jth expression at time t, and T_{ij} denotes the contraction rate of ith muscle while the person is displaying the jth expression.

The assignment of the matrix T can be simplified using the Facial Action Coding System (FACS) proposed by Ekman and Friesen [6]. FACS is a coding system for facial expressions based on so called Action Units (AU) which are atomic units of facial motion. Each AU corresponds to a certain facial muscle or a set of facial muscles which contract together. FACS has been applied in the animation of the human face[7] utilizing its higher capacity to discriminate facial motion patterns.

In the framework of FACS, the universality of some facial expression was asserted, that is, facial expressions such as anger, disgust, fear, happiness, sadness and surprise are universally recognized across human racial and cultural divides. The FACS representations of these principal facial expressions are shown in Table 1. The primary action and the activated muscles for each AU are tabulated in Table 2.

By applying FACS, the number of nonzero components of the matrix T was reduced to 36 from 174 when the number of states N equaled 3. This reduction of the number of parameters makes it easy to assign the value of the matrix component.

2.3 Synthesis module

In the synthesis module, the 3D wireframe model is deformed according to the contraction rate of each muscle after which a textured surface is mapped on the 3D model. The mode of deformation of the 3D model is divided into the three categories. The first category corresponds to a linear muscle. One terminus of a linear muscle attaches to bone while the other attaches to skin. The contraction of a linear muscle will move the tissue surrounding the skin attached to the muscle toward the other end of the muscle. This motion of the skin was simulated by using a method proposed by Waters[8]. His method assumed that the displacement of skin is falling off as a cosine function from the attached point. But to generate a realistic expression, the parameters specifying the region to be effected must be appropriately selected. Therefore, we derived an equation for the displacement of skin from first principles in physics. The derivation of the

Expression	List of AUs
Anger	AU2+4+5+10+23+24+28
Disgust	AU4+9+10+17
Fear	AU1+2+4+5+7+15+20+25
Happiness	AU1+2+5+6+10+12+13+20+25
Sadness	AU1+4+7+15
Surprise	AU1+2+5+26

Table 1. AU coding for the principal facial expressions

AU1	Inner brow raiser	Frontalis, pars medialis
AU2	Outer brow raiser	Frontalis, pars lateralis
AU4	Brow lowerer	Depressor supercilii
AU5	Upper lid raiser	Levator palebrae superioris
AU9	Nose wrinkler	Levator labii superioris alaeque nasi
AU10	Upper lid raiser	Levator labii superioris major
AU11	Nasolabial furrow deepener	Zygomatic minor
AU12	Lip corner puller	Zygomatic major
AU13	Cheek puffer	Levetor anguli oris
AU15	Lip corner depressor	Depressor anguli oris
AU16	Lower lip depressor	Depressor labii inferioris
AU20	Lip stretcher	Risorius

Table 2. Facial motion for AU and the activated muscle

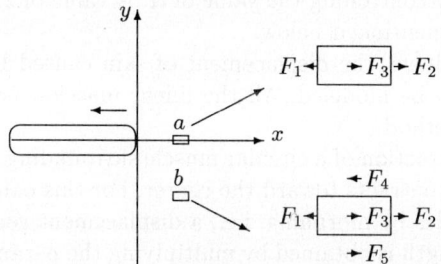

Fig. 3. Elastic model of displacement

equation was as follows. Suppose that xy is a plane on the surface of the skin and a muscle is on the negative x axis with one end attached to the skin at the origin as shown in Fig. 3. Consider the displacement of skin in the case when the muscle is contracting toward the negative infinity of the x axis. The forces acting on a small bar a at position $(x, 0)$ are expressed by the displacement in the x direction $u(x)$ as:

$$F_1 = -Yh\Delta y u(x) \qquad (3)$$
$$F_2 = Yh\Delta y u(x + \Delta x) \qquad (4)$$
$$F_3 = ku(x)\Delta x \Delta y \qquad (5)$$

where Δx, Δy, h are the length, width and height of the bar a, and Y, k are Young's modulus and the force constant of skin. As these forces are balanced, at the limit of Δx being zero, $u(x)$ satisfies the differential equation expressed as:

$$Yh\frac{\partial u(x)}{\partial x} = ku(x) \qquad (6)$$

The solution is derived by solving the differential equation for $u(x)$:

$$u(x) = u_0 \exp(-\frac{k|x|}{Yh}) \qquad (7)$$

In a similar way, displacement in the x direction at the general position (x, y) is derived as:

$$u(x,y) = u_0 \exp(-\frac{k|x|}{Yh} - \frac{k|y|}{Sh}) \qquad (8)$$

where S is the shear modulus. In this case, we assume that the displacement of skin is directed along the x axis. As generally the shear modulus is approximated to be one third of Young's modulus, Eq. (6) can be simplified as follows:

$$u(x,y) = u_0 \exp\{-\alpha(|x| + 3|y|)\} \qquad (9)$$

where α is defined as: $\alpha = \frac{k}{Yh}$. The coefficient α represents skin softness. As the coefficient increases, the effect of the muscle will be localized. The expression can be exaggerated or inhibited by controlling the value of α. A value of 2.0 will be used for α in the experiments mentioned below.

Using the method described above, the displacement of skin caused by the contraction of a linear muscle can be modeled. All the linear muscles listed in Table 2 are modeled using this method.

The second category is the contraction of a circular muscle surrounding either the eyes or the mouth and which contracts toward the center. For this category, we modeled the effect of the muscles by morphing, i.e., a displacement vector is assigned to each vertex and the length is obtained by multiplying the parameters to the vector.

The third category is jaw rotation. The deformation of 3D model for jaw rotation is performed as follows. First, the vertices attached to the mandible (jaw bone) are rotated. Then, the displacement of neighboring vertices are smoothed continuously. Finally, the boundary of the mouth is corrected by taking into account the effect of the tensile force using Eq. 9 as shown in Fig. 5.

3 Experimental results and discussion

To evaluate the proposed method, we compared the proposed method with the previous method implemented in the Virtual Kabuki Theater by converting the subject's facial expression into a Kabuki actor's. As the facial expressions of a real actor have not been analyzed yet, the original expression was mapped to the standard appearance of each expression, that is, all the facial actions in the Table 1 for each expression are to be displayed.

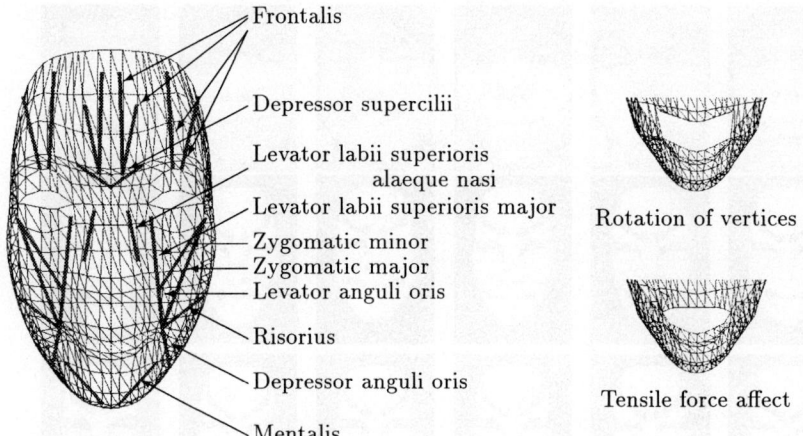

Fig. 4. Linear muscles   **Fig. 5.** Rendering of jaw rotation

In the previous method, the 3D model was deformed by the feature vector directly. The relation between the feature vector and the amount of deformation for each vertex was learned using the genetic algorithm. Several of the results are shown in Fig. 6. In Fig. 6, the last frame of the input sequences is shown at the top, that of the synthesized sequence using the previous method is shown in the middle, and that of the synthesized sequence using the proposed method is listed at the bottom. Although the mouth shapes generated by the previous method for anger, disgust, and sadness are quite similar, the proposed method generated easily distinguishable mouth shapes for these expressions. As the edge of the mouth is almost horizontal, it is difficult to detect the horizontal motion of the mouth accurately. But the proposed method can combine other features such as the motion of the nasolabial furrow to improve the accuracy for estimation of the facial motion.

4 Conclusion

A method for converting one person's facial expression into another person's has been proposed. In this method, the sequence of the feature vector for each expression is modeled by using HMM with the hidden states corresponding to the different muscle conditions (relaxed, contracting, and the end of contraction). The probability of each state is evaluated for each frame and the contraction rate of each muscle is obtained from the probability of each state using a matrix representing the characteristics of other people's expressions. The experiments showed the superior realism of the expression generated by our proposed method especially in the mouth motion of anger, disgust and sadness. More realistic synthesis of facial expressions such as the mouth shape is one of future work.

Anger   Disgust   Fear   Happiness   Sadness   Surprise

Fig. 6. Examples of the last frame (top: input, middle: synthesized by the previous method, bottom: synthesized by the proposed method)

References

1. J. Ohya, K. Ebihara, J. Kurumisawa, and R. Nakatsu, "Virtual Kabuki Theater: Towards the Realization of Human Metamorphosis Systems," *Proc. IEEE International Workshop on Robot and Human Communication*, pp. 416-421, Nov. 1996.
2. D. Terzopoulos, and K. Waters, "Analysis and Synthesis of Facial Image Sequences Using Physical and Anatomical Models," *IEEE Trans. PAMI*, vol. 15, no. 6, pp. 569-579, 1993.
3. I. A. Essa, "Analysis, Interpretation, and Synthesis of Facial Expressions," M.I.T. Media Laboratory, Perceptual Computing Group Report no. 303, 1994.
4. J. Ohya, Y. Kitamura, H. Ishii, H. Takemura, F. Kishino, and N. Terashima, "Virtual Space Teleconferencing: Real-time Reproduction of 3-D Human Image," Journal of Visual Communication and Image Representation (VCIR), vol. 6, no. 1, pp. 1-25, March 1995.
5. T. Otsuka, and J. Ohya, "Recognition of Racial Expressions Using HMM with Continuous Output Probabilities," *Proc. IEEE International Workshop on Robot and Human Communication*, pp. 323-328, Nov. 1996.
6. P. Ekman, and W. V. Friesen, "The Facial Action Coding System," Consulting Psychologists Press, Inc., 1978.
7. S. M. Platt, and N. I. Badler, "Animating Facial Expressions," *Computer Graphics*, vol. 15, no. 3, pp. 245-252, 1981.
8. K. Waters, "A Muscle Model for Animating Three-Dimensional Facial Expression," *Comput. Graphics*, vol. 22, no. 4, pp. 17-24, 1987.

Muscle-Based Feature Models for Analyzing Facial Expressions

Hiroshi OHTA, Hitoshi SAJI, and Hiromasa NAKATANI

Department of Computer Science, Shizuoka University
Johoku 3-5-1, Hamamatsu 432-8011, Japan

Abstract. We propose deformable models for tracking the continual motions of facial features, such as the eyebrows and mouth, in facial images. Directions and ranges of deformations of each facial feature are physically constrained by facial muscles. By using the directions and locations of facial muscles, feature models can be constructed with a few parameters and can be deformed only in the proper range and to the proper direction. The model parameters are obtained from the deformations of facial muscles, hence the proposed model can be easily applied to the anatomical analysis of facial expressions.

1 Introduction

With recent developments in man-machine interface, emotions and thoughts have been studied by analyzing human faces. Many techniques for facial image processing have been proposed, such as extraction of facial regions and facial features. To extract facial features which change their shapes variously, deformable models should be used, and active contour models, such as snakes [1] and deformable template models [2], have been investigated. The results of those methods may have anatomically impossible deformations of facial features. Since facial features are physically constrained by facial muscles [3], such a model that is consistent with anatomical knowledge should be constructed for extracting facial features. For synthesis of facial expressions, models based on anatomical structures were used [4]. In this paper, we use anatomical knowledge for recognition of facial expressions.

We have developed feature models for eyebrows, eyes and mouth [5]. They are deformable models that take into account the locations of muscular attachments and the directions of muscular contractions. Since our model parameters reflect muscular actions, both movements of facial features and changes of facial expressions can be tracked at the same time. Thus, facial expressions can be recognized by analyzing parameters obtained by tracking the movements.

In this paper, we propose to improve those models by selecting the locations of control points and muscular directions more precisely. This paper describes improved models of an eyebrow and a mouth which present distinctive motions of facial expressions.

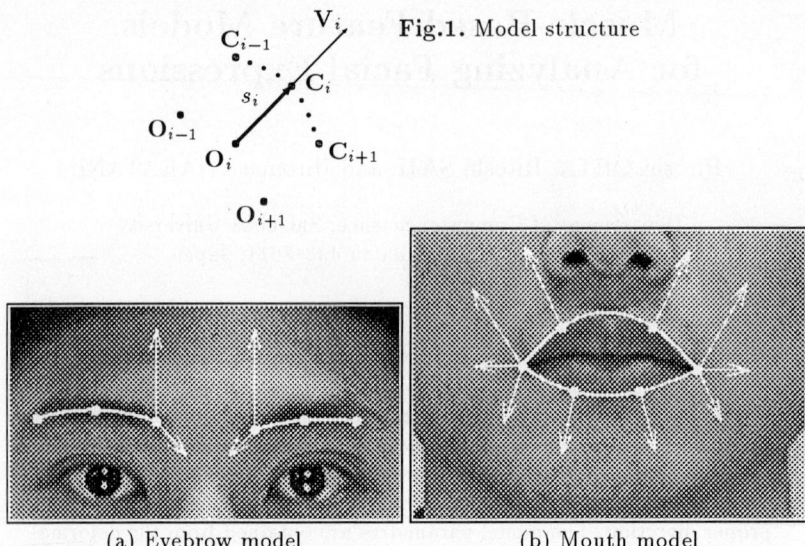

Fig.1. Model structure

(a) Eyebrow model (b) Mouth model

Fig.2. Models based on anatomical knowledge

2 Facial muscles and models

We propose facial feature models based on muscular functions. In this section, we describe relations between facial features and facial muscles, their related facial expressions, and structures of the proposed models.

The fundamental structure of the model is shown in Fig.1. The model consists of a few control points **C** and interpolated points between them. It also consists of starting points **O** and muscular contractile vectors **V**. Points **O** are detected manually from a facial image of a neutral expression. A muscular contractile vector **V** indicates the direction to which a control point **C** moves from a starting point **O**. The model is deformed by a parameter s. The value of s indicates a contractile degree of the corresponding muscle. We normalize the value of s to let $s = 1.0$ on the position of wrinkles appearing in the direction of the corresponding muscle when the muscle contracts to the limit. The heads of arrows in Fig.2 correspond to the positions where $s = 1.0$.

We construct those models individually and apply them to normalized images.

2.1 Eyebrow model

Eyebrows are deformed when facial expressions change, and they clearly represent changes of facial expressions. The muscles, Venter frontalis, M. depressor supercilii, M. corrugator supercilii, and M. orbicularis oculi, surround an eyebrow (Fig.3). We construct an eyebrow model by considering actions of those muscles.

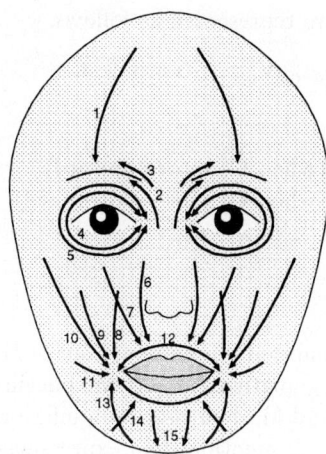

1. Venter frontalis
2. M. depressor supercilii
3. M. corrugator supercilii
4. M. orbicularis oculi (Pars palpebralis)
5. M. orbicularis oculi (Pars orbitalis)
6. M. levator labii superioris alaeque nasi
7. M. levator labii superioris
8. M. levator anguli oris
9. M. zygomaticus minor
10. M. zygomaticus major
11. M. buccinator
12. M. orbicularis oris
13. M. depressor anguli oris
14. M. depressor labii inferioris
15. M. mentalis

Fig. 3. Facial muscles

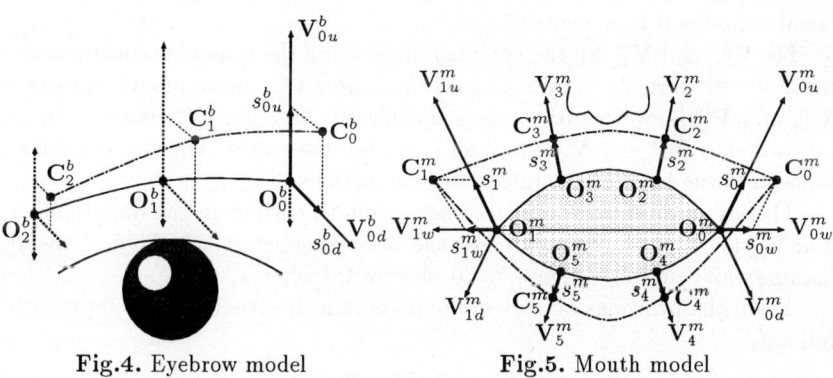

Fig. 4. Eyebrow model   Fig. 5. Mouth model

We locate three control points to control model deformations. They are the inside point $\mathbf{C}_0^b$, the center point $\mathbf{C}_1^b$, and the outside point $\mathbf{C}_2^b$. An eyebrow is quadratically interpolated between those control points. Points $\mathbf{O}_0^b$, $\mathbf{O}_1^b$, $\mathbf{O}_2^b$ are starting positions of control points $\mathbf{C}_0^b$, $\mathbf{C}_1^b$, $\mathbf{C}_2^b$, respectively, when the facial expression is at neutral.

Let $\mathbf{V}_{0u}^b$ be the vector representing the muscular contraction that moves an eyebrow upward, and $\mathbf{V}_{0d}^b$ downward. The size of each vector represents the limit of the contraction. The parameters s_{0u}^b and s_{0d}^b represent the strength of muscular contraction in the direction of $\mathbf{V}_{0u}^b$ and $\mathbf{V}_{0d}^b$, respectively.

Then, the locations of control points $\mathbf{C}_i^b$ are represented as follows:

$$\mathbf{C}_i^b = \mathbf{O}_i^b + w_{iu}s_{0u}^b\mathbf{V}_{0u}^b + w_{id}s_{0d}^b\mathbf{V}_{0d}^b \qquad i = 1, 2, 3, \qquad (1)$$

where we set the weights w as follows:

$$\begin{cases} w_{0u} = 1.0, w_{1u} = 0.9, w_{2u} = 0.4 & \text{(when } \mathbf{C}_i^b \text{ moves upward)} \\ w_{0u} = w_{1u} = w_{2u} = 1.0 & \text{(when } \mathbf{C}_i^b \text{ moves downward).} \end{cases}$$
$$w_{0d} = w_{1d} = w_{2d} = 1.0$$

2.2 Mouth model

We construct a mouth model by taking account of muscular actions of M. levator anguli oris, M. zygomaticus minor, M. zygomaticus major, M. buccinator, M. orbicularis oris, M. depressor anguli oris, and M. depressor labii inferioris. Since the movements of mouth corners reflect one's emotions and expressions, such a mouth model that is suitable for tracking mouth corners is constructed.

Figure 5 shows our mouth model, where we locate control points $\mathbf{C}_0^m$, $\mathbf{C}_1^m$ on mouth corners, $\mathbf{C}_2^m$, $\mathbf{C}_3^m$ on an edge of an upper lip, and $\mathbf{C}_4^m$, $\mathbf{C}_5^m$ on an edge of a lower lip. A mouth is cubically interpolated between those control points. Points $\mathbf{O}_i^m (i = 0, 1, \cdots, 5)$ are starting positions of control points $\mathbf{C}_i^m$ when the facial expression is at neutral.

Let $\mathbf{V}_{0u}^m$ and $\mathbf{V}_{1u}^m$ be the vectors representing the muscular contractions that move mouth corners outside upward; $\mathbf{V}_{0w}^m$ and $\mathbf{V}_{1w}^m$ move mouth corners wide; $\mathbf{V}_{0d}^m$ and $\mathbf{V}_{1d}^m$ move mouth corners downward; $\mathbf{V}_2^m$ and $\mathbf{V}_3^m$ move an upper lip upward; and $\mathbf{V}_4^m$ and $\mathbf{V}_5^m$ move an lower lip downward. The size of each vector represents the limit of the muscular contractions.

The strength of the force that moves mouth corners in the directions of $\mathbf{V}_{iu}^m$ and $\mathbf{V}_{id}^m$ ($i = 0, 1$) is represented by the same parameters, s_i^m. That is, the mouth corners move upward when $s_i^m \geq 0$, downward when $s_i^m < 0$.

Then, the positions of control points of mouth corners can be represented as follows:

$$\mathbf{C}_i^m = \begin{cases} \mathbf{O}_i^m + s_{iw}^m \mathbf{V}_{iw}^m + s_i^m \mathbf{V}_{iu}^m & (s_i^m \geq 0 \text{ :upward}) \\ \mathbf{O}_i^m + s_{iw}^m \mathbf{V}_{iw}^m - s_i^m \mathbf{V}_{id}^m & (s_i^m < 0 \text{ :downward}) \end{cases} \qquad i = 0, 1. \qquad (2)$$

The positions of control points of an upper and lower lip can be represented by

$$\mathbf{C}_i^m = \mathbf{O}_i^m + s_i^m \mathbf{V}_i^m \qquad i = 2, 3, 4, 5. \qquad (3)$$

The values of the strength parameters $s_i^m, (i = 0w, 1w, 2, 3, 4, 5)$ are positive when the corresponding muscles contract, while they are negative when M. orbicularis oris contracts.

3 Matching between facial features and models

A set of parameters s, two for an eyebrow and eight for a mouth, determines a shape of a feature. By changing those parameter values and finding the best-fit shape of a feature, we match the models to facial images.

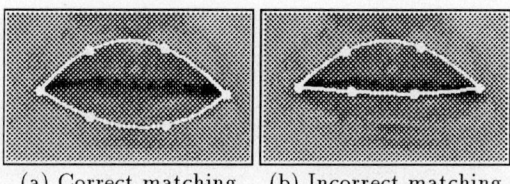
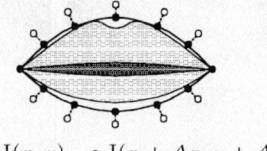

(a) Correct matching   (b) Incorrect matching   • $I(x,y)$ ○ $I(x+\Delta x, y+\Delta y)$

Fig.6. Matching of mouth model   **Fig.7.** Outside edge

3.1 Matching with the eyebrow model

We represent the eyebrow model of s_{0u}^b, s_{0d}^b by a set of x-y coordinates:

$$\mathbf{M}^b(s_{0u}^b, s_{0d}^b) = \{(x_i, y_i)\}, \tag{4}$$

where i represents the number of interpolated points between control points.

In the facial image, an eyebrow has low intensity generally. Therefore, we must detect proper parameter values that give such (x_i, y_i)'s that minimize the following total sum in the neighborhood of the model.

$$\sum_i \sum_{\Delta x} \sum_{\Delta y} I(x_i + \Delta x, y_i + \Delta y), \tag{5}$$

where Δx, Δy are ranges of model's neighborhood, and $I(x,y)$ is an intensity of pixel (x,y).

3.2 Matching with the mouth model

We represent the mouth model of s_{0w}^m, s_{1w}^m, s_0^m, s_1^m, s_2^m, s_3^m, s_4^m, and s_5^m by a set of x-y coordinates:

$$\mathbf{M}^m(s_{0w}^m, s_{1w}^m, s_0^m, s_1^m, s_2^m, s_3^m, s_4^m, s_5^m) = \{(x_i, y_i)\}. \tag{6}$$

Since the mouth model is a contour model, we must detect proper parameter values that give such (x_i, y_i)'s that maximize the following total sum.

$$\sum_i \left(I'(x_i, y_i) + I(x_i + \Delta x, y_i + \Delta y) \right), \tag{7}$$

where $I'(x,y)$ is the derivative of an image $I(x,y)$, and $I(x_i + \Delta x, y_i + \Delta y)$ is an intensity of an out point at a distance from (x_i, y_i) (Fig.7). The mouth model must match an outside edge of lips (Fig.6(a)). Therefore, we use the exterior intensities $I(x_i + \Delta x, y_i + \Delta y)$ not to match the model with an inside edge of lips (Fig.6(b)).

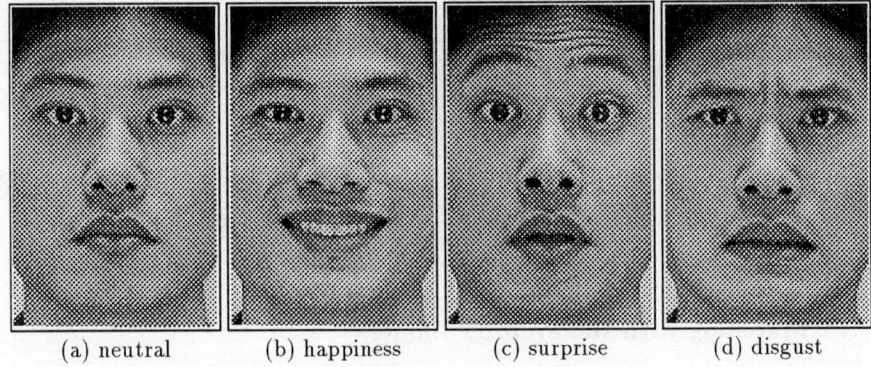

(a) neutral (b) happiness (c) surprise (d) disgust

Fig. 8. Example of facial expressions

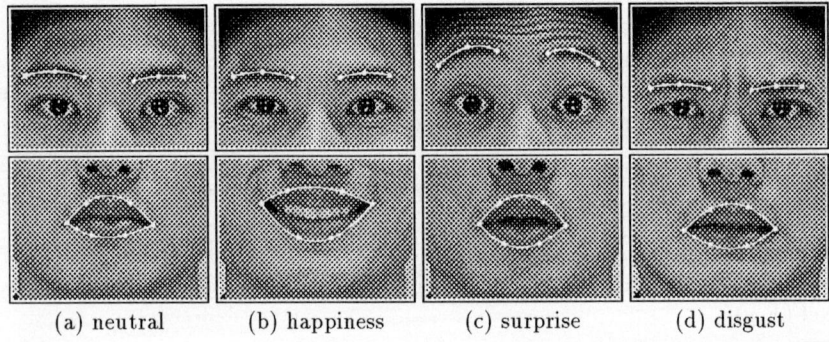

(a) neutral (b) happiness (c) surprise (d) disgust

Fig. 9. Results of tracking the eyebrows and mouth

4 Experiment

We perform an experiment for matching the proposed models with facial image sequences. The size of an image is 200 × 300 pixels. The number of interpolated points is 30 in an eyebrow and 120 in a mouth. The size of neighborhood in Eq. (5) is set to 5×5 pixels. The distance between (x_i, y_i) and $(x_i+\Delta x, y_i+\Delta y)$ in Eq. (7) is 4 pixels.

We take the typical facial expressions [6], "happiness", "surprise", and "disgust", in sequence using a video camera (Fig.8), and track motions of facial features. The image sequences are recorded every 1/30 seconds and 120 images are recorded. Images are then normalized with respect to position, scale, and orientation by measuring the distance and angle between two pupils in a face.

Before matching, we locate in the first frame starting points O_i^b of the eyebrow model and O_i^m of the mouth model. The models are also given the muscular contractile vectors, $\mathbf{V}_i^b$ and $\mathbf{V}_i^m$, from the locations of wrinkles in a face. Such contractile vectors are shown as arrows in Fig.2. Then, points between the control points are interpolated and the model is completed for the first frame. For the consecutive frames, the system repeatedly detects the most suitable parameter values and track the facial features. Figure 9 shows the facial features tracked by our method.

Transitions of parameter values are shown in Fig.4. The horizontal axis is the frame number of images, and the vertical axis is the parameter value. In Fig.4, we can see that eyebrow movements influence facial expressions. In sequences of "surprise", the parameter s_{0u}^b which moves eyebrow upward is increasing. In sequences of "disgust", the parameter s_{0d}^b which moves eyebrow inside downward is increasing while the parameter s_{0u}^b is decreasing. Transitions of parameter values are consistent with facial expressions.

In the mouth model, the parameters s_0^m and s_1^m which move mouth corners upward show expressions of "happiness" and "disgust" clearly. We can see that the model can track movements of the mouth corners correctly. The parameters s_2^m and s_3^m which move an upper lip upward are consistent with lip movements in "happiness".

The parameters s_{0w}^m and s_{1w}^m which move the mouth corners wide have no large values in "happiness" and "disgust," although the mouth corners are stretched wide. This is because those movements are caused mainly by contraction of M. zygomaticus major and M. depressor anguli oris rather than M. buccinator. The parameters s_{0w}^m and s_{1w}^m finely adjust the position of control points which move in the directions of s_0^m and s_1^m. Therefore, s_{0w}^m and s_{1w}^m do not represent exactly the degree of the contraction of M. buccinator by themselves. We can still distinguish the facial expressions by the parameter set.

5 Conclusions

In this paper, we have proposed facial feature models based on the facial muscles, and we have tracked the facial features by restricting the model deformations. Experimental results show that the parameters concerning eyebrows, mouth corners and an upper lip represent the changes of facial expressions effectively.

Since the proposed models are constructed by the parameters based on the muscular functions, both the movements of facial features and changes of facial expressions can be tracked at the same time. By giving initial values to the models individually, it has become easy to track features that vary from person to person. However, facial muscles are complex, and the movements interfere with each other. Future work will focus on the construction of such models that can reflect the mutual interference between muscles.

The emphasis of this paper was on tracking the facial features. Therefore we manually set the initial positions of the feature points, leaving model generalization techniques aside. Such problems of early processing stages are also left for our future work.

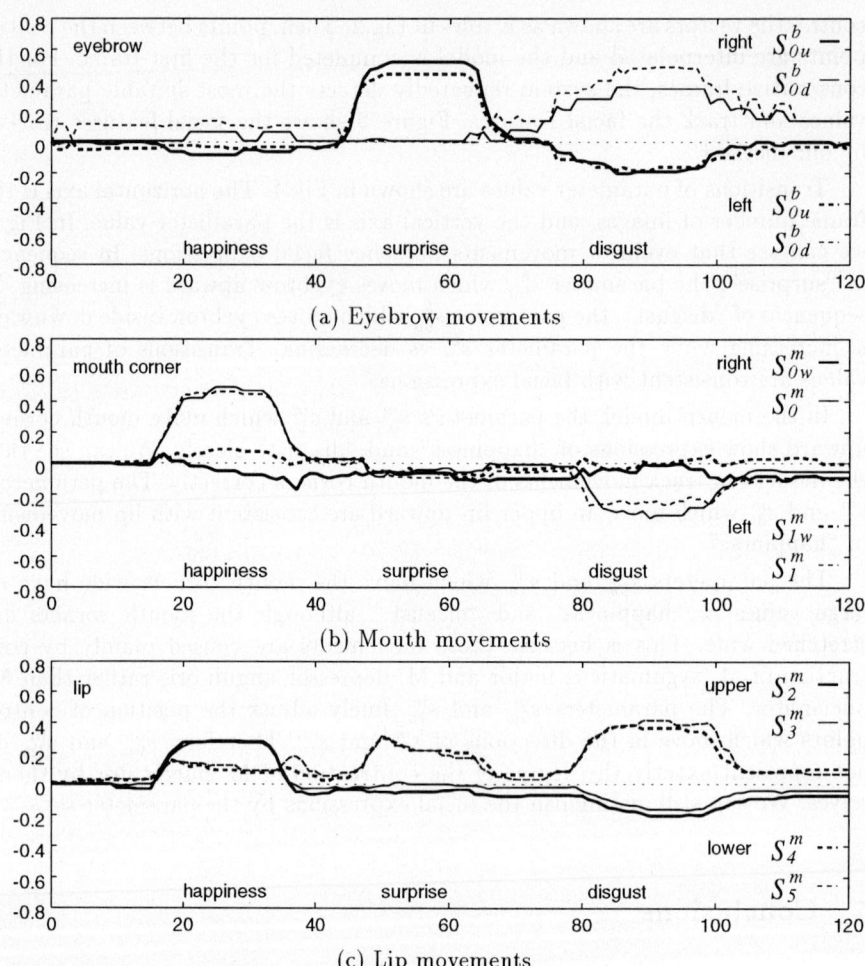

Fig.10. Transitions of parameter values

References

1. Kass, M., Witkin, A. and Terzopoulos, D.: Snakes: Active Contour Models. Int'l J. Computer Vision. **1** (1988) 321-331
2. Yuille, A. L., Hallinan, P. W. and Cohen, D. S.: Feature Extraction from Faces Using Deformable Templates. Int'l J. Computer Vision. **8** (1992) 99–111
3. Parke, F. I. and Waters, K.: Computer Facial Animation. A K Peters, Ltd. (1996) 19–54
4. Terzopoulos, D. and Waters, K.: Analysis and Synthesis of Facial Image Sequences Using Physical and Anatomical Models. IEEE Trans. Patt. Anal. Machine Intell. **15,6** (1993) 569-579
5. Ohta, H., Kimura, A., Saji, H. and Nakatani, H.: Tracking the Motions of Facial Components Using the Muscle-Based Model. Proc. of 3rd Korea-Japan Joint Workshop on Computer Vision. (1997) 133–138
6. Ekman, P. and Friesen, W. V.: Unmasking the Face. Prentice-Hall Inc. (1975)

A Morphological Method for Moving Object Segmentation and Posture Recognition

Yi Li*, Songde Ma, Hanqing Lu
National Lab. of Pattern Recognition
Automation Institute, Chinese Academy of Sciences
P.O.Box 2728,BeiJing,100080, P.R.China
*e-mail:liyi@prlsun8.ia.ac.cn

Abstract
Based on a parallel application dealing with the segmentation and posture recognition of 2-D figure moving over a relatively stationary background from a monocular video frame flow, this paper presents a general description and some important details of a coarse-to-fine morphological segmentation and matching method, which uses a multiscale morphological image filter and a "spore" element. Due to insensitivity to noise, flexibility and symmetricity of morphological operations, this method is robust enough and can be easily parallelized.

1. Introduction.

Motion segmentation is by no means a simple problem. Although there are a lot of documents in the literature[1,3,6,7,9,14], a robust, stable and real-time solution on this issue has not yet been presented. A chaotic approach using the Markov Random Fields[3,11] has provided a general framework, but often it is much slow to solve this kind of optimization problems, and, while the solution is very sensitive to the choice of chaotic model and parameters, many unacceptable premises such as the rigidity assumption must be met. Also there are many approaches[1,6,7] applying to partial aspects of this issue, and interesting results have been reached , but the combination of these approaches seems impossible.

Here we do not intend to pose another framework, but provide a simple but effective method for the most comprehensive application of motion segmentation and recognition: the segmentation and recognition of 2-D human figure from a relatively stationary background, which is the common situation of many security systems. The goal of our method is to segment the moving figure from the monocular video frame flow as accurately as possible and then recognize the actual meaning of action of the figure such as running, crouching, walking, moving closely or far away, etc. Many research efforts have been concentrated on this question. Methods based on adaptive extraction of background[6] are suitable for keeping track of slow changes of illumination, but their efficiency is impaired in the presence of abrupt changes [2] (due to reflection, shadow, shock, etc.). The detection of the changes of edges information in image frame flow [2] are more robust under practical conditions, however from the detected moving edge set which is incomplete to form a shape, at the same time contains many trivial lines, action recognition work seems difficult. An interesting approach based on a BP neutral network [15] gives consideration on many important points, but its segmentation condition is too ideal. [8] provides a good summary of the present framework research work, nevertheless there are too many uncertainties in this kind of methods for them to be applied.

The 2-D image of a moving person is typically not a rigid, but rather a morphological problem. This is the intrinsic reason why definite segmentation methods make a bad show when process this kind of problem. Furthermore, it is well-known that under many circumstances the failure of segmentation stems from the lack of knowledge of the objects been segmented. So, here we present a motion segmentation method based on morphological compensation and a posture matching method based on multiscale morphological filter[12,13]. To simplify our discussion, we make the following assumptions:

1. Only one human movement will be considered.
2. We only discuss gray-level video frames.
3. Human movement is the only reason of the conspicuous image change in the scene.
4. All motion of the body parts, such as legs and hands, can be approximated by planar motions

2. Morphological Compensation Based Motion Segmentation

To estimate the motion of objects and then segment them from a sequence of consecutive video frames, the traditional method is firstly carrying out the local or global matching of gray level or color data, through which the optical flow field can be ascertained; then calculating the motion parameters and segmenting the moving objects.[7,14]. Obviously the matching operation is an ill-pose problem whose performance and stability can not be ensured. Furthermore, this method is time consuming, and can not be parallelized efficiently.

The most remarkable factors effecting the correctness of segmentation include quantitative noise, shock, objects' shadow, which will cause false result of global or local optical flow fields; and the overlapping of regions containing same kinds of gray level in background and objects, which will cause the lost of optical fields. If the rigidity assumption can be met, we can get better result, but usually most moving objects are non-rigid. To deal with all these obstacles, here we provide a simple approach.

Suppose I_{t1}, I_{t2} are two frames of a video flow which contain a same object and relatively stationary background; [t1,t2] denotes a time period, during which the object makes detectable motion. If we compare I_{t1} and I_{t2} by simply subtract them, under ideal condition we can get a pro-set of the adumbration of the object(See Figure 1, we call it the 1-order differential image), and as the period of [t1,t2] is shorter, the difference between the pro-set and the real adumbration is less. Furthermore, two consecutive pro-set, if needed, can help us determine the direction of the motion. This fact indicates that, if we can process the comparison and the analysis fast enough, we can accurately do the segmentation.

Considering the uncertainties discussed above, as dt=t2-t1 is very small, the effects of shock and shadow to the comparison express as high-frequency components, which, together with the quantitative noise, can be effectively eliminated by band-limited smoothing filter. We can also use a adaptive threshold mechanism[6] which, takes in account of local condition to determine the local comparison threshold, to ameliorate our result. To ensure the correctness of the detected motion region(we do not insist on the completity of them), we can use a relatively high threshold. The final result of above processing is shown in Figure 2.

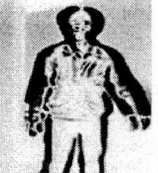

Figure 1. One-Order Differential Image.

Figure 2. 0-1 Adumbration Mask after Adaptive Band-Limited Smoothing Filtering.

Furthermore, we can use the erosion operation[4] with a fixed-size ball element to eliminate susceptible isolated flecks, as well as erode the superfluous edge region caused by the difference operation. Now the main problem we face is how to compensate the adumbration mask. Naturally this work need the 2-D adumbration knowledge of the object. Here we bring forth a "spore" element operator. It is the basic morphological element of the segmentation compensating and multiscale matching.

Define two finite integer sequence { r_i, i=0,...k-1} and { c_i, i=0,...k-1} as:

$$r_i = 2^i ; c_0 = 0, c_i = \sum_{j=1}^{i} r_{k-j};$$

the 2-D power-2 k-level (k≥1) "spore" element S_k^2 is defined as a 2-D integer dot set {(x, y)}, where,

$$\forall (x, y) \in S_k^2, \exists c_i \in C, i \in [0,1, ..., k-1], (x \pm c_i)^2 + y^2 <= (r_i)^2 \text{ or}$$
$$x^2 + (y \pm c_i)^2 <= (r_i)^2.$$

Figure 3 shows a 2-D power-2 3-level "spore" element S_3^2. Although the performance of this element is under further theoretical discussion, clearly when using the "spore" element to do morphological operation such as dilating or closing, it can on the one hand compensating and expanding the given image, on the other hand maintain enough details of the image. Using S_3^2 and the logical OR and AND operation as the morphological pixel operation to do the closing operation on the raw adumbration mask acquired in Figure 2, then using edge conditions to fill the holes in the mask, the result is shown in Figure 4. As we can see, this mask has roughly traced out the accurate adumbration of the object with little invalidate regions. Certainly more complex "spore" elements S_k^p can be designed, to achieve better result, the choice of parameters k and p is related to the parameters used by the image filter.

Conclusively we can denote the above motion segmentation approach as

$$S_3^2 \; close_m \; (B_2 \; erode_m \; F(I_{t1}, I_{t2})),$$

where S_3^2 denotes the "spore" element; $close_m$ denotes the closing morphological operation; B_2 denotes the 2-level "ball" element; $erode_m$ denotes the "erosion" morphological operation; F denotes the adaptive differential band-limited smoothing filter.

Figure 3. A S_3^2 element.   Figure 4. Segmentation Mask after Compensation by S_3^2

Conceptually, the use of this closing operation is a kind of knowledge about the adumbration of the object, i.e. continuity and smoothness. The morphological closing operation using the "spore" element is an intuitive and simple way to apply these knowledge.

3. Multiscale Morphological Matching

To recognize the posture of the adumbration mask acquired above, we introduce a round-corner rectangle and joint based 2-D model[16] of human posture, and a multiscale morphological matching method, which, in fact, is a coarse-to-fine relaxation method using definite morphological error constraint, but is much easier to handle.

Under the assumptions mentioned in Section 1, we can model the human body structure by 14 round-corner rectangles with some of them ocluded in some postures. Formally we can represent this model in an object-centered normalized coordinate system, as shown in Figure 5.

A round-corner rectangle **R** can be exactly described as a 9-elements array:

R =(x_0, y_0, w, h, $r_1, r_2, r_3, r_4, \alpha$),

where its components are:

(x_0, y_0) is the position of center point of the rectangle, defined in unit foursquare;

w is the width, $0 \leq w \leq 1$;

h is the height, $0 \leq h \leq 1$;

r_1, r_2, r_3, r_4 are radius of the four quarter-circle at corners, $0 \leq r_1, r_2, r_3, r_4 \leq 0.5$;

α is the angle between the axes of the rectangle and the x-axis, $0 \leq \alpha \leq \pi$.

A human body H can be represented as 14 round-corner rectangles, with constraints on the rectangle parameters and between them, i.e.:

H=(R_i, i∈ [0,1, ...,13]), where

R_0 is the trunk. The whole model is centered around R_0 because it should never disappear in any posture. Main constraints on its parameters include that, its length should be less than 0.5, width less than 0.25, angle between $\frac{1}{6}\pi$ and $\frac{5}{6}\pi$, etc.; R_1 is the head; R_2, R_3, R_4, R_5 are the four parts of two arms; R_6, R_7, R_8, R_9 are the four parts of two legs; R_{10}, R_{11} are two hands; R_{12}, R_{13} are two feet;respectively have constraints on their parameters.

The constraints between the rectangle parameters include location and adjacent relations, angles between adjacent rectangles such as the angle between two parts of a arms should between 0 to $\frac{5}{6}\pi$, and so on. All possible values of the parameters, restricted by all the constraint conditions, construct the search space of matching; a special parameter set, together with the relationship between them, constructs a kind of posture; a serial of consecutive posture constructs an action.

Based on this model, the procedure of our multiscale morphological relaxation matching method is:

among large parameter ranges, using a large scale morphological filter[12,13] to process both the models and the adumbration mask acquired in section 2; selecting the most similar one; dividing its parameter ranges, and doing the matching using smaller scale filter; until the model having desired precision is acquired.

The morphological filter we select is the dilation operation using the "spore" element S_k^p, whose parameter k is defined as the scale of the filter. The reason is the fact that, when using a monotonic similarity operator, under some loose conditions, there exists a causality of maximum similarity in morphological scale space using dilation, i.e.:

if S denotes a monotone similarity operator; $0 \leq S(I_1, I_2) \leq 1$ denotes the similarity between two image I_1, I_2 of same dimension; E_1, E_2 are same element in different scale; $dilate_m$ denotes the dilation operation, under some loose conditions, there will be

$S((E_1\ dilate_m\ I_1), (E_1\ dilate_m\ I_2)) \leq S((E_1\ dilate_m\ I_1), (E_1\ dilate_m\ I_3))$
provided
$S((E_2\ dilate_m\ I_1), (E_2\ dilate_m\ I_2)) \leq S((E_2\ dilate_m\ I_1), (E_2\ dilate_m\ I_3))$.

If this causality of maximum similarity does not maintain between different scales, we can not ensure the method we discussed here can get the optimal matching result. The "spore" element is the just element possessing many good attributes.

The similarity operator is designed under following considerations: first, it should be a monotonic operator; next, because we have choose the segmentation method in such a way that the acquired adumbration mask will have much less regions which the object does not contain(we call them positive error) than the lost ones(we call them negative error), the positive error should cause greater penalty on the similarity than negative ones.

When we use a set of parameters and constraints to construct a model, it will also form a 0-1 adumbration mask. If $card_{non-zero}$ denotes the operator counting non-zero elements of its argument set, we can simply design the similarity operator under a certain scale as:

$S(I_1, I_2) = e^{-4p-n} \bullet (1-p-n)$; where

I_1 is the 0-1 image of the acquired adumbration mask dilated by "spore" element of specific scale;

I_2 is the 0-1 image of the model adumbration mask using a specific parameter set and dilated by the "spore" element of specific scale;

p= $card_{non-zero}(I_1 \cap I_2^c) / card_{non-zero}(I_1 \cup I_2)$ denotes the positive error,

n= $card_{non-zero}(I_2 \cap I_1^c) / card_{non-zero}(I_1 \cup I_2)$ denotes the negative error.

This definition has the following attributes, which should be met due to common sense:

1. $S(I_1, I_2)=0 \Leftrightarrow p+n=1$, $S(I_1, I_2)=1 \Leftrightarrow p+n=0 \Leftrightarrow$ (p=0) and (n=0);

2. p has more effect on $S(I_1, I_2)$ than n;
3. $S(I_1, I_2) \neq S(I_2, I_1)$.

Using these definitions, we can do the model searching work in a effective way: First, zooming the standard posture models(facade, flank, crouching, etc.) according to the size of the acquired adumbration mask, dilating both the models and the acquired mask using "spore" element of largest scale, selecting the most similar model, giving proper relaxation(less in smaller scale) to this solution which construct the searching space in next scale, and doing the matching again in next scale, until the scale reach 0, at this time we can get the final solution. The matching algorithm can be, e.g., metropolis algorithm. The matching sequence is, firstly the trunk(R_0), then the head, upper parts of arms, legs, which connected directly to the trunk, finally the other parts.

Because the morphological operation and the similarity operator transform the structure matching problem to a linear parameters matching problem usually has only one local and global maximum point, we can use the 3-division method to accelerate matching speed. Detailedly, when we have calculated the similarity of four point p_0, p_1, p_2, p_3 ($p_0 < p_1 < p_2 < p_3$, $[p_0, p_3]$ is the current parameter interval) which divide $[p_0, p_3]$ into 3 subranges $[p_0, p_1]$, $[p_1, p_2]$, $[p_2, p_3]$, the desired point with maximum similarity will always in the subranges whose left or right endpoint has the maximum similarity among p_0, p_1, p_2, p_3. So at least one subrange can be discarded. Figure 6 shows the matching procedure of the h parameter of R_0 using S_3^2, with the other parts of the body in their initial states. Using the 3-division method, only 7 points(0.2, 0.8, 0.4, 0.6, 0.55, 0.575) need be calculated before the solution with enough precision can be gotten.

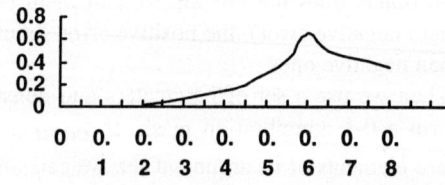

Figure 5. A Round-Corner 2-D Model of Human Body.

Figure 6. Matching of the h parameter of R_0

4. The Parallelizaion of the Method

From above discussion we can see that almost all the computation work focuses on morphological dilation and the calculation of the similarity between 0-1 images. Fortunately, these operations are all one-step operations, and can apply to every parts of the image simultaneously, so through dividing the image according to the number of processors and symmetrically distributing computation tasks, this approach can be easily parallelized and high acceleration ratio can be gotten. Also we can use vector architecture to do these computation by the design of specific instructions.

5. Future Work

Although tentative experiments show that our method is robust and stable enough to work under practical conditions, more theoretical analysis about the performance of the method should be done, in order that some important parameters used by the method can be chosen more rationally. Besides, there are still many works for us to do. The first one is how to ameliorate our method to recognize more than one object, the second is how to extend this method to work under 3-D condition. Furthermore, how to use the color information is also a valuable issue to discuss.

References

[1] C. Fuh, P. Maragos, and L. Vincent, "Visual motion correspondence by region-based approaches," Proc. ACCV'93, 784-789, Osaka, Japan, 1993.

[2] F. Bartollini and V. Cappellini, "Automatic Intrusion Detection Using Image Processing," Proceedings of the International Conference on Digital Signal Processing, Vol. 2, 468-471, July 1993.

[3] F. Heitz and P. Bouthemy, "Multi-Modal Motion Estimation and Segmentation Using Markov Random Fields," Proc. of 10th ICPR, 1990, Atlantic City, USA.

[4] J. Serra, "Image Analysis and Mathematical Morphology," Academic Press, 1982, Frlando, Florida, USA.

[5] J. Serra, "Image Analysis and Mathematical Morphology, Vol. 2: Theoretical Advances," Academic Press, 1988, Frlando, Florida, USA.

[6] K. P. Karmann, Achim von Brandt, and R. Gerl, "Moving Object Segmentation Based on Adaptive Reference Images," Signal Processing V: Theories and Applications, 951-954, 1990.

[7] K. Price and R. Reddy, "Matching Segments of Images," IEEE Trans. On Pattern Anal. And Machine Intell., PAMI-1(1):110-116, January 1979.

[8] K. Rohr, "Towards Model-Based Recognition of Human Movements in Image Sequences," CVGIP: Image Understanding, Vol. 59, No. 1, 94-115, January 1994.

[9] M. Bober and J. Kittle, "General Motion Estimation and Segmentation," V Siposium Nacional de Reconocimiento de Formas y Analisis de Imagenes, 21-25, September 1992, Valencia, Spain

[10] M. Haralick, S. R. Stemberg, and X. Zhuang, "Image Analysis Using Mathematical Morphology," IEEE Trans. On Pattern Anal. And Machine Intell., Vol. 9(4), 532-550, 1987.

[11] P. Bouthemy and F. Lalande, "Motion Detection in an Image Sequence Using Gibbs Distribution," Proceedings ICASSP 89, Vol. 3, 1651-1654, May 1989.

[12] R. Loce and E. Dougherty, "The Morphological Filter Mean-Absolute Error Theorem," Proc. SPIE, Vol. 1, 1658, San Jose, February 1992.

[13] R. Loce and E. Dougherty, "Facilitation of Optimal Morphological Filter Design via Structuring Element Libraries and Design Constraints," SPIE Journal of Optical Engineering, Vol. 31, No. 5, 1008-1025, May 1992.

[14] T. Huang and R. Tsai, "Image Sequence Analysis: Motion Estimation," In T. Huang, editor, Image Sequence Analysis, Springer-Verlag, 1981.

[15] Y. Guo, G. Xu, and S. Jsuji, "Understanding Human Motion Patterns," Proc. of 12th ICPR, Vol. 2, 325-330, January 1994.

[16] J. O'Rourke and N. I. Badler, "Model-Based Image Analysis of Human Motion Using Constraint Propagation," IEEE Trans. On PAMI, Vol. 2, No. 6, 522-536, November 1980.

Detection of Glasses in Facial Images

X. Jiang, M. Binkert, B. Achermann, H. Bunke

Department of Computer Science
University of Bern, CH-3012 Bern, Switzerland

Abstract. In this paper we consider the automatic detection of the presence of glasses. Using the eye locations detected in facial images, a region below the eyes is investigated where glasses are potentially expected. We introduce three measures for the likelihood of glasses. Experimental results have demonstrated a good separation of the two classes of facial images with and without glasses, based on these measures. Information about the presence of glasses will be useful to support automatic annotation of large image databases and indexing for fast face recognition.

1 Introduction

In dealing with large face databases, indexing is an important aspect of face recognition. In this context a number of salient facial features may be useful. Examples are gender, race, and approximate age. So far the annotation of image databases for such salient features has been done manually [5]. As a matter of fact, there exist only few works in the literature describing the automatic detection of salient facial features. The task of gender determination was tackled in [1]. O'Toole et al. [6] explored the race classification problem. The first work in the Image Understanding community on age classification was reported in [4]. An automatic feature detection will support both automatic annotation of large face databases and indexing for fast face recognition. In this paper we consider another potential feature for image database indexing: the presence of glasses, which possesses a considerable discrimination power as well.

Our goal is the determination of the presence of glasses, but not a precise localization of glasses. The main interest in the current work concerns facial images where a person looks straight into the camera, called frontal views in the following. For completeness we will also show experimental results on other views and discuss the limitations of the current method in dealing with arbitrary views and potential improvements. Our approach to the detection of glasses starts with a localization of the eyes. Given the eye locations, a region below the eyes is considered where glasses are potentially expected. We introduce three measures for the likelihood of glasses. Experimental results have demonstrated a good separation of the two classes of facial images with and without glasses, based on these measures.

2 Detection of eye locations

In the literature, methods are known for a precise localization of the iris contour [7]. In our application, however, approximate eye locations suffice. Therefore, we have adapted a Hough-transform approach introduced in [3] with some modifications. This approach is able to deal with complex background. It makes use of the fact that the most significant feature of eyes are the iris' circular shape and the strong contrast between the iris and the sclera. In this case the lines drawn at edge points along the gradient direction (from light to dark) must intersect at the iris' center. The detection of this center is based on an accumulation array. For all edge points with a significant gradient magnitude, the line along the gradient direction is considered and all cells in the accumulation array corresponding to points on this line are incremented. Then, a set of the largest accumulation values are selected as eye candidates, some of which may be caused by patterns on other parts of the face, the subject's clothes, or the background. All pairs of eye candidates are evaluated by a cost function. Finally, the pair with the smallest cost is chosen as the eyes.

Basically, we have made two modifications. First, the computation of gradients and lines along the gradients is replaced by a simpler strategy. For each pixel we consider the pixel in the 3×3 neighborhood with the smallest greylevel. If this minimum is significantly smaller than the current pixel, then a line is drawn from the current pixel to the minimum pixel and the accumulation cells on this line are incremented. Now there are merely four predefined line directions (horizontal, vertical, and diagonal) and the determination of involved accumulation cells becomes trivial. The second modification concerns the cost function for evaluating pairs of eye candidates. Given a pair of eye candidates at location p_{ij} and $p_{i'j'}$ with accumulation values H_{ij} and $H_{i'j'}$, respectively, the cost function in [3] is defined by:

$$C = (H_{ij} - H_{i'j'})^2 + \lambda(i-i')^2.$$

It requires that the two candidates have similar accumulation values and be at similar height in the image. The coefficient λ controls the relative weighting of the two criteria. In our implementation we have incorporated two additional criteria into the cost function. The accumulation values H_{ij} and $H_{i'j'}$ themselves should be high. Otherwise, a pair of eye candidates with similar but low accumulation values would receive a favorable evaluation. Moreover, the neighbors of the eye candidates should have high accumulation values as well. Considering all four criteria, we get a new cost function:

$$C^* = (H_{ij} - H_{i'j'})^2 + \lambda_1(i-i')^2 - \lambda_2(H_{ij} + H_{i'j'}) - \lambda_3 \sum_{p_{kl} \in N(p_{ij}) \cup N(p_{i'j'})} H_{kl},$$

where $N(p_{ij})$ denotes the set of eight direct neighbors of p_{ij}. In our experience the modified version of the eye detection method works faster at a slightly better performance.

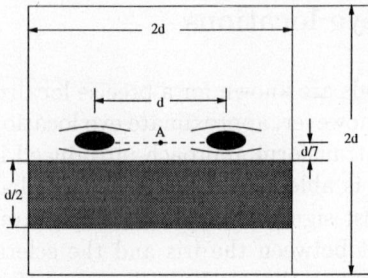

Fig. 1. Definition of the frame region F (the shaded region) and the region for automatic threshold determination.

3 Detection of glasses

Given the approximate location of the eyes, we propose in this section an approach to the detection of glasses based on a very simple idea. The glasses manifest themselves through high greylevel discontinuities of the frame against the facial part. The glasses frame is visible around the eyes (above, below, and between the eyes). In the area above the eyes, however, it is difficult to distinguish between greylevel discontinuities caused by glasses and those by the eyebrows. The area between the eyes may be too small to be suitably defined. It remains the region below the eyes that is generally free of foreign influences. We assume that in this area the majority of (high) greylevel discontinuities are caused by the glasses frame and its shadow on the face. Therefore, the edgeness measure of this region, referred to as *frame region* in the following discussion, provides a cue of the presence of glasses.

The frame region F is defined as a rectangular area below the eyes, see Figure 1. It has a length of $2d$ and a height of $d/2$, where d represents the distance between the eyes. The upper side of the frame region is located at a distance $d/7$ from the middle A of the eyes. Finally, the frame region is centered at A.

Within the frame region, cues of greylevel discontinuities are gathered. Note that we are only interested in the presence and magnitude of greylevel discontinuities, but not in their precise localization. Therefore, we apply a simple edgeness measure:

$$\text{edgeness}_{ij} = |g_{i+1,j} - g_{i-1,j}| + |g_{i,j+1} - g_{i,j-1}|,$$

where g_{ij} represents the greylevel at pixel p_{ij}. This edge operator may be iterated k times. This way the edgeness responses of the frame border will expand to the inner pixels of the frame so that not only the frame border, but all pixels of the frame contribute to the accumulation of edgeness cues. The parameter k is dependent on the expected (average) thickness of the frame. In our current implementation we have chosen $k = 2$. The high edgeness values can be propagated into their neighborhood by replacing the edgeness value of each pixel with the maximum in the 3×3 local neighborhood. Again, this maximum operation can be iterated to produce a final edgeness map. Currently, we perform three iterations.

After the computation of edge measures, a simple cue of the presence of glasses is the average edgeness measure of all pixels in the frame region:

$$M_1 = [\sum_{p_{ij} \in F} \text{edgeness}_{ij}] / |F|,$$

where $|F|$ represents the number of pixels in the frame region. Alternatively, we can perform a thresholding operation to generate a binary edge map of F and count the number of edge points. There exist a large number of methods for an automatic determination of thresholds. In our current implementation we have used the discriminant analysis technique proposed by Otsu [2]. The threshold determination is not applied to the frame region F. The reason is that in case of no glasses there are only few edge points in F and the computed threshold is not useful. Instead, we consider a quadratic region of side length $2d$ centered at the middle A of the eyes; see the larger unshaded region in Figure 1. For this region the edge detection and the maximum operation are applied. The histogram of the edge detection results is the input to Otsu' algorithm for threshold determination. Now we have a second cue of the presence of glasses:

$$M_2 = [\sum_{p_{ij} \in F} \text{if edgeness}_{ij} \geq T \text{ then } 1 \text{ else } 0] / |F|,$$

where T is the determined threshold. This represents the percentage of edge points within F. In the measure M_2 all edge points contribute equally to the cue accumulation. This way, however, weak edge points in the facial part of the frame region provide unnecessarily high support and may thus bias the cue accumulation. Instead, we can take the magnitude of edge points into account and formulate a third cue of the presence of glasses:

$$M_3 = [\sum_{p_{ij} \in F} \text{if edgeness}_{ij} \geq T \text{ then edgeness}_{ij} \text{ else } 0] / |F|.$$

In general we expect that the presence of glasses will give abundant support to all three cues defined above. Thus, facial images with and without glasses should result in a bimodal distribution, making a separation of the two kinds of facial images possible.

4   Experimental results

We have tested our algorithm on an image set consisting of 30 persons (mainly Europeans and a few Asians). Exactly half of the persons wear glasses and the other half doesn't. Available for each person are 10 images with varying viewing directions (2 frontal views, 2 looking to the left, 2 looking to the right, 2 downwards, and 2 upwards). All images are of the same resolution 512×342 and were taken under normal lighting conditions with a homogeneous or complex background.

	(a)	(b)	(c)	(d)	(e)	(f)
M_1	41	48	72	88	90	105
M_2	1	14	22	25	31	40
M_3	3	19	36	48	59	74

Table 1. The three measures of the presence of glasses for the six images in Figure 2.

Since our main interest in this work concerns the frontal view, we first look at the results on this subset S_f of 60 images. Figure 2 shows the results for six persons of S_f. The detected eyes are depicted by two dark crosses. The evaluation regions (the small frame region and the overall region for automatic threshold determination, see Figure 1) are drawn, in which the edgeness measures are visualized. In all 60 images, the eye position has been precisely localized. The eye detection works well even in case of inhomogeneous background, as demonstrated by two examples in Figure 2. Since the size of the evaluation region is dependent on the eye separation, our approach is able to adapt itself to different head sizes in the images. The head size in Figure 2(d) and (e), for instance, is much larger than that in Figure 2(f). Intuitively, the edgeness measures of the frame region visualized in Figure 2 confirm the expectation that the majority of significant edgeness values is caused by the glasses frame. On the entire subset S_f, the values of the three measures are histogrammed in Figure 3, where samples with and without glasses are denoted by the plus and circle symbol, respectively. The two classes of facial images are well separated. Here the three measures reveal a similar separability, although for M_1 and M_3, the positive samples (with glasses) demonstrate a broader distribution. For the six images in Figure 2, the three measures M_1, M_2, and M_3 have the values tabulated in Table 1. Actually, the image in Figure 2(b) is among the worst cases of all samples without glasses with regard to the three measures. On the other side, the worst case of samples with glasses is exemplified by the image in Figure 2(c). Certainly, our test basis is still too small to draw final conclusions on the ability of the proposed measures to separate the two image classes with and without glasses. But our experiments definitely demonstrated the potential of our approach.

We have also evaluated the behavior of our measures on the 240 images of our test set that have other viewing directions. In nine images the eye localization method made a false decision and we have therefore excluded them from further consideration. For our evaluation the 231 remaining images were divided into subsets S_d (downwards), S_u (upperwards), and S_s (sideviews). Because of the symmetry between looking to the left and right, these two subgroups were put together into the subset S_s. A representative from each of the three subsets is shown in Figure 4 and the measures M_1, M_2, and M_3 on each subset are histogrammed in Figures 5-7. For the downward view, the measures reveal a similar distribution and thus separability as in the case of S_f. For the other two views, however, the measures demonstrate overlaps, although the global

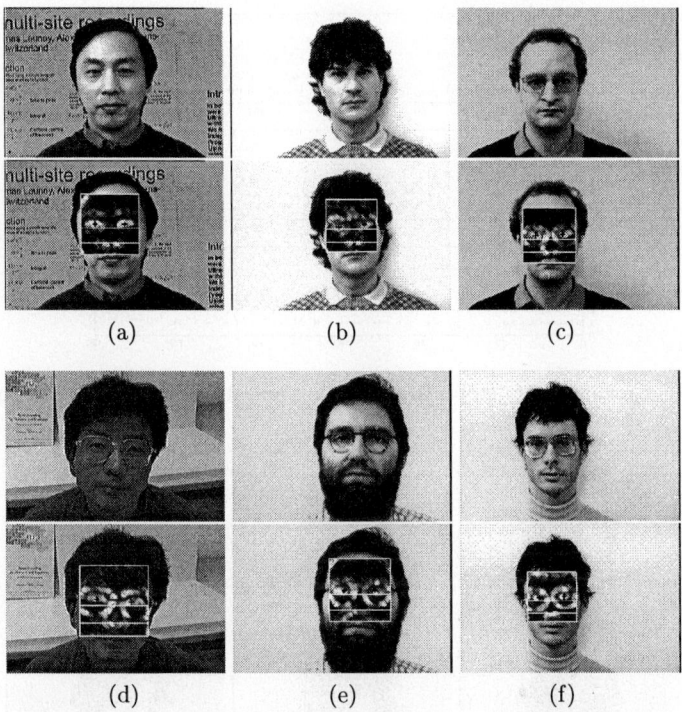

Fig. 2. Six facial images with the detected eye locations and the edgeness measures.

separation is still given. The reason for this phenomenon is that our definition of the evaluation area assumes a viewing direction more or less straight into the camera. Otherwise, unexpected facial parts with strong greylevel discontinuities may fall into this area and generate false support to the measures. As can be easily seen in Figure 4, this problem can be caused by the nostrils in the upward view and by the facial border against the background in the sideviews. In order to handle arbitrary views, therefore, we need a more sophisticated cue accumulation to minimize the influences of undesired edge responses.

5 Conclusions

In this paper we have presented an approach towards detecting the presence of glasses in facial images with unconstrained background. Based on the automatically detected edge locations, evaluation functions have been defined for areas where glasses are expected. Experimental results demonstrated the usefulness of this approach.

Acknowledgments

T. Perroud helped us implement and test the eye detection algorithm.

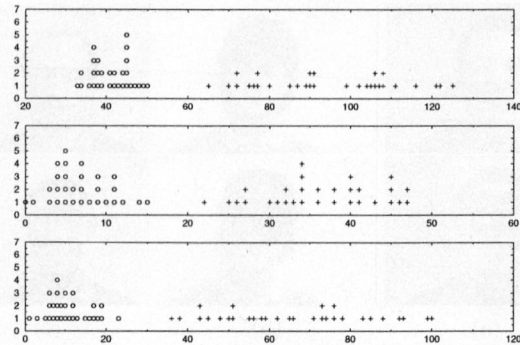

Fig. 3. Histograms of M_1 (top), M_2 (middle), and M_3 (bottom) for the subset S_f.

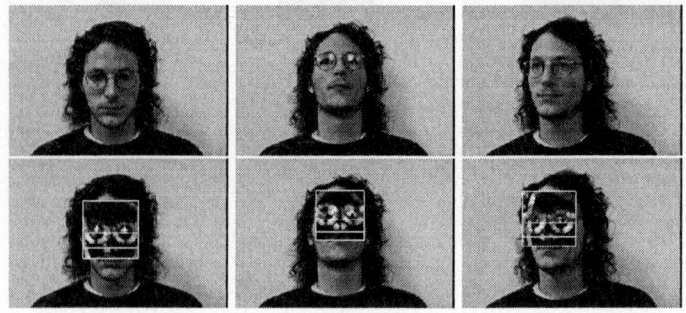

Fig. 4. Examples of other viewing directions.

References

1. B. Golomb, T. Sejnowski, Sex recognition from faces using neural networks, in A.F. Murray (Ed.), Applications of neural networks, Kluwer Academic Publishers, 71–92, 1995.
2. N. Otsu, A threshold selection method from gray-level histograms, IEEE Trans. on SMC, Vol. 9, 62–66, 1979.
3. R. Kothari, J. Mitchell, Detection of eye locations in unconstrained visual images, Proc. of Int. Conf. on Image Processing, 519–522, Lausanne, 1996.
4. Y.H. Kwon, N. Lobo, Age classification from facial images, Proc. of CVPR, 762–767, 1994.
5. A. Pentland, B. Moghaddam, T. Starner, View-based and modular eigenspaces for face recognition, Proc. of CVPR, 84–91, 1994.
6. A.J. O'Toole, *et al.*, Classifying faces by race and sex using autoassociative memory trained for recognition, Proc. of the Annual Meeting of the Cognitive Science Society, 847–851, 1991.
7. A.L. Yuille, P. Hallinan, D. Cohen, Feature extraction from faces using deformable templates, Int. Journal on Computer Vision, Vol. 8, No. 2, 99–111, 1992.

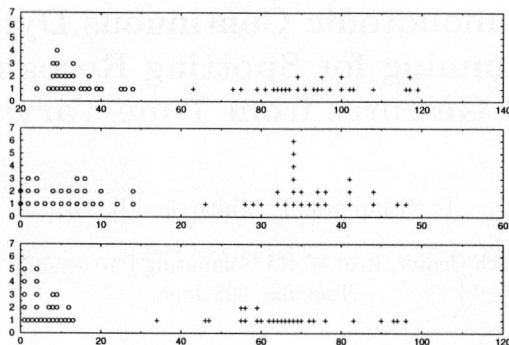

Fig. 5. Histograms of M_1 (top), M_2 (middle), and M_3 (bottom) for the subset S_d.

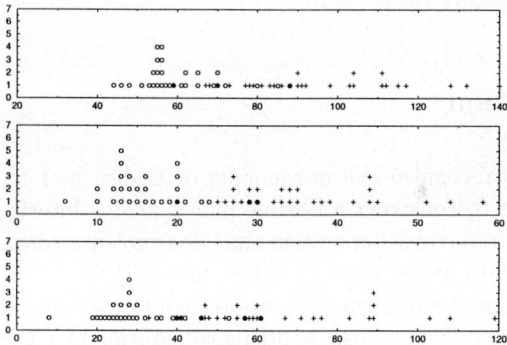

Fig. 6. Histograms of M_1 (top), M_2 (middle), and M_3 (bottom) for the subset S_u.

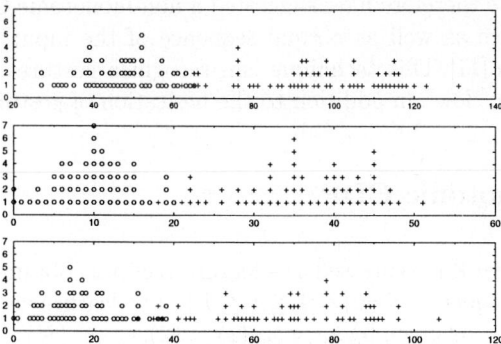

Fig. 7. Histograms of M_1 (top), M_2 (middle), and M_3 (bottom) for the subset S_s.

Non-monotonic Continuous Dynamic Programming for Spotting Recognition of Hesitated Gestures from Time-Varying Images

T. Nishimura, T. Mukai, and R. Oka[1]

Tsukuba Research Center, Real World Computing Partnership, 1-6-1, Takezono, Tsukuba, 305 Japan

Abstract. Continuous Dynamic Programming (CDP) has been proposed to recognize the meanings of human gestures from motion images. And this CDP has been extended to Non-monotonic CDP in order to recognize hesitated gestures. In this paper, we show the character of Non-monotonic CDP in detail.

1 Introduction

A technology to recognize the movements of bodies and hands of humans is essential in order to construct a flexible man-machine interface system. We are studying a gesture recognition system that does not use contact type sensors or marker such as data gloves worn by a test subject[1-9].

We have proposed a real-time gesture recognition system with low resolution feature[9][10] This system realises spotting recognition by CDP[8]. The CDP recognizes gestures continuously with automatic segmentation and it outputs the recognition results frame-wisely. Humans sometimes hesitate when doing gestures, and the hesitation pattern may change greatly. The recognition of such gestures by continuous DP is inefficient since it requires a large number of standard patterns. We therefore have suggested a non-monotonic CDP that matches a reverse sequence as well as normal sequence of the input sequence and the standard pattern.[11] This technique can recognize gestures indicating degree such as "high" or "low" in addition to the hesitation of gestures.

2 Non-monotonic CDP

A standard pattern Z is expressed as a feature vector z_τ obtained from a T frame standard image sequence $Z = \{z_\tau | 1 \leq \tau \leq T\}$. The feature vector z_τ, where N is the dimension of z_τ is given by $z_\tau = (z_\tau(1), z_\tau(2), \cdots, z_\tau(N))$. The input feature vector sequence $u_t (0 \leq t < \infty)$ are obtained from the input images. $d(t, \tau)$ is the local distance between u_t and z_τ .

The cumulative distance between the standard pattern and the input sequence with the end point (t, τ) is expressed as $S(t, \tau)$. In the non-monotonic continuous DP, $S(t,\tau)$ is calculated frame-wisely using the followin equations. The initial condition $(t = 0)$ is:

$$S(0,\tau) = d(0,\tau). \quad (1 \leq \tau \leq T) \tag{1}$$

Equations$(1 \leq t)$

$$S(t,\tau) = \alpha \cdot d(t,\tau) + (1-\alpha) \cdot \min_{m \in \{-1,0,1\}} S(t-1, \tau+m). \quad (1 \leq \tau \leq T) \tag{2}$$

where α is a normalizing factor $(0 \leq \alpha \leq 1)$. To simplify the equation, matching is assumed to be possible for the input series with -1 to 1 compression or expansion against a standard pattern asumming the local path shown in Figure 1(a). Various local paths can be defined as Figure 1(b). Expressions (1) and (2) are

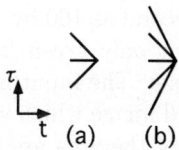

Fig. 1. Examples of local path of Non-monotonic CDP.

(a) $m \in \{-1, 0, 1\}$, (b) $m \in \{-2, -1, 0, 1, 2\}$

solved as follows:

$$S(t,\tau) = \min_{\{p_0, p_1, \cdots, p_t\}} \{\sum_{k=1}^{t} \alpha(1-\alpha)^{t-k} \cdot d(k, p_k) + (1-\alpha)^t \cdot d(0, p_0)\}.$$

$$(1 \leq t) \tag{3}$$

where the integer $p_0, p_1, \cdots, p_t$ is defined as:

$$p_t = \tau, |p_k - p_{k-1}| \leq 1 (k = t, t-1, \cdots, 1)$$

and $\tag{4}$

$$1 \leq p_k \leq T (k = 0, 1, \cdots, t)$$

When there are a number L of standard patterns, the cumulative distance is $S_\ell(t,\tau)(1 \leq \ell \leq L)$, the threshold value is h_ℓ and the number of frames is T_ℓ. The output of the non-monotonic continuous DP is the category No. $\ell^*(t)$ of the matched standard pattern and the number of frames $\tau^*(t)$ in that standard pattern may be expressed as:

$$\{\ell^*(t), \tau^*(t)\} =$$

$$\begin{cases} \text{Arg}[\min_\ell \{\min_{1 \leq \tau \leq T_\ell}(S_\ell(t,\tau) - h_\ell)\}] & \text{if } \exists \ell, \exists \tau \text{ so that } S_\ell(t,\tau) \leq h_\ell \\ \text{null} & \text{otherwise} \end{cases} \tag{5}$$

where Arg is a factor that returns an argument $\{\ell(t), \tau(t)\}$ and null is an empty category. From the standard pattern of these outputs, various motions may be distinguished from the traces of points $(t, \tau^*(t))$.

The normalization factor α can be changed as follows in order to match with longer past information when a gesture moves slowly.

$$\alpha(t) = \beta' \cdot \sum_{k=1}^{N} \{u_t(k) - u_{t-1}(k)\}^2 \tag{6}$$

3 Evaluation experiments

3.1 Experimental procedure

We used a Silicon Graphic's Indy (R4400 200MHz) and the IndyCom camera. The camera was set to capture the gestures of a person on a chair. The output images are taken 15 times per second as 160 by 120 pixel RGB images with each pixel having 256 gradations, but only green images were used. Those images were compressed to 64 by 64 pixel. The input image is divided into 3 by 3 area and compared with a background image which was taken beforehand. The ratios of value-changed pixels in each of the area are the nine-dimensional feature for gesture recognition.

The gestures used in the experiment were: (1) raise hand (raise right hand), (2) banzai(raise both hands), (3) small-large(spread both hands), (4) low-high(raise right hand in front of the face), and (5) right-left (swing right hand). Gestures (3), (4) and (5) show degrees. Figure 2 shows each gesture. The standard patterns were made from the start point to the end point of the arrow shown in Figure 3. For example, the standard pattern of the gesture "raise hand" is taken by raising the right hand from the lower position to the upper position as shown in Figure 4(a). The frame length T of the standard patterns were 10 to 27.

Shown in Figure 5, there are two similar frames among the five gestures. Frame#1 of Figure 5(a) "right-left" is similar to Frame#3 of Figure 5(b) "raise hand". This means the system has the possiblity to make mistakes if you stop hand in similar position to those two frames. The input image sequences were taken as follows. The five gestures were done in the reverse direction to the standard pattern as well as in the normal directions. And the similar pose shown in Figure 5 was also done when doing two gestures in order to know if the system mistakes the gestures or not. The total frame number of the input image sequences was 517, and we changed the normalizing factor α as $\{1, 0.1, variable\}$ and obtain recognition rate as follows.

$$\text{Recognition Rate} = \frac{\text{Number of correct } \{\ell^*(t), \tau^*(t)\}}{\text{Number of input frames}} \tag{7}$$

The permissible error of the matched number of frames $\tau^*(t)$ in the standard pattern is within ± 2 frames. When α is variable, the value β' of equation (6) is 1 to make the maximum value of a equal to 1.0.

$$\alpha(t) = \min\{1.0, U(t)\}. \tag{8}$$

3.2 Experimental results

The results of the recognition experiment for various normalization factors α are shown in Table 1. T The effectiveness of this method was confirmed with a recognition rate of 90% when α is variable. The major cause of recognition errors when α is variable was that a gesture was recognized even after the gesture had already been completed. This was because the standard patterns included the frames immediately before the hand moved out of the image, thus retaining the very last recognition when the feature vector stopped changing when the hand moved out of the image. When $\alpha = 1$, recognition became unstable with skipped τ^* and when the gesture stopped, incorrect detection occurred which lowered the recognition rate. When α was 0.1, the recognition was stable, but there were a delay in recognizing the start of gestures.

Table 1. Recognition results.

α	1	0.1	$\alpha(t)$
Recognition rate2(%)	72	87	92

Figure 6(b) shows the result of spotting recognition when α is variable. The horizontal axis is the input frame number and the vertical axis is the frame number τ^* of the standard pattern. Recognition results of the five gestures are marked with different marks shown in the Figure 6(b). The solid line and Figure 6(a) shows the ideal recognition by human. Figure 6(b) shows that the spotting recognition with the non-monotonic CDP can recognize gestures indicating degrees continuously in addition to hesitating gestures.

Figure 7 shows the recognition results when the similar pose shown in Figure 5 was done when doing gesture "right-left". The right hand was moved from right to left and then to right again and stopped. The white circle in Figure 7 is the result of the incorrect recognition of "right-left" and the rhomboid is recognized as "raise hand". The solid line indicates the ideal recognition by human.

As shown in Figure 7(a), errors occur when no past date is used (α=1). The previous trace information as shown in Figure 7(b) was retained with α=0.1 for a short period of time and the use of many past data resulted in a delay in recognizing the start of a gesture. We could reduce the delay and errors as shown in equation (6) by making a time variable as shown in Figure 7(c).

4 Conclusion

In this report we suggested non-monotonic continuous DP for recognizing gestures in the reverse direction and the stopping of gestures in partial sections of the standard pattern, and demonstrated its effectiveness through evaluation experiments. This technique can recognize hesitated gestures and gestures showing degree. The effectiveness of this method, was confirmed by some experiments.

One problem is that the recognition rate decreases when the position of the subject changes. Our future task is to develop a more robust technique of feature extraction using distance images or color information. We propose to verify the effectiveness of the system by applying the method to global position estimation with visual sensors of an autonomous mobile robot.

References

1. H.H. Baker and R.C. Bolles, "Generalizing Epipolar-Plane Image Analysis on The Spatio-temporal Surface", Proc. CVPR, pp. 2-9, 1988.
2. R.C. Bolles, H.H. Baker and D.H. Marimont, "Epipolar-Plane Image Analysis: An Approach to Determining Structure from Motion", International Journal of Computer Vision,1,pp. 7-55, 1987.
3. A.P. Pentland : "Visually Guided Graphics", International AI Symposium 92 Nagoya Proceedings, pp.37-44, 1922
4. M.A. Turk, A.P. Pentland : "Face Recognition Using Eigenfaces", Proc. CVPR, pp.586-590, 1991
5. J. Yamato, J. Ohya, K. Ishii: "Recognizing Human Action in Time-Sequential Images Using Hidden Markov Model", Proc. CVPR, pp.379-385, 1992
6. T. J. Darell and A. P. Pentland: "Space-Time Gestures", Proc.IJCAI'93 Looking at People Workshop(Aug. 1993)
7. H. Ishii, K. Mochizuki and F. Kishino, "A Motion Recognition Method from Stereo Images for Human Image Synthesis", The Trans. of the EIC, J76-D-II,8,pp.1805-1812,(1993-08)
8. R. Oka: "Continuous Word Recognition with Continuous DP", Report of the Acoustic Society of Japan, S78, 20, 1978 [in Japanese]
9. S. Seki, K. Takahashi and R. Oka: "Gesture Recognition from Motion Images by Spotting Algorithm", Proc. ACCV'93, pp.759-762(Nov.1993)
10. T. Nishimura and R. Oka, "Spotting Recognition of Human Gestures from Time-Varying Images", International Conf. on Automatic Face and Gesture Recognition, pp.318-322(1996-10)
11. T. Nishimura and R. Oka, "Towards the Integration of Spontaneous Speech and Gesture based on Spotting Method", International Conf. on Multisensor Fusion and Integration for Intelligent Systems, pp.433-437(1996)

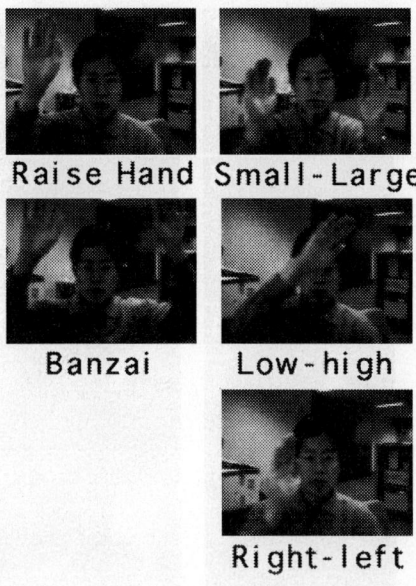

Fig. 2. Snapshots of five gestures.

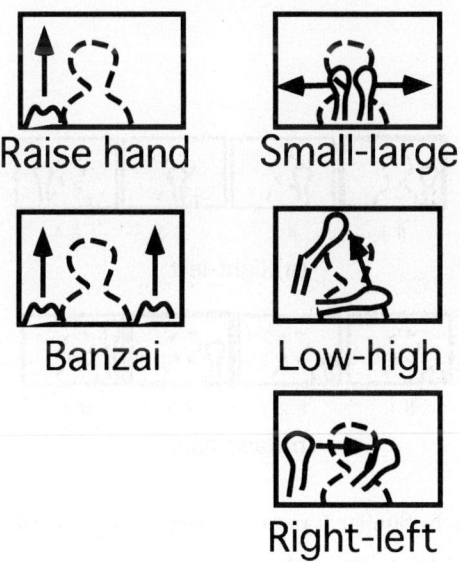

Fig. 3. Standard pattern of five gestures.
(From the start point to the end point of the arrow.)

Fig. 4. Standard pattern of gesture (a)"Raise hand",(b)"Small-large".

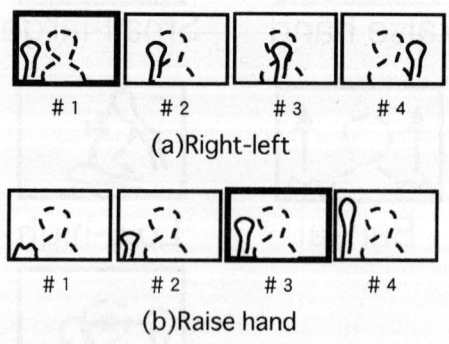

Fig. 5. Similar frames between standard patterns
Frame#1 of (a)"Right-left" and frame#3 of (b)"Raise hand".

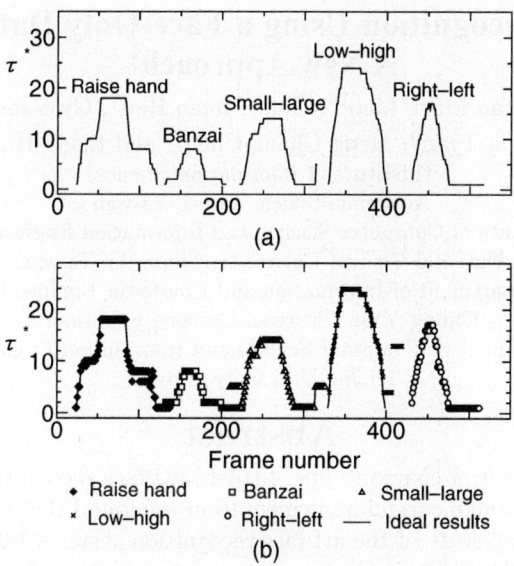

Fig. 6. Recognition results using Non-monotonic CDP.
(a)Recognition results of human, (b)Recognition results using Non-monotonic CDP.

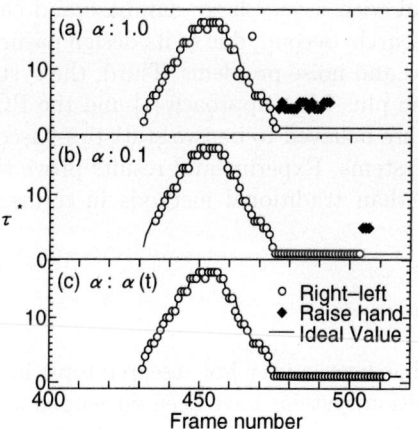

Fig. 7. Influence of normalization factor α.

Face Recognition Using a Face-Only Database: A New Approach[1]

Hong-Yuan Mark Liao[†][2], Chin-Chuan Han[†], Gwo-Jong Yu[‡],
Hsiao-Rong Tyan[§], Meng Chang Chen[†], and Liang-Hua Chen[¶]

†Institute of Information Science,
Academia Sinica, Taipei, Taiwan
‡Institute of Computer Science and Information Engineering,
National Central University, Chung-Li, Taiwan
§Department of Information and Computer Engineering,
Chung Yuan Christian University, Taiwan
¶Department of Computer Science and Information Engineering,
Fu Jen University, Taiwan

Abstract

In this paper, a coarse-to-fine, LDA-based face recognition system is proposed. Through careful implementation, we found that the databases adopted by two state-of-the-art face recognition systems[1,2] were incorrect because they mistakenly use some non-face portions for face recognition. Hence, a face-only database is used in the proposed system. Since the facial organs on a human face only differ slightly from person to person, the decision-boundary determination process is tougher in this system than it is in conventional approaches. Therefore, in order to avoid the above mentioned ambiguity problem, we propose to retrieve a closest subset of database samples instead of retrieving a single sample. The proposed face recognition system has several advantages. First, the system is able to deal with a very large database and can thus provide a basis for efficient search. Second, due to its design nature, the system can handle the defocus and noise problems. Third, the system is faster than the autocorrelation plus LDA approach [1] and the PCA plus LDA approach [2], which are believed to be two statistics-based, state-of-the-art face recognition systems. Experimental results prove that the proposed method is better than traditional methods in terms of efficiency and accuracy.

1 Introduction

Face recognition has been a very hot research topic in recent years[3]. Some successful face recognition systems have been developed and reported in the literature[1,2,4]. Among these works, the systems developed by Goudail *et al.*[1] and Swets and Weng[2] represent two state-of-the-art face recognition systems. In [1], Goudail *et al.* investigated the performance of a technique for face recognition based on the computation of 25 local autocorrelation coefficients. In the

---

[1] This work was partially supported by National Science Council of Taiwan under grants NSC86-2745-E-001-004 and NSC86-2213-E-001-023.
[2] To whom correspondence should be addressed.

autocorrelation coefficient computation stage, they used 25 3x3 filters to convolve with the whole image and came up with 25 numerical coefficients. This process was used to reduce the dimensionality of the raw images. They used the set of transformed 25-dimensional database samples to determine the set of most discriminating projection axes and then calculated each sample's projective feature vector. When an unknown image appeared, its corresponding projective feature vector was calculated and compared with those of the database samples. There is a major problem associated with their approach. That is, although they keep the color of the background and cloth dark, their "face" image is actually a combination of face, hair, shoulders and background. It was proven in [5] that this kind of database is incorrect in terms of "face" recognition.

Another state-of-the-art system was proposed by Swets and Weng[2]. In this work, they applied the PCA technique to reduce the dimensionality of the original images. They selected the top 15 principal axes and used them to derive a 15-dimensional feature vector for every sample. These transformed samples are then used as bases to execute LDA, and they reported a peak recognition rate of more than 90%. Again, we found that their face image contained face, hair, shoulders and background, not solely face.

In this paper, we propose a coarse-to-fine, LDA-based approach to solve the face recognition problems. In order to make the recognition process base its recognition on the "real" face portion, we build a face database with every sample containing only facial organs. In the feature extraction process, each sample image is equally partitioned into 2^{2i} regions at resolution $i(i = 1, 2, ...)$. The value of a partitioned region is obtained by averaging the gray values of all pixels in that region. Hence, the coarse-to-fine feature extraction process transforms all the database samples into 2^{2i}-dimensional feature vectors at resolution i (i=1,2,...). Based on the extracted feature vectors, LDA can be executed at different resolutions, and the 2^{2i} most discriminating projection axes can be determined at resolutions $i(i = 1, 2, ...)$. For resolution i, the transformed 2^{2i}-dimensional database samples are projected onto their corresponding most discriminating projection axes, and a set of 2^{2i}-dimensional projective feature vectors of database samples is obtained. When an unknown image comes in, the same coarse-to-fine feature extraction process is performed. Then, the set of different dimensional feature vectors is projected onto the most discriminating projection axes corresponding to their own resolutions, and a set of different dimensional projective feature vectors for the unknown image is obtained. Based on these projective feature vectors, a coarse-to-fine comparison can be executed, and the final results will be obtained at the finest resolution. There are several advantages associated with the proposed approach. First, the coarse-to-fine search structure can help deal with a very large database and provide a good basis for efficient search. Second, using an average gray level to represent a region can suppress the effect caused by defocus or noise. Third, the design nature makes the database extensible. Further, the search speed of this approach is faster than that of the approaches developed by Goudail et al.[1] and Swets and Weng[2]. As to accuracy, this approach is better than that of Goudail et al.. Finally and most importantly, our approach uses a face-only database to execute the face recognition task. Since the facial organs on a human face differ only slightly from person to person, the decision boundary set is hard to decide upon. However, experimental results have shown that our approach is still robust under these circumstances.

2 Why a Face-only Database?

In this section, we shall describe a series of experiments and show that the databases used in [1] and [2] were incorrect. In order to prove that our above statement is true, we have built a 128-person face database. In the database, 6 face images were taken of each person as shown in Figure 1. The 6 face images of each person included two frontal views, two 3/4 fontal views with the right side, and two 3/4 frontal views with the left side. Therefore, the total number of training samples was 768 face images. We then implemented the methods proposed in [1] and [2], respectively. Figure 2 shows a series of experimental results. Figure 2(a) is the results obtained by applying the autocorrelation plus LDA method proposed by Goudail et al.[1]. The upper-left face image of each experiment was a query (test) image which had to be different from all the training images. The remaining images were the retrieved database images that had the 11 Euclidean distances closest to the query image. The retrieved database images are ordered from left to right and top to bottom. Figure 2(b) shows another set of results obtained by applying the PCA plus LDA method proposed by Swets and Weng[2]. We found that, in both [1] and [2], the training images used to build the databases took hair, shoulders, and background into account. However, since both methods are statistics-based approaches, we wondered whether those non-face portions played a role in the face recognition process. In order to follow up on this suspicion, we cut out the face portion of another face image and attached it to the face portion of the query image shown in Figure 2(a). Then, we used this synthesized image as a query image to retrieve database images. Figures 3(a) and (b) are the results obtained by autocorrelation plus LDA [1] and PCA plus LDA [2], respectively. From the results shown in Figure 3, it is not difficult to find that, no matter how the method proposed in [1] or [2] was applied, the non-face part dominated the face recognition process. Based on these experiments, we conclude that face images used to train a face database should not include non-face portions.

3 Building a Face-only Database and Feature Extraction

3.1 Building a Face-only Database

A "correct" face here is defined as a face portion which includes the eyes, mouth, and nose. In the image acquisition process, it is very difficult to control the camera so as to photograph only the "face" portion. In this stage, we apply a previously developed face detection algorithm [6] to perform the task. Since we already had a face database containing 768 face images from 128 persons, the algorithm in [6] was used to "cut" out the face portion of each image. Figure 4 is an example showing this step. After performing the face detection algorithm proposed in [6], the detected face portions of the images in Figure 1 were those shown in Figure 5. In what follows, we shall discuss how to perform coarse-to-fine feature extraction based on the new database.

3.2 Coarse-to-fine Feature Extraction

The purpose of feature extraction is to reduce the dimensionality of the original images so that the small sample size problem[4] will not happen. In this paper, we propose a coarse-to-fine feature extraction mechanism. At the coarsest resolution, a face image is equally partitioned into 4 (2x2) regions as shown in Figure 6. For each region, the gray values of all the pixels are averaged, and the value is then used as the region's representative value. Therefore, at this resolution, every sample image is converted from its original size to a 4-dimensional feature vector. For the second resolution, every database image is partitioned into 16 (4x4) regions. Again, each region of this partition is represented by its corresponding average gray value(Figure 6). At this resolution, every sample image is represented as a 16-dimensional feature vector. Using a similar partition, every sample image at the third resolution should be represented by a 64-dimensional feature vector. Basically, the number of resolutions adopted in the coarse-to-fine feature extraction process should be provided in advance. Furthermore, this number is dependent on the scale of the face database. Figure 6 shows the details of the coarse-to-fine feature extraction mechanism. The proposed coarse-to-fine feature extraction strategy can guarantee a fast search process, which is a basic requirement in most nationwide face-based ID checking systems. In the next section, we shall report how, based on the coarse-to-fine features, LDA-based classification can be performed.

4 Classification Using Coarse-to-fine LDA

4.1 Linear Discriminant Analysis (LDA)

In the coarse-to-fine feature extraction stage, different dimensional feature vectors are extracted from resolution to resolution. The LDA technique discussed here is a general description which fits feature vectors extracted at different resolutions.

Let the training sample set be comprised of M classes, and let M_k be the number of samples contained in class k. X_j^k denotes a feature vector representing the jth sample of the kth class. $\bar{X}^k$ denotes the mean feature vector of the kth class, and $\bar{X}$ denotes the mean feature vector of the whole population. The physical meaning of LDA is to determine the mapping

$$Y_j^k = A' * X_j^k$$

that simultaneously maximizes the between-class scatter while minimizing the within-class scatter of all Y_j^k (where $k = 1, \ldots, M, j = 1, \ldots, M_k$) in the projective feature vector space. The within-class scatter in the projective feature space can be calculated as follows:

$$WS = \sum_{k=1}^{M} \sum_{j=1}^{M_k} (Y_j^k - \bar{Y}^k) * (Y_j^k - \bar{Y}^k)'.$$

The between-class scatter can be calculated as follows:

$$BS = \sum_{k=1}^{M} (\bar{Y}^k - \bar{Y}) * (\bar{Y}^k - \bar{Y})'.$$

The way to find the required mapping A is to maximize the following quantity:

$$\mathbf{tr}(WS^{-1} \times BS).$$

An algorithm which can solve the mapping matrix A can be found in [7].

4.2 LDA-based Coarse-to-fine Classification

In this section, we shall discuss how LDA can be applied to form coarse-to-fine clustering at different resolutions. Assume that the current resolution is i. In Section 3, we described how to reduce the dimensionality of all training samples to 2^{2i} dimensions. These 2^{2i} dimensional samples are then used as bases to calculate the set of most discriminating projection axes thru LDA. In order to simplify the process, at each resolution i, we calculate the 2^{2i} projection axes, i.e., the upper bound of allowable projection axes. After finding the 2^{2i} most discriminating projection axes at resolution i, the next step is to project all the training samples onto these axes and to obtain their corresponding 2^{2i}-dimensional projective feature vectors. When an unknown query image comes in, the coarse-to-fine feature extraction process will first be executed. This process will reduce the dimensionality of the unknown image to a 2^{2i}-dimensional feature vector at resolution i ($i = 1, 2, \ldots$). Then, for each resolution i, the transformed 2^{2i}-dimensional unknown data are projected onto the corresponding 2^{2i} most discriminating projection axes, and a 2^{2i}-dimensional projective feature vector for the unknown image at resolution i is obtained. The unknown projective feature vector obtained at resolution i can be compared with all the qualified database projective feature vectors at resolution i. If the Euclidean distance between the unknown projective feature vector and that of a database sample is larger than a statistically predetermined threshold, then at the next resolution (a finer resolution), this sample will not be considered. Figure 7 shows the coarse-to-fine search mechanism.

5 Experimental Results

In order to demonstrate the efficiency and accuracy of our method, a series of experiments was conducted. First of all, we used the morphology-based face detection algorithm proposed in [6] to detect the face portions of 768 database images. The detected face portions of each old database sample were put together to form a new face-only database, which contained 768 new database images. The original database was retained, and an index table was established to relate the old database to the new one. In order to avoid the small sample size problem, the number of resolutions selected was four (i.e., 2^1 to 2^4). In the recognition process, a set of test images which was different from the training set was used. In every experiment, a query image (test image) was presented to the system. The recognition system first performed face detection on the query image and then executed a normalization process on the detected face portion. The normalized face portion was then used as an input to retrieve the 11 closest samples (out of 768 samples) in the new database. The 11 retrieved closest samples were arranged in left to right, top to bottom order. Figure 8 shows one set of experimental results. Figure 8(a) shows the detected face portions. After normalization, the detected portion was used as a query to retrieve the new database. The 11 closest retrieved face-only samples together with the query image are shown in Figure 8(b). Figure 8(c) shows 11 original database images corresponding to the 11 retrieved new database samples. In this experiment, it is obvious that, although only the face portion was used to perform recognition, the top 6 of the 11 closest retrieved samples were correct identifications. We ran 1000 sets of experiments; among these experiments about 95.02% of the first ranked retrieved results were correct idenfications. If the scope was extended to the retrieved subset (11 closest samples), then the success rate could be further increased to 99.15%.

6 Conclusion and Future Work

In this paper, we have proposed a new face recognition system which bases its recognition process on a face-only database. In the feature extraction process, the average gray level technique is employed to reduce the dimensionality of the original image at different resolutions. Using this process, the effect caused by noises or defocus can be significantly suppressed. Unlike the PCA-based approach, our approach does not require recalculation of the projection axes, which makes the database extensible. Furthermore, based on a coarse-to-fine database architecture, our approach can handle an extremely large face database. As to the efficiency problem, our method is faster than the approaches proposed by Swets and Weng[2] and Goudail *et al.*[1]. As to the recognition rate, our approach is superior to Goudail *et al.*'s approach.

References

1. F. Goudail, E. Lange, T. Iwamoto, K. Kyuma, and N. Otsu, "Face recognition system using local autocorrelations and multiscale integration," *IEEE Trans. on Pattern Analysis and Machine Intelligence*, vol. 18, pp. 1024–1028, October 1996.

2. D. Swets and J. Weng, "Using discriminant eigenfeatures for image retrieval," *IEEE Trans. on Pattern Analysis and Machine Intelligence*, vol. 18, pp. 831–836, August 1996.
3. R. Chellappa, C. Wilson, and S. Sirohey, "Human and machine recognition of faces: A survey," *Proceedings of the IEEE*, vol. 83, no. 5, pp. 705–740, 1995.
4. Z.-Q. Hong and J.-Y. Yang, "Optimal discriminant plane for a small number of samples and design method of classifier on the plane," *Pattern Recognition*, vol. 24, no. 4, pp. 317–324, 1991.
5. H. Y. M. Liao, C. C. Han, and G. J. Yu, "Face + hair + shoulders + background ≠ face," in *Proc. Workshop on 3D Computer Vision '97*, pp. 91–96, 1997(Invited paper).
6. C. C. Han, H. Y. M. Liao, G. J. Yu, and L. H. Chen, "Fast face detection via morphology-based pre-processing," in *Proc. 9th International Conference on Image Analysis and Processing*, pp. 469–476, 1997.
7. K. Liu, Y. Cheng, and J. Yang, "Algebraic feature extraction for image recognition based on an optimal discriminant criterion," *Pattern Recognition*, vol. 26, no. 6, pp. 903–911, 1993.

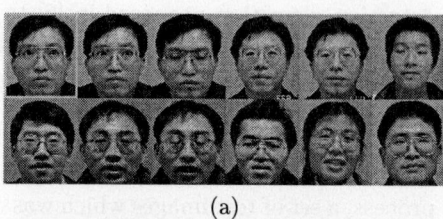

(a)

(b)

Fig. 2. (a) A set of retrieved results obtained by applying the autocorrelation plus LDA method. (b) A set of retrieved results obtained by applying the PCA plus LDA method. The upper-left face images of both cases are synthesized query images.

(a)

(b)

Fig. 3. (a) A set of retrieved results obtained by applying the autocorrelation plus LDA method. (b) A set of retrieved results obtained by applying the PCA plus LDA method. The upper-left face images of both cases are synthesized query images.

Fig. 1. Part of the original face database.

Fig. 4. The detected face portions obtained by using the algorithm proposed in [6].

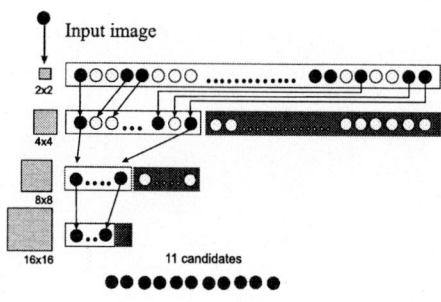

Fig. 7. The coarse-to-fine search mechanism.

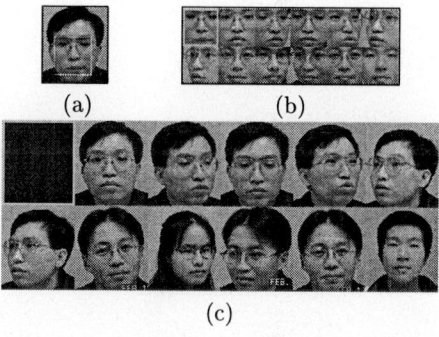

Fig. 8. (a) The detected face portion; (b) the 11 retrieved face-only samples and the query image; (c) the 11 original database samples which correspond to the 11 retrieved face-only database samples.

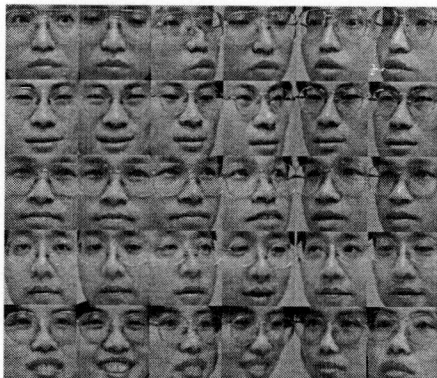

Fig. 5. The detected face portions of the face images in Fig. 1.

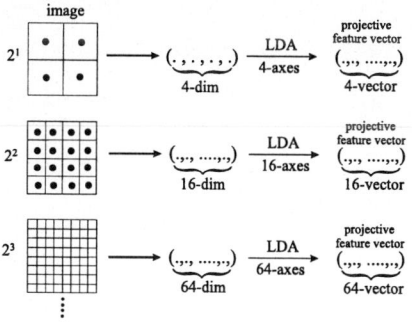

Fig. 6. Coarse-to-fine feature extraction and multi-scale LDA.

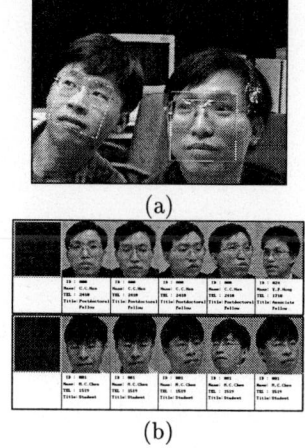

Fig. 9. Integration of the face detection and the face recognition subsystems.

Author Index

A

Abdallah, Samer M. I-386
Abdellatif, Mohamed.........I-208
Abe, Keiichi........................I-450
Abe, Norihiro.....................II-487
Achermann, B. II-726
Adachi, Hiroaki I-688
Adán, Antonio.....................I-482
Aggarwal, J.K. II-275
Ahuja, Narendra I-623
  II-33
  II-291
  II-323
Akutsu, Takashi I-321
Alexandrov, V.V. I-426
Alferez, Ronald-Bryan....I-542
Aloimonos, Yiannis..........II-283
Anderson, J.A.D.W. II-177
Arakawa, Kenichi...............I-321
Ariki, Yasuo II-695
Åström, Kalle....................II-169

B

Bab-Hadiashar, Alireza..I-566
  II-599
Bai, Xuesheng.......................I-240
Bajcsy, P. II-291
Beigi, Homayoon S.M. I-531
Bennet, Karen....................II-575
Berger, Marie-Odile II-360
Bhattacharyya, P..............I-590
Bhuiyan, Md. Shoaib II-25
Bilbro, Griff L. I-722
Binkert, M. II-726
Black, Michael S. II-267
Blum, Stefan A. I-128
Bolle, Ruud............................I-2
  II-283
Booth, William I-176
  II-623
Bottino, A. II-416
Bouvin, Christophe I-640

Brady, Michael.......................I-120
  I-152
  II-41
Brown, Michael S. I-558
Bryll, Robert K. II-591
Bunke, Horst II-299
  II-726
Burschka, Darius................I-128
Buxton, Hilary I-523
Byne, J.H.M. II-177

C

Caelli, Terry II-551
Califano, Andrea I-32
Cerrada, Carlos I-482
Cha, Sung-Hyuk..................I-370
Chaen, Atsushi I-288
Chai, Jinxiang I-272
Chan, Chorkin II-121
Chan, Kap Luk...................I-746
  II-225
Chan, S.L. I-714
Chan, Syin I-402
Chang, Chun-Ming...............I-192
Chaumette, François.........I-648
Chen, Dongwei I-192
Chen, Fang........................II-49
Chen, Liang-Hua II-742
Chen, Meng Chang II-742
Chen, Ming I-168
Chen, Tie Qi II-631
Chen, Youbin......................I-160
  I-168
Chen, Yunqiang II-9
Chen, Zen I-632
Chow, S.K. I-714
Chuang, Jen-Hui II-535
Chung, Jae-Moon I-346
Cipolla, Roberto...............I-515
Collina, Costantino.........I-338
Colville, Scott I-32
Cui, Yuntao.......................I-418

Cumani, Aldo I-582

D

Daum, M. I-72
Davis, Larry S. II-201
　　　　　　　　　　　　　II-267
Deguchi, Koichiro................I-48
　　　　　　　　　　　　　I-56
Dennis, Tim J. I-264
Deriche, Rachid..................I-640
Ding, Xiaoqing I-160
　　　　　　　　　　　　　I-168
　　　　　　　　　　　　　II-145
Dudek, G. I-72
Duric, Zoran I-305

E

Eason, Richard O. I-112
Endoh, Toshio II-1

F

Faugeras, Olivier..............I-640
Fejes, Sandor......................II-267
Feliu, Vicente I-482
Fermüller, Cornelia II-283
Fernandes, João L. I-64
Ferri, Massimo I-329
　　　　　　　　　　　　　I-338
Fountain, S.R. II-57
Frenkel, B.E. I-426
Frosini, Patrizio I-329
　　　　　　　　　　　　　I-338
Fu, J.S. I-746
Funahashi, Tatsushi........I-474
Funt, Brian II-257

G

Germain, Bob..........................I-32
Gillies, M. I-714
Gofuku, Akio I-208
Gong, Shaogang I-507
　　　　　　　　　　　　　I-607
　　　　　　　　　　　　　I-615
　　　　　　　　　　　　　II-679
Goss, Simon II-551

Greiffenhagen, Michael..I-418
Grimson, Eric I-498
Gross, Ari I-184
Gu, Jin...................................II-455
Gu, Weikang...........................I-144
　　　　　　　　　　　　　II-511
Guibas, Leonidas J. I-104
Guo, Fanxia II-145

H

Han, Chin-Chuan.................II-742
Hanmandlu, M. I-458
　　　　　　　　　　　　　I-466
Hariatoglu, Ismail..........II-267
Harley, Eric II-575
Harwood, David II-267
Hasegawa, Tsutomu..............I-442
Hata, Tadashi.....................II-639
Hauta-Kasari, M. I-248
Hayashi, Jun-ichiro II-1
Hedley, Mark II-376
　　　　　　　　　　　　　II-392
Heyden, Anders II-169
Hiura, Shinsaku.................II-495
Hong, Lin..................................I-16
Hoshide, Tsuyoshi..............I-442
Hotta, Kazuhiro II-89
Howell, A. Jonathan.........I-523
Hu, Wei.....................................I-362
Hu, Zhencheng.......................II-519
Huang, Chung-Lin................I-706
Huang, Qian...........................I-418
Huang, Thomas S. II-455
　　　　　　　　　　　　　II-463
Huang, Y.H. I-746
Hügli, H. I-490
Hung, Yi-Ping......................I-632
Hwang, Wey-Shiuan............II-503

I

Igi, Seiji I-698
Ikeuchi, Katsushi............II-209
　　　　　　　　　　　　　II-350
　　　　　　　　　　　　　II-408
　　　　　　　　　　　　　II-424

Imagawa, Kazuyuki I-698
Inokuchi, Seiji II-495
Ipson, Stanley S. I-176
Ishii, Naohiro....................I-40
Ishikawa, Noriyuki..........II-695
Ishikawa, Seiji I-354
Ishikawa, Takahiro..........II-671
Ishizawa, Akira I-688
Itoh, Hidenori I-474
Iwahori, Yuji I-40
         II-25
Iwai, Yoshio II-639
         II-647
Iwata, Akira II-25

J

Jain, Anil I-16
Jenkin, Michael R.M. I-656
Jenkinson, Mark...............II-41
Jia, Jin I-450
Jiang, Xiaoyi....................II-299
         II-726
Jo, Kang-Hyun II-368
Jojic, Nebojsa II-455
         II-463
Jost, T. I-490
Jurie, Frederic................II-440

K

Katafuchi, Norifumi.........I-200
Kato, Hirokazu II-495
Kato, Kunihito II-1
Kato, Zoltan I-738
Katsuyama, Yutaka II-137
Kaveti, Satish...................I-378
Kawaguchi, Eiji I-112
Kim, Chang-Hun II-241
Kim, Eun Yi......................I-730
Kim, Hang Joon................I-730
Kim, Hyoung Seop...........I-354
Kim, Jin Wook..................I-730
Kimmel, R. I-88
         I-574
King, Irwin........................I-410
         II-559

Kirihara, Satoshi II-448
Kittler, J. II-307
Koshimizu, Hiroyasu II-1
         II-663
Krishnamoorthi, R. I-590
Kuno, Yoshinori I-672
         II-368
Kurita, Takio......................II-89

L

Lai, Kok F. I-402
Laikov, E.V. I-426
Laine, Andrew.................I-192
Lan, Zhong-Dan I-313
Lang, J. I-656
Lang, S. I-136
Latecki, Longin I-184
Lau, Tak Kan I-410
Laurentini, A. II-416
Lee, John Chung-Mong I-738
Lee, Mi-Suen II-315
Lee, Sang Uk I-96
Lenz, R. I-248
Leow, Wee Kheng II-17
Li, Fuxing I-120
Li, Stan Ziqing I-746
         II-225
Li, Yi...............................II-719
Liang, Ping.....................II-400
Liao, H.Y. Mark...............II-742
Liao, Simon X. I-394
Lien, Cheng-Chang..........I-706
Lin, Xiaofan....................I-160
Liu, Jinhui I-160
Liu, Peilin II-209
Liu, Tianrong..................II-225
Liu, Yongmei II-153
Loft, Peter J. II-679
Lourakis, Manolis II-527
Lourens, Tino II-193
Lovato, Alberto I-329
Lu, G.Z. I-136
Lu, Hanqing II-719
Lu, Shan I-698
Lu, Weier II-687

Lu, Yi II-631
Luk, W.S. II-559

M

Ma, SongDe I-136
                               I-272
                               II-9
                               II-607
                               II-719
Maes, Stéphane H. I-531
Malladi, R. I-574
Marchand, Éric I-648
Maru, Noriaki I-256
Masai, Hiroyuki II-105
Masegawa, Tsutomu II-185
McIvor, Alan M. I-434
McKenna, Stephen J. I-507
                               I-607
                               I-615
Medioni, Gérard II-315
Melhi, Muhammed I-176
Menard, Christian..............I-550
Michalski, Ryszard S. I-305
Miller, Jason II-631
Minoh, Michihiko II-332
Mirmehdi, M. II-307
Mishima, Taketoshi............II-89
Mita, Takeshi.......................II-495
Miura, Jun I-672
Miyazaki, Fumio I-256
Miyazaki, Tsuyoshi I-474
Mohr, Roger..........................I-313
Mokhtarian, Farzin...........II-73
Mori, Hiroki II-655
Mori, Takeaki.......................II-471
Morishima, Shigeo............II-671
Morooka, Ken'ichi II-185
Motomura, Nachi I-354
Mukai, Toshiharu II-583
                               II-734
Mukaigawa, Yasuhiro.........I-680
Murakami, Kazuhito..............II-1
                               II-663
Murakami, Masamitsu...........I-40
Murase, Hiroshi I-321

Murata, Akio II-487

N

Nagai, Isaku.........................I-208
Nakamura, Yuichi................I-680
Nakatani, Hiromasa..........II-711
Nalwa, Vishvjit I-10
Naoi, Satoshi......................II-137
Nathan, Krishna I-1
Nebot, Eduardo M. I-386
Nevatia, Ram........................II-259
Ngan, Phillip M. II-615
Niimi, Michiharu.................I-112
Nishijima, Masakazu.......II-217
Nishikawa, Atsushi I-256
Nishimura, T. II-734
Noras, James M. II-623
Noronha, Sanjay.................II-259
Nozaki, Koichi.....................I-112

O

Ohara, Shuichi.....................I-200
Ohba, Kohtaro II-424
Ohnishi, Noboru I-232
                               I-346
                               II-153
                               II-583
Ohta, Hiroshi II-711
Ohta, Yuichi I-680
Ohya, Jun II-655
                               II-703
Oka, R. II-734
Okatani, Takayuki................I-48
                               I-56
Okudaira, Masashi I-200
Ong, Eng-Jon II-679
Or, S.H. II-559
Orphanoudakis, Stelios II-527
Oshiro, Naoki I-256
Osumi, Noriyoshi................II-471
Otsuka, Takahiro................II-703

P

Pagliari, Carla L. I-264
Palmer, P.L. II-307

Pankanti, S. I-2
Park, In Kyu I-96
Park, Se Hyun I-730
Parkkinen, J. I-248
Pawlak, Miroslaw I-394
Peake, G.S. II-97
Pearce, Adrian R. II-551
Perrin, Benoit II-323
Pong, Ting-Chuen I-738
Porcellini, Eleonora I-338
Pramadihanto, Dadet II-647
Priese, Lutz I-598
Psarrou, Alexandra I-664

Q

Qi, Hairong I-722
Qiu, G. II-81
Quan, Long I-313
Quek, Francis K.H. II-591

R

Raja, Yogesh I-607
                                          I-615
Ramakrishna, R.S. I-296
Ratakonda, Krishna............II-33
Ratha, Nalini I-2
Rehrmann, Volker I-598
Rubner, Yossi I-104
Rye, David C. I-386

S

Saito, Hideo II-448
Saito, Takafumi II-471
Saji, Hitoshi II-711
Sakauchi, Masao II-209
                                        II-408
Salden, Alfons II-65
Samarasekera, Supun I-418
Sano, Mutsuo I-200
Sato, Yoichi II-350
                                        II-424
Schütz, C. I-490
Se, Stephen I-152

Seales, W. Brent...............I-362
                                          I-558
                                        II-233
Seki, Hirohisa I-474
Sera, Hajime II-671
Shah, Shishir II-275
Shantaram, V. I-458
                                          I-466
Shen, Helen C. II-455
Shen, Weicheng I-24
Shimada, Nobutaka I-672
Shinya, Mikio II-471
Shirai, Yoshiaki................I-672
                                        II-368
                                        II-432
Simon, Gilles II-360
Snyder, Wesley E...............I-722
Sochen, N. I-574
Srinivasa, Narayan..........I-623
                                        II-323
Sugie, Noboru I-232
                                        II-153
Sugimoto, Akihiro...........II-161
Sugimoto, Noriko..............II-543
Sugiyama, Yoshiaki..........II-695
Suh, Tae-Jung II-241
Sumi, Yasushi II-249
Suter, David I-566
                                          II-49
                                        II-599

T

Takai, Mikiko II-663
Takebe, Hiroaki................II-137
Takemura, Haruo I-288
Tan, T.N. II-57
                                          II-97
Tanaka, Hiromi T...............I-688
Tanaka, Yutaka I-208
Tang, Cheng-Yuan I-632
Taniguchi, Rin-ichiro...II-479
Taniguchi, Yasuhiro.......II-432
Tanner, Jonathan...............I-664
Teoh, Eam Khwang................I-378

Terzopoulos, Demetri......II-671
Tian, Ying-li........................I-216
                                   I-224
Tomasi, Carlo.......................I-104
Tominaga, Shoji....................I-80
                                   II-258
Tomita, Fumiaki................II-249
Tong, W.B. I-136
Toriu, Takashi.......................II-1
Torreão, José R.A. I-64
Toyooka, S. I-248
Trajković, Miroslav.......II-376
                                   II-392
Tsai, Chi-Hao II-535
Tsai, Wei-Hsin...................II-535
Tso, S.K. I-136
Tsotsos, John K................II-575
Tsui, Hung-Tat I-216
                                   I-224
                                   II-384
Tsuruta, Naoyuki...............II-479
Tyan, Hsiao-Rong..............II-742

U
Uchimura, Keiichi............II-519
Utsumi, Akira II-655

V
Vaidyanathan, B. I-296

W
Waibel, Alex II-687
Waltenberg, Peter T. I-434
Wang, Han I-378
Wang, Pingtao II-408
Wang, W. I-248
Wang, Xinli II-129
Wang, Yuan-Fang I-542
                                   II-400
Watanabe, Toyohide II-105
                                   II-113
                                   II-217

Weng, John (Juyang) II-503
Westling, Mark F. II-201
Wheeler, Mark D................II-350
Wirtz, Brigitte I-499
Wong, K.H. II-559
Wong, Pak-Kwong II-121
Woodham, Robert J. I-40
Wu, Haiyuan II-647
Wu, Wei II-209
Wu, Youshou...........................I-160
                                   I-168
                                   II-145
Würtz, Rolf P. II-193

X
Xie, M. I-280
Xu, Donglai..........................II-623
Xu, Gang II-543
Xu, Guangyou I-240

Y
Yachida, Masahiko II-639
                                   II-647
                                   II-655
Yacoob, Yaser.....................II-267
Yamada, Masashi I-474
Yamamura, Tsuyoshi..........II-153
Yamazawa, Kazumasa...........I-288
Yang, Chuei-Yaw II-535
Yang, Jie II-687
Yang, Jun I-232
Yang, Qing II-607
Ye, Xiuqing..........................II-511
Ye, Yimimg II-575
Yeung, S.Y. I-216
                                   I-224
Yokokura, Naoko.................II-113
Yokoya, Naokazu I-288
Yonemoto, Satoshi II-479
Yow, Kin Choong I-515
Yu, Gwo-Jong II-742
Yuan, Cheng Jiun...............II-233
Yun, Il Dong I-96

Z

Zambelli, Chiara……………I-329
Zha, Hongbin …………………I-442
  II-185
Zhang, Dili ………………………I-232
Zhang, Nan …………………………II-17
Zhang, Qi ……………………………I-144
  I-305
  II-511
Zhang, Zhengyou ……………II-340
  II-567
Zhang, Zhong-Ying …………II-384
Zheng, Jiang Yu ……………II-487
Zheng, Jing …………………………II-145
Zhou, Hong …………………………I-402
Zuccone, P.…………………………II-416

Lecture Notes in Computer Science

For information about Vols. 1–1278

please contact your bookseller or Springer-Verlag

Vol. 1279: B. S. Chlebus, L. Czaja (Eds.), Fundamentals of Computation Theory. Proceedings, 1997. XI, 475 pages. 1997.

Vol. 1280: X. Liu, P. Cohen, M. Berthold (Eds.), Advances in Intelligent Data Analysis. Proceedings, 1997. XII, 621 pages. 1997.

Vol. 1281: M. Abadi, T. Ito (Eds.), Theoretical Aspects of Computer Software. Proceedings, 1997. XI, 639 pages. 1997.

Vol. 1282: D. Garlan, D. Le Métayer (Eds.), Coordination Languages and Models. Proceedings, 1997. X, 435 pages. 1997.

Vol. 1283: M. Müller-Olm, Modular Compiler Verification. XV, 250 pages. 1997.

Vol. 1284: R. Burkard, G. Woeginger (Eds.), Algorithms — ESA '97. Proceedings, 1997. XI, 515 pages. 1997.

Vol. 1285: X. Jao, J.-H. Kim, T. Furuhashi (Eds.), Simulated Evolution and Learning. Proceedings, 1996. VIII, 231 pages. 1997. (Subseries LNAI).

Vol. 1286: C. Zhang, D. Lukose (Eds.), Multi-Agent Systems. Proceedings, 1996. VII, 195 pages. 1997. (Subseries LNAI).

Vol. 1287: T. Kropf (Ed.), Formal Hardware Verification. XII, 367 pages. 1997.

Vol. 1288: M. Schneider, Spatial Data Types for Database Systems. XIII, 275 pages. 1997.

Vol. 1289: G. Gottlob, A. Leitsch, D. Mundici (Eds.), Computational Logic and Proof Theory. Proceedings, 1997. VIII, 348 pages. 1997.

Vol. 1290: E. Moggi, G. Rosolini (Eds.), Category Theory and Computer Science. Proceedings, 1997. VII, 313 pages. 1997.

Vol. 1291: D.G. Feitelson, L. Rudolph (Eds.), Job Scheduling Strategies for Parallel Processing. Proceedings, 1997. VII, 299 pages. 1997.

Vol. 1292: H. Glaser, P. Hartel, H. Kuchen (Eds.), Programming Languages: Implementations, Logigs, and Programs. Proceedings, 1997. XI, 425 pages. 1997.

Vol. 1293: C. Nicholas, D. Wood (Eds.), Principles of Document Processing. Proceedings, 1996. XI, 195 pages. 1997.

Vol. 1294: B.S. Kaliski Jr. (Ed.), Advances in Cryptology — CRYPTO '97. Proceedings, 1997. XII, 539 pages. 1997.

Vol. 1295: I. Prívara, P. Ružička (Eds.), Mathematical Foundations of Computer Science 1997. Proceedings, 1997. X, 519 pages. 1997.

Vol. 1296: G. Sommer, K. Daniilidis, J. Pauli (Eds.), Computer Analysis of Images and Patterns. Proceedings, 1997. XIII, 737 pages. 1997.

Vol. 1297: N. Lavrač, S. Džeroski (Eds.), Inductive Logic Programming. Proceedings, 1997. VIII, 309 pages. 1997. (Subseries LNAI).

Vol. 1298: M. Hanus, J. Heering, K. Meinke (Eds.), Algebraic and Logic Programming. Proceedings, 1997. X, 286 pages. 1997.

Vol. 1299: M.T. Pazienza (Ed.), Information Extraction. Proceedings, 1997. IX, 213 pages. 1997. (Subseries LNAI).

Vol. 1300: C. Lengauer, M. Griebl, S. Gorlatch (Eds.), Euro-Par'97 Parallel Processing. Proceedings, 1997. XXX, 1379 pages. 1997.

Vol. 1301: M. Jazayeri, H. Schauer (Eds.), Software Engineering - ESEC/FSE'97. Proceedings, 1997. XIII, 532 pages. 1997.

Vol. 1302: P. Van Hentenryck (Ed.), Static Analysis. Proceedings, 1997. X, 413 pages. 1997.

Vol. 1303: G. Brewka, C. Habel, B. Nebel (Eds.), KI-97: Advances in Artificial Intelligence. Proceedings, 1997. XI, 413 pages. 1997. (Subseries LNAI).

Vol. 1304: W. Luk, P.Y.K. Cheung, M. Glesner (Eds.), Field-Programmable Logic and Applications. Proceedings, 1997. XI, 503 pages. 1997.

Vol. 1305: D. Corne, J.L. Shapiro (Eds.), Evolutionary Computing. Proceedings, 1997. X, 307 pages. 1997.

Vol. 1306: C. Leung (Ed.), Visual Information Systems. X, 274 pages. 1997.

Vol. 1307: R. Kompe, Prosody in Speech Understanding Systems. XIX, 357 pages. 1997. (Subseries LNAI).

Vol. 1308: A. Hameurlain, A M. Tjoa (Eds.), Database and Expert Systems Applications. Proceedings, 1997. XVII, 688 pages. 1997.

Vol. 1309: R. Steinmetz, L.C. Wolf (Eds.), Interactive Distributed Multimedia Systems and Telecommunication Services. Proceedings, 1997. XIII, 466 pages. 1997.

Vol. 1310: A. Del Bimbo (Ed.), Image Analysis and Processing. Proceedings, 1997. Volume I. XXII, 722 pages. 1997.

Vol. 1311: A. Del Bimbo (Ed.), Image Analysis and Processing. Proceedings, 1997. Volume II. XXII, 794 pages. 1997.

Vol. 1312: A. Geppert, M. Berndtsson (Eds.), Rules in Database Systems. Proceedings, 1997. VII, 214 pages. 1997.

Vol. 1313: J. Fitzgerald, C.B. Jones, P. Lucas (Eds.), FME '97: Industrial Applications and Strengthened Foundations of Formal Methods. Proceedings, 1997. XIII, 685 pages. 1997.

Vol. 1314: S. Muggleton (Ed.), Inductive Logic Programming. Proceedings, 1996. VIII, 397 pages. 1997. (Subseries LNAI).

Vol. 1315: G. Sommer, J.J. Koenderink (Eds.), Algebraic Frames for the Perception-Action Cycle. Proceedings, 1997. VIII, 395 pages. 1997.

Vol. 1316: M. Li, A. Maruoka (Eds.), Algorithmic Learning Theory. Proceedings, 1997. XI, 461 pages. 1997. (Subseries LNAI).

Vol. 1317: M. Leman (Ed.), Music, Gestalt, and Computing. IX, 524 pages. 1997. (Subseries LNAI).

Vol. 1318: R. Hirschfeld (Ed.), Financial Cryptography. Proceedings, 1997. XI, 409 pages. 1997.

Vol. 1319: E. Plaza, R. Benjamins (Eds.), Knowledge Acquisition, Modeling and Management. Proceedings, 1997. XI, 389 pages. 1997. (Subseries LNAI).

Vol. 1320: M. Mavronicolas, P. Tsigas (Eds.), Distributed Algorithms. Proceedings, 1997. X, 333 pages. 1997.

Vol. 1321: M. Lenzerini (Ed.), AI*IA 97: Advances in Artificial Intelligence. Proceedings, 1997. XII, 459 pages. 1997. (Subseries LNAI).

Vol. 1322: H. Hußmann, Formal Foundations for Software Engineering Methods. X, 286 pages. 1997.

Vol. 1323: E. Costa, A. Cardoso (Eds.), Progress in Artificial Intelligence. Proceedings, 1997. XIV, 393 pages. 1997. (Subseries LNAI).

Vol. 1324: C. Peters, C. Thanos (Eds.), Research and Advanced Technology for Digital Libraries. Proceedings, 1997. X, 423 pages. 1997.

Vol. 1325: Z.W. Raś, A. Skowron (Eds.), Foundations of Intelligent Systems. Proceedings, 1997. XI, 630 pages. 1997. (Subseries LNAI).

Vol. 1326: C. Nicholas, J. Mayfield (Eds.), Intelligent Hypertext. XIV, 182 pages. 1997.

Vol. 1327: W. Gerstner, A. Germond, M. Hasler, J.-D. Nicoud (Eds.), Artificial Neural Networks – ICANN '97. Proceedings, 1997. XIX, 1274 pages. 1997.

Vol. 1328: C. Retoré (Ed.), Logical Aspects of Computational Linguistics. Proceedings, 1996. VIII, 435 pages. 1997. (Subseries LNAI).

Vol. 1329: S.C. Hirtle, A.U. Frank (Eds.), Spatial Information Theory. Proceedings, 1997. XIV, 511 pages. 1997.

Vol. 1330: G. Smolka (Ed.), Principles and Practice of Constraint Programming – CP 97. Proceedings, 1997. XII, 563 pages. 1997.

Vol. 1331: D. W. Embley, R. C. Goldstein (Eds.), Conceptual Modeling – ER '97. Proceedings, 1997. XV, 479 pages. 1997.

Vol. 1332: M. Bubak, J. Dongarra, J. Waśniewski (Eds.), Recent Advances in Parallel Virtual Machine and Message Passing Interface. Proceedings, 1997. XV, 518 pages. 1997.

Vol. 1333: F. Pichler. R.Moreno-Díaz (Eds.), Computer Aided Systems Theory – EUROCAST'97. Proceedings, 1997. XII, 626 pages. 1997.

Vol. 1334: Y. Han, T. Okamoto, S. Qing (Eds.), Information and Communications Security. Proceedings, 1997. X, 484 pages. 1997.

Vol. 1335: R.H. Möhring (Ed.), Graph-Theoretic Concepts in Computer Science. Proceedings, 1997. X, 376 pages. 1997.

Vol. 1336: C. Polychronopoulos, K. Joe, K. Araki, M. Amamiya (Eds.), High Performance Computing. Proceedings, 1997. XII, 416 pages. 1997.

Vol. 1337: C. Freksa, M. Jantzen, R. Valk (Eds.), Foundations of Computer Science. XII, 515 pages. 1997.

Vol. 1338: F. Plášil, K.G. Jeffery (Eds.), SOFSEM'97: Theory and Practice of Informatics. Proceedings, 1997. XIV, 571 pages. 1997.

Vol. 1339: N.A. Murshed, F. Bortolozzi (Eds.), Advances in Document Image Analysis. Proceedings, 1997. IX, 345 pages. 1997.

Vol. 1340: M. van Kreveld, J. Nievergelt, T. Roos, P. Widmayer (Eds.), Algorithmic Foundations of Geographic Information Systems. XIV, 287 pages. 1997.

Vol. 1341: F. Bry, R. Ramakrishnan, K. Ramamohanarao (Eds.), Deductive and Object-Oriented Databases. Proceedings, 1997. XIV, 430 pages. 1997.

Vol. 1342: A. Sattar (Ed.), Advanced Topics in Artificial Intelligence. Proceedings, 1997. XVII, 516 pages. 1997. (Subseries LNAI).

Vol. 1343: Y. Ishikawa, R.R. Oldehoeft, J.V.W. Reynders, M. Tholburn (Eds.), Scientific Computing in Object-Oriented Parallel Environments. Proceedings, 1997. XI, 295 pages. 1997.

Vol. 1344: C. Ausnit-Hood, K.A. Johnson, R.G. Pettit, IV, S.B. Opdahl (Eds.), Ada 95 – Quality and Style. XV, 292 pages. 1997.

Vol. 1345: R.K. Shyamasundar, K. Ueda (Eds.), Advances in Computing Science - ASIAN'97. Proceedings, 1997. XIII, 387 pages. 1997.

Vol. 1346: S. Ramesh, G. Sivakumar (Eds.), Foundations of Software Technology and Theoretical Computer Science. Proceedings, 1997. XI, 343 pages. 1997.

Vol. 1347: E. Ahronovitz, C. Fiorio (Eds.), Discrete Geometry for Computer Imagery. Proceedings, 1997. X, 255 pages. 1997.

Vol. 1348: S. Steel, R. Alami (Eds.), Recent Advances in AI Planning. Proceedings, 1997. IX, 454 pages. 1997. (Subseries LNAI).

Vol. 1349: M. Johnson (Ed.), Algebraic Methodology and Software Technology. Proceedings, 1997. X, 594 pages. 1997.

Vol. 1350: H.W. Leong, H. Imai, S. Jain (Eds.), Algorithms and Computation. Proceedings, 1997. XV, 426 pages. 1997.

Vol. 1351: R. Chin, T.-C. Pong (Eds.), Computer Vision – ACCV'98. Proceedings Vol. I, 1998. XXIV, 761 pages. 1997.

Vol. 1352: R. Chin, T.-C. Pong (Eds.), Computer Vision – ACCV'98. Proceedings Vol. II, 1998. XXIV, 757 pages. 1997.

Vol. 1353: G. BiBattista (Ed.), Graph Drawing. Proceedings, 1997. XII, 448 pages. 1997.

Vol. 1354: O. Burkart, Automatic Verification of Sequential Infinite-State Processes. X, 163 pages. 1997.

Vol. 1355: M. Darnell (Ed.), Cryptography and Coding. Proceedings, 1997. IX, 335 pages. 1997.

Vol. 1356: A. Danthine, Ch. Diot (Eds.), From Multimedia Services to Network Services. Proceedings, 1997. XII, 180 pages. 1997.